D0161386

North American
Terrestrial Vegetation

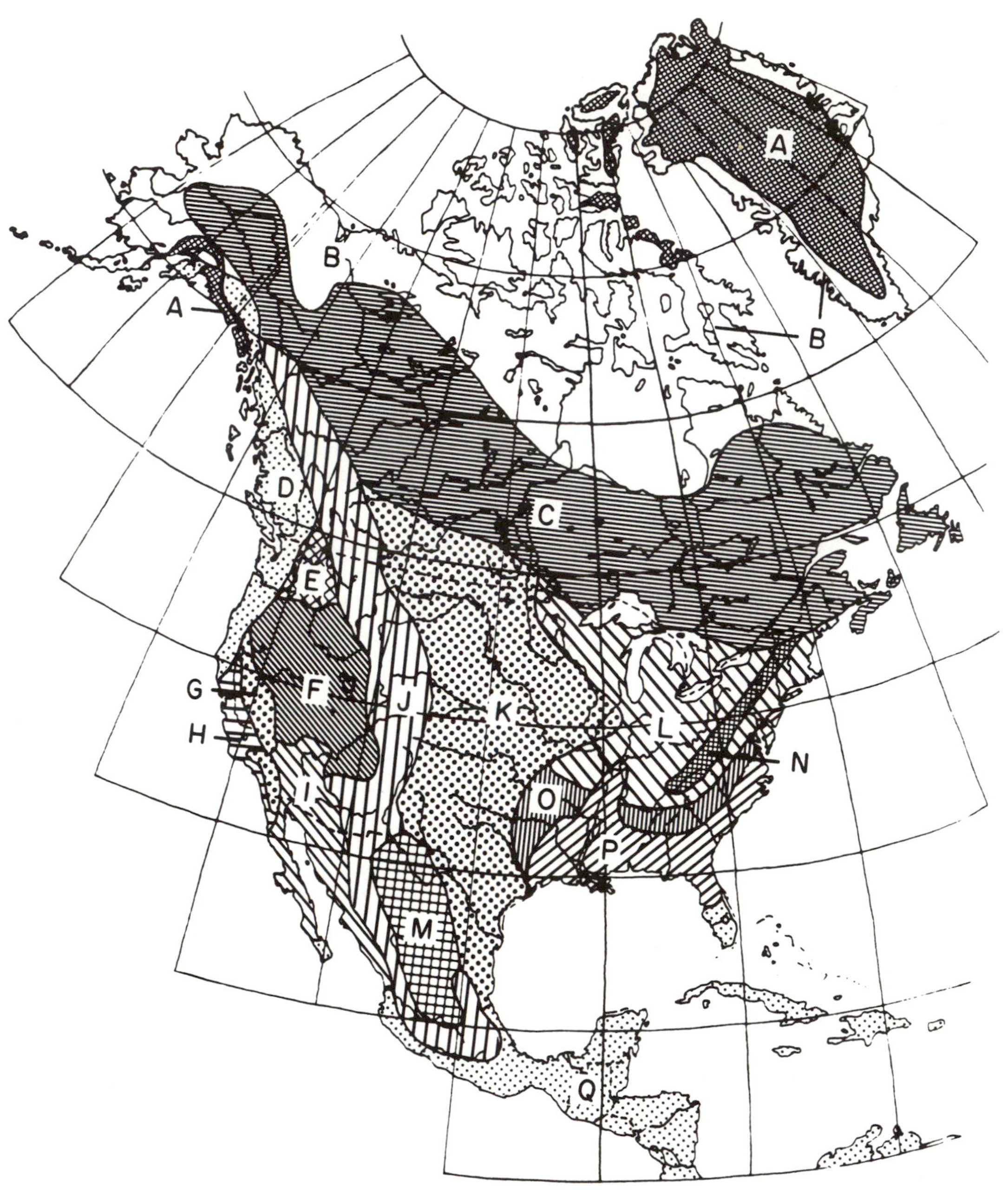

Generalized vegetation formations of North America. A = ice; B = arctic tundra; C = taiga; D = Pacific coastal/Cascadian forests; E = Palouse prairies; F = Intermountain deserts, shrub steppes, woodlands, and forests; G = Californian forests and alpine vegetation; H = Californian grasslands, chaparral, and woodlands; I = Mojave and Sonoran deserts; J - Rocky Mountain forests and alpine vegetation; K = central prairies and plains; L = mixed deciduous forests; M = Chihuahuan deserts and woodlands; N = Appalachian forests; O - piedmont oak-pine forests; P = coastal plain forests, bogs, swamps, marshes, and strand; Q = tropical forests. Boundaries and formation names according to W.D. Billings; art by Christina Weber-Johnson.

North American Terrestrial Vegetation

Edited by

Michael G. Barbour
William Dwight Billings

CAMBRIDGE UNIVERSITY PRESS

Cambridge
New York New Rochelle Melbourne Sydney

Published by the Press Syndicate of the University of Cambridge
The Pitt Building, Trumpington Street, Cambridge CB2 1RP
32 East 57th Street, New York, NY 10022, USA
10 Stamford Road, Oakleigh, Melbourne 3166, Australia

First published 1988

Printed in the United States of America

Library of Congress Cataloging-in-Publication Data

North American terrestrial vegetation.

Includes bibliographies and index.

1. Botany–North America–Ecology. 2. Plant communities–North America. 3. Phytogeography–North America. I. Barbour, Michael G. II. Billings, W. D. (William Dwight), 1910–
QK110.N854 1987 581.5'264'097 86-31769

British Library Cataloguing in Publication Data

North American terrestrial vegetation.

1. Botany–North America–Ecology
I. Barbour, Michael G. II. Billings, W. Dwight
581.5'097 QK110

ISBN 0 521 26198 8

Contents

Contributors

Michael G. Barbour
Botany Department
University of California
Davis, California

William Dwight Billings
Botany Department
Duke University
Durham, North Carolina

L. C. Bliss
Botany Department
University of Washington
Seattle, Washington

Norman L. Christensen
Botany Department
Duke University
Durham, North Carolina

Deborah L. Elliott-Fisk
Geography Department
University of California
Davis, California

Jerry F. Franklin
College of Forest Resources
University of Washington
Seattle, Washington

Andrew M. Greller
Biology Department
Queens College
Flushing, New York

Gary S. Hartshorn
Tropical Science Center
San Jose
Costa Rica

Jon E. Keeley
Biology Department
Occidental College
Los Angeles, California

Sterling C. Keeley
Biology Department
Whittier College
Los Angeles, California

James A. MacMahon
Biology Department
Utah State University
Logan, Utah

Robert K. Peet
Biology Department
University of North Carolina
Chapel Hill, North Carolina

Phillip L. Sims
USDA Field Station
Woodward, Oklahoma

Neil E. West
Range Science Department
and Ecology Center
Utah State University
Logan, Utah

Preface

Rather naively, back in the fall of 1982, when this book first began to be organized, we expected to have a 300-page publication in hand three years later. The complexity of the subject and the intricate schedules of fourteen contributing authors and two editors lengthened the preparation period by 100% and increased the size of the book by 50%. We were helped in our early discussions about the book's content, format, and contributors by the hard work and encouragement of Richard Ziemacki, then editor at the New York office of Cambridge University Press. We thank him now for his vision when ours was still in the "what if . . ." stage.

We reached final decisions about book objectives, format, and the list of contributors by mid-1983. But the distribution of fine-scale topics was in a state of dynamic equilibrium well into 1986. Book writing, it seems, is a seral procedure, with many analogies to succession in vegetation. We trust that the climax presented to the reader is an accurate and challenging summary of what is currently understood about North American vegetation.

Chapters focus on the major plant formations of North America, but they also include information on many other, more local, vegetation types. The authors have devoted enough attention to each vegetation type discussed to give the reader details on vegetation structure, response to disturbance, community/environment relations, nutrient cycling and productivity, and autecological behavior of dominant species. We have selected contributors who are active researchers in extensive regions or vegetation types, and we have invited them to flavor their reviews with their own research biases (many chapters include previously unpublished data, analyses, or models). The same standard topical outline was given to each author to follow, but each region or vegetation type has a peculiarly skewed literature that reflects the environmental factors long considered to be important by regional ecologists. In one area, those factors may be fire and soil moisture; in another, wind storms and soil nutrient availability; in another, paleoecological events of the Pleistocene or Holocene; in another, human actions that have introduced grazing animals or weeds or modified the scale of natural disturbances. Consequently, any two chapters may have little in common beyond starting with a frontispiece map and introduction and concluding with suggestions for future research. Although the sequences of topics vary from chapter to chapter, each chapter includes sections on paleobotany, the modern environment, human-caused vegetation changes, successional patterns following disturbance, and a quantitative description of major vegetation types.

There is some overlap, in terms of vegetation types discussed, between pairs of chapters. Petran chaparral, for example, is described in Chapter 3 (Rocky Mountains), Chapter 6 (chaparral), and Chapter 7 (intermountain deserts, shrub steppes, and woodlands); the New Jersey pine barrens are considered in both Chapter 10 (deciduous forest) and Chapter 11 (southeastern coastal plain); the forest-tundra ecotone is discussed in both Chapter 1 (arctic tundra and polar desert biome) and Chapter 2 (boreal forest). In the interest of efficient use of space, only one chapter of such a pair describes an overlapping vegetation unit in detail, but the units are such logical inclusions within each chapter that the remaining degree of overlap is important to the book's overall objective and to the reader's understanding of vegetation gradients.

Our intended audience includes knowledgeable laypeople, advanced undergraduates, graduate students, and professional ecologists in both basic and applied fields. Every chapter addresses management questions and problems, and every chapter includes considerable information on basic ecological topics, quite apart from immediate management considerations. Common names of dominant plants are sometimes used, but text, tables, and figures emphasize scientific names. The metric system is adopted exclusively. Figures and tables are added when they can efficiently convey a sense of the vegetation better than can the text. Space limitations prevent us from illustrating every vegetation type included in the text, and some types simply do not lend themselves well to black-and-white photography. Some of the graphs in the text were drawn by Christina Weber-Johnson.

There is a rich heritage of texts on North American vegetation, and we stand on the shoulders of their authors in order to see just a bit farther. Our predecessors include Clements, Daubenmire, Harshberger, Knapp, Küchler, and Vankat; their works span the past 75 years. Most recently, Chabot and Mooney have edited a volume on the physiological ecology of North American vegetation, and we consider their work to be an appropriate complementary companion to this book.

Many colleagues have served as reviewers for portions of this book, and they are acknowledged at the end of each chapter. We, as editors, take this opportunity now to thank them as well. We also congratulate the authors on the care and effort with which they have created their chapters, and we anticipate that these chapters will stand as the best summaries for some years to come. Quite simply, we believe that the authors have made this text a balanced, modern, detailed reference work that has had no equal in the past and that certainly has no current equal.

Finally, we deeply appreciate the help of the following people for their work in indexing: Julie Barbour, Frederika Bowcutt, Robert Boyd, Sonia Cook, Ronnie James, Steven Jennings, Scott Martens, Tisa Owen, Wayne Owen, Oren Pollack, Robert Rhode, George Robinson, Diane Ryerson, Jane Sakauye, Judy Sindel-Dorsey, Glenn Turner, and Debbie Woodward. Mrs. Judi Steinig was invaluable to us in the production aspects of this book; we are indebted to her for the exceptional care she consistently applied to our project.

M.G.B. and W.D.B.

Chapter 1

Arctic Tundra and Polar Desert Biome

L. C. BLISS

INTRODUCTION

The Arctic is often viewed as a monotonous landscape with a limited number of vascular plants and an abundance of cryptogams. In reality, the arctic tundra and polar desert biome is as diverse in its vegetation types and soils as are the grassland biome and the coniferous biomes of western mountains and the taiga. The Arctic constitutes about 20% of North America (Fig. 1.1), about 2.5 million km^2 in Canada, 2.0 million km^2 in Greenland, and 0.3 million km^2 in Alaska. Over this large area there are considerable variations in climate, ice cover, soils, sizes of the flora (cryptogam and vascular), and plant communities.

As used here, "Arctic" refers to those areas beyond the climatic limit of the boreal forest and treeline. There often are small pockets of trees, usually *Picea glauca*, in protected mesohabitats (south-facing slopes, river terraces) beyond the treeline, but the uplands are dominated by arctic tundra vegetation. Throughout these cold-dominated landscapes, soils are permanently frozen (permafrost), with only the upper portion (20–60 cm, but 100–200 cm along rivers) thawing in summer, the active layer. The Circumpolar Arctic is divided into the Low Arctic and High Arctic (Fig. 1.2), based on many ecological characteristics (Table 1.1). Tundras predominate in the Low Arctic, whereas polar semideserts and polar deserts dominate the High Arctic. "Tundra" is a generic term that includes vegetation types that range from tall shrub (2–5 m high) to dwarf shrub heath (5–20 cm high) and graminoids and mosses. These landscapes have a total plant cover of 80–100%, including an abundance of cryptogams in most sites.

PHYSICAL ENVIRONMENT

Climate

The macroclimate of the Arctic is characterized by continuous darkness in midwinter, with a nearly continuous cover of snow and ice for 8–10 mo, and by continuous light in summer, with its short growing season of 1.5–4 mo. In winter, a large semipermanent high-pressure system of intensely cold dry air occurs over the Yukon Territory and the western Northwest Territories. The average southern position of these anticyclonic systems parallels the southern boundary of the taiga in winter. However, outbreaks of cold arctic air extend into the midwestern and southern United States during severe storms. In summer, the arctic high-pressure system is weaker, with its southern limit parallel-

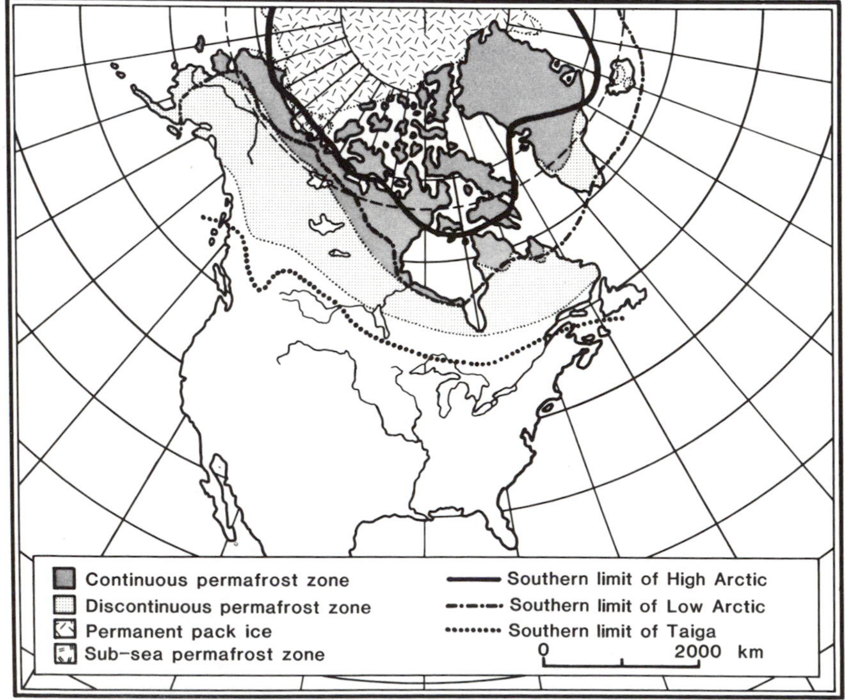

Figure 1.1. Major subdivisions of the North American and Greenland Arctic.

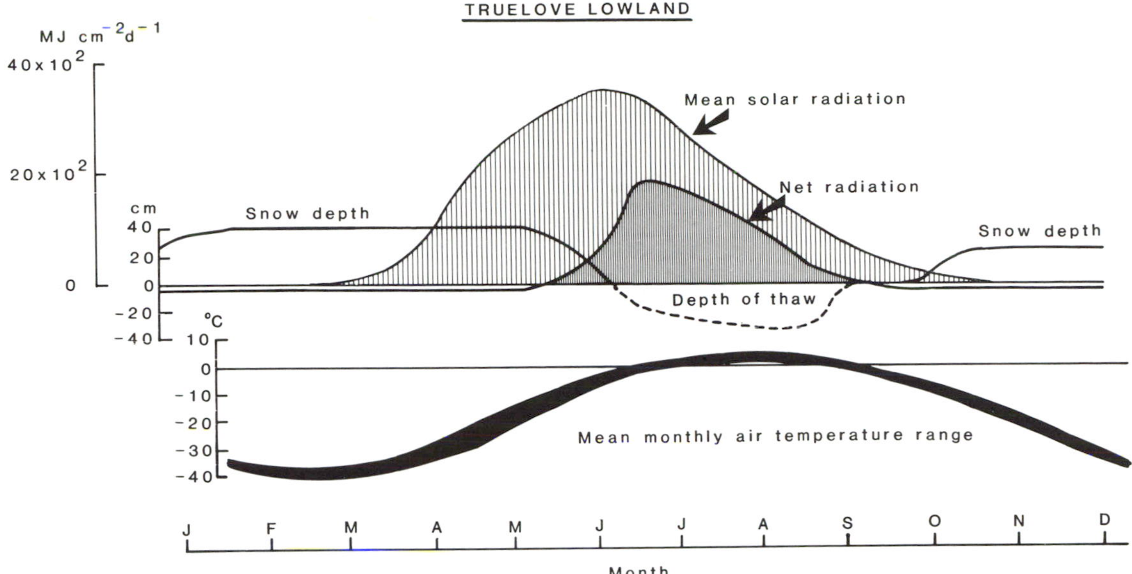

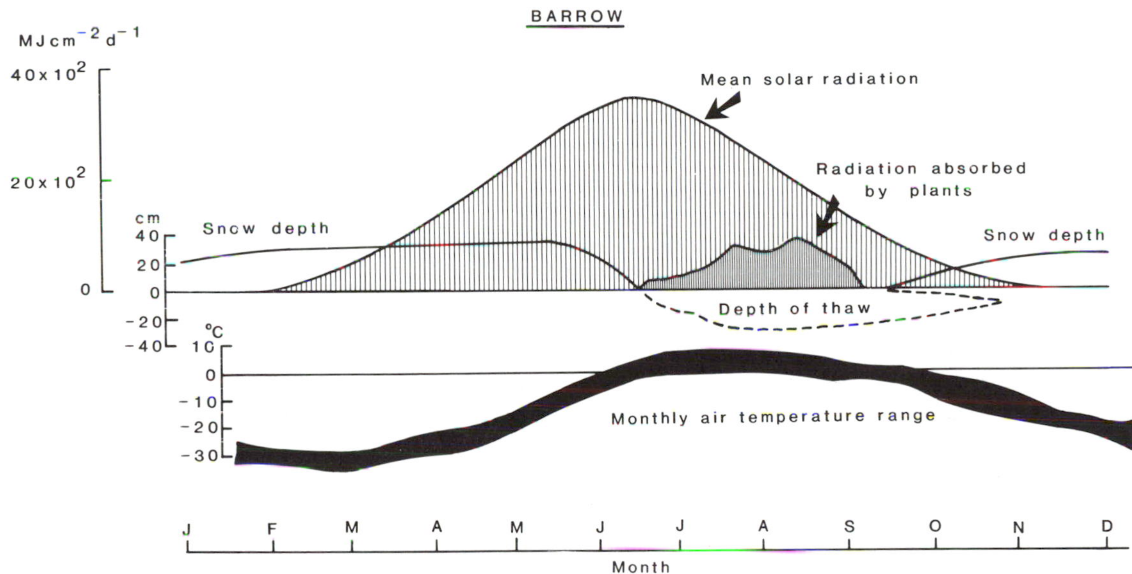

Figure 1.2. Diagram of mean maximum and minimum temperatures, snow depth, active-layer depth, and solar radiation at Barrow (from Chapin and Shaver 1985a) and Truelove Lowland, both in the High Arctic.

ing the arctic–boreal forest boundary over 50% of the time (Bryson 1966). Low-pressure systems in the Gulf of Alaska and between Baffin Island and Greenland bring increased precipitation to the Arctic in summer, when these storm systems move across Alaska into the Canadian Arctic Archipelago and Greenland. Winter penetration of low-pressure systems into the Arctic seldom occurs except in southern Baffin Island and southern Greenland, regions with much higher annual precipitation.

Whether the positioning of the Arctic Front (leading edge of polar air) in summer along the treeline results from changes in surface albedo, evapotranspiration, and surface roughness, as suggested by Hare (1968), or from an assemblage of climatic variables, as suggested by Bryson (1966), is not known. However, both the vegetation limits of

Table 1.1. *Comparison of environmental and biotic characteristics of the Low and High Arctic in North America*

Characteristics	Low Arctic	High Arctic
Environmental		
Length of growing season (mo)	3–4	1.5–2.5
Mean July temperature (°C)	8–12	3–6
Mean July soil temperature (°C) at −10 cm	5–8	2–5
Accumulated degree-days above 0°C	600–1400	150–600
Active-layer depth		
Fine-textured soils (cm)	30–50	30–50
Coarse-textured soils (cm)	100–300	70–150
Botanical/vegetational		
Total plant cover (%)		
Tundra	80–100	80–100
Polar semidesert	20–80	20–80
Polar desert	1–5	1–5
Vascular plant flora (species)	700	350
Bryophytes	Common	Abundant
Lichens	Fruticose and foliose growth forms common	Fruticose growth form minor, crustose and foliose common
Growth-form types	Woody and graminoid common	Graminoid, cushion, and rosette common
Plant height (cm)		
Shrubs	10–500	5–100
Forbs	5–30	2–10
Sedges	10–50	5–20
Shoot:root ratios (alive)		
Shrubs	1 : 1	1 : 1
Forbs	1 : 1–2	1 : 0.5–1
Sedges	1 : 3–5	1 : 2–3

forest and tundra and the air-mass systems are responding to the basic solar radiation control of net radiation (Hare and Ritchie 1972). Annual net radiation averages 1000–1200 MJ m^{-2} yr^{-1} in the northern boreal forest, 750–800 MJ m^{-2} yr^{-1} across Canada, and about 670 MJ m^{-2} yr^{-1} across Alaska at the treeline. In the High Arctic, it averages 200–400 MJ m^{-2} yr^{-1} over much of the land and drops to near zero in northern Ellesmere Island and northern Greenland in the High Arctic. Similar patterns of reduced summer temperature, reduced mean annual temperature, and reduced precipitation occur northward (Table 1.2). These dramatic shifts in summer climate and length of the growing season explain much of the change in vegetation from the Low Arctic to the High Arctic. Winter snow cover averages 20–50 cm over many upland areas, whereas in the Low Arctic the lee slopes and depressions may have snowbanks of 1–5+ m. In the High Arctic, with greatly reduced winter precipitation, many uplands have only 10–20 cm of snow, and some areas even less. Deep snowbanks (1-3+ m) occur in protected slopes and in depressions. With a lower sun angle and lower summer temperatures, many of these snowbanks do not melt until late July or August, whereas in the Low Arctic nearly all the snow is melted by mid-June. The time of snowmelt greatly influences the distribution of species, especially north of latitude 74° N. There is also a strong correlation between shrub height and average snow depth, especially in the Low Arctic. This explains why tall shrubs (2–5 m) are confined to river bottoms, steep banks, and drainages in uplands.

Permafrost

"Permafrost" refers to those areas of soil, rock, and sea floor where the temperature remains below 0°C for 2+ yr. In coarse gravels and frozen rock there is little ice; a dry permafrost predominates. Ice, however, is generally an important component of permafrost, reaching 80–100% by volume in massive ice wedges and ice lenses. Permafrost underlies all of the Arctic and extends into the boreal forest, tapering off as discontinuous permafrost where the

Table 1.2. *Climatic data for select stations in the Low and High Arctic*

Station and latitude	Temperature (0°C)					Precipitation (mm)			
	Mean monthly			Mean annual	Degree-days	Mean monthly			Mean annual
	June	July	August		(>0°C)	June	July	August	
Low Arctic									
Frobisher, Bay, N.W.T., 64°	0.4	3.9	3.6	−12.7	610	38	53	58	425
Baker Lake, N.W.T. 65°	3.2	10.8	9.8	−12.2	1251	16	36	34	213
Tuktoyaktuk, N.W.T., 70°	4.7	10.3	8.8	−10.4	903	13	22	29	130
Kotzebue,AK, 67°	8.2	12.4	8.5	− 4.3	1462	22	44	37	246
Umiat, AK, 69°	9.1	11.7	7.3	−11.7	993	43	24	20	119
Angmagssalik, Greenl., 65°	5.8	7.4	6.6	− 0.4	793	44	35	62	770
Godthaab, Greenl., 64°	5.7	7.6	6.9	− 0.7	809	46	59	69	515
Umanak, Greenl., 71°	4.8	7.8	7.0	− 4.0	682	12	12	12	201
High Arctic									
Barter Island, AK, 70°	1.8	4.6	3.7	−11.6	368	10	18	25	124
Barrow, AK, 71°	1.3	3.8	2.2	−12.1	288	8	22	20	100
Cambridge Bay, N.W.T., 69°	1.5	8.1	7.0	−14.8	579	13	22	26	137
Sacks Harbour, N.W.T., 72°	2.2	5.5	4.4	−13.6	458	8	18	22	102
Resolute, N.W.T., 75°	−0.3	4.3	2.8	−16.4	222	13	26	30	136
Isachsen, N.W.T., 79°	−0.9	3.3	1.1	−19.1	161	8	21	22	102
Eureka, N.W.T., 80°	2.2	5.5	3.6	−19.3	318	4	13	9	58
Alert, N.W.T., 80°	−0.6	3.9	0.9	−18.1	167	13	18	27	156
Scoresbysund, Greenl., 70°	2.4	4.7	3.7	− 6.7	333	26	38	33	428
Nord, Greenl., 81°	−0.4	4.2	1.6	−11.1	202	5	12	19	204

Source: Data from various sources.

mean annual temperature in the atmosphere is about −8°C. The depth of permafrost depends on the mean annual temperature, the thermal conductivities of soil and rock, proximity to the sea, and topographic position. Permafrost is about 400 m thick at Barrow, 500−600 m in the Mackenzie River Delta region, and 400−650+ m throughout the High Arctic.

Permafrost and the dynamic processes of the annual freeze−thaw cycle result in many surface features. Ice-cored hills (pingos) (Fig. 1.3) and small ice mounds (palsas) occur in the Low Arctic, but are

Figure 1.3. Pingo (ice-cored hill) east of the Mackenzie River Delta. These form in old lake basins as permafrost invades formerly unfrozen soils.

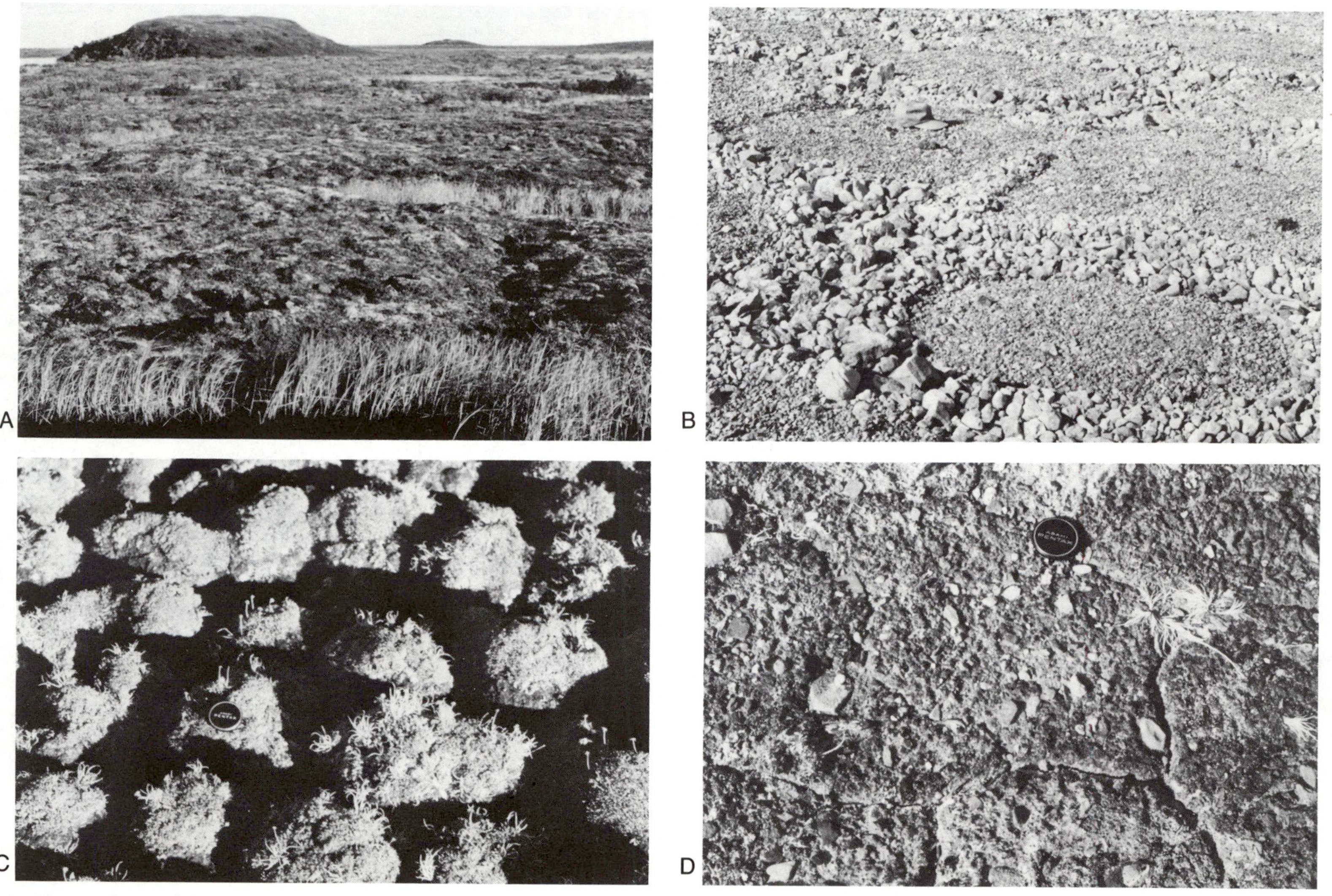

Figure 1.4. *Patterned ground features in the Arctic: (A) raised-center polygonsin the MacKenzie River Delta area with sedges in the troughs and heath shrubs and lichens on top; note the pingo beyond; (B) sorted polygons on Cornwallis Island within a polar desert landscape; (C) soil hummocks with* Dryas integrifolia, *Banks Island; (D) desication cracks, with the development of mosses and "safe sites" for vascular plants to grow, King Christian Island.*

more limited farther north. The most common features, which are not restricted to the Arctic, are sorted and nonsorted circles and stripes, polygons, earth hummocks, and solifluction steps. The circles, stripes, and polygons may be 10–25 cm across each unit, but many are 5–10 m, even 50–100 m in diameter. Raised and depressed-center polygons, sorted and nonsorted circles, soil hummocks, and terraces all influence the distribution of plants and the development of soils (Fig. 1.4A–C). Large soil polygons (>1 m) are common in lowlands, whereas stone nets and stripes are typical of uplands. Polygonal patterns occur on lands that are level or have gentle slopes (1–3°), and elongated polygons and stripes are common on slopes greater than 3–5° (Washburn 1956, 1980).

Not all patterned ground features result from ice formation. In the High Arctic, many small polygonal patterns (10–30 cm) result from desiccation cracks (Fig. 1.4D) that form as the silty to sandy soils dry each summer. Soil hummocks probably result from intense soil churning during the spring and fall freezing cycles. Needle ice is another important feature in finer-textured soils. This forms in many soils each spring and fall causing lifting of surface soils which greatly inhibits seedling establishment in some sites.

Soils

Plant communities and soils are interrelated in the Arctic as they are in all biomes. Because of reduced plant cover, a short growing season, and the short time since ice retreat (12,000 yr) or emergence from the sea, the soils are less well developed than in temperate regions. The presence of permafrost, which limits the vertical movement of water and the churning of some soils (cryoturbation), further restricts soil and plant development. Consequently, chemical decomposition, release of nutrients, and synthesis of minerals from weathering of clay all progress very slowly.

The early concepts that arctic soils form under processes quite different from those in temperate regions have been replaced by the realization that the same soil-forming processes occur, but at greatly reduced rates. The process of podzolization is limited in the Arctic to well-drained soils with a deep active layer. Where dwarf shrub heath species and dwarf bitch predominate, weakly developed podzols (Spodosols) are found. Less well developed soils of uplands and dry ridges are the arctic brown soils (Inceptisols). Cushion plant and heath shrub communities occur in these areas. The most common group of soils in the Low Arctic includes the tundra soil (Inceptisols) of cottongrass–dwarf shrub heath and some sedge communities of imperfectly drained habitats. These soils form under the process of gleization. Poorly drained lowlands where soils remain saturated all summer accumulate peats, both sedge and moss peats. These bog and half-bog soils (Histosols) are dominated by sedge-moss or grass-moss plant communities (Fig. 1.5).

The arctic podzols (Spodosols) and arctic browns (Inceptisols) show some translocation of humus and iron, with iron-enriched B_2 horizons and weakly eluviated A_2 horizons in the podzols. The surface layers tend to be quite acidic (pH 6–4) and low in available nutrients, but quite well drained above the permafrost layer. Inceptisols of imperfectly drained lands (arctic tundra soils) are acidic (pH 6.5–4.5), contain B horizons that have subangular to angular structures, are grayish in color, with iron oxide mottles, and are low in available nutrients. Histosols of poorly drained lands are acidic (pH 6.5–5.0) and are similar to the arctic tundra soils in having limited translocation of minerals into the B horizon.

Within the High Arctic, soil-forming processes are further reduced. The very limited lowlands of sedge-moss or grass-moss communities have thin peats 2–20 cm thick overlying gleyed (arctic tundra) soils. Many slopes, with their cushion plant–cryptogam communities, contain well-drained soils of the Arctic brown group (Inceptisols). Within the barren polar deserts, soils develop under the process of gleization or calcification. These soils are generally basic (pH 7.5–8.5), contain very little organic matter, and are base-saturated, but are very deficient in nitrogen and phosphorus. There are some areas where free carbonates are released from sedimentary rocks or from recently uplifted marine sediments along coastal areas. In these habitats, calcium and magnesium salts effervesce on the soil surface during the brief warm and drier periods in summer. Halophytic species predominate in these sites, as they often do in the extremely depauperate salt marshes in some coastal areas. For a detailed discussion of arctic soils, see Tedrow (1977).

PALEOBOTANY

On a geologic basis, arctic ecosystems are relatively young. In middle Miocene time (≃18–15 million yr BP), coniferous and mixed coniferous-deciduous forests occupied uplands in central Alaska (Wolfe 1969). By Pliocene time (≃6–3 million yr BP), coniferous forests still predominated, along with insects typical today of southern British Columbia and northern Washington (Hopkins et al. 1971). The Beaufort Formation, extending from Banks Island

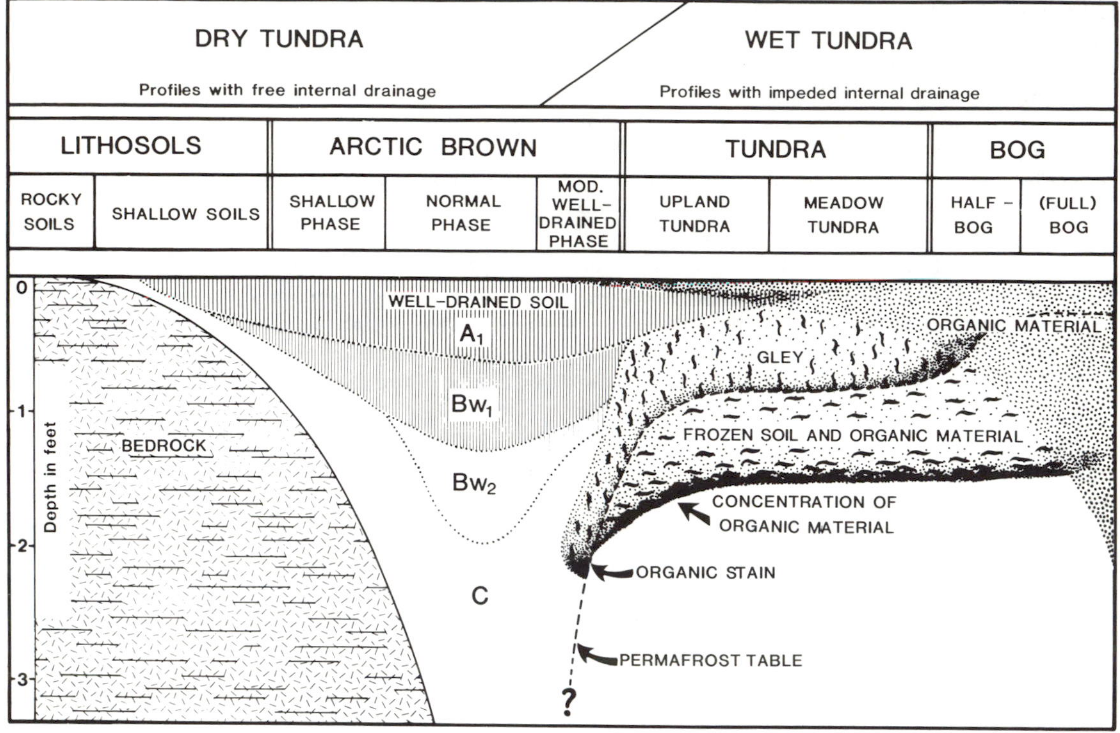

Figure 1.5. Generalized diagram of major soils in the Arctic (modified from Tedrow 1977).

(71° N) to Meighen Island (80° N) and estimated to be of middle Miocene to Pliocene age, contains fossils of mixed coniferous-deciduous forests (Hills et al. 1974). An amazing group of fossil seed plants, mosses, and invertebrates from Meighen Island indicates a forest-tundra vegetation in late Miocene to early Pliocene time on a land surface now polar desert. Tundra probably occurred farther north and at higher elevations on Axel Heiberg and Ellesmere islands and northern Greenland. Much of the arctic flora and fauna is believed to have evolved in the highlands of central Asia (Hoffman and Taber 1967; Yurtsev 1972) and to a more limited degree in the central and northern Rocky Mountains (Billings 1974; Packer 1974). From these centers of origin, plants and animals spread north and across Beringia to enrich the biota on both continents.

A circumpolar flora of perhaps 1500 species developed prior to the onset of Pleistocene glaciations (Löve and Löve 1974). Glacial advances and retreats, with accompanying climatic changes, reduced both the flora and fauna of these northern lands. The result is a truly circumpolar arctic vascular plant flora of 1000–1100 species, with about 700 of these species occurring in the North American Arctic.

The evolution of arctic ecosystems in Beringia (the land connection between Alaska and Siberia during the late Pleistocene) and the northern Yukon Territory is thoroughly discussed in recent books (Hopkins et al. 1982; Ritchie 1983). The pollen record from the Bering Strait region shows that during the Duvanny Yar Interval (25,000–15,000 BP), herbaceous tundra vegetation, similar to that at Barrow today, occupied the region (Colinvaux 1964, 1967). Other studies by Ager (1982) and Matthews (1974) in western Alaska and Ritchie (1977, 1983) and Ritchie and Cwynar (1982) in the Yukon Territory and Northwest Territories (N.W.T.) all indicate that herbaceous tundra was widespread and was dominated by grasses, sedges, and forbs. Ericaceous heath shrubs probably have been underestimated because of low pollen production. Spruce became extinct, and balsam poplar and aspen became more restricted during that period. Willow scrub was confined to river bottoms, as today. Cottongrass–dwarf shrub heath, so abundant across the foothills of Alaska and the Yukon today, was minor or absent at that time (Colinvaux 1967).

Although many fossil remains of large mammals have been found in central Alaska, most have been reworked by rivers, so that accurate carbon dates are difficult to determine. Earlier studies by Guthrie (1972) indicated that an arctic grassland must have predominated to support these mammals. However, more recent studies and summaries of various research indicate little basis for this belief (Ritchie 1983). Ritchie has presented new data from the Bluefish Cave site in the northern Yukon indicating that there were major changes in vegetation from herb tundra (15,000–12,000 BP) to closed and open woodlands at low elevations and to herb tundra in the uplands (9000 BP to the present). The late Pleistocene fauna of woolly mammoth, horse, Dall sheep, caribou, bison, musk-ox, and elk shifted to caribou and moose with these major changes in vegetation. This is one of the few records in the north that conclusively documents simultaneous changes in mammal and vegetation patterns. The reduction and extinction of large mammal species over a 4000-yr period probably resulted from a combination of two factors: climatic changes that induced major changes in vegetation, and overhunting by primitive humans (Martin 1974, 1982). Undoubtedly, more much will be written on the saga of Beringia and the important ecological role the region played in the evolution and modification of tundra ecosystems as we know them today.

VEGETATION

Given the diversity of arctic landscapes (mountains, Precambrian Shield, low-elevation lands) and their great latitudinal extent (55° N along Hudson Bay to 83° N at northern Ellesmere Island and Greenland) and associated climatic changes, it is not surprising that there are diverse patterns of plant communities. Although arctic vegetation has a general structure and pattern of herbaceous species and (often) low scattered shrubs, not all of it has the same appearance. The major types include (1) tall shrub tundra (*Salix, Alnus, Betula* 2–5 m high) along river terraces, stream banks, steep slopes, and lake shores, (2) low shrub tundra (*Salix, Betula* 40–60 cm high) on slopes and uplands beyond the forest tundra, (3) dwarf shrub heath tundra (5–20 cm high) and cottongrass–dwarf shrub heath tundra (tussock tundra) on rolling terrain with soils of intermediate drainage, (4) graminoid-moss tundra (20–40 cm high) on poorly drained soils, and (5) cushion plant–herb–cryptogam polar semidesert (2–5 cm high) on wind-exposed slopes and ridges with limited snow cover. This last vegetation type is more typical of the High Arctic.

There are general reductions in total plant cover and plant height, reductions in woody and vascular plant species, and increases in lichens and mosses from the Low Arctic to the High Arctic. At the final limit of plant growth in the polar desert barrens, the ultimate restriction to plant growth appears to be that the soils become saturated in the spring and bake hard later in the summer. These soils lack nitrogen and have few "safe sites" for vascular plants to become established because of lack of moss and lichen mats that provide sites with more water and nutrients, and the bare soils favor an abundance of needle ice during spring and fall. Where soil moisture is greater and the summers have more clear days, vascular plants grow to within a few meters of glaciers, at elevations of 900–1200 m, on Ellesmere Island.

Forest Tundra

Throughout northern Alaska and Canada there is a relatively narrow zone, 10–50 km wide, of forest tundra. This vegetation consists of scattered clumps of trees in more protected sites where snow accumulates and with low shrub tundra elsewhere. In many places, the transition from forest tundra to shrub tundra consists only in loss of stunted trees. Because of this patterning, Soviet ecologists consider shrub tundra a part of the subarctic rather than the Arctic (Aleksandrova 1980), for they believe that fire eliminated the trees in many places. Repeated fires within these subarctic communities have little effect on the understory vascular plants, for they easily resprout. However, lichens and mosses are slow to return. In the Mackenzie River Delta region, it takes about 200+ yr for an extensive lichen cover to develop on the fine-textured soils (Black and Bliss 1978). On sandy soils to the southeast, where shrubs are minor, a lichen cover developed in 60–120+ yr in the Abitan Lake region (Maikawa and Kershaw 1976). Studies conducted near Inuvik, N.W.T., indicate that there is a period of only 5–8 yr following fire during which seedlings of *Picea mariana* have a chance to become established, because of a short period of seed viability, rapid seed release from surviving cones, and seed destruction by rodents and insects (Black and Bliss 1980). Massive fires within the forest tundra could drive the treeline 50–100 km south should they occur during a cool climatic period, for the seeds germinate at surface temperatures above 15°C. Chapter 2 describes this vegetation and the role of fire in more detail.

Low Arctic Tundra

As stated earlier, most landscapes within the Low Arctic are covered with vascular plants, lichens, and mosses. Their general appearance is that of a

Figure 1.6. Tall shrub tundra dominated by Salix alexensis, *with lesser amounts of* S. glauca *ssp.* desertorum, S. arbusculoides, *and herbaceous plants including* Lupinus arcticus, Hedysarum mackenzii, Deschampsia caespitosa, *and* Agropyron sericeum, *on river gravels along the Colville River, Umiat, Alaska.*

grassland in which low to dwarf shrubs are common, except in the wetter habitats. The vascular plant flora for an area of 100–200 km^2 may have 100–150 species, although any given vegetation type may have only 10–50 species of vascular plants and 20–30 species of mosses and lichens.

Tall shrub tundra. The rivers that flow across the tundra have sandbars and gravel bars, islands, and terraces with well-drained soils that are relatively warm, have a deep active layer (1–1.5 m), and contain higher nutrient levels than adjacent uplands. These habitats and steeper slopes above lakes and rivers that have a deep snow cover in winter (2–5 m) are generally covered with various mixtures of *Salix, Betula,* and *Alnus.* In arctic Alaska, the Yukon, and the northwestern Northwest Territories, *Salix alaxensis* predominates on the coarser-textured alluvium, dune sands, and slopes (Fig. 1.6). Associated shrubs, 1–2 m high, of lesser importance and of varying compositions, site to site, include *S. arbusculoides, S. glauca* ssp. *richardsonii, S. pulchra,* and *Alnus crispa* (Table 1.3) (Bliss and Cantlon 1957; Drew and Shanks 1965; Johnson et al. 1966; Gill 1973; Komarkova and Webber 1980). There is often a rich understory of grasses and forbs in these shrub communities.

Depauperate outliers of the *Salix alaxensis, S. lanata* ssp. *richardsonii,* and *S. pulchra* shrub community, with sedges and forbs, occur along rivers and steep slopes on southern Banks and Victoria islands (Kuc 1974), within the southern High Arctic. *Salix alaxensis* is usually only 0.5–1.5 m tall, and the stands look similar to those in the upper river drainages that are on the north slope of the Brooks Range.

Low shrub tundra. Plant communities dominated by varying combinations of dwarf birch, low willows, heath species, scattered forbs, and graminoids are common in rolling uplands beyond the forest tundra in Alaska and northwestern Canada (Hanson 1953; Corns 1974). It is believed that open forest and forest tundra occupied these lands in northwestern Canada in the past and that fires and changing climate have forced the treeline south (Ritchie and Hare 1971; Ritchie 1977).

The open canopy of shrubs is 40–60 cm high and is dominated by *Betula nana* ssp. *exilis, Salix glauca* ssp. *acutifolia, S. planifolia* ssp. *pulchra,* and *S. lanata* ssp. *richardsonii,* with the combinations of species varying from place to place (Fig. 1.7). The ground cover includes *Carex lugens* and *C. bigelowii* in Alaska (*C. bigelowii* alone in the Northwestern Territories), *Eriophorum vaginatum,* and numerous forbs, grasses, and heath shrubs (10–20 cm high). Varying combinations of *Vaccinium uliginosum, V. vitis-idaea* ssp. *minus, Empetrum nigrum* ssp. *hermaphroditum, Ledum palustre* ssp. *decumbens, Arctostaphylos alpina, A. rubra, Rubus chamaemorus,* and

Table. 1.3. *Prominence values (cover × square root of frequency) for plant communities in the Low Artic at Umiat, Alaska*

Species	Tall shrub	Low shrub	Cottongrass-heath	Cushion plant	Graminoid-moss
	Plant communities				
Eriophorum vaginatum	—	—	45	—	—
Arctagrostis latifolia	29	27	19	—	27
Carex lugens	—	30	27	—	—
Carex glacialis	—	—	—	27	—
Luzula confusa	—	—	19	—	—
Dryas integrifolia	—	—	—	90	—
Ledum palustre ssp. decumbens	9	30	30	—	—
Vaccinium vitis-idaea	9	29	45	—	—
Vaccinium uliginosum	14	23	—	—	—
Arctostaphylos alpina	—	21	27	19	—
Rhododendron lapponicum	—	—	—	9	—
Cassiope tetragona	—	14	29	9	—
Empetrum nigrum ssp. hermaphroditum	—	21	27	—	—
Rubus chamaemorus	—	—	25	—	—
Salix planifolia ssp. pulchra	80	14	9	—	—
Salix glauca ssp. acutifolia	9	29	—	25	—
Betula nana. ssp. exilis	9	30	30	—	—
Alnus crispa	63	23	—	—	—
Pedicularis lanata	—	—	—	25	—
Lupinus arcticus	—	—	—	29	—
Sussurea angustifolia	—	14	—	—	—
Saxifraga punctata	—	14	19	—	—
Saxifraga cernua	9	—	—	9	29
Polemonium acutiflorum	—	14	—	—	—
Petasites frigidus	9	—	19	—	—
Polygonum bistorta	—	23	16	—	—
Pyrola grandiflora	21	25	—	—	—
Stellaria laxmannii	—	—	—	—	27
Caltha palustris	—	—	—	—	23
Carex aquatilis	—	—	—	—	375
Eriophorum angustifolium	—	—	—	—	21
Cardamine pratensis	—	—	—	—	16
Hedysarum alpinum	—	—	—	19	29
Mosses	29	190	625	30	90
Lichens	—	90	375	25	—
Total vascular species	13	20	17	14	9

Note: Sampling area for each stand was 10m^2.
Source: Data from Churchill (1955).

dwarf species of *Salix* occur along with an abundant ground cover of lichens and mosses (Table 1.3). Where snow lies late into June, the heaths are often dominated by *Cassiope tetragona*. The most common mosses include *Aulacomnium turgidum, Hylocomium splendens*, and *Polytrichum juniperinum*. Important fruticose lichens are *Cetraria nivalis, C. cucullata, Cladonia gracilis, Cladina mitis, C. rangiferina*, and *Thamnolia vermicularis*.

In much of the northeastern mainland of Canada, low shrub tundra is minor, probably because of limited winter snow and abrasive winter winds (Savile 1972). In the Chesterfield Inlet of Hudson Bay, northern Quebec, and southern Baffin Island, the dominant shrubs are *Betula glandulosa, Salix glauca* ssp. *callicarpaea*, and the aforementioned heath species, graminoids, and cryptogams (Polunin 1948).

In Greenland, low shrub tundras occur in the inner fjord regions that are warmer in summer. These shrub communities, dominated by *Salix glauca* ssp. *callicarpaea, Betula nana*, and *B. glandulosa*, predominate on steep slopes and along streams and rivers from Disko Island (69° N) and Sønder Strom-

Figure 1.7. Low shrub tundra dominated by Salix planifolia *ssp.* pulchra, Salix glauca *ssp.* acutifolia, Betula nana *ssp.* exilis, Vaccinium uliginosum, V. vitis-idaea, Empetrum nigrum *ssp.* hermaphroditum, Ledum palustre *ssp.* decumbens, *and* Carex lugens *in the Brooks Range.*

fjord (67° N) southward. The shrubs (>1 m in height) are associated with herbs and dwarf heath shrubs (Böcher 1954, 1959; Hansen 1969).

Dwarf shrub heath tundra. As used here, "heath" refers to species within the Ericaceae, Empetraceae, and Diapensiaceae. Some authors have broadened the concept to include *Dryas* and dwarf species of *Salix*. A characteristic feature of these heath plants is the evergreen leaf, although deciduous-leaved species also occur (*Vaccinium uliginosum, Arctostaphylos alpina, A. rubra*). Heath-dominated communities occur on well-drained soils of river terraces, slopes, and uplands where winter snows are at least 20–30 cm deep. Heath tundra occupies relatively small areas, a few hundred meters square, rather than the many hectares or square kilometers of other tundra vegetation types.

In western Alaska (Hanson 1953; Churchill 1955; Johnson et al. 1966), northern Yukon (Hettinger et al. 1973), and the Mackenzie River Delta region (Corns 1974), communities of heath species (10–20 cm high) are quite common. The species occur in various combinations, including *Ledum palustre* ssp. *decumbens, Vaccinium uliginosum, V. vitis-idaea, Empetrum nigrum* ssp. *hermaphroditum, Loiseleuria procumbens, Rhododendron lapponicum, R. kamschaticum,* and *Cassiope tetragona. Betula nana* ssp. *exilis* and one or more species of dwarf *Salix* are commonly associated, along with abundant lichens and mosses.

In the Keewatin District, N.W.T., west of Hudson Bay, dwarf shrub heath predominates on north exposures, fellfields, and gravel summits. Here, the heaths are less rich floristically, with *Ledum palustre* ssp. *decumbens, V. uliginosum, V. vitis-idaea, Empetrum nigrum* ssp. *hermaphroditum,* and *Cassiope tetragona* dominating (Larsen 1965, 1972). Where snow is deep, *Betula glandulosa* and highly deformed *Picea mariana* (60–90 cm tall) occur near the treeline at Ennadai.

At Chesterfield Inlet along the west coast of Hudson Bay, at Wakeham Bay in northern Quebec, and at Lake Harbour, Baffin Island, heath vegetation is common. The dominant species include *V. uliginosum, Ledum palustre* ssp. *decumbens, Empetrum nigrum* ssp. *hermaphroditum, Phyllodoce caerulea,* and the sedge *Carex bigelowii*. Where snow lies into July, *Cassiope tetragona* dominates, with *Salix reticulata, S. herbacea, Dryas integrifolia, Carex misandra, Luzula nivalis,* and numerous lichens, and mosses (Polunin 1948).

In southeast and west Greenland, heath vegetation is found on steep, moist slopes that are warmer. Many of these habitats are in the inland valleys, which have a warmer and drier continental climate, as compared with the moist maritime climate along the coast and outer fjords (Böcher 1954). Species dominating in various combinations include *Cassiope tetragona, Vaccinium uliginosum* ssp. *microphyllum, V. vitis-idaea* ssp. *minus, Ledum palustre* ssp. *decumbens, Rhododendron lapponicum, Phyllodocae cae-*

Figure 1.8. Cottongrass–dwarf shrub heath tundra in the Caribou Hills, northwest of Inuvik, N.W.T.

rulea, Empetrum nigrum ssp. *hermaphroditum,* and *Loiseleuria procumbens* (Sørensen 1943; Böcher 1954, 1959, 1963; Böcher and Laegaard 1962; Hansen 1969; Daniels 1982). *Betula nana* is a component of the more southern heaths. *Ledum, Phyllodocae,* and *Loiseleuria* drop out in the Melville Bugt region (72–75° n). At Thule (78° N), the dominant species of the limited heaths include *Cassiope tetragona, Vaccinium uliginosum* ssp. *microphyllum,* and *Salix arctica* (Sørensen 1943), as they do in Ellesmere in the High Arctic.

Cottongrass-dwarf shrub heath tundra. Large areas of rolling uplands between the mountains and the wet coastal plain in Alaska and the Yukon Territory are dominated by tussocks of *Eriophorum vaginatum,* dwarf shrubs, lichens, and mosses (Hanson 1953; Churchill 1955; Britton 1957; Johnson et al. 1966; Hettinger et al. 1973; Wein and Bliss 1974; Komarkova and Webber 1980). Much more limited areas of this vegetation type occur in the Northwest Territories, except in the Mackenzie River Delta region (Corns 1974), where it is common (Fig. 1.8). This type of tundra is absent from most of the eastern Canadian Arctic, with the exception of southern Baffin Island and northern Quebec (Polunin 1948).

Historically, these landscapes have been called cottongrass-tussock tundra or tussock-heath tundra because of the conspicuousness of the cottongrass tussocks. However, the phytomass and the net annual production of heath and low shrub species often are greater than for cottongrass, and plant cover is also often greater for the shrubs. The Alaskan and western Canadian cottongrass is *E. vaginatum* ssp. *vaginatum,* whereas the subspecies *spissum* occurs in the eastern Arctic.

The predominant heath species are *V. vitis-idaea* ssp. *minus, V. uliginosum* ssp. *alpinum, Ledum palustre* ssp. *decumbens,* and *Empetrum nigrum* ssp. *hermaphroditum.* Scattered low shrubs of *Betula nana* ssp. *exilis* and *Salix pulchra* are common, along with *Carex bigelowii, C. lugens,* and several species of forbs (Table 1.3). The cryptogam layer is well developed and includes the mosses *Dicranum elongatum, Aulacomnium turgidum, A. palustre, Rhacomitrium lanuginosum, Hylocomium splendens, Tomenthypnum nitens,* and several species of *Sphagnum.* The most common lichens include *Cetraria cucullata, C. nivalis, Cladina rangiferina, C. mitis, Cladonia sylvatica, Dactylina arctica,* and *Thamnolia vermicularis* (Bliss 1956; Johnson et al. 1966).

Graminoid-moss tundra. The concept of arctic tundra is often associated with treeless wetlands in which species of *Carex, Eriophorum,* and sometimes the grasses *Arctagrostis, Dupontia, Alopecurus,* and *Arctophila* predominate, along with an abundance of bryophytes, but few lichens. In northern and western Alaska, arctic wetland meadows are minor in the mountain valleys, increase in importance in the Foothill Province, and dominate the Coastal

Plain Province (Churchill 1955; Britton 1957; Webber 1978; Komarkova and Webber 1980). Arctic wetlands are again minor in the Yukon mountains, but increase in importance on the coastal plain and eastward to the Mackenzie River region (Hettinger et al. 1973; Corns 1974).

The dominant sedges are *Carex aquatilis, C. rariflora, C. rotundata, C. membranaceae, Eriophorum angustifolium, E. scheuchzeri*, and *E. russeolum* (Table 1.3). The grasses *Arctagrostis latifolia, Dupontia fisheri, Alopecurus alpinus*, and *Arctophila fulva* occur along a gradient from drier sites to saturated soils and standing water. In shallow waters of lakes and ponds (50–60 cm deep), *Menyanthes trifoliata, Equisetum variegatum*, and *Arctophila fulva* predominate, with *Potentilla palustris* and *Hippuris vulgaris* in water 20–30 cm deep. The various species of *Carex* and *Eriophorum* and *Dupontia fisheri* occur with little (10–20 cm deep) or no standing water. *Carex aquatilis, Eriophorum angustifolium*, and *E. scheuchzeri* are common in low-center polygons and in the troughs of high-center polygons (Britton 1957; Hettinger et al. 1973; Corns 1974). The nearly continuous cover of mosses includes species of *Aulacomnium, Ditrichum, Calliergon, Drepanocladus, Sphagnum, Hylocomium splendens*, and *Tomenthypnum nitens* (Britton 1957; Johnson et al. 1966).

Studies of plant succession in the thaw-lake cycle of northern Alaska have shown that high-center polygons form in drained lake basins from the meeting of ice wedges and post-drainage thermokarst erosion. Pioneer plants include the moss *Psilopilum cavifolium* on dry peaty sites and the graminoids *Arctophila fulva* in wet sites and *Eriophorum scheuchzeri* and *Dupontia fisheri* in moist swales. *Eriophorum angustifolium* and *Carex aquatilis* enter somewhat later, but in time they predominate (Britton 1957; Billings and Peterson 1980).

Sedge-dominated wet meadows are found in the Ennadai area (Larsen 1965), Chesterfield Inlet, northern Quebec, and Lake Harbour (Polunin 1948). Common species include *Eriophorum angustifolium, E. scheuchzeri, Carex rariflora, C. membranaceae*, and *C. stans* (Polunin 1948).

In Greenland, graminoid-moss tundra is present, but as with large areas of eastern Canada, the marshes are less common. Important species include *Carex rariflora, C. vaginata, C. holostoma, Eriophorum angustifolium*, and *E. triste* (Böcher 1954, 1959).

Other graminoid vegetation. With thousands of kilometers of arctic shoreline, one might assume that salt marshes, coastal dune complexes, and other coastal vegetation would be common. Such is not the case, for favorable habitats are limited. Factors that limit arctic salt marshes are the limited areas of fine sands and silts, the annual reworking of shorelines by sea ice, a limited tidal amplitude, low salinity of coastal waters, a very short growing season, and low soil temperatures. Coastal salt marshes have been described from northern Alaska (Jeffries 1977; Taylor 1981), Tuktoyaktuk (Jeffries 1977) and Hudson Bay, N.W.T. (Kershaw 1976), the west central part of southern Greenland (Sørensen 1943; Vestergaard 1978), and Devon, Ellesmere, and Baffin islands (Polunin 1948; Jeffries 1977).

In many of these marshes the plant cover is only 15–25%, with plant heights of 1–5 cm. Floristically, most salt marshes have only three to five herbaceous species, none of which are woody or belong to the Chenopodiaceae, a family of plants so common in temperate and tropical latitudes. The most common species are *Puccinellia phryganodes* on mud, with scattered clumps of *Carex ursina, C. ramenskii, C. subspathacea, Stellaria humifusa*, and *Cochlearia officinalis*. In many ways it is quite amazing that there are any species at all adapted to living in these harsh coastal environments, in contrast with the extensive and highly productive salt marshes of temperate regions.

Grasslands are also minor features in the Arctic. In coastal sandy areas, *Elymus arenarius* ssp. *mollis* often dominates, with small amounts of the semisucculent *Honckenya peploides, Mertensia maritima*, and *Cochlearia officinalis* ssp. *groenlandica*, along with *Festuca rubra* and *Matricaria ambigua*.

Small areas of grassland are found in the inner fjords of east central Greenland, dominated by *Calamagrostis purpurascens, Arctagrostis latifolia, Poa arctica*, and *P. glauca* (Seidenfaden and Sørensen 1937; Oosting 1948). In west Greenland, larger areas (1–5 ha) of grassland occur. These are dominated by *Calamagrostis neglecta, Poa pratensis*, and *P. arctica* (Böcher 1959). Similar small grasslands are found in the eastern Canadian Arctic of Baffin Island (Polunin 1948) and northern Keewatin (Larsen 1972).

Low Arctic Semidesert: Cushion Plant–Cryptogam

From the Rocky Mountains of Montana north to the Yukon Territory and Alaska, wind-swept slopes and ridges are often dominated by cushions or mats of *Dryas integrifolia* and *D. octopetala*. This vegetation type, often called *Dryas* fellfield or *Dryas* tundra, is limited in areal extent within the western Low Arctic but increases in importance in the large areas of barren rock and lag gravel surfaces in the eastern Arctic. In the Mackenzie District and Keewatin District, N.W.T., where acidic soils derived from granites predominate, *Dryas* is less prominent. Other cushion plants include *Silene acaulis*, *Saxifraga oppositifolia*, and *S. tricuspidata*. The lichens *Alectoria nitidula*, *A. ochroleuca*, *Cetraria cucullata*, and *C. nivalis*, along with the moss *Rhacomitrium lanuginosum*, are common in the Repulse Bay, Peely Lake, and Snow Bunting Lake areas (Larsen 1971, 1972).

Within the ecotone between forest and tundra in the Campbell-Dolomite uplands near Inuvik, N.W.T., mats of *Dryas integrifolia* and *Cladonia alpestris* cover the limestone and dolomite rocks. Where small pockets of soil occur, open stands of *Picea glauca* predominate (Ritchie 1977). In the Brooks Range (Spetzman 1959) and northward along exposed ridges within the Foothill Province, *Dryas*-lichen vegetation is common. *Dryas integrifolia* or *D. octopetala* and their hybrids often comprise 80–90% of the vascular plant cover. Associated species include *Silene acaulis*, *Carex rupestris*, *C. capillaris*, *Kobresia myosuroides*, *Anemone parviflora*, and *Polygonum viviparum*. Cushions of *Dryas* are often 0.5–1.0 m across, and they seldom reach a height greater than 2–5 cm. The ground cover of herbs, lichens, and mosses is generally sparse. As a result, there is often much bare rock and soil, although crustose lichens are often common.

High Arctic Tundra

In contrast with the Low Arctic, the High Arctic is characterized by herbaceous rather than woody species, and there is generally much less plant cover, especially of vascular plants. Lichens and mosses contribute a much larger percentage of total cover and biomass than in the Low Arctic. Vast areas are dominated by lichens and mosses, with only 5–25% cover of flowering plants. Equally large areas of polar deserts have almost no cryptogams (0–3% cover) and only 0.1–4% cover of vascular plants (Table 1.1). Vascular plants are much smaller in size (3–10 cm) than those of the same species (10–30 cm) in the Low Arctic, and their root systems are also smaller.

Based on vegetation types, the High Arctic can be divided into small areas of tundra (tall shrub, dwarf shrub heath, cottongrass tussock, and graminoid-moss), vast areas of cushion plant–cryptogam and cryptogam-herb polar semidesert, and the herb barrens and very limited snowflush herb–moss vegetation of the polar deserts. With the exception of a few locations, woody species other than dwarf willows and semiwoody *Dryas* and *Cassiope* are very minor components.

Graminoid-moss tundra. Of the common vegetation types in the Low Arctic, only the graminoid-moss tundra is ecologically important farther north. It occupies 5–40% of the lands on the southern islands, with the exception of Baffin Island, which is mountainous. In the Queen Elizabeth Islands, less than 2% of the area contains this vegetation; yet it provides the major grazing habitat for musk-ox and breeding grounds for waterfowl and shore birds.

In the eastern and southern islands, *Carex stans* and *C. membranaceae* are the dominants, with lesser amounts of *Eriophorum scheuchzeri*, *E. triste*, *Dupontia fisheri*, and *Alopecurus alpinus*. Small clumps of *Salix arctica* and *Dryas integrifolia* are restricted to moss hummocks or rocky areas within the wet meadows–the best-drained and best-aerated microsites. Plant communities of this vegetation type occur where drainage is impeded along river terraces, small valleys, and coastal lowlands (Beschel 1970; Bird 1975; Muc 1977; Thompson 1980; Sheard and Geale 1983; Freedman et al. 1983; Bliss and Svoboda 1984). In the northwestern islands, including northern Melville Island, the grass *Dupontia fisheri* dominates (Fig. 1.9), with lesser amounts of *Juncus biglumis* and *Eriophorum triste* in wetlands. Where there are shallow ponds, *Pleuropogon sabinei* occurs–the ecological equivalent of *Arctophila fulva* to the south (Savile 1961; Bliss and Svoboda 1984).

Bryophytes are abundant in the various wetland plant communities, including *Orthothecium chryseum*, *Campylium arcticum*, *Tomenthypnum nitens*, *Drepanocladus revolvens*, *Ditrichum flexicaule*, and *Cinclidium arcticum*. Blue-green algae are abundant, including species of *Nostoc* and *Oscillatoria*. Lichens are minor in these wetlands.

Dwarf shrub heath tundra. Compared with the Low Arctic, these heaths have few species, and they are almost always in snowbed sites that melt by early July. The heaths of central and northern Baffin Island are richer floristically than elsewhere. *Cassiope tetragona* is the dominant and characteristic species, although *Vaccinium uliginosum* ssp. *microphyllum*, *Salix herbacea*, *S. arctica*, *Carex bigelowii*, *Luzula nivalis*, and *L. confusa* are common. Important lichens include *Cladina mitis*, *Dactylina arctica*, *Stereocaulon turgidum*, *Drepanocladus uncinatus*, *Hylocomium splendens*, and *Rhacomitrium lanuginosum* (Polunin 1948).

Heath vegetation is sparse in west Greenland, in terms of species and areal extent. *Cassiope* domi-

Figure 1.9. Graminoid-moss tundra in the High Arctic dominated by Dupontia fisheri *on Melville Island.*

Figure 1.10. Dwarf shrub heath tundra dominated by Cassiope tetragona, *with lesser amounts of* Dryas integrifolia, Vaccinium uliginosum *ssp.* microphyllum, *and* Luzula parviflora. *Note the wet sedge-moss meadows beyond. Truelove Lowland, Devon Island.*

nates, with lesser amounts of *Salix arctica, Dryas integrifolia,* and *Luzula confusa* (Sørensen 1943). Small areas of heath with *Cassiope, Dryas,* and *Salix* occur in northern Greenland at 81–83° N (Holmen 1957).

Farther north and west in the Queen Elizabeth Islands, the depauperate heaths are dominated by *Cassiope, Dryas,* and *Luzula*. Only in the warmer eastern High Arctic is *Vaccinium uliginosum* ssp. *microphyllum* present (Fig. 1.10) (Beschel 1970; Brassard and Longton 1970; Bliss et al. 1977; Reznicek and Svoboda 1982). Lichens and mosses are also common in these heaths.

Other tundra vegetation types. Cottongrass tussock and tall shrub tundra occupy small areas on Banks and Victoria islands, the northern limits for these common vegetation types of the Low Arctic. Tussock tundra is dominated by *Eriophorum vaginatum* ssp. *vaginatum*, with scattered plants of *Vaccinium vitis-idaea* ssp. *minus*, the only heath species, and a few other vascular species and cryptogams. Along some of the rivers and near the lakes of these same southern islands there are small willow thickets (0.5–1.5 m), with *Salix alaxensis* and *S. pulchra* (Kuc 1974).

High Arctic Semidesert

The arctic vegetation and plant communities included within this landscape unit have generally been considered polar desert (Aleksandrova 1980; Andreyev and Aleksandrova 1981). However, in terms of plant cover, plant and animal biomass, species richness, and soil development, the High Arctic semidesert is so different from barren polar deserts that the two have been separated for North America (Bliss 1975, 1981). Areas of polar semidesert cover about 50–55% of the southern islands and about 25% of the northern islands. The two major vegetation types include cushion plant–cryptogam and cryptogam-herb.

Cushion plant–cryptogam vegetation. Large areas of the southern islands and the Boothia and Melville peninsulas are covered with large mats of *Dryas integrifolia*. Associated species include *Salix arctica*, *Saxifraga oppositifolia*, *S. caespitosa*, *S. cernua*, *Draba corymbosa*, *Papaver radicatum*, and two to three species each of *Minuartia* and *Stellaria*. Graminoids are always present as scattered plants, including *Carex rupestris*, *C. nardina*, *Luzula confusa*, and *Alopecurus alpinus* (Table 1.4). Lichens and mosses provide 30–60% of the total plant cover; vascular plants provide 5–25% of the cover (Fig. 1.11).

This vegetation is common on rolling uplands and gravelly raised beaches that are drier and warmer than surrounding lands in summer. Plant communities of this type have been described from Devon Island (Svoboda 1977), Ellesmere Island (Brassard and Longton 1970; Freedman et al. 1983), Bathurst Island (Sheard and Geale 1983; Bliss et al. 1984), Cornwallis and Somerset islands (Bliss et al. 1984), and South Hampton Island (Reznicek

Table 1.4. *Prominence values (cover × square root of frequency) for plant communities in the High Artic*

	Plant communities				
Species	Cushion plant	Cryptogam-herb	Graminoid-steppe	Snowflush	Herb barrens
Dryas integrifolia	28	—	—	—	—
Saxifraga oppositifolia	8	—	—	4	—
Saxifraga caespitosa	—	11	—	1	—
Saxifraga hieracifolia	—	9	—	—	—
Saxifraga cernua	—	10	—	1	—
Saxifraga flagellaris	—	17	—	9	—
Phippsia algida	—	—	—	—	—
Papaver radicatum	—	14	—	6	2
Cerastium alpinum	—	4	—	—	—
Oxyria digyna	—	7	—	—	—
Ranunculus sulphureus	—	4	—	—	—
Minuartia rubella	—	3	—	—	—
Stellaria crassipes	—	6	—	—	—
Draba corymbosa	—	2	—	5	9
Draba subcapitata	—	1	—	—	—
Festuca brachyphylla	—	14	—	—	—
Alopecurus alpinus	—	37	20	—	—
Puccinellia angustata	—	—	—	—	6
Luzula confusa	—	6	81	—	—
Luzula nivalis	—	5	2	—	—
Mosses	—	116	76	66	—
Lichens	22	41	24	14	—
Total vascular species	6	19	4	14	5

Note: Sampling area for each stand was 40 m^2.
Source: Data from Bliss and Svoboda (1984) and Bliss et al. (1984).

Figure 1.11. Cushion plant–cryptogam vegetation of the polar semideserts. Dryas *dominated community with lesser amounts of* Saxifraga oppositifolia, *and* Salix arctica *with* Cassiope tetragona *in the small depressions. Thompson River, Banks Island.*

and Svoboda 1982). These landscapes are the major habitats for Peary's caribou, collared lemming, ptarmigan, and several species of passerine birds.

Cryptogam-herb vegetation. This vegetation type is common on the western Queen Elizabeth Islands, where the previous vegetation type is minor or totally lacking. Cryptogam-herb vegetation covers low rolling uplands with sandy to salty and clay loam soils. Vascular plants contribute 5–20% cover, and cryptogams 50–80%.

Alopecurus alpinus, Luzula confusa, and *L. nivalis* are the common graminoids, along with three to five species of *Saxifraga, Papaver radicatum, Draba corymbosa, Cerastium alpinum,* and *Juncus albescens* (Table 1.4). Abundant bryophytes include the mosses *Rhacomitrium lanuginosum, R. sudeticum, Aulacomnium turgidum, Polytrichum juniperinum, Pogonatum alpinum, Ditrichum flexicaule, Dicranoweisia crispa, Tomenthypnum nitens,* and *Schistidium holmenianum* and, in wetter soils, the liverwort *Gymnomitrion corallioides.* Common lichens include crustose *Lecanora epibryon, Lepraria neglecta,* and *Dermatocarpon hepaticum* and fruticose *Cladonia delisei, C. gracilis, Parmelia omphalodes, Cetraria cucullata, C. nivalis,* and *Dactylina ramulosa.* There is often a black crust of lichens and blue-green algae called patena on soils (Bliss and Svoboda 1984). Seedlings of vascular plants germinate at random on a variety of microsite surfaces, but their survival to adult plants is strongly favored by moss mats or desiccation cracks where mosses are also important (Sohlberg and Bliss 1984). Plant communities of this type have been described from Axel Heiberg Island (Beschel 1970), Melville, Cameron, King Christian, and Ellef Ringnes islands (Bliss and Svoboda 1984), Laugheed Island (Edlund 1980), and Prince Patrick Island (Bird 1975).

Graminoid steppe. Large areas of fine sand to clay loam soils are dominated by *Luzula confusa* and *Alopecurus alpinus* on Ellef Ringnes and King Christian islands (Fig. 1.12). Other areas on Melville, Laugheed, King Christian, and Ellef Ringnes islands are dominated by *A. alpinus* on black soils derived from shales of Lower Cretaceous age. There are very few herbs or cryptogams associated with *Alopecurus* (Bliss and Svoboda 1984). This great reduction in species richness may result from soil churning, which is so common in these dark soils. *Alopercurus* is rhizomatous and can tolerate soil movement.

High Arctic Polar Desert

Herb barrens. Landscapes that are almost totally devoid of plants occupy thousands of square kilometers, especially in the Queen Elizabeth Islands (Fig. 1.13A). These landscapes occur from near sea level (10–20 m) to upland plateaus at over 200–300 m. The basic control appears to be geologic substrate, as this influences soil development and the

Figure 1.12. Luzula confusa *and* Alopecurus alpinus *dominate this graminoid steppe on Ellef Ringnes Island.*

ability of soils to remain moist throughout much of the short growing season (1–1.5 mo). Soil texture ranges from medium-grained sands to clay loams. In many areas there is a thick veneer of rocks that covers much of the surface, reducing the area of "safe sites" for plant establishment. In other areas there is at least 30–50% open soil.

There are no woody or semiwoody species in these barren landscapes, only rosette (*Draba*), cushion (*Saxifraga*), and mat-forming (*Puccinellia*) species. In a study of 23 barren sites on six islands, a total of 17 vascular species and 14 species of cryptogams were found, with a mean species richness of 9 per site (60 m^2). Vascular plant cover averaged 1.8%, and cryptogams 0.7%; the remaining 98% was bare soil, frost-shattered rocks, and pebbles (Fig. 1.13B). Most vascular plants were found in desiccation cracks within the polygonal patterns.

A

Figure 1.13. A: Aerial view of a polar desert landscape on Prince Patrick Island. The darker areas are snowflush communities with an abundance of mosses and a few vascular plants. B (p.20): Polar desert barrens on Ellef Ringnes Island, with 1–2% cover of Papaver radicatum *and* Puccinella angustata.

Figure 1.13. (cont.) Polar desert-barrens.

Cryptogams were seldom present in close association with vascular plants. This is in contrast to polar semidesert and snowflush habitats within this barren landscape, where moss mats are preferred sites for vascular plants. Of the vascular plants recorded, *Draba corymbosa, D. subcapitata, Papaver radicatum, Minuartia rubella, Saxifraga oppositifolia,* and *Puccinellia angustata* were most commonly present (Table 1.4). The most important cryptogams were *Hypnum bambergeri, Tortula ruralis, Thamnolia subuliformis, Dermatocarpon hepaticum,* and *Lecanora epibryon* (Bliss et al. 1984). The lack of lichens on rocks and the very depauperate flora and plant cover suggest that these lands, especially uplands, may have become vegetated only since the Little Ice Age (130–430 yr BP) (Svoboda 1982). This hypothesis is more important for high-elevation areas of polar barrens.

Snowflush communities. Within the polar barrens, the only habitats that have significant plant cover are those below large snowbanks. The meltwaters, present all summer, enable the development of moss mats in which flowering plants become established. Although snowflush communities occupy only 3–5% of the landscape, they are very conspicuous features when present. Species richness was found to be much greater than in barrens (30 species of vascular and 27 species of cryptogams) at 12 sites (60 m^2) on three islands (Bliss et al. 1984). Plant cover was also much greater in these habitats (vascular plants 9.5%, bryophytes 18.5%, lichens 8.2%) than in the herb barrens (Table 1.4).

Plant composition can be quite variable, with graminoid-dominated meadows in some sites (*Eriophorum triste, Alopecurus alpinus*) and herbs (*Saxifraga oppositifolia, S. cernua, Papaver radicatum, Draba corymbosa, Minuartia rubella,* and *Stellaria longipes*) in others. The polygonal troughs and stripes contain *Orthothecium chryseum, Ditrichum flexicaule,* and *Drepanocladus revolvens,* and nearly all vascular plants are restricted to these moss mats (Fig. 1.14). These microsites also support blue-green algae and thus the ability to fix nitrogen. Few species other than *Phippsia algida* and *Alopecurus alpinus* occur in the large areas of bare soil.

PLANT COMMUNITY DYNAMICS

Succession

Plant succession is seldom discussed in the arctic literature, for it is not a conspicuous feature. The number of pioneer species (r-selection or ruderal stress tolerators) is quite small, and until recently there were few disturbed soil areas of sufficient size to warrant study. Following disturbance, an increase in plant cover involves relative shifts in the covering by certain species rather than total replacement of species as in temperate regions. *Arctagrostis latifolia* often provides only 1–2% cover in undisturbed low shrub tundra, but with surface disturbance, its cover increases to 5–10%, even in 5 to 8 yr.

Pioneer species are often graminoids, with some species reproducing by seed (*Arctagrostis latifolia,*

Figure 1.14. Snowflush habitats within the polar desert barrens. Snowflush site dominated by mosses, with a few scattered vascular plants of Alopecurus alpinus, Luzula nivalis, *and* Draba alpina *on Cornwallis Island.*

Calamagrostis canadensis, Eriophorum vaginatum, Poa arctica) and others by rhizomes (*Carex aquatilis, C. stans, Arctophila fulva*). These same species are part of the stable, climax vegetation.

In the Low Arctic of the Mackenzie River Delta region and the North Slope of Alaska, where petroleum exploration began in the 1960s, human-induced surface soil disturbance and succession are most evident. Six years following exposure of peats and mineral soils, plant cover was usually 30–50% in low shrub–heath and cottongrass-heath tundra. In wet sedge–moss tundra, *Carex aquatilis* and *C. rariflora* provided 95% cover within 1 yr because of rhizomatous growth (Hernandez 1973). At Eagle Creek, Alaska, *Eriophorum vaginatum* provided nearly 100% cover from seedlings 10 yr after disturbance (Shaver et al. 1983). Shrubs of heath, birch, and willow are slow to establish unless rootstalks remain.

In the High Arctic, pioneer species are almost lacking. *Phippsia algida* and *Arctagrostis latifolia*, reproducing by seed, and *Alopecurus alpinus* and *Carex stans*, reproducing by rhizomes, perform this role in limited areas. The time required for recovery of vascular plant cover is probably 50–100+ yr, but for recovery of cryptogams, probably 150+ yr. In the Low Arctic, recovery of vascular plants generally requires less than 25 yr, unless areas greater than 1 km^2 are denuded.

Natural succession in the Arctic is best exemplified by plant development on Alaskan river gravel and sandbars (Fig. 1.15). Plant communities dominated by herbs (grasses, legumes, composites) commonly pioneer on river alluvium. With increases in deposition of sands and silts, *Salix alaxensis* becomes established, often the result of masses of willows stranded with receding floodwaters in spring. There is a well-developed understory of grasses and forbs where the active soil layer is 1–2 m deep in July. With time, *Salix arbusculoides, S. glauca* ssp. *desertorum, S. niphoclada* var. *ferrae, S. pulchra, S. lanata* ssp. *richardsonii*, and *Alnus crispa* become established and provide greater shade and ground cover. Here the active layer is 0.5–0.7 m deep in late July, and a thin organic layer develops on the surface. On the first terrace above the river, *S. alaxensis* and several other willows drop out, and *Alnus* becomes dominant. Farther back, where the active layer is shallower (0.25–0.4 m in late August) and soils are colder and wetter, cottongrass-heath tundra and wet sedge tundra dominate (Bliss and Cantlon 1957).

Successional patterns have also been described for sand dunes along the Meade River 100 km south of Barrow. Here, *Elymus arenarius* and *Salix alaxensis* are important pioneer species, followed by *Dryas integrifolia* and *Arctostaphylos rubra*. With time, lichen-heath tundra develops on dry sites. In poorly drained sites, *Dupontia fisheri, Eriophorum scheuchzeri, E. angustifolium*, and *Carex aquatilis* dominate, with a shift to the latter species as organic matter accumulates (Peterson and Billings 1978, 1980).

A cyclic pattern of succession occurs with the development and eventual drainage of thaw lakes in the Coastal Plain of Alaska (Fig. 1.16). The cycle is driven by physical, rather than biological, pro-

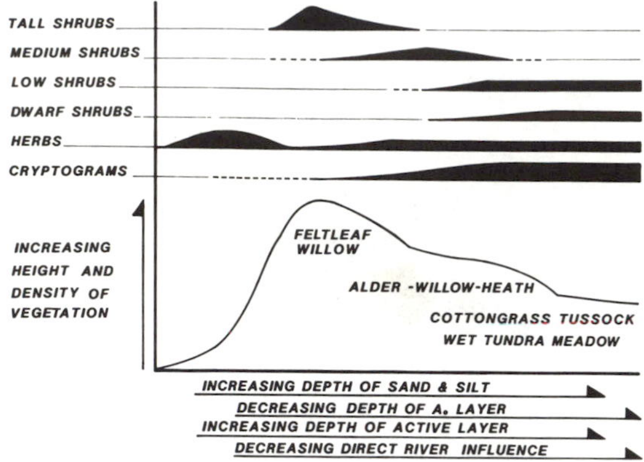

Figure 1.15. Pattern of plant succession along the Covalle River, nothern Alaska. Note the changes in plant cover, organic matter, and depth of the active layer (from Bliss and Cantlon 1957).

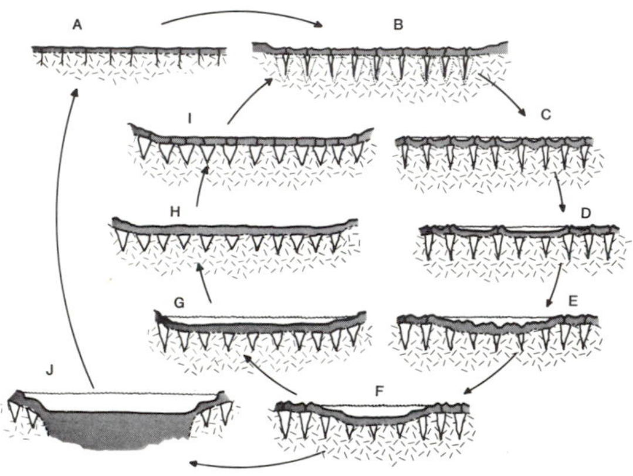

Figure 1.16. Sequence of events in the thaw-lake cycle, coastal plain, Alaska. On newly exposed marine sediments, contraction cracks form (A), followed by ice wedges (B). Low-center polygons form (C) and, with further melt, coalesce to form a shallow thaw lake (D–G). A shallow thaw lake may drain and the old ice wedges become rejuvenated to initiate a new cycle (H,I,B), or a deep thaw lake may form, with permafrost melt (J), followed by lake drainage and the initiation of a new cycle with the development of contraction cracks (A), followed by ice wedges (from Billings and Peterson 1980).

cesses, in contrast to most examples of succession. The thaw-lake cycle of succession was studied by Britton (1957), who proposed the concept, and more recently by Billings and Peterson (1980), who studied the physical and biological changes associated with the artificial drainage of four shallow lakes in 1950 and succession on older naturally drained lake basins. Soon after drainage, the surface consists of relatively low, high-center polygons with wide troughs above the ice wedges (Fig. 1.16A). The troughs are invaded by mosses, *Dupontia fisheri, Eriophorum scheuchzeri,* and *Saxifraga cernua,* and also by *Arctophila fulva*. In a matter of a century or two, *Carex aquatilis* becomes established, and a mixed graminoid-moss vegetation develops. With time, a thousand years or so, the ice wedges increase in size, the polygonal pattern becomes more prominent, and the polygon centers with their high ice content (ice lenses) melt, forming mini-ponds, in which *Carex aquatilis* dominates (Fig. 1.16C). As the ice lenses and wedges continue to melt, the low-center polygons coalesce, forming thaw ponds. As these increase in size and water depth, *Arctophila fulva* becomes dominant (Fig. 1.16E–G). The thaw ponds continue to increase in area and become oriented NE-SW because of differential erosion from prevailing NE-SW winds (Black 1969).

Fire

Fire is common to many biomes, but until the past decade, tundra fires received little attention. They have been reported from the District of Keewatin (Wein and Shilts 1976), east of the Mackenzie River Delta (Wein and Bliss 1973; Wein 1976), arctic Alaska, especially the Seward Peninsula (Wein 1976; Racine 1981; Racine et al. 1983), and northeastern Greenland (Oosting 1948). Most fires burn 1–10 ha, but some have covered 25 to more than 1000 km^2. These fires remove litter and biomass, darken the soil surface, increase the depth of summer thaw 30–50%, and release nutrients.

Plant recovery following fire is quite rapid in cottongrass–dwarf shrub heath and low shrub vegetation (the most commonly burned type, for it has a standing crop of 400 g m^{-2} or more). There is abundant flowering of *Eriophorum vaginatum* and rapid resprouting of *Betula, Salix,* and *Ledum* shrubs. Most heath species, lichens, and some bryophytes show much slower rates of recovery. In some sites, seedlings of *Eriophorum, Carex bigelowii,* and *Calamagrostis canadensis* are abundant after a burn (Bliss and Wein 1972; Racine et al. 1983), presumably because of the large seed bank (McGraw 1980).

Aboveground plant production in cottongrass-heath tundra reached 51–115% of that in control plots in only 1–2 yr at sites in Alaska and northwestern Canada (Wein and Bliss 1973). Assuming a decomposition rate of 20% yr^{-1}, the accumulation rate of 60 g m^{-2} yr^{-1} would lead to recovery of standing crop in 7–17 yr, a rapid rate of recovery for cold-dominated systems. Production measurements at an Elliott, Alaska, site 13 yr after a burn showed a total aboveground production 145%

Table 1.5. *Summary data from various sources on annual plant production, root : shoot ratios, and leaf area index (LAI) for various Arctic vegetation types*

	Plant production (g m^{-2})		Root : shoot		
Vegetation type	Vascular	Cryptogam	Phytomass	Net production	LAI
Low arctic tundra					
Tall shrub	250–400	5–25	1 : 2	1 : 2	1–2
Low shrub	125–175	25–50	1 : 2	1 : 2	1–2
Cottongrass-herb	150–200	25–100	3 : 1	1 : 1	0.5–1
Wet sedge–moss	150–200	5–25	20 : 1	3 : 1	0.5–1
High arctic tundra					
Wet sedge–moss	100–175	10–40	20 : 1	3 : 1	0.5–1
High arctic polar semidesert					
Cushion plant–cryptogam	5–25	2–5	1 : 1	1 : 5	0.1–0.2
Cryptogam-herb	10–20	5–30	2 : 1	1 : 2	0.1–0.2
High arctic polar desert					
Herb barrens	0.1–2	T[a]–1	1 : 3	1 : 2	—[b]

[a]T = trace.
[b]No data.

greater on the old burn than on the adjacent control; *Eriophorum*, *Carex*, and *Ledum* contributed 78% of the total. Although the total plant growth had recovered, the species composition had not for two species of *Vaccinium* (Fetcher et al. 1984).

Studies of nutrient regimes within cottongrass-heath tundra show a higher nutrient content in new plant shoots than in controls 1–2 yr after a burn. Nutrient release from the burn, deeper thawed and warmer soils, and increased microbial activity all may contribute to the rapid uptake of nutrients by graminoid and heath species. However, 6 yr after the Inuvik, N.W.T., fire, there were still net losses of nitrogen and phosphorus from tundra and forest-tundra ecosystems in terms of kilograms of nutrients per unit area (MacLean et al. 1983).

Where permafrost is ice-rich, soil slumpage and erosion gullies may form. This is generally a minor problem, in part because the thick organic mat seldom burns. Near the treeline, fires may result in eliminating tree seedling establishment unless the following 3–6 yr are unusually warm and moist (Black and Bliss 1980). As a result, tundra advances south. Soviet scientists have referred to this as "pyrogenic tundra" and "subarctic tundra." Fires rarely, if ever, occur in high arctic vegetation.

STANDING CROP AND PLANT PRODUCTION

It is known that arctic plants are small in stature, that they grow slowly, and that a standing crop of plant and animal material accumulates slowly. Decomposition is also very slow. However, there is great variation in carbon allocation and in dry-matter production when comparing plant communities from different vegetation types in the Arctic. Unfortunately, many studies lack estimates for root production. As used here, "standing crop" refers to living and dead plant material, "phytomass" refers to only living material, and "plant production" refers to net production over time, usually per year.

Table 1.5 summarizes data on major arctic vegetation types. These data show the dramatic shifts in plant production, root : shoot ratios, and the relative roles of cryptogams in major low and high arctic systems.

Tundras

Standing crop and net annual production are greatest in tall shrub tundra dominated by *Salix alaxensis*, *S. lanata* ssp. *richardsonii*, and *S. glauca* (Komarkova and Webber 1980). Total net annual production is nearer 250–400 g m^{-2} if it is assumed that root production is 50–100% of aboveground production. These values will be somewhat higher south of the Meade River (ca. 70° 15′ N) in the warmer sector of the Foothills Province. Low shrub tundra dominated by *Betula nana* ssp. *exilis*, *S. glauca*, *S. pulchra*, various species of heath, and *Carex bigelowii* or *C. lugens* has an aboveground standing crop similar to that of cottongrass–dwarf shrub heath and of wet sedge–moss tundras (Table 1.6). Total

Table 1.6. *Standing crop (g m⁻²) for select Arctic vegetation types*

Location	Vascular plants		Litter	Cryptogams	Total
	Aboveground	Belowground			
Low arctic tundra					
Atkasook, AK[1]					
Tall shrub	1496	—[a]	142	82	—
Low shrub	342	—[a]	427	137	—
Dwarf shrub heath	421	—[a]	423	403	—
Wet sedge–moss	257	—[a]	24	499	—
Cottongrass-heath	418	—[a]	323	577	—
Eagle Creek, AK[2]					
Cottongrass-heath	419	6563	77	89	7148
Dempster Highway, Yukon[2]					
Heath-cottongrass	168	6724	39	302	7233
High arctic tundra					
Barrow, AK[3]					
Wet sedge–moss	83	4366	67	45	4561
Devon Island, N.W.T.[4]					
Wet sedge–moss	273	2230	—[b]	908	3411
Barrow, AK[3]					
Herb–cushion plant	75	3376	41	31	3524
High arctic polar semidesert					
Devon Island, N.W.T.[4]					
Cushion plant–cryptogram	387	57	—[b]	79	523
Victoria Island, N.W.T.[5]					
Cushion plant–cryptogam	1078	300	—[b]	327	1705
Melville Island, N.W.T.[6]					
Cryptogam-herb	127	21	138	782	920
King Christian Island, N.W.T.[6]					
Cryptogam–herb	41	23	—[b]	2146	2210
High arctic polar desert					
Devon Island, N.W.T.[7]					
Herb barrens	11	1	12	2	26
Somerset Island, N.W.T.[7]					
Herb barrens	8	1	15	1	25

[a]No data.
[b]Litter included in above ground data.
Data sources: 1, Komarkova and Webber (1980); 2, Wein and Bliss (1974); 3, Webber (1978); 4, Bliss (1977); 5, Bliss and Svoboda (unpublished); 6, Bliss and Svoboda (1984); 7, Bliss et al. (1984).

annual production for these three types is also similar, 125–250 g m^{-2} if root production is included. Based on root : shoot ratios for annual production of 1 : 2 for low shrub, 2 : 1 for cottongrass-heath, and 3 : 1 for wet sedge–moss tundras, estimates of total production of vascular plants are approximately 125–175 g m^{-2}, 150–200 g m^{-2}, and 150–250 g m^{-2}, respectively. Cryptogams play a significant role in the low shrub, dwarf shrub heath, some cottongrass-heath, and the wet sedge vegetation types (Table 1.6). In the latter communities, biomass root: shoot ratios are generally 15:1 to 21:1, and net annual production is 2:1 to 3:1. Note that plant production in the wet sedge–moss communities in the High Arctic of Devon Island is similar to that at Barrow, Alaska (Table 1.5).

From very limited data, low shrub- and graminoid-dominated communities have a leaf area index (LAI) of 0.6–2. LAI and chlorophyll content are generally well correlated with aboveground vascular plant production (Wielgolaski et al. 1981).

Polar Semideserts and Polar Deserts

There are significant shifts in the magnitudes of standing crop and net annual production of plant communities in the High Arctic (Table 1.5 and 1.6). This reflects the shorter growing season, reduced degree-days, and colder soils, with less available nitrogen and phosphorus. The cushion plant–cryptogam vegetation has a large aboveground standing crop because of the large amount of dead material associated with *Dryas integrifolia* and *Salix arctica*. Elsewhere, the cryptogam-herb vegetation has a considerable standing crop and biomass of bryophytes and lichens, and vascular plants contribute a relatively small percentage. Belowground biomass and annual production are minor contributors to carbon allocation; root : shoot ratios are generally 1 : 1 to 1 : 2 for phytomass and annual production, indicating that root development is generally much more restricted in these cold soil environments that range from moist to dry. Net annual production is generally 25–50 g m^{-2}. The LAI based on only two examples of the cushion plant–cryptogam vegetation is only 0.2. The LAI may be 0.1–0.2 in the cryptogam-herb vegetation.

In the polar barrens, plant production and standing crop are extremely low. In contrast with the tundras and polar semideserts, where cryptogams are almost always large carbon contributors, in the polar barrens they are nearly absent. In 18 stands sampled on five islands, annual plant production averaged only 0.8 g m^{-2} (Bliss et al. 1984).

Controlling Parameters

Why are phytomass and annual production so low in these diverse Arctic landscapes? In the polar deserts, the reason is lack of available soil water once surface soils dry in midsummer, and also because of extremely low levels of nitrogen and phosphorus. Within the polar semideserts there is considerably greater cover of cryptogams. These surfaces hold more moisture in summer and provide "safe sites" for vascular plants to develop (Sohlberg and Bliss 1984), and they contain blue-green algae that fix limited amounts of nitrogen. In these high arctic ecosystems with their limited vascular plant cover, low air and soil temperatures are further limitations. Soil temperature 5–10 cm below the surface (where most roots are found) during the summer are generally 2–4°C. At higher temperatures, root growth increases dramatically (Bell and Bliss 1978). Plant production is significantly higher in the wet meadows, which contain more available water. Soil nitrogen is also higher because blue-green algae are more abundant in these moist soils. These habitats also have cold soils with a shallower active layer (20–30 cm, compared with 40–60 cm in the semideserts and deserts).

In the Low Arctic, these same parameters are central to the control of plant growth, although the environmental extremes are not as severe. Greater production of the shrub complexes and the wet sedge or wet grass–moss vegetation reflects the higher nutrient content of these systems. The growing season is longer, and the number of degree-days is greater (Table 1.1). Recent simulation modeling within the cottongrass–dwarf shrub heath tundra clearly shows the interaction of nutrients (N and P), soil water, and thermal regime in controlling plant growth. When fertilizer is added, growth is stimulated in the graminoids, but growth is depressed in heath species (Miller et al. 1984). The same phenomenon occurs in the High Arctic when populations of *Carex stans* and *Dryas integrifolia* are fertilized (Babb and Whitfield 1977).

LIFE HISTORY PATTERNS AND PHYSIOLOGICAL ADAPTATIONS

Arctic plants grow in some of the most extreme environments in the world. Consequently, they have been intensively studied for three decades. These plants are subjected to low air and soil temperatures, to low nutrient availability, and often to extremes in water availability. Plants of some species grow in cold, water-saturated soils with low oxygen tensions; others grow in soils that are saturated after snowmelt, but may dry considerably by midsummer. These environmental constraints are further controlled by a short growing season that limits carbon gain and sexual reproduction.

Reproduction and Plant Establishment

Arctic plants are long-lived; there are very few annuals or short-lived perennials, and their populations are maintained and dispersed by seeds or by vegetative reproduction. Within plant communities of the Low Arctic, many perennials depend on tillering (*Carex, Dupontia, Eriophorum, Luzula, Petasites*) and branch layering (*Betula, Empetrum, Salix, Vaccinium*); yet these same species flower, fruit, and produce viable seeds. Seedlings are not common in undisturbed tundra, but with surface disturbance (fire, vehicle tracks, animal burrows), seedling establishment of graminoids is common (Bliss and Wein 1972; Hernandez 1973; Chapin and Chapin 1980; Chester and Shaver 1982; McGraw and Shaver 1982). Seed dormancy mechanisms are generally absent or minor in low arctic (Bliss 1971) and in high arctic species (Bell and Bliss 1980). The optimum temperature for germination is quite high (15–30°C) for most species (Billings and Mooney 1968; Bell and Bliss 1980). The presence of *Eriophorum vaginatum* and *Carex bigelowii* seedlings on disturbed peat or mineral soils, versus their near absence in undisturbed vegetation, suggests that warmer surfaces provide the stimulus for enhanced germination. In the High Arctic, seed germination is generally low, with seedling survival more successful on bryophyte or desiccation-crack microsites (Bell and Bliss

1980; Sohlberg and Bliss 1984). These microsites are more moist and have higher nutrient content and less needle-ice formation than bare soil or lichen surfaces. Seed rain and seed germination are at random, indicating the importance of microsites (safe sites) in the establishment and maintenance of species. Seedling mortality is much higher in summer than in winter, indicating the importance of soil moisture in seedling survival. Estimates of seed bank sizes including 150 seeds m^{-2} at Barrow (Leck 1980) in a sedge-moss tundra, 430–820 seeds m^{-2} in upland shrub and lowland wet sedge tundra in northern Alaska (Roach 1983), 570–1600 seeds m^{-2} in graminoid barren and cryptogam-herb vegetation on King Christian Island (Sohlberg and Bliss 1984), 1–131 seeds m^{-2} in wet sedge and cushion plant and 450–7800 seeds m^{-2} near animal dens and disturbed soils on Ellesmere Island (Freedman et al. 1982), and 3400 seeds m^{-2} in tussock tundra in central Alaska (McGraw 1980). The abundance of flowering and the presence of seed banks and seedlings confirm that arctic plants allocate rather large amounts of carbon to sexual reproduction, even though the cold, short, and sometimes dry summer environments are not conducive to seedling establishment. We now know that sexual and asexual reproduction are important in low and high arctic habitats.

Plant Growth and Phenological Development

The general patterns of plant growth and development differ little from those in temperate regions. There is, however, a telescoping of these events into a shorter time frame. Early season growth of deciduous forbs and shrubs is rapid because of their near-synchronous leaf initiation and leaf expansion from preformed buds of the previous year (Sørensen 1941; Bliss 1956; Johnson and Tieszen 1976; Tieszen 1978). In contrast, graminoids continue to grow until shorter days and lower temperatures supress leaf and stem elongation. Many species carry some green leaves over the winter and initiate growth as soon as the snow melts in June. In the High Arctic, of 27 species examined, only *Oxyria digyna* failed to retain wintergreen or evergreen leaves over winter. Eight of these species were grown under controlled summer environmental conditions (continuous light, alternating "day" and "night" temperature conditions, and high humidity), and all entered dormancy in spite of favorable growing conditions (Bell and Bliss 1977), indicating a periodic growth pattern. Field studies of *Luzula confusa* showed that leaf senescence began in 45–50 d, and root growth dropped to near zero in 49 d. Under controlled environment conditions, the growing period was 55 d (Addision and Bliss 1984). Nonperiodic growth patterns have been observed in snowbed forbs in Greenland (Sørensen 1941) and in Alaska (Murray and Miller 1982). Species may flower and fruit synchronously in one microenvironment and be asynchronous in other environments (Bliss 1956; Billings 1974; Murray and Miller 1982).

Root growth often begins as soon as the active layer thickens and soils warm above 0°C, but often lateral root growth is more active the latter half of the summer (Shaver and Billings 1975; Bell and Bliss 1978). The temporal partitioning of root and shoot growth may reflect adjustments in plant growth to reduce the demand for limited carbon and nutrient resources and to take advantage of the most favorable environmental conditions aboveground (early) and belowground (late). Both shoots and roots are capable of growth at low temperatures (0–5°C) (Billings et al. 1978; Bell and Bliss 1978), but experimental studies in field minigreenhouses and plant growth chambers show that growth is enhanced in different growth forms (graminoids, forbs, deciduous shrubs) by higher temperatures (10–20°C) (Chapin and Shaver 1985b).

Plant growth and phenological development are more rapid in warmer microenvironments for vascular plants (Sørensen 1941; Bliss 1956; Warren Wilson 1959) and for bryophytes (Oechel and Sveinbjörnsson 1978). In general, the height of arctic plants conforms to the height aboveground where the air is warmest in summer and to the mean depth of snow cover in winter. This is especially true for shrubs, cushion- and mat-forming plants such as *Dryas*, *Saxifraga caespitosa*, and *S. oppositifolia*, the tussock form of *Eriophorum vaginatum*, and the small rosette form of species of *Draba*, *Saxifraga*, *Papaver*, and *Minuartia*. Leaf temperatures within the boundary layer are often 5–10°C higher than ambient and may reach 25–30°C above ambient in *Dryas* leaves on clear, calm days. The parabola-shaped flowers of some arctic species serve as basking sites for insects and raise the temperature of the developing ovary 5–10°C, which enhances pollination and speeds seed development (Kevan 1972).

The growth rate of arctic graminoids on a daily basis is often as high as that for temperate-region graminoids, although arctic temperatures at plant height are 10–15°C rather than the 20–35°C in grasslands to the south. Total plant production is less in the Low Arctic than in comparable temperate grasslands and herblands because of the shorter growing season.

Physiological Characteristics of Plant Growth Forms

It is now known that there are strong correlations between the physiological characteristics of taxonomically diverse species with the same growth form and their adaptation to certain habitats. Later I shall discuss deciduous shrubs, evergreen shrubs, graminoids, tussock graminoids, and cushion plants.

Deciduous shrubs require large carbon gains each summer to produce an entire new leaf surface. Consequently, this group of species has high photosynthetic rates (10–30 mg CO_2 dm^{-2} hr^{-1}) and relatively high leaf conductances (Tieszen 1978; Limbach et al. 1982) (Table 1.7). Root growth, root respiration, and rates of phosphorus and nitrogen uptake are relatively high in *Salix pulchra* and *Betula nana* as compared with evergreen species. These shrubs also have higher root and shoot growth rates, which place greater demands on nutrient uptake. Mycorrhizal effects on their roots further facilitate nutrient uptake.

The evergreen-leaved heath species are conservative plants in a number of ways (Table 1.7). Their photosynthetic rates are low, leaf growth and shoot growth are slow, their leaves function for 2–5 yr, they have low rates of water conductance, the plants maintain small root systems and limited annual root production, they are conservative in use nutrients, they have low rates of nutrient uptake, and they show limited response to added nutrients. Respiration rates are relatively low, in part because growth rates are slow and leaves are maintained for more than 1 yr (Tieszen 1978; Chapin and Tryon 1982; Limbach et al. 1982). These plants have a further energy drain to the production of tannins, alkaloids, terpenes and anthraquinones, especially in *Empetrum hermaphroditum*, *Ledum palustre* ssp. *decumbens*, and *Cassiope tetragona* (Batzli and Jung 1980).

Single-stemmed graminoids such as the sedges *Carex aquatilis*, *C. stans*, cottongrasses, *Eriophorum angustifolium*, and *E. scheuchzeri* and the grasses *Dupontia fisheri* and *Alopecurus alpinus* are, in various combinations or in monoculture, the dominant growth form in large areas of the Arctic. These species carry some green stem and leaf tissue over winter, they grow rapidly as soon as snow melts and soils warm in June, and they continue growth until dieback in August and September. These graminoids have relatively high photosynthetic rates (15–20 mg CO_2 dm^{-2} hr^{-1}) and high rates of leaf conductance, and they are adapted to living in cold soils that are often water-saturated (Mayo et al. 1977; Tieszen 1978). As a group, these species produce two to three times more root biomass than shoot biomass per year.

Nutrients are taken up in smaller quantities by single-stemmed graminoids, in part because of the lack of mycorrhizae. There is considerable retranslocation of nutrients out of senescing leaves, an adaptation of plants growing in nutrient-poor soil systems (Chapin et al. 1975). Leave of *Dupontia* and *Carex aquatilis* are mature in terms of high photosynthetic rates for only 10 and 27 d, respectively (Tieszen 1978). In the High Arctic, the two or three leaves produced per shoot apparently function efficiently for a longer time.

The tussock graminoid *Eriophorum vaginatum*

Table 1.7. *General physiological characteristics of different plant growth forms in the Arctic*

	Plant growth form				
Characteristic	Deciduous shrubs	Evergreen shrubs	Single-stem graminoids	Tussock graminoids	Cushion plants
Resource allocation					
Reproduction	Low	Low	Low	Low	Low
Shoots	High	Medium	Medium	Medium	Low
Belowground	Low	Low	High	High	Low
Photosynthetic rate	High	Low	Medium	Low	Low
Respiration	High	Low	Medium	High	High
Leaf conductance	High	Low	High	High	Low
Leaf longevity (yr)	1	2–5	1+	1+	2–3
Plant or shoot age (yr)	50–100+	20–50+	5–7	50–100+	20–100+
Carbohydrate reserves	Medium	Low	High	Medium	Medium
Habitat preference	Snow-protected, moist	Late-snow, moist	Wet to dry	Moist	Dry

has been studied in more detail than any other arctic species (Miller et al. 1984). This species is more physiologically conservative than single-stemmed graminoids (Table 1.7). Net photosynthesis averages only 2–3 mg CO_2 dm^{-2} hr^{-1}, but root and stem respiration rates are relatively high compared with those for other species (Limbach et al. 1982). Leaves of cottongrass typically live for 1+ yr, but all roots are new each year, a very large energy drain, because root production is two to three times that for shoots. Tussocks are long-lived, ranging in age from 122 to 187 yr (Mark et al. 1985). Most roots (75%) are found in the elevated and organic-rich tussock (Chapin et al. 1979), although some roots grow into the mineral soil, where they grow within 1 cm of the permafrost table at a temperature near 0°C (Bliss 1956; Chapin et al. 1979). The freeze–thaw cycle that occurs at the top of the permafrost table (bottom of the active layer) may make phosphorus and cations more available, and water movement along this surface may also facilitate water uptake. This species also retranslocates nutrients into the tussock each fall, conserving the limited nutrient pool. The fact that most roots remain in this organic layer enables maximum nutrient uptake each year. Rates of nutrient uptake are as high in *Eriophorum* as they are in *Carex aquatilis* (Chapin and Tryon 1982).

Plants with the cushion growth form predominate in the High Arctic, where *Dryas integrifolia* has been studied in considerable detail. This growth form is near the ultimate in physiological conservatism. It occupies windy sites that are also droughty in summer and have limited snow cover in winter. This is also a long-lived species, 100–150 yr (Svoboda 1977). *Dryas* has low photosynthetic rates (2–3 mg CO_2 g^{-1} hr^{-1}), low leaf conductances, and very low growth rates of shoots and roots. Leaves are evergreen and function for 2 yr, but remain on the plant for many years before decomposing (Mayo et al. 1977; Svoboda 1977). Each clump is a self-contained unit that conserves the very limited nutrient pool at the base of the shoots. Most active roots are confined to this zone. Respiration rates are high for both roots and shoots. Root : shoot ratios are low, 0.4–0.6, as they are for rosette species in the High Arctic. This is in strong contrast to the patterns for sedges and grasses of wet sites throughout the Arctic, for which ratios are often 10–20 for living root systems.

General Physiological Characteristics

Although arctic plants live in some of the world's most severe environments, adaptations are not especially well developed. Seeds of most species germinate on only warm microsites (15–20°C), and, in general, germination is a slow process, especially in the High Arctic. On the other hand, arctic plants initiate spring growth with a burst, some while there is still a thin snow cover. Plants in warmer microsites grow larger and often flower and fruit more abundantly than do plants in such nutrient-rich sites as animal burrows and bird cliffs.

Arctic plants utilize only the C_3 carbon photosynthetic pathway, and their maximum rates of photosynthesis are very similar to those for plants of the same growth form in temperate regions. However, the temperature optima for photosynthesis are on the lower end of the range for temperate region species. Most species can maintain positive carbon accumulation over a wide temperature range, some to −2 to −4°C. In general, arctic plants have a high Q_{10} of 2–3 for both photosynthesis and respiration. Arctic plants have higher rates of mitochondrial activity, shoot respiration, and phosphate absorption than temperate climate plants. These high reaction rates may result from higher rates of enzyme activity and probably are necessary to maintain rapid growth rates at low temperature.

Although many species in the Low Arctic do not experience water stress, many in the High Arctic are periodically drought-stressed. Cushion plants, some grasses, and rosette species in sites with bare and well-drained soils commonly have leaf water potentials of −2.0 to −3.0 MPa and sometimes less than −4.0 MPa, especially in summers with less precipitation (Mayo et al. 1977; Grulke 1983). Leaf water potentials of plants in wet sites seldom exceed −1.0 to 1.5 MPa. In cold soils, root resistance to water movement can be an important factor. Arctic plants control water loss through osmoregulation and stomatal movement.

Nutrients are limiting in arctic ecosystems, and plants have evolved various mechanisms to cope with these situations. For some species, this entails the ability to get by with low rates of uptake, because the leaves and roots function for 2 yr or more. Other species confine most of their roots to the nutrient-rich organic mass associated with tussock or cushion plant development. Retranslocation of nutrients to shoot bases and belowground organs prior to leaf senescence characterizes most species. When nutrients are added, shoot growth increases, but not photosynthetic rates (Tieszen 1978).

AREAS FOR FUTURE RESEARCH

Plant communities within the Low Arctic have been well studied, with the exception of those in the Brooks Range, the mountains of the northern Yukon, and the barren lands of Keewatin District and northeastern Mackenzie District, N.W.T. Shrub communities of riparian habitats deserve more emphasis in terms of successional patterns, nutrient relationships, and their roles as wildlife habitats.

There have been few detailed studies of plant communities in the High Arctic, with vast areas having received little botanical research. The historical and environmental reasons for the barrenness of the polar deserts need detailed study, especially the mechanisms by which cryptogam mats develop in the conversion of desert to semidesert landscapes.

Our understanding of the patterning of species within mesotopographic landscapes will be enhanced by ecophysiological studies of dominant species. Research on soil–plant water relations, mineral nutrition, and cold hardiness of dominant species in relation to plant life forms is especially needed. *Eriophorum vaginatum* has received a disproportionate amount of research, in part because of its abundance and ease of growth within controlled environments. Arctic fungi and the roles they play in decomposition, the importance of mycorrhizae, and floristic relationships are worthy of more study.

Arctic gardening is just beginning to receive the attention it deserves in Alaska (Dearborn 1979), as well as through the studies of Josef Svoboda and his colleagues in the Canadian Arctic (Romer et al. 1983). Their studies at Rankin Inlet on Hudson Bay and Alexandra Fiord on eastern Ellesmere Island show that cold temperate vegetables can be grown in the Arctic. Raising crops in small portable plastic greenhouses can provide much-needed fresh vegetables at a relatively low cost. This will be more successful in those parts of the Arctic with greater amounts of summer sunshine.

REFERENCES

Addison, P. A., and L. C. Bliss. 1984. Adaptations of *Luzula confusa* to the polar semi-desert environment. Arctic 37:121–132.

Ager, T. A. 1982. Vegetational history of western Alaska during the Wisconsin Glacial Interval and the Holocene, pp. 75–93 in D. M. Hopkins, J. V. Matthews, C. E. Schweger, and S. B. Young (eds.), Paleoecology of Beringia. Academic, New York.

Aleksandrova, V. D. 1980. The Arctic and Antarctic: their division into geobotanical areas. Cambridge University Press.

Andreyev, V. N. and V. D. Aleksandrova. 1981. Geobotanical division of the Soviet arctic, pp. 25–37 in L. C. Bliss, D. W. Heal, and J. J. Moore (eds.), Tundra ecosystems: a comparative analysis. Cambridge University Press.

Babb, T. A., and D. W. A. Whitfield. 1977. Mineral nutrient cycling and limitation of plant growth in the Truelove Lowland ecosystem, pp. 587–606 in L. C. Bliss (ed.), Truelove Lowland, Devon Island, Canada: a high arctic ecosystem. University of Alberta Press, Edmonton.

Batzli, G. O., and H. G. Jung. 1980. Nutritional ecology of microtene rodents: resource utilization near Atkasook, Alaska. Arct. Alp. Res. 12:483–499.

Bell, K. L., and K. C. Bliss. 1977. Overwinter phenology of plants in a polar semi-desert. Arctic 30:118–121.

Bell, K. L., and L. C. Bliss. 1978. Root growth in a polar semidesert environment. Can. J. Bot. 56:2470–2490.

Bell, K. L. and L. C. Bliss. 1980. Plant reproduction in a high Arctic environment. Arct. Alp. Res. 12:1–10.

Beschel, R. E. 1970. The diversity of tundra vegetation, pp. 85–92 in W. A. Fuller and P. G. Kevan (eds.), Productivity and conservation in northern circumpolar lands. IUCN new series 16. Morges, Switzerland.

Billings, W. D. 1974. Adaptations and origins of alpine plants. Arct. Alp. Res. 6:129–142.

Billings, W. D., and H. A. Mooney. 1968. The ecology of arctic and alpine plants. Biol. Rev. 43:481–530.

Billings, W. D., and K. M. Peterson. 1980. Vegetation change and ice-wedge polygons through the thaw-lake cycle in arctic Alaska. Arct. Alp. Res. 12:413–432.

Billings, W. D., K. M. Peterson, and G. R. Shaver. 1978. Growth turnover and respiration rates of roots and tillers in tundra graminoids, pp. 415–434 in L. L. Tieszen (ed.), Vegetation and production ecology of an Alaskan arctic tundra. Ecological Studies, vol. 29. Springer-Verlag, New York.

Bird, C. D. 1975. The lichen, bryophyte, and vascular plant flora and vegetation of the Landing Lake area, Prince Patrick Island, Arctic, Canada. Can. J. Bot. 53:719–744.

Black, R. A., and L. C. Bliss. 1978. Recovery sequence of *Picea mariana–Vaccinium uliginosum* forests after fire near Inuvik, Northwest Territories, Canada. Can. J. Bot. 56:2020–2030.

Black, R. F. 1969. Thaw depressions and thaw lakes–a review. Biuletyn Peryglacjalny 19:131–150.

Bliss, L. C. 1956. A comparison of plant development in microenvironments of arctic and alpine tundras. Ecol. Monogr. 26:303–337.

Bliss, L. C. 1971. Arctic and alpine plant life cycles. Ann. Rev. Ecol. Syst. 2:405–438.

Bliss, L. C. 1975. Tundra grasslands, herblands, and shrublands and the role of herbivores. Geosci. Man 10:51–79.

Bliss, L. C. 1977. General summary: Truelove Lowland ecosystem, pp. 657–675, in L. C. Bliss (ed.), Truelove Lowland, Devon Island, Canada: a high arctic ecosystem. University of Alberta Press, Edmonton.

Bliss, L. C. 1981. North American and Scandinavian tundras and polar deserts, pp. 8–24 in L. C. Bliss, O. W. Heal, and J. J. Moore (eds.), Tundra ecosystems: A comparative analysis. Cambridge University Press.

Bliss, L. C., and J. E. Cantlon. 1957. Succession on river alluvium in northern Alaska. Amer. Midl. Nat. 58:452–569.

Bliss, L. C., J. Kerik, and W. Peterson. 1977. Primary production of dwarf shrub heath communities, Truelove Lowland, pp. 217–244 in L. C. Bliss (ed.), Truelove Lowland, Devon Island, Canada: a High Arctic ecosystem. University of Alberta Press, Edmonton.

Bliss, L. C., and J. Svoboda. 1984. Plant communities and plant production in the western Queen Elizabeth Islands. Holarctic Ecol. 7:324–344.

Bliss, L. C., J. Svoboda, and D. I. Bliss. 1984. Polar deserts, their plant cover and plant production in the Canadian High Arctic. Holarctic Ecol. 7:304–324.

Bliss, L. C., and R. W. Wein. 1972. Plant community responses to disturbances in the western Canadian arctic. Can. J. Bot. 50:1097–1109.

Böcher, T. W. 1954. Oceanic and continental vegetational complexes in southwest Greenland. Medd. om Grønland, 148(1):1–336.

Böcher, T. W. 1959. Floristics and ecological studies in middle west Greenland. Medd. om Grønland, 156(5):1–68.

Böcher, T. W. 1963. Phytogeography of middle west Greenland. Medd. om Grønland, 148(3):1–289.

Böcher, T. W., and S. Laegaard. 1962. Botanical studies along the Arfersiorfik Fiord, west Greenland. Botanisk Tidsskrift 58:168–190.

Brassard, G. R., and R. E. Longton. 1970. The flora and vegetation of Van Hauen Pass, northwestern Ellesmere Island. Can. Field-Nat. 84:357–364.

Britton, M. E. 1957. Vegetation of the arctic tundra, pp. 22–61 in H. P. Hansen (ed.), Arctic biology. Oregon State University Press. Corvallis.

Bryson, R. A. 1966. Airmasses, streamlines and the boreal forest. Geogr. Bull. 8:228–269.

Chapin, F. S., and M. C. Chapin. 1980. Revegetation of an arctic disturbed site by native tundra species. J. Appl. Ecol. 17:449–456.

Chapin, F. S., and G. R. Shaver. 1985a. Arctic, pp. 16–40 in G. F. Chabot and H. A. Mooney (eds.), Physiological ecology of North American plant communities. Chapman & Hall, New York.

Chapin, F. S., and G. R. Shaver. 1985b. Individualistic growth response of tundra plant species to environmental manipulations in the field. Ecology 66:564–576.

Chapin, F. S., and P. R. Tryon. 1982. Phosphate absorption and root respiration of different plant growth forms from northern Alaska. Holarctic Ecol. 5:164–171.

Chapin, F. S., K. Van Cleve, and M. C. Chapin. 1979. Soil temperature and nutrient cycling in the tussock growth form of *Eriophorum vaginatum*. J. Ecol. 67:169–189.

Chapin, F. S., K. Van Cleve, and L. L. Tieszen. 1975. Seasonal nutrient dynamics of tundra vegetation at Barrow, Alaska. Arct. Alp. Res. 7:209–226.

Chester, A. L., and G. R. Shaver. 1982. Reproductive effort in cotton grass tussock tundra. Holarctic Ecol. 5:200–206.

Churchill, E. D. 1955. Phytosociological and environmental characteristics of some plant communities in the Umiat region of Alaska. Ecology 36:606–627.

Colinvaux, P. A. 1964. The environment of the Bering land bridge. Ecol. Monogr. 34:297–329.

Colinvaux, P. A. 1967. Quaternary vegetation history of arctic Alaska, pp. 207–231 in D. M. Hopkins (ed.), The Bering land bridge. Stanford University Press.

Corns, I. G. W. 1974. Arctic plant communities east of the Mackenzie Delta. Can. J. Bot. 52:1730–1745.

Daniels, F. J. A. 1982. Vegetation of the Angmagssalik District, Southeast Greenland. IV. Shrub, dwarf shrub and terricolous lichens. Medd. om Grønland, Biosci. 10:1–78.

Dearborn, C. H. 1979. Horticultural limitations and potentials of Alaska's Arctic, particularly the Kobuk River region. Arctic 32:248–263.

Drew, J. V., and R. E. Shanks. 1965. Landscape relationships of soils and vegetation in the forest-tundra, Upper Firth River Valley, Alaska-Canada. Ecol. Monogr. 35:285–306.

Edlund, S. A. 1980. Vegetation of Lougheed Island, District of Franklin. Geol. Surv. Can. 80-1A:329–333.

Fetcher, N., T. F. Beatly, B. Mullinax, and D. S. Winkler. 1984. Changes in arctic tussock tundra thirteen years after fire. Ecology 65:1332–1333.

Freedman, B., N. Hill, J. Svoboda, and G. Henry. 1982. Seed banks and seedling occurrence in a high arctic oasis at Alexandra Fiord, Ellesmere Island, Canada. Can. J. Bot. 60:2112–2118.

Freedman, B., J. Svoboda, C. Labine, M. Muc, G. Henry, M. Nams, J. Stewart, and E. Woodley. 1983. Physical and ecological characteristics of Alexandra Fiord, a high arctic oasis on Ellesmere Island, Canada, pp. 301–304 in Permafrost: Fourth International Conference Proceedings. National Academy Press, Washington, D.C.

Gill, D. 1973. Ecological modifications caused by the removal of tree and shrub canopies in the Mackenzie Delta. Arctic 26:95–111.

Grulke, N. E. 1983. Comparative morphology, ecophysiology and life history characteristics of two high arctic grasses, N.W.T. Ph.D. thesis, University of Washington, Seattle.

Guthrie, R. D. 1972. Mammals of the mammoth steppe as paleoenvironmental indicators, pp. 307–326 in D. H. Hopkins, J. V. Matthews, C. E. Schweger, and S. B. Young (eds.), Paleoecology of Beringia. Academic, New York.

Hansen, K. 1969. Analyses of soil profiles in dwarf-shrub vegetation in south Greenland. Medd. om Grønland 178(5):1–33.

Hanson, H. C. 1953. Vegetation types in northwestern Alaska and comparisons with communities in other arctic regions. Ecology 34:111–140.

Hare, F. K. 1968. The Arctic. Q. J. Roy. Meteor. Soc. 94:439–459.

Hare, F. K., and J. C. Ritchie. 1972. The boreal bioclimates. Geogr. Rev. 62:333–365.

Hernandez, H. 1973. Natural plant recolonization of surficial disturbances, Tuktoyaktuk Peninsula region, Northwest Territories. Can. J. Bot. 51:2177–2196.

Hettinger, L., A. Janz, and R. W. Wein. 1973. Vegetation of the northern Yukon Territory. Arctic Gas Biol. Rept. Serv. Vol. 1, Canadian Arctic Gas Study Ltd., Calgary.

Hills, L. V., J. E. Klovan, and A. R. Sweet. 1974. *Juglans eocinerea* n. sp., Beaufort Formation (Tertiary), southwestern Banks Island, Arctic Canada. Can. J. Bot. 52:65–90.

Hoffman, R. S., and R. D. Taber. 1967. Origin and history of holarctic tundra ecosystems, with special reference to vertebrate faunas, pp. 143–170 in W. H. Osburn and H. E. Wright, Jr. (eds.), Arctic and alpine environments. Indiana University Press, Bloomington.

Holmen, K. 1957. The vascular plants of Peary Land, north Greenland. Medd. om Grønland 124(9):1–149.

Hopkins, D. M., J. V. Matthews, J. A. Wolfe, and M. L. Silberman. 1971. A Pliocene flora and insect fauna from the Bering Strait region. Palaeogeogr. Palaeoclimatol. Palaeoecol. 9:211–213.

Hopkins, D. M., J. V. Matthews, C. E. Schwegner, and S. B. Young (eds.). 1982. Paleoecology of Beringia. Academic, New York.

Jeffries, R. L. 1977. The vegetation of salt marshes at some coastal sites in arctic North America. J. Ecol. 65:661–672.

Johnson, A. W., L. A. Viereck, R. E. Johnson, and H. Melchior. 1966. Vegetation and flora, pp. 277–354 in N. J. Wilimovsky and J. N. Wolfe (eds.), Environment of the Cape Thompson region, Alaska. U.S.A.E.C., Div. Tech. Info., Washington, D.C.

Johnson, D. A., and L. L. Tieszen. 1976. Aboveground biomass allocation, leaf growth, and photosynthesis patterns in tundra plant forms in Arctic Alaska. Oecologia 24:159–173.

Kershaw, K. A. 1976. The vegetation zonation of the East Pen Island salt marshes, Hudson Bay. Can. J. Bot. 54:5–13.

Kevan, P. 1972. Insect pollination of high arctic flowers. J. Ecol. 60:831–847.

Komarkova, V., and P. J. Webber. 1980. Two low arctic vegetation maps near Atkasook, Alaska. Arct. Alp. Res. 12:447–472.

Kuc, M. 1974. Noteworthy vascular plants collected in southwestern Banks Island, N.W.T. Arctic 26:146–150.

Larsen, J. A. 1965. The vegetation of the Ennadai Lake area, N.W.T.: studies in subarctic and arctic bioclimatology. Ecol. Monogr. 35:37–59.

Larsen, J. A. 1971. Vegetation of Fort Reliance, Northwest Territories. Can. Field-Nat. 85:147–178.

Larsen, J. A. 1972. The vegetation of northern Keewatin. Can. Field-Nat. 86:45–72.

Leck, M. A. 1980. Germination in Barrow, Alaska tundra soil cores. Arct. Alp. Res. 12:343–349.

Limbach, W. E., W. C. Oechel, and W. Lowell. 1982. Photosynthetic and respiratory responses to temperature and light of three tundra growth forms. Holarctic Ecol. 5:150–157.

Löve, A., and D. Löve. 1974. Origin and evaluation of the arctic and alpine floras, pp. 571–603 in J. D. Ives and R. G. Barry (eds.), Arctic and alpine environments. Methuen, London.

McGraw, J. B. 1980. Seed bank size and distribution of seeds in cottongrass tussock tundra, Eagle Creek, Alaska. Can. J. Bot. 58:1607–1611.

McGraw, J. B., and G. L. Shaver. 1982. Seedling density and seedling survival in Alaskan cotton grass tussock tundra. Holarctic Ecol. 5:212–217.

MacLean, D. A., S. J. Woodley, M. G. Weber, and R. W. Wein. 1983. Fire and nutrient cycling, pp. 111–132 in R. W. Wein and D. A. MacLean (eds.), The role of fire in northern circumpolar ecosystems. Wiley, New York.

Maikawa, E., and K. A. Kershaw. 1976. Studies on lichen-dominated systems. XIX. The postfire recovery sequence of black spruce–lichen woodland in the Abitan Lake Region, N.W.T. Can. J. Bot. 54:2679–2689.

Mark, A. F., N. Fetcher, G. R. Shaver, and F. S. Chapin. 1985. Estimated ages of mature tussocks of *Eriophorum vaginatum* along a latitudinal gradient in central Alaska, U.S.A. Arct. Alp. Res. 17:1–5.

Martin, P. S. 1974. Paleolithic plays on the American stage, pp. 669–700 in J. D. Ives and R. G. Barry (eds.), Arctic and alpine environments. Methuen, London.

Martin, P. S. 1982. The pattern and meaning of Holarctic mammoth extinction, pp. 399–408 in D. M. Hopkins, J. V. Matthews, C. E. Schweger, and S. B. Young (eds.), Paleoecology of Beringia. Academic, New York.

Matthews, J. V. 1974. A preliminary list of insect fossils from the Beaufort Formation, Meighen Island, District of Franklin. Geol. Surv. Can. 74-1A:203–206.

Mayo, J. M., A. P. Hartgerink, D. G. Despain, D. G. Thompson, R. G. Thompson, E. M. van Zinderin Bakker, and S. D. Nelson. 1977. Gas exchange studies of *Carex* and *Dryas*, Truelove Lowland, pp. 265–280 in L. C. Bliss (ed.), Truelove Lowland, Devon Island, Canada: a high arctic ecosystem. University of Alberta Press, Edmonton.

Miller, P. C., P. M. Miller, M. Blake-Jacobson, F. S. Chapin, K. R. Everett, D. W. Hilbert, J. Kummerow, A. E. Linkins, G. M. Marion, W. C. Oechel, S. W. Roberts, and L. Stuart. 1984. Plant-soil processes in *Eriophorum vaginatum* tussock tundra in Alaska: systems modeling approach. Ecol. Monogr. 54:361–405.

Muc, M. 1977. Ecology and primary production of the Truelove Lowland sedge-moss meadow communities, pp. 157–184 in L. C. Bliss (ed.), Truelove Lowland, Devon Island, Canada: a high arctic ecosystem. University of Alberta Press, Edmonton.

Murray, C., and P. C. Miller. 1982. Phenological observations of major plant growth forms and species in montane and *Eriophorum vaginatum* tussock tundra in central Alaska. Holarctic Ecol. 5:109–116.

Oechel, W. C., and B. Sveinbjörnsson. 1978. Primary production processes in arctic bryophytes at Barrow, Alaska, pp. 269–298 in L. L. Tieszen (ed.), Vegetation and production ecology of an Alaskan arctic tundra. Springer-Verlag, New York.

Oosting, H. J. 1948. Ecological notes on the flora of east Greenland and Jan Mayen, pp. 225–269 in L. A. Boyd et al. (eds.), The coast of northeast Greenland. American Geographical Society special publication 30, Washington, D.C.

Packer, J. G. 1974. Differentiation and dispersal in alpine floras. Arct. Alp. Res. 6:117–128.

Peterson, K. M., and W. D. Billings. 1978. Geomorphic processes and vegetational change along the Meade River sand bluffs, northern Alaska. Arctic 31:7–23.

Peterson, K. M., and W. D. Billings. 1980. Tundra vegetational patterns and succession in relation to microtopography near Atkasook, Alaska. Arct. Alp. Res. 12:473–482.

Polunin, N. 1948. Botany of the Canadian eastern Arctic. Part III. Vegetation and ecology. National Museum of Canada Bulletin 104, Ottawa.

Racine, C. H. 1981. Tundra fire effects on soils and three plant communities along a hill-slope gradient in the Seward Peninsula, Alaska. Arctic 34:71–85.

Racine, C. H., W. A. Patterson, and J. G. Dennis. 1983. Permafrost thaw associated with tundra fires in northwest Alaska, pp. 1024–1029 in Permafrost: Fourth international conference proceedings. National Academy Press, Washington, D.C.

Reznicek, S. A. and J. Svoboda. 1982. Tundra communities along a microenvironmental gradient at Coral Harbour, Southhampton Island, N.W.T. Naturalist Can. 109:585–595.

Ritchie, J. C. 1977. The modern and late Quaternary vegetation of the Campbell-Dolomite uplands, near Inuvik, N.W.T., Canada. Ecol. Monogr. 47:401–423.

Ritchie, J. C. 1983. Past and present vegetation of the far northwest of Canada. Toronto University Press.

Ritchie, J. C., and L. C. Cwynar. 1982. The late Quaternary vegetation of the North Yukon, pp. 113–126 in D. M. Hopkins, J. V. Matthews, C. E. Schweger, and S. B. Young (eds.), Paleoecology of Beringia. Academic, New York.

Ritchie, J. C. and F. K. Hare. 1971. Late Quaternary vegetation and climate near the arctic line of northwestern North America. Quat. Res. 1:331–342.

Roach, D. A. 1983. Buried seed and standing vegetation in two adjacent tundra habitats, northern Alaska. Oecologia 60:359–364.

Romer, M. J., W. R. Cummins, and J. Svoboda. 1983. Productivity of native and temperate "crop" plants in the Keewatin District, N.W.T. Naturalist Can. 110:85–93.

Savile, D. B. O. 1961. The botany of the northwestern Queen Elizabeth Islands. Can. J. Bot. 39:909–942.

Savile, D. B. O. 1972. Arctic adaptations in plants. Canadian Department of Agriculture Monograph 6., Ottawa.

Seidenfaden, G., and T. Sørensen. 1937. The vascular plants of northeast Greenland from 74° 30′ to 79° 00′ N lat. Medd. om Grønland 101(4):1–215.

Shaver, G. R., and W. D. Billings. 1975. Root production and root turnover in a wet tundra ecosystem, Barrow, Alaska. Ecology 56:401–409.

Shaver, G. R., B. L. Gartner, F. S. Chapin, and A. E. Linkins. 1983. Revegetation of arctic disturbed sites by native tundra plants, pp. 1133–1138 in Permafrost: fourth international conference proceedings. National Academy Press, Washington, D.C.

Sheard, J. W., and D. W. Geale. 1983. Vegetation studies of Polar Bear Pass, Bathurst Island, N.W.T. I. Classification of plant communities. Can. J. Bot. 61:1618–1636.

Sohlberg, E., and L. C. Bliss. 1984. Microscale pattern of vascular plant distribution in two high arctic communities. Can. J. Bot. 62:2033–2042.

Sørensen, T. 1941. Temperature relations and phenology of the northeast Greenland flowering plants. Medd. om Grønland 125:1–305.

Sørensen, T. 1943. The flora of Melville Bugt. Medd. om Grønland 124(5):1–70.

Spetzman, L. A. 1959. Vegetation of the Arctic Slope of Alaska. U.S. Geological Survey professional paper 302-B. Washington, D.C.

Svoboda, J. 1977. Ecology and primary production of raised beach communities, Truelove Lowland, pp. 185–216 in L. C. Bliss (ed.), Truelove Lowland, Devon Island, Canada: a high arctic ecosystem. University of Alberta Press, Edmonton.

Svoboda, J. 1982. Due to the Little Ice Age climatic impact most of the vegetation cover in the Canadian High Arctic is of recent origin. A hypothesis. Proc. 33rd Alaskan Science Conference 33:206.

Taylor, R. J. 1981. Shoreline vegetation of the Arctic Alaskan coast. Arctic 34:37–42.

Tedrow, J. C. F. 1977. Soils of the polar landscapes. Rutgers University Press, New Brunswick, N.J.

Thompson, D. C. 1980. A classification of the vegetation of Boothia Peninsula and the northern district of Keewatin, N.W.T. Arctic 33:73–99.

Tieszen, L. L. 1978. Photosynthesis in the principal Barrow, Alaska species: a summary of field and laboratory responses, pp. 241–268 in L. L. Tieszen (ed.), Vegetation and production ecology of an Alaskan arctic tundra. Springer-Verlag, New York.

Vestergaard, P. 1978. Studies in vegetation and soil of coastal salt marshes in the Disko area, west Greenland. Medd. om Grønland 204(2):1–51.

Warren Wilson, J. 1959. Notes on wind and its effects on arctic-alpine vegetation. J. Ecol. 47:415–427.

Washburn, A. L. 1956. Classification of patterned ground and review of suggested origins. Geol. Soc. Amer. Bull. 67:823–865.

Washburn, A. L. 1980. Geocryology: a survey of periglacial processes and environments. Wiley, New York.

Webber, P. J. 1978. Spatial and temporal variation in the vegetation and its production, Barrow, Alaska, pp. 37–112 in L. L. Tieszen (ed.), Vegetation and production ecology of an Alaskan arctic tundra. Springer-Verlag, New York.

Wein, R. W. 1976. Frequency and characteristics of arctic tundra fires. Arctic 29:213–222.

Wein, R. W., and L. C. Bliss. 1973. Changes in arctic *Eriophorum* tussock communities following fire. Ecology 54:845–852.

Wein, R. W., and L. C. Bliss. 1974. Primary production in arctic cottongrass tussock tundra communities. Arct. Alp. Res. 6:261–274.

Wein, R. W., and W. W. Shilts. 1976. Tundra fires in the District of Keewatin. Can. Geol. Surv. 76-1A:511–515.

Wielgolaski, F. E., L. C. Bliss, J. Svoboda, and G. Doyle. 1981. Primary production of tundra, pp. 187–225 in L. C. Bliss, O. W. Heal, and J. J. Moore (eds.), Tundra ecosystems: a comparative analysis. Cambridge University Press.

Wolfe, J. A. 1969. Neogene floristic and vegetational history of the Pacific Northwest. Madroño 20:83–110.

Yurtsev, B. A. 1972. Phytogeography of northeastern Asia and the problems of transberingian floristic interrelations, pp. 19–54 in A. Graham (ed.), Floristics and paleofloristics of Asia and eastern North America. Elsevier, New York.

Chapter 2

The Boreal Forest

DEBORAH L. ELLIOTT-FISK

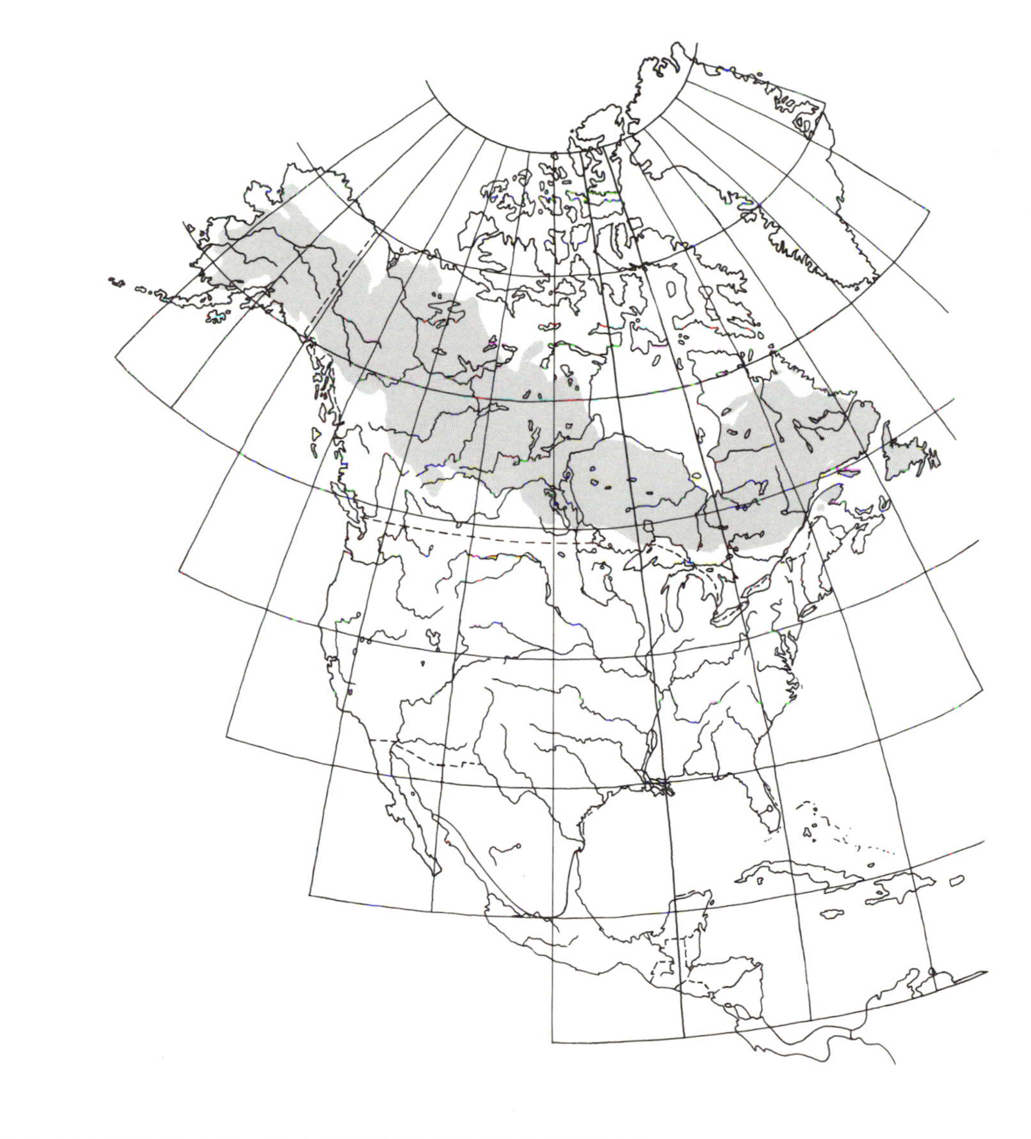

INTRODUCTION

The boreal forest of North America is a continuous vegetation belt at high latitudes stretching across the continent from the Atlantic shoreline of central Labrador westward across Canada to the mountains and interior and central coastal plains of Alaska. Larsen (1980) has referred to this major vegetation formation as the North American boreal continuum. This northern coniferous forest (or taiga) is part of the larger Northern Hemisphere circumpolar boreal forest belt.

In North America (Fig. 2.1), the boreal forest spans more than 10° of latitude in places, with transitions to the Arctic tundra to the north (Chapter 1), the subalpine forest of mountainous Alberta and British Columbia to the west (Chapter 3), the prairie grasslands to the southern interior (Chapter 9), and the Great Lakes–St. Lawrence mixed forest to the southeast (Chapter 10) (Rowe 1972). This large area contains many ecotonal communities, with boreal species ranging into the Arctic tundra, tundra species ranging southward into the boreal forest, and so forth. The forest-tundra ecotone in particular will be discussed later in this chapter in regard to its physiognomy, phytosociology, and history.

Delimitation of the boreal forest has confused many scientists because of the often broad transitions into other vegetation formations. In western Canada, the Rocky Mountain forests mingle with the boreal forest, with some species, such as *Picea glauca* (white spruce), *Pinus contorta* (lodgepole pine), and *Populus tremuloides* (quaking or trembling aspen), being components of both. Even more confusion stems from the many and differing definitions of treeline and tree limit (Hustich 1966, 1979; Elliott 1979b; Elliott-Fisk 1983).

Although the boreal forest is composed of diverse plant communities, its physiognomy is quite uniform throughout its large geographical range (Larsen 1980). This is largely because of the dominance of conifers in the formation. Despite the large size of the forest, its diversity is low compared with more temperate and tropical communities. *Picea glauca* and *P. mariana* (black spruce) are almost ubiquitous species, with *Larix laricina* (larch or tamarack), *Abies balsamea* (balsam fir), *Pinus banksiana* (jack pine), *P. contorta, Populus tremuloides, P. balsamifera* (balsam poplar), and *Betula papyrifera* (white or paper birch) also common forest components. Intermediate forms, subspecies, varieties, and ecotypes are present (Larsen 1980). Many shrubs and

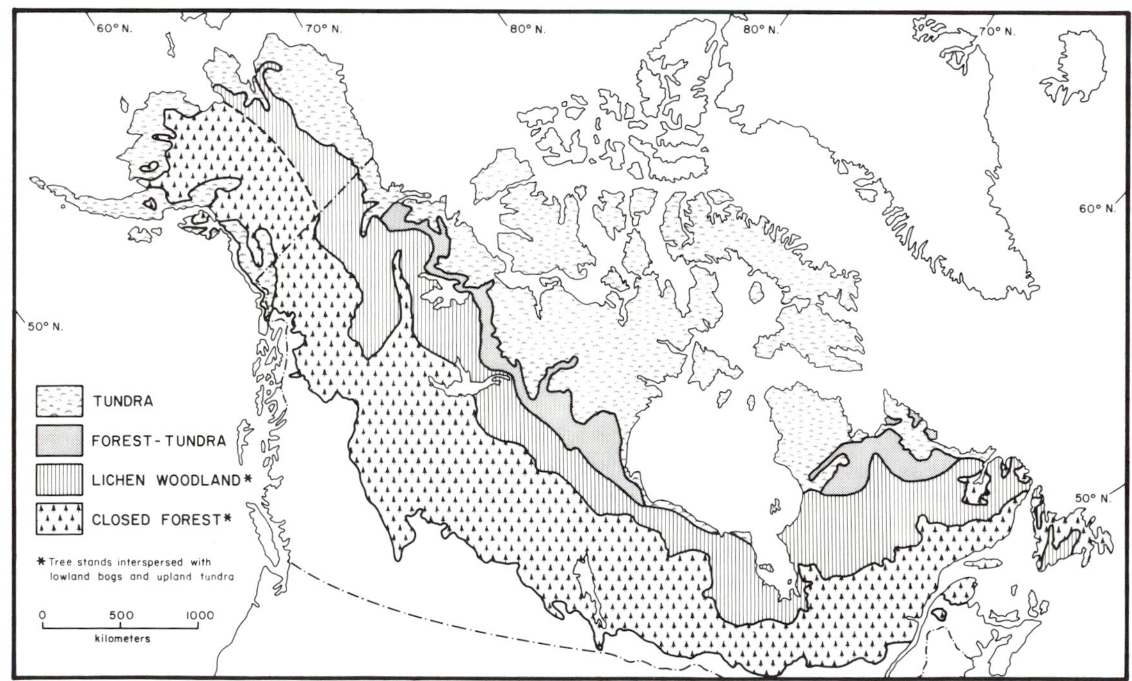

Figure 2.1. The boreal forest of North America can be broadly subdivided into three formations: closed forest, lichen woodland, and forest-tundra ecotone. The northern treeline separates the lichen woodland from the forest-tundra ecotone (or, occasionally, the tundra proper). The northern tree limit is most often found at the northern edge of the forest-tundra ecotone immediately south of the Low Arctic tundra.

Table 2.1. *Characteristics of the abiotic environment*

Parameter	West	Central	East
Geology	Precambrian shield: basement complex of granitoid-gneissic rocks rimmed by sedimentary Paleozoic rocks (outcropping along Hudson Bay and in the mountainous region of the west)		
Surficial deposits	Till and outwash predominate; lacustral and fluviatile sediments locally important; occasional bed-rock outcrops and unglaciated regions		
Topography	Mountainous (elev. > 5000 m)	Gentle plain (elev. < 300 m)	Dissected, tilted plateau (elev. 600–1000 + m)
Glaciation	Extensive coverage by Cordilleran and western edge of Laurentide ice sheets; some ice-free area in Richardson and Mackenzie mountains	Inundated by the Laurentide ice sheet	Mostly inundated by the Laurentide ice sheet; coastal and possibly alpine nunataks
Soils	Shallow Spodosols/podzols (forest and woodland), Entisols/regosols, and Histosols/fibrisols (tundra)		
Climate	Humid, cool microthermal (see Table 2.2 for details)		

herbs are also characteristic of boreal communities and will be discussed later. In particular, lichens and feathermosses are well-known understory components, playing important roles in successional processes and tree regeneration.

The North American boreal forest covers such a large geographical area that it is important to discuss the variations in physical environments of the region. In particular, climate, surficial geology, glacial history, topography, and soils influence forest structure (Table 2.1).

Climate

Climate encompasses the major group of limiting factors for the boreal forest. Although the forest spans large ranges of latitude and longitude, the regional climate may be classified as cool, humid microthermal (Hare 1950), with very cold winters of 7–9 mo allowing persistence of snow cover during all but the brief, relatively cool summer season. Its physiological processes (translocation, water absorption, rates of photosynthesis, rate of respiration) (Warren Wilson 1967; Oechel and Lawrence 1985) are often temperature-controlled and temperature-limited, though soil drought and waterlogging may play roles in limiting growth at certain sites.

The northern and southern limits of the boreal forest appear to be thermally established, though the causal relationships are as yet largely unknown (Vowinckel et al. 1975; Black and Bliss 1980; Larsen 1980). According to Larsen (1980), the northern boundary of the forest corresponds roughly to the July isotherm of 13°C, with departures due to montane or marine influences. The southern forest limit in central and eastern Canada approximates the 18°C July isotherm, though in the west the forest border shifts to slightly cooler regions with higher precipitation.

Radiation has been well investigated by Ritchie and Hare (1971) as a control over boreal forest limits. Net radiation, in particular, as a measure of available energy, may partly control the position of the northern forest border. Other climatic influences and climate summaries for the North American boreal forest have been presented by Hare (1950, 1954), Bryson (1966), Barry (1967), Larsen (1971, 1980), Hare and Ritchie (1972), Rowe (1972), Haag and Bliss (1974), Hare and Hay (1974), Miller and Auclair (1974), Streten (1974), Lettau and Lettau (1975), and Vowinckel and associates (1975).

Climatological gradients through the boreal forest reflect latitude (with temperature and net radiation generally decreasing to the north), air-mass trajectories and sources (influencing cloud cover and precipitation patterns), topography (with an altitudinal decrease in temperature), and maritime/continental position. Table 2.2 presents select climatological data for several sites west to east across the boreal forest.

Topoclimate (mesoclimatic) and microclimatic variations may be more important than microclimatic gradients in controlling forest existence. Topoclimates are especially important in mountainous regions, such as Labrador, northwestern Canada, and interior Alaska, where altitudinal forest limits may be established. The presence of a mountain range may also restrict the forest to sites south of its true climatic latitudinal limit (Elliott 1979b; Elliott and Short 1979; Elliott-Fisk 1983; Short and Elliott-Fisk in press). A forest canopy efficiently modifies the microclimate by acting as a windbreak and energy trap (Larsen 1965, 1980), providing an ameliorated environment for understory species. However, the microclimate is so modified as the

Table 2.2. *Climatological data for northern boreal forest sites in Alaska and western, central, and eastern Canada*

Parameter	Fairbanks, Alaska	Doll Creek, Yukon	Ennadai Lake, NWT	Napaktok Bay, Labrador
Mean daily temperature, January (°C)	−24	−35	−31.5	−20
Mean daily temperature, July (°C)	15	14	13	7.5
Mean annual temperature (°C)	−3	−10	−9.3	−5
Mean annual precipitation (mm)	290	240	290	500
Mean growing-season length (d)	100	120	100	75
Annual growing degree-days (5.6°C base)	1200	600	1000	550
Mean annual net radiation (kly yr^{-1})[b]	22	20	19	18
Net radiation, July (1y yr^{-1})	—[a]	—[a]	225	200

[a]Data not available.
[b]Kilolangleys per year.
Sources: Canada (1957, 1978), Wilson (1971), Hare and Hay (1974), Ritchie (1982), and Elliott-Fisk (1983).

stand develops that stand degeneration (tree exclusion) may occur in certain settings.

Geology and Glacial History

Bedrock and structural geology through the boreal forest region does not strongly affect forest existence or composition. Bedrock varies from Precambrian granitic and gneissic rocks of the Canadian Shield in eastern and central Canada to Paleozoic and Cretaceous sedimentary rocks in the region southwest of Hudson Bay, the south-central prairie−forest border, western Canada, and Alaska. Hustich (1949) and others have noted differences in forest communities dominated by either *Picea glauca* or *P. mariana* on acidic-versus-basic parent materials in eastern Canada and elsewhere, but it appears that surficial deposits may account for more of these differences than bedrock geology. As discussed briefly later, *P. mariana* is better able to tolerate shallow, poorly drained soils than is *P. glauca,* which does best on alluvial substrates. Therefore, a patterning is often seen in a forest region, with communities dominated by white spruce on the alluvial deposits along river valleys, and black spruce communities on upland bedrock sites, but also on the lowland peats.

Almost all of the region now occupied by the boreal forest of North America experienced intensive glaciation as recently as the late Pleistocene (maximum 18,000 BP). Forest communities were generally forced south with the expansion of the Laurentide and Cordilleran ice sheets, though refugia existed along the Labrador coast, the Old Crow region of the Yukon Territory, in interior Alaska, and elsewhere. It is unlikely that many of these displaced communities remained intact as boreal and Arctic species were faced with competition by taxa from more temperate regions.

Soils

Soil and vegetation types are highly correlated, with boreal forest communities most frequently associated with podzol (Spodosol) soils (Canadian system of soil classification and U.S. Seventh Approximation System, respectively). These soils are the result of podzolization, which is a consequence of low temperatures and excess precipitation above that needed for evapotranspiration. In this process, iron and aluminum (with organic materials) are leached from the upper A horizon and illuviated (deposited) in the lower B horizon. Nutrients are removed from the upper soil, which is low in important bases such as calcium. With low temperatures, soil microorganisms may be unable to decompose organic matter effectively, resulting in acidic soil conditions and low nitrogen and mineral levels (Larsen 1980).

Soil and permafrost development may thus inhibit what is thought of as the normal succession of seral stages in more temperate or tropical settings. Instead, conditions become more inimical to tree growth, nutrient cycling is limited, permafrost (frozen ground) may form, with the active layer becoming shallow as the forest canopy closes, and the site may become unfavorable for trees.

Although podzols are the dominant soils of the taiga, other soil types are found, with variations in parent material, plant cover, and other factors. Podzols are most easily derived from coarse-grained granitic rocks or sandy deposits, with regosols (entisols), fibrisols (histosols), and gray- and brown-wooded soils developing in other settings (Larsen 1980).

Minor soil changes are seen latitudinally through the boreal forest, with podzols generally shallower to the north. A transition to gray-wooded and gray-brown podzols soils and, more southerly, brown forest soils of the temperate deciduous forest proper is seen at the southern limit of the boreal forest (boreal forest–eastern deciduous forest ecotone). In this ameliorated environment, organic decomposition is more rapid, and the deciduous litter is higher in bases, with more rapid recycling of nutrients. As a result, the upper soil horizons tend to be enriched in nutrients rather than leached of nutrients (Tedrow 1970, 1977; Larsen 1980).

MAJOR VEGETATION TYPES

Although the boreal forest formation is often depicted as a monotonous coniferous forest of uniform composition and physiognomy, in reality it is a complex mosaic of different plant communities, though almost all are dominated by (principally coniferous) trees. Bogs and meadows of varying sizes are found throughout the region. Changes in dominant tree, shrub, and herb taxa can be seen latitudinally through the forest, though there are no dramatic discontinuities (Rowe 1972; Larsen 1980).

The term "taiga" is broadly used in ecological literature and may be defined as a coniferous northern forest with no admixture of nonconiferous species except *Betula* and *Populus* (Ritchie 1962; Larsen 1980). The forest canopy becomes more open and generally shorter northward, with a transition from closed forest to open lichen woodland to scattered trees of the forest-tundra ecotone.

Rowe (1972) has divided what he terms the "boreal forest region" of Canada into 35 sections. Neiland and Viereck (1978) have prepared a detailed and very workable subdivision of the "forest types and ecosystems" of Alaska, with basic divisions into bottomland, lowland, and upland forests. Alaska's boreal forest is perhaps more diverse in both type and pattern than that over all of Canada, because of the complex topography of the state. Neiland and Viereck point out that the mosaic of forest communities is a response to (1) topography, microtopography, climate, microclimate, river flooding, permafrost occurrence and depth, organic matter, and fires, (2) chance, and (3) variations in reproductive abilities, productivity, and distribution patterns of the various species.

Boreal Forest Regions

Larsen's (1980) division of the North American boreal forest zone (continuum) into seven regions is most easily used and will be outlined briefly here. The Alaskan taiga (region 1) is a mixture of upland and lowland forests dominated by *Picea glauca, P. mariana, Betula papyrifera,* and *Populus tremuloides* (Fig. 2.2). Fire and permafrost contribute to successional processes and vegetation patterning.

The boreal forests of the Cordillera or northern Rocky Mountains occupy large sections of the Mackenzie and Richardson mountains of the Yukon Territory and the northwestern Mackenzie District of the Northwest Territories (region 2). Ritchie and his colleagues (Ritchie 1962, 1977, 1982; Ritchie and Hare 1971; Cwynar and Ritchie 1980; Ritchie and Cwynar 1982; Spear 1982; MacDonald 1983) have contributed a great deal to our understanding of forest dynamics in this region. The forest reaches its northernmost location here along the Mackenzie River delta, extending almost to the Arctic Ocean (Larsen 1980).

In southwestern Mackenzie and northern Alberta, the interior (region 3) forest is composed predominantly of *Picea mariana, P. glauca, Pinus banksiana, P. contorta* var. *latifolia, Populus tremuloides,* and *P. balsamifera,* with occasional stands of *Larix laricina* and *Abies balsamea*. This forest extends from the Cordilleran foothills eastward to the western edge of the Canadian Shield.

On the Canadian Shield proper (region 4), both west and east of Hudson Bay, the forest is relatively uniform in appearance, with topographic and microclimatic factors affecting vegetation structure at certain sites. The relative importances of the tree species change from south to north, with richer forests in the south, *Picea mariana, P. glauca,* and *Pinus banksiana* increasing in dominance northward, and *P. mariana* being dominant near the northern forest border. Boreal forest outliers to the south exist in northern Minnesota, Wisconsin, and Michigan, with communities transitional to the northern conifer-hardwood forest (Curtis 1959), but disturbance by agriculture and logging has severely altered many of these stands (Larsen 1980).

In the Gaspe-Maritime region of eastern Canada (region 5), a wide variety of forest communities is found, as best illustrated by the finding (Rowe 1972) of 18 major divisions of the boreal forest there. Many of these forests are closely related to those of the Great Lakes–St. Lawrence region, as is apparent from their understory vegetation. Again, the boreal forest of eastern Canada is predominantly coniferous, but with *Abies balsamea* playing a more dominant role there than elsewhere, especially on well-drained uplands. In contrast, monospecific stands of *Picea mariana* are rare (Larsen 1980).

The forest of Labrador-Ungava (region 6) has been intensively studied by Hustich (1949, 1950, 1954) and others. Although *Picea mariana* and *P. glauca* are often dominants, *Abies balsamea, Larix laricina, Betula paprifera, Populus tremuloides, P. bal-*

Figure 2.2. Boreal forest along the Klutina River west of the Wrangell Mountains in interior Alaska (elev. 500 m). The closed forest here is dominated by Picea glauca *on the uplands and* P. glauca, Populus balsamifera, P. tremuloides, *and* Betula papyrifera *on the terrace slopes and lowlands. The understory vegetation varies with the degree of canopy closure, with mosses dominating in closed sites, and woody shrubs (*Vaccinium, Rosa, *and* Salix *species) under canopy openings.*

samifera, Pinus banksiana, and *Thuja occidentalis* (white cedar) are also found. The complex topography of this dissected, faulted, and glaciated plateau has led to the development of a mixture of community types partitioned along moisture and geologic gradients. Black spruce communities become increasingly important to the north, with understory vegetation changing in species frequency and composition while the arborescent stratum remains essentially unchanged (Larsen 1980). Elevational treelines are also encountered, reaching progressively lower elevations northward.

Much research has been concentrated along the northward forest border (region 7), which is also variously termed the continental Arctic treeline, the northern tree limit, the northern forest–tundra ecotone and so forth. The forest usually extends northward along river lowlands, with tundra found to the south along upland surfaces (Larsen 1980). *Picea mariana* again dominates, but *P. glauca* is present on well-drained sites, and *Larix laricina* in boggy situations. Occasionally, all three species are found together in outlier stands.

Closed forest. The boreal forest can be broadly divided into three structural units (formations): closed forest, lichen woodland, and forest-tundra ecotone. These will be briefly discussed, along with shrublands, whose ecology is as yet poorly understood.

Over large areas, closed forest communities dominate the southern boreal forest zone on a variety of topographic sites, soils, and lithologies. Though the species composition of these stands may change, the structural appearance of the forest is almost unchanged throughout.

The characteristic forest community here is the "spruce-feathermoss forest," with either *Picea mariana* or *P. glauca* the dominant (Fig. 2.3). La Roi (1967; La Roi and Stringer 1976) has classified these forests as follows: (1) western, black spruce–feathermoss forest (mixed wood); (2) eastern, black spruce–feathermoss forest and white spruce–fir–feathermoss forest (mixed wood).

The transition from western to eastern forest occurs along a longitudinal axis from western Lake Superior to Lake Winnipeg to Churchill (South-

Figure 2.3. Eastern white spruce–feathermoss forest at Napaktok Bay, Labrador. The tree stratum here is composed solely of Picea glauca. *This stand is at the northern tree limit; yet the population is vigorously reproducing (see Fig. 2.6). The bryophyte stratum interfingers with an irregular, patchy, dwarf shrub stratum of principally* Betula glandulosa. *The maximum age of trees in this stand is ~400 yr.*

western Hudson Bay). The semipermanent trough of the upper (westerly) circumpolar vortex is situated over Hudson Bay and this transitional area. These western and eastern regions experience different temperature, precipitation, and radiation regimes, especially as witnessed through changes in humidity and potential evapotranspiration.

La Roi and Stringer (1976) classified 60 boreal spruce-fir forest stands across North Amercia. Their stands occur in 24 of Rowe's (1972) 38 boreal forest sections. Both the vascular flora and bryophyte assemblages proved useful in this classification, with the complex patterns of both gradual and abrupt changes in species composition giving a geographic basis for the two primary stand types (Figs. 2.4 and 2.5).

The black spruce–feathermoss forests have a uniform tree stratum that is moderately dense, with an understory of almost continuous bryophyte cover. In contrast, the tree stratum of the white spruce (mixed) forests is more irregular and open, with strata of broadleaf shrubs and herb–dwarf shrubs that are both dense and species-rich, with a patchy bryophyte stratum. The white spruce–fir forests are 50% richer in number of bryophyte species than the black spruce forests, even though the latter have a much higher bryophyte cover (La Roi and Stringer 1976). The vascular flora is also richer on the average (Table 2.3). The higher overall diversity may be a function of the higher productivity and greater mean age of the white spruce–fir stands or of a greater habitat diversity. The microclimates of these stands should be studied, because the physiognomy of white spruce–feathermoss forests across Canada appears to be much more diverse than that of the closed black spruce–feathermoss stands.

Hylocomium splendens and *Pleurozium schreberi* are the characteristic cover dominants in bryophyte strata of the spruce-feathermoss forests, though *H. splendens* dominates in the mixed woods, with *P. schreberi* being the primary species in the black spruce forests. Though the two species can be considered ubiquitous, their distributions may reflect different microclimatic conditions in the two types of communities.

Although La Roi and Stringer's community analysis shows discrete clusters of *Picea mariana* versus *P. glauca* communities in the southern (closed) boreal forest, Larsen (1972, 1974, 1980) has found that *P. mariana* forest stands are almost uniform in structure throughout their latitudinal range. This is most likely a function of similar habitat conditions, with the trees exerting a considerable influence on their own environment. The growth and closure of the forest canopy result in warmer winters, cooler summers under calm conditions, warmer summers in windy sites, higher moisture, encroachment of the permafrost table, and poor drainage.

To the north, lichen woodlands dominate the landscape. The feathermosses require a high precipitation input, with summer humidity especially critical; thus, their importance decreases northward as climates become cooler and more xeric, with lichens then dominating the lower forest strata (La Roi and Stringer 1976). The partitioning of tree, shrub, and herb taxa along moisture gradients is obvious in the southern closed forest (Table 2.4) and contributes to community development (Larsen 1980).

Lichen woodland. The lichen woodlands of northern Canada, in particular, have intrigued scientists for years. These woodlands have been variously regarded as a successional sere following fire in the progression to closed boreal forest or muskeg, as a fire climax, and as a true climax community. These woodlands dominate the northern region of the boreal forest zone (Hare 1959; Ritchie 1962; Kershaw 1977; Larsen 1980; Elliott-Fisk 1983). The transition from southern closed spruce-feathermoss forest to lichen woodland is sharp in many areas (such as northern Manitoba and Saskatchewan), but transitional in others, with stands of lichen woodland alternating with closed forest and mus-

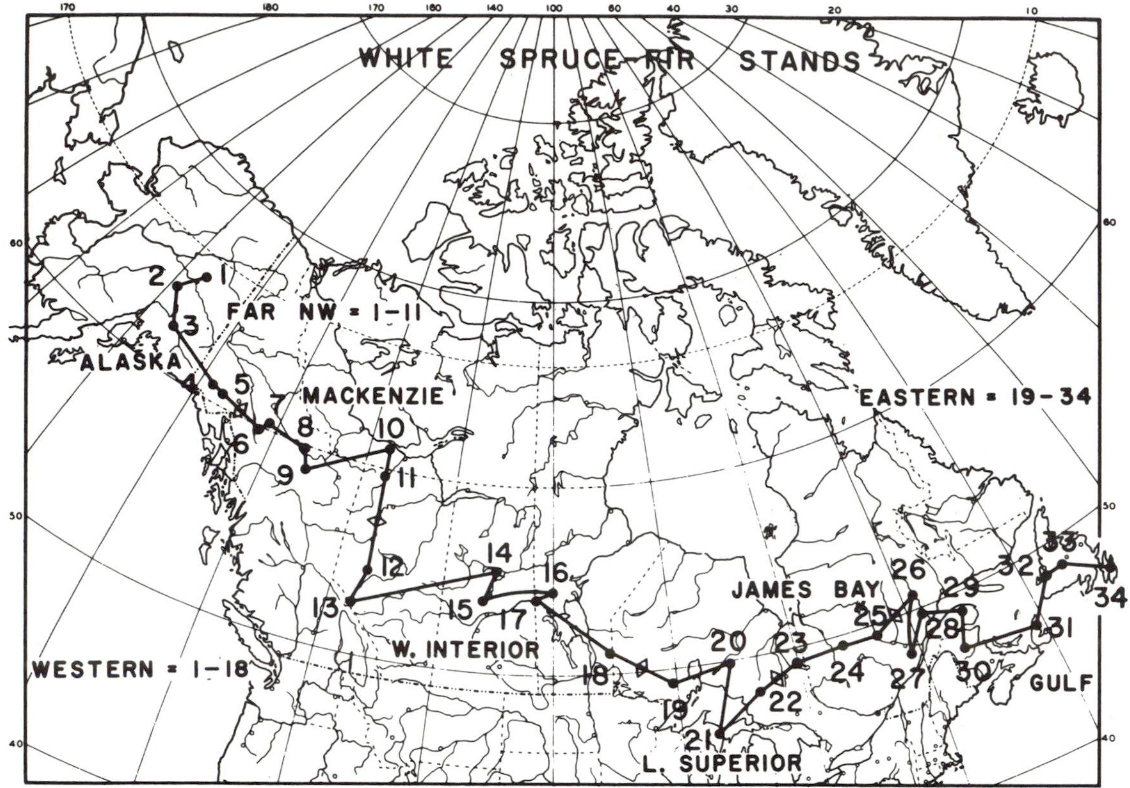

<table>
<tr><td colspan="7">Classification of White Spruce - Fir Stands on Bryofloristic Criteria</td></tr>
<tr><td colspan="7">HYLOCOMIUM SPLENDENS – PLEUROZIUM SCHREBERI Order</td></tr>
<tr><td colspan="4">CERATODON PURPUREUS – PYLAISIELLA POLYANTHA Alliance</td><td colspan="3">RHYNCHOSTEGIUM SERRULATUM – TETRAPHIS PELLUCIDA Alliance</td></tr>
<tr><td>RHYTIDIUM RUGOSUM – DICRANUM ACUTIFOLIUM Association</td><td>AULACOMNIUM PALUSTRE – LOPHOZIA SPP Association</td><td colspan="2">MNIUM CUSPIDATUM – ONCOPHORUS WAHLENBERGII Association</td><td>RHYTIDIADELPHUS TRIQUETRUS Association</td><td colspan="2">BAZZANIA TRILOBATA – CEPHALOZIA MEDIA Association</td></tr>
<tr><td>No. 1 2 3 4 5</td><td>6 7 8 9</td><td>10 11 12 13</td><td>14 15 16 17</td><td>18 19 20 21 22 23 24</td><td>25 26 27 28 29 30 31</td><td>32 33 34</td></tr>
<tr><td colspan="2">Solidago multiradiata – Empetrum nigrum Stand Group</td><td>Lonicera dioica / Rubus pubescens – Lathyrus ochroleucus Stand Group</td><td>Lonicera dioica / Vaccinium vitis-idaea – Geocaulon lividum Stand Group</td><td>Corylus cornuta / Diervilla lonicera / Aster macrophyllus – Anemone quinquefolia Stand Group</td><td>Oxalis montana – Gaultheria hispidula Stand Group</td><td>Dryopteris austriaca – Trientalis borealis Stand Group</td></tr>
<tr><td colspan="3">Populus spp./Salix spp./Shepherdia canadensis</td><td colspan="2">Abies balsamea - Populus spp.// Rosa acicularis / Mertensia paniculata</td><td colspan="2">Abies balsamea / Acer spicatum – Pyrus spp.// Clintonia borealis</td></tr>
<tr><td colspan="7">PICEA GLAUCA - ABIES Community Type</td></tr>
<tr><td colspan="7">Classification of White Spruce - Fir Stands on Vascular Floristic Criteria</td></tr>
</table>

B

Figure 2.4. Picea glauca – Abies stands in the North American boreal forest (A) have been classified by La Roi and Stringer (1976) (part B) using both vascular and bryophyte floristic criteria.

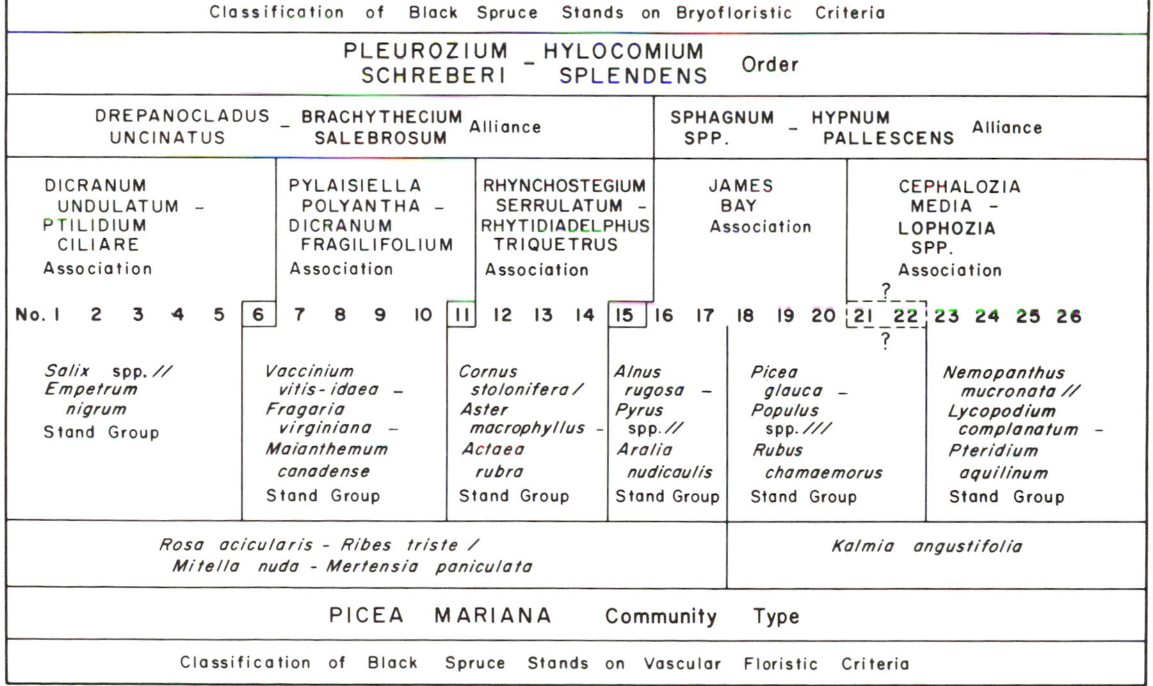

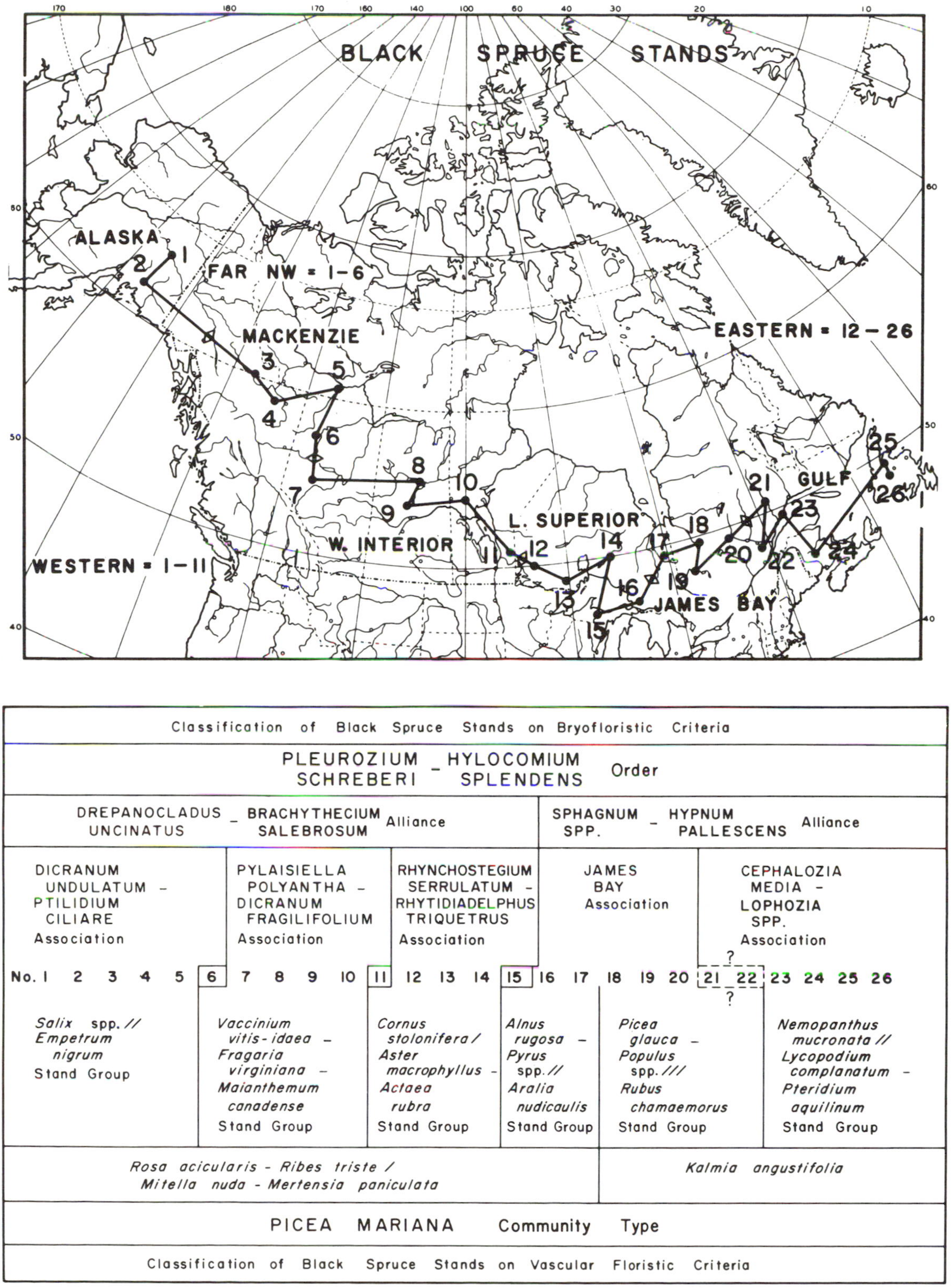

Figure 2.5. Picea mariana *stands in the North American boreal forest (A) have been classified by La Roi and Stringer (1976) (part B) using both vascular and bryophyte floristic criteria.*

Table 2.3. *Vascular flora of spruce-feathermoss forests*

Stratum	Presence class	White spruce–fir	$\bar{X}$ spp.	Black spruce	$\bar{X}$ spp.
Tree	61–100%	*Picea glauca*	4.8	*Picea mariana*	4.3
		Betula papyrifera		*Betula papyrifera*	
		Abies balsamea			
		Picea mariana			
	41–60%	*Populus tremuloides*		*Picea glauca*	
		Populus balsamifera		*Abies balsamea*	
				Populus tremuloides	
Low tree & tall shrub	41–60%	*Alnus crispa*	3.3	*Alnus crispa*	2.2
		Pyrus decora			
Medium & low shrub	61–100%	*Viburnum edule*	6.4	*Ledum groenlandicum*	6.4
		Rosa acicularis		*Rosa acicularis*	
		Ribes triste			
	41–60%	*Rubus idaeus*		*Viburnum edule*	
Herb-dwarf	61–100%	*Linnaea borealis*	28.4	*Cornus canadensis*	22.9
		Cornus canadensis		*Linnaea borealis*	
		Pyrola secunda		*Maianthemum canadense*	
		Mitella nuda		*Pyrola secunda*	
		Maianthemum canadense		*Gaultheria hispidula*	
		Rubus pubescens		*Coptis groenlandica*	
		Moneses uniflora		*Geocaulon lividum*	
				Vaccinium myrtilloides	
	41–60%	*Trientalis borealis*		*Clintonia borealis*	
		Goodyera repens		*Trientalis borealis*	
		Lycopodium annotinum		*Vaccinium angustifolium*	
		Aralia nudicaulis		*Goodyera repens*	
		Clintonia borealis		*Epilobium angustifolium*	
		Epilobium angustifolium		*Equisetum sylvaticum*	
		Mertensia paniculata		*Mitella nuda*	
		Coptis groenlandica		*Petasites palmatus*	
		Dryopteris austriaca		*Rubus pubescens*	
		Viola renifolia		*Equisetum arvense*	
		Gymnocarpium dryopteris		*Vaccinium vitis-idaea*	
		Pyrola asarifolia			
		Streptopus roseus			
Total for stand			42.9		35.8

Source: La Roi (1967).

keg according to topographic position. This zonal pattern is especially disrupted in Alaska and the Yukon Territory because of the mountainous terrain (Larsen 1980).

Kershaw (1977) has classified lichen woodlands into two broad groups (1) a western *Stereocaulon paschale* woodland and (2) eastern *Cladonia stellaris* woodland (Table 2.5). Although knowledge of the environmental controls of these formations is incomplete, the transition of woodland types occurs southwest of Hudson Bay (in the same area as the western–eastern spruce-feathermoss forest transition) and may be a function of summer moisture regime (Kershaw 1977). Edaphic conditions may be important as well, but they are intimately tied to microclimate. Undoubtedly, the mosaic of vegetation types seen in many areas is a function of soil characteristics, as well as the fire history of the region.

Picea mariana and *P. glauca* are the dominant trees of the lichen woodlands, with *P. mariana* increasing in importance to the north. Although tree density is obviously less in these woodlands than in the closed forest, tree stature is often equal to, or greater than, that in the closed forests. Branching also occurs to ground level because of higher light levels in the open canopy.

Forest-tundra ecotone. The location of the forest-tundra ecotone reflects the general circulation of the atmosphere, with irregularities due to differences in topography, glacial history, bedrock, fire history, and topoclimatic and microclimatic influences. The ecotone is found north of the treeline, which is defined as the boundary separating arboreal and nonarboreal vegetation, with forest/woodland communities dominating at least 50% of the terrestrial landscape at, and south of, the treeline

Table 2.4. *Distribution of select tree, shrub, and herb species along moisture gradients in the southern boreal forest*

Strata	Very dry/dry	Fresh	Moist	Very moist	Wet
Trees		*Pinus banksiana*			
				Larix laricina	
			Picea glauca		
		Betula papyrifera			
			Populus tremuloides		
			Populus balsamifera		
			Abies balsamea		
			Picea mariana		
Tall shrubs	*Alnus crispa*	*Amelanchier alnifolia*	*Acer spicatum*	*Acer negundo*	*Alnus rugosa*
Medium shrubs	*Juniperus communis*	*Vaccinium myrtilloides*	*Lonicera dioica*	*Ledum groenlandicum*	*Betula glandulosa*
Herbs	*Arctostaphylos uva-ursi*	*Viola rugulosa*	*Petasites palmatus*	*Geocaulon lividum*	*Petasites sagittatus*
	Lycopodium complanatum	*Lycopodium obscurum*	*Mertensia paniculata*	*Equisetum scirpoides*	*Equisetum arvense*
	Potentilla tridentata	*Pyrola asarifolia*	*Vaccinium vitis-idaea*	*Mitella nuda*	*Rubus chamaemorus*

Source: Larsen (1980).

Table 2.5. *Arboreal density and understory cover and composition of western and eastern lichen woodlands*

Parameter	Western	Eastern	Parameter	Western	Eastern
Mean density:			Club moss and moss		
(trees ha^{1})			*Pleurozium* spp.	19.3	0
Larix laricina	16	10	*Lycopodium annotinum*	0.1	0.6
Picea glauca	39	124	*Dicranum* spp.	0.1	2.3
Picea mariana	445	422	*Hylocomium splendens*	0	1.0
Total	500	556	*Polytrichum* spp.	5.8	1.8
Composition			Total	5.7	30.4
and mean cover of:			Lichens		
Shrub layer			*Cetraria* spp.	0.4	0.5
Betula glandulosa	9.5	5.5	*Cladonia alpestris*	—	78.1
Empetrum nigrum	5.0	1.7	*C. mitis*	—	2.1
Ledum groenlandicum	1.9	4.3	*C. rangiferina*	—	–
Vaccinium angustifolium	0	4.0	Other *Cladonia* spp.	16.9	9.3
V. uliginosum	0.6	1.0	*Stereocaulon paschale*	15.8	5.0
V. vitis idaea	9.6	0.5	Total	34.0	97.4
Total	27.2	17.0	Grand total (shrub and ground layer)	91.6	120.4
Grass layer					
Festuca ovina	0	0.3			
Total	0	0.3			

Sources: Western, Small Tree Island, N.W.T. (Maini 1966); eastern, Schefferville, Nouveau Quebec (Rencz and Auclair 1978).

(Larsen 1974; Elliott-Fisk 1983). Physiological controls over the treeline have been discussed by Billings and Mooney (1968) and Oechel and Lawrence (1985). Tranquillini (1979) has given a physiological definition for the treeline, stating that it is the limit beyond which forest existence is no longer possible because damage halts development somewhere in the life cycle from seed set to maturity. Unsuccessful or irregular sexual regeneration has been well documented for these northernmost tree stands (Elliott 1979a, 1979b, 1979c; Black and Bliss 1980; Elliott-Fisk 1983) (Fig. 2.6).

North of the treeline, woodland patches, clones, and individual trees are scattered over the tundra

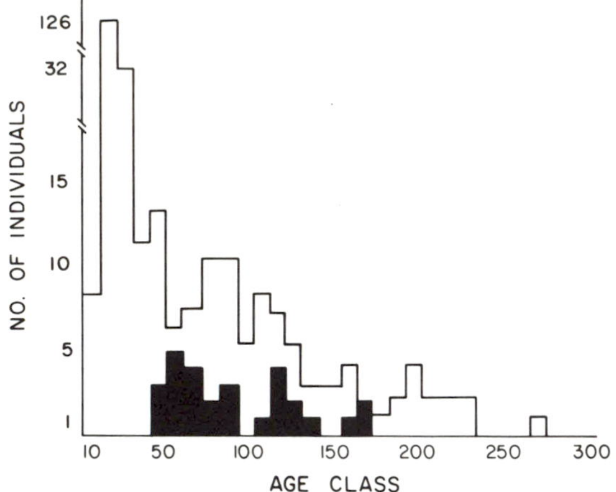

Figure 2.6. The age structures of isolated stands of Picea glauca *in the forest-tundra ecotone at Ennadai Lake, N.W.T. (black bars) and Napaktok Bay, Labrador (open bars) show distinct differences. The Ennadai Lake stand is not in equilibrium (note lack of juveniles), and it is maintaining itself through layering, whereas the Napaktok Bay stand is in equilibrium and is maintaining itself through seed production.*

Figure 2.7. This mixed stand of Picea glauca, P. mariana, *and* Larix laricina *is found in a ravine in the forest–tundra ecotone at Ennadai Lake, N.W.T. Many typical boreal forest taxa are found in the understory, and paleoecological data suggest that this stand is a relict of a formerly more extensive forest cover. Source: Elliott (1979a), permission of Arctic and Alpine Research and Regents, Univ. Colorado.*

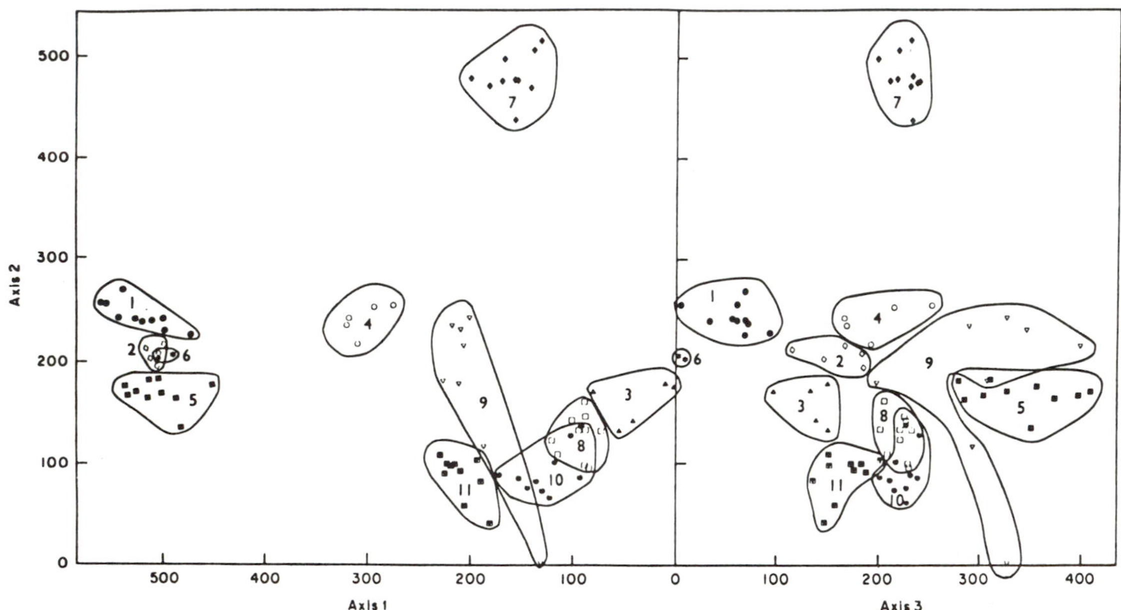

Figure 2.8. DECORANA ordination analysis of vegetation plots from the Doll Creek area of the northern Yukon Territory illustrates the diversity of plant communities in the forest-tundra ecotone (Ritchie 1982). The subjectively determined vegetation types are (1) tundra, (2) black spruce woodlands on moraine surfaces, (3) snowpatch, (4) limestone cliff and crag, (5) black spruce–larch on bottomland mires, (6) open white spruce forest on steep shale-sandstone surfaces, (7) tundra, (8) tundra, (9) white spruce woodlands on south- and southwest-facing limestone slopes, (10) white spruce woodland on north-facng limestone slopes, and (11) spruce-larch woodland at treeline on north-facing limestone slopes.

landscape (Fig. 2.7), eventually reaching the northern tree limit (limit of tree growth). Occasionally, tree species may be found assuming a shrub growth form even north of this, where they form the northern tree species limit (Elliott-Fisk 1983). The diversity of plant communities in the ecotone is high and is related to environmental gradients and site history. This is well illustrated by Ritchie's (1982) ordi-

Figure 2.9. Dwarf clones of Picea mariana, *as shown here in the Northwest Territories, are frequently seen scattered through the forest-tundra ecotone. This pattern is most pronounced over flat terrain that was inhabited by forest/woodland during the Hypsithermal.* Alnus crispa *shrubs are found in the same stratum, with an understory of various low ericaceous shrubs and lichens. An* Eriophorum *species bog is visible between spruce clones in the foreground and along the lake shore.*

nation of communities in the Doll Creek region of the northern Yukon Territory (Fig. 2.8).

The width of the forest-tundra ecotone varies across North America, narrower at both its western and eastern ends and up to 235 km in width in central Canada. In Quebec, the ecotone may exceed 300 km in width, with isolated woodland patches found in river lowlands where the rest of the landscape is tundra-covered. The decrease in elevation from Schefferville north toward Ungava Bay has effectively widened the vegetation zones by lessening the apparent latitudinal decrease in temperature northward (Hare 1950; Elliott-Fisk 1983). The western Cordillera may serve to compress latitudinal temperature variation, effectively anchoring air-mass trajectories (Sorenson 1977). The dissection of the landscape has also decreased the width of the ecotone, with severe climatic conditions on ridge tops blocking migration routes. Historical fluctuations of the ecotone, as reflective of climate changes, have been greatest in the relatively flat interior plains of Canada and least along its western and eastern margins.

The physiognomy of the ecotone vegetation is extremely variable across North America, though species compositions remain similar. In certain areas, such as southwestern Keewatin, small clonal spruce may extend up river valleys as outliers, perhaps 50 km from their nearest neighbors. Trees here are typically dwarfed (<2 m tall) and only occasionally krummholzed (Fig. 2.9). Marr (1948) has documented the dramatic changes in tree growth forms both away from the coast and to the north for the Great Whale River region east of Hudson Bay. Other areas of the ecotone have very large woodland patches that often have an understory of true boreal species, as related to the history of the site (Larsen 1965, 1972; Elliott 1979a, 1979c; Elliott and Short 1979; Elliott-Fisk 1983).

Larsen (1965, 1972, 1974, 1980) has investigated the plant communities of the forest-tundra ecotone in western and central Canada. Many of the species found here have large geographic ranges, occurring in both the northern boreal forest and Low Arctic tundra. However, the majority of ecotonal communities are depauperate, because neither true Arctic nor boreal species can occupy this zone of variable climatic (frontal) conditions. Larsen found that only the more ubiquitous species occurred in sufficient abundance that they appeared regularly with high frequencies in transects. The physiological characteristics of the plants that account for this are unknown.

Shrublands

Shrublands of varying composition are found throughout the boreal forest. Although these communities may in some instances be successional to forest or woodland, at other sites they appear to be the "climax" vegetation.

Dense thickets 2–6 m tall of *Alnus crispa* (green alder), *Betula glandulosa* (dwarf birch), and various *Salix* species (willow) have been documented across Canada and Alaska (Fig. 2.10). These stands are most frequently seen near the northern elevational treeline and frequently are classified as Arctic tundra rather than boreal in nature. Neiland and Viereck (1978), Oswald and Senyk (1977), Short (1978a, 1978b), Elliott and Short (1979), and Short and Elliott-Fisk (in press) have described these communities at upland sites in Alaska, the Yukon Territory, and Labrador-Ungava. Some of these stands show little or no sign of tree invasion, though site conditions appear amiable for tree growth. Further research on these shrublands and their history is needed. See also Chapter 1, pages 10-11.

Bogs

Various wetland plant communities are found in the boreal forest region. Bogs are the most common of these and have been defined by Curtis (1959) as plant communities of specialized shrubs and herbs growing on a wet, acidic, peat substrate. Trees are found as a minor component of bog vegetation in many areas. Bogs can be contrasted with fens, which form under more alkaline conditions on wet peat. Heinselman (1963, 1970) and others have devised more elaborate schemes for classifying northern

Figure 2.10. Extensive tall shrublands are associated with forest/woodland vegetation in certain regions of the northern boreal forest. Dark, dense thickets of Alnus crispa, Betula glandulosa, *and* Salix planifolia *are shown here just north of Hebron Fiord along the central Labrador coast.*

wetlands based on topography, hydrology, water chemistry, and landscape evolution (Larsen 1982).

Few studies have been conducted on northern bogs, largely because of their poor accessibility. These communities are widespread in regions that have undergone continental glaciation. Bogs have formed in kettle lakes and shallow ponds, on stable lake margins, in marshland, along stream margins, and over dense shrublands. Many of them are apparently long-lived, whereas others are ephemeral in sites that are frequently disturbed (Larsen 1982).

Bog taxa are predominantly members of the Orchidaceae or Ericaceae, though Cyperaceae, Asteraceae, and Poaceae species are not uncommon. Bog ericads and sedges generally develop on a layer of *Sphagnum* species, with black spruce a nearly ubiquitous component of treed bogs. *Larix laricina* is frequently found in bogs possessing an open tree canopy. In southern Alaska, *Picea mariana* is largely replaced by coastal forest species such as *Tsuga heterophylla, Thuja plicata,* and *Chamaecyparis nootkatensis,* and also by *Pinus contorta,* which possesses a taproot and may survive hygric conditions longer than other conifers. All of these taxa are ombrotrophic and adapted to at least seasonally saturated, acidic, nutrient-poor, organic soils (Larsen 1982). Input of nutrients into the system occurs principally through wet or dry atmospheric deposition.

Bog vegetation is uniform in physiognomy throughout the boreal forest (Larsen 1982). The component species are evergreen xeromorphs because of the short growing season, the fluctuating groundwater table, and the nutrient-deficient substrate. Nutrients are largely retained in the biomass.

Bog succession has interested scientists for several decades. Trees of several species may be found around the margins of bogs and their meadows, and in many instances they appear to be invading the bog vegetation. However, in many cases that is not an invasion. Many of these trees are dwarfed adults several decades in age that have had slow growth rates because of the low nutrient availability and saturation of the bog soils (Lawrence 1958). Long-term survival under these conditions is tenuous unless the tree can extend its roots into a more favorable substrate or unless the climate changes and the site becomes more mesic. Many bogs serve as valuable recorders of climate and other environmental changes; consequently, bog deposits have been investigated by paleoecologists interested in both succession and climatic history in various regions of the boreal forest.

COMMUNITY DYNAMICS

Succession

The successional status of North America's boreal forests has intrigued scientists for many decades, especially regarding the role that fire plays in natural ecosystem dynamics. Researchers have focused on the question whether or not the closed spruce forest is the ultimate (climax) stage in boreal succession (Larsen 1980). Various workers have argued the question pro (Raup and Denny 1950; Lutz 1956; Kershaw 1977; Oswald and Senyk 1977; Black and Bliss 1980; Larsen 1980) and con (Rowe 1961; Strang 1973; Viereck 1973; Black and Bliss 1978; Larsen 1980) with regard to particular forest communities at certain sites. In regard to the entire boreal forest, the best answer undoubtedly is "perhaps" (Larsen 1980), because site conditions exert a considerable influence on the successional process.

In this light, it is important to understand that the boreal forest is a "disturbance forest," as argued by Rowe (1961): "In short, there are no species in the Canadian western boreal forest possessing in full the silvical characteristics appropriate to participation in a self-perpetuating climax. The boreal forest is a disturbance forest, usually maintained in youth and health by frequent fires to which all species, with the probable exception of fir, are nicely adapted."

Larsen (1980) has argued that traditional concepts of forest succession are inapplicable here: Instead of succession resulting in more mesophytic conditions (as occurs in the most temperate regions), tree survival becomes more difficult as the canopy closes. Some of the degradative environmental changes acompanying succession in the boreal forest are (1) an increase in thickness of the organic mat, (2) a decrease in nutrient availability, (3) a decrease in summer soil temperatures, (4) an increase in the level of the permafrost table (i.e., a decrease in active-layer depth), (5) a decrease in soil drainage, resulting in waterlogging, anaerobic conditions, and gleization, and (6) an increase in frost heave and thurst. Patchiness of the microenvironments (substrate, fire, flooding, logging, and other disturbances) contributes to the mosaic of seral communities seen in the region and has led Larsen (1980) to state that succession does not progress to a single regional climax. The "climax" could be a closed forest on eskers, floodplains, and other well-drained sites, or it may be an open muskeg or bog in lowlands and other poorly drained sites. Carleton and Maycock (1978), in their work in the James Bay region of Ontario, found through ordination studies that species did not show any tendency to progress toward a single climax type, though certain species tended to associate in seral stages.

Succession in bogs and other wetland areas in the boreal forest region is poorly understood. As Heinselman (1963, 1970) stated, the only trend in bog succession is toward landscape diversity, not mesophytism. Topography, the biogeochemistry of the site, climate, and disturbance all alter the successional process such that there is a wide variety of bog/peatland communities. Nevertheless, Gates (1942) has outlined a general pattern of boreal bog succession, with *Chamaedaphne* species pioneering succession around the edge of a depression in to open water; a mat soon forms from roots and rhizomes, growing over the pool, with organic matter deposited on the lake bed; this surface is then invaded by shrubs and moss (generally *Sphagnum*), forming a dense thicket conducive to black spruce and larch establishment. *Carex* species typically pioneer secondary succession on peatlands following disturbance (Larsen 1982). Fire, drought, and other disturbances alter the moisture and nutritional status of the peatland, adding to the diversity of peatland plant communities throughout the boreal forest.

Fire has been documented as the primary disturbance factor disrupting successional processes in the boreal forest (Rowe and Scotter 1973; Viereck 1973; Kershaw 1977; Larsen 1980). Many tree species of the boreal forest are adapted to fire, as shown by their morphologic and reproductive characteristics (Table 2.6) (Rowe and Scotter 1973).

Populus balsamifera (Fig. 2.11) is the most resistant of all the boreal forest trees to destruction by fire, a function of its thick bark (>10 cm near the ground) and regeneration by root suckers as well as seeds (Larsen 1980). Neiland and Viereck (1978) have rated balsam poplar alluvial forests, along with those composed of *Populus trichocarpa* (black cottonwood), as the most productive of all the Alaskan taiga communities.

Although the conifers are not able to stump-sprout or root-sucker, all but *Abies* retain some of their seeds on the tree. Thus, a seed source (especially for species that are serotinous) is readily available following a fire. Layering may also be adaptive. Layering has been documented for *Picea mariana* in many regions, resulting in stand stability (maintenance) in the absence of fire or major climatic deterioration. Black and Bliss (1980), in their studies of black spruce communities at the northern treeline and tree limit near Inuvik (Northwest Territories), found layering able to replace the loss

Table 2.6. *A rating of regenerative successes of conifers following fire*[a]

Parameter	Pine	Black spruce	Larch	White spruce	Balsam fir
Seed retention on tree	10	6	2	3	1
Early seed production	5	4	2	3	1
Seed mobility	2	5	5	3	1
Seedling frost-hardiness	5	3	4	2	1
Seedling growth rate	5	2	4	3	1
Seedling palatability	3	5	2	4	1
Seedling response to full sun	4	3	5	2	1
Totals	34	28	24	20	7

[a]Relative scale from high (10) to low (1).
Source: Rowe and Scotter (1973).

Figure 2.11. Populus balsamifera *is the most fire-resistant tree of the boreal forest. Here individuals 20 m or greater in height are important components of the vegetation at the southern margin of the boreal forest in Elk Island National Park, Alberta.*

of dominant canopy trees during periods of fire exclusion. If sexual reproduction is rare or infrequent, however, as is often the case near the northern treeline (Larsen 1965; Elliott 1979a, 1979c; Black and Bliss 1980; Elliott-Fisk 1983), fire can destroy stands that are normally maintained through layering, and a rapid retreat of the northern forest border can occur. This is believed to be the explanation for very rapid, large-scale retreats of the northern Canadian treeline with neoglacial climatic deterioration.

Lastly, the question of the seral versus climax status of lichen woodland needs to be addressed. There are compositional (if not structural) variations in lichen woodlands across North America, with the most important being understory dominance by *Stereocaulon paschale* in the west versus *Cladonia stellaris* in the east. Lichens can dominate the ground vegetation rapidly if an area is cleared by fire, especially if the substrate is such that water is in physically short supply (Kershaw 1977). Under these conditions, tree cover is inhibited long enough for a complete cover of lichens to develop. The lichen mat will then serve to decrease summer soil temperatures, further retarding tree growth and, for at least a period of time, conserving the open nature of the woodland. Brown and Mikola (1974) have also found that *C. stellaris* inhibits the growth of pine and spruce seedlings, though the mechanisms involved in this process are not known (Kershaw 1977).

If fires are excluded from a woodland site for at least 200 yr, a closed-canopy spruce-feathermoss forest is likely to develop, especially in the west (Kershaw 1977). Layering is prevalent in lichen woodlands in these open-light situations; hence, trees eventually are able to dominate the site, as lichens are largely shade-intolerant. However, fires are rarely excluded from any given area for 200 yr; thus, lichen woodlands are perpetuated.

Productivity

The productivity of boreal forest communities is a function of latitude, maritime/continental position,

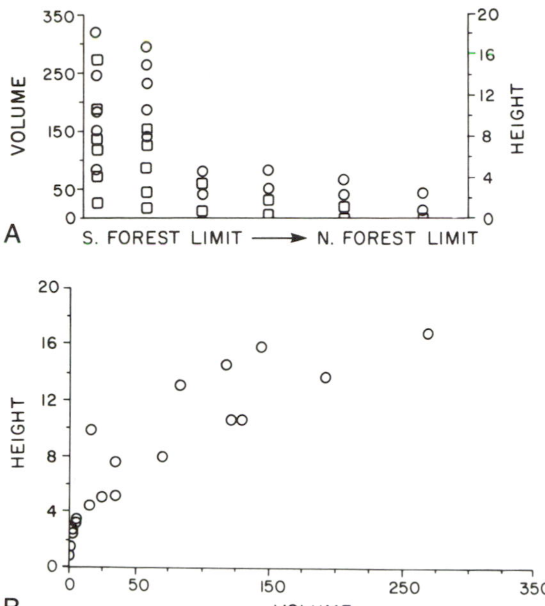

Figure 2.12. A: The decreasing productivity of the boreal forest with increasing latitude is reflected by decreases in stand volume (m^3 ha^{-1}, squares) and mean tree height (m, circles). B: Stand volume and tree height are positively correlated ($r^2 = 0.781$ at the 0.01 confidence level). Data from Black and Bliss (1978).

substrate, topographic position, and seral status. Net primary productivity varies between 400 and 2000 g m^{-2} yr^{-1}, with an average value of 800 g m^{-2} yr^{-1}, relatively low compared with those for other global biomes (Whittaker and Likens 1975). Decreased productivity to the continental north is well documented by smaller annual growth increments, a reduction in aboveground biomass, and the lower stature of trees (Larsen 1965, 1972; Zasada et al. 1978), with a decreased length of the growing season (Fig. 2.12). (In Alaska, however, the most productive forests are found north of the Alaska Range, because cooler summers due to marine influence exist south of the range.) Tree biomass ranges from 9590 to 163,360 kg ha^{-1} for spruce communities and from 21,500 to 51,200 kg ha^{-1} for aspen communities (Oechel and Lawrence 1985).

In reference to topographic position, tremendous differences in biomass production are seen along a gradient from uplands to lowland river bottoms. Uplands are dominated by evergreens, which tend to be nutrient-conservative (retaining foliage for several years) and hence often are adapted to nutrient-poor conditions. As a thick organic (peat) mat builds up during succession, the thickness of the 0 horizon increases, with nutrients largely found in the 0 and A horizons. Organics, largely in an undecomposed state, are stored on the forest floor. Decomposition activity is slow because of low temperatures, resulting in increased insulation of the soil, lower soil temperatures, decreased active-layer thickness, and impeded drainage. The shallow root systems of *Picea* and other boreal conifers (such as *Larix*) are well adapted to these conditions, up to a point.

Nutrient availability in the soil generally decreases over time, though the forest biomass may continue to increase. In early seral stages, most of the biomass (~99%) is found in the overstory, whereas understory vegetation (especially the mosses) plays a dominant role in later stages (tree biomass < 50%). The partitioning of nutrients in the system then changes. For example, nitrogen availability will decrease in the soil, but nitrogen will accumulate in the phytomass. Living feathermosses act as a nutrient "sponge," such that any nitrogenous or ionic nutrient that is deposited on the moss surface is absorbed by the moss and immobilized (Oechel and Lawrence 1985). This may, however, limit growth on some sites, because not all species are well adapted to low nitrogen levels (Zasada et al. 1978). The upland forest community may then operate as a relatively closed system, unless the succession progresses to an even more nutrient-poor state, that of the peatland. The partitioning of biomass between the overstory, forest floor, and rhizosphere is shown in Table 2.7 for select forest communities in Alaska (Zasada et al. 1978).

A peatland (bog or muskeg) may develop such that nutrients are largely unavailable for tree growth. This successional pathway does not usually lead to increased nutrient availability and ameliorated conditions, in contrast to the situation in more temperate regions, unless thermokarst collapse and erosion occur (Luken and Billings 1983).

Picea mariana communities in northernmost latitudes and some maritime settings are then the least productive forest communities of the boreal region. This is partly because of their tolerance of poor site conditions. *Picea glauca*, which is more site-demanding, but may be competitively excluded from poorer sites, is more productive.

In contrast, lowland floodplain communities are very productive. Hardwoods, such as *Populus balsamifera*, dominate these sites and tend to be less nutrient-conservative than the conifers. These floodplain communities function more as open systems, because nutrients are periodically added to the system by flooding. Flooding also erodes the mat of insulative organic matter; hence, soil temperatures are higher at these sites. The high groundwater table not only promotes maintenance of a deep active layer but also provides additional nutrients to the rooting zone. However, water infiltration promotes loss of soluble constituents through

Table 2.7. *Distribution of biomass for select boreal forest communities in Alaska*

Community dominant	Stand age (yr)	Biomass ($g\ m^{-2}$, dry wt.) Overstory	Forest floor	Roots
Alnus incana	20	4790.4	2078.2	2473.0
Populus tremuloides	50	4872.0	3728.4	1804.4
Betula papyrifera	60	11,894.8	6877.2	4429.7
Populus balsamifera	60	18,031.0	2222.7	6678.2
Picea mariana	61	2416.3	13,325.9	1040.1
Betula papyrifera	120	14,763.0	4374.3	5467.8
Picea mariana	130	10,398.7	11,923.5	5169.7
Picea glauca	165	24,945.0	19,786.0	12,401.4

Source: Zasada et al. (1978).

leaching, with these coarse soils exhibiting a low cation-exchange capacity. Organic matter turns over relatively rapidly in these systems (Zasada et al. 1978). Such communities are often kept in a state of disequilibrium, with lower seral stages maintained by disturbance. Nonetheless, nutrient retention in the biomass will increase through the successional cycle.

AUTECOLOGY OF TREE SPECIES

The trees of the boreal forest are relatively diverse in terms of their floristic affinities and histories, life histories, and ecologies. The common trees of boreal North America number approximately 18 species. The cold-hardy Pinaceae dominate this list, with 5 genera and 13 species. *Picea glauca* and *P. mariana* are the most widespread members of this family, ranging from Alaska to Labrador and from the northern to southern regions of the boreal forest. The deciduous guild in the boreal forest is dominated by the Salicaceae. The genus *Populus* has four common species in the boreal forest, with one of these, *P. tremuloides*, being the most wide-ranging tree species in all of North America.

Important ecological attributes of the eight most common dominant tree species of the boreal forest are shown in Tables 2.8 and 2.9. Unfortunately, many studies of these species have been done in only the southern portions of their ranges, because of their greater economic significance in this area. Inferences can generally be made for northern populations, and research on these populations is increasing. The reader is referred to Fowells (1965) and the U.S. Department of Agriculture (1974) for information on all the common tree species of the boreal forest.

The boundaries of the range of any species are determined by the genetically controlled tolerance ranges of its populations and the biotic and abiotic factors of the environments of its individuals. These include competition and drought stress. At both the northern and southern boreal forest limits, trees are exposed to both severe and highly fluctuating environmental conditions (Vowinckel et al. 1975; Larsen 1980). Researchers generally agree that plants survive at the limits of their ecological amplitude (range) through conservative allocation of their resources, especially energy reserves.

The ecophysiology of boreal tree species is poorly understood. Although no one has studied carbon balance (photosynthetic rates) for all the tree species (or their understory flora) at a particular site, species-specific measurements have been taken at a few sites. The photosynthetic period has been shown to be longer for evergreen than for deciduous taxa because of leaf senescence in the late summer for the deciduous plants (Prudhomme 1983; Oechel and Lawrence 1985). Maximum photosynthetic rates (P_{max}) are also consistently higher for deciduous hardwoods versus conifers, though the maximum rates reached during the growing season are similar (Oechel and Lawrence 1985). Temperature does not generally limit photosynthesis during the growing season. Photosynthesis remains positive from temperatures less than 0°C to temperatures greater than 40°C. The temperature optimum within this range for black spruce in central Quebec is ~15°C (Vowinckel et al. 1975). Light intensity has been proposed as a possible factor limiting summer photosynthesis in areas with frequent cloud cover, though the low light compensation and light saturation points of many boreal forest species allow daily carbon gains under low light intensities (Vowinckel et al. 1975; Skre et al. 1983; Oechel and Lawrence 1985).

The study of water stress on boreal tree taxa has been confined to *Picea mariana*. Field measurements of water potential indicate little water stress, with minimum water potentials of −1.5 MPa at Scheffer-

Table 2.8. *Ecological attributes of common boreal tree species*

	Balsam fir	Paper birch	Larch	White spruce	Black spruce	Jack pine	Balsam poplar	Aspen
Evergreen	Yes	No	No	Yes	Yes	Yes	No	No
Deciduous	No	Yes	Yes	No	No	No	Yes	Yes
Needleleaf	Yes	No	Yes	Yes	Yes	Yes	No	No
Broadleaf	No	Yes	No	No	No	No	Yes	Yes
Growth rate	Moderate	Rapid	Rapid	Moderate	Moderate	Rapid	Rapid	Rapid
Max. age (yr)	200	200	350	600	350	230	200	200
Plasticity	Moderate	High	Low	High	High	High	High	Low
Disturbance tolerance	Low	High	Moderate	Low	High	High	High	Moderate
Competitive ability	Moderate	Moderate	Low	High	High	Moderate	Moderate	Low
Frost-free period (d)	80	75	75	60	60	80	75	80
Site requirements								
Substrate preference	Acidic loam	None	Acidic	Basic loam	Acidic	Acidic	Alluvium	Basic
Permafrost table	Moderate	Low	High	Low	High	Low	Low	Low
Soil moisture	Mesic	Xeric-mesic	Mesic-hygric	Xeric-mesic	Xeric-hygric	Xeric-mesic	Mesic	Mesic
Shade tolerance	High	Low	Low	High	High	Low	Low	Low
Nutrient requirements	Moderate	High	Low	Moderate	Low	Moderate	Moderate	High
Seral status	Late	Early	Subclimax	Late	Early-late	Early	Early	Early

Table 2.9. *Reproductive attributes of common boreal tree species*

	Balsam fir	Paper birch	Larch	White spruce	Black spruce	Jack pine	Balsam poplar	Aspen
Vegetative growth								
Layering	Yes	No	Yes	Yes	Yes	No	No	No
Root suckers	No	No	No	No	No	No	Yes	Yes
Crown/stump sprouts	No	Yes	No	No	No	No	Yes	Yes
Sexual reproduction								
Age of maturation (yr)	15	15	12–15	13	10	5–10	15–20	20
Period of seed set (yr)	2	1	3	3	3	3	1	1
Seed crop frequency (yr)	2–4	1	2–6	2–6	4	3–4	3–4	4–5
Germination requirements								
Light	Low	High	High	Low–high	High	Moderate–high	High	High
Temperature (°C)	21	?	18–21	15	15	21	?	?
Moisture	Moderate	High	Moderate	Moderate	Moderate	Moderate	High	Moderate
Cone retention	No	No	A few yr	No	Several yr	Many yr	No	No
Dispersal distance	Low	High	Low	Moderate	Low	Low	High	High
Optimal seedbed	Org./mineral	Mineral	Org./mineral	Mineral	Org./mineral	Mineral	Mineral	Mineral

ville and −2.1 MPa in the Mackenzie delta (Black and Bliss 1980; Oechel and Lawrence 1985). However, water stress measurements by Black and Bliss (1980) on greenhouse-grown seedlings show a 50% reduction in net photosynthesis under a water potential of −1.5 MPa, and a decrease in leaf water potential to −2.5 MPa, further reducing net photosynthesis to the compensation point. In the field, trees are able to maintain a constant turgor of −1 MPa by osmoregulation, allowing continuation of growth and cell expansion (Black and Bliss 1980).

Long-lived trees must first allocate resources to trunk and foliage maintenance ("self-maintenance") and may therefore exhibit sporadic sexual reproduction in marginal environments. Studies have shown that allocation of resources to seed production decreases the growth of the individual (Harper 1977). The successful marginal trees will then be of species that also have the ability to reproduce vegetatively (through layering, root suckers, or stump/crown sprouts), because environmental conditions necessary for the development of vegetative tissues are not as exacting as those for the development of sexual organs and propagules (Elliott 1979c; Elliott-Fisk 1983). Whether or not the processes of clonal growth are competitive with those involved in producing seeds is not known (Harper 1977), but, nonetheless, vegetative reproduction (clonal growth) is a valuable adaptive attribute in the boreal forest because of natural cyclic disturbances by fire, flooding, windthrow, and rises in the permafrost table.

All the dominant tree species of the boreal forest are able to reproduce vegetatively to some extent, with the exception of *Pinus banksiana* (Table 2.9). Species in the genus *Pinus* rarely layer, but members of this taxon in the boreal forest are adapted to fire because they possess serotinous or semiserotinous cones, retaining a viable seed store on the tree for several years. *Picea mariana* also possesses this valuable trait.

Deciduous trees are adapted to coping with site disturbances by the production of large crops of light wind-dispersed seeds. Populations may be maintained on site through rapid root suckering and stump/crown spreading. Many of these species are "r-strategists," though some are able to remain the climax vegetation on special sites. Deciduous trees also frequently occur along floodplains.

True mixed (evergreen and deciduous) forests occur in only special situations, and thus a comparison of their competitive abilities is warranted. Trees of many of the deciduous species (such as *Larix laricina*) are able to outgrow the evergreens at the seedling and sapling stages, thus attaining dominance in the canopy. The eventual demise of these deciduous trees in many stands may be a function of the evergreen's photosynthetic capabilities. The evergreens have an advantage where the growing season is either short or unpredictable, and they are able to retain their needles for as many as 15 yr (Black and Bliss 1980).

Insect pests and biological agents may damage or kill individual trees and occasionally entire stands. Defoliation by budworms (*Choristoneura fumiferana* and *C. pinus*), tent caterpillars (*Malacosoma disstria*), and the larch sawfly (*Pristiphora erichsonii*) is an episodic problem in certain forest regions. Spruce beetle (*Dendroctonus rufipennis*) infestation can lead to death of mature and overmature white spruce, but this is easily controlled by salvage cutting. Fungal pests are of minor importance, as is dwarf mistletoe (*Arceuthobium americanum* and *A. pusillum*), which triggers the formation of witches' brooms in spruce and pine (Moody and Cerezke 1984).

From Table 2.8 it appears that tree species of the boreal forest have relatively narrow site requirements. It should be remembered that these are *optimal* requirements. The majority of the species are able to tolerate other conditions, though populations on these sites may be less vigorous and eventually become competitively excluded. Many of these species are also phenotypically plastic, with ecotypes and local genetic populations (Halliday and Brown 1943; Elliott 1979c). Hybridization, which is frequently introgressive, has also been documented between many congenerics (Gordon 1976; Elliott 1979c).

Picea Glauca and *Picea Mariana*

The genus *Picea* A. Dietr. is composed of ~40 species of evergreen trees ranging through cool, temperate regions of the Northern Hemisphere. *Picea glauca*, geographically, is the most widespread member of this genus and of the Coniferales as a whole in North America (Nienstaedt and Teich 1972; Elliott 1979c). However, *P. mariana* is the most frequently encountered tree of the boreal forest, because it is able to occupy a wider range of sites than *P. glauca*.

The phenology of *Picea* has been outlined by Zasada and Gregory (1969), Rowe (1970), and Owens and Molder (1976, 1977; Owens et al. 1977). These monoecious trees possess distinct male and female cones, though bisexual cones are occasionally found (Zasada et al. 1978; Elliott 1979a, 1979b). Three years are required for completion of the sexual reproductive cycle from axillary bud scale initiation to seed maturation.

Trees of *P. mariana* reach sexual maturity slightly earlier than those of *P. glauca*; so black spruce populations may be able to invade new habitats more rapidly. Although seed crop intervals in the two species are similar, crops usually are more frequent for *P. mariana* in northern latitudes (Zasada 1971), though *P. glauca* produces more cones (and seeds) in a typical crop. Seedfall peaks in late summer and autumn for *P. glauca*, whereas *P. mariana* releases most of its seeds immediately following a fire. Some seeds are released in both species throughout the year, though this is minimal. Because the fire frequency is highest in the north in June and July (with frontal activity), *P. mariana* has a competitive advantage. Trees of *P. mariana* retain their seed in semiserotinous cones on the tree, an appropriate strategy, because a thin seed coat and small endosperm make the embryo vulnerable to destruction on the forest floor. Rowe and Scotter (1973) stated that *P. mariana* is the most successful regenerator of all boreal trees following a fire; this species appears to be the most opportunistic of the boreal trees.

Although the sizes and morphologies of the seeds of these two spruces differ only slightly, *P. glauca* has proved to be a better disperser of its seed, a useful strategy, because it is susceptible to destruction by fire (Zasada and Gregory 1969; Elliott 1979c). Squirrels are heavy predators on *P. glauca* cones and are effective agents for seed dispersal, though they may reduce the available seed crop (Rowe 1952; Elliott 1979b).

Seedling establishment is optimum for both species on mesic sites, though seedlings of *P. mariana* can tolerate more moisture than those of *P. glauca*. Growth of both species, however, is poorest on wet sites (Maini 1966). *Picea mariana* seedlings are also able to grow a little faster than those of *P. glauca*, though adults are not as long-lived.

Layering is common in both spruces, though *P. mariana* layers vigorously in almost all environments, whereas *P. glauca* layers only occasionally and in open, marginal sites (Elliott 1979b, 1979c; Elliott-Fisk 1983). Layering allows populations to persist during times of relatively minor (but perhaps prolonged) climatic deterioration.

Both *P. mariana* and *P. glauca* are phenotypically plastic, with numerous growth forms found in different settings. White spruce is typically either an upright tree or simply a dwarfed (short) tree; it does not usually take on a twisted or mat krummholz form. Black spruce, on the other hand, assumes many growth forms, both in individuals and in clones. This has triggered discussion of which of the species is hardier. Hustich (1953) stated that *P. glauca* is hardier, keeping an erect form on sites where *P. mariana* is prostrate. Yet *P. mariana* has a much easier time propagating vegetatively through adventitious rooting than does *P. glauca*. These decumbent forms seem well adapted to the environment, with their height equaling that of the "average" snow cover; thus, shoots are not exposed to desicating, abrasive winter winds and low temperatures.

Occasionally there is confusion in identifying particular specimens as either *P. glauca* or *P. mariana*, especially in marginal locations where cone production is infrequent. This may be due to introgressive hybridization of the two species (Little and Pauley 1958; Larsen 1965, 1980; Dugle and Bols 1971; Elliott 1979c). Species of *Picea* are known to hybridize readily (Owens and Molder 1977). Hybrids have been reported from Minnesota (the Rosendahl spruce of Little and Pauley 1958), Manitoba, Ontario, British Columbia, and the Northwest Territories (Larsen 1965; Roche 1969; Dugle and Bols 1971; Parker and McLachlan 1978). Unfortunately, the techniques used in identifying putative hybrids have not been consistent. Von Rudloff and Holst (1968), for example, identified the Rosendahl spruce on the basis of leaf oil terpenes, but other studies have focused on morphology. Some individuals that have been studied have exhibited intermediate morphologies for some characters but not others, and this may be a function of whether the individual is a true F_1 hybrid or whether the hybrid population has back-crossed with a parental type (Abercrombie et al. 1973; Parker and McLachlan 1978). Further work is needed, especially on introgressive hybridization of these species at the northern tree limit, an area where stress may induce rare and unusual reproductive events (Anderson 1948; Elliott 1979c). In such an area, species that would otherwise maintain quite separate identities occasionally will hybridize, with introgression resulting in an intermingling of morphological traits of each species.

Although both *P. glauca* and *P. mariana* are wide-ranging through the boreal forest, occasionally even found as codominants in a stand, the species have distinct site preferences that have no doubt promoted their maintenance as separate species. *Picea glauca* is the more site-demanding of the two species, strongly preferring well-drained, basic mineral soils and intolerant of high permafrost tables. *Picea mariana*, on the other hand, is more tolerant of poor environmental conditions and is able to survive in poorly drained areas and on various substrates (though, like *P. glauca*, it does best on well-drained sites). Black spruce is frequently found on poorly drained acidic soils (i.e., peat) that are unsuitable for most tree species. It is not as nutrient-demanding as white spruce, which according to Fowells (1965) is one of the most exacting conifers in terms of nutrient requirements. The very shallow root system of *P. mariana* allows it to exist in areas where there is a high permafrost table, but this also makes it susceptible to windthrow; its prostrate growth form at exposed tree-limit sites may compensate for the shallow root system (Elliott 1979c). Black spruce is not well adapted to coastal fog (Payette and Gagnon 1979) nor to salt-laden winds. This is thought to be the explanation for its limited occurrence in maritime settings, in both Labrador-Ungava and Alaska, where *P. glauca* dominates and is the northern treeline species (Elliott and Short 1979; Short and Elliott-Fisk in press).

The two species form mixed stands, sometimes in association with *Larix laricina*, on well-drained sites near the northern forest border (Larsen 1965; Elliott 1979b). *Picea glauca* is frequently the larger, older tree, but because both species are shade-tolerant climax species, neither is excluded. Stands such as this are hypothesized to have escaped fire for several hundred to perhaps several thousand years (Elliott 1979a, 1979c; Elliott-Fisk 1983), in agreement with the finding by Johnson and Rowe (1975) that the frequency of fires is lower north of the forest border, outside of frequent frontal-zone activity.

HISTORICAL ASPECTS

The history of the boreal forest is complex and is not understood in its details, because of the scarcity of Tertiary fossiliferous materials from boreal associations and the magnitude and frequency of late Cenozoic glacials and interglacials. Our best picture of historical forest dynamics is from late glacial and particularly Holocene deposits, as these deposits are typically better preserved, more abundant, and more readily dated than older materials. We must also remember that each species has its own unique floristic history. Variations in genetic plasticity, dispersal strategy, and competitive ability allow floras to evolve through time, experiencing environmental changes, such that present communities may no longer resemble those of even the recent past.

Both the boreal and Arctic floras of North America are believed to be derived from the Arcto-Tertiary geoflora. Conifers and deciduous hardwoods were dominants in this geoflora, which ranged as much as 20° north of the present boreal and Low Arctic vegetation zones (Larsen 1980). Wolfe (1971, 1972,

1978) and others have debated the circumpolar extent of this geoflora, contesting that it did not exist in all northern regions (e.g., Alaska). This is certainly possible, because North American geofloras are known to have intermingled, much as our present floras do, giving the flora of a particular region multiple origins.

The earth's Cenozoic climatic history (spanning the last 70 million yr) is well known on the global scale. The first part of this period saw a general warming trend (with more tropical conditions globally) lasting to the Eocene-Oligocene boundary about 38 million yr ago, followed by a rapid cooling triggered by the opening of Drake Passage between Antarctica and Australia. This allowed the southern circumpolar current (Westerly Wind Drift) to develop, thermally isolating the high latitudes, and resulted in the spread of cool-temperate taxa globally and, possibly, the evolution of polar and subpolar floras. A gradual cooling trend continued through the Oligocene to the Miocene, when temperatures once again dropped in the higher latitudes. This led to the development in the middle to late Miocene of the East Antarctic Ice Sheet and to mountain glaciation in the Northern Hemisphere. By late Pliocene time (~3 million yr ago) near the end of the Tertiary, ice sheets had also developed in the Northern Hemisphere, especially adjacent to the north Atlantic Ocean. Boreal forest species and communities continued to evolve throughout this time, with many of the modern tree species coming into existence by the late Pliocene.

Late Tertiary cooling resulted in elimination of many "temperate" species from the northern floras, leaving evergreen conifers such as *Picea* and *Pinus* dominant (Larsen 1980). Changes in understory vegetation must have accompanied these changes in the forest canopy, though our knowledge of these is poor.

Research on the Quaternary history of the boreal forest has concentrated on the last 15,000 yr, focusing on (1) the presence of possible glacial refugia of boreal species, (2) the southernmost extent of boreal forest taxa during glacials, (3) rates of northward migration of the forest/woodland accompanying deglaciation, (4) fluctuations of the northern forest border, and (5) vegetation dynamics related to succession, climate change, and anthropogenic disturbance.

It is important to note that many boreal species occurred in associations different from those that exist today. Paleoecological data show that each species behaved individualistically, as proposed long ago by Shreve (1915) and Gleason (1926), because of differences in phenotypic plasticity, reproductive strategy, and competitive ability (Hultén 1937; Larsen 1980; Davis 1983; Webb et al. 1983). Thus, many of the present boreal forest associations are recent in the geologic sense, having existed for only the last 3000–6000 yr (Ritchie and Yarranton 1978a). This youth may account for the relative impoverishment of the boreal flora and the small number of associations and communities present. Many species can be considered generalists, not occupying tight ecological niches, but instead occurring in a wide array of sites (Larsen 1980).

Phytogeographers have long thought that boreal forest species or entire associations may have persisted in isolated glacial refugia during the Wisconsin or previous glacials. Matthews (1974) and Rampton (1971) have reported that their studies support the existence of local spruce communities in Alaska during the mid-Wisconsin glacial. Limited areas of spruce woodland are also documented for east central Alaska during the late Wisconsin glacial, when the elevational treeline was 400–600 m lower than today (Colinvaux 1967; Hopkins 1967; Matthews 1970, 1979; Péwé 1975). Ager (1975) and Hopkins and associates (1981) have presented pollen and macrofossil evidence for a refugium in interior Alaska, with *Picea, Betula papyrifera, Larix laricina*, and an *Alnus* species present. This refugium was eliminated from Alaska by low temperatures during the last glacial; however, it is probable that some *Alnus* species, *Populus balsamifera*, and possibly *P. tremuloides* survived in unglaciated Alaska during the Wisconsin glacial, perhaps in thermal belts on hill slopes (Hopkins et al. 1981; Ager 1982). The existence of refugia on nunataks (unglaciated highlands and lowlands) in eastern North America is more controversial (Ives 1974; Short 1978a).

The glacial and postglacial history of the vegetation in Alaska is extremely complex because of topographic variability, multiple glaciations of mountainous areas, and repeated land-bridge connections with Siberia (Ager 1983). In contrast, most of the Canadian Shield and surrounding planated rock formations were covered by the Laurentide and Cordilleran ice sheets. Far southern positions of boreal species have been found in the southeastern United States (Watts 1970) some 1200 km south of their present limits. Various tundra formations occupied much of the area immediately south of the ice sheets, regions now vegetated by various hardwood and mixed evergreen-deciduous forest associations. Vegetation changes in subarctic and temperate North America have been exceedingly rapid in the last geologic epoch.

With the deterioration of the late glacial ice sheets, tundra and forest components migrated

Table 2.10. *Reconstructed Holocene treeline migrations for Canada*

Years BP	Western Canada	Central Canada	Eastern Canada
	RETREAT	RETREAT	(further cooling?)
1000	(major cooling)	(cooling) forest-	
	dwarf birch/heath tundra	tundra	
		ADVANCE	CONTRACTION
2000		(minor warming)	(cooling)
		forest	decreased productivity
3000		?(minor warming)?	
		RETREAT	ADVANCE
		(cool) tundra	(warming)
4000		?RETREAT (onset of	spruce woodland
		(cooling)?	
	RETREAT		
5000	(cooling)		
	tall shrub tundra	ADVANCE	ADVANCE
		(warming)	(warming)
6000		forest	tall shrub tundra
	ADVANCE	(cool tundra)	(peat initiation)
7000	(warming)		
	closed spruce/birch	DEGLACIATION	
8000	forest		DELAY IN MIGRATION
			(cold, dry?)
9000			low (herb) tundra
	ADVANCE		
10,000	(warming)		
	forest-tundra		
11,000			DEGLACIATION (?)
12,000	(cool)		
	low tundra		
13,000	DEGLACIATION		

Source: Elliot-Fisk (1983).

rapidly northward, colonizing what is now the area occupied by the boreal forest. Nichols (1975, 1976) found that in central Canada, the early Holocene forest species moved about 480 km north in 2000 yr. If we analyze the Holocene history of the entire North American boreal forest, we see a general transgressive pattern of deglaciation, followed by progressive establishment of an herbaceous tundra, shrub tundra, and spruce-dominated woodland or forest. *Picea* generally appeared in southern sites between 14,000 and 10,000 BP and in northern sites by 8000–6500 BP (Ritchie 1976). These site histories have not been summarized for the entire boreal forest of North America because there were great differences in the timing of these vegetation changes dependent on the latitudinal location of the site, among other factors. Some regional summaries, such as that for select localities in Alaska by Ager (1983), are available.

Sites at the margins of boreal species ranges have recorded the most detailed history of Holocene changes, which were principally a function of climate, but also were influenced by soil and community development. Along the southern boreal forest border, shifts in grassland, open parkland, and the southern transitional boreal forest have been detected for the Holocene. This southern border has remained in the same position for the last 2000 yr (Ritchie and Yarranton 1978a, 1978b), though anthropogenic changes have occurred in many regions. Shifts in the northern boreal forest border have been more thoroughly documented, because the northern treeline has been correlated with several atmospheric parameters, and changes in the treeline position have then been used to record climatic changes and aid in their explanation (Bryson 1966; Sorenson and Knox 1974; Elliott 1979c; Elliott-Fisk 1983). Tables 2.10 and 2.11 illustrate climatically induced fluctuations of the entire northern Canadian treeline, with the accompanying vegetation changes shown for Labrador-Ungava (Elliott-Fisk 1983; Short and Elliott-Fisk, in press). Although Holocene climatic changes have been both directional and synchronous across Canada, deteriorations/ameliorations perhaps (1) lagged from west to east, (2) were buffered by local geographic factors, or (3) were not registered in the fossil record owing to the inherent persistence of

Table 2.11. *Holocene vegetation changes, years BP, in Labrador-Ungava*

Site[a]	Low tundra	Shrub tundra	Spruce woodland	Contraction
Hebron Lake	10,200	6900/6400	absent	none
Napaktok Lake/Bay	8700	5500	4800/4300	2200
Ublik Pond	10,500	6500	4300	3000/2400
Umiakoviarusek Lake bog	absent	absent	-2650-	1000
Nain Pond	8000(?)	6500	4500	2500
Pyramid Hills Pond	7000	6500	4500	2500
Kogaluk Plateau Lake	9000	6700	4400	3000(?)

[a]Sites arranged from north to south and from coast to interior.
Source: Short and Elliot-Fisk (in press).

the vegetation at some sites (Elliott-Fisk 1983). Some portions of the north central and northwestern Canadian treelines also are relictual (from hypsithermal warmer climates), with low sexual regeneration capacities, and these are susceptible to destruction by further climatic deterioration or anthropogenic disruption (Elliott-Fisk 1983).

Although human disruption of the boreal forest ecosystem typically results in successive occupancy of seral communities (like those described previously associated with fire), particular disturbances can greatly alter not only the vegetation but also the substrate. The water budget for a site is readily altered by mining, hydropower development, and destruction of the surface vegetation. In northern regions, this can lead to destruction of frozen ground (permafrost) and rapid landscape degradation. In certain marginal settings, trees may be unable to regenerate following cutting, reducing the vegetation to a more or less permanent tundra subclimax (Larsen 1965, 1980; Stang 1973).

The majority of the boreal forest, however, is quite vigorous and suitable for sustained harvest; 25% of Canada's land ($\sim 10^9$ ha) is covered by forest suitable for regular harvesting (Larsen 1980). About 20% ($\sim 220 \times 10^6$ ha) of this is being harvested (Love and Overend 1978). Reforestation and management of these ecosystems are the primary concerns of several governmental agencies, with the Institute of Northern Forestry of the USDA Forest Service (Fairbanks, Alaska) a prime example. A great deal of attention has been paid to genetic improvement of breeding stock and management of plantations to reduce rotation time between harvests.

Acid deposition by both wet and dry mechanisms has been proposed as a potential threat to northern forests. "Acid rain" is the product of long-distance transport and transformation of sulfur and nitrogen oxides from combustion of fossil fuels (Shugart 1984). At the present time, only the southeastern portions of the boreal forest (Ontario, southern Quebec, and southern Labrador), with an average annual precipitation pH of 4.5–5.0, appear to be vulnerable to acid deposition, because they are (at least seasonally) downwind of major sulfur sources in the industrialized Northeast and Great Lakes region (Turk 1983). With the dominant westerly circulation over North American, the majority of the boreal forest in Canada and Alaska is not susceptible to regional acid deposition (Barrie 1982; Turk 1983; Shugart 1984).

The future impact of acid deposition on the North American boreal forest cannot be predicted at the present time, for the following reasons:

Site-specific studies near large smelters show acute effects immediately downwind, but chronic exposure to multiple pollutants from multiple sources in a regional airshed has not been documented (Shugart 1984).

The region lacks a historical sampling network and data base on the acidification of natural waters and soils; however, a network of precipitation sampling sites (CANSAP) for monitoring acid deposition has been established in the last 10 yr (Barrie 1982).

The pH of natural precipitation can vary between 4.5 and 5.6, making trends in acidification difficult to detect (Turk 1983).

The presence of a winter background of sulfur in Arctic air masses makes it difficult to detect anthropogenic sulfur; this is coupled with a general lack of careful evaluation of natural versus anthropogenic sources of acidity (Barrie 1982; Turk 1983).

Field studies have not documented the impact of SO_2 and other regional-scale pollutants on forest-stand dynamics (Shugart 1984).

The results of natural soil formation (leaching of nutrients, release of aluminum, and acidification of soil and water) are the same as those attributed to acid deposition (Krug and Frink 1983).

The ability of hydrogen ions to remove nutrient cations from soils more acidic than pH 5 is low (Krug and Frink 1983).

The proportion of bases to acids in precipitation is usually greater than that in strongly acidic forest soils (Krug and Frink 1983).

Changes in land use and subsequent vegetation recovery and soil recovery following disturbance play an important role in the area's susceptibility to acidification (Krug and Frink 1983).

It is safe to state that the boreal forest is much easier to manage than many other natural vegetation types. The boreal ecosystem is adapted to widely fluctuating environmental conditions and thus can better resist exploitation than can ecosystems adapted to more stable environments (Larsen 1980).

AREAS FOR FUTURE RESEARCH

The boreal forest remains relatively unexplored and poorly understood. Physical factors exert important influences on the natural vegetation; yet the physiological tolerances of even the most dominant species are not well known. Environmental conditions in a stand can change markedly through the successional process, such that conditions that favor maintenance of mature trees are unfavorable for tree seedling establishment. Bioclimatological studies are desperately needed to decipher (1) the extension of forest along river bottoms, (2) the ecology of understory lichen and bryophyte species and associations, (3) the physiology of ecotonal shrubs and herbs that occupy both boreal and Arctic vegetation formations, and (4) the existence of shrublands in sites apparently suitable for forest vegetation.

Though many researchers have tried to link the boreal forest to a particular climatic type, this has not yet been done successfully. Atmospheric scientists need to work with boreal ecologists to link the vegetation to the large-scale upper-level dynamics, as well as the surface dynamics, of the atmosphere. Atmospheric chemistry needs to be fully linked to boreal biogeochemical cycling so that we can evaluate the impact of acid deposition on northern ecosystems.

The genetic and phenotypic plasticity of dominant boreal species needs to be investigated if we are to fully understand the evolution and ecology of the vegetation. How can a given tree species (such as *Picea mariana* or *Populus tremuloides*) occupy such a large and apparently diverse geographical area? Future documentation of genotype variation, ecotypes, and introgressive hybridization may more clearly explain this.

The physiology of boreal trees also needs to be investigated, especially in regard to their short life spans. In temperate subalpine regions, many conifers live to great ages in stressful habitats. Trees at the northern forest border and in the forest-tundra ecotone are under severe ecological stress; yet trees more than 300 yr old are rarely found. Many boreal stands can maintain populations for thousands of years during (minor) climatic deterioration. Although several good ecological studies of treeline communities have been done, the subject of long-term stand dynamics deserves a great deal more effort.

The boreal and boreal-Arctic shrublands are vegetation types that have largely been uninvestigated. Shrub communities are important colonizers following deglaciation and prior to forest establishment at many sites. The current distribution and maintenance of these communities, in what appear to be some of the best environments for tree establishment, remain mysteries. Little is known about productivity and nutrient cycling in these systems. These stands provide valuable wildlife habitat and browse. In addition, the belowground component in all boreal forest communities warrants further research.

The structure of boreal forest vegetation is well documented. Its history is not completely understood, though great advances are being made on the late Quaternary history of the forest. We must now turn our research efforts toward physiological ecology and bioclimatology. Human disruption and destruction of this ecosystem do not seem imminent in this or the next century. Preservation of "undisturbed" communities is still necessary, however, to ensure ecological stability of the landscape.

REFERENCES

Abercrombie, M., C. J. Hickman, and M. L. Johnson. 1973. A dictionary of biology. Penguin, Middlesex, England.

Ager, T. A. 1975. Late Quaternary environmental history of the Tanana Valley, Alaska. Ohio State University, Institute of Polar Studies Report 54. Columbus.

Ager, T. A. 1982. Vegetational history of western Alaska during the Wisconsin glacial interval and Holocene, pp. 75–93 in D. M. Hopkins, J. V. Matthews, Jr., C. E. Schweger, and S. B. Young (eds.), Paleoecology of Beringia. Academic, New York.

Ager, T. A. 1983. Holocene vegetation history of

Alaska, pp. 128–141 in H. E. Wright, Jr. (ed.), The Holocene, vol. 2, Late-Quaternary environments of the United States. University of Minnesota Press, Minneapolis.

Anderson, E. 1948. Hybridization of the habitat. Evolution 2:1–9.

Barrie, L. A. 1982. Environment Canada's long range transport of atmospheric pollutants program: atmospheric studies, pp. 141–161 in F. M. D'itri (ed.), Acid precipitation: effects on ecological systems. Ann Arbor Science, Ann Arbor, Mich.

Barry, R. G. 1967. Seasonal location of the arctic front over North America. Geogr. Bull. 9:79–95.

Billings, W. D., and H. A. Mooney. 1968. The ecology of arctic and alpine plants. Biol. Rev. 43:481–530.

Black, R. A. 1977. Reproductive biology of *Picea mariana* (Mill.) B.S.P. at treeline. Dissertation, University of Alberta, Edmonton.

Black, R. A., and L. C. Bliss. 1978. Recovery sequence of *Picea mariana/Vaccinium uliginosum* forests after burning near Inuvik, Northwest Territories, Canada. Can. J. Bot. 56:2020–2030.

Black, R. A., and L. C. Bliss. 1980. Reproductive ecology of *Picea mariana* (Mill.) B.S.P. at treeline near Inuvik, Northwest Territories, Canada. Ecol. Monogr. 50:331–354.

Brown, R. T., and P. Mikola. 1974. The influence of fruticose soil lichens upon the mycorrhizal and seedling growth of forest trees. Acta For. Fennica 141:5–22.

Bryson, R. A. 1966. Air masses, streamlines, and the boreal forest. Geogr. Bull. 8:228–260.

Canada. 1957. Atlas of Canada. Department of Mines and Technical Surveys, Geographical Branch, Ottawa.

Canada. 1978. Miscellaneous climatic data obtained from Atmospheric Environment Service, Fisheries and Environment Canada, Ottawa.

Carleton, T. J., and P. F. Maycock. 1978. Dynamics of boreal forest south of James Bay. Can. J. Bot. 56:1157–1173.

Colinvaux, P. A. 1967. Quaternary vegetational history of Arctic Alaska, pp. 207–231 in D. M. Hopkins (ed.), The Bering land bridge. Stanford University Press.

Curtis, J. T. 1959. The vegetation of Wisconsin. University of Wisconsin Press, Madison.

Cwynar, L. C., and J. C. Ritchie. 1980. Arctic steppe-tundra: a Yukon perspective. Science 208:1375–1377.

Davis, M. B. 1983. Holocene vegetational history of the eastern United States, pp. 166–181 in H. E. Wright, Jr. (ed.), The Holocene, vol. 2, Late-Quaternary environments of the United States. University of Minnesota Press, Minneapolis.

Dugle, J. R., and N. Bols. 1971. Variation in *Picea glauca* and *P. mariana* in Manitoba and adjacent areas. Publication AECL-3681. Atomic Energy Commission, Canada Limited, Whiteshell Nuclear Research Establishment, Pinawa, Manitoba.

Elliott, D. L. 1979a. The current regenerative capacity of the northern Canadian trees, Keewatin, N.W.T., Canada: some preliminary observations. Arct. Alp. Res. 11:243–251.

Elliott, D. L. 1979b. The occurrence of bisexual strobiles on black spruce [*Picea mariana* (Mill.) B.S.P.] in the forest-tundra ecotone: Keewatin, N.W.T. Can. J. Forest Res. 9:284–286.

Elliott, D. L. 1979c. The stability of the northern Canadian tree limit: current regenerative capacity. Dissertation, University of Colorado, Boulder.

Elliott, D. L., and S. K. Short. 1979. The northern limit of trees in Labrador: a discussion. Arctic 32:201–206.

Elliott-Fisk, D. L. 1983. The stability of the northern Canadian tree limit. Ann. Assoc. Amer. Geogr. 73:560–576.

Fowells, H. A. 1965. Silvics of forest trees of the United States. USDA Forest Service agricultural handbook 271.

Gates, F. C. 1942. The bogs of northern lower Michigan. Ecol. Monogr. 12:216–254.

Gleason, H. A. 1926. The individualistic concept of the plant association, Bull. Torrey Botanical Club 53:7–26.

Gordon, A. G. 1976. The taxonomy and genetics of *Picea rubens* and its relationship to *Picea mariana*. Can. J. Bot. 54:781–813.

Haag, R. W., and L. C. Bliss. 1974. Functional effects of vegetation on the radiant energy budget of boreal forest. Can. Geotech. J. 11:374–379.

Halliday, W. E. D., and A. W. A. Brown. 1943. The distribution of some important forest trees in Canada. Ecology 24:353–373.

Hare, F. K. 1950. Climate and zonal divisions of the boreal forest formation in eastern Canada. Geogr. Rev. 40:615–635.

Hare, F. K. 1954. The boreal conifer zone. Geogr. Studies 1:4–18.

Hare, F. K. 1959. A photo-reconnaissance survey of Labrador-Ungava. Geographical Survey of Canada Memoir 6:1–83.

Hare, F. K., and J. E. Hay. 1974. The climate of Canada and Alaska, pp. 49–192 in R. A. Bryson and F. K. Hare (eds.), World survey of climatology, vol. 2, Climates of North America. Elsevier, Amsterdam.

Hare, F. K., and J. C. Ritchie. 1972. The boreal bioclimates. Geogr. Rev. 62:333–365.

Harper, H. L. 1977. Population biology of plants. Academic, New York.

Heinselman, M. L. 1963. Forest sites, bog processes, and peatland types in the glacial Lake Agassiz region, Minnesota. Ecol. Monogr. 33:327–374.

Heinselman, M. L. 1970. Landscape evolution, peatland types, and the environment in the Lake Agassiz peatlands natural area, Minnesota. Ecol. Monogr. 40:235–261.

Hopkins, D. M. 1967. The Bering land bridge. Stanford University Press.

Hopkins, D. M., P. A. Smith, and J. V. Matthews, Jr. 1981. Dated wood from Alaska: implications for forest refugia in Beringia. Quat. Res. 15:217–249.

Hultén, E. 1937. Outline of the history of arctic and boreal biota during the Quaternary period. Stockholm.

Hustich, I. 1949. On the forest geography of the Labrador peninsula. A preliminary synthesis. Acta Geogr. 10:3–63.

Hustich, I. 1950. Notes on the forest on the east coast of Hudson Bay and James Bay. Acta Geogr. 11:1–83.

Hustich, I. 1953. The boreal limits of conifers. Arctic 6:149–162.

Hustich, I. 1954. On forests and tree growth in the Knob Lake area, Quebec-Labrador peninsula. Acta Geogr. 13:1–60.

Hustich, I. 1966. On the forest-tundra and the northern tree-lines. Ann. Univ. Turku [Series A2] 36:7–47.

Hustich, I. 1979. Ecological concepts and biogeographical zonation in the north: the need for a generally accepted terminology. Holarctic Ecol. 2:208–217.

Ives, J. D. 1974. Biological refugia and the nunatak hypothesis, pp. 605–636 in J. D. Ives and R. G. Barry (eds.), Arctic and alpine environments. Methuen, London.

Johnson, E. A., and J. S. Rowe. 1975. Fire in the subarctic wintering ground of the Beverely Caribou herd. Amer. Midl. Nat. 94:1–14.

Kershaw, K. A. 1977. Studies on lichen-dominated systems. An examination of some aspects of the northern boreal lichen woodlands in Canada. Can. J. Bot. 55:393–410.

Krug, E. C., and C. R. Frink. 1983. Acid rain on acid soil: a new perspective. Science 221:520–525.

La Roi, G. H. 1967. Ecological studies in the boreal spruce-fir forests of the North American taiga. Ecol. Monogr. 37:229–253.

La Roi, G. H., and M. H. L. Stringer. 1976. Ecological studies in the boreal spruce-fir forests of the North American taiga. II. Analysis of the bryophyte flora. Can. J. Bot. 54:619–643.

Larsen, J. A. 1965. The vegetation of the Ennadai Lake area, N.W.T.: studies in subarctic and arctic bioclimatology. Ecol. Monogr. 35:37–59.

Larsen, J. A. 1971. Vegetation of Fort Reliance, Northwest Territories. Can. Field-Nat. 85:147–178.

Larsen, J. A. 1972. Vegetation and terrain (environment): Canadian boreal forest and tundra. University of Wisconsin Report UW-G1128, Madison.

Larsen, J. A. 1974. Ecology of the northern continental forest border, pp. 341–369 in J. D. Ives and R. G. Barry (eds.), Arctic and alpine environments. Methuen, London.

Larsen, J. A. 1980. The boreal ecosystem. Academic, New York.

Larsen, J. A. 1982. Ecology of the northern lowland bogs and conifer forests. Academic, New York.

Lawrence, D. B. 1958. Glaciers and vegetation in southeast Alaska. Amer. Sci. 46:89–122.

Lettau, H., and K. Lettau. 1975. Regional climatonomy of tundra and boreal forest in Canada, pp. 210–221 in G. Weller and S. Bowling, (eds.), Climate of the Arctic. University of Alaska, Fairbanks.

Little, E. L., and S. S. Pauley. 1958. A natural hybrid between black and white spruce in Minnesota. Amer. Midl. Nat. 60:202–211.

Love, P., and R. Overend. 1978. Tree power: an assessment of the energy potential of forest biomass in Canada. Energy, Mines and Resources Canada report ER 78-1.

Luken, J. O., and W. D. Billings. 1983. Changes in bryophyte production associated with a thermokarst erosion cycle in a subarctic bog. Lindbergia 9:163–168.

Lutz, H. J. 1956. Ecological effects of forest fires in the interior of Alaska. USDA technical bulletin 1133.

MacDonald, G. M. 1983. Holocene vegetation history of the upper Natla River area, Northwest Territories, Canada. Arct. Alp. Res. 15:157–168.

Maini, J. S. 1966. Phytoecological study of sylvotundra at Small Tree Lake, N.W.T. Arctic 19:220–243.

Marr, J. W. 1948. Ecology of the forest-tundra ecotone on the east coast of Hudson Bay. Ecol. Monogr. 18:117–144.

Matthews, J. V., Jr. 1970. Quaternary environmental history of interior Alaska: pollen samples from organic colluvium and peats. Arct. Alp. Res. 2:241–251.

Matthews, J. V., Jr. 1974. Wisconsin environment of interior Alaska: pollen and macrofossil analysis of a 27-meter core from the Isabella Basin (Fairbanks, Alaska). Can. J. Earth Sci. 11:828–841.

Matthews, J. V., Jr. 1979. Beringia during the late Pleistocene: arctic-steppe or discontinuous herb-tundra? A review of the paleontological evidence. Geological Survey of Canada open-file report 649. GSC, Ottawa.

Miller, W. S., and A. N. Auclair. 1974. Factor analytic models of bioclimate for Canadian forest regions. Can. J. Forest Res. 4:536–548.

Moody, B. H., and H. F. Cerezke. 1984. Forest insect and disease conditions in Alberta, Saskatchewan, Manitoba, and the Northwest Territories in 1983 and predictions for 1984. Canadian Forestry Service, Northern Forest Research Centre, information report NOR-X-261.

Neiland, B., and L. A. Viereck. 1978. Forest types and ecosystems, pp. 109–136 in North American forest lands at latitudes north of 60 degrees. Proceedings of a symposium, September 19–22, 1977, University of Alaska, Fairbanks.

Nichols, H. 1975. Palynological and paleoclimatic study of the late Quaternary displacement of the boreal forest-tundra ecotone in Keewatin and Mackenzie, N.W.T., Canada. Institute of Arctic and Alpine Research, occasional paper 15. University of Colorado, Boulder.

Nichols, H. 1976. Historical aspects of the northern Canadian treeline. Arctic 29:38–47.

Nienstaedt, H., and A. Teich. 1972. The genetics of white spruce. USDA Forest Service research paper WO-15.

Oechel, W. C., and W. T. Lawrence. 1985. Taiga, pp. 66–94 in B. F. Chabot and H. A. Mooney (eds.), Physiological ecology of North American plant communities. Chapman & Hall, New York.

Oswald, E. T., and J. P. Senyk. 1977. Ecoregions of Yukon Territory. Canadian Forest Service, Pacific Forest Research Centre, Victoria, British Columbia.

Owens, J. N., and M. Molder. 1976. Bud development in Sitka spruce. II. Cone differentiation and early development. Can. J. Bot. 54:766–779.

Owens, J. N., and M. Molder. 1977. Bud development in *Picea glauca*. II. Cone differentiation and early development. Can. J. Bot. 55:2746–2760.

Owens, J. N., M. Molder, and H. Langer. 1977. Bud development in *Picea glauca*. I. Annual growth cycle of vegetation buds and shoot elongation as they relate to date and temperature sums. Can. J. Bot. 55:2728–2745.

Parker, W. H., and D. G. McLachlan. 1978. Morphological variation in white and black spruce: investigation of natural hybridization between *Picea glauca* and *P. mariana*. Can. J. Bot. 56:2512–2520.

Payette, S., and R. Gagnon. 1979. Tree-line dynamics in Ungava Peninsula, northern Quebec. Holarctic Ecol. 2:239–248.

Péwé, T. L. 1975. Quaternary geology of Alaska. U.S. Geological Survey professional paper 835.

Prudhomme, T. I. 1983. Carbon allocation to antiherbivore compounds in a deciduous and an evergreen subarctic shrub species. Oikos 40:344–356.

Rampton, V. 1971. Late Quaternary vegetational and climatic history of the Snag-Klutlan area, southwestern Yukon Territory, Canada. Geol. Soc. Amer. Bull. 82:959–978.

Raup, H. M., and C. S. Denny. 1950. Photo interpretation of the terrain along the southern part of the Alaska Highway. U.S. Geol. Surv. Bull. 963-D:95–135.

Rencz, A. N., and A. N. D. Auclair. 1978. Biomass distribution in a subarctic *Picea mariana–Cladonia alpestris* woodland. Can. J. Forest Res. 8:168–176.

Ritchie, J. C. 1962. A geobotanical survey of northern Manitoba. Arctic Institute of North America technical paper 9. AINA, Montreal.

Ritchie, J. C. 1976. The late Quaternary vegetational history of the western interior of Canada. Can. J. Bot. 54:1793–1818.

Ritchie, J. C. 1977. The modern and late Quaternary vegetation of the Campbell-Dolomite uplands near Inuvik, N.W.T., Canada. Ecol. Mongr. 47:401–423.

Ritchie, J. C. 1982. The modern and late-Quaternary vegetation of the Doll Creek area, north Yukon, Canada. New Phytol. 90:563–603.

Ritchie, J. C., and L. C. Cwynar. 1982. The late-Quaternary vegetation of the north Yukon, pp. 113–126 in D. M. Hopkins, J. V. Matthews, Jr., C. E. Schweger, and S. B. Young (eds.), Paleoecology of Beringia. Academic, New York.

Ritchie, J. C., and F. K. Hare. 1971. Late Quaternary vegetation and climate near the arctic treeline of northwestern North America. Quat. Res. 1:331–342.

Ritchie, J. C., and G. A. Yarranton. 1978a. Patterns of change in the late-Quaternary vegetation of the western interior of Canada. Can. J. Bot. 56:2177–2183.

Ritchie, J. C., and G. A. Yarranton. 1978b. The late-Quaternary history of the boreal forest of central Canada, based on standard pollen stratigraphy and principal components analysis. J. Ecol. 66:199–212.

Roche, L. 1969. A genecological study of the genus *Picea* in British Columbia. New Phytol. 68:505–554.

Rowe, J. S. 1952. Squirrel damage to white spruce. Canada Department of Resource Development, Division of Forest Research silvic leaflet 61.

Rowe, J. S. 1961. Critique of some vegetational concepts as applied to forests of northwestern Alberta. Can. J. Bot. 39:1007–1017.

Rowe, J. S. 1970. Spruce and fire in northwest Canada and Alaska, pp. 245–254 in Proceedings of the annual Tall Timbers fire ecology conference. Tall Timbers Research Station, Tallahasse, Fla.

Rowe, J. S. 1972. Forest regions of Canada. Department of the Environment, Canadian Forestry Service, publication number 1300.

Rowe, J. S., and G. W. Scotter. 1973. Fire in the boreal forest. Quat. Res. 3:444–464.

Short, S. K. 1978a. Holocene palynology in Labrador-Ungava: climatic history and culture change in the central coast. Dissertation, University of Colorado, Boulder.

Short, S. K. 1978b. Palynology: a Holocene environmental perspective for archaeology in Labrador-Ungava. Arctic Anthropol. 15:9–35.

Short, S. K., and D. L. Elliott-Fisk. in press. Holocene history of the northern Labrador tree limit. Rev. Palaeobot. Palynol.

Shreve, F. 1915. The vegetation of a desert mountain range as conditioned by climatic factors. Publication 217. Carnegie Institution of Washington, Washington, D.C.

Shugart, H. H. 1984. A theory of forest dynamics: the ecological implications of forest succession models. Springer-Verlag, New York.

Skre, O., W. C. Oechel, and P. M. Miller. 1983. Moss leaf water content and solar radiation at the moss surface in a mature black spruce forest in central Alaska. Can. J. Forest Res. 13:860–868.

Sorenson, C. J. 1977. Reconstructed Holocene bioclimates. Ann. Assoc. Amer. Geogr. 67:214–222.

Sorenson, C. J., and J. C. Knox. 1974. Paleosols and paleoclimates related to late Holocene forest/tundra border migrations: Mackenzie and Keewatin, N.W.T., Canada, pp. 187–203 in International conference on prehistory and paleoecology of western North American Arctic and subarctic. Archeological Association, University of Calgary, Calgary, Alberta.

Spear, R. W. 1982. The late Quaternary distribution of spruce (*Picea glauca, P. mariana*) in the Mackenzie Delta, Northwest Territories, Canada, p. 45 in Abstracts of the eleventh annual Arctic Workshop, Boulder, Colorado.

Strang, R. M. 1973. Succession in unburned sub-Arctic woodlands. Can. J. Forest Res. 3:140–142.

Streten, N. A. 1974. Some features of the summer climate of interior Alaska. Arctic 27:273–286.

Tedrow, J. C. F. 1970. Soils of the subarctic regions, pp. 189–206 in Ecology of the subarctic regime. Proceedings of the Helsinki symposium. UNESCO, New York.

Tedrow, J. C. F. 1977. Soils of the polar landscapes. Rutgers University Press, New Brunswick, N.J.

Tranquillini, W. 1979. Physiological ecology of the alpine timberline. Springer-Verlag, New York.

Turk, J. T. 1983. An evaluation of trends in the acidity of precipitation and the related acidification of surface water in North America. U.S. Geological Survey water-supply paper 2249.

U.S. Department of Agriculture. 1974. Seeds of woody plants in the United States. USDA agricultural handbook 450.

Viereck, L. A. 1973. Wildfire in the taiga of Alaska. Quat. Res. 3:465–495.

Von Rudloff, E., and M. J. Holst. 1968. Chemosystematic studies in the genus *Picea* (Pinaceae). III. The leaf oil of a *Picea glauca* × *mariana* (Rosendahl spruce). Can. J. Bot. 46:1–4.

Vowinckel, T., W. C. Oechel, and W. G. Boll. 1975. The effect of climate on the photosynthesis of *Picea mariana* at the sub-arctic tree line. I. Field measurements. Can. J. Bot. 53:604–620.

Warren Wilson, J. 1967. Ecological data on dry-matter production by plants and plant communities, pp. 77–127 in E. F. Bradley and O. T. Denmead (eds.), The collection of processing of field data. Wiley, New York.

Watts, W. A. 1970. The full glacial vegetation of northwestern Georgia. Ecology 51:631–642.

Webb III, T., E. J. Cushing, and H. E. Wright, Jr. 1983. Holocene changes in the vegetation of the Midwest, pp. 142–165 in H. E. Wright, Jr. (ed.), The

Holocene, vol. 2, Late-Quaternary environments of the United States. University of Minnesota Press, Minneapolis.

Whittaker, R. H., and G. E. Likens. 1975. The biosphere and man, pp. 305–328 in H. Leith and R. H. Whittaker (eds.), The primary production of the biosphere. Springer-Verlag, New York.

Wilson, C. V. 1971. The climate of Quebec. I. Climatic Atlas. Climatological Studies 11, Canadian Meterological Service, Ottawa.

Wolfe, J. A. 1971. Tertiary climatic fluctuations and methods of analysis of Tertiary floras. Palaeogeogr. Palaeoclimatol. Palaeoecol. 9:27–57.

Wolfe, J. A. 1972. An interpretation of Alaskan Tertiary floras, pp. 225–233 in A. Graham (ed.), Floristics and paleofloristics of Asia and eastern North America. Elsevier, New York.

Wolfe, J. A. 1978. A paleobotanical interpretation of Tertiary climates in the northern hemisphere. Amer. Sci. 66:694–703.

Zasada, J. C. 1971. Natural regeneration of interior Alaska forests–seed, seedbed, and vegetative reproduction considerations, pp. 231–246, in Proceedings, fire in the northern environment, a symposium. Pacific Northwest Forest and Range Experiment Station, Portland, Ore.

Zasada, J. C., and R. A. Gregory. 1969. Regeneration of white spruce with reference to interior Alaska: a literature review. U.S. Forest Service research paper PNW-79. Portland, Ore.

Zasada, J. C., K. Van Cleve, R. A. Werner, J. A. McQueen, and E. Nyland. 1978. Forest biology and management in high-latitude North American forests, pp. 137–195 in North American forest lands at latitudes north of 60 degrees. Proceedings of a symposium, September 19–22, 1977, University of Alaska, Fairbanks.

Chapter 3

Forests of the Rocky Mountains

ROBERT K. PEET

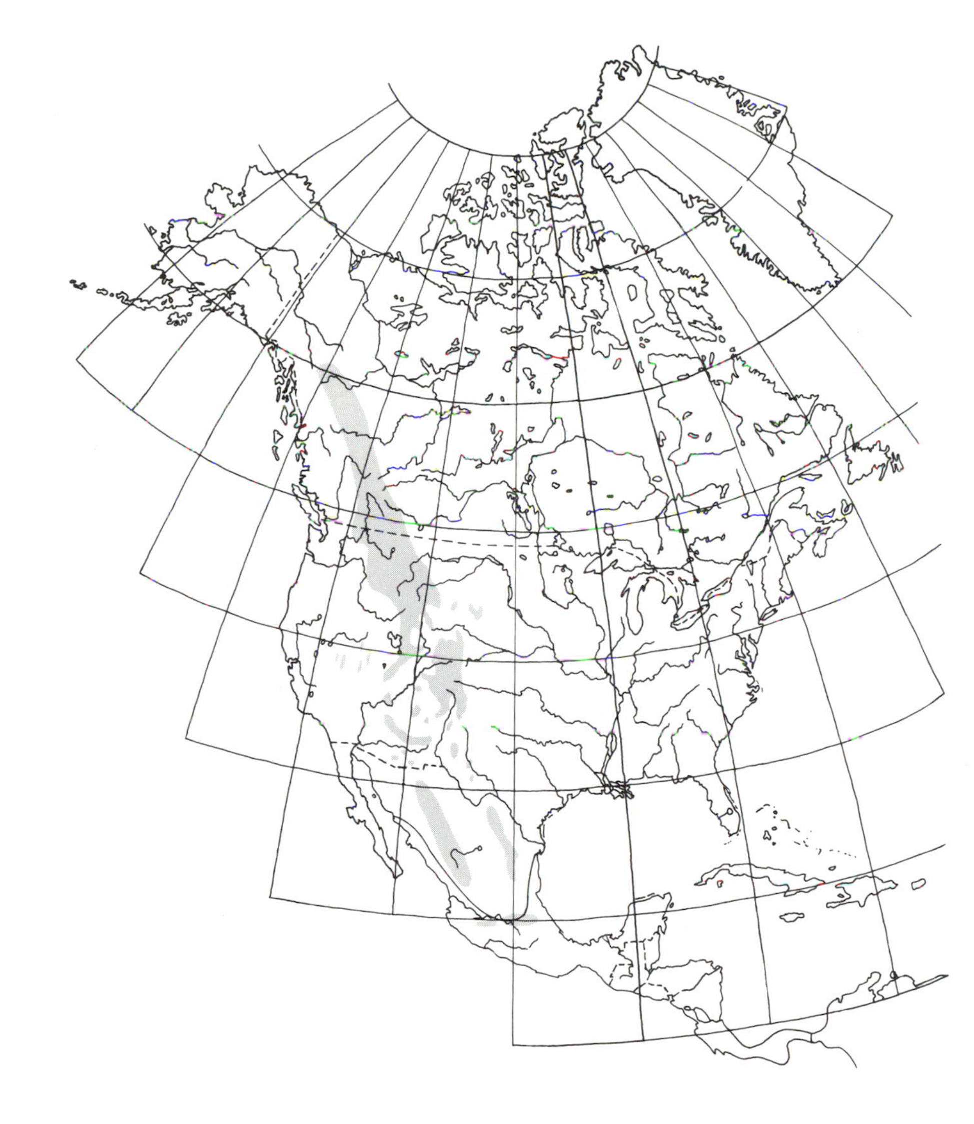

INTRODUCTION

The east slope of the Rocky Mountains can be an impressive sight to the traveler who has just crossed the flat expanses of the Great Plains. By way of example, the Colorado Front Range rises abruptly from 1500 m at the base of the foothills to peaks reaching 4400 m and gives the impression of an impassable wall. The slopes of these mountains are accented by a distinct dark band of forest bordered both above and below by lighter-colored grassland vegetation, the abrupt topography making the vertical zonation of vegetation particularly dramatic. Closer examination shows the forests to be dominated almost exclusively by conifers, and to have both physiognomic and floristic similarities with forests of the boreal region.

The Rocky Mountains are massive chains of mountains that form the backbone of the North American continent. In the broad sense, Rocky Mountain forests can be defined as extending from north of the 65° N latitude in the boreal regions of Alaska and Canada south to the towering peaks of the Mexican volcanic belt at 19° N latitude (Fig. 3.1), and perhaps even to the western highlands of Guatemala (Steyermark 1950). Given the length of the cordillera, the surprise is not that latitudinal variation exists in Rocky Mountain forests but that major vegetation types, often representing elevational zones, remain essentially constant over great distances.

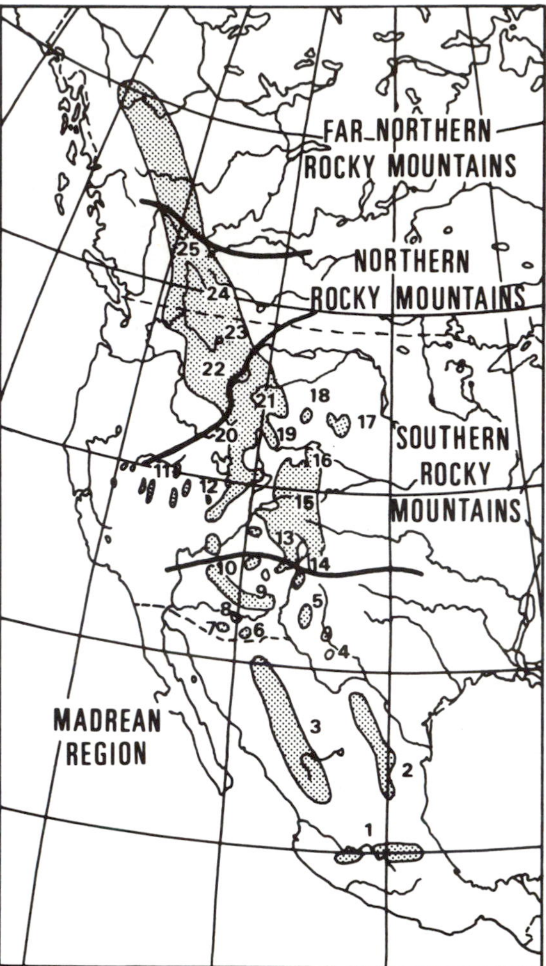

Figure 3.1. Distribution of Rocky Mountain vegetation in North America. Approximate boundaries of the four major floristic provinces are indicated by solid lines. Major mountain ranges and locations mentioned in the text include: (1) Trans-Mexican volcanic belt, (2) Sierra Madre Oriental, (3) Sierra Madre Occidental, (4) Davis Mountains, (5) Sierra Blanca, (6) Chiricahua Mountains, (7) Santa Catalina Mountains, (8) Pinaleno Mountains, (9) Mogollon Mesa, (10) San Francisco Peaks, (11) Great Basin ranges, (12) Wasatch Range, (13) San Juan Mountains, (14) Sangre de Cristo Mountains, (15) Front Range, (16) Medicine Bow Mountains, (17) Black Hills, (18) Bighorn Mountains, (19) Wind River Range, (20) Teton Mountains, (21) Yellowstone National Park, (22) Bitterroot Mountains, (23) Glacier National Park, (24) Banff National Park, (25) Jasper National Park.

Floristic Regions and Latitudinal Trends

Four relatively distinct Rocky Mountain floristic regions are often recognized (Fig. 3.1) (Daubenmire 1943). The northern region extends from northwest of Yellowstone Park in southern Montana north to Jasper National Park in Alberta and contains numerous species typical of the western slope of the Cascade Mountains. As one example, sheltered valley bottoms can contain forests of *Tsuga heterophylla*, *Thuja plicata*, and *Taxus brevifolia* (common names of major tree species are listed in Table 3.1), species otherwise not found in the Rockies. Similarly, *Tsuga mertensiana* and *Larix lyallii* are subalpine species of the Far West that occur in the Rockies only in this area of more moderate climate.

The far northern region – the boreal mountain region of Arno and Hammerly (1984) – occurs beyond the zone of intrusion of the Cascadian species. Here a few cordilleran species such as *Abies lasiocarpa* and *Pinus contorta* retain dominance, but other species such as *Picea engelmanii* and *Pseudotsuga* are largely replaced by boreal species such as *Picea glauca*.

The southern Rockies – Central Rockies of Daubenmire (1943), Southern and Central Rocky Mountain Physiographic Provinces of Thornbury (1965),

I thank R. B. Allen, S. F. Arno, W. D. Billings, D. H. Knight, W. H. Romme, and T. T. Veblen for numerous helpful comments on the manuscript.

Table 3.1. *Distributions and common names of tree species in the four Rocky Mountain floristic regions*

Species	Far north	Northern	Southern	Madrean
Conifers				
Abies concolor concolor (white fir)			×	×
Abies grandis (grand fir)		×		
Albies lasiocarpa lasiocarpa (subalpine fir)	×	×	×	
Albies lasiocarpa arizonica (corkbark fir)			+	×
Cupressus arizonica (Arizona cypress)				×
Juniperus communis (common juniper)	×	×	×	+
Juniperus deppeana (alligator juniper)				×
Juniperus flaccida (drooping juniper)				×
Juniperus monosperma (one-seed juniper)			×	×
Juniperus osteosperma (Utah juniper)			×	×
Juniperus scopulorum (Rocky Mountain juniper)		×	×	×
Larix laricina (tamarack)	+			
Larix lyallii (subalpine larch)		×		
Larix occidentalis (western larch)		×		
Picea engelmannii (Engelmann spruce)	+	×	×	×
Picea glauca (white spruce)	×	+		
Picea mariana (black spruce)	×			
Picea pungens (blue spruce)			×	×
Pinus albicaulis (whitebark pine)	+	×	+	
Pinus aristata (bristlecone pine)			×	
Pinus cembroides (Mexican pinyon)				×
Pinus contorta latifolia (lodgepole pine)	×	×	×	
Pinus edulis (pinyon)			×	×
Pinus engelmannii (Apache pine)				×
Pinus flexilis (limber pine)		×	×	+
Pinus leiophylla (Chihuahua pine)				×
Pinus monophylla (singleleaf pine)			×	
Pinus monticola (western white pine)		×		
Pinus ponderosa ponderosa (ponderosa pine)		×		
Pinus ponderosa scopulorum (ponderosa pine)		×	×	×
Pinus ponderosa arizonica (Arizona pine)				×
Pinus strobiformis (Mexican white pine)				×
Pseudotsuga menziesii (Douglas-fir)	+	×	×	×
Taxus brevifolia (Pacific yew)		×		
Thuja plicata (western redcedar)	+	×		
Tsuga heterophylla (western hemlock)	+	×		
Tsuga mertensiana (mountain hemlock)		×		
Selected angiosperms				
Acer glabrum (Rocky Mountain maple)	×	×	×	×
Acer grandidentatum (bigtooth maple)			+	+
Alnus oblongifolia (Arizona alder)				+
Alnus rubra (red alder)		+		
Alnus sinuata (Sitka alder)	×	×		
Alnus tenuifolia (thinleaf alder)	×	×	×	+
Arbutus arizonica (Arizona madrone)				×
Arbutus texana (Texas madrone)				×
Betula occidentalis (water birch)		×	×	
Betula papyrifera (paper birch)	×	×		
Ostrya species (hop hornbeam)			+	+
Populus angustifolia (narrowleaf cottonwood)		+	×	×
Populus balsamifera (balsam poplar)		×	+	

(cont.)

Table 3.1. *(cont.)*

Species	Far north	Northern	Southern	Madrean
Populus fremontii (Fremont cottonwood)			+	+
Populus sargentii (plains cottonwood)		×	×	
Populus tremuloides (quaking aspen)	×	×	×	×
Populus trichocarpa (black cottonwood)	×	×	+	
Prunus serotina (black cherry)				×
Quercus chrysolepis (canyon live oak)				+
Quercus emoryi (Emory oak)				×
Quercus gambelii (Gambel oak)			×	×
Rhamnus purshiana (cascara buckthorn)		×		
Robinia neomexicana (New Mexican locust)			×	×
Salix scouleriana (Scouler willow)	×	×	×	+
Sorbus scopulina (mountain-ash)	×	×	×	+
Sorbus sitchensis (Sitka mountain-ash)	×	×		

Note: Not included are the numerous southern Rocky Mountain species found only in Mexico. A minor presence on the margin of the region is indicated by a plus (+) sign.
Sources: Data from Little (1971, 1976).

and, in part, the Southern Rocky Mountain Region of Axelrod and Raven (1985) – which extend south from southern Montana through the Sangre de Cristo Mountains of northern New Mexico and the San Francisco Peaks of Arizona, have few species peculiar to them. Among the conifers, only *Pinus aristata* appears restricted to this region.

The madrean region (Axelrod and Raven 1985) – the Southern Rockies of Daubenmire (1943) – in contrast, has numerous distinctive species. Pines and oaks are particularly well represented and probably are components of a distinct flora that evolved in the Sierra Madre of Mexico (Axelrod 1958, 1979) and, following glacial retreats, expanded northward. The madrean region is sufficiently large that virtually no species are in common between the northern and southern extremes. As additional information becomes available on the floristics and vegetation of the Mexican mountains, recognition of additional floristic regions likely will become desirable.

Daubenmire (1943, 1975) hypothesized the location of the northern Rocky Mountain floristic region to be largely a consequence of a major storm track along which Pacific air penetrates to the Rockies between northwestern Oregon and southwestern British Columbia. Mitchell (1976) has shown that from June through September the Pacific air mass typically penetrates to the Rockies, the southern boundary of this area of intrusion running northeast from northern California into western Montana. South of this line, little Pacific air penetrates, and warmer continental air dominates. Thus, Cascadian species in the Rockies are likely limited on the south by summer heat and drought.

The border between the southern Rocky Mountain and madrean floristic regions is less well defined. Mitchell (1976) delimited an area southeast of a line from southwestern Arizona to Salt Lake and across to the Front Range as characterized by summer rains (the "Arizona monsoon") generated by the Bermuda high. This region has its precipitation concentrated during the summer months, in contrast to a region to the west, running from the Mojave Desert to central Idaho, in which winter rains predominate. Building on Mitchell's work, Neilson and Wullstein (1983) proposed that the northern limit of *Quercus gambelii*, one of the most northerly madrean species, occurs where its lower elevation limit set by drought intersects its upper elevation limit set by cold temperature. If this species is typical, it could be that a combination of summer drought and winter cold defines the northern limit of the madrean region. Because the specific drought and temperature tolerances of various species differ, this boundary will likely be diffuse.

Elevational Zonation and Environmental Gradients

Because of their apparent geographical constancy, elevational vegetation zones provide the oldest and simplest means of classifying mountain vegetation. Such zones have been identified typically on the basis of either climate or dominant species. Ramaley (1907, 1908), familiar with the efforts of Merriam (1890) and Schimper (1898) to delimit life zones, recognized four major Rocky Mountain climatic zones – Foothill, Montane, Subalpine, and Alpine – corresponding roughly to Merriam's Transition, Canadian, Hudsonian, and Arctic-Alpine zones. In contrast, Daubenmire (1943) used dominant species to designate zones. Between the basal plains and the alpine he recognized (1) the oak–mountain mahogany (*Quercus-Cercoparpus*) zone, (2) the juniper-pinyon (*Juniperus-Pinus edulis*) zone, (3) the ponderosa pine (*Pinus ponderosa*), zone, (4) the Douglas fir (*Pseudotsuga menziesii*) zone, and (5) the spruce-fir (*Picea engelmannii–Abies lasiocarpa*) zone. The first and second correspond to the Foothill zone of Ramaley, whereas the third and fourth have often been designated as the Lower and Upper Montane, respectively (Marr 1961). The spruce-fir zone is synonymous with the Subalpine zone of Ramaley.

The vegetation zones recognized by Daubenmire and others portray vegetation on an idealized mountain. In actuality, all zones are not necessarily present in a region, nor do they occur at consistent elevations. A northerly aspect, high moisture availablility, and increased latitude all tend to lower the elevation at which a formation occurs.

The primary difficulty with an elevational zonation approach to vegetation classification is that vegetation usually is not composed of discrete bands. Rather, composition varies continuously along environmental gradients, with the consequence that no two investigators are likely to recognize the same zones. In addition, environmental factors other than elevation influence vegetation. In this chapter I take two complementary approaches: I recognize 11 major groups of communities as foci for discussion, but I also use environmental gradients to present major vegetation patterns.

The single most important environmental gradient controlling vegetation composition in the Rockies is undoubtedly elevation. Elevation is a complex gradient that combines several environmental variables important for plant growth. Generally, with increasing elevation, temperature drops, precipitation increases (though perhaps with a secondary decline at highest elevations), solar radiation and particularly ultraviolet radiation increase, wind increases, snow depth and duration may increase, and so on (Greenland et al. 1985).

Local topographic variation makes climatic information difficult to obtain or summarize for the mountainous regions of western North America (Baker 1944). Most weather stations are located at low elevations, and few long-term records are available from sites above the foothills. One notable exception to the dearth of climatic data is a result of work at the Mountain Research Station of the University of Colorado by John Marr and colleagues (Marr 1961, 1967; Marr et al. 1968a, 1968b; Barry 1972, 1973). They collected data from weather stations situated at 2195 m (lower montane), 2580 m (upper montane), 3050 m (subalpine), and 3750 m (alpine). Supplemented with a National Oceanic and Atmospheric Administration (NOAA) station in the nearby foothills (1603 m), these stations provide a detailed view of an elevational gradient in climate (Fig. 3.2). Along this transect, the mean annual temperature drops from 8.8°C at 1603 m to −3.3°C at 3750 m. The mean daily minimum for January decreases from −7.8°C at 2195 m to −16.1°C at 3750 m, and the mean July maximum drops from 30.7°C at 1603 m to 19.4°C at 3750 m. Monthly precipitation is greatest in May and increases with elevation from 395 mm per year at the base of the foothills to 1050 mm in the alpine zone.

The topographic-moisture gradient is a second major complex gradient determining vegetation composition. Sites situated so as to received high levels of incident solar radiation, such as south-facing slopes and ridge tops, are warmer and drier than north-facing slopes and sheltered valley bottoms. In addition, upper slopes and ridges lose water to downslope flow, whereas lower slopes and bottoms often have a net gain from runoff. The topographic moisture gradient combines these factors and others that influence temperature and moisture within a given elevational belt (Whittaker 1967, 1973).

When placed as orthogonal axes, the elevational and topographic moisture gradients provide a useful frame of reference for studying mountain vegetation. An idealized representation for the central Rockies is shown in Fig. 3.3. The interaction of the two gradients is clearly evident in the predominantly diagonal orientation of the vegetation zones: A given vegetation type tends to be found at higher elevations on drier sites. This trend is particularly marked in sheltered valleys, where high moisture content, low incident radiation, and cold-air drainage combine to produce an environment and vegetation more characteristic of that nearly 500 m higher on open slopes.

Soil provides a third complex environmental variable important for interpretation of vegetation composition. Much of the Rocky Mountain region has young soils derived from Precambrian granites and chemically similar gneisses and schists. In areas dominated by these rocks, the major soil factors influencing vegetation are texture and depth, both of which in turn have direct impacts on water availability (Smith 1985). Trees almost always dominate sites with thin or rocky soils, but where deep, fine-textured soils occur, grasses and forbs can form a dense sod that inhibits tree regeneration.

The conspicuous importance of elevation and moisture appears to have obscured the importance of substrate variation for most Rocky Mountain

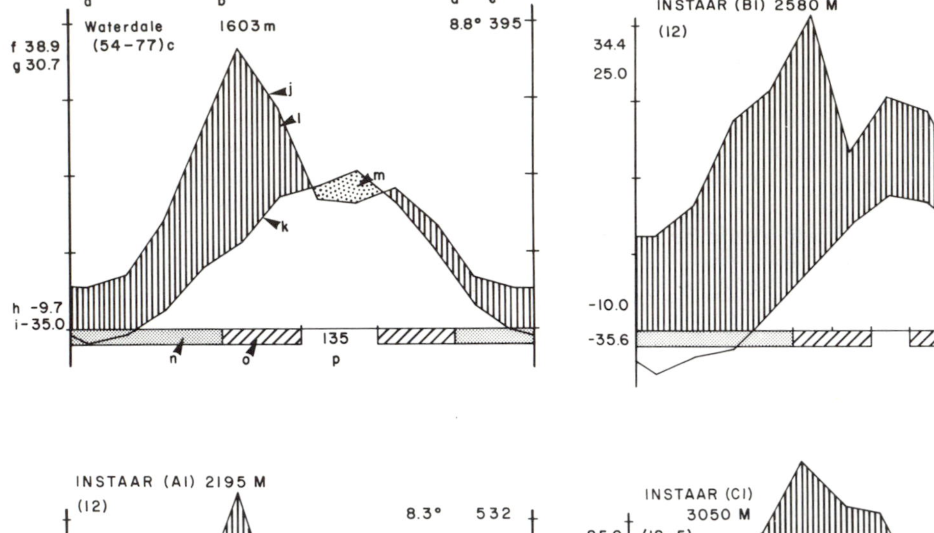

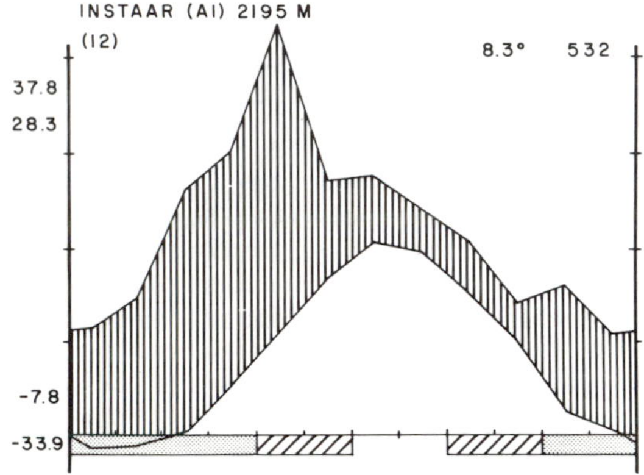

Figure 3.2. An elevational sequence of climate diagrams based on data collected at Waterdale and the Mountain Research Station west of Boulder Colorado. The diagrams follow the format of Walter and Lieth (1967). Abcissa: months, starting in January. Ordinate: one division = 10°C or 20 mm precipitation; a = station, b = elevation in meters above sea level, c = years of temperature and precipitation records, d = mean annual temperature, e = mean annual precipitation, f = highest temperature on record, g = mean daily maximum of warmest month, h = mean daily minimum of coldest month, i = lowest temperature on record, j = mean monthly precipitation curve, k = mean monthly temperature curve, l = relative humidity season (vertical shading), m = relative period of drought (dotted shading), n = months with mean daily minimum below 0°C (neutral shading), o = months with absolute minimum below 0°C (diagonal shading), p = mean duration of frost-free period in days (from Peet 1981).

vegetation research. Limestone and granitic soils, when in close proximity to each other, provide a clear exception. Such dissimiliar ranges as the Bighorn Mountains of Wyoming (Despain 1973) and the Santa Catalina Mountains of Arizona (Whittaker and Niering 1968) have much more xeric vegetation at a given site on limestone than on granitic soils, though this may be more a consequence of

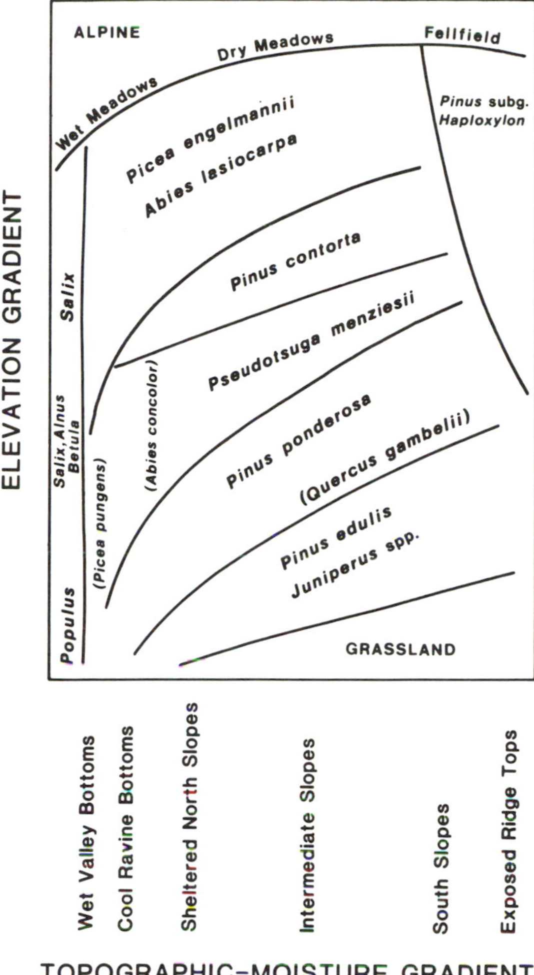

Figure 3.3. Major vegetation zones of the central Rocky Mountains as related to elevation and topographic moisture gradients. Parenthetical species are not consistent dominants.

differences in soil textures (Wentworth 1981). The influence of soil chemistry on Rocky Mountain vegetation remains largely unstudied. In one of the few Rocky Mountain vegetation studies to emphasize the importance of parent material, Despain (1973) demonstrated that vegetation in the Bighorn Mountains of north central Wyoming is strongly dependent on whether the soil parent material is granite or limestone. Subsequently, Despain (pers. commun., 1984) documented divergent patterns of vegetation for rhyolite- and andesite-derived soils in Yellowstone National Park, where rhyolite-derived soils are more sterile and the resulting vegetation is more xeromorphic in character than on otherwise equivalent but chemically more base-rich andesite-derived soils. In other studies (Billings, 1950; Salisbury 1964), highly sterile, acidic soils derived from hydrothermally altered rocks have been shown to support conifer woodland in regions where the primary vegetation is semidesert shrubland.

DISTURBANCE, SUCCESSION, AND ECOSYSTEM DEVELOPMENT

Disturbance Events

Nearly all Rocky Mountain forests are in some stage of recovery from prior disturbance. Fire, wind, insects, disease, ungulate browsing, avalanches, landslides, extreme weather, volcanism, and, of course, humans all have major impacts on the landscape. As a consequence, the vegetation is perhaps best thought of not as a uniform and stable cover but rather as a mosaic, with the character of each tessera frequently changing and the borders being periodically redefined.

For most of the Rocky Mountain landscape, fire has historically been the most important form of natural disturbance. Charcoal can be found in the soil of virtually any forest. Mean fire intervals have now been calculated for a sufficient number of localities that general patterns can be hypothesized. However, the available data must be interpreted with caution, as fire intervals likely vary with local conditions within a forest type, and the forests cover such huge areas that a few studies are not sufficient for general characterization.

Current fire regimes give little indication of fire regimes of the past. The great majority of the Rocky Mountain region has experienced a major reduction in fire frequency since the beginning of the century, owing to cessation of the periodic burning practiced by Indians, fire suppression, reduced fuel levels due to grazing, public education, and the continuing dissection of the landscape into smaller units separated by artificial firebreaks such as roads. In contrast, for much of the region, the activities of early European visitors, such as prospecting, land clearing, and simple arson, led to a major increase in fire frequency during the later part of the nineteenth century (Veblen and Lorenz 1986). However, portions of the ponderosa pine forest generally showed a decline in fire frequency during this same period owing to an increase in the grazing of domestic stock and the subjugation and declining populations of the native peoples, who originally used fire as a common management practice (Dieterich 1980; Arno 1985; Gruell 1985).

The Rocky Mountain forests with the highest fire frequencies were the low elevation *Pinus ponderosa* woodlands with their abundance of flammable grasses and forbs. A combination of lightning fires and aboriginal burning led in some areas to virtually annual fires (Dieterich 1980, 1983), with a mean fire interval of 5−12 yr being typical of the region as a whole (Weaver 1951; Arno 1980; Gruell 1985). Such frequent fires were mostly of low intensity, burning accumulated dead grass and surface litter, but rarely causing major damage to the thick-barked dominant trees. The low elevation *Pseudotsuga* woodlands of Wyoming and Montana had fire frequencies slightly lower than those for the more southern *P. ponderosa* woodlands. Arno and Gruell (1983) reported pre-1910 intervals of 35−40 yr, and Gruell (1985) suggested 20−40 yr to be typical. The lower elevation pinyon-juniper woodlands also burned regularly with low-intensity surface fires, but with lower frequency than the *P. ponderosa* woodlands, owing to lower rates of fuel accumulation (Arno 1985).

Whereas lightning likely caused many of the presettlement fires, current evidence suggests that the native Indians were responsible for the bulk of the fires (Arno 1985; Gruell 1985). Barrett and Arno (1982) were able to document the role of Indian fires in the Bitterroot forests of Montana by determining fire frequencies in matched areas, some heavily used by Indians and others remote. Mean fire intervals in the isolated areas were roughly twice those found in the heavily used areas. One implication of this finding is that maintenance of presettlement vegetation may require prescribed burns in addition to natural lightning fires.

Pinus contorta and *Picea-Abies* forests burned less frequently than the *Pinus ponderosa* woodlands, but the fires that did occur were more commonly severe, stand-replacing crown fires (Wellner 1970). Romme and Knight (1982; Romme 1982) reported a pattern for the *Pinus contorta*-dominated subalpine forests of Yellowstone National Park that may be typical for higher elevations in the Rocky Mountain region. They reported a 300−400-yr mean fire interval during which large areas burn within a relatively short interval. This is followed by a relatively long fire-free period during which fuel accumulates. Stand ages in the Rockies south of Yellowstone largely lend support to a 200−400-yr mean fire interval for subalpine *Pinus contorta* and *Picea-Abies* forests (Peet 1981; Romme and Knight 1981), but an interval of 50−150 yr is likely more typical of lower-elevation *Pinus contorta* forests (Clements 1910; Loope and Gruell 1973; Peet 1981). Studies conducted farther north by Tande (1979) in Jasper Park, Alberta, and by Arno (1980) in Montana also support the pattern of a few widespread fires, but suggest a shorter (65−100 yr) mean fire interval. Hawkes (1980), working in Kananaskis Park, Alberta, found a similar fire interval (90 yr) for low-elevation forests (1830 m), but a somewhat longer interval (153 yr) for high-elevation forests (>1830 m). Throughout the Rockies, *Pinus contorta* and *Picea-Abies* forests in proximity to frequently burned communities such as *Pinus ponderosa* or *Pseudotsuga* woodlands have experienced frequent, low-intensity surface fires at a shorter fire interval, typically 15−50 yr (Houston 1973; Heinselman 1981).

Other major forms of forest disturbance can also be characterized by mean intervals, but only for fire have such intervals been determined often enough for patterns to be clear. Wind storms can destroy large patches of subalpine forest, with major damage usually being concentrated in old stands that have a high incidence of trunk rot. No regional determinations of windthrow frequency are currently available for Rocky Mountain forests.

Outbreaks of insect pests also destroy large areas of forest. For example, in the 1940s, an outbreak of spruce beetle (*Dendroctonus rufipennis*) killed virtually all of the *Picea* and most of the *Abies* trees greater than 10 cm in diameter on the White River Plateau of northwestern Colorado, destroying an estimated timber volume of 10^7 m^3 of *Picea* alone (Miller 1970; Alexander 1974). Mountain pine beetles launch periodic outbreaks that cause extensive damage to such species as *Pinus contorta* and *P. ponderosa* (Amman 1977, 1978; Romme et al. 1986), and budworms similarly engage in frequent epidemics affecting mainly *Pseudotsuga* (McKnight 1968). Fungal pathogens such as *Armillaria mellea* and *Phellinus weirii* are less well known but can also be significant. James and associates (1984) reported that about 35% of the annual tree mortality in two national forests in Idaho is associated with root diseases caused by such fungi. Further, there is widespread support for the hypothesis that the reduced fire frequency during the past 50−100 yr is largely responsible for conditions that have allowed major outbreaks of several insects (e.g., western spruce budworm, Douglas fir tussock moth, mountain pine beetle) and root diseases (e.g., *Armillaria*), (Arno, pers. commun.). Insects and fungal pathogens typically are far less devastating than fire. Indeed, loss of 50% or more of the mature trees in a *Pinus contorta* stand over a 2-yr period, though causing an immediate drop in production, can be compensated for within 5 yr by increased production by the remaining trees (Romme et al. 1986).

Disturbance types of seemingly limited importance, such as avalanches and volcanism, also play major roles in particular regions. For example, in a study of the influence of avalanches on vegetation in the central third of Glacier National Park, Butler (1979) identified more than 800 avalanche paths. Because avalanches occur annually or nearly so in many such tracks, the variances in velocity and distance of flow of avalanches within a track become critical variables in determining the impact on vegetation (Johnson et al. 1985). As might be expected, the most common woody plants in tracks with high avalanche frequencies are short-lived species with flexible stems, such as *Alnus* species, *Acer glabrum*, *Betula pumila*, and *Salix* species (Butler 1979; Malanson and Butler 1984; Johnson et al. 1985).

Stand Development

Just as community composition varies with position on environmental gradients such as elevation and moisture, the course of stand development following a major disturbance such as fire is a function of gradient position. Three models characteristic of low, middle, and high elevations serve to illustrate the range of stand development patterns seen.

Stands in the middle portions of the elevation and moisture gradients typically exhibit a developmental sequence widely encountered in montane and boreal conifer forests. Four-stage sequences have been envisaged by several workers (Bloomberg 1950; Daubenmire and Daubenmire 1968; Day 1972; Bormann and Likens 1979; Oliver 1980; Peet 1981), with the details varying, but the main outline being consistent. The first stage is one of little competition and extensive tree establishment. During the second stage there is intense competition among the established trees such that little, if any, new establishment is possible. This is primarily a period of decreasing tree density and increasing tree size. In the third stage, mortality finally exceeds the ability of established trees to fill canopy gaps, with the result that resources again become available for new establishment. The fourth stage is a form of steady state in which tree mortality is balanced by tree recruitment. This basic sequence can be found in forests dominated by *Pinus contorta*, *Picea engelmannii*, *Pseudotsuga*, and various other species.

The synchrony of tree establishment following disturbance can be highly variable and depends on site conditions and the intensity of competition by herbaceous species. On the most favorable sites, establishment can be confined to a 10-yr window. More typically, one encounters a 40–70-yr range of establishment dates, though the window can be significantly wider on severe sites (Franklin and Hemstrom 1981; Peet 1981). Age distributions (determined by increment corings at ground level) for *Picea* and *Abies* on a typical site in Rocky Mountain National Park are shown in Fig. 3.4. In this stand, *Picea* largely became established in the 80 yr following an extensive forest fire 260 yr prior to sampling,

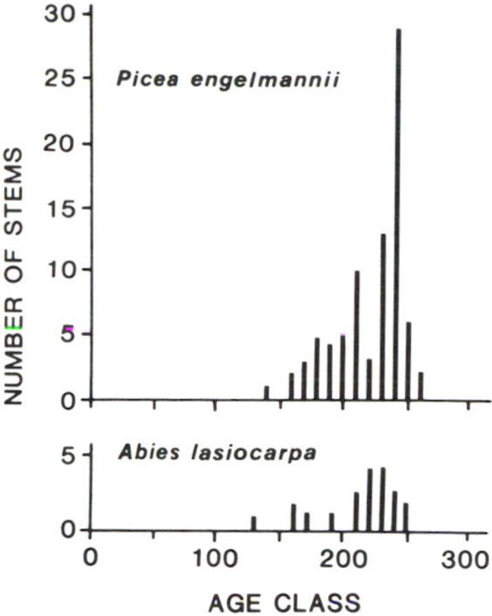

Figure 3.4. Age distributions in 10-yr age classes for Picea engelmannii *and* Abies lasiocarpa *in a 260-yr-old even-aged stand in Rocky Mountain National Park. Based on stems >2.5 cm diameter at 2 cm above the ground, cored at the ground surface.*

and very little establishment has taken place since. Notice also that *Abies*, which is primarily confined to the subcanopy of this stand, became established during the same interval as *Picea*. *Pseudotsuga* often plays the same role in *Pinus contorta*- and *Pinus ponderosa*-dominated stands, appearing to be younger than the primary canopy species owing to its smaller size, but having actually become established during the same interval.

The steady-state stage of stand development is infrequent in Rocky Mountain forests, owing to the high frequency of disturbances. Usually it is recognized by a reverse-J or negative-exponential size distribution of the dominant species (Leak 1965; Parker and Peet 1984), though this can also occur in broadly even-aged stands. Figure 3.5 shows age distributions for an old stand in Rocky Mountain National Park that comes as close to a steady state as can normally be found. Even this stand fails to

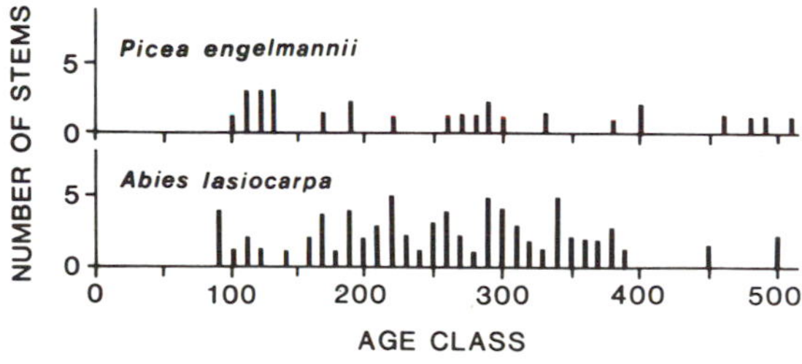

Figure 3.5. Age distributions in 10-yr age classes for Picea engelmannii *and* Abies lasiocarpa *in a steady-state stand in Rocky Mountain National Park. Based on stems >2.5 cm diameter at 2 cm above the ground, cored at the ground surface.*

have a true negative-exponential age distribution and probably represents the second generation of trees following a disturbance, with most of the trees having become established during a 250-yr interval of canopy breakup.

On extreme, high-elevation sites or xeric sites, tree establishment is often very slow as a consequence of the harsh environment and sometimes the more rapid establishment of a highly competitive herbaceous stratum (Stahelin 1943; Bollinger 1973; Noble and Ronco 1978; Peet 1981). Establishment can be sufficiently slow that by the time the canopy is nearly complete, trees are dying from senescence-related causes. Consequently, the second stage of stand development with its intense tree competition and no recruitment is bypassed, as is the third stage, with pronounced canopy breakup and renewed establishment. Instead, one finds a stage in which both stand density and tree size slowly increase toward steady-state conditions. Later in the chapter, Fig. 3.12B show a 70-yr-old stand of such a type in which canopy closure is nowhere near complete and establishment is likely to continue for another century or more (p.87).

Establishment on some extreme, high-elevation sites can be episodic, reflecting variation in climatic or other environmental variables. For example, several ecologists working in the Pacific Northwest have reported tree invasions of subalpine meadows to occur only during extended periods of drought and associated long snow-free periods (Franklin et al. 1971; Agee and Smith 1984). Dunwiddie (1977) has described expansion of trees into subalpine meadows of the Wind River Mountains of Wyoming as being correlated with periods of grazing by domestic stock.

In open, low-elevation communities, such as *Pinus ponderosa* woodlands, episodic establishment can be particularly important. On these sites, fires generally are not detrimental to the canopy trees, but occur frequently (though less so than prior to 1900) at low intensity, killing young woody plants and removing accumulated dead grass and litter. Here, tree establishment occurs in pulses associated with the co-occurrence of a good seed year, favorable weather (e.g., good spring and summer moisture), and absence of fire, and perhaps there had been an earlier fire that removed competition. Such strict requirements can result in long intervals between pulses of establishment (Potter and Green 1964; Peet 1961; White 1985; cf. Cooper 1960).

All intermediates between the three development patterns described earlier can be found, with stand dynamics varying continuously along environmental gradients. A typical example has been reported by Tesch (1981) from Montana: On a mesic, north-slope *Pseudotsuga* site, even-aged development occurred. In contrast, on the opposite, south-facing slope, recruitment was slow, and an uneven-aged stand had developed, with the shade-intolerant pioneer species *Pinus contorta* continuing to reproduce beneath the open *Pseudotsuga* upper canopy.

Biomass, Production, and Species Diversity

As with stand development, three general patterns of successional changes in biomass, production, and diversity can be associated with low-, middle-, and high-elevation forests (Peet 1978b, 1981). Two of these general patterns are illustrated in Fig. 3.6.

Following a major disturbance on a middle-elevation site, the biomass is low, but it increases steadily to an asymptote late in the self-thinning (second) phase of forest development (Fig. 3.6A). With canopy breakup (middle to late stage 3), the biomass declines rapidly, only to again increase with the renewal of regeneration. Production similarly starts low but increases rapidly to reach a peak early in the self-thinning phase, shortly after canopy closure. [Epidemics such as those of pine beetles simply mimic natural thinning, albeit at an accelerated rate, so that no long-term reduction in production is observed (Romme et al. 1986).] Production then appears to decline slowly until canopy breakup frees resources and again stimulates production. Species diversity (vascular plant species per 0.1 ha) tends to track resource availability. Early in the sequence, diversity is quite high, but diversity drops dramatically with the onset of self-thinning and

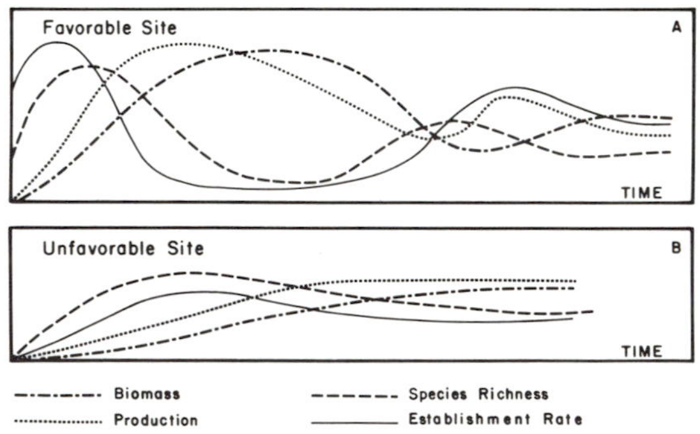

Figure 3.6. Generalized development patterns for forests of the Colorado Front Range. The favorable site (A) is typical of forests dominated by Pinus contorta, Pseudotsuga menziesii, *and* Picea engelmannii. *Time scales vary with site, the secondary low in biomass occurring at ~300 yr for* Pinus contorta *and ~450 yr for* Picea engelmannii. *The unfavorable site (B) is typical of extreme, high-elevation* Picea-Abies *forests and xeric* Pinus flexilis *forests (from Peet 1981).*

stays low until canopy breakup again provides resources for understory growth (see Table 3.4 for specific examples, p.88).

On extreme, high-elevation sites or exposed sites where tree establishment is slow, biomass and production do not show the pronounced cycles characteristic of middle-elevation stands, but rather increase asymptotically (Fig. 3.6B). Variation in species diversity is also reduced, though again the variation tracks available resources and thus shows an initial high followed by a modest decline to steady-state values. In low-elevation *Pinus ponderosa* woodlands characterized by episodic regeneration, production is relatively constant, and the biomass tracks the oscillations in numbers of large trees.

Biomass and production also vary with site and regional climate, but detailed data are sufficiently sparse that only very coarse patterns can be demonstrated (Table 3.2). An elevation sequence in the Santa Catalina Mountains of Arizona (first 10 stands in Table 3.2) shows that almost the entire ranges of biomass and production normally found in Rocky Mountain forests can be found in a single, environmentally heterogeneous locality. Indeed, Weaver and Forcella (1977), studying data form some 50 varied forest stands in Montana, Idaho, and eastern Oregon, reported a range of biomass values contained within the range Whittaker and Niering (1975) found in the Santa Catalina Mountains. Although the changes in biomass and production with elevation seen in Whittaker and Niering's data do not show as clear a relationship as might be hoped, this could in part result from the fact that those authors did not control for stand age or successional status.

The final 12 stands in Table 3.2 illustrate variation within forests dominated by a single species, *Pinus contorta*. In these stands, the biomass ranges from 52 to 245 t ha^{-1}, and production from 2.2 to 8.4 t ha^{-1} yr^{-1}. After study of a subset of these data, including two very different but adjacent stands (18 and 19), Pearson and associates (1984) concluded that for *Pinus contorta*, the biomass is at least as tightly linked to stand density as to site conditions.

Species diversity can also vary greatly within a small but environmentally heterogeneous area. In Rocky Mountain National Park and the adjacent foothills, species richness (per 0.1 ha) takes the form of a cup-shaped response on a plot of elevation versus moisture gradient. Sites near the middle of the gradients for moisture and elevation generally have low numbers of species, with the bottom of the "cup" falling near an average of 13 species per 0.1 ha in dry *Pinus contorta* forests. At both the high- and low-elevation transitions from forest to grassland, diversity is much higher, typically averaging 40 species per 0.1 ha. The highest-diversity sites are mesic ravine forests at moderately low elevations, where deciduous trees often dominate, and diversity averages up to 60 species per 0.1 ha.

The complex diversity findings for Rocky Mountain National Park show the futility of modeling diversity as a simple function of elevation, moisture, or successional development; strong interactions are involved, and diversity can be understood only as a multidimensional phenomenon. Perhaps the only clear pattern to emerge is that dense conifer forest depresses diversity, and this depression can be ameliorated by increased levels of moisture or light (Daubenmire and Daubenmire 1968; del Moral 1972; Peet 1978b; McCune and Antos 1981).

MAJOR VEGETATION TYPES

The combined influences of elevation, moisture, soil, and latitude make for complex vegetation patterns. The necessary addition of communities recovering from various forms of disturbances yields, at a minimum, a five-dimensional vegetation model. Such continuous, multidimensional variation is difficult to present graphically or in words. Figure 3.7 represents an attempt to display several dimensions of variation through a series of gradient diagrams. Each diagram has a vertical gradient of elevation and a horizontal gradient of moisture and exposure. Two of these diagrams have major seral species indicated by shading. The latitudinal gradient can be deduced by comparison of the diagrams. To simplify discussion, I recognize 11 major vegetation types in the Rocky Mountain forest region. These are discussed individually in later sections. The reader should keep in mind that although not always explicitly discussed, many of the relationships between the recognized types can be seen by examination of Fig. 3.7.

Riparian and Canyon Forests

Low-elevation stream-side forests of the Rockies are at once the first forests encountered by the traveler and the least typical of the region as a whole. Broad-leaved deciduous species, mostly cottonwoods (e.g., *Populus angustifolia*) and willows (*Salix* species), line the major streams of the foothills and adjacent semiarid lowlands (Fig. 3.8). The result is a series of fingers of mesophytic forest in an otherwise semiarid landscape of low grass or desert scrub. The dominant species of these low-elevation forests shift geographically, with *Populus trichocarpa* occurring from the northern Rockies to

Table 3.2. *Selected aboveground biomass and production estimates for Rocky Mountain forests*

Dominant species[a]	Elevation (m)	Age (yr)	Density (ha^{-1})	BA[b] ($m^2\ ha^{-1}$)	Biomass[c] ($t\ ha^{-1}$)	LAI[d] ($m^2\ m^{-2}$)	Production[e] ($t\ ha^{-1}\ yr^{-1}$)	Location	Ref.[f]
Abies lasiocarpa	2720	106	590	57.8	357	14.7	8.6	Arizona	1
Abies concolor	2340	124	1510	58.6	361	15.5	11.1	Arizona	1
Pseudotsuga menziesii						16.7	10.7	Arizona	1
Abies concolor	2640	321	400	118.1	790				
Pseudotsuga menziesii	2650	252	340	70.5	438	15.5	8.3	Arizona	1
Pinus ponderosa *Pinus strobiformis*	2740	93	2700	39.4	161	7.6	6.1	Arizona	1
Pinus ponderosa	2470	142	1100	46.3	250	5.9	5.7	Arizona	1
Pinus ponderosa	2180	150	1280	34.9	163				
Quercus hypoleucoides						4.7	4.9	Arizona	1
Pinus leiophylla *Quercus arizonica*	2040	101	2780	26.0	114	3.7	4.3	Arizona	1
Pinus cembroides *Juniperus deppeana*	2040	115	570	4.3	19	2.0	0.65	Arizona	1
Quercus oblongifolia *Quercus emoryi*	1310	117	190	4.0	11	1.8	0.72	Arizona	1
Pinus edulis *Juniperus osteosperma*	2135	270	1111	—	119	—	5.6	Arizona	2
Populus tremuloides	3200	80	3070	31.8	157	—	11.7	New Mexico	3
Juniperus occidentalis	1356	350+	246	—	21	2.0	1.1	Oregon	4

Thuja plicata *Pinus monticola* (3)	—	105	798	56.8	290	—	8.7	Idaho	5
Tsuga heterophylla *Pinus monticola*	—	250+	105	49.8	316	—	5.5	Idaho	5
Pinus monticola	—	103	710	63.7	446	—	13.4	Idaho	5
Abies grandis *Larix occidentalis*	—	105	1127	53.5	290	—	9.2	Idaho	5
Pinus monticola *Abies grandis* (5)	—	103	314	66.4	504	—	13.8	Idaho	5
Pinus contorta	2800	110	2217	42	142	7.3	—	Wyoming	6
Pinus contorta	2800	110	14640	50	101	7.1	—	Wyoming	6
Pinus contorta	-2800	110	9700	55	124	8.8	—	Wyoming	6
Pinus contorta	2900	110	1850	64	144	9.9	—	Wyoming	6
Pinus contorta	3050	75	1280	26	96	9.0	—	Wyoming	6
Pinus contorta	2950	240	420	37	132	4.5	—	Wyoming	6
Pinus contorta	—	72	8600	—	122	4.5	4.3	Colorado	7
Pinus contorta	—	71	1650	—	52	4.5	2.2	Colorado	7
Pinus contorta	—	77	3800	—	275	14.0	8.4	Colorado	7
Pinus contorta	1400	100	2520	52.3	245	—	—	Alberta	8
Pinus contorta	1400	100	717	34.9	194	—	—	Alberta	8
Pinus contorta	1400	100	12256	35.9	92	—	—	Alberta	8

[a]Numbers of stands averaged together shown in parentheses.
[b]Basal area.
[c]Aboveground biomass of trees.
[d]Leaf area index, two-sided.
[e]Aboveground primary production of trees.
[f]References: 1, Whittaker and Niering (1975); 2, Darling (1966); 3, Gosz (1980); 4, Gholz (1980); 5, Hanley (1976); 6, Pearson, et al. (1984); 7, Moir (1972); Moir and Francis (1972); 8, Johnstone (1971).

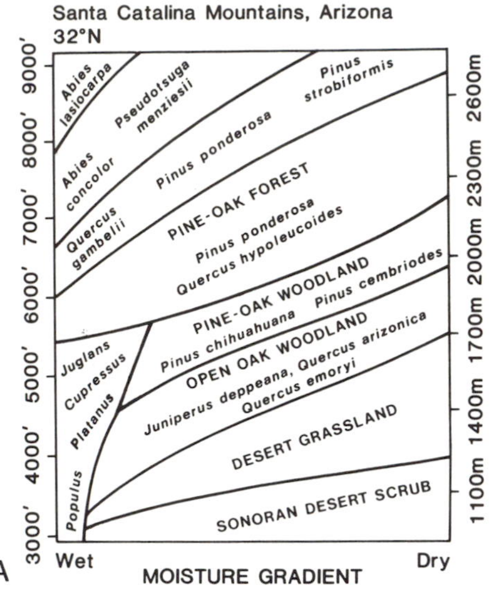

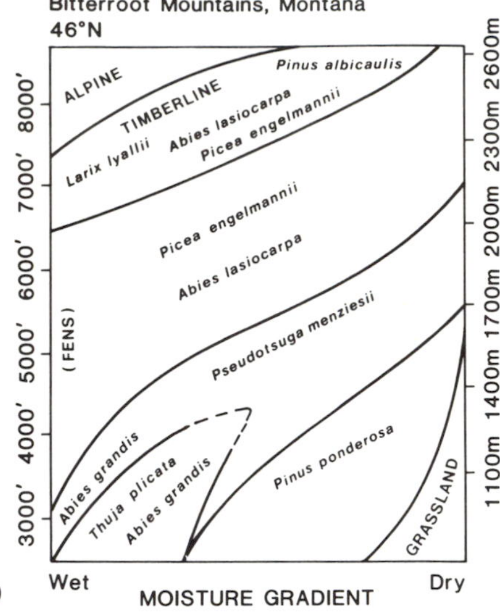

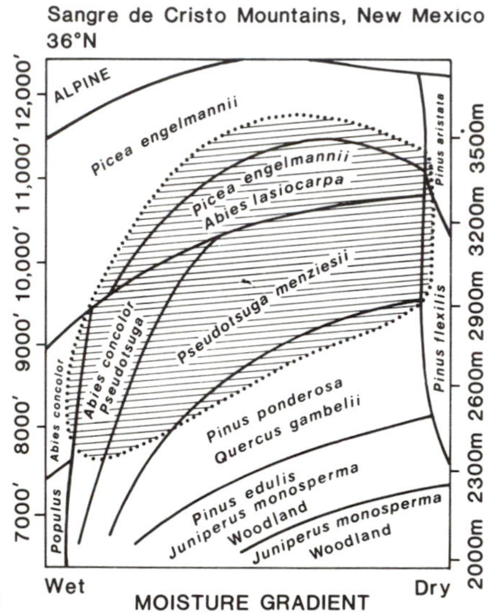

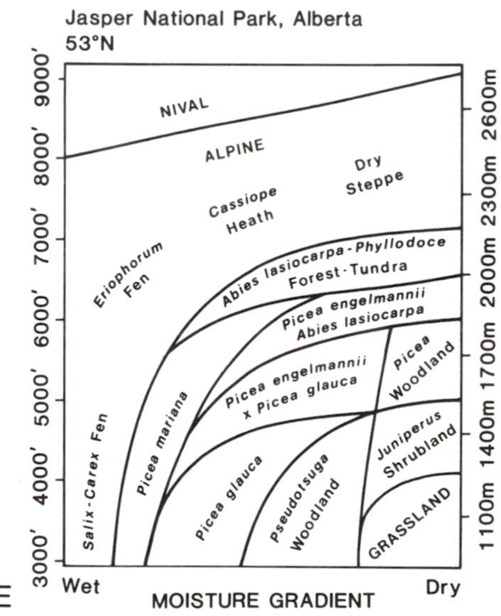

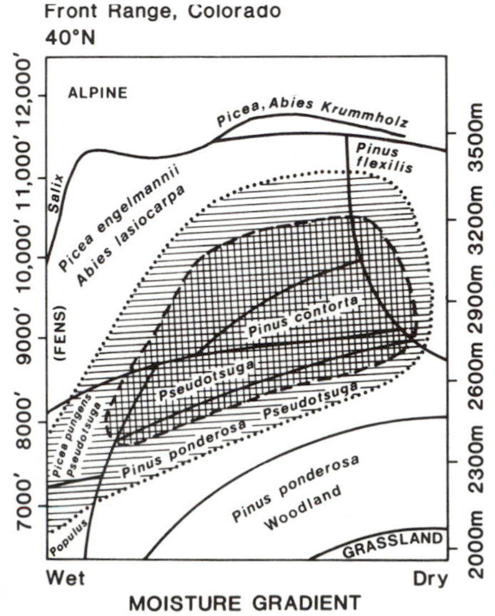

Figure 3.7. Gradient mosaic diagrams illustrating variations in vegetation compositions with elevations and topographic positions for seven sites along a latitudinal sequence. In B and C, the shading down to the left indicates the range of Populus tremuloides *as an important post-disturbance species, whereas shading down to the right indicates the range of* Pinus contorta *as an important post-disturbance species. Successional* P. contorta *and* P. tremuloides *communities may also occur at locations D and E, but they are not indicated because of insufficient data on the exact distributions. A: Santa Catalina Mountains, Arizona (from Whittaker and Niering 1965). B: Southern Sangre de Cristo Mountains near Santa Fe, New Mexico (redrawn from Peet 1978a). C: Northern Front Range, northern Colorado (redrawn from Peet 1978a, 1981). D: Bitterroot Mountains, central western Montana (adapted from Habeck 1972, Arno 1979, and Arno, pers. commun.) E: Jasper National Park, Alberta (adapted from La Roi and Hnatiuk 1980).*

Figure 3.8. Riparian Populus angustifolia *(cottonwood) forest bordering the Wind River in western Wyoming. Riparian forests regularly extend down into the grass- and shrub-dominated foothills and plains.*

coastal Alaska, *Populus sargentii* dominant on the eastern fringe of the southern Rockies, and *Populus fremontii* dominant in the western portion of the southern Rockies and much of the northern madrean region.

Riparian communities, like Rocky Mountain forests in general, exhibit elevational variation. With increasing elevation, forests of the wide-leaved *Populus sargentii* or *P. fremontii* give way to forests of the narrower-leaved *Populus angustifolia*. Still higher, *Alnus tenuifolia* and *Betula occidentalis*, along with numerous species of *Salix*, replace *Populus* on the alluvial flats and moist meadow edges. *Salix* thickets continue to dominate many stream sides and valley bottoms up to the timberline. Where middle-elevation streams pass through sheltered valleys or canyons, the dominant deciduous species are often replaced by evergreen conifers such as *Picea pungens, Pseudotsuga,* and *Abies concolor* (Pace and Layser 1977; Peet 1978a; Romme and Knight 1981). In the madrean region, the low-elevation riparian vegetation can be further divided into a lower *Platanus wrightii, Juglans major,* and *Fraxinus velutina* community and a higher *Alnus oblongifolia* community.

The similarities of the foothill riparian forests to the deciduous forests of eastern North America do not stop with the deciduous habit of the dominant trees. Numerous species of herbs typical of eastern forests occur in a narrow band along the foothills of the east slope, where they are confined to mesophytic riparian habitats. Included are species such as *Carex sprengelii, Ranunculus abortivus,* and *Aralia nudicaulis*. Other species typical of the East, such as *Ratibida pinnata*, are widespread in the central and northern Rockies, but are largely confined to similar riparian habitats or to the deciduous *Populus tremuloides* forests.

In the semiarid madrean region, cool, moist canyons take on added importance as a habitat for rare or relict mesophytic species. Trees typical of eastern forests, such as *Ostrya virginiana, Cercis canadensis,* and *Quercus muhlenbergii*, occur as isolated populations in canyons, and some predominantly Mexican species, such as *Cupressus arizonica*, are mostly confined to such canyons at their northern limits. Except for the uncommonly high importance of *Cupressus*, Niering and Lowe's (1985) description of Bear Canyon in the Santa Catalina Mountains of southern Arizona serves well to represent this type. They reported the common tree species (>2.5 cm dbh) to be, in descending order of abundance, *Cupressus arizonica, Alnus oblongifolia, Quercus rugosa,*

A

B

Figure 3.9. Lower-treeline forests. A: Pygmy conifer woodland of Pinus edulis *and* Juniperus monosperma *east of Taos, New Mexico. B:* Quercus gambelii *chaparral with scattered* Pinus ponderosa *at the mountain front near Colorado Springs, Colorado.*

Fraxinus velutina, *Q. hypoleucoides*, *Q. oblongifolia*, *Platanus wrightii*, *Q. emoryi*, *Pinus cembroides*, and *Juglans major*.

Pygmy Conifer Woodland

Over much of the southern and madrean regions, pygmy conifer woodland (Fig. 3.9A) forms the transition from grassland or desert on the basal plains to montane conifer forest. The dominant junipers (*Juniperus* species) and pinyon pines (*Pinus* subsection *Cembroides*) generally have a low stature seldom exceeding 7 m, rounded crowns, and multiple stems that give the trees a shrub-like appearance. The lowest, driest sites are dominated by *Juniperus* alone, with pinyon assuming dominance at higher elevations. Pygmy conifer woodlands are best developed in the semiarid lands west of the main cordillera and are therefore more fully described in Chapter 7 on cold deserts.

On the east slope of the Rockies, from the Texas border north to central Colorado, *Pinus edulis* and *Juniperus monosperma* are the dominant trees of foothill woodlands. Southward, *Pinus cembroides* and

Table 3.3. *Compositions of typical pinyon-juniper woodlands*

	Basal area (m^2 ha^{-1})		Density (stems ha^{-1})	
Species	Texas	New Mexico	Texas	New Mexico
Cercoparpus montanus				5.9
Juniperus deppeana	2.22		180.6	
Juniperus erythrocarpa	0.17		21.0	
Juniperus monosperma		9.81		346.3
Juniperus scopulorum		0.60		18.3
Pinus cembroides	1.00		127.5	
Pinus edulis		2.91		161.5
Quercus emoryi	0.26		40.0	
Quercus grisea	0.85		94.3	
Quercus undulatus				23.7
Rhus trilobata				9.4

Note: Average basal area and density are shown for four sites in the Manzano Mountains of central New Mexico and three sites in the Davis Mountains of southwest Texas; all stems > 10 cm in diameter at 0.3 m height are included.
Source: Data from Woodin and Lindsey (1954)

Juniperus deppeana assume dominance, though with a progressively greater intermixing of madrean forest elements (Table 3.3). *Pinus cembroides* continues south to the Mexican volcanic belt, but throughout the woodlands of Mexico the pinyon shares dominance with various oak species, except on the most highly disturbed sites (Robert 1977; Passini 1982). In north central Colorado, both *Juniperus monosperma* and *Pinus edulis* are absent from the east slope, except for a disjunct population of pinyon northwest of Fort Collins, Colorado, likely established in the last few hundred years (Wright 1952). In a region stretching from Idaho on the west to the Black Hills of South Dakota on the east, and north into Alberta, *Juniperus scopulorum* dominates low-elevation woodlands. This expansion of *J. scopulorum* from its typical forest margin habitat in the south might be a consequence of the absence of competing pinyon and juniper species, though other environmental and biotic factors cannot yet be discounted. In some locations in the northern Rockies, *Pinus flexilis* occurs with *J. scopulorum*, filling the niche of the codominant pine (Daubenmire 1943; Arno 1979). West of the main Rocky Mountain massif, from the Mogolon Mesa of Arizona north to the Wyoming border, *Juniperus osteosperma* and *Pinus edulis* dominate vast areas. Westward, in the Great Basin region of Nevada and western Utah, *Pinus monophylla* replaces *P. edulis* as the dominant pine. On the east side of the cordillera, *J. osteosperma* occurs in the Bighorn Mountains of Wyoming, but strictly in association with limestone substrate (Despain 1973).

Numerous sources of information, including old photographs, age-structure analyses, and personal observations of long-time residents, clearly document a recent but major expansion of juniper woodlands into adjacent grasslands, and to a lesser extent an increase in density of the existing woodlands (Springfield 1976; Tausch et al. 1981; Gruell 1983). Several studies, such as those of Burkhardt and Tisdale (1969, 1976), have shown older *Juniperus* to be associated with rocky sites, whereas the expansion has been largely onto finer-textured soils. This, together with the correlation between heavy grazing and juniper expansion (Johnsen 1962; West et al. 1975), suggests a scenario wherein the introduction of intense grazing by domestic cattle reduced the cover and vigor of the dominant grasses. This both allowed establishment of *Juniperus* and reduced the frequency of grass fires owing to fuel reduction. With increased establishment opportunities and reduced fire frequency and intensity, *Juniperus* could have quickly spread out of its traditional rocky refugia (Young and Evans 1981). However, other explanations are plausible. Aboriginal exploitation of this pygmy woodland for fuel may have reduced the importance of juniper and pinyon in at least some areas (Samuels and Betancourt 1982) such that part of the current expansion of these species might simply be a result of recovery following this earlier land use. Also, a marked shift in climate toward decreased severity of drought that occurred around 1900 has been shown to have had a significant impact on establishment of grasses in the semiarid Southwest (Neilson 1986). This climatic shift could also have been responsible for the coincident increase in woody plant establishment.

Ponderosa Pine Woodland

The vegetation type most Americans associate with western mountains is *Pinus ponderosa* woodland. Novels, films, and television programs have romanticized this landscape of tall but sparse trees growing over grassy rangeland (Fig. 3.10). Although best developed in the southwestern states, *P. ponderosa* woodland extends from the Sierra Madre Occidental of Mexico north into the dry interior valleys of southern British Columbia. East of the continental divide, from the Colorado-Wyoming border northward, the low-elevation habitat of ponderosa pine occurs only along the far eastern fringe of the Rocky Mountain region. Thus, the more interior Medicine Bow, Wind River, and Teton ranges of Wyoming have little, if any, *P. ponderosa*, whereas outlying ranges like the Laramie and Bighorn, plus the Black Hills of South Dakota and numerous rocky scarps of the western Great Plains, have extensive *P. ponderosa* forests and woodlands (Wells 1965; Alexander and Edminster 1981).

Near its upper elevational limit, *P. ponderosa* increases in density to form well-developed forests, though often with the pine being only successional to *Pseudotsuga* (Peet 1981). In the madrean region, low-elevation *P. ponderosa* woodland grades downward into either pygmy conifer woodland or encinal (an oak-dominated chaparral-like community). Where pygmy conifers are absent, the forest of *P. ponderosa* becomes progressively more open with decreasing elevation, until only scattered individuals remain in the most rocky areas.

The lush grass understory of a *P. ponderosa* woodland is highly flammable during the dry summer months. As a consequence, both lightning fires and aboriginal fires were common in the years before European settlement. Studies of multiple fire scars have shown that in many areas of the Southwest, fires occurred almost annually during the eighteenth and nineteenth centuries (Cooper 1960; Dieterich 1980, 1983; Dieterich and Swetnam 1984). Farther north, along the east slope of the Colorado Front Range, where the landscape is more dissected, the fire "return time" was longer, generally 25–40 yr (Rowdabaugh 1978; Laven et al. 1980).

The frequent occurrence of fire not only kept the understory of *P. ponderosa* woodlands free of invading species but also limited pine regeneration. In addition, the semiarid climate of the pinelands led to low establishment rates (Pearson 1923), and the pronounced year-to-year variation in seed production (Schubert 1974) limited most regeneration to only occasional years. In short, for significant levels of regeneration, the pines required a good seed year (preferably preceded by a fire to prepare the seedbed), followed by a period free of summer drought or excessive winter cold, combined with several years free of fire. The age structures of old *P. ponderosa* stands still reflect the episodic nature of regeneration (Cooper 1960; Potter and Green 1964; Larsen and Schubert 1969; Schubert 1974; White 1985).

Original *P. ponderosa* forests of the Southwest have been reported to have had a patchy appearance, with most trees occurring in small, family-like groups, with considerable grass and forb development between, but not within, the groups. Cooper (1960, 1961) described this two-phase mosaic and reported that the patches were even-aged and represented simultaneous regeneration following death of an earlier patch of trees. Specifically, the earlier occupants supplied fuel that produced a particularly hot fire, which in turn provided openings in the grass for tree invasion. Subsequently, White (1985) has shown that some sites with such a patch-like appearance have an uneven-aged within-patch structure. He suggested that there was periodic regeneration of trees in the patches, usually following a fire that had burned accumulated fuel beneath the living trees and thus provided regeneration sites safe from competition with grasses. The possibility exists that Cooper and White were describing essentially the same phenomenon, with differences in fire intensities being responsible for the death or survival of the original patch dominants.

Following suppression of wildfires and the introduction of domestic cattle, the *P. ponderosa* woodlands underwent a dramatic transformation. The first change was establishment of numerous new trees, which caused a shift from woodland to forest physiognomy (Veblen and Lorenz 1986). This increased regeneration could have been the result of either reduced grass density, owing to grazing, or fire suppression, which would allow greater post-establishment survival (Marr 1961). These dense stands of young trees, lacking recurrent fires, readily built up high levels of fuel, such that present-day fires frequently are catastrophic, killing virtually all the trees in a forest (Weaver 1959; Cooper 1960; Kallander 1969).

Madrean Pine-Oak Woodland

With decreasing latitude and more moderate climate, *P. ponderosa* woodlands are replaced by a diverse assemblage of pines and broad-leaved, sclerophyllous angiosperms. The northern limit for this community can be defined approximately by the Santa Catalina and Chiricahua mountains of Arizona on the west and the Davis Mountains of

Figure 3.10. Ponderosa pine forest. A: Pinus ponderosa *with* Quercus gambelii *understory near Santa Fe, New Mexico. With decreasing latitude, low-elevation pine stands gain madrean elements, especially oaks. B:* Pinus ponderosa *woodland with a shrub layer of* Artemisia tridentata *and a herb stratum dominated by* Muhlenbergia montana *in Rocky Mountain National Park, Colorado. Beyond the trees, open-park vegetation is visible.*

Texas on the east. Depauperate variants of this type occur farther north, such as the *P. ponderosa* forests of the southern Sangre de Cristo Mountains of northern New Mexico and southern Colorado, where *Quercus gambelii* is an important understory shrub species (Fig. 3.10A) (Peet 1978a).

At elevations between roughly 1800 and 2100 m, the Santa Catalina Mountains support a pine-oak woodland composed of such species as *Pinus leiophylla, P. cembroides, P. ponderosa arizonica, Quercus hypoleucoides, Q. arizonica, Q. emoryi, Q. rugosa, Juniperus deppeana,* and *Arctostaphylos pungens* (Whittaker and Niering 1965; Niering and Lowe 1985), several of which are important over extensive areas in Mexico. Similar vegetation has been described from the mountains of Chihuahua (Shreve 1939) and Coahuila (Muller 1947). Pine-oak woodland at 2250 m in the Davis Mountains of Texas is not as diverse as the Santa Catalina woodlands, but it contains such species as *Pinus edulis, P. ponderosa, P. strobiformis, Juniperus deppeana, Quercus grisea, Q. hypoleucoides, Q. gravesii, Q. gambelii, Populus tremuloides,* and *Arbutus texana* (Hinckley 1944). South of the Davis Mountains near the Texas-Mexico border, the Chisos Mountains have a more diverse madrean flora, with numerous broad-leaved tree species (Whitson 1965). The increasing diversity southward is perhaps most dramatically illustrated by the report of Leopold (1950) that approximately 112 species of *Quercus* and 39 species of *Pinus* can be found in the pine-oak woodlands that form the dominant vegetation over much of the Mexican Plateau.

With progressively more moderate winters, broad-leaved, sclerophyllous, often shrub-like species largely replace the conifers. The resultant encinal communities, dominated by genera such as *Rhus, Ceanothus, Quercus, Cercocarpus, Arctostaphylos,* and *Arbutus,* have clear floristic affinities with the chaparral of California, despite a very different climatic regime (Muller 1939; Axelrod and Raven 1985). They can be found at the lower limit of the forest on the west slope of the Sierra Madre Oriental, as well as in more western ranges, and even on the shallow soils of exposed ridge tops in the Santa Catalina and Chiricahua mountains of southern Arizona (Whittaker and Niering 1965).

Species-poor encinal communities, dominated almost exclusively by *Quercus gambelii,* with various amounts of *Rhus* and *Cercocarpus,* can be found as far north as Colorado Springs, Colorado, on the east slope (Fig. 3.9B), and north to the Wyoming and Idaho borders on the west (Ream 1964). On the east slope beyond the range of *Quercus,* an attenuated form of the encinal community, composed of low shrubs such as *Rhus trilobata, Cercocarpus montanus,* and *Purshia tridentata,* occurs between the grasslands of the plains and the foothill *Pinus ponderosa* woodlands as far north as the Laramie Mountains in Wyoming (Ramaley 1931; Peet 1978a).

In Utah and western Colorado, *Quercus gambelii* dominates extensive areas along the lower slopes of the mountains. The sites appear similar to those typically dominated by either *Pinus edulis* or *Pinus ponderosa* elsewhere in the Rockies, but no explanation for the dominance of oak over pine has received widespread acceptance (Harper et al. 1985). The successional status of oak varies with site conditions. In some locations, *Quercus* is clearly successional to *Pinus ponderosa* (Dixon 1935; Cronquist et al. 1972), whereas in others it is successional to *Pinus edulis* (Floyd 1982), *Acer grandidentatum* (Nixon 1967), or *Abies concolor* and *Pseudotsuga* (Harper et al. 1985). Stable stands have also been described (Brown 1958).

Douglas-Fir Forest

Pseudotsuga menziesii (Douglas-fir) is found on appropriate sites throughout the montane zone of the Rocky Mountains, ranging from central British Columbia south deep into Mexico. Only in northern British Columbia and the Yukon is *Pseudotsuga* absent, though the Rocky Mountain montane forests of which it is a dominant are essentially absent north of Jasper National Park, Alberta (Stringer and La Roi 1970). Over its range, this species is the potential climax dominant for a broad range of sites. In addition, *Pseudotsuga* functions as an important seral species in many areas.

In the Front Range of Colorado, *Pseudotsuga* is the dominant tree species of north-facing slopes and steep ravines form the lower treeline at 1650 m up to 2700 m. However, on open slopes it is more restricted and is confined mostly to sites between 2300 and 2800 m. Northward, in western Montana, *Pseudotsuga* is more broadly distributed, ranging from either the lower treeline or *Pinus ponderosa* woodland upward to *Picea-Abies* forest.

Pseudotsuga is often associated with shade-intolerant seral species such as *Pinus contorta, Pinus ponderosa,* and, in the northwest, *Larix occidentalis.* All of these species regenerate well following fire, and all except *Pinus contorta* are tolerant of repeated, low-intensity surface fires.

In the madrean region, *Pseudotsuga* is more a species of high peaks, north slopes, and mesic sites. Reports from the Huachuca Mountains (Brady and Bonham 1976) and Santa Catalina Mountains (Niering and Lowe 1985) indicate dominance on peaks above 2450 m. On mesic slopes, *Pseudotsuga* often codominates with, or is successional to, *Abies con-*

color (Moir and Ludwig 1979), a species that extends north along the east slope to central Colorado, but that in Utah, where minimum temperatures are not as extreme, ranges north to the Idaho border. *Picea pungens* can codominate in moist canyon bottoms of the southern Rockies and northern madrean region.

Cascadian Forests

Several tree species that dominate extensive areas in the Cascade Mountains of Oregon, Washington, and British Columbia reach eastward to the Rockies only in ranges near the Columbian Plateau and adjacent Canada. This is an area where Pacific air penetrates the Cascades, bringing heavy rains and cool temperatures to the western slopes of the Rockies. Near the eastern limits of *Tsuga heterophylla* (western hemlock) and *Thuja plicata* (western redcedar), outbreaks of arctic air occur periodically, damaging trees not in sheltered localities. Among the Cascadian species, *Tsuga heterophylla, Thuja plicata, Abies grandis, Taxus brevifolia, Tsuga mertensiana,* and *Larex lyallii* (at treeline) are locally climax species of the northern Rockies. Important understory species with similar distribution patterns include *Menziesia ferruginea, Oplopanax horridum, Philadelphus lewisii, Rhododendron albiflorum, Sorbus sitchensis, Vaccinium globulare, V. membranaceum,* and *Xerophyllum tenax* (Aller 1969; Pfister et al. 1977) (see Chapter 4 for similar vegetation in the Cascades).

On mesic sites at moderately low elevations (ca. 1200 m), particularly in sheltered valley bottoms, luxuriant forests of *Thuja plicata* and *Tsuga heterophylla* dominate and appear essentially equivalent to forests in the Cascade Mountains to the west (see Chapter 4). Basal areas of 100 m^2 ha^{-1} are not rare, and values in excess of 200 have been reported (Daubenmire and Daubenmire 1968). *Tsuga* is ubiquitous in those forests, and like *Abies* in the *Picea-Abies* forests, it clearly dominates the seedling and sapling strata. The abundance of seedlings suggests that *Tsuga* will slowly replace *Thuja* (Daubenmire and Daubenmire 1968; Habeck 1968), though the evidence is equivocal, and *Thuja* may out perform *Tsuga* on somewhat drier slopes.

Between the *Tsuga heterophylla–Thuja plicata* forests on moist sites and the *Psudeotsuga* forests on drier sites, *Abies grandis* is the potential climax dominant. In addition, *Abies grandis* dominates between the warm, low-elevation *Pseudotsuga* forests and the *Picea-Abies* forests of cool, high-elevation sites. These forests are best developed in northern Idaho (Steele et al. 1981), but also are important in western Montana (Pfister et al. 1977; Antos and Habeck 1981).

Like most Rocky Mountain forest types, *Tsuga, Thuja,* and *Abies grandis* forests are periodically subject to severe forest fires (Habeck 1968; Antos and Habeck 1981). The principal successional species are *Pinus contorta, P. monticola,* and *Larix occidentalis*. On moister sites, *Abies grandis* can occur as a successional species (Daubenmire and Daubenmire 1968; Antos and Habeck 1981).

Montane Seral Forests

The ubiquity of disturbances in Rocky Mountain forests, particularly fire, has resulted in seral species dominating a large proportion of the landscape. Two such species are *Populus tremuloides* (quaking aspen) and *Pinus contorta* (lodgepole pine), both of which are widespread species that can form steady-state forests under certain conditions, though they are much more important as post-fire invaders (Fig. 3.11). *Pinus contorta* dominates many post-fire forests, from the northern end of the Rockies in the Yukon Territory of Canada south to the Sangre de Cristo Mountains of southern Colorado. The species is absent, however, from the madrean region and most of the more isolated and xeric ranges of the Great Basin. *Populus tremuloides* is the most widely distributed tree species in North America and is the only major deciduous tree species in the Rocky Mountains. It can be found on post-fire sites as well as forest margins from near the Artic treeline south into Mexico, though south of 32° N latitude it occurs only as isolated, apparently relict, populations. Trees of both species grow rapidly following fire and form extensive, even-aged stands.

Pinus contorta is often viewed as the archetypal post-fire species. The pattern described long ago by Clements (1910) and others (Mason 1915) is that, following fire, the serotinous cones of *P. contorta* release large quantities of seeds that produce a dense, even-aged stand. Such a forest undergoes initial rapid growth, followed by slower growth and natural thinning until the next fire comes along to reset the cycle. Though correct in broad outline, that scenario has had to be revised because of recent studies of regeneration and age structure (R.K. Peet unpublished data; Veblen 1986a, 1986b). Seedling establishment often is not particularly rapid, with the consequence that a typical lodgepole stand has an initial cohort with a broad range of establishment ages, often upward of 30–50 yr (Peet 1981; Veblen and Lorenz 1986) and sometimes reaching 100 yr. In those situations in which numerous seedlings do become established within a few years following fire, "dog-hair" stands develop, with little size hierarchy being apparent. The result often is slow growth for all the trees.

Figure 3.11. Middle-elevation seral forests. A: Populus tremuloides *on a middle-elevation open slope in the southern Sangre de Cristo near Sante Fe, New Mexico. The location is south of the range of* Pinus contorta *on a site that the species might otherwise dominate. B: A senescent, even-aged* Pinus contorta *stand in Yellowstone Park, Wyoming.*

Although *Pinus contorta* does produce serotinous cones, the level of serotiny varies geographically and with stand history. In areas with extensive gentle topography over which fires can spread readily with little interruption, serotiny is the rule (Lotan 1975). However, in areas of rugged topography where bare rocky ridges break up the landscape, keeping fires small, and where patches of pines can be certain to escape fire, many trees are not serotinous. In intermediate habitats in Montana, where the average return time for fire approaches the longevity of the species, the level of serotiny has been shown to correlate with stand history. Stands originating following fire have a high percentage of serotinous trees, whereas stands originating following blowdown or insect damage have a high percentage of trees with nonserotinous cones (Muir and Lotan 1985). In addition, Mutch (1970) has proposed that selection has favored flammability in the shade-intolerant *P. contorta*, with high flammability leading to catastrophic fires that kill invading understory climax species, thus ensuring continued success for the serotinous phenotypes of *P. contorta*.

Populus tremuloides contrasts greatly, in terms of ecology, with *Pinus contorta*. Unlike *Pinus contorta*, *Populus* in the Rocky Mountains regenerates almost totally from root sprouts instead of seeds (Ellison 1943; Larson 1944). Virtually all forest stands between 2000 and 3200 m in the Colorado Front Range contain *Populus* sprouts, even where no mature trees are found. Cottam (1954) found that although aspen does produce viable seed, establishment is almost nonexistent. He suggested that nearly all of the aspen in the extensive aspen woodlands of western Colorado and eastern Utah are of sprout origin and that the clones largely date from an earlier period of greater and more seasonally even precipitation. To illustrate the clonal nature of the aspen woodlands, he described distinct varieties that can be identified from leaf phenology. In either spring or autumn, one can look at a hillside from a distance and pick out large clonal groves of aspen by their distinctive phenologies.

Following fire, *Populus* sprouts rapidly from previously established root systems, often developing complete canopy coverage in only 3–5 yr (Jones and Trujillo 1975). Extremely hot fires can damage or kill *Populus* roots. However, although *Populus* stands may burn frequently, intense fires are uncommon in *Populus* stands, which typically have a well-developed understory of mesophytic forbs and grasses, in contrast to most conifer stands with sparse understories and considerable accumulated woody litter. The undergrowth is to some extent a consequence of the more mesic sites that *Populus* occupies, but must also be viewed as a consequence of the biology of the species. This is particularly evident south of the range of *Pinus contorta* in the Sangre de Cristo Mountains, where sites that in the north probably would be dominated by *Pinus contorta*, with little undergrowth, are occupied by *Populus*, with a typical luxuriant herbaceous understory. Apparently, the rapidly decomposing wood and nutrient-rich deciduous leaves of *Populus* lead to low levels of woody fuel accumulation and rapid nutrient cycling, all of which encourages growth of mesophytic understory species (Vitousek et al. 1982; Parker and Parker 1983). The end result is what appears to be virtually the opposite of the flammability selection proposed by Mutch for *Pinus contorta: Populus* maintains its root-sprouting potential by maintaining low flammability of the forest understory. At least for some regions, the possibility then exists that *Pinus contorta* and *Populus tremuloides* represent, in the presence of a natural fire regime, alternative stable states for the same site (Peterson 1984).

The ecological relationships between *Pinus contorta* and *Populus tremuloides* have been sources of considerable discussion and confusion. *Populus* has been variously described to grow on moister, drier, finer-textured, rockier, and less acidic sites than *Pinus contorta*. The relative dominance of *Pinus contorta* on coarse, granitic soils and *Populus* on finer, more calcareous soils is well documented (Langenheim 1962; Patten 1963; Reed 1971; Despain 1973). The distributional pattern is further clarified by noting that in the Sangre de Cristo Mountains of northern New Mexico, south of the range of *Pinus contorta, Populus* occupies a broad range of habitat types. However, as one moves northward, *Pinus contorta* occupies an increasing portion of the range of habitats, such that, in the Front Range, the distribution of *Populus tremuloides* takes on the appearance of a doughnut when viewed on a plot of elevation versus moisture gradient (Fig. 3.7B-C). *Pinus contorta* appears to be the better competitor in the middle of the gradient mosaic, but *Populus* appears to have the broader range of tolerance, which allows it to win on the periphery.

Both *Pinus contorta* and *Populus tremuloides* stands are typically replaced by stands of more shade-tolerant species. *Abies lasiocarpa* is the most important invader of high-elevation successional stands, though *Picea engelmannii* can also be important. At lower elevations, *Pseudotsuga* is typically the most important invader.

Despite the similarity of invading species, the actual patterns of growth and establishment are typically quite different in *Pinus contorta* stands and *Populus tremuloides* stands. In *P. contorta* stands, the understory herb layer is usually sparse, offering little competition for new tree seedlings. Thus, seedling establishment is frequent. However, intense competition from the established trees, and perhaps low availability of nitrogen in the soil, results in very slow growth rate for the new seedlings. As a consequence, a typical 200-yr-old *P. contorta* stand has numerous stunted seedlings, or "Oscars," in the terminology of Silvertown (1982), of *Abies*, and to a lesser extent of *Picea*. These seedlings can remain in the understory indefinitely, "waiting" for the death of nearby canopy trees. It is common to find such *Abies* seedlings under a meter in height but in excess of 100 or even 150 yr of age. Following a catastrophic windstorm in which the canopy is largely destroyed, a seemingly even-aged *Abies* stand can develop from the released seedlings, though the stand will be truly even-aged only for ages collected at heights a meter or more above the ground. In contrast, in *Populus tremuloides* stands there is usually a well-developed herb layer that appears to interfere competitively with seedling establishment. Thus, seedlings are much less common than in *P. contorta* stands. However, once established, conifer seedlings in *P. tremuloides* stands appear to grow rather steadily, either because *Populus* does not offer as much competition for resources or because the associated soil is not as nitrogen-limiting (Vitousek et al. 1982).

Although usually viewed as seral species, *Pinus contorta* and *Populus tremuloides* do form stable, self-maintaining stands. In the Colorado Front Range, occasional old, stable *P. contorta* stands occur (Fig. 3.11) (Whipple and Dix 1979; Peet 1981), though not as commonly as sometimes suggested (Moir 1969). Northward, they appear rather frequently (Despain 1973, 1983; La Roi and Hnatiuk 1980), though usually in situations in which there is no nearby seed source for other potential climax species like *Pseudotsuga* or *Abies lasiocarpa*, or where soils are distinctly infertile. In the southern Rockies, stable *Populus tremuloides* stands are best developed on the western slopes, where *Populus* often forms a low-elevation belt transitional from steppe or shrubland to forest (Langenheim 1962; Morgan 1969; Reed 1971; Harniss and Harper 1982). A similar but less extensive fringing aspen woodland can be found along the east slope in Montana and southern Alberta (Lynch 1955).

Populus tremuloides and *Pinus contorta* are not the only important seral species of the Rockies. *Larix occidentalis* is a widespread post-fire species in the

Cascade forests of northern Idaho and adjacent portions of Montana, British Columbia, Washington, and Oregon. The species is shade-intolerant and requires a mineral seedbed, but in contrast to other seral species of the region, its mature trees, with their thick bark, lack of resins, and open branching pattern, are quite fire-resistant (Schmidt et al. 1976). Davis (1980) reported moderate-intensity surface fires that did not kill *Larix* to have occurred at 10–30-yr intervals, whereas destructive crown fires occurred at intervals of around 140 yr. *Pinus monticola* is another seral species of the northern Rockies. It is most characteristic of the mesic sites where *Tsuga heterophylla* or *Thuja plicata* is the potential climax species. Several additional species play important seral roles, even though they are best known as climax species in the Rocky Mountain region. As examples, *Pseudotsuga menziesii, Pinus flexilis,* and *Pinus ponderosa* all act as successional species on sites more mesic than those on which they are typically climax (Peet 1981).

Spruce-Fir Forest

Spruce-fir (*Picea-Abies*) forest characterizes the subalpine portion of the Rocky Mountains from the Yukon Territory south to the Mexican border (Fig. 3.12). These forests represent a southern extension and modification of the boreal conifer forests to which they are similar both floristically and structurally. The primary tree species of the mountains are *Picea engelmannii* (Engelmann spruce) and *Abies lasiocarpa* (subalpine fir), both of which are genetically similar to, and sometimes hybridize with, the boreal dominants *Picea glauca* and *Abies balsamea*. Indeed, as in the case for *Pinus contorta* and *P. banksiana*, the mountain species of each pair represents a genetically heterogeneous species from which the more homogeneous boreal species may well have have been derived following one of the Pleistocene glacial retreats (Boivin 1959; Taylor 1959; Parker et al. 1981, 1984: Critchfield 1985).

Picea engelmannii is the dominant high-elevation spruce south of about 54° N latitude. *Picea glauca* replaces *P. engelmannii* in the northernmost Rockies and occurs at lower elevations throughout the Canadian Rockies and parts of northern Montana; hybrids not uncommonly link populations of the two species (Moss 1955; Daubenmire 1974). In the Black Hills of western South Dakota, *Picea glauca* occurs some 1000 km disjunct from the main part of its range. Otherwise, *Picea engelmannii* and *Abies lasiocarpa* are the dominant subalpine species of the Rockies.

On isolated peaks and ranges, particularly in the semiarid madrean regions, the occurrence of *Picea* and *Abies* appears in part to reflect random, postglacial events of colonization and extinction. The Chiricahua Mountains of southeastern Arizona contain *Picea engelmannii* but *Abies lasiocarpa* is absent (Sawyer and Kinraide 1980), whereas the Santa Catalina Mountains only 160 km to the west contain *A. lasiocarpa* but lack *P. engelmannii* (Whittaker and Niering 1965). A similar situation exists with respect to several mountain ranges of the Great Basin of western Utah and eastern Nevada (Loope 1969; Harper et al. 1978; Wells 1983; Arno and Hammerly 1984). Pike's Peak, located slightly to the east of the main body of the Rockies in Colorado, lacks both *Abies* and *Pinus contorta*, perhaps as a consequence of a drier, more continental environment (Peet 1978a; Diaz et al. 1982).

Spruce-fir forests are only poorly developed in the mountains of Mexico. Probably neither *Picea engelmannii* nor *Abies lasiocarpa arizonica* occurs south of the U.S. border (Little 1971). Spruce is ecologically unimportant in Mexican forests, occurring only as two narrow endemics on high northern peaks, *Picea chihuahuana* in the Sierra Madre Occidental (Gordon 1968) and *Picea mexicana* in the Sierra Madre Oriental (Martinez 1961) [*P. engelmannii* var. *mexicana* of Taylor and Patterson (1980)]. *Abies* is represented by some eight species, with *A. religiosa* of the high southern peaks and *A. durangensis* of the northwest, the most important. Nowhere do these forests cover a major portion of the Mexican landscape (Leopold 1950; Lauer 1973; Rzedowski 1981).

Picea-Abies forests are remarkably consistent over the length of the Rockies. On open slopes, the understory often consists of continuous cover of a low *Vaccinium* (e.g., *V. scoparium, V. myrtillus*) and often a dense layer of moss. Typically, few herbaceous species are present, often not more than 10 in a 0.1-ha plot. Among the more widespread species are *Arnica cordifolia, Pyrola secunda,* and *Polemonium delicatum*. *Abies* regeneration provides the bulk of additional woody understory vegetation. On moist sites with seeps, or adjacent to running water, a lush herbaceous stratum can be expected (e.g., *Senecio, Mertensia, Erigeron, Cardamine, Saxifraga, Veratrum*), together with taller shrubs (e.g., *Lonicera involucrata, Viburnum edule*). Forests such as these occur from central Canada south at least to central New Mexico (Oosting and Reed 1952; Pfister et al. 1977; Peet 1978a, 1981; Moir and Ludwig 1979). The major latitudinal pattern is higher diversity of spruce-fir communities at higher latitudes. In Arizona, Nevada, and New Mexico, *Picea-Abies* forest covers only a narrow range of sites, and consequently only a few species combinations are found in any one locality. In contrast, in the northern Rockies, *Picea-Abies* forests dominate a large por-

A

B

Figure 3.12. *Subalpine* Picea engelmannii – Abies lasiocarpa *forest, Rocky Mountain National Park, Colorado. A: Typical dense stand in a site sheltered from fire behind a cirque lake. B: High-elevation* Picea-Abies *forest (~3400 m) slowly recovering from a fire that occurred ~70 yr earlier.*

Table 3.4. *Development sequences for* Picea engelmannii–Abies lasiocarpa *forests*

Site and stage	Basal area (m^2 ha^{-1})	Relative importance		Species number (per 1000 m^2)
		Picea	*Abies*	
Wet				
1	9.8	27.6	72.4	49.0
2	73.9	66.9	33.1	25.5
3	51.9	65.8	34.2	34.2
4	49.7	59.5	40.5	34.8
Mesic				
1	12.2	37.0	67.3	23.7
2	41.2	76.5	22.5	10.3
3	53.8	63.1	34.0	15.7
4	40.5	56.6	40.8	28.7
Xeric				
1	12.2	60.4	39.6	35.0
2	29.7	52.1	36.9	22.3
3	45.3	59.2	39.7	8.7
4	37.5	56.1	32.5	15.7

Note: Changes in average basal area and relative importance (average of relative density and relative basal area) of *Picea* and *Abies* are shown for three contrasting site conditions in the Colorado Front Range. Stage 1 contains even-aged stands (as judged by their bell-shaped diameter distributions) with an average diameter for *Picea* <12.5 cm. Stages 2 and 3 contain similar stands, with average diameters 13–25 and >25 cm, respectively. Stage 4 contains stands with diameter distributions indicative of mature, steady-state forests.
Source: Data from Peet (1981).

tion of the landscape and exhibit numerous variations (Pfister et al. 1977).

Because forest structure varies with site conditions and disturbance history, it is not possible to provide a simple structural characterization. Table 3.4. illustrates this variation with averages for sets of stands along successional gradients in wet, mesic, and dry *Picea-Abies* forests in the northern Front Range of Colorado. For these stands, peak basal areas range form 45 to 74 m^2 ha^{-1}, and species numbers per 0.1 ha range from 10 to 49.

Picea and *Abies* have somewhat different ecological characteristics. Perhaps most striking is that *Picea* is more tolerant of extreme conditions than is *Abies*. For example, in the forests of the Front Range, *Picea* is the dominant species on very wet or boggy sites, as well as on drier sites (Peet 1981). *Abies* is numerically dominant on more mesic sites, although *Picea* still usually dominates in basal area. *Abies* is absent altogether from the dry subalpine zone of Pike's Peak. Farther south in the Sangre de Cristo Range, *Picea* dominates the upper slopes, with *Abies* largely absent from the upper 300 m of forest. However, this may reflect physiological characteristics of *Abies lasiocarpa* var. *arizonica*, which replaces the typical form of *Abies* in the Sangre de Cristo Mountains and southward.

The relative ecological roles of *Picea* and *Abies* in mixed stands have long been a puzzle to ecologists. The most common pattern is for *Picea* to have the greater number of large trees and the greater basal area, but for the greater portion of seedlings and saplings to be *Abies*. For the Rocky Mountains as a whole, Hobson and Foster (1910) found 50–90% of the small stems to be *Abies*, but 75% of the mature timber to be *Picea*. Oosting and Reed (1952) reported a seven-to-one dominance of *Abies* in the forest understory in spruce-fir stands in the Medicine Bow Mountains of southern Wyoming, but a dominance of *Picea* in the overstory. One interpretation is that the mixed stands are successional and that *Abies* will, in the absence of large-scale disturbance, largely replace *Picea* (Hansen 1940; Bloomberg 1950; Daubenmire and Daubenmire 1968; Loope and Gruell 1973). A smaller number of ecologists expect *Picea* to increase relative to *Abies* owing to greater longevity (Miller 1970; Alexander 1974; Schmidt and Hinds 1974). Still others favor the hypothesis that these forests are near equilibrium and that the structure will remain relatively constant owing to the greater longevity of *Picea* (Oosting and Reed 1952; Marr 1961; Veblen 1986a). Fox (1977) has proposed the novel hypothesis that a cyle occurs wherein *Abies* is more frequent under *Picea*, and *Picea* is more frequent under *Abies*. It is probable that versions of all these viewpoints are true under appropriate conditions.

In the debate concerning the relative roles of

Picea and *Abies*, successional status of stands has largely been ignored. *Picea* appears to be a far more successful species at establishing on mineral soil following fire (Day 1972; Alexander 1974; Whipple and Dix 1979; Peet 1981). *Abies*, in contrast, is the better of the two at establishing in the shade and on organic substrate (Knapp and Smith 1982). Following fire on mesic sites in the Front Range, *Picea* usually dominates. Where a new stand with a dense canopy develops, *Picea* regeneration is virtually absent, and *Abies* regeneration occurs mostly as stunted seedlings or Oscars. However, with thinning of the canopy, regeneration increases, and *Abies* seedlings are released. Gradually, over a period of perhaps 500 yr, *Abies* largely, but never completely, replaces the canopy *Picea*. Alternatively, pecularities of climate and seed rain can lead to an initial dominance of *Abies*. In this situation, *Picea* may slowly increase in importance, but cannot be expected to eventually dominate.

Subalpine White Pine Forests

On dry ridges and exposed southern slopes of the subalpine zone, instead of the dense forests of tall, conical *Picea* and *Abies*, one finds forests of shorter, round-crowned, more widely spaced trees of the white pine group (*Pinus* subgenus *Haploxylon*) (Fig. 3.13). In the northern Rockies, south through the Wind River Range of Wyoming, the dominant white pine is *Pinus albicaulis* (whitebark pine). In the southern Rockies, from central Colorado south through the Sangre de Cristo Mountains of northern New Mexico and westward to the San Francisco Peaks in northeastern Arizona, *Pinus aristata* (bristlecone pine) dominates these sites. Northwestward on the high peaks of the Great Basin, the closely related *Pinus longaeva* replaces *P. aristata*. *Pinus longaeva* is well known for attaining the greatest longevity of any known tree species (Currey 1965).

Pinus flexilis, another member of this subgenus, ranges across most of both the southern and northern Rockies, including the higher mountains of the Great Basin. Where *Pinus aristata* or *Pinus albicaulis* occur, *P. flexilis* is mostly confined to dry, exposed ridges at lower elevations, though sometimes dominance is shared in subalpine sites. In the northern Rockies, in particular, *P. flexilis* is not a forest species, but is confined to especially droughty sites at low and occasionally middle elevations (Pfister et al. 1977). In some Great Basin ranges such as the Ruby Mountains of Nevada, *Pinus flexilis*, *P. albicaulis*, and *P. longaeva* are the only important montane and subalpine conifers (Loope 1969; Lewis 1971; Arno and Hammerly 1984). Along the Front Range, between the geographical ranges of the southern *P. aristata* and the northern *P. albicaulis*, *P. flexilis* shows what appears to be competitive release, its range having expanded upward to dominate exposed ridges near treeline (Peet 1978a).

The understories of the various white pine forests show only gradual compositional changes with latitude. *Pinus albicaulis* sites described from Montana (Weaver and Dale 1974) and Wyoming (Reed 1976) are essentially equivalent in physiognomy and species composition to *P. flexilis* sites of the Colorado Front Range (Peet 1981), and to a lesser extent to sites in northern New Mexico (Peet 1978a).

At roughly the latitude of central New Mexico and Arizona, *P. flexilis* is replaced by *Pinus strobiformis*. In its extreme form, *P. strobiformis* is highly distinctive, but over much of the northern portion of its range the species appears to intergrade genetically with *P. flexilis*. Like *P. flexilis*, *P. strobiformis* grows on dry, exposed ridges, though not at treeline, because the mountains do not reach to treeline elevation in the areas where it occurs.

The white pines, because of their tolerance for extreme environmental conditions, can colonize high-elevation sites unavailable to *Picea* and *Abies*. On the less extreme of these sites, *Picea* and *Abies* can subsequently become established in the understory and eventually grow up through and shade out the white pines (Beasley and Klemmedson 1980; Peet 1981; Veblen 1986b). The typical pattern is one of a broad environmental range over which white pines are potentially dominant, but with a more restricted core region where they are potential climax species. Because of its broad elevational range, *Pinus flexilis* can be an important post-fire seral species, replacing *Pinus contorta* in this role on drier sites in the southern Rockies.

Treeline Vegetation

At the northern end of the Rockies, in the southern Yukon, the elevation of treeline, or the upper limit of forest growth, is at about 1400 m on favorable sites. With decreasing latitude, treeline rises steadily at a rate of about 100 m per degree of latitude to an elevation of over 3600 m in northern New Mexico. Farther south, timberline does not rise as rapidly, the highest treelines occurring near 4000 m in the tropical mountains of the Mexican volcanic belt (Fig. 3.14). The rate of decline in treeline with increasing latitude is similar to the −83 m per degree of latitude reported for eastern North America by C.V. Cogbill and P.S. White (pers. commun.), and the continent-wide value of −110 m per degree of latitude reported by Daubenmire (1954).

Two species, *Picea engelmannii* and *Abies lasio-*

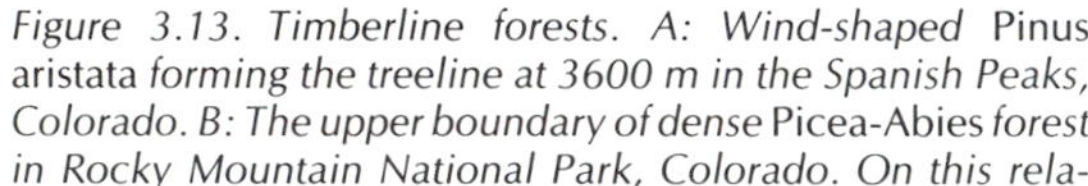
Figure 3.13. Timberline forests. A: Wind-shaped Pinus aristata *forming the treeline at 3600 m in the Spanish Peaks, Colorado. B: The upper boundary of dense* Picea-Abies *forest in Rocky Mountain National Park, Colorado. On this relatively sheltered site, the trees are well formed. Regeneration is nearly absent, perhaps as a consequence of high snow accumulation in the glades.*

carpa, are the dominant treeline species throughout much of the Rocky Mountain region (Fig. 3.13B). Nonetheless, treeline vegetation patterns can be extraordinarily complex. Environmental factors such as wind, snow depth, and snowmelt take on new importance at high elevations.

Where wind is strong, abrasion and winter desiccation can cause trees to assume stunted, often cushion-like growth forms (Fig. 3.15A). On extreme sites, the shrub-like trees occur only where sheltered behind rocks or undulations in the topography (Holtmeier 1985). In similar but less exposed sites, tree islands can form. These often consist of a clone of trees produced vegetatively from one parent through layering (i.e., rooting of branches pressed against the ground by the snow). In other cases, multiple seedlings become established in the shelter of a single original plant. In either case, the appearance is that of a clump of shrubs. Where the wind is somewhat less strong, emergent trees rise from the shrub island, often with flag-shaped crowns (Holtmeier 1980, 1985). Marr (1977) and Benedict (1984) have documented the movement of tree islands at rates of roughly 2 cm yr^{-1}, with wind abrading the windward side while vegetative regeneration extends the plant on the lee side.

Snow accumulation and duration can be critical at treeline. Where no snow accumulates, winter desiccation can be severe, and tree establishment and growth can fail altogether. Where accumulation is sufficient for snow to persist through much of the growing season, tree establishment can be greatly inhibited. Once established, trees can trap blowing snow, and their shade can slow snowmelt.

One of the most intriguing forms of interaction between wind and snow is the formation of "ribbon forests" (Fig. 3.15B). These alternating parallel strips of forest and intervening "snow glades" of moist alpine meadow have been described from the subalpine regions of Wyoming and Montana (Billings 1969) and Colorado (Buckner 1977; Holtmeier 1978, 1985), with the trees apparently functioning as snow fences. Snow accumulation inhibits seedling establishment in a band on the lee side of the ribbon. Adjacent to the drift, however, tree growth is improved because of continual summer water from snowmelt, plus increased protection from desiccation during the winter. In many cases, the forest

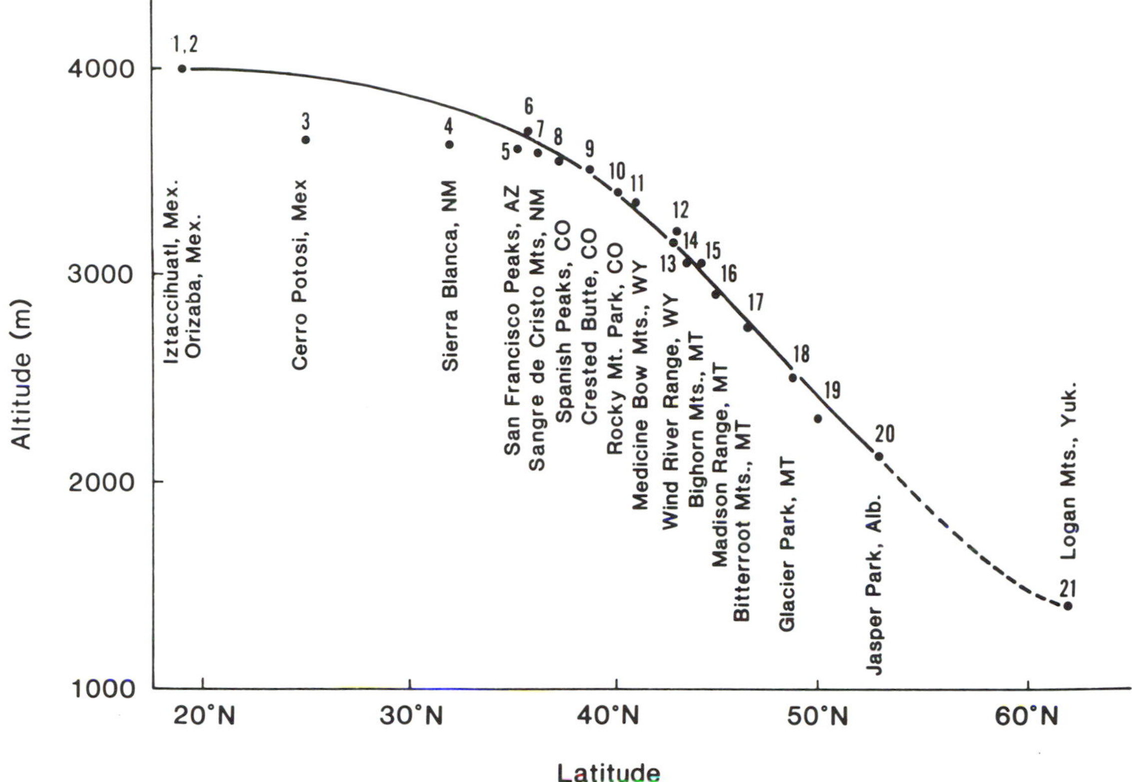

Figure 3.14. Treeline as a function of elevation and latitude. Numbers in parentheses following the references below indicate uppermost limits for trees, as opposed to the prevailing treeline indicated in the figure. The rate of decline of the treeline between 40°N and 55°N latitude is approximately 100 m elevation per degree of latitude. 1, Beaman (1962) (4100 m); 2, Lauer and Klaus (1975); 3, Beaman and Andresen (1966); 4, Dye and Moir (1977); 5, Rominger and Paulik (1983); 6, Peet (1978a) (3800 m); 7, Baker (1983); 8, Peet (1978a) (3650 m); 9, Langenheim (1962); 10, Peet (1978a, 1981) (3550 m); 11, Oosting and Reed (1952), Peet (1978a) (3500 m); 12, Griggs (1938); 13, Reed (1976); 14, Griggs (1938); 15, Despain (1973); 16, Patten (1963); 17, Arno and Habeck (1972), Habeck (1972); 18, Kessell (1979) (2600 m); 19, Wardle (1965); 20, Griggs (1938); 21, Oswald and Senyk (1977).

ribbons appear to have established along solifluction terraces, where seedlings would have had modest protection from blowing snow and ice (Holtmeier 1985).

Forest fires that start in lower *Picea-Abies* forests or montane *Pinus contorta* forests frequently burn uphill until they reach the forest boundary. Because both decomposition and tree seedling establishment are extremely slow, skeleton-like remains of old forests can remain for centuries after a fire has passed (Fig. 3.12B) (Stahelin 1943; Peet 1981). The picture this gives is one of a dynamic treeline, with fire periodically destroying the forest, and only very slowly does the forest creep back up the mountain, frequently taking in excess of 500 yr to regain lost ground (Peet 1981; Arno and Hammerly 1984). Alternatively, it has been suggested that climatic cooling has lowered the zone of natural forest regeneration, leaving the present timberline as a remnant of past conditions (Ives and Hansen-Bristow 1983; Hansen-Bristow and Ives 1984). However, Shankman (1984) and Daly and Shankman (1985) have documented sufficient tree regeneration at the treeline and in high-elevation burns to cast considerable doubt on this hypothesis.

Picea and *Abies* are not the only treeline species in the Rockies. The important roles of *Pinus aristata*, *P. flexilis*, and *P. albicaulis* on dry or exposed ridges have already been described. Other pines dominate at treeline in the high mountains of central Mexico, the most important being *Pinus hartwegii*, one of the only two tree species to reach above 4000 m elevation in North America (*Juniperus standleyi* reaches 4100 m in Guatemala). Most of these pines do not assume the krummholz shape characteristic of treeline *Picea* and *Abies*, though krummholz forms

Figure 3.15. Influences of wind and snow at treeline. A: Islands of predominantly Abies lasiocarpa *krummholz with conspicuous layering in Rocky Mountain National Park, Colorado. B: Ribbon forest in Montana (from Billings 1969). The interaction of snow, wind, and perhaps fire has resulted in bands of forest perpendicular to the prevailing wind.*

of *P. albicaulis* and *P. flexilis* do occur. Further, the low density and the dearth of understory vegetation in the white pine stands result in low levels of fuel, and thus lower fire frequencies than are typical for other treeline forests. *Pinus hartwegii* stands, in contrast, are often grassy and savanna-like in appearance.

Two additional tree species characteristic of treeline habitats, but confined in the Rockies strictly to the northern region, are *Larix lyallii* and *Tsuga mertensiana*. Both of these species occur only where the climate is moderated by periodic incursions of Pacific air. *Larix* grows largely on cold, north slopes where snow lies much of the summer, often above the treeline for *Picea* and *Abies* (Arno and Habeck 1972). *Tsuga mertensiana* is of even more limited range in the Rockies, its distribution being confined primarily to the Cascades and the Sierra Nevada (see Chapters 4 and 5). Habeck (1967) reported *T. mertensiana* to occur in Montana on a few moist ridges along the western border of the state. It is more common at the headwaters of the Saint Joe River in northern Idaho and in the Selkirks of British Columbia's Glacier National Park (Shaw 1909; Arno and Hammerly 1984).

Meadows and Parks

Scattered through the forestlands of the Rockies are treeless patches of various sizes dominated by grasses, sedges, and forbs (Fig. 3.16). The larger, low-elevation examples are locally called parks and include such well-known examples as South Park and Estes Park in Colorado. Smaller, high-elevation, moist examples are usually called meadows. Although the reason for the treeless state of mountain parks and meadows has often been discussed in the ecological literature, little agreement exists, probably because no one explanation fits all examples, and often a variety of factors and their interactions are involved.

Soil texture is one of the most critical factors for explaining the existence of low-elevation parks (Dunnewald 1930; Ives 1942; Daubenmire 1943; Peet 1981; Veblen and Lorenz 1986). The parks typically occupy rounded valley bottoms, sites with predominantly fine-textured alluvial or colluvial soils. This is in marked contrast to the coarse, rocky material of the adjacent forested slopes (Fig. 3.16A). The dense fibrous roots of the dominant grasses can form a thick sod in the fine-textured soils that trees appear rarely able to penetrate to the more moisture-rich soil below. Where trees such as *Pinus ponderosa* or *Juniperus scopulorum* occur, they often are immediately adjacent to large rocks that provide entry for roots to the deeper soil horizons (Robbins and Dodds 1908; Peet 1981). Similar mechanisms have been invoked for the wooded rocky scarps of the Great Plains, with their disjunct populations of Rocky Mountain tree species (Wells 1965), as well as for various tropical savannas (Knoop and Walker 1985; Walter 1985). However, this explanation does not preclude gradual invasion of parkland by trees, the control of which must be attributed to some other factor, such as periodic fire, soil drought, or grazing (Veblen and Lorenz 1986).

Some parks doubtless result from factors other than soil texture. Simple soil drought may often be the explanation. Steep, south-facing slopes in the *Pinus ponderosa* and lower *Pseudotsuga* zones of the Front Range, particularly where soil is thin, with little water-holding capacity, frequently support open, park-like vegetation dominated by *Muhlenbergia montana* or other graminoids. Daubenmire (1968) has reported similar thin, droughty soils in northern Idaho to be dominated by *Festuca idahoensis–Agropyron spicatum* steppe at an elevation that otherwise would support mesic *Tsuga*, *Thuja*, and *Abies* forest.

The high-elevation meadows of the Rockies have been attributed to an even greater range of possible causal mechanisms. Like low-elevation sites, the subalpine valley bottoms often are treeless, here giving the appearance of alpine vegetation extending as narrow fingers sometimes kilometers down into the *Picea-Abies* zone (Fig. 3.16B). Although soil texture likely plays a role on drier sites at high elevations, it probably is not as pivotal in this elevational zone as excess soil moisture. In addition, cold-air drainage and frost pockets, high snow accumulation, slow post-fire regrowth, lake filling, paludification, beavers, and avalanches have all been suggested to be important.

Excess soil moisture is the explanation most often suggested for montane and subalpine meadows. Sites with soils saturated or nearly so occur along streams or valley bottoms, on lake margins, at the bases of slopes, behind clayey morainal deposits, and even on open slopes where an impermeable substrate brings water to the surface. Ives (1942) suggested that valley bottoms often undergo paludification, with peat and silt accumulation resulting in lateral spread of meadow or bog into the adjacent forest. Beavers (*Castor*) have often accelerated the paludification process. For millennia, beavers have repeatedly dammed streams, only to have the impoundments fill with peat and silt and the old dams become buried beneath the bog surface. Today, many of these old beaver meadows give the appearance of a terraced landscape.

A

B

Figure 3.16. Treeless areas in forested landscapes, Rocky Mountain National Park, Colorado. A: Two types of parks can be seen in the photograph. The valley bottom of fine-textured alluvial and colluvial soils has an extensive treeless area. In addition, treeless areas can be seen on the south slope above the Pinus flexilis *in the foreground. The mountain across the valley is largely dominated by even-aged* Pinus contorta. *B: Subalpine valley bottom dominated by* Salix *shrubs (mostly* S. brachycarpa*), forbs, and grasses. The treeless condition is probably due to a combination of saturated soils, high snow accumulation, cold-air drainage, and fine-textured soils.*

Grassy balds similar to those of the southern Appalachians have been described from high open slopes and mountaintops, particularly in Montana (Korterba and Habeck 1971; Root and Habeck 1972). Drought resulting from limited soil depth and high winds provides the most likely explanation (Daubenmire 1968; Root and Habeck 1972).

Slow recovery from disturbance is critically important in the maintenance of subalpine meadows and balds (Stahelin 1943). On sites where tree regeneration is slow and forests require up to 500 yr to recover from fire, post-fire meadow communities can cover a significant portion of the landscape. Meadows also can dominate avalanche tracks, where the frequency of disturbances precludes development of mature forest (Butler 1979). In both cases, graminoids and forbs, once established, competitively inhibit the establishment of trees and thus further reduce the rate of forest recovery (Robbins 1918; Daubenmire 1943; Bollinger 1973; Peet 1981).

Some subalpine treeless areas defy simple explanation. Cinnabar Park, a 44-ha upland meadow at 2926 m in the Medicine Bow Mountains of Wyoming, has been shown to have been treeless for a long period. Despite extensive study (Miles and Singleton 1975; Vale 1978), no explanation for its origin or maintenance has yet received wide acceptance.

Relatively little has been written on the vegetation of the meadows and parks of the Rockies. Working in the wet meadows on the west slope of the Front Range, Wilson (1969) reported two meadow sequences. For typically ponded or floodplain habitats, mostly at middle elevations, he described a sequence with *Carex utriculata* dominant on the wettest sites, *Carex aquatilis* dominant on the moderately wet sites, and *Calamagrostis canadensis* dominant on the moist sites. In some valley-bottom sites, *Salix* thickets dominate in place of graminoids, but the alternation of dominance between the growth forms is not well understood. A second series, more characteristic of higher elevations and often occurring on seepage slopes, has *Eleocharis pauciflora* dominant on the wettest sites, a mixed community with species such as *Caltha leptosepala*, *Deschampsia caespitosa*, *Carex illota*, and *Podagrostis humilis* on the moderately wet sites, and again a diverse community with *Phleum alpinum*, *Poa reflexa*, *Erigeron peregrinus*, *Bistorta bistortoides*, and others on the more mesic sites (cf. Rydberg 1915; Reed 1917; Robbins 1918; Langenheim 1962). Some high-elevation seepage slopes and occasional shallow-lake margins develop a mat of *Sphagnum* and assume a character not unlike bogs of the boreal zone. In the Front Range, the ericad component of these include *Vaccinium myrtillus*, *Kalmia polifolia*, and *Gaultheria humifusa* (Peet 1981). Such bogs are relatively uncommon in the southern Rockies, but increase in frequency northward.

AREAS FOR FUTURE RESEARCH

Rocky Mountain vegetation, stretching from Alaska and the Yukon to northern Central America, is both varied and complex. The depth of our current understanding of this vegetation is equally varied, but additional research is needed in virtually all areas of ecological inquiry.

The first stage in ecological analysis should almost always be descriptive. Few modern descriptive studies have been published for Rocky Mountain areas north of Jasper Park, Alberta, or south of the Mexico–United States border. Studies are particularly needed for Mexico because of the floristic complexity of the region and because the high intensity of land use and rapidly growing population are rapidly altering the remaining remnants of natural vegetation. As a consequence of the land classification activities of the USDA Forest Service, considerable published information is available for the central portion of the Rockies (Pfister et al. 1977; Steele et al. 1981, 1983), though there remains ample opportunity for synthetic analysis of these and other data sets. Only a few studies have organized compositional information relative to environmental or geographical gradients, and none of these has yet carefully examined the interactions involved in correlations of vegetation with soil chemistry and with other environmental gradients.

Rocky Mountain forests are disturbance forests, with climax stands being less common than seral communities. Quantitative assessment of disturbance regimes has only just begun, and the impacts of various forms of disturbances on tree population dynamics and various ecosystem processes are just beginning to be studied. Only for fire are reliable estimates of mean return intervals in either presettlement or contemporary forests available, and even for fire the data are still scant. The magnitude of the impact of modern fire suppression on forest communities throughout the Rockies needs to be investigated.

Few in-depth, integrated studies of ecosystem processes, including such factors as primary production, nutrient cycling, and water fluxes, have been undertaken. The most notable exceptions have been studies in the southern Sangre de Cristo Mountains of New Mexico (Gosz 1980) and the Medicine Bow Mountains of Wyoming (Fahey 1983; Pearson et al. 1984; Fahey et al. 1985; Knight and

Fahey 1985). In no case are we yet able to superimpose isopleths for rates of ecological processes on a gradient representation of vegetation.

REFERENCES

Agee, J. K., and L. Smith. 1984. Subalpine tree establishment in the Olympic Mountains, Washington. Ecology 65:810–819.

Alexander, R. R. 1974. Silviculture of subalpine forests in the central and southern Rocky Mountains: the status of our knowledge. USDA Forest Service research paper RM-121.

Alexander, R. R., and C. B. Edminster. 1981. Management of ponderosa pine in even-aged stands in the Black Hills. USDA Forest Service research paper RM-228.

Aller, A. R. 1960. The composition of the Lake McDonald forest, Glacier National Park. Ecology 41:29–33.

Amman, G. D. 1977. The role of mountain pine beetle in lodgepole pine ecosystems: impact on succession, pp. 3–18 in W. J. Mattson (ed.), The role of arthropods in forest ecosystems. Springer-Verlag, New York.

Amman, G. D. 1978. The biology, ecology, and causes of outbreaks of mountain pine beetle in lodgepole pine forests, pp. 39–53 in M. A. Berryman, G. D. Amman, and R. W. Stark (eds.), Theory and practice of mountain pine beetle management in lodgepole pine forests. University of Idaho Experiment Station, Moscow.

Antos, J. A., and J. R. Habeck, 1981. Successional development in *Abies grandis* (Dougl.) forbes forests in the Swan Valley, western Montana. Northwest Sci. 55:26–39.

Arno, S. F. 1979. Forest regions of Montana. USDA Forest Service research paper INT-218.

Arno, S. F. 1980. Forest fire history in the northern Rockies. J. Forestry 78:460–465.

Arno, S. F. 1985. Ecological effects and management implications of Indian fires. USDA Forest Service general technical report INT-182, pp. 81–86.

Arno, S. F., and G. E. Gruell. 1983. Fire history of the forest-grassland ecotone in southwestern Montana. J. Range Management 36:332–336.

Arno, S. F., and J. R. Habeck. 1972. Ecology of alpine larch (*Larix lyallii* Parl.) in the Pacific Northwest. Ecol. Monogr. 42:417–450.

Arno, S. F., and R. P. Hammerly. 1984. Timberline: mountain and artic forest frontiers. The Mountaineers, Seattle.

Axelrod, D. I. 1958. Evolution of the Madro-Tertiary geoflora. Bot. Rev. 24:433–509.

Axelrod, D. I. 1979. Age and origin of Sonoran Desert vegetation. California Academy of Sciences Occasional Papers 132:1–74.

Axelrod, D. I., and P. H. Raven. 1985. Origins of the Cordilleran flora. J. Biogeography 12:21–47.

Baker, F. S. 1944. Mountain climates of the western United States. Ecol. Monogr. 14:223–254.

Baker, W. L. 1983. Alpine vegetation of Wheeler Peak, New Mexico, U.S.A.: Gradient analysis, classification, and biogeography. Arct. Alp. Res. 15:223–240.

Barrett, S. W., and S. F. Arno. 1982. Indian fires as an ecological influence in the northern Rockies. J. Forestry 10:647–651.

Barry, R. G. 1972. Climatic environment of the east slope of the Front Range, Colorado. Institute of Arctic and Alpine Research Occasional Papers 3.

Barry, R. G. 1973. A climatological transect on the east slope of the Front Range, Colorado. Arct. Alp. Res. 5:89–110.

Beaman, J. H. 1962. The timberlines of Iztaccihautl and Popocatepetl, Mexico, Ecology 43:377–385.

Beaman, J. H., and J. W. Andresen. 1966. The vegetation, floristics, and phytogeography of the summit of Cerro Potosi, Mexico. Amer. Midl. Nat. 75:1–33.

Beasley, R. S., and J. O. Klemmedson. 1980. Ecological relationships of bristlecone pine. Amer. Midl. Nat. 104:242–252.

Benedict, J. B. 1984. Rates of tree-island migration, Colorado Rocky Mountains, USA. Ecology 65:820–823.

Billings, W. D. 1950. Vegetation and plant growth as affected by chemically altered rocks in the western Great Basin. Ecology 31:62–74.

Billings, W. D. 1969. Vegetational pattern near alpine timberline as affected by fire-snowdrift interactions. Vegetation 19:192–207.

Bloomberg, W. G. 1950. Fire and spruce. Forest Chron. 26:157–161.

Boivin, B. 1959. *Abies balsamea* (Linne) Miller et ses variations. Naturaliste Canadiene 86:219–223.

Bollinger, W. H. 1973. The vegetation patterns after fire at the alpine forest-tundra ecotone in the Colorado Front Range. Dissertation, University of Colorado, Boulder.

Bormann, F. H., and G. E. Likens. 1979. Pattern and process in a forested ecosystem. Springer-Verlag, New York.

Brady, W., and C. D. Bonham. 1976. Vegetation patterns on an altitudinal gradient, Huachuca Mountains, Arizona. Southwestern Naturalist 21:55–66.

Brown, H. E. 1958. Gambel oak in west central Colorado. Ecology 39:317–327.

Buckner, D. L. 1977. Ribbon forest development and maintenance in the central Rocky Mountains of Colorado. Dissertation, University of Colorado, Boulder.

Burkhardt, J. W., and E. W. Tisdale. 1969. Nature and successional status of western juniper vegetation in Idaho. J. Range Management 22:264–270.

Burkhardt, J. W., and E. W. Tisdale. 1976. Causes of juniper invasion in southwestern Idaho. Ecology 57:472–484.

Butler, D. R. 1979. Snow avalanche path terrain and vegetation, Glacier National Park, Montana. Arct. Alp. Res. 11:17–32.

Butler, D. R. 1985. Vegetation and geomorphic change on snow avalanche paths, Glacier National Park, Montana, USA. Great Basin Naturalist 45:313–317.

Clements, F. E. 1910. The life history of lodgepole burn forests. USDA Forest Service bulletin 79.

Cooper, C. F. 1960. Changes in vegetation, structure, and growth of southwestern pine forests since white settlement. Ecol. Monogr. 30:129–164.

Cooper, C. F. 1961. Pattern in ponderosa pine forests. Ecology 42:493–499.

Cottam, W. P. 1954. Prevernal leafing of aspen in Utah Mountains. J. Arnold Arboretum 35:239–250.
Critchfield, W. B. 1985. The late Quaternary history of lodgepole and jack pines. Can. J. Forest Res. 15:749–772.
Cronquist, A. A., A. H. Holmgren, N. H. Holmgren, and J. L. Reveal. 1972. Intermountain flora—vascular plants of the intermountain west, U.S.A., Vol. 1. Hafner, New York.
Currey, D. R. 1965. An ancient bristlecone pine stand in eastern Nevada. Ecology 46:564–566.
Daly, C., and D. Shankman. 1985. Seedling establishment by conifers above tree limit on Niwot Ridge, Front Range, Colorado, USA. Arct. Alp. Res. 17:389–400.
Darling, M. S. 1966. Structure and productivity of a pinyon-juniper woodland in northern Arizona. Dissertation, Duke University, Durham, N.C.
Daubenmire, R. 1943. Vegetation zonation in the Rocky Mountains. Bot. Rev. 9:325–393.
Daubenmire, R. 1954. Alpine timberlines in the Americas and their interpretation. Butler University Botanical Studies 11:119–136.
Daubenmire, R. 1968. Soil moisture in relation to vegetation distribution in the mountains of northern Idaho. Ecology 49:431–438.
Daubenmire, R. 1974. Taxonomic and ecologic relationships between *Picea glauca* and *Picea engelmannii*. Can. J. Bot. 52:1545–1560.
Daubenmire, R. 1975. Floristic plant geography of eastern Washington and northern Idaho. J. Biogeography 2:1–18.
Daubenmire, R., and J. B. Daubenmire. 1968. Forest vegetation of eastern Washington and northern Idaho. Washington Agricultural Experiment Station technical bulletin 60.
Davis, K. M. 1980. Fire history of a Western Larch/Douglas-fir forest type in northwestern Montana. USDA Forest Service general technical report RM-81, pp. 69–74.
Day, R. J. 1972. Stand structure, succession, and use of southern Alberta's Rocky Mountain forest. Ecology 53:474–478.
Despain, D. G. 1973. Vegetation of the Big Horn Mountains, Wyoming, in relation to substrate and climate. Ecol. Monogr. 43:329–355.
Despain, D. G. 1983. Nonpyrogenous climax lodgepole pine communities in Yellowstone National Park. Ecology 64:231–234.
Diaz, H. F., R. G. Berry, and G. Kiladis, 1982. Climatic characteristics of Pike's Peak, Colorado (1874–1888) and comparisons with other Colorado stations. Mountain Res. Dev. 2:359–371.
Dieterich, J. H. 1980. Chimney Spring forest fire history. USDA Forest Service research paper RM-220.
Dieterich, J. H. 1983. Fire history of southwestern mixed conifers: a case study. Forest Ecol. Management 6:13–31.
Dieterich, J. H., and T. W. Swetnam. 1984. Dendrochronology of a fire-scarred ponderosa pine. Forest Sci. 30:238–247.
Dixon, H. 1935. Ecological studies on the high plateaus of Utah. Bot. Gaz. 97:272–320.
Dunnewald, T. J. 1930. Grass and timber soils distribution in the Big Horn Mountains. J. Amer. Soc. Agronomy 22:577–586.
Dunwiddie, P. W. 1977. Recent tree invasion of subalpine meadows in the Wind River Mountains, Wyoming. Arct. Alp. Res. 9:393–399.
Dye, A. J., and W. H. Moir. 1977. Spruce-fir forest at its southern distribution in the Rocky Mountains, New Mexico. Amer. Midl. Nat. 97:133–146.
Ellison, L. 1943. A natural seedling of western aspen. J. Forestry 41:767–768.
Fahey, T. J. 1983. Nutrient dynamics of aboveground detritus in lodgepole pine (*Pinus contorta* ssp. *latifolia*) ecosystems, southeastern Wyoming. Ecol. Monogr. 53:51–72.
Fahey, T. J., J. B. Yavitt, J. A. Pearson, and D. H. Knight. 1985. The nitrogen cycle in lodgepole pine forests, southeastern Wyoming. Biogeochemistry 1:257–275.
Floyd, M. E. 1982. The interaction of pinyon pine and gambel oak in plant succession near Dolores, Colorado. Southwestern Naturalist 27:143–147.
Fox, J.F. 1977. Alternation and coexistence of tree species. American Naturalist 111:69–89.
Franklin, J. F., and M. A. Hemstrom. 1981. Aspects of succession in the coniferous forests of the Pacific Northwest, pp. 212–229 in D. C. West, H. H. Shugart, and D. B. Botkin (eds.), Forest succession: concepts and application. Springer-Verlag, New York.
Franklin, J. F., W. H. Moir, G. W. Douglas, and C. Wiberg. 1971. Invasion of subalpine meadows by trees in the Cascade Range, Washington and Oregon. Arct. Alp. Res. 3:215–224.
Gholz, H. L. 1980. Structure and productivity of *Juniperus occidentalis* in central Oregon. Amer. Midl. Nat. 103:251–261.
Gordon, A. G. 1968. Ecology of *Picea chihuahuana* Martinez. Ecology 49:880–896.
Gosz, J. R. 1980. Biomass distribution and production budget for a nonaggrading forest ecosystem. Ecology 61:507–514.
Greenland, D., J. Burbank, J. Key, L. Klinger, J. Moorehouse, S. Oaks, and D. Shankman, 1985. The bioclimates of the Colorado Front Range. Mountain Res. Dev. 5:251–262.
Griggs, R. F. 1938. Timberlines in the northern Rocky Mountains. Ecology 19:548–564.
Gruell, G. E. 1983. Fire and vegetative trends in the Northern Rockies: interpretations from 1871–1982 photographs. USDA Forest Service general technical report INT-158.
Gruell, G. E. 1985. Fire on the early western landscape: an annotated record of wildland fires 1776–1900. Northwest Sci. 59:97–107.
Habeck, J. R. 1967. Mountain hemlock communities in Western Montana. Northwest Sci. 41:169–177.
Habeck, J. R. 1968. Forest succession in the Glacier Park cedar-hemlock forests. Ecology 49:872–880.
Habeck, J. R. 1972. Fire ecology investigations in Selway-Bitterroot Wilderness: historical considerations and current observations. University of Montana publication R1-72-001.
Hanley, D. P. 1976. Tree biomass and production estimated for three habitat types in northern Idaho. College of Forestry, Wildlife and Range Sciences, University of Idaho, bulletin 14.
Hansen, H. P. 1940. Ring growth and dominance in a spruce-fir association in southern Wyoming. Amer. Midl. Nat. 23:442–448.
Hansen-Bristow, K. J., and J. D. Ives. 1984. Changes in

the forest-alpine tundra ecotone: Colorado Front Range. Phys. Geogr. 5:186–197.

Harniss, R. O., and K. T. Harper. 1982. Tree dynamics in seral and stable aspen stands of central Utah. USDA Forest Service research paper INT-297.

Harper, K. T., D. C. Freeman, W. K. Ostler, and L. G. Klikoff. 1978. The flora of Great Basin mountain ranges: diversity, sources, and dispersal ecology. Great Basin Naturalist Memoirs 2:81–103.

Harper, K. T., F. J. Wagstaff, and L. M. Kunzler. 1985. Biology and management of the Gambel oak vegetation type: a literature review. USDA Forest Service general technical report INT-179.

Hawkes, B. C. 1980. Fire history of Kananaskis Provincial Park–mean fire return intervals. USDA Forest Service general technical report RM-81, pp. 42–45.

Heinselman, M. L. 1981. Fire intensity and frequency as factors in the distribution and structure of northern ecosystems, pp. 7–57 in H. A. Mooney, T. M. Bonnicksen, N. L. Christensen, J. E. Lotan, and W. A. Reinsers (eds.), Fire regimes and ecosystems properties. USDA Forest Service general technical report WO-26.

Hinckley, L. C. 1944. The vegetation of the Mount Livermore area in Texas. Amer. Midl. Nat. 32:236–250.

Hobson, E. R., and J. H. Foster. 1910. Engelmann spruce in the Rocky Mountains. USDA Forest Service circular 170.

Holtmeier, F. K. 1978. Die bodennahen Winde in den Hochlagen der Indian Peaks section (Colorado Front Range). Muenstersche Geographische Arbeiten 3:7–47.

Holtmeier, F. K. 1980. Influence of wind on tree-physiognomy at the upper timberline in the Colorado Front Range. New Zealand Forest Service technical paper 70, pp. 247–261.

Holtmeier, F. K. 1982. "Ribbon-forest" und "Hecken"; streifenartige Verbreitungsmuster des Baumwuchses an der oberen Waldgrenze in den Rocky Mountains. Erdkunde 36:142–153.

Holtmeier, F. K. 1985. Climatic stress influencing the physiognomy of trees at the polar and mountain timerline. Eidgenoessische Anstalt für das forstliche Versuchswesen, Berichte 270:31–40.

Houston, D. B. 1973. Wildfires in northern Yellowstone National Park. Ecology 54:1111–1117.

Ives, J. D., and K. J. Hansen-Bristow. 1983. Stability and instability of natural and modified upper timberline landscapes in the Colorado Rocky Mountains, USA. Mountain Res. Dev. 3:149–155.

Ives, R. L. 1942. Atypical subalpine environments. Ecology 23:89–96.

James, R. L., C. A. Stewart, and R. E. Williams. 1984. Estimating root disease losses in northern Rocky Mountain national forests. Can. J. Forest Res. 14:652–655.

Johnsen, T. N., Jr. 1962. One-seed juniper invasion of North Arizona grasslands. Ecol. Monogr. 32:187–207.

Johnson, E. A., L. Hogg, and C. S. Carlson. 1985. Snow avalanche frequency and velocity for the Kananaskis Valley in the Canadian Rockies. Cold Regions Sci. Technol. 10:141–151.

Johnstone, W. D. 1971. Total standing crop and tree component distributions in three stands of 100-year-old lodgepole pine, pp. 81–89 in H. E. Young (ed.), Forest biomass studies. University of Maine, Orono.

Jones, J. R., and D. P. Trujillo. 1975. Development of some young aspen stands in Arizona. USDA Forest Service research paper RM-151.

Kallander, H. 1969. Controlled burning on the Fort Apache Indian Reservation, Arizona. Tall Timbers Fire Ecology Conference Proceedings 9:241–249.

Kessell, S. R. 1979. Gradient modeling: resource and fire management. Springer-Verlag, New York.

Knapp, A. K., and W. K. Smith. 1982. Factors influencing understory seedling establishment of Engelmann spruce (*Picea engelmannii*) and subalpine fir (*Abies lasiocarpa*) in southeast Wyoming. Can. J. Bot. 60:2753–2761.

Knight, D. H., and T. J. Fahey. 1985. Water and nutrient outflow from contrasting lodgepole pine forests in Wyoming. Ecol. Monogr. 55:29–48.

Knoop, W. T., and B. H. Walker. 1985. Interactions of woody and herbaceous vegetation in a southern African savanna. J. Ecol. 73:235–154.

Koterba, W. D., and J. R. Habeck. 1971. Grasslands of the North Fork Valley, Glacier National Park, Montana. Can. J. Bot. 49:1627–1636.

Langenheim, J. H. 1962. Vegetation and environmental patterns in the Crested Butte area, Gunnison County, Colorado. Ecol. Monogr. 32:249–285.

La Roi, G. H., and R. J. Hnatiuk. 1980. The *Pinus contorta* forests of Banff and Jasper national parks: a study in comparative synecology and syntaxonomy. Ecol. Monogr. 50:1–29.

Larsen, M. M., and G. H. Schubert. 1969. Root competition between ponderosa pine seedlings and grass. USDA Forest Service research paper RM-54.

Larson, G. C. 1944. More on seedlings of western aspen. J. Forestry 42:452.

Lauer, W. 1973. The altitudinal belts of the vegetation in the central Mexican highlands and their climatic conditions. Arct. Alp. Res. 5:A99–A113.

Lauer, W., and D. Klaus. 1975. Geoecological investigations on the timerline of Pico de Orizaba, Mexico. Arct. Alp. Res. 7:315–330.

Laven, R. D., P. N. Omi, J. G.Wyant, and A. S. Pinkerton. 1980. Interpretation of fire scar data from a ponderosa pine ecosystem in the central Rocky Mountains, Colorado, pp. 46–49 in M. A. Stokes and J. H. Dieterich (eds.), Proceedings of the fire history workshop, October 20–24, 1980, Tucson, Arizona. USDA Forest Service general technical report RM-81.

Leak, W. B. 1965. The J-shaped probability distribution. Forest Sci. 11:405–409.

Leopold, A. S. 1950. Vegetation zones of Mexico. Ecology 31:507–518.

Lewis, M. E. 1971. Flora and major plant communities of the Ruby–East Humboldt Mountains. USDA Forest Service, Humboldt National Forest.

Little, E. L., Jr. 1971. Atlas of United States trees. Vol. 1. Conifers and important hardwoods. USDA miscellaneous publication 1146.

Little, E. L., Jr. 1976. Atlas of United States trees. Vol. 3. Minor western hardwoods. USDA miscellaneous publication 1314.

Loope, L. L. 1969. Subalpine and alpine vegetation of northeastern Nevada. Dissertation, Duke University, Durham, N.C.

Loope, L. L., and G. E. Gruell, 1973. The ecological

role of fire in the Jackson Hole area, northwestern Wyoming. Quat. Res. 3:425–443.

Lotan, J. E. 1975. The role of cone serotiny in lodgepole pine forests, pp. 471–495 in D. M. Baumgartner (ed.), Management of lodgepole pine ecosystems. Washington State University Cooperative Extension Service, Pullman.

Lynch, D. 1955. Ecology of the aspen groveland in Glacier County, Montana. Ecol. Monogr. 25:321–344.

McCune, B., and J. A. Antos. 1981. Diversity relationships of forest layers in the Swan Valley, Montana. Bull. Torrey Botanical Club 108:354–361.

McKnight, M. E. 1968. A literature review of the spruce, western, and 2-year cycle budworms. USDA Forest Service research paper RM-44.

Malanson, G. P., and D. R. Butler. 1984. Transverse pattern of vegetation on avalanche paths in northern Rocky Mountains, Montana. Great Basin Naturalist 44:453–458.

Marr, J. W. 1961. Ecosystems of the east slope of the Front Range in Colorado. University of Colorado Studies, Biology 8.

Marr, J. W. 1967. Data on mountain environments. I. Front Range, Colorado, sixteen sites, 1952–1953. University of Colorado Studies, Biology 27.

Marr, J. W. 1977. The development and movement of tree islands near the upper limit of tree growth in the southern Rocky Mountains. Ecology 58:1159–1164.

Marr, J. W., J. M. Clark, W. S. Osburn, and M. W. Paddock. 1968a. Data on mountain environments. III. Front Range, Colorado, four climax regions, 1959–1964. University of Colorado Studies, Biology 29.

Marr, J. W., A. W. Johnson, W. S. Osburn, and O. A. Knorr. 1968b. Data on mountain environments. II. Front Range, Colorado, four climax regions, 1953–1958. University of Colorado Studies, Biology 28.

Martinez, M. 1961. Una nueva especie de *Picea* en Mexico. Anales del Instituto de Biologia, Universidad Nacional Mexico 32:137–142.

Mason, D. T. 1915. The life history of lodgepole pine in the Rocky Mountains. USDA bulletin 154.

Merriam, C. H. 1890. Life zones and crop zones of the United States. USDA Division of Biological Survey Bulletin 10:9–79.

Miles, S. R., and P. C. Singleton, 1975. Vegetative history of Cinnebar Park in Medicine Bow National Forest, Wyoming. Soil Science Society of America Proceedings 39:1204–1208.

Miller, P. C. 1970. Age distributions of spruce and fir in beetle-killed forests on the White River Plateau, Colorado. Amer. Midl. Nat. 83:206–212.

Mitchell, V. L. 1976. The regionalization of climate in the western United States. J. Appl. Meteorol. 15:920–927.

Moir, W. H. 1969. The lodgepole pine zone in Colorado. Amer. Midl. Nat. 81:87–98.

Moir, W. H. 1972. Litter, foliage, branch, and stem production in contrasting lodgepole pine habitats of the Colorado Front Range, pp. 189–198 in J. F. Franklin, L. J. Dempster, and R. H. Waring (eds.), Research on coniferous forest ecosystems—a symposium, USDA Forest Service, Portland, Ore.

Moir, W. H., and R. Francis. 1972. Foliage biomass and surface area in three *Pinus contorta* plots in Colorado. Forest Sci. 18:41–45.

Moir, W. H., and J. A. Ludwig. 1979. A classification of spruce-fir and mixed conifer habitat types of Arizona and New Mexico. USDA Forest Service research paper RM-207.

Moral, R. del. 1972. Diversity patterns in forest vegetation of the Wenatchee Mountains, Washington. Bull. Torrey Botanical Club 99:57–64.

Morgan, M. D. 1969. Ecology of aspen in Gunnison County, Colorado. Amer. Midl. Nat. 82:204–228.

Moss, E. H. 1955. The vegetation of Alberta. Bot. Rev. 21:493–567.

Muir, P. S., and J. E. Lotan, 1985. Disturbance history and serotiny of *Pinus contorta* in western Montana. Ecology 66:1658–1668.

Muller, C. H. 1939. Relations in the vegetation and climatic types in Nuevo Leon, Mexico. Amer. Midl. Nat. 21:687–729.

Muller, C. H. 1947. Vegetation and climate of Coahuila, Mexico. Madroño 9:33–57.

Mutch, R. W. 1970. Wildland fires and ecosystems–a hypothesis. Ecology 51:1046–1051.

National Oceanic and Atmospheric Administration, Environmental Data Service. Climatological data, Colorado. Asheville, N.C.

Neilson, R. P. 1986. High resolution climatic analysis and southwest biogeography. Science 232:27–34.

Neilson, R. P., and L. H. Wullstein. 1983. Biogeography of two southwestern American oaks in relation to atmospheric dynamics. J. Biogeography 10:275–297.

Niering, W. A., and C. H. Lowe. 1984. Vegetation of the Santa Catalina Mountains: community types and dynamics. Vegetatio 58:3–28.

Nixon, E. S. 1967. A comparative study of the mountain brush vegetation in Utah. Great Basin Naturalist 27:59–66.

Noble, D. L., and F. Ronco. 1978. Seedfall and establishment of Engelmann spruce and subalpine fir in clearcut openings in Colorado. USDA Forest Service research paper RM-200.

Oliver, C. D. 1980. Forest development in North America following major disturbances. Forest Ecol. Management 3:153–168.

Oosting, H. J., and J. F. Reed. 1952. Virgin spruce-fir forest of the Medicine Bow Mountains, Wyoming. Ecol. Monogr. 22:69–91.

Oswald, E. T., and J. P. Senyk. 1977. Ecoregions of Yukon Territories. Canadian Forest Service publication BC-X-164.

Pace, C. P., and C. E. Layser. 1977. Classification of riparian habitat in the southwest. USDA Forest Service general technical report RM-43:5–9.

Parker, A. J., and K. C. Parker. 1983. Comparative successional roles of trembling aspen and lodgepole pine in the southern Rocky Mountains. Great Basin Naturalist 43:447–455.

Parker, A. J., and R. K. Peet. 1984. Size and age structure of conifer forests. Ecology 65:1685–1689.

Parker, W. H., J. Maze, F. E. Bennett, T. A. Cleveland, and W. G. McLachlan. 1984. Needle flavinoid variation in *Abies balsamea* and *A. lasiocarpa* from western Canada. Taxon 33:1–12.

Parker, W. H., J. Maze, and G. E. Bradfield. 1981. Implications of morphological and anatomical variation in *Abies balsamea* and *A. lasiocarpa* (Pinaceae) from western Canada. Amer. J. Bot. 68:843–854.

Passini, M. 1982. Les forets de *Pinus cembroides* au Mexique: Etude phytogeographique et ecologique. Etudes Mesoamericaines II-5. Mission Archeologique et ethnologique Francaise au Mexique. Paris.

Patten, D. T. 1963. Vegetational pattern in relation to environments in the Madison Range, Montana. Ecol. Monogr. 33:375–406.

Pearson, G. A. 1923. Natural reproduction of western yellow pine in the Southwest. USDA bulletin 1105.

Pearson, J. A., T. J. Fahey, and D. H. Knight. 1984. Biomass and leaf area in contrasting lodgepole pine forests. Can. J. Forest Res. 14:259–265.

Peet, R. K. 1978a. Latitudinal variation in southern Rocky Mountain forests. J. Biogeography 5:275–289.

Peet, R. K. 1978b. Forest vegetation of the Colorado Front Range: patterns of species diversity. Vegetatio 37:65–78.

Peet, R. K. 1981. Forest vegetation of the Colorado Front Range: composition and dynamics. Vegetatio 45:3–75.

Peterson, C. H. 1984. Does a rigorous criterion for environment identity preclude the existence of multiple stable points? American Naturalist 124:127–133.

Pfister, R. D., B. L. Kovalchik, S. F. Arno, and R. C. Presby. 1977. Forest habitat types of Montana. USDA Forest Service general technical report INT-34.

Potter, L. D., and D. L. Green. 1964. Ecology of ponderosa pine in western North Dakota. Ecology 45:10–23.

Ramaley, F. 1907. Plant zones in the Rocky Mountains of Colorado. Science 26:642–643.

Ramaley, F. 1908. Botany of northeastern Larimer County, Colorado. University of Colorado Studies 5:119–131.

Ramaley, F. 1931. Vegetation of chaparral-covered foothills, southwest of Denver, Colorado. University of Colorado Studies 18:231–237.

Ream, R. R. 1964. The vegetation of the Wasatch Mountains, Utah and Idaho. Dissertation, University of Wisconsin, Madison.

Reed, E. L. 1917. Meadow vegetation in the montane region of northern Colorado. Bull. Torrey Botanical Club 44:97–109.

Reed, R. M. 1971. Aspen forests of the Wind River Mountains, Wyoming. Amer. Midl. Nat. 86:327–343.

Reed, R. M. 1976. Coniferous forest habitat types of the Wind River Mountains, Wyoming. Amer. Midl. Nat. 95:159–173.

Robbins, W. W. 1918. Successions of vegetation in Boulder Park, Colorado. Bot. Gaz. 65:493–525.

Robbins, W. W., and G. S. Dodds. 1908. Distributions of conifers on the mesas. University of Colorado Studies 6:37–49.

Robert, M. 1977. Aspects phytogeographiques et ecologiques des forets de *Pinus cembroides*. I. Les forets de l'est et du nord-est du Mexique. Bulletin de la Societe Botanique de France 124:197–216.

Rominger, J. M., and L. A. Paulik. 1983. A floristic inventory of the plant communities of the San Francisco Peaks Research Natural Area. USDA Forest Service general technical report RM-96.

Romme, W. H. 1982. Fire and landscape diversity in subalpine forests of Yellowstone National Park. Ecol. Monogr. 52:199–221.

Romme, W. H., and D. H. Knight. 1981. Fire frequency and subalpine forest succession along a topographic gradient in Wyoming. Ecology 62:319–326.

Romme, W. H., and D. H. Knight. 1982. Lanscape diversity: the concept applied to Yellowstone National Park. Bioscience 32:664–670.

Romme, W. H., D. H. Knight, and J. B. Yavitt. 1986. Mountain pine beetle outbreaks in the central Rocky Mountains: regulators of primary production. American Naturalist 127.

Root, R. A., and J. R. Habeck. 1972. A study of high elevation grassland communities in western Montana. Amer. Midl. Nat. 87:109–121.

Rowdabaugh, K. M. 1978. The role of fire in the ponderosa pine–mixed conifer ecosystems. Thesis, Colorado State University, Fort Collins.

Rydberg, P. A. 1915. Phytogeographical notes on the Rocky Mountain region. V. Grasslands of the subalpine and montane zones. Bull. Torrey Botanical Club 42:629–642.

Rzedowski, J. 1981. Vegetacion de Mexico. Limusa, Mexico, D. F.

Salisbury, F. B. 1964. Soil formation and vegetation on hydrothermally altered rock material in Utah. Ecology 45:1–9.

Samuels, M. L., and J. L. Betancourt. 1982. Modeling the long-term effects of fuelwood harvests on pinyon-juniper woodlands. Environ. Mangement 6:505–515.

Sawyer, D. A., and T. B. Kinraide. 1980. The forest vegetation at higher altitudes in the Chiricahua Mountains, Arizona. Amer. Midl. Nat. 104:224–241.

Schimper, A. F. W. 1898. Pflanzen-Geographie auf physiologischer Grundlage. Fischer, Jena.

Schmidt, J. M., and T. E. Hinds. 1974. Development of spruce-fir stands following spruce beetle outbreaks. USDA Forest Service research paper RM-131.

Schmidt, W. C., R. C. Shearer, and A. L. Roe. 1976. Ecology and silviculture of western larch forests. USDA Forest Service technical bulletin 1520.

Schubert, G. H. 1974. Silviculture of southwestern ponderosa pine: the status of our knowledge. USDA Forest Service research paper RM-123.

Shankman, D. 1984. Tree regeneration following fire as evidence of timberline stability in the Colorado Front Range, U.S.A. Arct. Alp. Res. 16:413–417.

Shaw, C. H. 1909. The causes of timberlines on mountains: the role of snow. Plant World 12:169–181.

Shreve, F. 1939. Observations on the vegetation of Chihuahua. Madroño. 5:1–13.

Silvertown, J. W. 1982. Introduction to plant population ecology. Longman, London.

Smith, W. K. 1985. Western montane forests, pp. 95–126 in B. F. Chabot and H. A. Mooney (eds.), Physiological ecology of North American plant communities. Chapman & Hall, New York.

Springfield, H. W. 1976. Characteristics and management of southwestern pinyon-juniper ranges: the status of our knowledge. USDA Forest Service research paper RM-160.

Stahelin, P. 1943. Factors influencing the natural restocking of high altitude burns by coniferous trees in the central Rocky Mountains. Ecology 24:19–30.

Steele, R., R. D. Pfister, R. A. Ryker, and J. A. Kittams. 1981. Forest habitat types of central Idaho. USDA Forest Service general technical report INT-114.

Steele, R., S. V. Cooper, D. M. Ondov, D. W. Roberts, and R. D. Pfister. 1983. Forest habitat types of eastern Idaho–western Wyoming. USDA Forest Service general technical report INT-144.

Steyermark, J. A. 1950. Flora of Guatemala. Ecology 31:368–372.

Stringer, P. W., and G. H. La Roi. 1970. The Douglas-fir forests of Banff and Jasper National Parks, Canada. Can. J. Bot. 48:1703–1726.

Tande, G. F. 1979. Fire history and vegetation pattern of coniferous forests in Jasper National Park, Alberta. Can. J. Bot. 57:1912–1931.

Tausch, R. J., N. E. West, and A. A. Nabi. 1981. Tree age and dominance in Great Basin pinyon-juniper woodlands. J. Range Management 34:259–264.

Taylor, R. J., and T. F. Patterson. 1980. Biosystematics of Mexican spruce species and populations. Taxon 29:421–440.

Taylor, T. M. C. 1959. The taxonomic relationship between *Picea glauca* (Moench) Voss and *Picea engelmannii* Parry. Madroño 15:111–115.

Tesch, S. D. 1981. Comparative stand development in an old-growth Douglas-fir (*Pseudotsuga menziesii* var. *glauca*) forest in western Montana. Can. J. Forest Res. 11:82–89.

Thornbury, W. D. 1965. Regional geomorphology of the United States. Wiley, New York.

Vale, T. R. 1978. Tree invasion of Cinnabar Park in Wyoming. Amer. Midl. Nat. 100:277–284.

Veblen, T. T. 1986a. Treefalls and the coexistence of conifers in subalpine forests in the Central Rockies. Ecology.

Veblen, T. T. 1986b. Size and age structure of subalpine forests in the Colorado Front Range. Bull. Torrey Botanical Club.

Veblen, T. T., and D. C. Lorenz, 1986. Anthropogenic disturbance and patterns of recovery in montane forets of the Colorado Front Range. Phys. Geogr.

Vitousek, P. M., J. R. Gosz, C. C. Grier, J. M. Melillo, and W. A. Reiners. 1982. A comparative analysis of potential nitrification and nitrate mobility in forest ecosystems. Ecol. Monogr. 52:155–177.

Walter, H. 1985. Vegetation of the earth and ecological systems of the geo-biosphere, 3rd ed. Springer-Verlag, New York.

Walter, H., and H. Lieth. 1967. Klimadiagramm–Weltatlas. Gustav Fischer, Jena.

Wardle, P. 1965. A comparison of alpine timberlines in New Zealand and North America. New Zealand Journal of Botany 3:113–135.

Weaver, H. 1951. Fire as an ecological factor in the southwestern ponderosa pine forests. J. Forestry 49:93–98.

Weaver, H. 1959. Ecological changes in the ponderosa pine forest of the Warm Springs Indian Reservation in Oregon. J. Forestry 57:15–20.

Weaver, T., and D. Dale. 1974. *Pinus albicaulis* in central Montana: environment, vegetation and production. Amer. Midl. Nat. 92:222–230.

Weaver, T., and F. Forcella. 1977. Biomass of fifty conifer forests and nutrient exports associated with their harvest. Great Basin Naturalist 37:395–401.

Wellner, C. A. 1970. Fire history in the northern Rocky Mountains, pp. 42–64 in The role of fire in the intermountain West. University of Montana, Missoula.

Wells, P. V. 1965. Scarp woodlands, transported grassland soils, and concept of grassland climate in the Great Plains region. Science 148:246–249.

Wells, P. V. 1983. Paleobiogeography of montane islands in the Great Basin since the last glacio-pluvial. Ecol. Monogr. 53:341–382.

Wentworth, T. R. 1981. Vegetation on limestone and granite in the Mule Mountains, Arizona. Ecology 62:469–482.

West, N. E., K. H. Rea, and R. J. Tausch. 1975. Basic synecological relationships in juniper-pinyon woodlands, pp. 41–53 in G. F. Gifford and F. E. Busby (eds.), The pinyon-juniper ecosystem: a symposium. Utah Agricultural Experiment Station, Utah State University, Logan.

Whipple, S. A., and R. L. Dix. 1979. Age structure and successional dynamics of a Colorado subalpine forest. Amer. Midl. Nat. 101:142–158.

White, A. S. 1985. Presettlement regeneration patterns in a southwestern ponderosa pine stand. Ecology 66:589–594.

Whitson, P. D. 1965. Phytocoenology of Boot Canyon Woodland, Chisos Mountains, Big Bend National Park, Texas. Thesis, Baylor University, Waco. Tex.

Whittaker, R. H. 1967. Gradient analysis of vegetation. Biol. Rev. 42:207–264.

Whittaker, R. H. 1973. Direct gradient analysis: techniques, pp. 9–31 in R. H. Whittaker (ed.), Ordination and classification of communities. Handbook of vegetation science 5. Junk, The Hague.

Whittaker, R. H., and W. A. Niering. 1965. Vegetation of the Santa Catalina Mountains, Arizona: a gradient analysis of the south slope. Ecology 46:429–452.

Whittaker, R. H., and W. A. Niering. 1968. Vegetation of the Santa Catalina Mountains, Arizona. IV. Limestone and acid soils. J. Ecology 56:523–544.

Whittaker, R. H., and W. A. Niering. 1975. Vegetation of the Santa Catalina Mountains, Arizona. V. Biomass, production, and diversity along the elevation gradient. Ecology 56:771–790.

Wilson, H. C. 1969. Ecology and successional patterns of wet meadows, Rocky Mountain National Park, Colorado. Dissertation, University of Utah, Salt Lake City.

Woodin, H. E., and A. A. Lindsey. 1954. Juniper-pinyon east of the continental divide, as analyzed by the line-strip method. Ecology 35:473–489.

Wright, C. W. 1952. An ecological description of an isolated pinon pine grove. Thesis, University of Colorado, Boulder.

Young, J. A., and R. Evans. 1981. Demography and fire history of a western juniper stand. J. Range Management 34:501–506.

Chapter 4

Pacific Northwest Forests

JERRY F. FRANKLIN

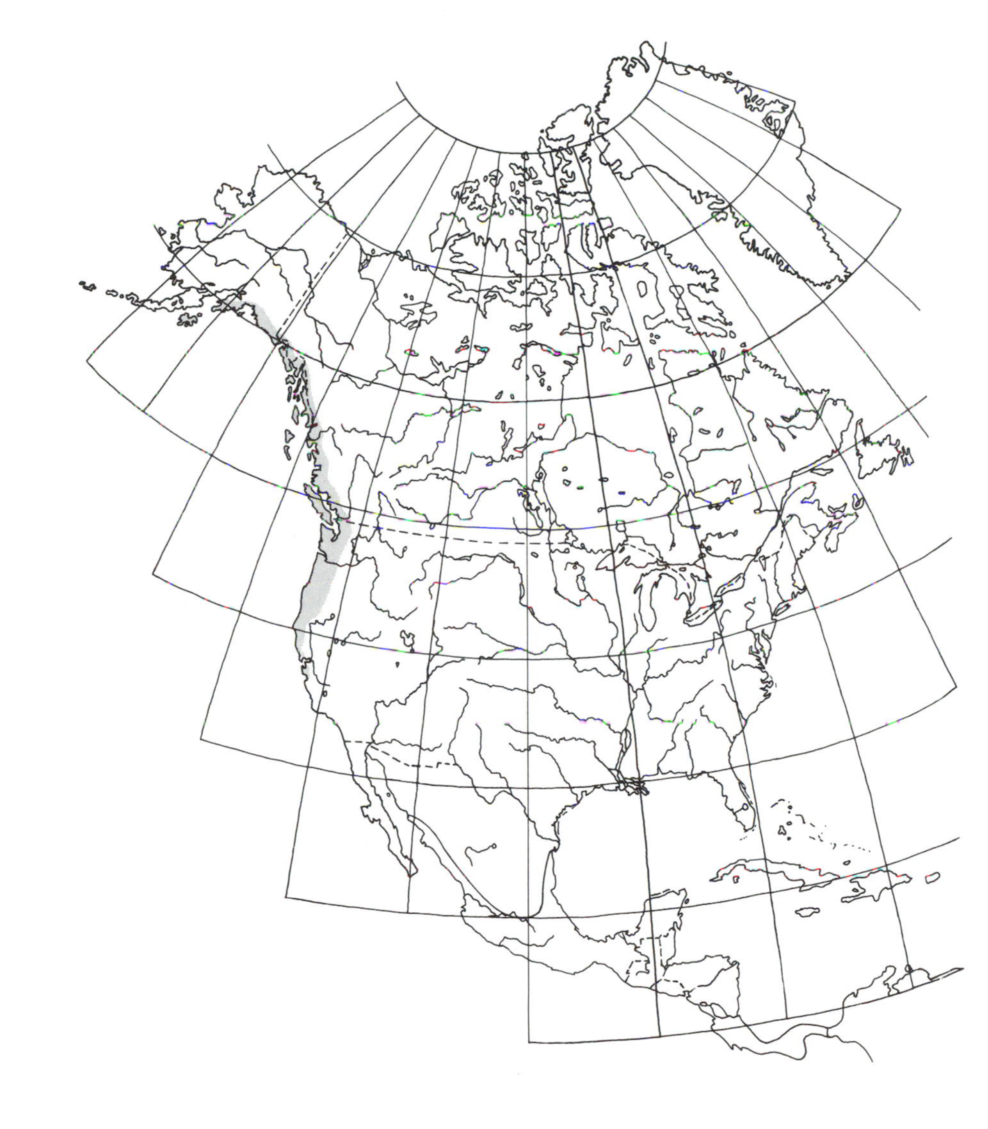

INTRODUCTION

The forests and associated vegetation of coastal northwestern North America cover nearly 20 degrees of latitude, extending from the Gulf of Alaska to northern California, but within 60–120 km of the Pacific Ocean (Frontispiece). Mild, moist maritime conditions characterize the region, producing expanses of forest dominated by massive evergreen conifers, including *Pseudotsuga menziesii* (Douglas fir), *Tsuga heterophylla* (western hemlock), *Thuja plicata* (western red cedar), *Picea sitchensis* (Sitka spruce), and *Sequoia sempervirens* (coast redwood). The outstanding ecological features of these northwestern forests and the contrasts that they offer with forests in other temperate regions of the world include the following:

1. Dominance by numerous coniferous species, species that are the largest and longest-lived representatives of their genera (Table 4.1). Hardwoods typically dominate other mesic, temperate forest regions of the world, but in the Pacific Northwest the ratio of hardwoods to conifers has been estimated at 1 : 1000 (Kuchler 1946). Hardwood species are few in number and generally are confined to specialized habitats (*Quercus qarryana*, Oregon white oak, and *Populus trichocarpa*, black cottonwood) or to early successional stages (*Alnus rubra*, red alder). *Lithocarpus densiflorus* (tan oak) is an evergreen hardwood with broad environmental and successional amplitude within the Klamath Mountain region.

2. Forests that exhibit the greatest biomass accumulations and some of the highest productivity levels of any in the world, temperate or tropical.

3. A climate of wet, mild winters and relatively dry summers. These conditions favor evergreen life forms and needle-leaved conifers by permitting extensive photosynthesis outside of the growing season and reducing net photosynthesis during the summer months (Waring and Franklin 1979).

4. Strong climatic gradients associated with latitude, longitude, and elevation, as well as strong localized gradients associated with slope, aspect, and soil moisture conditions. Vegetation gradients follow climatic gradients closely: thus, primary and secondary environmental factors in ordinations are indices of moisture and temperature.

5. Disturbance regimes that, within the heart of the northwestern forests, are dominated by infrequent catastrophic events, such as wildfires, at intervals of several hundred years. This contrasts with the patterns of frequent, noncatastrophic fires that dominate disturbance regimes in many other forested regions of western North America, including California and the Rocky Mountains (see Chapters 3 and 5).

6. Extended forest-alpine transition zones with attractive and diverse parklands of forest, tree patches, and meadowlands. Deep, persistent winter snowpacks within a relatively warm subalpine zone are believed to be key factors in the formation and maintenance of these parklands.

Terminology in this chapter reflects a strong phytosociological tradition of the Pacific Northwest in which Professor R. Daubenmire has been very influential. "Succession" refers to a directional change in community structure or composition or both, either observed or inferred. "Climax" refers to the ability of a species or community to perpetuate its structure and composition without a catastrophic disturbance. "Series" refers to a group of community types that have the same reproducing tree species, such as *Pinus ponderosa* (ponderosa pine) or *Tsuga heterophylla*. The series level is becoming the most commonly used level in the phytosociological hierarchy, above habitat type or association, because of its importance in the USDA Forest Service's western regional ecology programs.

ENVIRONMENTAL CONDITIONS

The Pacific Northwest is characterized by north-south-trending mountain ranges, including the Cascade Range, the Coastal Ranges of Oregon, the Olympic Mountains, and the British Columbia Coastal Range. Lowland areas, such as the Willamette Valley and Puget Trough, separate the coastal and Cascadian mountain systems for most of their length. The major topographic and climatic divide of the region is the Cascade Range, which is bisected by only three river systems: the Fraser, Columbia, and Klamath. Geologic conditions are highly varied, with sedimentary rock types typical of the Oregon Coast Ranges and metamorphic rocks dominating much of the northern Cascade Range and Olympic Mountains. Volcanic rocks of Miocene, Pliocene, and Pleistocene age are typical of the southern two-thirds of the Cascade Range. Glaciation has been an important process at higher elevations in the Cascade Range and Olympic Mountains, as well as in the ranges of British Columbia and Alaska. Continental glaciation extended a short distance south of Puget Sound.

Forest soils are highly varied, reflecting the diverse parent materials and topography of the region. Haplumbrepts, Haplohumults, Haplorthods, Xerochrepts, Cryorthods, and Vitrandepts are most characteristic. Deposition of parent materials by

Table 4.1. *Ages and dimensions typically attained by forest trees on better sites in the Pacific Northwest and their relative shade tolerances and fire sensitivities*[a]

Species	Age (yr)	Diameter (cm)	Height (m)	Shade tolerance[b]	Fire sensitivity
Abies amabilis	400+	90−110	45−55	VTOL	HIGH
Abies concolor	300+	100−150	40−55	TOL	INTER
Abies grandis	300+	75−125	40−60	TOL	INTER
Abies lasiocarpa	250+	50−60	25−35	TOL	HIGH
Abies magnifica	300+	100−125	40−50	INTER	INTER
Abies procera	400+	100−150	45−70	INTOL	HIGH
Chamaecyparis lawsoniana	500+	120−180	60	TOL	INTER
Chamaecyparis nootkatensis	1000+	100−150	30−40	TOL	INTER
Larix occidentalis	700+	140	50	INTOL	LOW
Libocedrus decurrens	500+	90−120	45	INTER	INTER
Picea engelmannii	400+	100+	45−50	TOL	HIGH
Picea sitchensis	500+	180−230	70−75	TOL	HIGH
Pinus contorta	250+	50	25−35	INTOL	INTER
Pinus lambertiana	400+	100−125	45−55	INTER	INTER
Pinus monticola	400+	110	60	INTER	INTER
Pinus ponderosa	600+	75−125	30−60	INTOL	LOW
Pseudotsuga menziesii	750+	150−220	70−80	INTOL	LOW
Sequoia sempervirens	1250+	150−380	75−100	TOL	LOW
Thuja plicata	1000+	150−300	60+	TOL	INTER
Tsuga heterophylla	400+	90−120	50−65	VTOL	HIGH
Tsuga mertensiana	500+	75−100	35+	TOL	HIGH
Acer macrophyllum	300+	50	15	TOL	INTER
Alnus rubra	100	55−75	30−40	INTOL	HIGH
Castanopsis chrysophylla	150	30+	20+	INTER	INTER
Lithocarpus densiflorus	180	25−125	15−30	TOL	INTER
Populus trichocarpa	200+	75−90	25−35	INTOL	HIGH
Prunus emarginata	50	15−30	15	INTOL	HIGH
Quercus garryana	500	60−90	15−25	INTOL	LOW

[a]Developed from a variety of sources, the most important being Fowells (1965). Maximum ages and sizes for species are generally much greater than those indicated here.
[b]Tolerance scale: VTOL = very tolerant of shade; TOL = tolerant; INTER = intermediate shade tolerance (greater in youth, lesser at maturity); INTOL = intolerant.

alluvial, colluvial, glacial, or eolian action is an important soil-forming process. Soils in the Cascade Range, for example, are much deeper than might be expected, because of aerial deposits of various volcanic ejecta.

A maritime climate characterizes the Pacific Northwest. In coastal regions, temperatures are mild, with prolonged cloudy periods, muted extremes, and narrow diurnal fluctuations (6−10°C). Winters are mild, with precipitation of 800−3000 mm, 75−85% of which occurs between October 1 and March 31, mostly as rain, or as snow at higher elevations. Summers are cool and can be relatively dry. Most precipitation is the result of low-pressure systems that approach from the Pacific Ocean.

There are major variations in the climate of the region associated with latitude, elevation, and positions of mountains ranges (Fig. 4.1). There is a general latitudinal increase in precipitation and decrease in temperature. Mountain masses block maritime air masses, creating rainshadows in their lee; for example, the coastal mountains are responsible for the drier and less muted climates of the Willamette Valley and Puget Trough. Mountain ranges also produce local increases in precipitation (because of orographic effects) and in the proportion of precipitation that falls as snow.

MAJOR TEMPERATE FORESTS

Pseudotsuga Menziesii−*Tsuga Heterophylla* Forests

This is the major forest complex of the Pacific Northwest, encompassing seral forests dominated by *Pseudotsuga menziesii* and massive old-growth forests of *Pseudotsuga*, *Tsuga heterophylla*, *Thuja plicata*, and other species (Fig. 4.2). These forests occur from sea level up to elevations of 700−1000 m in the Coast Ranges, Olympic Mountains, and Cas-

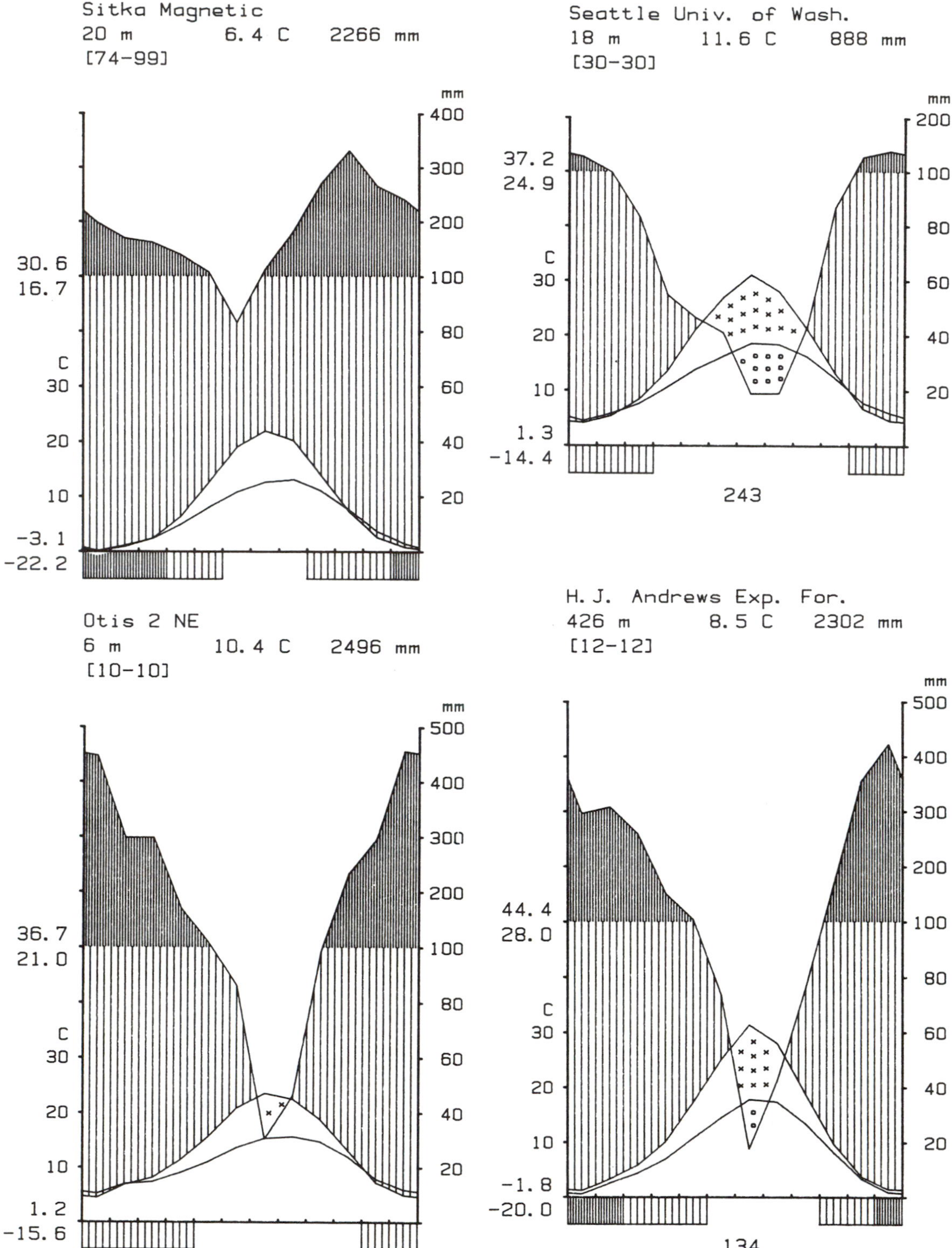

Figure 4.1. Climatic diagrams illustrating temperature and moisture regimes for selected locations in the Pacific Northwest. The Sitka, Alaska, and Otis, Oregon (Cascade Head Experimental Forest), stations are near opposite latitudinal ends of the Picea sitchensis *zone. Seattle, Washington, and H. J. Andrews Experimental Forest, Oregon, are in the* Tsuga heterophylla *zone in the Puget Trough and western Oregon Cascade Range, respectively. Longmire and Paradise, Washington, are located in the lower* Abies amabilis *and* Tsuga mertensiana *zones, respectively, within Mount Rainier National Park (diagrams produced by Bradley Smith from U.S. Weather Bureau data).*

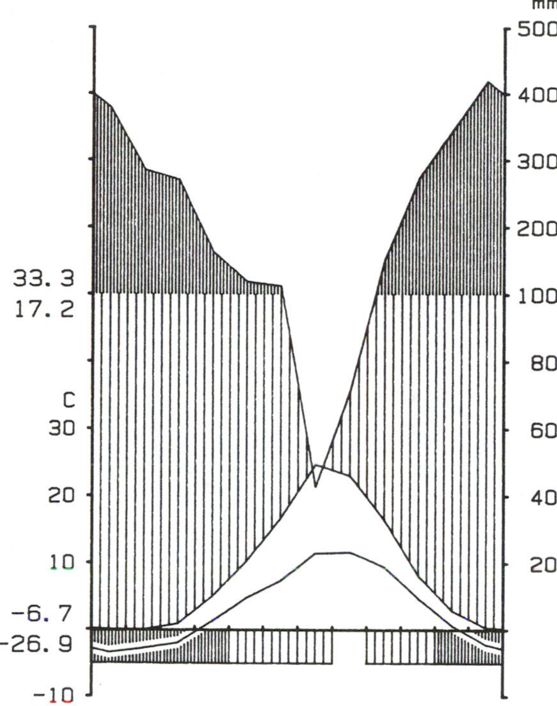

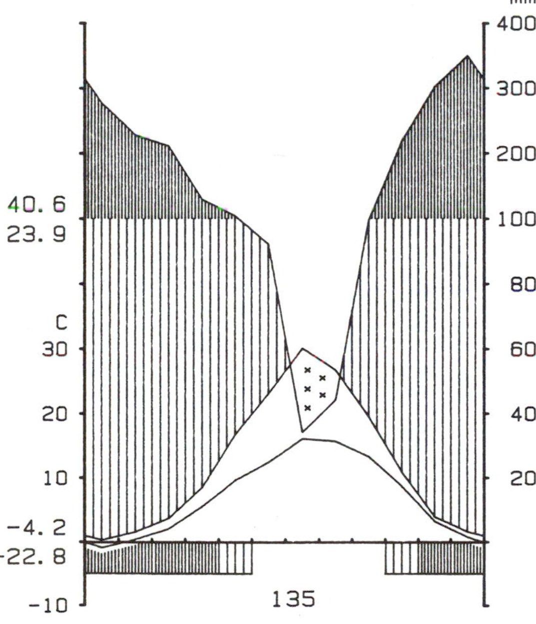

Figure 4.1. (Cont.)

cade Range north of latitude 43° 15′ N. They are equivalent to the *Tsuga heterophylla* zone defined by Franklin and Dyrness (1973) (Table 4.2).

Pseudotsuga menziesii–Tsuga heterophylla forests occupy a wide range of environments and are highly variable in composition and structure, depending on local conditions, especially the moisture regime. Major forest trees are *Pseudotsuga menziesii*, *Tsuga heterophylla*, and *Thuja plicata*; *Abies grandis* (grand fir), *Picea sitchensis*, and *Pinus monticola* (western white pine) occur sporadically. In central Oregon, *Calocedrus decurrens* (incense cedar), *Pinus lambertiana* (sugar pine), and even *Pinus ponderosa* may be encountered. *Abies amabilis* (Pacific silver fir) is common near the upper altitudinal limits of the *Pseudotsuga* forests and even at lower elevations in the Olympic Mountains, northern Cascade Range, and British Columbia Coast Ranges. *Taxus brevifolia* (western yew) is ubiquitous but is always a subordinate tree. Hardwoods are not common except on recently disturbed sites or in specialized habitats, such as riparian zones. Typical species are *Alnus rubra*, *Acer macrophyllum* (bigleaf maple), and *Prunus emarginata* (bitter cherry). *Populus trichocarpa* and *Fraxinus latifolia* (Oregon ash) also occur along waterways. *Arbutus menziesii* (Pacific madrone), *Castanopsis chrysophylla* (golden chinkapin), and *Quercus garryana* are found on warmer, drier sites.

In forest communities, composition, structure, and productivity vary markedly along moisture gradients, as has been demonstrated by numerous studies (Spilsbury and Smith 1947; McMinn 1960; Dyrness et al. 1974; Zobel et al. 1976). Less productive communities dominated by *Gaultheria shallon* or *Holodiscus discolor* characterize the hot, dry end of the gradient; *Pseudotsuga menziesii* may even be a major reproducing tree species on such sites. At the moist end of the gradient, herbaceous understories dominated by *Polystichum munitum* and *Oxalis oregana* are characteristic. Intermediate, mesic sites have understories dominated by a variety of evergreen shrubs, subshrubs, and herbs, of which *Berberis nervosa* and (in Oregon) *Rhododendron macrophyllum* are most common.

Although the details of community compositions vary with locales, the basic pattern–*Gaultheria* at the dry end and *Polystichum* at the wet end of the scale–is repeated throughout the *Pseudotsuga menziesii–Tsuga heterophylla* type. This spectrum is illustrated in Table 4.3, using data from five old-growth associations on the western slope of the Oregon Cascade Range at about 45° N (Dyrness et al. 1974). The same types of understories are found in much younger forests, including stands solely dominated by *Pseudotsuga menziesii*.

The early stages of successional development

Figure 4.2. Typical old-growth forest stand of Pseudotsuga menziesii *and* Tsuga heterophylla *in the H.J. Andrews Experimental Forest, western Oregon Cascade Range (photo courtesy USDA Forest Service).*

following logging and burning are well known (Dyrness 1973). Invading herbaceous species, such as *Senecio sylvaticus* and *Eplobium angustifolium*, are early dominants. *Senecio* typically dominates for only the second year, because of its high nutrient requirements (West and Chilcote 1968). After 5–7 yr, invading perennial herbs begin to give way to shrubs, including residual species such as *Acer circinatum, Rubus ursinus, Berberis nervosa, Rhododendron macrophyllum*, and *Ceanothus velutinus*. *Ceanothus* can regenerate from seed stored in the forest floor for up to several centuries. The vegetational composition is heterogeneous in the preforest stages of succession, reflecting variability in logging and fire disturbances. Invading species (and *Ceanothus*) show a marked preference for burned microsites, for example, whereas residual species are more important on less impacted habitats. The diversity and biomass of the shrubs and herbs decline dramatically once the canopy of trees closes (Long and Turner 1975).

Tsuga heterophylla and *Pseudotsuga menziesii* are the major climax and seral species, respectively. *Pseudotsuga* typically dominates young forests, often forming pure stands because of its relatively large and hardy seedlings and rapid growth rate. *Tsuga heterophylla* or *Thuja plicata* may establish themselves early in succession or later under the

Table 4.2. *Major series and Society of American Foresters (SAF) cover types and vegetation zones associated with Pacific Northwest forest types*

Type	Series[a]	Zone[b]	Major SAF types[c]
Pseudotsuga menziesii–Tsuga heterophylla	*T. hererophylla*	*T. heterophylla*	Pacific Douglas fir–western hemlock, western hemlock, western red cedar–western hemlock, western red cedar
Picea sitchensis–Tsuga heterophylla	*T. heterophylla, P. sitchensis*	*P. sitchensis*	Western hemlock–Sitka spruce
Sequoia sempervirens	*Lithocarpus densiflorus, T. heterophylla, S. sempervirens*	*P. sitchensis*	Redwood
Sierra-type mixed conifer	*Abies concolor, P. menziesii*	Mixed conifer	Sierra Nevada mixed conifer, white fir, Pacific ponderosa pine–Douglas fir
Klamath Mountain mixed evergreen	*L. densiflorus, P. menziesii, A. concolor*		Douglas fir–tan oak–Pacific madrone, Pacific ponderosa pine–Douglas fir
Abies amabilis–T. heterophylla	*A. amabilis*	*A. amabilis*	Coastal true fir–hemlock
Abies magnifica var. *shastensis*	*A. magnifica* var. *shastensis, A. concolor*	*A. magnifica* var. *shastensis*	Red fir
Tsuga mertensiana	*A. amabilis, T. mertensiana*	*T. mertensiana*	Mountain hemlock, coastal true fir–hemlock

[a] "Series" refers to a group of community types that have the same climax tree species; it is a category widely used in the USDA Forest Service Area Ecology Program.
[b] From Franklin and Dyrness (1973).
[c] From Eyre (1980).

Table 4.3. *Percentages of constancy and cover of important species in five associations found in the* Tsuga heterophylla *zone of the western Oregon Cascade Range (continues on pp.110, 111)*

	PSME/HODI[a]		TSHE/RHMA/GASH		TSHE/RHMA/BENE		TSHE/POMU		TSHE/POMU/OXOR	
Species and stratum	Cons.	Cover	Cons.	Cover	Cons.	Cover	Cons.	Cover	Cons.	Cover
Overstory tree										
Tsuga heterophylla	—[b]	—	76	20	100	43	100	44	100	29
Pseudotsuga menziesii	100	41	100	45	100	45	100	42	100	38
Thuja plicata	—	—	47	3	72	13	80	16	75	13
Libocedrus decurrens	50	6	12	1	—	—	—	—	—	—

(Cont.)

Table 4.3. *(Cont.)*

	PSME/HODI[a]		TSHE/RHMA/GASH		TSHE/RHMA/BENE		TSHE/POMU		TSHE/POMU/OXOR	
Species and stratum	Cons.	Cover	Cons.	Cover	Cons.	Cover	Cons.	Cover	Cons.	Cover
Pinus lambertiana	50	1	12	2	—	—	—	—	—	—
Acer macrophyllum	25	2	12	1	—	—	47	2	62	7
Arbutus menziesii	38	2	—	—	—	—	—	—	—	—
Understory tree										
Tsuga heterophylla	25	T	100	8	100	8	100	9	100	11
Taxus brevifolia	38	4	59	6	78	7	47	4	50	1
Cornus nuttallii	50	2	53	3	50	2	33	1	38	1
Castanopsis chrysophylla	50	1	82	2	78	2	13	T	25	T
Pseudotsuga menziesii	100	8	—	—	—	—	—	—	—	—
Thuja plicata	—	—	29	1	56	2	53	3	38	2
Libocedrus decurrens	38	3	—	—	—	—	—	—	—	—
Pinus lambertiana	38	T	6	T	—	—	—	—	—	—
Acer macrophyllum	12	T	6	T	11	T	13	T	—	—
Arbutus menziesii	12	T	—	—	—	—	—	—	—	—
Shrub										
Acer circinatum	88	19	88	21	83	9	87	2	88	6
Rhododendron macrophyllum	12	T	100	40	89	13	67	1	12	T
Holodiscus discolor	88	5	—	—	—	—	—	—	—	—
Corylus cornuta var. *californica*	88	7	12	T	11	T	33	T	12	T
Vaccinium parvifolium	62	1	71	2	83	1	87	2	100	3
Beberis nervosa	100	16	100	14	100	11	100	8	100	13
Gaultheria shallon	62	7	100	40	89	4	53	2	75	4
Rubus ursinus	75	1	65	1	83	2	67	1	75	1
Symphoricarpos mollis	88	2	—	—	17	2	—	—	—	—
Herb										
Achlys triphylla	50	T	29	T	33	T	47	1	75	2
Viola sempervirens	12	T	65	1	83	2	73	2	62	1
Trillum ovatum	—	—	29	T	56	1	87	1	62	1
Polystichum munitum	100	4	65	1	94	4	100	26	100	27
Linnaea borealis	75	3	82	5	100	13	80	13	50	11
Vancouveria hexandra	25	T	6	T	39	T	27	T	88	4
Galium triflorum	38	T	18	T	11	T	60	1	38	1
Trientalis latifolia	100	1	29	T	44	T	27	T	12	T
Lathyrus polyphyllus	38	3	—	—	—	—	—	—	—	—
Madia gracilis	50	1	—	—	—	—	—	—	—	—
Collomia heterophylla	38	1	—	—	—	—	—	—	—	—
Hieracium abliflorum	62	1	12	T	28	T	20	T	25	T
Synthyris reniformis	75	4	12	T	17	T	—	—	—	—
Xerophyllum tenax	25	2	53	2	50	2	7	T	—	—
Iris tenax	62	1	6	T	—	—	—	—	—	—
Festuca occidentalis	88	1	6	T	—	—	7	T	—	—

(Cont.)

Table 4.3. *(Cont.)*

Species and stratum	PSME/HODI[a] Cons.	PSME/HODI[a] Cover	TSHE/RHMA/GASH Cons.	TSHE/RHMA/GASH Cover	TSHE/RHMA/BENE Cons.	TSHE/RHMA/BENE Cover	TSHE/POMU Cons.	TSHE/POMU Cover	TSHE/POMU/OXOR Cons.	TSHE/POMU/OXOR Cover
Whipplea modesta	100	8	29	1	17	1	20	T	—	—
Chimaphila umbellata	88	1	82	2	83	4	53	T	38	1
Coptis laciniata	—	—	53	1	89	4	73	3	25	1
Tiarella unifoliata	—	—	12	T	28	T	73	4	12	T
Disporum hookeri	—	—	18	T	17	T	27	T	50	1
Asarum caudatum	—	—	—	—	—	—	27	T	25	1
Blechnum spicant	—	—	—	—	—	—	27	1	38	1
Oxalis oregana	—	—	—	—	—	—	7	T	100	38

[a]PSME/HDI = *Pseudotsuga menziesii–Holodiscus discolor;* TSHE/RHMA/GASH = *Tsuga heterophylla–Rhododendron macrophyllum–Gaultheria shallon;* TSHE/RHMA/BENE = *Tsuga heterophylla–Rhododendron macrophyllum–Berberis nervosa;* TSHE/POMU = *Tsuga heterophylla–Polystichum munitum;* TSHE/POMU/OXOR = *Tsuga heterophylla–Polystichum munitum–Oxalis oregana.*
[b]T = trace (less than 0.5% cover); a dash (–) means that the species was not found.
Source: Based on Table 8 of Franklin and Dyrness (1973).

canopy of *Pseudotsuga.* Stand basal areas shift toward *Tsuga–Thuja* dominance after 400–600 yr. *Pseudotsuga menziesii* may persist as individuals for more than 1000 yr. *Thuja plicata* is a minor climax species; *Thuja* reproduction occurs at low levels, but it apparently has a high survival ability and appears capable of perpetuating the species within many stands on moist to wet habitats. *Pseudotsuga menziesii* can be a climax species on dry sites within the *Pseudotsuga–Tsuga* forest complex; such habitats actually belong to the *Pseudotsuga* series, not the *Tsuga heterophylla* series. Several other species, such as *Acer macrophyllum* and *Calocedrus decurrens*, appear capable of playing minor climax roles. *Alnus rubra* is the most conspicuous of the seral hardwoods, although the short-lived *Prunus emarginata* may have a wider ecological amplitude.

There are many variants of the widespread *Pseudotsuga–Tsuga* forests. A notable contrast between Oregon and Washington concerns the importance of *Rhododendron macrophyllum.* Forests dominated by *Thuja plicata,* with understories of *Oplopanax horridum* and *Athyrium filix-femina,* or even *Lysichiton americanum* and *Carex obnupta,* occur on very wet sites. *Chamaecyparis lawsoniana* (Port Oxford cedar) is a localized endemic often associated with these forests in southwestern Oregon (Zobel et al. 1985). The Puget lowlands have a number of notable features, including prairies, Mima mounds, and common occurrence of *Pinus contorta* and *P. monticola* (Franklin and Dyrness 1973). Talus communities of *Acer circinatum* and *A. macrophyllum* are often associated with the forests and are particularly noticeable in the Columbia Gorge.

Picea Sitchensis–Tsuga Heterophylla Forests

These forests characterize a relatively narrow band adjacent to the Pacific Ocean, extending from northern California to the Gulf of Alaska. Maritime influences are maximum, with cool, muted temperatures, high precipitation, and frequent fogs. *Picea-Tsuga* forests typically extend only a few kilometers inland, except up river valleys or where the coastal plain is unusually broad. These forests are found at elevations from sea level up to several hundred meters or, in Alaska, essentially to the timberline.

Picea sitchensis and *Tsuga heterophylla* are the major tree species. *Tsuga* may dominate numerically by a factor of two or more, but *Picea* distinguishes these forests from types found in less humid inland regions. In areas immediately adjacent to the ocean, *Picea sitchensis* may form nearly pure forests or codominate with *Pinus contorta* (lodgepole pine), because of the spruce's high tolerance for salt spray. Typical forest associates vary with latitude: *Pseudotsuga menziesii, Thuja plicata, Abies grandis, Acer macrophyllium,* and *Abies amabilis* are typical over much of its range. *Alnus rubra* has come to occupy large areas following logging; prolific seeding and rapid early growth make it an aggressive competitor. *Sequoia sempervirens* occurs in the south. *Tsuga mertensiana* (mountain hemlock) and *Chamaecyparis nootkatensis* (Alaska cedar) are associates in southeastern Alaska and parts of British Columbia. *Pinus contorta* occurs on swampy or boggy habitats, in ocean-front areas, and on sand dunes.

Understories in mature forests typically include

Figure 4.3. Alluvial Picea sitchensis–Tsuga heterophylla *forest in the Hoh River valley of Olympic National Park, Washington (photo courtesy USDA Forest Service).*

substantial shrub and herb coverage and a well-developed moss layer. One widespread community type in coastal Washington and Oregon is the *Picea sitchensis–Oxalis oregana* type (Henderson and Peter 1982; Hemstrom and Logan 1984). Typical herbs include *Polystichum munitum, Oxalis oregana, Rubus pedatus, Blechnum spicant, Tiarella trifoliata, Montia sibirica,* and *Maianthemum dilatatum*. Typical shrubs include *Acer circinatum, Rubus spectabilis, Vaccinium parvifolium, Menziesia ferruginea,* and *Vaccinium alaskaense*. Wetter habitats may incorporate an even greater variety of herbs, as well as *Oplopanax horridum*, a spiny shrub. The understory composition is similar in southeast Alaska, with well-developed ericaceous shrub layers, but an absence of *Oxalis oregana* (Alaback 1982). The *Picea sitchensis–Gaultheria shallon* community type is characteristic of sand dunes and areas of salt spray (Hemstrom and Logan 1984). Its shrubby understory of *Gaultheria shallon* and depauperate herb layer contrast sharply with the other types.

Revegetation is very rapid following disturbances on the highly productive habitats characteristically occupied by *Picea sitchensis–Tsuga heterophylla* forests. As mentioned, *Alnus rubra* is an aggressive colonizer and often forms pure or nearly pure stands following a disturbance. *Alnus* is typically replaced by coniferous species such as *Tsuga, Picea,* and *Thuja*. Conifer stands may also regenerate directly following disturbances, resulting in dense, even-aged stands. The high leaf areas in young and fully stocked stands can prevent development of understory plants; reestablishment of a significant shrub and herb community may take several centuries, a matter of concern where such forests are used as wintering grounds by herbivores (Alaback 1982). *Tsuga* is the major climax species based on size-class distributions. Under old-growth conditions, canopies often are sufficiently open for reproduction of *Picea*.

Alluvial rain forests of the western Olympic Peninsula are outstanding examples of *Picea-Tsuga* forest (Franklin and Dyrness 1973) (Fig. 4.3). Forests have relatively low tree densities, with open canopies. Epiphytes, mostly cryptogams, are conspicuous. *Selaginella oregana* is particularly abundant. Branches covered with such epiphytes may sprout roots in a process of air-layering (Nadkarni 1983). Rotting logs are important seedbeds for tree reproduction throughout the *Picea sitchensis* zone, but nurse-log phenomena are especially conspicuous in the alluvial forests (McKee et al. 1982). Competition from understory plants is the major factor restricting tree seedlings to rotten-wood seedbeds (Harmon 1986). Finally, *Cervis canadensis* var. *roosevelti* (Roosevelt elk) significantly influence the composition and structure of the forest understory by grazing; such preferred species as *Oplopanax horridum* and *Rubus spectabilis* are found only in protected microsites.

Sequoia Sempervirens Forests

The coastal forests of northern California and southern Oregon are distinguished by the presence of *Sequoia sempervirens*. This is the world's tallest tree (112 m); with its long life span and massive growth form, it results in forests that have the greatest biomass accumulations known. The primary belt of *S. sempervirens* is about 16 km in width, with its western boundary often several kilometers inland from the ocean, and its eastern limits 35+ km or more inland (Zinke 1977). A wide variety of environments and community types with *Sequoia* are represented along the coast-inland gradient (Waring and Major 1964; Zinke 1977).

Associates of *Sequoia sempervirens* vary substantially with local environmental conditions (Waring and Major 1964). *Pseudotsuga menziesii* and *Lithocarpus densiflorus* are typically present on a wide variety of sites. *Tsuga heterophylla*, *Umbellularia californica*, and *Alnus rubra* are common on moist habitats. *Picea sitchensis* is associated near the ocean. Dry-site species include *Arbutus menziesii*, *Calocedrus decurrens*, and even *Pinus attenuata*.

Major climax species on *Sequoia sempervirens* sites include *Lithocarpus densiflorus*, *Tsuga heterophylla*, and *Sequoia sempervirens*. Regeneration of *Sequoia* was once believed to be dependent on disturbance, specifically fire or flood (Stone and Vasey 1968), but age-structure analyses of cutover stands have shown that *Sequoia* can reproduce in sufficient numbers to perpetuate its population, even in the absence of catastrophic disturbance (Viers 1982).

Waring and Major (1964) have provided detailed information on understory compositions in various *Sequoia* forest types. Forests on alluvial flats and moist lower slopes have the lushest understories: *Polystichum munitum*, *Oxalis oregana*, *Disporum smithii*, *Anemone deltoidea*, *Tiarella unifoliata*, *Trillium ovatum*, *Asarum caudatum*, *Viola glabella*, and *Hierochloe occidentalis* are characteristic. Stands in middle- and upper-slope positions are often characterized by evergreen shrubs, such as *Gaultheria shallon*, *Rhododendron macrophyllum*, and *Vaccinium ovatum*, and by reduced diversity and coverage of herbs.

Klamath Mountains Mixed Evergreen Forests

Mixed forests of conifers and evergreen hardwoods characterize the majority of the forested landscape within the Klamath Mountains of northwestern California and southwestern Oregon. They are bounded by mesic coastal forests on the west, subalpine forests of *Abies* species at high elevations, and dry *Quercus* woodlands and *Pinus* forests in the interior valleys to the east. The Klamath Mountains contain regions of complex and strongly contrasting geology, including ultrabasic rock types, such as serpentine; consequently, soil conditions are highly variable. There are also sharp gradients in moisture and temperature regimes between the coastal and inland valleys. Complex vegetational patterns are therefore characteristic of the region (Whittaker 1960). Furthermore, the region contains a large number of endemic or relictual species that increase the vegetational complexity.

The mixed evergreen forests typically include conifers, primarily *Pseudotsuga menziesii*, and one or more evergreen hardwoods, such as *Lithocarpus densiflorus*, *Quercus chrysolepis*, *Arbutus menziesii*, and *Castanopsis chrysophylla* (Franklin and Dyrness 1973; Sawyer et al. 1977; Atzet and Wheeler 1984) (Fig. 4.4). Stands are often two-storied, with *Pseudotsuga* forming a canopy up to 65 m in height emergent above a hardwood canopy up to 35 m in height (Thornburgh 1982). Three series comprise most of this formation on normal soils: *Lithocarpus densiflorus*, *Pseudotsuga menziesii*, and *Abies concolor* (Thornburgh 1982; Atzet and Wheeler 1984). The *Pinus jeffreyi* series is most characteristic on ultrabasic parent material.

Pseudotsuga-Lithocarpus communities are widespread on the windward or coastal side of the Klamath Mountains, regions with relatively high summer humidities (Atzet and Wheeler 1984). Twenty-one *Lithocarpus densiflorus* community types have been recognized in the Siskiyou National Forest (Atzet and Wheeler 1984), and 11 of 16 habitat types listed for northwestern Californian mixed evergreen forests are also *Pseudotsuga-Lithocarpus* (Thornburgh 1982). Characteristic tree species are *Pseudotsuga* and *Pinus lambertiana* in the overstory

Figure 4.4 Mixed evergreen forests in southwestern Oregon consist of a mixture of evergreen hardwood and conifer species that include Arbutus menziesii *and* Psuedotsuga menziesii *in this stand (photo courtesy of USDA Forest Service).*

Table 4.4. *Percentages of constancy and coverage (cons./cover.) of selected species in stands belonging to the four major series in the Siskiyou National Forest in southwestern Oregon*

Layer and species	Community type[a] ABCO	LIDE	PSME	PIJE
Tree upper story				
Pseudotsuga menziesii	87/40	99/59	94/48	36/4
Abies concolor	83/34	3/9	9/9	T/T
Pinus lambertiana	32/6	37/10	34/14	10/7
Abies magnifica var. *shastensis*	21/13	—	1/1	—
Calocedrus decurrens	17/18	4/12	15/13	44/5
Pinus ponderosa	18/10	8/8	35/22	—
Pinus jeffreyi	1/6	1/5	7/15	95/18
Chamaecyparis lawsoniana	414	6/22	3/8	—
Picea breweriana	1/4	—	—	—
Tsuga heterophylla	T/8	—	1/20	—
Pinus monticola	2/2	2/15		15/4
Tree lower story				
Abies concolor	100/29	13/6	28/5	10/2
Quercus sadleriana	21/16	4/17	8/18	—
Pseudotsuga menziesii	53/7	70/7	95/27	49/4
Calocedrus decurrens	25/4	11/6	22/11	62/8
Abies magnifica var. *shastensis*	17/5	—	1/1	—
Pinus lambertiana	13/2	28/3	33/5	5/7
Pinus ponderosa	5/2	1/3	21/10	—
Lithocarpus densiflorus	11/7	100/50	28/13	—

(Cont.)

Table 4.4. *(Cont.)*

Layer and species	Community type[a] ABCO	LIDE	PSME	PIJE
Arbutus menziesii	17/10	48/11	63/12	3/1
Castanopsis chrysophylla	36/8	29/16	27/16	—
Quercus chrysolepis	19/9	46/13	49/14	8/5
Quercus kelloggii	2/7	5/4	30/10	10/3
Chamaecyparis lawsoniana	8/6	11/7	4/4	8/3
Pinus jeffreyi	T/1	—	6/8	95/8
Pinus monticola	2/3	3/8	4/8	23/9
Umbellularia californica	—	13/18	3/12	26/10
Shrub				
Rosa gymnocarpa	77/4	45/3	56/3	3/1
Symphoricarpos mollis	66/5	22/3	44/7	—
Berberis nervosa	65/16	62/14	38/15	—
Gaultheria shallon	2/29	38/42	—	—
Vaccinium ovatum	—	32/36	—	—
Rhododendron macrophyllum	1/40	30/36	—	—
Rhus diversiloba	1/2	25/9	38/13	—
Lonicera hispidula	3/4	17/3	25/3	—
Berberis piperiana	6/2	6/3	25/5	3/1
Holodiscus discolor	35/4	6/6	29/6	8/9
Archostaphylos nevadensis	5/7	2/7	5/22	31/18
Archostaphylos viscida	—	1/12	7/17	54/14
Herb				
Grasses	100/1	100/1	100/7	100/39
Achillea millifolium	5/2	T/1	7/5	36/3
Chimaphila umbellata	56/5	31/4	27/4	—
Goodyera oblongifolia	52/1	55/1	43/1	3/1
Anemone deltoidea	50/2	3/1	7/2	—
Disporum hookeri	44/2	29/1	34/2	—
Achlys triphylla	41/16	33/6	21/7	—
Smilacina stellata	39/4	1/2	9/5	5/1
Linnaea boreallis	33/17	20/11	15/13	—
Polystichum munitum	20/3	55/4	36/5	3/1
Xerophyllum tenax	8/7	25/6	37/9	26/8
Pteridium aquilinum	21/3	39/3	34/4	—
Whipplea modesta	27/7	36/6	29/8	8/1
Galium aparine	2/2	11/1	27/4	—
Hieracium albiflorum	38/2	20/2	39/2	3/1
Trientalis latifolia	60/3	25/2	37/3	10/1

[a]*Abies concolor* series (ABCO, 331 plots in 24 community types), *Lithocarpus densiflorus* series (LIDE, 251 plots in 21 community types), *Pseudotsuga menziesii* series (PSME, 188 plots in 20 community types), and *Pinus jeffreyi* series (PIJE, 39 plots in community types). T=trace, dash (—) = absent.
Source: Based primarily on Atzet and Wheeler (1984).

and *Pseudotsuga*, *Lithocarpus*, *Arbutus menziesii*, *Castanopsis chrysophylla*, *Quercus chrysolepis*, and *Abies concolor* in the lower tree layers (Table 4.4). Dense evergreen ericaceous shrub layers occur in the *Lithocarpus* series, with dominants such as *Rhododendron macrophyllum*, *Vaccinium ovatum*, *Gaultheria shallon*, and *Berberis nervosa* (Table 4.4). *Polystichum munitum* is the most common herb.

The *Pseudotsuga* series characterizes drier environments on the interior slopes of the Klamath Mountains (Atzet and Wheeler 1984). *Pseudotsuga*, *Pinus lambertiana*, and *Pinus ponderosa* are overstory dominants, and *Castanopsis chrysophylla*, *Quercus chrysolepis*, *Arbutus menziesii*, and *Quercus kelloggii* form a lower tree canopy. *Lithocarpus densiflorus* and *Abies concolor* are associates in some *Pseudotsuga* community types, but they are considerably less important than in their respective series (Table 4.4). *Rhus diversiloba* and other deciduous species are typical components of the shrub layer, along with *Berberis* species, and grasses are a conspicuous element of the herbaceous layer.

The *Abies concolor* series generally occurs at elevations above the *Lithocarpus* series on coastal slopes and on moister and cooler habitats than the *Pseudotsuga* series that is on interior slopes. The *Abies concolor* community types are really conifer types rather than mixed evergreen types and are analogous to some of the Sierran mixed conifer types. They occur in the Klamath Mountains in close association with mixed evergreen types and have some distinctive compositional features (Sawyer et al. 1977). *Picea breweriana* is associated with this series (Atzet and Wheeler 1984). The most common evergreen hardwood associate is *Quercus sadleriana* (Table 4.4).

Forest communities found on ultrabasic rock types, with their poorly developed and chemically unique soils, vary greatly with temperature and moisture regime (Whittaker 1960). *Chamaecyparis lawsoniana* is a characteristic species on wet sites (Zobel et al. 1985). The *Pinus jeffreyi* series is confined to ultrabasic types and occurs across a broad elevational range. Open woodlands of *Pinus jeffreyi* are characteristic, with *Pinus monticola* commonly associated at higher elevations and *Umbellularia californica* on moist sites (Table 4.4). *Arctostaphylos* species and grasses are characteristic understory species.

Successional relationships in mixed evergreen forests are complex. Fire, grazing, and logging have been important in creating the current community mosaic, which includes many multi-aged stands. Major climax species (and community series) are *Pseudotsuga*, *Abies concolor*, and *Lithocarpus* (Atzet and Wheeler 1984). There are many minor climax associates, including *Arbutus*, *Castanopsis*, *Quercus chrysolepis*, *Pinus lambertiana*, and *Chamaecyparis lawsoniana* on various habitats (Thornburgh 1982). The successional status of *Calocedrus* is problematic on many sites; it appears at least as shade-tolerant as *Pseudotsuga* and is represented in smaller size classes within stands that may represent either older suppressed trees or relatively recent reproduction.

On *Lithocarpus* habitat types, two-storied stands of *Pseudotsuga* and *Lithocarpus* are the expected climax. Even-aged stands of *Lithocarpus* and *Pseudotsuga*, either pure or in mixture, typically develop following disturbances. *Lithocarpus* has a distinct advantage because of its ability to sprout from the root crown. *Pseudotsuga*, however is less likely to be killed by ground fires,and it has a faster growth rate which allows it to overtop *Lithocarpus* at 15–30 yr if they become established simultaneously (Thornburgh 1982). Where pure stands of *Lithocarpus* are established, canopy gaps develop after 60–100 yr that permit *Pseudotsuga* establishment. Brushfields of *Ceanothus*, *Arctostaphylos*, and other species may develop in mixed evergreen habitats following fire or other disturbances and can retard establishment of any conifers for several decades.

More southerly versions of mixed evergreen forest, generally with associated conifers other than *Pseudotsuga*, are discussed in Chapter 5.

Sierran-Type Mixed Conifer Forests

Mixed stands of *Pseudotsuga menziesii*, *Pinus lambertiana*, *Abies concolor*, *Pinus ponderosa*, and *Calocedrus decurrens* characterize the montane forests of the southern Cascade Range and eastern Siskiyou Mountains (Franklin and Dyrness 1973). These forests are northern extensions of the Sierran mixed conifer type, which is described in Chapter 5. The transition between the Sierran mixed conifer forest and *Pseudotsuga menziesii–Tsuga heterophylla* forest is gradual, with southern species extending into northern Oregon (e.g., *Pinus lambertiana* and *Calocedrus decurrens* into the Mount Hood region), and northern species extending well into southern Oregon. The McKenzie River drainage, at about 45° N, is often identified as a major transition point.

There is considerable variability in the Sierran mixed conifer forest associated with latitude, elevation, and local site conditions (Griffin 1967; Franklin and Dyrness 1973). *Pinus ponderosa* and *Quercus garryana* are most common at lower elevations, especially adjacent to major valleys. *Quercus chrysolepis* may be a major associate of *Pseudotsuga menziesii* on some sites, producing stands resembling the mixed evergreen forests of the Klamath Mountains. Highly varied forests of *Pseudotsuga*, *Abies*, *Pinus lambertiana*, *Calocedrus decurrens*, and *Quercus kelloggii* characterize middle elevations. Stands at higher elevations, adjacent to subalpine *Abies magnifica* var. *shastensis* forests, are often totally dominated by *Abies concolor*. *Pinus jeffreyi* and *Pinus ponderosa* forests occur at high elevations on the eastern slope of the southern Cascade Range (Rundel et al. 1977), but *Pinus jeffreyi* is absent from the western slope mixed conifer forest in the Cascade Range, in contrast to the situation in the Sierra Nevada.

The *Pseudotsuga menziesii* and *Abies concolor* series characterize most of the Sierra-type mixed conifer forest. The *Pseudotsuga* series typifies hotter, drier habitats. Shrubby understories are common. The *Abies concolor* series dominates the bulk of the mixed conifer region, including modal and moist sites. Understories associated with the *Abies* series vary from shrubby types (with shrubs of either northern or southern origin) to those with a rich array of herbs (e.g., *Abies concolor–Linnaea borealis* community) (Franklin and Dyrness 1973).

Successional relationships in the mixed conifer forests resemble those in the mixed evergreen types discussed previously. Fire, grazing, and logging have been important in creating the current community mosaic, which includes many multi-aged stands. The major climax species are *Abies concolor* and *Pseudotsuga menziesii*. There are many minor associates, because often there are sufficient openings or gaps in the forest to allow for regeneration of less shade-tolerant associates such as *Pinus lambertiana* and *Quercus kelloggii*. *Calocedrus decurrens* appears to be a climax species on many sites, as noted earlier; *Calocedrus* also is bimodally distributed, typically occurring on very dry forest sites and, with *Abies concolor*, on moist to wet sites (and often invading meadows) at relatively high elevations.

SUBALPINE FORESTS AND PARKLANDS

Abies Amabilis–Tsuga Heterophylla Forests

These forests characterize montane regions from the central Oregon Cascades north though the mountains of southern British Columbia and in the Olympic Mountains (Krajina 1965; Fonda and Bliss 1969; Franklin and Dyrness 1973). They are a mixture of temperate-zone and subalpine species (Fig. 4.5). Cooler temperatures and permanent winter snowpacks distinguish the environment of the *Abies amabilis* zone from that of the lower-elevation *Tsuga heterophylla* zone.

Forest compositions are highly variable, depending on stand age, history, and locale (Franklin 1965). *Abies amabilis*, *Tsuga heterophylla*, *Abies procera* (noble fir), *Pseudotsuga menziesii*, *Thuja plicata*, and *Pinus monticola* are typical tree species. In the High Cascades and more continental environments, *Pinus contorta*, *Abies lasiocarpa* (subalpine fir), *Picea engelmannii* (Engelmann spruce), *Abies grandis*, and *Larix occidentalis* (western larch) may occur. *Tsuga mertensiana* and *Chamaecyparis nootkatensis* can be important associates at higher elevations and in more northerly latitudes.

Understories are typically dominated by ericaceous genera, such as *Vaccinium*, *Menziesia*, *Gaultheria*, *Chimaphila*, *Rhododendron*, and *Pyrola*. *Cornus canadensis*, *Clintonia uniflora*, *Rubus pedatus*, *Rubus lasiococcus*, *Linnaea borealis*, and *Xerophyllum tenax* are also common species. The four major groups of

Figure 4.5. Mixed forest of Abies amabilis, Abies procera, *and* Tsuga heterophylla *in the Cascade Range of Oregon, Wildcat Mountain Research Natural Area, Oregon (photo courtesy USDA Forest Service).*

communities are (1) *Vaccinium alaskaense*, (2) herb-rich communities, (3) *Vaccinium membranaceum–Xerophyllum tenax*, and (4) *Gaultheria shallon* (Table 4.5). *Abies amabilis–Vaccinium alaskaense* communities are widely distributed and occur over large areas, from northern Oregon to southern British Columbia and in the Olympic Mountains (Franklin 1966; Henderson and Peter 1982). The herb-rich communities typically have high diversity and coverage of a variety of ferns and dicotyledonous herbs; shrubs may be abundant, as in the *Abies amabilis–Oplopanax horridum* type, or essentially absent. Community types with understories dominated by *Vaccinium membranaceum* or *Xerophyllum tenax* or both and by *Gaultheria shallon* typify, respectively, the snowier and drier *Abies amabilis* types.

Succession typically leads from more diverse mixed forests toward old-growth stands of *Abies*

Table 4.5. *Constancy (%) and average basal area (m^2 ha^{-1}) of trees and constancy and cover (%) of shrubs and herbs in five associations found in the* Abies amabilis *zone of Mount Rainier National Park, Washington*

Growth form and species	ABAM/OPHO[a]		ABAM/TIUN		ABAM/MEFE		ABAM/VAAL		ABAM/GASH	
	Cons.	BA	Cons.	BA	Cons.	BA	Cons.	BA	Cons.	BA
Trees										
Abies amabilis	94	25.6	100	34.4	100	46.1	98	23.4	79	2.5
Tsuga heterophylla	94	40.2	75	20.1	79	21.1	98	33.4	100	40.3
Pseudotsuga menziesii	35	28.0	31	6.4	21	8.0	49	15.9	79	14.9
Chamaecyparis nootkatensis	6	2.7	25	3.9	84	14.1	28	3.2	29	1.0
Thuja plicata	47	16.6	—	—	11	0.1	38	4.9	86	7.5
Tsuga mertensiana	—[b]	—	16	1.3	68	9.3	15	1.0	—	—
Abies lasiocarpa	—	—	1	0.5	—	—	—	—	—	—
	Cons.	Cover.	Cons.	Cover.	Cons.	Cover.	Cons.	Cover.	Cons.	Cover.
Shrubs										
Vaccinium alaskaense	100	11	33	T	54	7	97	28	77	9
Vaccinium membranceum	28	T	80	2	96	12	83	3	85	2
Vaccinium ovalifolium	83	4	68	2	96	8	71	7	46	1
Menziesia ferruginea	56	2	20	1	100	16	65	1	54	1
Berberis nervosa	11	1	6	T	4	T	38	1	100	7
Rubus spectabilis	72	2	40	T	10	T	7	T	7	T
Oplopanax horridum	100	12	33	T	5	T	14	T	—	—
Acer circinatum	22	1	30	3	—	—	28	2	—	—
Ribes lacustre	44	1	13	T	—	—	5	T	—	—
Vaccinium parvifolium	61	1	7	T	—	—	66	33	100	5
Gaultheria shallon	11	T	—	—	—	—	12	T	100	25
Rhododendron albiflorum	—		—	—	63	3	—	—	—	—
Herbs										
Rubus pedatus	94	7	72	1	79	4	87	4	40	T
Clintonia uniflora	94	3	74	3	63	2	78	2	15	T
Rubus lastococcus	44	1	85	5	96	2	68	1	23	T
Linnaea borealis	72	5	13	T	8	T	64	4	92	3
Cornus canadensis	72	3	13	T	10	T	46	1	69	1
Chimaphila umbellata	6	T	20	T	5	T	53	1	100	2
Blechnum spicant	67	5	20	1	15	T	68	4	20	T
Pyrola secunda	50	T	68	1	79	T	72	T	31	T
Viola sempervirens	61	1	78	2	54	T	26	T	46	T
Trillium ovatum	94	T	78	T	42	T	32	T	7	T
Achlys triphylla	92	14	84	14	25	1	38	1	—	—
Streptopus roseus	47	3	72	7	46	1	37	1	—	—
Valeriana sitchensis	22	1	72	4	29	T	5	T	—	—
Athyrium filix-femina	83	3	33	1	5	T	12	T	—	—
Gymnocarpium dryopteris	45	7	50	3	17	T	7	T	—	—
Smilacina stellata	50	2	59	4	13	T	23	T	—	—
Viola glabella	56	1	53	2	13	T	2	T	—	—
Streptopus streptopoides	28	T	48	4	17	T	9	T	—	—
Tiarella nifoliata	83	6	87	15	58	1	47	T	—	—
Adenocaulon bicolor	78	11	—	—	8	T	5	T	—	—
Tiarella trifoliata	67	5	53	1	—	—	9	T	—	—
Majanthemum dilatatum	50	1	20	T	—	—	12	T	—	—
Xerophyllyum tenax	—	—	—	—	54	3	29	1	100	15

[a]ABAM/OPHO = *Abies amabilis–Oplopanax horridum*, ABAM/TIUN = *Abies amabilis–Tiarella unifoliata;* ABAM/MEFE = *Abies amabilis–Menziesia ferruginea;* ABAM/VAAL = *Abies amabilis–Vaccinium alaskaense;* ABAM/GASH = *Abies amabilis–Gaultheria shallon.*

[b]T = trace (less than 0.5% cover); dash (—) indicates species not found.

Source: Data on file with Research Work Unit 1251, Pacific Northwest Forest and Range Experiment Station, Forestry Sciences Laboratory, 3200 Jefferson Way, Corvallis, OR 97331.

amabilis and *Tsuga heterophylla*. *Abies procera*, *Pseudotsuga menziesii*, and *Pinus monticola* are typical seral species. Any species may regenerate directly following a disturbance, although the heavy-seeded, fire-sensitive *Abies amabilis* is often the last to invade a site (Schmidt 1957). *Abies amabilis* is the major climax species, as shown by size- and age-class analyses. Although *Tsuga heterophylla* often dominates stands, it reproduces poorly within the *Abies amabilis* zone, perhaps because of its small, fragile

Figure 4.6. Shrub communities typified by Alnus sinuata *typify snow avalanche tracks and other disturbed habitats within the* Abies amabilis *zone, H.J. Andrews Experimental Forest, Oregon (photo courtesy of USDA Forest Service).*

seedlings that are buried under snow-compressed accumulations of litter (Thornburgh 1969).

Dense shrub communities of *Alnus sinuata* (Sitka alder) (Fig. 4.6) are often associated with *Abies amabilis–Tsuga heterophylla* forests (Franklin and Dyrness 1973). These 3–5-m-tall communities occupy sites subject to deep winter snow accumulations and extensive snow creep; recurrent snow avalanches are common.

Abies Magnifica var. *Shastensis* Forests

Forests characterized by *Abies magnifica* var. *shastensis* (Shasta red fir) dominate the lower portion of the subalpine zone in the southern Cascade Range (to about 44° N) and the Klamath Mountains. Elevations are typically 1600–2000 m in the Cascades and 1800–2200 m in the Klamath Mountains. These are extensions of Sierra Nevada *Abies magnifica* forests (see Chapter 5).

Typical associated tree species are *Abies concolor* and *Tsuga mertensiana* lower and higher in the zone, respectively. *Pinus monticola* and *P. contorta* are other common associates. Many other tree species may occur sporadically. The Klamath endemic, *Picea breweriana*, is noteworthy and may be a local dominant.

The composition and density of understory vegetation in *Abies magnifica* var. *shastensis* stands vary widely. Many stands have relatively depauperate understories; under dense canopies the understory may consist solely of small ericads and orchids, many of which are nearly or completely achlorophyllous–the "*Pirola-Corallorrhiza* union" of Oosting and Billings (1943). On moist sites, Shasta red fir forests typically have luxuriant herbaceous understories incorporating grasses, sedges, and forbs, many of which are also found in associated mountain meadows. Tall ericaceous shrubs, such as *Vaccinium ovalifolium*, may also be present. On colder, more rigorous sites, mixed forests of Shasta red fir and *Tsuga mertensiana* are characterized by depauperate understories of *Vaccinium scoparium*.

Abies magnifica var. *shastensis* appears to be the major climax species. *Abies concolor* is a climax associate on better sites and replaces red fir at lower elevations. *Tsuga mertensiana* is a climax associate on colder sites. *Pinus contorta* and *P. monticola* are primarily seral species.

Tsuga Mertensiana Forests

This is the coldest and snowiest forest zone in the Pacific Northwest. It is a true subalpine environment, with deep, persistent winter snowpacks: 400–1400 cm of snowfall are typical, accumulating in snowpacks up to 7.5 m in depth. The sharp increase in snow accumulation reflects the typical elevation of the freezing isotherm during the winter (Brooke et al. 1970). Temperatures are cool, but much warmer during winter months than in comparable zones in more continental regions.

The *Tsuga mertensiana* zone extends throughout the Cascade Range, the British Columbia Coast Ranges, the Olympic Mountains, the Klamath Mountains, and two-thirds the length of the Sierra Nevada (see Chapter 5). Typical elevations are 1250–1850 m in central Washington and 1700–2000 m in southern Oregon. Throughout much of its distribution, the zone can be divided into a lower subzone of closed forest and an upper parkland subzone. The parkland subzone in considered in the following section.

Tsuga mertensiana provides the continuity for

Figure 4.7. Pure stands of Tsuga mertensiana *are very common in the High Cascades of Oregon, but often have very depauperate understories typified by* Vaccinium scoparium; *proposed Torrey-Charlton Research Natural Area (photo courtesy USDA Forest Service).*

this latitudinally extensive forest formation. Forests of essentially pure *Tsuga* are best developed in the central and southern Oregon Cascade Range, where extensive, undulating topography occurs at appropriate elevations. *Abies amabilis* codominates stands from central Oregon to central British Columbia and in the Olympic Mountains, except at the highest elevations. Other common associates are *Pinus contorta, P. monticola, P. albicaulis, Picea engelmannii,* and *Abies lasiocarpa. Chamaecyparis nootkatensis* is a distinctive associate from northern Oregon northward. *Abies procera* and *Abies magnifica* var. *shastensis* occur with *Tsuga* north and south, respectively, of the McKenzie River in Oregon.

Communities vary dramatically in their diversity. Forests with *Abies amabilis* as an associate typically have greater species and structural diversity (Dyrness et al. 1974; Franklin et al. 1987). Communities dominated by understories of *Xerophyllum tenax* or *Vaccinium membranaceum* or both are common in northern Oregon and southern Washington. Well-developed tall shrub layers composed of species such as *Menziesia ferruginea, Vaccinium ovalifolium, Vaccinium alaskaense,* and *Rhododendron albiflorum* increase in abundance to the north, as do forests with relatively lush understories of dicotyledonous herbs, ferns, and shrubs such as *Oplopanax horridum.*

Tsuga mertensiana stands in the southern Cascade Range are typically depauperate (Fig. 4.7). *Tsuga mertensiana–Vaccinium scoparium* is the most common community type. Stands near the forest line often have a nearly monospecific understory of *Luzula.*

Succession can be very slow following wildfire or other disturbance in the *Tsuga mertensiana* zone. Early successional communities usually are dominated by surviving species, such as *Xerophyllum tenax* or *Vaccinium* species. Reestablishment of closed forest may take a century or more. Successional sequences of tree species vary geographically, with *Pinus* species common in the southern half of the zone. *Abies lasiocarpa*, a major climax species in the Rocky Mountains, is apparently seral when associated with *Tsuga mertensiana* (Franklin and Mitchell 1967). *Tsuga mertensiana* and *T. heterophylla* ranges do overlap, but there is no convincing evidence of hybridization.

Abies amabilis appears to be the major climax species wherever it occurs within the closed forest subzone of the *Tsuga mertensiana* zone; consequently, the *Abies amabilis* series dominates much of the zone. *Tsuga* reproduction is typically much less abundant than that of *Abies amabilis*. *Tsuga* and *Chamaecyparis nootkatensis* can be minor climax associates. In the southern Cascade Range and Klamath Mountains, where more shade-tolerant tree species are absent, *Tsuga* is the major climax species, and the *Tsuga mertensiana* series is characteristic.

Subalpine Parkland

The subalpine meadow-forest mosaic, or parkland, is one to the most distinctive features of the mountains of the Pacific Northwest (Fig. 4.8). As mentioned earlier, deep, late-lying snowpacks are believed to be responsible for the 300–400-m-wide ecotonal band that is recognized here as the upper subzone of the *Tsuga mertensiana* zone. The variety and richness of the meadow flora and communities make the parkland attractive to scientists and laymen alike (Brooke et al. 1970; Douglas 1970, 1972; Kuramoto and Bliss 1970).

Subalpine parklands attain maximal development in the Olympic Mountains and Cascade Range of Washington. Forest patches and tree groups present within the parkland may have a composition different from that of the closed forest subzone (Franklin and Dyrness 1973). The nonforested communities are numerous and varied and largely reflect snowpack duration (Fig. 4.8). Meadows can typically be assigned to one of five broad groups: (1) *Phyllodoce-Cassiope-Vaccinium*, the heath shrub or heather-huckleberry group; (2) *Valeriana sitchensis–Carex spectabilis*, the lush herbaceous group; (3) *Carex nigricans*, the dwarf sedge group; (4) the pioneer and low herbaceous group, and (5) *Festuca viridula*, the grass or dry grass group.

Exemplary data are provided for communities belonging to each of these types in Table 4.6. Heather-huckleberry communities are typically dominated by some combination of *Phyllodoce empetriformis*, *Cassiope mertensiana*, and *Vaccinium deliciosum*. These communities are closely related to heather types of the lower alpine zone. Lush herbaceous types are typically dominated by showy herbs greater than 1 m tall, such as *Valeriana sitchensis* and *Veratrum viride*, as well as the robust *Carex spectabilis*. *Carex nigricans* occupies sites with late-lying snowpacks and cold, wet soils. The simple communities are characterized by nearly pure mats of this short sedge. A series of diverse, poorly developed pioneer communities compose the early pioneer group. *Saxifraga tolmei*, *Luetkea pectinata*, and *Antennaria lanata* are typical dominants, but total plant cover is low. Grassy meadows dominated by *Festuca viridula* or *F. idahoensis* characterize interior subalpine parklands, but they also occur in rainshadow regions of the coastal mountains. Such communities typically contain significant forbs (e.g., *Lupinus latifolius*, *Potentilla flabellifolia*, *Polygonum bistortoides*, *Ligusticum grayi*, *Anemone occidentalis*, and *Aster ledophyllus*). Successional patterns have been hypothesized for both forest groups and meadow communities (Franklin and Mitchell 1967; Douglas 1970; Henderson 1973).

DISTURBANCE PATTERNS

Wildfire and wind are the major natural disturbances in the forests of the Pacific Northwest. The tendency toward infrequent catastrophic disturbances in these forests contrasts with patterns of more frequent, noncatastrophic wildfires typical of coniferous forests in the Sierra Nevada and Rocky Mountains. Insect outbreaks are also of lesser importance in the northwestern coastal forest types. There are some broad gradients in disturbance patterns within the region. Wind increases in relative importance from interior to coastal regions. Fire increases in frequency from north to south.

Large, intense, and infrequent forest fires appear to be the most important natural agent of forest destruction in the Pacific Northwest. Evidence includes the extensive acreages of comparable forest age classes and records of forest fires dating from the early 1800s. An analysis of fire patterns in Mount Rainier National Park, Washington (Hemstrom and Franklin 1982) identified 16 important fire events since the year 1230 and a natural fire rotation of 434 yr. The largest episode was in 1230 and affected over 47% of the park. The Yacholt Burn in southwestern Washington (initial fire in 1902) and Tillamook Burn in coastal Oregon (initial fire in 1933) are historical demonstrations that fires can cover thousands of hectares in a very short time when conditions are appropriate. Trees frequently survive within such burns as individuals, groups and small stands, and so revegetation, including reestablishment of trees, can be rapid. Such burned sites have a tendency to reburn in the century following the initial fire, however, and repeatedly burned sites reforest slowly because of reduced seed supplies (fewer survivors), increased

Figure 4.8. An attractive parkland, or mosaic of meadow and tree communities, typifies the upper portion of the Tsuga mertensiana *zone in the Olympic Mountains and Cascade Range; Butler Creek Research Natural Area, Mount Rainier National Park, Washington (photo courtesy USDA Forest Service).*

competing vegetation, and a more severe physical environment for tree seedlings. (Large-scale deforestation can drastically modify macroclimate, as well as microclimate, such as by reducing precipitation levels in regions where condensation in tree crowns is a significant process (Harr 1982). Repeated wildfires are believed to be one factor contributing to the wide range in age classes typical

Table 4.6. *Mean prominence values (average percentage cover multiplied by square root of frequency) for selected shrubs and herbs in subalpine meadow communities in the northern Washington Cascade Range*

	Communities[a]					
Species	Came-Phem	Vasi-Vevi	Casp	Cani	Lupe	Sato
Cassiope mertensiana	441	—[b]	—	11	11	T
Phyllodoce empetriformis	386	—	—	2	15	1
Vaccinium deliciosum	92	—	T	T	21	1
Luetkea pectinata	73	—	2	21	502	—
Lycopodium sitchense	16	—	—	T	16	—
Deschampsia atropurpurea	9	T	T	15	44	1
Polygonum bistortoides	1	8	30	T	9	—
Valeriana sitchensis	5	305	15	—	47	T
Carex spectabilis	1	52	782	6	41	2
Veratrum viride	—	290	2	—	T	—
Lupinus latifolius	4	59	42	—	—	—
Carex nigricans	2	—	16	803	13	1
Epilobium alpinum	—	T	T	32	10	3
Hieracium gracile	3	—	1	3	36	4
Luzula wahlenbergii	—	—	—	2	10	47
Potentilla flabellifolia	—	4	7	1	13	—
Castilleja parviflora	T	—	—	T	16	4
Anemone occidentalis	T	—	2	—	11	T
Saxifraga tolmei	—	—	—	T	—	78

[a]Came-Phem = *Cassiope mertensiana–Phyllodoce empetriformis*; Vasi-Vevi = *Valeriana sitchensis–Veratrum viride*; Casp = *Carex spectabilis*; Cani = *Carex nigricans*; Lupe = *Luetkea pectinata*; Sato = *Saxifraga tolmei*.
[b]T = trace amounts; dash (—) indicates absence.
Source: Douglas (1972).

of old-growth stands of *Pseudotsuga* (Franklin and Hemstrom 1981).

The frequencies of wildfires, both light and catastrophic, vary both locally and regionally. Natural fire rotation is shorter, for example, in the central Oregon Cascade Range than at Mount Rainier. Noncatastrophic wildfires do occur, creating gaps and thinning forest canopies; these can result in very complex forest structures (Stewart 1984). More frequent, lower-intensity wildfires apparently increase with decreasing latitude, and fire regimes south of the Willamette-Umpqua River divide begin to approach those of the Sierra Nevada. Partial burns are important factors in maintaining *Pseudotsuga menziesii* as a part of coastal *Sequoia sempervirens* stands (Viers 1982).

Wind is important in catastrophic and chronic disturbances. Series in coastal *Picea sitchensis–Tsuga heterophylla* forests in southeastern Alaska and coastal British Columbia are typically initiated by major windstorms. Catastrophic windstorms occur in other areas as well. Henderson and Peter (1982) have reported that windthrow, rather than fire, has been the most important agent of disturbance on the western Olympic Peninsula; a particularly destructive storm occurred on the peninsula in 1934. Ruth and Yoder (1953) have documented catastrophic windstorms on the Oregon coast. The Columbus Day (October 12, 1962) windstorm affected both the Coast and Cascade ranges of Oregon and Washington, destroying several billion board-feet of timber. The chronic importance of wind varies regionally between the coastal *Picea sitchensis* forests and the *Pinus ponderosa* forests east of the Cascade Range (Franklin et al. 1986). Wind-related factors are responsible for around 80% of within-stand mortality in *Picea sitchensis–Tsuga heterophylla* forests, 40% in Cascadian *Pseudotsuga menziesii–Tsuga heterophylla* forests and less than 20% in *Pinus ponderosa* forests.

Pathogens can create significant disturbances in some situations, but generally they are not as important as they are in many other western conifer forests. Outbreaks of bark beetles occur, but they typically kill only individuals or groups of trees. Outbreaks of defoliators are uncommon and rarely appear to threaten stands. Several diseases may create patch-wise mortality (e.g., *Phellinus wierii*), but leave stands basically intact. Several introduced pathogens (a root rot on *Chamaecyparis lawsoniana*, an aphid on several species of *Abies*, and blister rust on five-needled pines) have seriously disrupted individual species, but, again, they have rarely eliminated entire stands. Thus, pathogens appear to be chronic, rather than catastrophic, agents of disturbance.

Disturbance types and intensities have significant effects on paths and rates of succession and on ecosystem processes. For example, fire initially produces large numbers of standing dead trees and tends to kill from below (i.e., killing smaller and less fire-resistant specimens), thereby favoring regeneration of *Pseudotsuga menziesii* and other pioneer species. Windthrow, on the other hand, generates downed logs, rather than snags, and tends to eliminate larger specimens, leaving most of the seedlings and saplings of shade-tolerant species untouched. Therefore, windthrow accelerates succession toward the climax tree species, whereas fire favors reestablishment of early successional tree species. Because snags and logs fulfill different wildlife functions and decay at different rates (snags decompose three to four times as fast as logs of comparable size in the Pacific Northwest forests), wildfire and wind storms will also have significantly different impacts on wildlife populations and nutrient and energy cycling.

Clear-cutting is, of course, the most common current agent of disturbance in Pacific Northwest forests, and it differs markedly from wildfire in its effects. Considerable research has been conducted on succession following clear-cutting (West and Chilcote 1968; Dyrness 1973), and on the effects on erosion and nutrient losses. Rates and paths of succession are altered by planting and other cultural practices, such as elimination of nonarboreal species. Furthermore, logging usually removes most snags and logs, eliminating their potential functional roles.

Generalized successional relationships have been discussed in the sections on individual forest types. Additional interpretations can be made for individual tree species utilizing relative shade tolerances (Table 4.1). Two cautionary notes are essential, however. First, shade tolerance is a physiological feature of each species, whereas successional role is dependent on a community context. For example, *Pinus ponderosa* is a shade-intolerant tree species, but it may play either a seral or climax role, depending on whether or not more shade-tolerant tree species are capable of growing on a specific site. In general, shade tolerance is required for climax status on sites that can develop closed canopies. Second, few climax tree species require that a seral species precede to ameliorate site conditions. Hence, most climax species can also function as pioneers, although they may be unable to compete with faster-growing seral species early in succession. As an example, *Tsuga heterophylla* can dominate young stands following fire or clear-cutting over much of the Pacific Northwest, but *Pseudotsuga menziesii* is more common.

Much attention is currently focused on relations between forest types and successional stages and usage by vertebrates (Brown 1985). Both early (before tree canopy closure) and late (old-growth) stages in succession have been identified as periods of special interest because of higher levels of diversity or of special-interest species, or both, than are found in young forest stands. Several bird and mammal species find optimum habitats in old-growth *Pseudotsuga* forests, for example, and require special management consideration (Franklin et al. 1981; Harris 1984). One key to the special role of old-growth forests is their structural complexity compared with younger forest ecosystems.

ECOSYSTEM CHARACTERISTICS

Biomass

The outstanding structural feature of forests in the Pacific Northwest is the huge biomass accumulation typically present (Franklin and Dyrness 1973), which results from long-lived species capable of growing to very large sizes and from high productivity. Values for aboveground live biomass are typically in the range of 500–1000 Mt ha^{-1} (Table 4.7), exceeding values for temperate deciduous forests and tropical rain forests by factors of 2 to 4 (Franklin and Waring 1981). *Sequoia sempervirens* forests hold the world record for maximum biomass, with a basal area of 343 m^2 ha^{-1} and a stem biomass of 3461 Mt ha^{-1} (Fujimori 1977); with addition of branch, leaf, and root biomass, the estimate of standing crop would probably approach Fujimori's (1972) estimate of 4525 Mt ha^{-1}. Although maximum values for *Pseudotsuga menziesii* and *Abies procera* are less than half those for *Sequoia* (Fujimori et al. 1976), they still greatly exceed maxima for any other forests.

Foliage is a particularly important biomass component. Leaf biomass and surface area require decades to recover to maximal levels following disturbances in northwestern conifer forests, a very slow rate compared with those for forests in the eastern United States, where equilibria may be reached in less than a decade. Ultimately, the levels of foliage mass and area are typically very large in mature and old-growth stands (Table 4.7). Values again exceed those for other temperate and tropical forests by a factor greater than 2 (Franklin and Waring 1981). The high values for leaf surface area reflect the large needle surface areas associated

Table 4.7. *Aboveground total biomass and leaf biomass and projected leaf area (one side only) for four forest types in the Pacific Northwest*

Forest type and age class	Number of stands	Aboveground biomass Average (Mt ha^{-1})	Aboveground biomass Range (Mt ha^{-1})	Leaf mass (Mt ha^{-1})	Projected leaf area (m^{2}m^{-2})
Pseudotsuga menziesii (70–170 yr)	10	604	422–792	19	9.7
Pseudotsuga menziesii–Tsuga heterophylla	19	868	317–1423	23	11.7
Picea sitchensis–Tsuga heterophylla	4	1163	916–1492	21	13.2
Abies procera	1	880	—	18	10

Source: Franklin and Waring (1981).

with individual old-growth trees: Nine old-growth *Pseudotsuga* trees averaged 2850 m^2 total needle surface per tree (Massman 1982). Leaf surface areas are strongly related to site water balances (Grier and Running 1977; Gholz 1982) and to temperature regimes (Waring et al. 1978). Maximum values are usually found on sites at middle elevations with favorable moisture regimes and moderate air and soil temperatures.

Coarse woody debris, primarily standing dead trees and downed logs, is increasingly recognized as an important organic structure in forests and streams, particularly in the Pacific Northwest (Franklin et al. 1981; Maser and Trappe 1984). This woody debris plays significant roles in energy and nutrient cycling, geomorphic processes, and provision of habitat for terrestrial and aquatic organisms (Harmon et al. 1986). Large masses of such material are typically present in natural forests of all ages, because few catastrophes consume or remove much wood from trees that are killed. Tonnages in old-growth stands average 75–100 Mt ha^{-1} and may range to more than 500 Mt ha^{-1} (Franklin and Waring 1981).

Productivity and Nutrient Cycling

Productivity of forest stands in the Pacific Northwest is generally comparable to that in other temperate forest regions (Table 4.8). Biomass in young stands accumulates at 15–25 Mt ha^{-1} yr^{-1} in fully stocked stands on better than average sites. Mature and old-growth stands have lower net productivities. On the best sites, particularly in *Sequoia* and *Tsuga-Picea* stands, annual net productivity is as high as at any place on earth (Fujimori 1971, 1977). A record net annual production of 36.2 Mt ha^{-1} yr^{-1} occurred in a 26-yr-old *Tsuga heterophylla* stand on the Oregon coast.

Such high productivities are exceptional, however. Annual productivity in many other mesic, temperate forests around the world equals or exceeds that for average sites in the Pacific Northwest. Greater biomass accumulations are primarily the result of sustained height growth and longevity of the dominant trees in the Northwest, not the result of superior annual productivity. This growth is aided by the trees' ability to accumulate and maintain a large amount of foliage. Trees of northwestern species continue to grow substantially in diameter and height, and stands continue to increase in biomass, long after forests in other temperate regions have reached equilibrium. This is well illustrated by comparing growth of *Pinus taeda* (loblolly pine) in the Southeast and *Pseudotsuga* in the Northwest. Wood production from a single 100-yr rotation of *Pseudotsuga* is about 22% greater than from two 50-yr rotations of pine (Worthington 1954). Recent studies of height growth patterns of several northwestern conifers show that height growth may be sustained into their second and third centuries.

Patterns of nutrient cycling have been described for several northwestern forest types (Sollins et al. 1980; Edmonds 1982). Typical and important features of these cycles are very large nutrient pools and the "tightness" of the forests, as indicated by low nutrient losses. Winter decomposition is important because of mild winter temperatures and moisture limitations on decomposition processes during the relatively dry summer season. Nitrogen is generally considered the limiting nutrient on most sites, and study of sources and losses of nitrogen is a major research topic. Numerous sources for nitrogen fixation have been identified in recent years, including (1) shrubs and trees with nitrogen-fixing symbionts, such as *Alnus* and *Ceanothus* species and *Purshia tridentata* (Tarrant et al. 1967; Jarmillo 1985), (2) canopy lichens with blue-green algal associates (Carroll 1980), (3) microbial organ-

Table 4.8. *Aboveground net primary production estimates for forests west of the Cascade Range crest in Oregon and Washington*

Forest type	Stand age (yr)	Biomass (Mt ha^{-1})	Net primary production (Mt ha^{-1} yr^{-1})	Source
Abies amabilis	23	49	6.4[a]	Grier et al. (1981)
Tsuga heterophylla	26	192	36.2	Fujimori (1971)
Pseudotsuga menziesii	40	248, 467	7.3, 13.7[b]	Keyes & Grier (1981)
Pseudotsuga menziesii–Tsuga heterophylla	100	661	12.7	Fujimori et al. (1976)
Tsuga heterophylla–Picea sitchensis	110	871	10.3	Fujimori et al. (1976)
Abies procera–Pseudotsuga menziesii	115	880	13.0	Fujimori et al. (1976)
Pseudotsuga menziesii	125	449	6.2	Gholz (1982)
Picea sitchensis–Tsuga heterophylla	130	1080, 1492	15, 13	Gholz (1982)
Pseudotsuga menziesii–Tsuga heterophylla	150	527, 865	9.5, 10.5	Gholz (1982)
Abies amabilis	180	446	4.6[a]	Grier et al. (1981)
Pseudotsuga menziesii–Tsuga heterophylla	450	718	10.8	Grier & Logan (1977)

[a]Total net primary production was 18.3 and 16.8 Mt ha^{-1}yr^{-1} in young and old stands, respectively; belowground accounted for 65% and 73%, respectively, of those totals.
[b]Total net primary production was 15.4 and 17.8 Mt ha^{-1} on high- and low-quality sites; belowground accounted for 65% and 73%, respectively, of those totals.

isms in rotting wood (Harmon et al. 1986), (4) organisms living in the rhizosphere, and (5) free-living organisms associated with decaying leaf litter.

Research on hydrologic cycling has focused on effects of forest cutting on stream flow and water quality (Rothacher 1970). Forest removal typically results in increased water yields, particularly during summer low-flow periods, as a consequence of reduced transpirational losses. Condensation of fog or cloud moisture in tree canopies results in substantial amounts of fog drip in some coastal and mountain forests, however, and forest cutting may reduce water yields under such circumstances. In one study in the Oregon Cascade Range, for example, fog drip added 30% or 88 cm of precipitation to the 216 cm received in the open (Harr 1982). The deep crowns and large surface areas of needles, lichens, twigs, and branches (Pike et al. 1977) found in old-growth forests make them particularly effective as condensing and precipitating surfaces for moisture, nutrients, and pollutants.

PALEOECOLOGICAL CONSIDERATIONS

Paleoecological research on the vegetation of the Pacific Northwest includes studies of fossil floras (Chaney 1956; Axelrod 1976), pollen profiles (Hansen 1947; Heusser 1960; Baker 1983; Heusser 1983), glacial records (Burke and Birkeland 1983; Porter et al. 1983; Waitt and Thorson 1983), volcanic-ash depositions (Mullineaux 1974; Sarna-Wojcicki et al. 1983), and tree-ring records (Brubaker and Cook 1983). These approaches have been particularly successful in reconstructing Quaternary vegetational history. Some interesting linkages between native peoples and vegetation development have been discovered, such as between *Thuja* expansion and evolution of a woodworking technology (Hebda and Mathewes 1984).

Daubenmire (1978) has summarized vegetational development up to the Quaternary. The Arcto-Tertiary geoflora was an important ancestral formation. This flora comprised a widespread and complex temperate forest in the warm period at the close of the Eocene. The mixed hardwood and coniferous forests included *Abies*, *Chamaecyparis*, *Calocedrus*, *Picea*, *Pinus*, *Pseudotsuga*, and *Tsuga*. Cooling and the rise of mountain ranges during the Oligocene, Miocene, and Pliocene resulted in development of a xerophytic flora, northward expansion and incorporation of some elements of the Madro-Tertiary geoflora, and loss of most of the hardwood tree species and genera.

Both continental and alpine glaciations were important during the Pleistocene. The continental ice sheet occupied the Puget Trough to a few kilometers south of Olympia, Washington, and affected additional areas by creating outwash plains and channels and lakes through damming of river valleys (Waitt and Thorson 1983). The maximum extent of the continental ice sheet during the Fraser Glaciation was achieved at 22,000–18,000 BP. Alpine glaciation was extensive in both the Olympic Mountains and Cascade Range (Burke and Birkeland 1983; Porter et al. 1983). The histories of the glaciations are complex, and the patterns and extents of recent glaciations vary substantially among mountain ranges (Burke and Birkeland 1983). There have been numerous studies of glacial fluctuations in the Pacific Northwest post-1800 CE.

Recent vegetational histories are based largely on pollen records (Baker 1983). In the lowlands, these indicate the occurrence of tundra and taiga-like vegetation associated with glaciation about 17,000 BP, development of subalpine-type forests at 12,500 BP, a warmer and drier period around 10,000–6000 BP, and subsequent development of typical coniferous forests. The "hypsithermal maximum" was 4000–6000 yr ago (Hansen 1947).

Palynological reconstructions are aided by the widespread occurrence of layers of tephra in soils and sediments (Sarna-Wojcicki et al. 1983). "Tephra" refers to the whole array of materials, solid and liquid, erupted from a volcanic vent and aerially deposited. Some tephras are distinctive petrographically and chemically, and that allows identification of their sources. Once identified and dated, such layers are used for temporal correlations and age dating (i.e., in "tephrochronology" and "tephrostratigraphy"). Twelve ash layers of moderate to large volumes have been identified in the Pacific Northwest during the last 13,000 yr. Seven layers are associated with Mount St. Helens, and three with Glacier Peak, and a single very extensive layer is from Mount Mazama (Crater Lake).

Tree-ring analyses have been most useful in reconstruction of very recent fluctuations in climate (Brubaker and Cook 1983). Tree ages have also dated glacial fluctuations (Sigafoos and Hendricks 1972) and meadow invasions (Franklin et al. 1971).

Interactions between geomorphic processes and vegetation are providing active areas for study in the Pacific Northwest. These include relations between forests and fluvial processes (Swanson and Lienkaemper 1982), landslides and earthflows, and, of course, volcanoes. The May 18, 1980, eruption of Mount St. Helens has provided an exceptional laboratory for studying recovery processes, including the effects of geomorphic processes such as erosion (Franklin et al. 1985).

AREAS FOR FUTURE RESEARCH

There are many exciting directions for ecological research in the forests of the Pacific Northwest. Development of an immense data base, through the USDA Forest Service's area ecology program, has provided an outstanding opportunity for major syntheses of forest community patterns within the region. Detailed studies on patterns, rates, and mechanisms in succession are needed, including considerations of small-scale disturbances and gaps. Research on ecosystem processes has only begun, with many topics needing attention, such as belowground and canopy processes, nitrogen dynamics, forest-stream interactions (including riparian vegetation), and effects of herbivores. The development of structural complexity in stands, as represented by dead wood structures and overall heterogeneity, is also an important and underexplored topic.

Development of the necessary research infrastructure is more critical than any specific research topic. Long-term data bases must be developed for a variety of population, community, and ecosystem processes, from the population dynamics of selected organisms on permanent sample plots to water chemistry of benchmark watersheds, as a basis for formulating and definitively testing important hypotheses. Development of these data bases is a collective responsibility of the scientific community. Such long-term research is under way at several sites in the region, including the Hoh River valley of Olympic National Park, Washington, and the H. J. Andrews Experimental Forest in the central western Cascades of Oregon, a site supported by the National Science Foundation as part of its long-term ecological research program.

Ecological reserves need to be identified and protected for scientific study of both natural and manipulated ecosystems. Many efforts are under way, such as those of federal and state agencies and The Nature Conservancy, but continued support and encouragement are needed. A major obligation of the scientific community is to use reserved areas for research whenever possible.

REFERENCES

Alaback, P. B. 1982. Forest community structural changes during secondary succession in southeast Alaska, pp. 70–79 in J. E. Means (ed.), Forest succession and stand development research in the northwest. Oregon State University Forest Research Laboratory, Corvallis.

Atzet, T., and D. L. Wheeler. 1984. Preliminary plant associations of the Siskiyou Mountain Province. USDA Forest Service, Pacific Northwest Region, Portland, Ore.

Axelrod, D. I. 1976. History of the coniferous forests, California and Nevada. University of California Press, Berkeley.

Baker, R. G. 1983. Holocene vegetational history of the western United States, pp. 109–127 in H. E. Wright, Jr. (ed.), Late-Quaternary environments of the United States, vol. 2, The Holocene.

University of Minnesota Press, Minneapolis.

Brooke, R. C., E. B. Peterson, and V. J. Krajina. 1970. The subalpine mountain hemlock zone. Ecol. West. North Am. 2:147–349.

Brown, E. R. (ed.). 1985. Management of wildlife and fish habitats in forests of western Oregon and Washington, Part I, chapter narratives. USDA Forest Service, Pacific Northwest Region, Portland, Ore.

Brubaker, L. B., and E. R. Cook. 1983. Tree-ring studies of Holocene environments, pp. 222–238 in H. E. Wright, Jr. (ed.), Late-Quaternary environments of the United States, vol. 2, The Holocene. University of Minnesota Press, Minneapolis.

Burke, R. M., and P. W. Birkeland. 1983. Holocene glaciation in the mountain ranges of the western United States, pp. 3–11 in H. E. Wright, Jr. (ed.), Late-Quaternary environments of the United States, vol. 2, The Holocene. University of Minnesota Press, Minneapolis.

Carroll, G. C. 1980. Forest canopies: complex and independent subsystems, pp. 87–107 in R. H. Waring (ed.), Forests: fresh perspectives from ecosystem research. Oregon State University Press, Corvallis.

Chaney, R. W. 1956. The ancient forests of Oregon. University of Oregon Press, Eugene.

Daubenmire, R. 1978. Plant geography. Academic Press, New York.

Douglas, G. W. 1970. A vegetation study in the subalpine zone of the western north Cascades, Washington. M.S. thesis, University of Washington, Seattle.

Douglas, G. W. 1972. Subalpine plant communities of the western north Cascades, Washington. Arct. Alp. Res. 4:147–166.

Dyrness, C. T. 1973. Early stages of plant succession following logging and burning in the western Cascades of Oregon. Ecology 54:57–69.

Dyrness, C. T., J. F. Franklin, and W. H. Moir. 1974. A preliminary classification of forest communities in the central portion of the western Cascades in Oregon. U.S. International Biological Program Coniferous Forest Biome Bull. 4:1–123.

Edmonds, R. L. (ed.). 1982. Analysis of coniferous forest ecosystems in the western United States. Hutchinson Ross Publishing, Stroudsburg, Pa.

Emmingham, W. H., and R. H. Waring. 1977. An index of photosynthesis for comparing forest sites in western Oregon. Can. J. Forest Res. 7:165–174.

Eyre, F. H. (ed.). 1980. Forest cover types of the United States and Canada. Society of American Foresters, Washington, D.C.

Fonda, R. W., and L. C. Bliss. 1969. Forest vegetation of the montane and subalpine zones, Olympic Mountains, Washington. Ecol. Monogr. 39:271–301.

Fowells, H. A. 1965. Silvics of forest trees of the United States. USDA handbook 271.

Franklin, J. F. 1965. Tentative ecological provinces within the true fir-hemlock forest areas of the Pacific Northwest. USDA Forest Service Research paper PNW-22.

Franklin, J. F. 1966. Vegetation and soils in the subalpine forests of the southern Washington Cascade Range. Ph.D. thesis, Washington State University, Pullman.

Franklin, J. F., K. Cromack, Jr., W. Denison, A. McKee, C. Maser, J. Sedell, F. Swanson, and G. Juday. 1981. Ecological characteristics of old-growth Douglas-fir forests. USDA Forest Service general technical report PNW-118.

Franklin, J. F., and C. T. Dyrness. 1973. Natural vegetation of Oregon and Washington. USDA Forest Service general technical report PNW-8.

Franklin, J. F., and M. A. Hemstrom. 1981. Aspects of succession in the coniferous forests of the Pacific Northwest, pp. 219–229 in D. C. West, H. H. Shugart, and D. B. Botkin (eds.), Forest succession: concepts and application. Springer-Verlag, New York.

Franklin, J. F., M. Klopsch, K. Luschessa, and M. Harmon. 1986. Tree mortality in some mature and old-growth forests in the Cascade Range of Oregon and Washington. Unpublished manuscript, Forestry Sciences Laboratory, Corvallis, Ore.

Franklin, J. F., J. A. MacMahon, F. J. Swanson, and J. R. Sedell. 1985. Ecosystem responses of Mount St. Helens. National Geogr. Res. 1:198–216.

Franklin, J. F., and R. G. Mitchell. 1967. Successional status of subalpine fir in the Cascade Range. USDA Forest Service research paper PNW-46.

Franklin, J. F., W. H. Moir, G. W. Douglas, and C. Wiberg. 1971. Invasion of subalpine meadows by trees in the Cascade Range, Washington and Oregon. Arct. Alp. Res. 3:215–224.

Franklin, J. F., W. H. Moir, M. A. Hemstrom, and S. Greene. 1987. Forest ecosystems of Mount Rainier National Park. National Park Service Scientific monograph.

Franklin, J. F., and R. H. Waring. 1981. Distinctive features of the northwestern coniferous forest: development, structure, and function, pp. 59–86 in R. H. Waring (ed.), Forests: fresh perspectives from ecosystem research. Oregon State University Press, Corvallis.

Fujimori, T. 1971. Primary productivity of a young *Tsuga heterophylla* stand and some speculations about biomass of forest communities on the Oregon coast. USDA Forest Service research paper PNW-123.

Fujimori, T. 1972. Discussion about the large forest biomasses on the Pacific Northwest in U.S.A. J. Jpn. Forestry Soc. 54:230–233.

Fujimori, T. 1977. Stem biomass and structure of a mature *Sequoia sempervirens* stand on the Pacific coast of northern California. J. Jpn. Forestry Soc. 59:435–441.

Fujimori, T., S. Kawanabe, H. Saito, C. C. Grier, and T. Shidei. 1976. Biomass and primary production in forests of three major vegetation zones of the northwestern United States. J. Jpn. Forestry Soc. 58:360–373.

Gholz, H. L. 1982. Environmental limits on aboveground net primary production, leaf area, and biomass in vegetation zones of the Pacific Northwest. Ecology 63:469–481.

Grier, C. C. 1978. A *Tsuga heterophylla–Picea sitchensis* ecosystem of coastal Oregon: decomposition and nutrient balance. Can. J. Forest Res. 8:198–206.

Grier, C. C., and R. S. Logan. 1977. Old-growth *Pseudotsuga menziesii* communities of a western Oregon watershed: biomass distribution and production budgets. Ecol. Monogr. 47:373–400.

Grier, C. C., and S. W. Running. 1977. Leaf area of mature northwestern coniferous forests: relation to site water balance. Ecology 58:893–899.

Grier, C. C., K. A. Vogt, M. R. Keyes, and R. L. Edmonds. 1981. Biomass distribution and above- and below-ground production in young and mature *Abies amabilis* zone ecosystems of the Washington Cascades. Can. J. Forest Res. 11:155–167.

Griffin, J. R. 1967. Soil moisture and vegetation patterns in northern California forests. USDA Forest Service research paper PSW-46.

Hansen, H. P. 1947. Postglacial forest succession, climate, and chronology in the Pacific Northwest. Trans. Amer. Phil. Soc. New Series 37:1–130.

Harmon, M. E. 1986. Logs as sites of tree regeneration in *Picea sitchensis–Tsuga heterophylla* forests of Washington and Oregon. Ph.D. thesis, Oregon State University, Corvallis.

Harmon, M. E., J. F. Franklin, F. J. Swanson, et al. 1986. Ecology of coarse woody debris in temperate ecosystems. Adv. Ecol. Res. 15:133–302.

Harr, R. D. 1982. Fog drip in the Bull Run municipal watershed. Water Resources Bull. 18:785–789.

Harris, L. D. 1984. The fragmented forest: island biogeography theory and the preservation of biotic diversity. University of Chicago Press.

Hebda, R. J., and R. W. Mathewes. 1984. Holocene history of cedar and native Indian cultures of the North American Pacific coast. Science 225:711–713.

Hemstrom, M. A., and J. F. Franklin. 1982. Fire and other disturbances of the forests in Mount Rainier National Park. Quat. Res. 18:32–51.

Hemstrom, M. A., and S. E. Logan. 1984. Preliminary plant association and management guide Siuslaw National Forest. Willamette National Forest, Eugene, Ore.

Henderson, J. A. 1973. Composition, distribution, and succession of subalpine meadows in Mount Rainier National Park, Washington. Ph.D. thesis, Oregon State University, Corvallis.

Henderson, J. A., and D. Peter. 1982. Preliminary plant associations and habitat types of the Soleduck Ranger District, Olympic National Forest. USDA Forest Service, Region 6, Portland, Ore.

Heusser, C. J. 1960. Late-Pleistocene environments of North Pacific North America. American Geographical Society special publication 35.

Heusser, C. J. 1983. Vegetational history of the northwestern United States including Alaska, pp. 239–258 in S. C. Porter (ed.), Late-Quaternary environments of the United States, vol. 1, The late Pleistocene. University of Minnesota Press, Minneapolis.

Jarmillo, A. (ed.). 1985. Biology of *Ceanothus*. USDA Forest Service general technical report PNW-182.

Keyes, M. R., and C. C. Grier. 1981. Above- and below-ground net production in 40-year-old Douglas-fir stands on low and high productivity sites. Can. J. Forest Res. 11:599–605.

Krajina, V. J. 1965. Biogeoclimatic zones and classification of British Columbia, pp. 1–17 in V. J. Krajina (ed.), Ecology of western North America, vol. 2. University of British Columbia Department of Botany, Vancouver.

Krajina, V. J. 1969. Ecology of forest trees in British Columbia. Ecol. West. North Am. 2:1–147.

Krajina, V. J., K. Klinka, and J. Worrall. 1982. Distribution and ecological characteristics of trees and shrubs of British Columbia. University of British Columbia Faculty of Forestry, Vancouver.

Kuchler, A. W. 1946. The broadleaf deciduous forests of the Pacific Northwest. Ann. Assoc. Amer. Geogr. 36:122–147.

Kuramoto, R. T., and L. C. Bliss. 1970. Ecology of subalpine meadows in the Olympic Mountains, Washington. Ecol. Monogr. 40:317–347.

Long, J. N., and J. Turner. 1975. Aboveground biomass of understory and overstory in an age sequence of four Douglas-fir stands. J. Appl. Ecol. 12:179–188.

McKee A., G. La Roi, and J. F. Franklin. 1982. Structure, composition, and reproductive behavior of terrace forests, South Fork Hoh River, Olympic National Park, pp. 19–29 in E. E. Starkey, J. F. Franklin, and J. W. Matthews (eds.), Ecological research in national parks of the Pacific Northwest. Oregon State University Forest Research Laboratory, Corvallis.

McMinn, R. G. 1960. Water relations and forest distribution in the Douglas-fir region on Vancouver Island. Canadian Department of Agriculture publication 1091.

Maser, C., and J. M. Trappe. 1984. The seen and unseen world of the fallen tree. USDA Forest Service general technical report PNW-164.

Massman, W. J. 1982. Foliage distribution in old-growth coniferous tree canopies. Can. J. Forest Res. 12:10–17.

Mullineaux, D. R. 1974. Pumice and other pyroclastic deposits in Mount Rainier National Park, Washington. U.S. Geol. Surv. Bull. 1326:1–83.

Nadkarni, N. 1983. The effects of epiphytes on nutrient cycles within temperate and tropical rainforest tree canopies. Ph.D. thesis, University of Washington, Seattle.

Oosting, H. J., and W. D. Billings. 1943. The red fir forest of the Sierra Nevada: *Abietum magnificae*. Ecol. Monogr. 13:259–274.

Pike, L. H., R. A. Rydell, and W. C. Denison. 1977. A 400-year-old Douglas fir tree and its epiphytes: biomass, surface area, and their distributions. Can. J. Forest Res. 7:680–699.

Porter, S. C., K. L. Pierce, and T. D. Hamilton. 1983. Late Wisconsin mountain glaciation in the western United States, pp. 71–111 in S. C. Porter (ed.), Late-Quaternary environments in the United States, vol. 1, The late Pleistocene. University of Minnesota Press, Minneapolis.

Rothacher, J. 1970. Increases in water yield following clear-cut logging in the Pacific Northwest. Water Resources Res. 6:653–658.

Rundel, P. W., D. J. Parsons, and D. T. Gordon. 1977. Montane and subalpine vegetation of the Sierra Nevada and Cascade Ranges, pp. 559–599 in M. G. Barbour and J. Major (eds.), Terrestrial vegetation of California. Wiley, New York.

Ruth, R. H., and R. A. Yoder. 1953. Reducing wind damage in the forests of the Oregon Coast Range. USDA Forest Service, Pacific Northwest, Forest and Range Experiment Station research paper 7.

Sarna-Wojcicki, A. M., D. E. Champion, and J. O. Davis. 1983. Holocene volcanism in the coterminous United States and the role of silicic volcanic ash layers in correlation of latest-Pleistocene and Holocene deposits, pp. 52–77 in H. E. Wright, Jr. (ed.), Late-Quaternary environments of the United States, vol. 2, The Holocene. University of Minnesota Press, Minneapolis.

Sawyer, J. O., and D. A. Thornburgh. 1977. Montane and subalpine vegetation of the Klamath Mountains, pp. 699–732 in M. G. Barbour and J. Major (eds.), Terrestrial vegetation of California. Wiley, New York.

Sawyer, J. O., D. A. Thornburgh, and J. R. Griffin. 1977. Mixed evergreen forest, pp. 359–381 in M. G. Barbour and J. Major (eds.), Terrestrial vegetation of California. Wiley, New York.

Schmidt, R. L. 1957. The silvics and plant geography of the genus *Abies* in the coastal forests of British Columbia. British Columbia Forest Service technical publication T46.

Sigafoos, R. S., and E. L. Hendricks. 1972. Recent activity of glaciers of Mount Rainier, Washington. U.S. Geological Survey professional paper 387-B.

Sollins, P., C. C. Grier, F. M. McCorison, K. Cromack, Jr., R. Fogel, and R. L. Fredricksen. 1980. The internal element cycles of an old-growth Douglas-fir ecosystem in western Oregon. Ecol. Monogr. 50:261–285.

Spilsbury, R. H., and D. S. Smith. 1947. Forest site types of the Pacific Northwest. British Columbia Forest Service publication T30.

Starkey, E. E., J. F. Franklin, and J. W. Matthews (eds.). 1982. Ecological research in national parks of the Pacific Northwest. Oregon State University Forest Research Laboratory, Corvallis.

Stewart, G. H. 1984. Forest structure and regeneration in the *Tsuga heterophylla–Abies amabilis* transition zone, central western Cascades, Oregon. Ph.D. thesis, Oregon State University, Corvallis.

Stone, E. C., and R. B. Vasey. 1968. Preservation of coast redwood on alluvial flats. Science 159:157–161.

Swanson, F. J., and G. W. Lienkaemper. 1982. Interactions among fluvial processes, forest vegetation, and aquatic ecosystems, South Fork Hoh River, Olympic National Park, pp. 30–34 in E. E. Starkey, J. F. Franklin, and J. W. Matthews (eds.), Ecological research in national parks of the Pacific Northwest. Oregon State University Forest Research Laboratory, Corvallis.

Tarrant, R. F., J. M. Trappe, and J. F. Franklin (eds.). 1967. Biology of alder. USDA Forest Service, Pacific Northwest, Forest and Range Experiment Station, Portland, Ore.

Thornburgh, D. A. 1969. Dynamics of the true fir-hemlock forests of the west slope of the Washington Cascade Range. Ph.D. thesis, University of Washington, Seattle.

Thornburgh, D. A. 1982. Succession in the mixed evergreen forests of northwestern California, pp. 87–91 in J. E. Means (ed.), Forest succession and stand development research in the northwest. Oregon State University Forest Research Laboratory, Corvallis.

Viers, S. D., Jr. 1982. Coast redwood forest: stand dynamics, successional status, and the role of fire, pp. 119–141 in J. E. Means (ed.), Forest succession and stand development research in the northwest. Oregon State University Forest Research Laboratory, Corvallis.

Waitt, R. B., Jr., and R. M. Thorson. 1983. The Cordilleran ice sheet in Washington, Idaho, and Montana, pp. 53–70 in S. C. Porter (ed.), Late-Quaternary environments of the United States, vol. 1, The late Pleistocene. University of Minnesota Press, Minneapolis.

Waring, R. H., W. H. Emmingham, H. L. Gholz, and C. C. Grier. 1978. Variation in maximum leaf area of coniferous forests in Oregon and its ecological significance. Forest Sci. 24:131–140.

Waring, R. H., and J. F. Franklin. 1979. Evergreen coniferous forests of the Pacific Northwest. Science 204:1380–1386.

Waring, R. H., and J. Major. 1964. Some vegetation of the California coastal redwood region in relation to gradients of moisture, nutrients, light, and temperature. Ecol. Monogr. 34:167–215.

West, N. E., and W. W. Chilcote. 1968. *Senecio sylvaticus* in relation to Douglas-fir clear-cut succession in the Oregon Coast Range. Ecology 49:1101–1107.

Westman, W. E., and R. H. Whittaker. 1975. The pygmy forest region of northern California: studies on biomass and primary productivity. J. Ecol. 63:493–520.

Whittaker, R. H. 1960. Vegetation of the Siskiyou Mountains, Oregon and California. Ecol. Monogr. 30:279–338.

Worthington, N. 1954. The loblolly pine of the south versus the Douglas fir of the Pacific Northwest. Pulp Paper 28:87–90.

Zinke, P. J. 1977. The redwood forest and associated north coast forests, pp. 679–698 in M. G. Barbour and J. Major (eds.), Terrestrial vegetation of California. Wiley, New York.

Zobel, D. B., A. McKee, G. M. Hawk, and C. T. Dyrness. 1976. Relationships of environment to composition, structure, and diversity of forest communities of the central western Cascades of Oregon. Ecol. Monogr. 46:135–156.

Zobel, D. B., L. F. Roth, and G. M. Hawk. 1985. Ecology and management of Port-Orford-cedar. USDA Forest Service general technical report PNW-184.

Chapter 5

Californian Upland Forests and Woodlands

MICHAEL G. BARBOUR

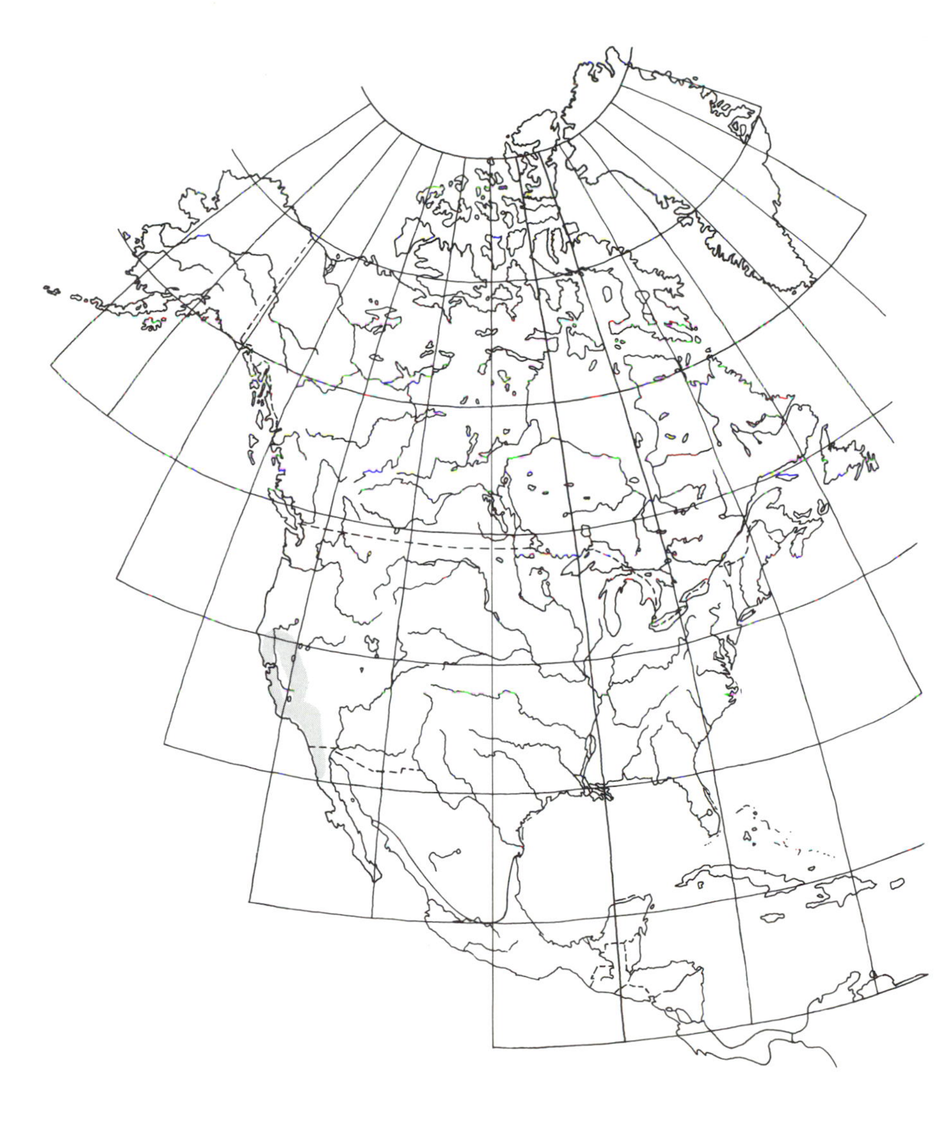

INTRODUCTION

"California" is used here to indicate the area within the "Californian Floristic Province," as defined and discussed by Stebbins and Major (1965) and Raven and Axelrod (1978). In the northern part of the state, the California province increasingly blends with an Oregonian province. For the sake of convenience, several vegetation types in that region are included in Chapter 4. The coastal low-elevation strip of redwood forest is also discussed in Chapter 4. "Upland," as used in the chapter title, means that low-elevation vegetation such as the riparian forest of the Central Valley is not included. Only woodland and forest vegetation is discussed, vegetation defined by UNESCO (1973) and Paysen and associates (1982) as dominated by trees at least 5 m tall and with a tree canopy covering at least 25% of the ground.

The part of California that is considered in this chapter (Fig. 5.1) totals approximately 170,000 km^2, or 42% of the state's area. In addition, the Californian Floristic Province extends south into Baja California, largely along the western flank of the Sierra

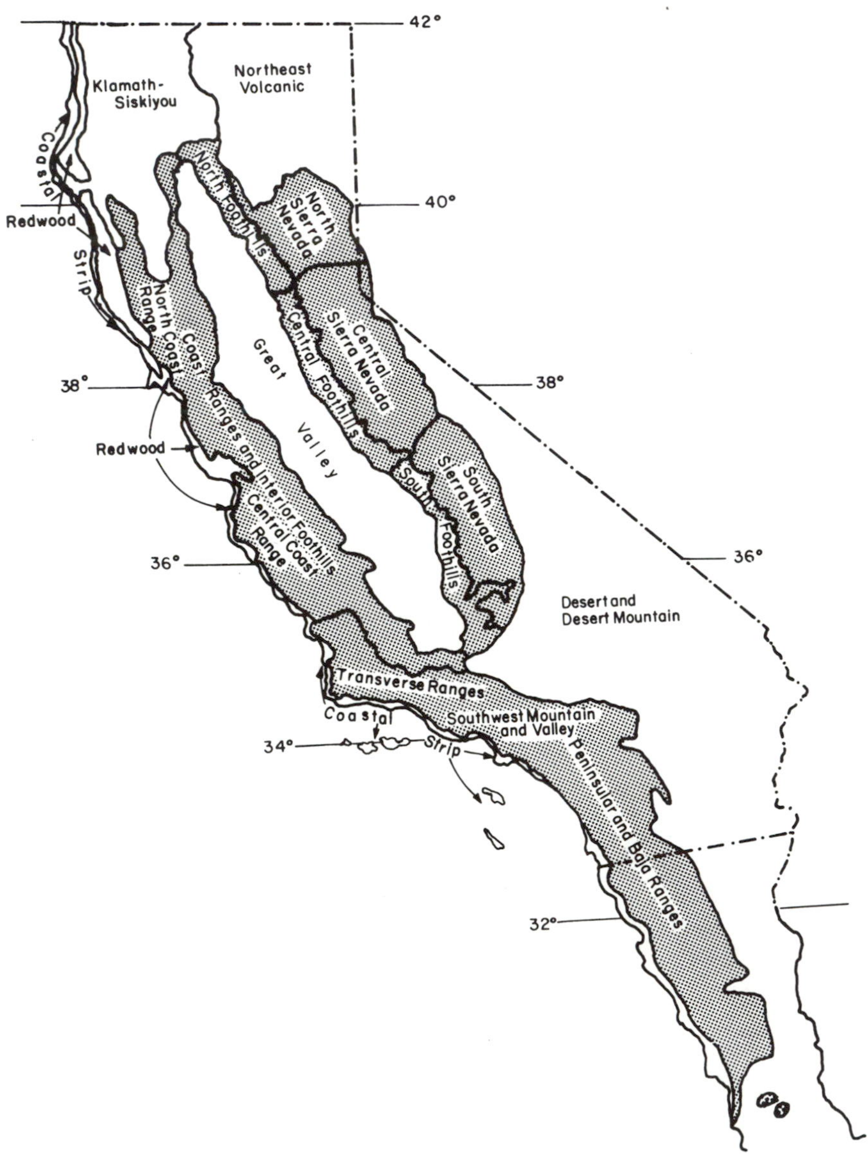

Figure 5.1. Regions of California and Baja California discussed in this chapter. Major landscape province names and boundaries are taken from Mason (1970) and Nelson (1922).

Juařez and the Sierra San Pedro Mártir, for an additional area of 27,000 km^2, and the vegetation of those areas is also discussed here.

I have divided the woodlands and forests into seven regional or elevational classes, and then into 20 communities that I call "phases," in order to stress the continuum that embraces them (Table 5.1). Eight taxa suffice to serve as threads to define the seven classes: *Quercus douglasii* for the blue oak woodland, *Q. agrifolia* for the southern oak woodland, *Q. chrysolepis* for the mixed evergreen forest, *Pinus ponderosa* and *P. jeffreyi* for the midmontane and eastside conifer forests, *P. contorta* var. *murrayana* for the upper montane conifer forest, and *P. albicaulis* and *P. flexilis* for the subalpine woodland. My class names (Table 5.1) could be modified to incorporate these species names, but I have tried to select names that already have widespread acceptance. Tree nomenclature follows Eyre (1980), and all other nomenclature follows Munz and Keck (1959). An elevational summary for some of the communities discussed appears in Fig. 5.2.

There have been several intensive efforts to classify California vegetation on a regional basis in the past decade, and Table 5.1 illustrates four such attempts. In general, the classification scheme used in this chapter agrees well with those four. I find less agreement with classification attempts by Küchler (1977), who defined and mapped "potential, natural formations," by W. J. Barry (unpublished), who developed an ecosystem classification in 1973 for use by the California Department of Parks and Recreation, and by Matyas and Parker (1980), who mapped 43 units that were a mixture of series and subformations. Matyas and Parker's CALVEG map is at the same 1 : 1-million scale as Küchler's map, but their emphasis on existing rather than potential vegetation gives a significantly different picture of the state.

The Nature Conservancy has recently published a checklist of 375 "natural communities" in cooperation with the state of California and unpublished work by Cheatham and Haller (Holland 1986). Communities are named and defined by the species that dominate the overstory and by words that relate to the habitat or the locale. An extension of this approach, with even more units, is the continuing USFS Soil-Vegetation Mapping Program (Colwell 1977). More than 4.5 million ha of upland vegetation have been mapped on over 300 quadrangle sheets. Units as small as 4 ha are delineated. Finally, a good summary of progress to date in mapping vegetation on the basis of Landsat imagery is available (CDFR 1979).

The geologic histories of California forests and woodlands have been especially well documented and interpreted. The space limitations of this chapter prohibit a full discussion of the fossil record, but three recent reviews can be recommended (Axelrod 1976, 1977; Raven and Axelrod 1978).

BLUE OAK WOODLAND

This community forms a nearly continuous ring around the Central Valley of California, generally between 100 and 1200 m in elevation. Common synonyms include digger pine–oak forest or woodland, pine-oak woodland, and foothill woodland. The name I have chosen emphasizes the dominant species (*Quercus douglasii*). Küchler's (1964) map of United States vegetation shows 9–10% of California to be dominated by this type (Barbour and Major 1977); however, his description is so broad that it includes part of the mixed evergreen forest and all of the southern oak woodland. Probably this type covers less than 8% of the state.

This is essentially a two-layered community (Fig. 5.3). An overstory canopy, 5–15 m tall, is 30–80% closed, and blue oak, with an importance value above 200 (Brooks 1971; Vankat and Major 1978), is dominant. Associated trees include coast and interior live oaks (*Q. agrifolia* toward the coast, *Q. wislizenii* toward the interior), digger pine (*Pinus sabiniana*), and two deciduous oaks, valley oak (*Q. lobata*) at lower elevations with shallow water tables and black oak (*Q. kelloggii*) at higher elevations or on mesic slopes. Sapling and tree densities combined usually total less than 200 ha^{-1}, but dense stands can reach 1000 ha^{-1} (Griffin 1977). A somewhat shorter, more spreading deciduous tree, *Aesculus californica*, occurs as scattered individuals or in small clumps. There are significant differences in tree compositions between Coast Range and Sierran phases of blue oak woodland (Table 5.2). Tree life spans (except for *Q. lobata*) are less than 300 yr, and tree girths are modest, averaging 20–30 cm dbh (diameter breast height).

Shrubs 1–2 m tall are regularly present, but cover is insignificant, usually 5%. Common genera include *Arctostaphylos, Ceanothus, Cercis, Heteromeles, Rhamnus*, and *Toxicodendron*. *Cercis* and *Toxicodendron* are deciduous. The herbaceous ground stratum, now composed mainly of introduced annual grasses and of native annual and perennial forbs, averages over 80% cover (Table 5.2).

Blue oak woodland is not significantly different from valley grassland, parkland, or chaparral in terms of temperature means or extremes. Based on my survey of 17 oak woodland sites near U.S. weather stations (USDC 1964), mean annual temperature is 16°C, mean annual amplitude (warmest month minus coldest month) is 19°C, and mean annual pre-

Table 5.1. *Community types discussed in this chapter and comparable names applied by other classification schemes*

This chapter	Cheatham & Haller (1973)	Paysen et al. (1980, 1982)	Kyre (1980) (SAF no.)	Thorne (1976)
Blue oak woodland				Foothill woodland
Blue oak phase	Blue oak woodland	*Quercus douglasii/ Pinus sabiniana*	Blue oak–digger pine (250)	
Coast live oak phase	Coast live oak forest	*Quercus agrifolia*	Coast live oak (255)	
Interior live oak phase		*Quercus wislizenii*		
Southern oak woodland				Southern oak woodland
Coast live oak phase	Southern coastal oak woodland	*Quercus agrifolia*	Coast live oak (255)	
Engelmann oak phase	Southern interior oak woodland	*Quercus engelmannii*	Coast live oak (255)	
Mixed evergreen forest				
Douglas fir–hardwood phase	Mixed evergreen forest	*Pseudotsuga/ hardwoods*	Douglas fir–tan oak–madrone (234)	Northern mixed evergreen forest
Coulter pine–hardwood phase	Coulter pine forest, big-cone spruce forest	*Quercus chrysolepis/ Pinus coulteri*		Coulter pine forest, southern mixed evergreen forest
Mixed hardwood phase	Mixed evergreen forest	*Lithocarpus/Arbutus*	California black oak (246)	
Canyon live oak phase	Canyon live oak forest		Canyon live oak (249)	
Black oak phase	Northern oak woodland	*Quercus kelloggii*	California black oak (246)	Northern oak woodland
Midmontane conifer forest				
Mixed conifer phase	Westwide Sierran ponderosa pine forest, Coast Range mixed conifer forest, Sierran mixed conifer forest	*Pseudotsuga/Pinus, Pinus ponderosa, Abies concolor/ Pinus*	Ponderosa pine–Douglas fir (244), ponderosa pine (245), Sierra Nevada mixed conifer (243)	Yellow pine forest, mixed conifer forest
White fir phase	Sierran white fir forest, southern California white fir forest	*Abies concolor*	California white fir (211)	White fir–sugarpine forest
Big-tree phase	Sierra big-tree forest			
Jeffry pine phase	Jeffrey pine forest, Jeffrey pine–fir forest	*Pinus jeffreyi*	Jeffrey pine (247)	Yellow pine forest
Upper montane conifer forest				
Lodgepole pine phase	Lodgepole pine forest	*Pinus contorta*	Sierra lodgepole pine (218)	Lodgepole pine forest
Red fir phase	Red fir forest	*Abies magnifica*	Red fir (207)	Red fir forest
Aspen parkland phase		*Populus tremuloides*		Aspen woodland
Mixed subalpine woodland				
Whitebark pine phase	Foxtail pine forest, Sierran mixed subalpine forest, whitebark pine–Lodgepole pine forest	*Pinus albicaulis, P. balfouriana, Tsuga mertensiana*	California mixed subalpine (256)	Whitebark pine–mountain hemlock forest
Limber pine phase	Southern California subalpine forest	*Pinus flexilis*	Limber pine (219)	Limberpine forest
Sierran east-side montane forest	Jeffrey pine forest, Jeffrey pine–fir forest	*Pinus jeffreyi*	Jeffrey pine (247)	Yellow pine forest

Note: A blank space indicates that the type was not specifically identified as a separate type, but it may have been included within another type.

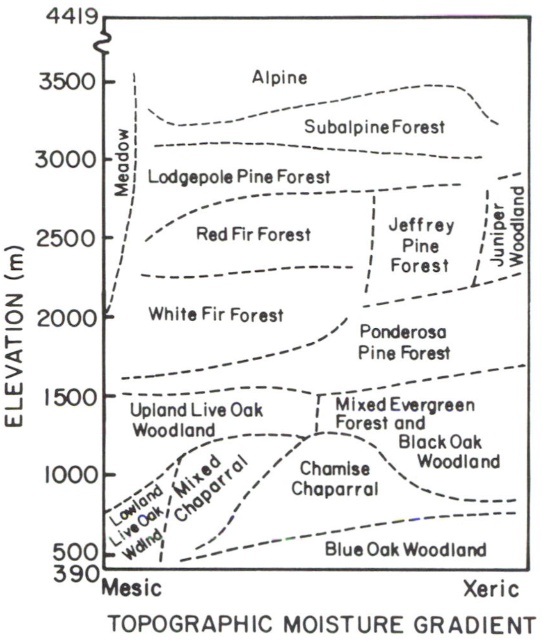

Figure 5.2. Diagrammatic representation of community positions along elevational, topographic, and moisture gradients for the Sierra Nevada (redrawn from Vankat 1982).

cipitation is 530 mm (range 280–1000). Oak woodland occurs on moderately rich, loamy, well-drained soils with neutral or slightly basic pH, in the Inceptisol, Alfisol, and Mollisol orders. Topography is often gently rolling to steep (10–30% slope). Oak woodland often occurs in a mosaic with grassland, parkland (<25% tree cover), and chaparral – a mosaic that reflects differences in slope, aspect, soil depth, and frequency of fire more than differences in climate.

Phases

The Coast Range phase of blue oak woodland, with *Q. agrifolia*, has already been mentioned in Table 5.2. Other phases may be adjacent to each other, as shown by recent studies in the Sierra Nevada (Brooks 1971; Vankat and Major 1978; Baker et al. 1981). An interior live oak phase may occur both below blue oak woodland, in somewhat riparian habitats, and above it, on steeper slopes. The tree canopy tends to be more closed, tall shrubs may increase, and herbaceous ground cover may decrease (Table 5.3). Increasing elevation may add canyon live oak (*Q. chrysolepis*), *Q. kelloggii*, and

Figure 5.3. Typical aspect of blue oak woodland, with Pinus sabiniana a prominent associate. Napa County, about 150 m elevation.

Table 5.2. *Relative tree densities in Coast Range and Sierran phases of blue oak woodland.*

Species	Coast Ranges	Sierran
Quercus douglasii	36	54
Q. agrifolia	36	0
Q. lobata	8	2
Pinus sabiniana	6	19
Q. kelloggii	5	4
Aesculus californica	2	18
Q. wislizenii	3	0.5
	Absolute density ha^{-1}	
All trees	159	171
	Herbaceous cover (%)	
All herbs	89	80

Note: Based on several hundred 800-m^2 plots placed in the VTM survey.
Source: Griffin (1977).

conifers (*Pinus ponderosa, Pseudotsuga menziesii*) to form a transition to mixed evergreen forest. On other slopes or on exposed ridges, a black oak phase of blue oak woodland may occur at 1000–1500 m elevation; in this phase, tree and shrub canopy covers are not greatly changed, but herb cover declines to 30–65% (Table 5.3). Formally, I include canyon live oak and black oak phases within the mixed evergreen forest.

Similarly, at increasing elevations in the Coast Range, blue oak woodland is replaced by a xeric mixed evergreen forest, dominated by *Pinus coulteri, Q. agrifolia, Q. chrysolepis*, and *Q. kelloggii* (Griffin 1977; Sawyer et al. 1977). Well within the blue oak woodland zone, but in riparian-like habitats, is a live oak–buckeye phase, dominated by *Q. agrifolia* or *Q. wislizenii* and *Aesculus californica*, and with *Q. douglasii* regularly present (Bowerman 1944).

Axelrod (1965), Major (1977), Myatt (1980), and Vankat (1982) have documented gradients in precipitation, mean annual temperature, and annual amplitude of temperature with elevation that correspond with the upper limit of blue oak woodland. There is also some evidence (Dunn 1980; Baker et al. 1981) that soil C : N ratios and contents of N, P, Ca, and organic matter rise as one moves upslope out of blue oak woodland. Within blue oak woodland, there is conflicting evidence for a positive or negative impact of blue oak on the growth and nutritional content of associated understory species (Holland 1980; Holland and Morton 1980; Kay and Leonard 1980).

Stand Dynamics

White (1966) made a detailed study of 38 ha of blue oak woodland in the Central Coast Range, Monterey County. Two blue oak age groups were evident: 60–100 yr old and 150–260 yr old. Successful establishment had been declining for the past 90 yr, and essentially no establishment had taken place in the last 30 yr. Establishment, then, appeared to be episodic, with the most recent flush having occurred in the 1870s. Vankat and Major (1978), working from historical records and photographs in the Sequoia–Kings Canyon area, also concluded that there had been a flush of blue oak establishment in the 1870s.

In a series of papers, Griffin (1971, 1976, 1977, 1980a) concluded that major causes of deciduous oak seedling mortality were inability of the root to penetrate compact soils, summer drought away from tree canopies on southern exposures, and browsing by pocket gophers, aboveground rodents, deer, and cattle. Thousands of seedlings were marked,

Table 5.3. *Stand structures for three phases of blue oak woodland in Sequoia National Park, 400–1500 m elevation*

	Blue oak woodland				Lowland live oak				Black oak woodland			
Species	C	D	BA	IV	C	D	BA	IV	C	D	BA	IV
Quercus douglasii	56	636	12	238	18	100	2	52	11	233	2	69
Aesculus californica	6	50	2	34	32	150	2	60	5	33	2	34
Q. wislizenii	2	14	0.4	14	73	400	12	188	1	33	0.6	26
Q. kelloggii	4	7	1	13	—	—	—	—	60	300	13	172
Total	68	707	15	300	123	650	16	300	77	599	18	300

Note: Based on 18 stand surveys of variable areas and on 20 transects, each 2 × 50 m, by Vankat and Major (1978). C = absolute cover (%), D = density ha^{-1}, BA = basal area (m^2 ha^{-1}); IV = importance value = relative frequency + relative basal area + relative density. *Pinus sabiniana* is absent from this part of the Sierra Nevada.

and only those on north-facing slopes, in partial shade, and protected from all grazing animals survived for 6 yr, at a density of 9 m^{-2}. Mortality was 100% in all other cases. The flush of establishment in the 1870s could have coincided with low herbivore population numbers and optimal fall germination conditions, followed by mild summers. Rodent, deer, and livestock populations currently are all too high to permit significant establishment. Deciduous oaks (*Q. douglasii* and *Q. lobata*) show less successful establishment than evergreen oaks (*Q. agrifolia*).

Stand structure may, of course, reflect episodic disturbances in the form of ground or crown fires. There has been surprisingly little research on assessing the role of natural fire frequency in oak woodlands. It is known that all oak tree species are capable of sprouting in California, though *Q. lobata* and *Q. douglasii* lose that ability once they reach a certain mass or age (Griffin 1980b; Plumb 1980).

In terms of water relations, *Q. douglasii* is much more xerophytic than *Q. lobata* or *Q. agrifolia*. In Coast Range woodlands, Griffin (1973) showed that dry-season xylem water potentials in *Q. douglasii* ranged down to −2.6 MPa (pre-dawn) and −4.0 MPa (midday), and the average summer pre-dawn water potential was −1.1 MPa over the course of three summers. At the same time, nearby *Q. lobata* and *Q. agrifolia* trees averaged only −0.4 MPa. In a Sierran woodland, Rundel (1980) and Baker and associates (1981) showed that peak late-summer pre-dawn potentials for *Q. douglasii*, *Q. wislizenii*, and *Q. chrysolepis* all averaged −2.0 MPa.

It is clear from Rundel's (1980) review of oak ecology that we have insufficient data to define or compare photosynthetic rates, productivity, or patterns of biomass allocation of California oak trees. We cannot yet assume that the generally held tenets about photosynthetic differences between evergreen and deciduous species (Mooney 1972; Larcher 1980; Chabot and Hicks 1982) apply to California oaks. These oaks provide an excellent test for such hypotheses, and they even offer hybrids between evergreen and deciduous species (e.g., *Q. × morehus*, a common hybrid of *Q. kelloggii* and *Q. wislizenii*) (Tucker (1980).

SOUTHERN OAK WOODLAND

This vegetation class occurs in the outer portion of the Central Coast Range, beginning near the northern border of San Luis Obispo County (35° 45′) and extending south into the Transverse Ranges, where it occupies north-facing and coast-facing slopes and ravines below 1200 m elevation. It also occurs in interior valleys and on gentle foothill slopes of the Peninsular Ranges, mainly at 150–1400 m elevation, continuing south to the Sierra San Pedro Mártir, 30° N, on western slopes below 2000 m.

Southern oak woodland has a physiognomy and stand architecture similar to those of blue oak woodland. The overstory is 9–22 m tall and incompletely closed, and the understory herbaceous layer approaches 80% cover. In the Coast Range, *Q. agrifolia* is the major dominant, but in the Transverse, Peninsular,and Baja California ranges, it is associated with (and sometimes subordinated to) two deciduous species: California walnut (*Juglans californica*, especially from Orange County to Santa Barbara County), and mesa or Engelmann oak (*Quercus engelmanii*, especially in an 80-km-wide belt running north-south about 30 km from the coast, from Los Angeles County to San Diego County) (Pavlik 1976). Synonyms for these phases within southern oak woodland include coast live oak woodland, walnut-oak woodland, and Engelmann oak woodland.

Axelrod's (1977) analysis of climatic relationships among California oak woodlands suggests that the southern California types differ from each other, and from blue oak woodland, in terms of warmth (ET, effective temperature) and equability (M, a measure of temperature amplitude during the year) (Fig. 5.4). Equability declines (that is, the amplitude of annual temperature increases) in the following order: coast live oak, walnut woodlands, Engelmann oak, blue oak. The mean annual temperature is still 16°C, but annual fluctuations are reduced in the south, because of a maritime influence and summer fog, to only 10°C (USDC 1964). Mean annual precipitation remains about 530 mm, as in blue oak woodland. However, my analysis of 10 oak woodland weather stations in Baja California at elevations 900–1400 m, using data from Hastings and Humphrey (1969), showed only 260 mm annual precipitation, mean annual temperature of 15°C, and annual amplitude of 14°C. The differences are due to higher elevation and more southern latitude.

According to Griffin's (1977) review, few descriptive data on stands for the southern oak woodland exist. Walnut woodlands have been largely disturbed, modified, or supplanted because of human activity. The best description of the interior phase comes from work in the Santa Ana mountains and from vegetation type map (VTM) surveys conducted by Wieslander in Riverside and San Diego counties in the 1930s, as summarized by Griffin (1977). Only 10% of VTM plots located in oak woodland contained *Q. engelmanii*; the rest were dominated by *Q. agrifolia*. Some Engelmann oak stands were parkland, with 27 trees ha^{-1}, but

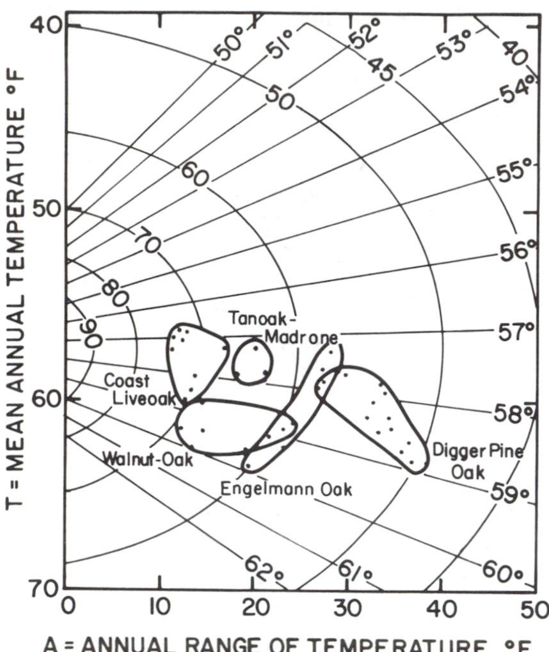

Figure 5.4. Thermal relations of five woodland/forest types in California. Dots represent meteorological stations for each type. Radii are lines of warmth, or effective temperature; arcs represent temperateness, or equability, reflecting departures from 100, a yearly constant 14°C (57°F) (redrawn from Axelrod 1977).

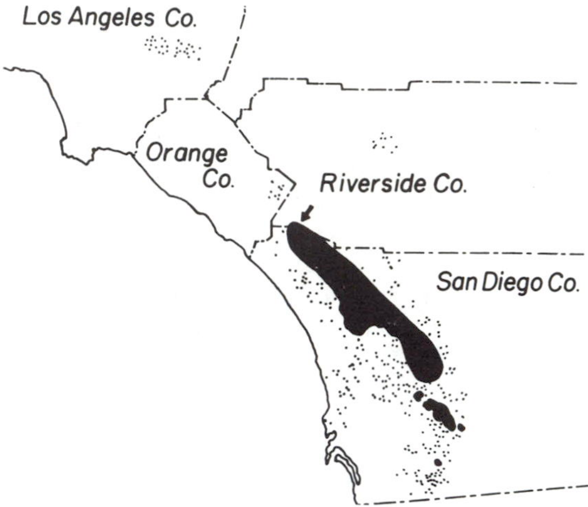

Figure 5.5. Range of Quercus engelmannii *in California. Arrow shows proposed area for acquisition by the state of California for a mesa oak reserve. The range of mesa oak continues south into the Sierra Juárez of Baja California to about 31° 40′ N (redrawn from Schmidle 1983).*

others were denser, with 50–150 trees ha^{-1}, and these were either with equal densities of Engelmann and coast live oak or with a mix of Engelmann, coast, and black oaks in a 63 : 25 : 12 ratio. The Engelmann oak diameter distribution is much like that for blue oak, with most trees being in the 20–30-cm dbh class.

In general, *Q. agrifolia* is more abundant on steeper or moister slopes, and *Q. engelmannii* on gentler, more arid slopes. This pattern continues into Baja California, where *Q. engelmannii* occupies dry slopes below 1200 m, whereas *Q. agrifolia* ranges up to 2000 m (Wiggins 1980). In the more open, parkland stands, 90% of the trees are *Q. engelmannii.* Both Engelmann and coast live oaks resprout following fire or cutting, *Q. engelmannii* doing so most vigorously (Snow 1980).

The Nature Conservancy, in cooperation with the state of California, has put both Engelmann oak and walnut woodlands in their highest-priority class for acquisition of vegetation/habitat data and for inclusion in protected natural areas (Holstein 1981). They are viewed as two of the state's 32 most endangered plant communities. The California Department of Parks and Recreation is considering acquisition of 400–800 ha of Engelmann oak woodland at the northern edge of the Santa Ana Mountains (Fig. 5.5). The woodland lies on the Santa Rosa Plateau, at about 600 m elevation. The oaks are 14 m tall and 60 cm dbh maximum; they occur in dense woodlands, open parkland stands, or in riparian corridors (Schmidle 1983). Other excellent Engelmann oak stands are on the slopes of Mt. Palomar near Lake Henshaw (D. I. Axelrod, pers. commun.).

MIXED EVERGREEN FOREST

This community is between oak woodland below and midmontane conifer forests above; consequently, the list of characteristic species can be long, taking in species that range considerably above and below the mixed evergreen forest. The floristic composition of the type and the distributional limits of its various phases are well known, but quantitative descriptions are few. The lack of stand data is surprising, given the forest's wide distribution, covering 3–4% of California's area (Barbour and Major 1977). Cooper (1922) included it in his "broad sclerophyll forest formation," and synonyms for some of its phases include canyon oak woodland, Coulter pine forest, big-cone spruce forest, mixed hardwood forest, Douglas fir–hardwood forest, tan oak–madrone forest, and Santa Lucia fir forest. To the north, this type grades into a Douglas fir forest, described in Chapter 4, pages 113–116.

Küchler's (1977) brief description of the forest's structure is an excellent beginning: "Low to medium tall, broad-leaved evergreen forest with an admixture of broad-leaved deciduous and needle-leaved evergreen trees; the latter may be towering above the canopy. The forest is more or less dense . . . ." A review by Sawyer and associates (1977) identified three trees as common throughout the forest's range—*Acer macrophyllum, Quercus chrysolepis,* and *Umbellularia californica*—but I would emphasize the oak as the most common thread, much as *Q. douglasii* is the identifying thread for the blue oak woodland and its northern phases. The maple and bay are more representative of mesic canyons and are not as widespread as the oak.

The coniferous overstory, when present, is generally scattered and 30–60 m in height. Beneath it is a more completely closed canopy, 15–30 m tall, of broad-leaved evergreen trees with scattered deciduous trees. Both canopies together may contribute 40–100% cover (Fig. 5.6). Shrub, moss, and perennial herb cover (5–25%) is greater than for oak woodlands. In some hardwood phases, the community is essentially one-layered, the ground covered with a thick mat of undecomposed leaf litter, and with shrubs and herbs largely absent.

As defined in this chapter, mixed evergreen forest extends in a broken ring around the Central Valley, facing the valley on the Sierran slopes at 600–1200 m elevation, but away from the valley in the Coast Ranges, and there expanding its zone to 300–1500 m, depending on proximity to the ocean. In the Transverse, Peninsular, and Baja ranges, the forest exists between 900 and 1400 m elevation (Wright 1968; Minnich 1976; Vogl 1976; Thorne 1977; Wiggins 1980).

Climatic data for 11 mixed evergreen forest stations (USDC 1964; Major 1967, 1977; Talley 1974; McDonald 1980; Wainwright and Barbour 1984) showed the mean annual temperature to be 14°C, significantly cooler than in blue oak woodland, and the mean annual precipitation to be 870 mm, nearly 40% above that for oak woodland. In Axelrod's (1977) thermal scheme, tan oak–madrone does not seem to differ significantly from the cluster of oak woodland phases (Fig. 5.4). Myatt (1980), however, found that the various Sierran phases of mixed evergreen forest fell out rather well between oak woodland and mixed conifer on an elevational/moisture gradient (Fig. 5.7).

Figure 5.6. Mixed evergreen forest, Santa Lucia Range, 950 m elevation. Dominants include Arbutus menziesii, Quercus chrysolepis, *and* Umbellularia californica.

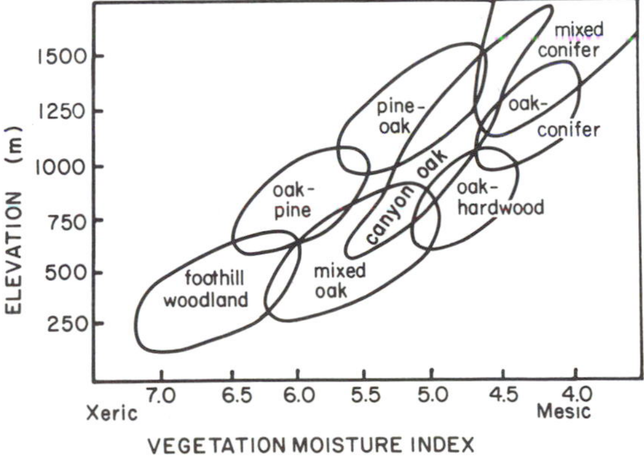

Figure 5.7. Vegetation relationships between various blue oak woodland phases and mixed conifer forest in the Stanislaus River region of the Sierra Nevada (redrawn from Myatt 1980).

Douglas Fir–Hardwood Phase

This phase occurs at relatively low elevations in Marin and Sonoma counties and at higher elevations in the Sierra Nevada. One detailed study was done at Annadel State Park, 300 m elevation, in the Sonoma Mountains of Sonoma County (Wainwright and Barbour 1984). *Pseudotsuga menziesii* and

Table 5.4. *Average characteristics of 10 stands of mixed evergreen forest at Annadel State Park, Sonoma County*

Species	D[a]	BA
Overstory species		
Pseudotsuga menziesii	521	27
Umbellularia californica	497	5
Quercus kelloggii	264	8
Heteromeles arbutifolia	248	<1
Arbutus menziesii	168	3
Quercus agrifolia	120	3
Q. garryana	54	2
Sequoia sempervirens	18	2
Woody understory species		
Amelanchier pallida	19	
Arctostaphylos manzanita	2	
Aesculus californica	2	
Fraxinus latifolia	<1	

Note: Total cover by all ground-stratum species = 23%.
[a]D = density (ha^{-1}), BA = basal area ($m^2\ ha^{-1}$)
Source: Wainwright and Barbour (1984).

Table 5.5. *Importance values (IV) of all woody species encountered in the mixed evergreen forests of Snow Mountain (Lake County, 39° 22′ N) and the west flank of the Sierra Nevada (Yuba and Sierra countries, ~39° 30′N)*

Species	Snow Mt.	Sierra
Pinus ponderosa	33	28
Quercus chrysolepis	19	10
Q. wislizenii	15	—
Pseudotsuga menziesii	1	21
Quercus kelloggii	—	15
Calocedrus decurrens	—	13
Quercus durata	8	—
Ceanothus integerrimus	7	1
Arctostaphylos viscida	1	6
A. glandulosa	5	—
Cercocarpus betuloides	5	—
Arctostaphylos canescens	4	—
Lithocarpus densiflora	—	3
Arbutus menziesii	—	3
Pinus lambertiana	2	1
Adenostoma fasciculatum	1	—
Pinus sabiniana	1	—
Elevational range (m)	1000–1400	600–1200

Note: I have used sites 5–8 on Gray's Snow Mountain transect and sites 1–8 on his Sierran transect. IV = 0.5 (relative density + relative cover).
Source: Gray (1978).

Umbellularia californica shared dominance in terms of stems per hectare (Table 5.4). Douglas firs in some stands were large, 30–170 cm dbh. Low woody vegetation (excluding juveniles of overstory species) contributed only 5% cover, ferns and mosses 9% cover, grasses and sedges 3% cover, and (largely perennial) forbs 6% cover.

Gray (1978) sampled elevational gradients along Snow Mountain in Lake County and along the northern Sierra Nevada. *Pseudotsuga menziesii* was uncommon on Snow Mountain, but it was a leading dominant in the Sierra Nevada (Table 5.5). Gray wrote of the Sierran stations abstracted in his Table 2 as belonging to the "yellow pine belt," but I choose to view them as mixed evergreen forest because of the high importance values (IV) for *Quercus chrysolepis* and *Q. kelloggii* and the presence of *Lithocarpus densiflora* and *Arbutus menziesii*. The coefficient of community or similarity index between the two forests in Table 5.5, weighted by IV, is 40, which is not relatively high (Sorensen/Motyka method of calculation) (Mueller-Dombois and Ellenberg 1974).

Coulter Pine–Hardwood Phase

Pinus coulteri is scattered from Mount Diablo (38° N) south through the Coast Ranges, then as large patches into the Transverse Ranges and down the Peninsular Ranges into Baja California. Within this area it becomes an important, sometimes dominant element in the mixed evergreen forest.

In the Santa Lucia Range it is a regular but minor part of forests on steep slopes at 1200–1600 m elevation (Talley and Griffin 1980). Shrub and herb cover is modest, the common genera being *Arctostaphylos, Carex, Galium, Gayophytum, Lupinus, Pyrola, Rhamnus,* and *Ribes*. Extending down to lower elevations, 250–1500 m, *Abies bracteata* can become an added element, but the other associated species do not change (Sawyer et al. 1977) (Table 5.6). Tree cover is 40–65%, and herb and shrub covers are each 5%. These steep, rocky sites are relatively fire-free (Griffin 1978). *Abies bracteata* has a very limited range of 1800 km^2 in the Central Coast Range (Griffin and Critchfield 1972); so it is today a minor element of mixed evergreen forest as a whole. Historically, it has been an associate of the forest for more than 13 million years (D.I. Axelrod, pers. commun.).

In the Transverse and Peninsular ranges, the Coulter pine phase is best developed between 1200 and 1800 m elevation (Minnich 1976; Thorne 1977), and it is associated with *Pinus ponderosa* (upper elevations), *Quercus chrysolepis, Q. kelloggii*, and a shrub stratum of 10% cover with *Arctostaphylos, Ceanothus,* and *Cercocarpus* species. Vogl (1976) pointed out the close spatial relationship between chaparral and Coulter pine forest–woodland, then concluded:

Table 5.6. *Basal areas (m^2 ha^{-1}) for all trees encountered in 45 plots within a Santa Lucia fir forest*

Species	Summit (N = 10)	Slope (N = 6)	Ravine (N = 29)
Quercus chrysolepis	12	25	13
Abies bracteata	6	13	12
Pinus lambertiana	1	1	<1
P. coulteri	1	—	<1
Lithocarpus densiflora	—	2	<1
Pinus ponderosa	—	1	<1
Arbutus menziesii	—	<1	1
Calocedrus decurrens	—	—	1
Umbellularia californica	—	—	1
Sequoia sempervirens	—	—	1
Acer macrophyllum	—	—	<1
Aesculus californica	—	—	<1
Quercus agrifolia	—	—	<1
Q. wislizenii	—	—	<1
Platanus racemosa	—	—	<1
Alnus rhombifolia	—	—	<1

Note: Plots are grouped into three topographic categories. The elevational range for summit plots is 1280–1560 m, for slope plots is 1160–1420 m, and for ravine plots the elevation extends down to 240 m.
Source: Talley (1974; pers. commun.).

> Since Coulter pine often exists amidst seas of dense chaparral, flourishes on sites with southern exposures, has semi-serotinous cones, readily reseeds burned sites, has a relatively short life span, and has growth characteristics conducive to crown fires, it may well have a life cycle requiring certain natural fire frequencies . . . in southern California mountains. It is interesting that this tree, abundant in the southern California coast ranges and reaching its maximum development there, has yet to be studied ecologically.

Borchert (1985) has shown that Coulter pinecones are serotinous only when the tree is associated with chaparral, *Quercus chrysolepis*, or *Cupressus sargentii*, but not when associated with *Q. agrifolia* forest, which burns less frequently.

On steep north-facing slopes and in ravines, big-cone spruce (*Pseudotsuga macrocarpa*) can be a common element (Fig. 5.8). At lowest elevations of 1100 m (Minnich 1976, 1980), *P. macrocarpa* trees are scattered individuals 15–30 m tall above a closed canopy of *Quercus chrysolepis*. At about 1500 m elevation, *Q. chrysolepis* thins and *P. macrocarpa* becomes increasingly abundant (80–190 trees ha^{-1}). Tree canopy cover may total 85% (about equally divided between oak and big-cone spruce), and shrub cover is 5–13%. As in *Abies bracteata* stands, these stands appear to be more fire-free than surrounding mixed evergreen phases such as canyon live oak and Coulter pine (Minnich 1980). A study in the Santa Ana Mountains by Littrell and McDonald (1974) showed a stable age structure for oak and big-cone spruce in mature stands.

Although all three species—*Quercus chrysolepis*, *Pinus coulteri*, and *Pseudotsuga macrocarpa*—occur sympatrically in the Transverse and Peninsular ranges, Coulter pine and big-cone spruce tend to dominate different sites. Coulter pine and canyon live oak occur on more xeric, frequently disturbed sites adjacent to chaparral, and big-cone spruce and canyon live oak occur on more mesic, protected, fire-free sites (Minnich 1980). A recent vegetation map of the San Bernardino Mountains (Minnich 1976) shows a mosaic of *Q. chrysolepis* + *P. macrocarpa*, *Q. chrysolepis* + *P. coulteri*, and *Q. kelloggii* types between 1200 and 2100 m elevation. Hanes (1976) called mixed evergreen forest with *P. macrocarpa* a "high elevation riparian woodland" to emphasize its mesic habitat in the San Gabriel Mountains. In moist, cool canyons between 600 and 1500 m, it is associated with *Acer macrophyllum*, *Populus trichocarpa*, *Quercus chrysolepis*, and *Umbellularia californica*.

Nonconiferous Phases

Within the mixed evergreen forest belt, the forest may be dominated exclusively by broad-leaved trees: *Aesculus californica*, *Arbutus menziesii*, *Lithocarpus densiflora*, *Quercus agrifolia*, *Q. chrysolepis*, *Q. kelloggii*, *Q. wislizenii*, and *Umbellularia californica*. In some cases this phase is seral following fire or logging that has removed a conifer element once present. The hardwood species typically recover by stump sprouting, resulting in a dense stand of polesize trees with little commercial value. McDonald (1980) described a forest such as this in the Sierra Nevada, Yuba County, at 650 m elevation: essentially a one-layered community dominated by *Arbutus menziesii*, *Lithocarpus densiflora*, and *Quercus kelloggii* in a 1 : 4 : 2 ratio, all about 20 m tall; stems greater than 5 cm dbh numbered 1628 ha^{-1}, and basal area totaled only 46 m^2 ha^{-1}.

Other hardwood phases are edaphic or topographic climaxes, on steep slopes, in canyons, or on poor soils; these are nearly pure stands of *Q. chrysolepis* or *Q. kelloggii* (Table 5.7). The USDA Forest Service map by Matyas and Parker (1980) shows that considerable areas of the west flank of the Sierra Nevada are in these two oak phases. Both may extend well into the montane conifer zone (Myatt 1980). Canyon live oak forest, as mentioned earlier in this chapter, is also common at all elevations within the mixed evergreen forest in southern California. In Baja California, *Q. chrysolepis* extends south to 29° N, on steep slopes below 2000 m (Wiggins 1980).

Figure 5.8. Mixed evergreen forest on Mt. Palomar, 1500 m elevation. Overstory species include Pinus coulteri, Pseudotsuga macrocarpa, Quercus chrysolepis, *and* Q. kelloggii.

Table 5.7. *Nonconiferous mixed evergreen forest, Sequoia–Kings Canyon National Forest, ~1000 m elevation, means of three stands*

Tree species	C[a]	D	BA	IV
Quercus kelloggii	77	60	70	207
Q. douglasii	13	29	16	58
Aesculus californica	7	6	10	23
Quercus wislizenii	2	6	4	12
Totals:				
Absolute C (%)				77
Absolute D (trees ha^{-1})				600
Absolute BA (m^2 ha^{-1})				17
Absolute shrub cover (%)				10

[a]C = relative cover; D = relative density; BA = relative basal area; IV = C + D + BA.
Source: Vankat (1970).

MIDMONTANE CONIFER FOREST

This is the most extensive forest in California, covering 13–14% of the state's area (Barbour and Major 1977). It is most commonly called the mixed conifer forest, but various phases of it are given equal ranking by some ecologists. Such phases have been called ponderosa pine forest, yellow pine forest, white fir forest, big tree forest, and Transition Life Zone, sometimes with north-south or east-west prefixes attached to those terms (Table 5.1). In general, *Pinus ponderosa* is the thread that holds the type and its phases together, much as *Quercus douglasii* is the matrix for oak woodland and *Q. chrysolepis* is the matrix for mixed evergreen forest. In more xeric or colder portions of the type's range – especially in the eastern and southern Sierra Nevada, and in the Transverse and Peninsular ranges – *Pinus jeffreyi* may replace ponderosa pine or range above it.

The mixed conifer forest is a four-layered community, though the cover contribution by each layer can be quite variable, and in the most xeric sites the herbaceous element is not significant. Overstory trees are needle-leaved conifers 30–60 m tall, commonly greater than 1 m dbh, with interlocking crowns that exhibit 50–80% cover (Fig. 5.9). In the relatively arid San Bernardino Mountains, trees are only 12–30 m tall and contribute 20–60% cover (Minnich 1976). Exceptionally, *Pinus lambertiana* crowns may reach to 78 m, and *Sequoiadendron giganteum* to 85 m.

A subdominant tree canopy, 5–10 m tall, is present, but individual members are so scattered that total cover is insignificant. Members include deciduous species (*Quercus kelloggii, Cornus nuttallii, Acer macrophyllum*) and broad-leaved evergreens (*Quercus chrysolepis*, tree forms of *Cercocarpus ledifolius*).

In contrast, a shrub canopy less than 1 m tall is regularly present and contributes 10–30% cover. Deciduous and evergreen species are common in the genera *Arctostaphylos, Ceanothus, Chamaebatia, Castanopis, Lithocarpus, Prunus, Quercus, Ribes, Symphoricarpus*, and *Vaccinium*. Many of these reproduce vegetatively or otherwise spread outward in clumps, and for this reason shrub cover is patchy and variable.

Herb cover may reach 20% (Vankat and Major 1978), but is more commonly 5–10%. Perennial forbs predominate, some of which are hemi-parasitic, and species richness is relatively high. The more common genera include *Adenocaulon, Clintonia, Disporum, Galium, Iris, Lupinus, Osmorhiza, Pteridium, Pyrola, Smilacina*, and *Viola*.

The general architecture of the undisturbed forest prior to fire-suppression policies gave the impression of openness. It was almost parklike, with loose clusters of large trees alternating with open-

Figure 5.9. Aspect of the midmontane mixed conifer forest. A: In the San Jacinto Mountains of the Transverse Ranges, 1600 m elevation courtesy of R.F. Thorne. B: In the central Sierra Nevada, Placer County, 1550 m elevation. Prominent species include Abies concolor, Calocedrus decurrens, Pinus lambertiana, Pinus ponderosa, *and* Quercus kelloggii.

Table 5.8. *Estimates of lower and upper elevational limits for midmontane conifer forest, as interpreted from information provided by various authors*

Region	North latitude	Lower	Upper	Authors[a]
N. Sierran	39° 30′	900	1700	Gray (1978)
Coast Range (Snow Mt.)	39° 29′	1300	1900	Gray (1978)
C. Sierran	38° 20′	800	2000	Yeaton et al. (1980), Yeaton (1981)
C. Sierran	37° 50′	1200	2000	Parker (1982)
S. Sierran	36° 30′	1350	2150	Rundel et al., (1977), Vankat (1982)
S. Sierran	35° 40′	1600	2300	Twisselman (1967)
Transverse Range (Mt. Pinos)	34° 40′	1750	2700	Twisselman (1967), Vogl and Miller (1968)
Transverse Range (San Gabriel Mts.)	34° 15′	1400	2400	Hanes (1976), Küchler (1977), Thorne (1977)
Transverse Range (San Bernardino Mts.)	34° 10′	1600	2600	Küchler (1977), Minnich (1976), Thorne (1977)
Peninsular Range (San Jacinto Mts.)	33° 50′	1450	2500	Thorne (1977), Vogl (1976)
Peninsular Range (Sierra San Pedro Mártir)	30° 30′	1400	2400	Wiggins (1980), Goldman (1916)

[a]Where more than one author is cited, I may have chosen average elevations.

ings, and relatively few trees of intermediate height. There is evidence for a change in shrub and tree cover in the past 100 yr, as described in a later section on fire ecology.

Physical Factors

The elevational limits of mixed conifer forests reflect latitude, proximity to humid air, slope aspect, and soil depth. Considering its entire range, the type may extend from 800 to 2600 m in elevation (Table 5.8). The easternmost portion of the Transverse Ranges, the San Bernardino Mountains, shows the maximum upward shift due to aridity, more so than in ranges farther south. The forest is absent from the western Transverse Ranges and the Central Coast Range, where peaks are below 1400 m (vegetation near the highest peaks in the Santa Lucia Range supports a depauperate mixed evergreen forest dominated by *Quercus chrysolepis*, *Pinus lambertiana*, and *Abies bracteata*). Several scattered peaks in the southern Yolla Bolly Mountains of Glen, Mendocino, and Lake counties do support mixed conifer forest: Snow Mountain and Black Butte are the best examples. More northerly examples are discussed in Chapter 4, pages 116–117.

Judging from Major's (1967, 1977) analyses of climatic gradients along the Sierra Nevada west flank, and from my own analysis of gradients in southern California, the lower elevational limit corresponds to a mean annual temperature of 13°C and annual precipitation of 850 mm. Well within the forest, the mean annual temperature is 11°C, and annual precipitation exceeds 1000 mm. Mount Pinos (430 mm per year, 2690 m elevation) is an exception, as are east-side forests along the Sierra Nevada and midmontane forests in the Sierra San Pedro Mártir (500 mm per year) (Hastings and Humphrey 1969). Major (1977) showed that temperature and precipitation lapse rates along the west flank of the Sierra Nevada are relatively similar north to south, averaging about −0.5°C per 100-m rise and +40 mm per 100-m rise. My own limited analysis of precipitation lapse rates for the Transverse and Peninsular ranges for elevations between 300 and 1800 m shows considerably shallower precipitation lapse rates: +11 mm per 100-m rise on Mount Pinos and +5 mm per 100-m rise elsewhere. The percentage of precipitation that falls as snow also increases with elevation. In the northern Sierra Nevada, about 33% falls as snow within the mixed conifer forest zone (Major 1977).

Soils at lower elevations and on moderate slopes are typically loamy Alfisols or Ultisols, often distinctively red in color, within pH 5–6, and a depth of 80–180 cm. Common soil series include Aiken, Josephine, Mariposa, and Sites (all Ultisols) and Atwell, Boomer, Cohasset, Holland, and Musick (all Alfisols). On steeper terrain, with less soil development, Inceptisols are common (Chaix, Corbett, Hugo, Masterson, McCarthy, Neuns, Sheetiron) (CDFR 1979; Laacke 1979).

Phases

Mixed conifer phase. The mixed conifer forest exhibits shared or shifting dominance by six conifer species: ponderosa pine (*Pinus ponderosa*), Jeffrey pine (*P. jeffreyi*), sugar pine (*P. lambertiana*), Douglas fir (*Pseudotsuga menziesii*), California white fir (*Abies concolor* var. [or ssp.] *lowiana*), and incense cedar (*Calocedrus decurrens*, synonymous with *Libocedrus decurrens*). Depending on latitude, region, stand history, and microenvironment, any one of the taxa may be dominant, or some number of the six may share dominance. A compilation of five stand samples from Snow Mountain in the Coast Range shows shared dominance by three of the five, with *Quercus chrysolepis* equally important (Table 5.9). The coefficient of community or similarity index between Coast Range and Sierran forests in Table 5.9, weighted by IV, is 57, which is higher than the similarity of their mixed evergreen belts just below.

Barbour and associates (1981) conducted a detailed survey of the mixed conifer forest at Calaveras Big Trees State Park in the central Sierra Nevada (Table 5.10). Four conifer species shared dominance. *Pseudotsuga menziesii* was not common in this forest, even though its southern limit is 100 km farther south. Four subdominant tree taxa were found in these stands, but only *Quercus kelloggii* had a significant presence and importance value. Notice that the densities of *Calocedrus* and *Abies* are two to six times those of the two *Pinus* species, but their basal areas (BA) are only 1.5 times as large. This indicates that the pines dominate the oldest cohorts and the *Calocedrus* and *Abies* overwhelmingly dominate the youngest–a shift reflecting fire-suppression policies for the past 80 yr.

Table 5.9. *Importance values for trees and shrubs in the midmontane mixed conifer forest at Snow Mountain (39°29 ') and in the northern Sierra Nevada (39° 30')*

Species	Snow Mt.	Sierra
Abies concolor var. *lowiana*	24	24
Pinus ponderosa	18	23
Quercus chrysolepis	33	6
Pinus lambertiana	15	2
Arctostaphylos canescens	6	9
Calocedrus decurrens	0	13
Psuedotsuga menziesii	0	10
Pinus jeffreyi	0	7
Ceanothus cordulatus	1	3
Arctostaphylos glandulosa	3	0
Ribes roezlii	0	3
Amelanchier alnifolia	0	2
Arctostaphylos patula	0	2
Garrya fremontii	0	2
Pinus contorta	0	1
Quercus kelloggii	0	<1
Ribes nevadensis	0	<1
Elevational range (m)	1500–2000	1300–1700

Note: IV = (0.5) × (relative density + relative cover). I have used Gray's sites 9–14 for Snow Mountain and sites 9–13 for the Sierra Nevada
Source: Gray (1978).

Table 5.10. *Tree composition of midmontane mixed conifer forest in the central Sierra Nevada, 1500 m, Calaveras Big Trees State Park, 38° 20' N latitude*

Species	P	D	BA	IV
Calocedrus decurrens	100	454	22	69
Abies concolor var. *lowiana*	94	311	25	60
Pinus ponderosa	78	158	17	37
P. lambertiana	83	72	14	24
Quercus kelloggii	56	19	3	6
Cornus nuttallii	17	21	<1	2
Quercus chrysolepis	22	9	<1	1
Acer macrophyllum	6	6	<1	1

Note: Means of 18 stand samples taken by Barbour et al. (1981). P = presence in 18 stands (%); D = trees ha^{-1}; BA = basal area (m^2 ha^{-1}); IV = relative density + relative basal area.

In southern California, Gemmill (1980) (Table 5.11) surveyed the midmontane forest of the San Bernardino Mountains. *Pinus jeffreyi* here is more a part of the mixed conifer forest than it is in the central northern Sierra Nevada, where it is primarily an upper montane species. The penetration of Jeffrey pine into the mixed conifer zone becomes more and more pronounced with declining latitude in the Sierra Nevada, and it becomes most pronounced in the Transverse and Peninsular ranges. At the arid extreme–Mount Pinos (Vogl and Miller 1968)–ponderosa pine is completely replaced by Jeffrey pine, with an importance value (relative BA + relative density) of 172. In the somewhat less arid Sierra San Pedro Mártir of Baja California, ponderosa pine is present, but is much less common than Jeffrey pine (Wiggins 1980). Haller (1959), Thorne (1977), and Yeaton (1981) have hypothesized from field observations that Jeffrey pine is more tolerant of drought, low temperatures, and smog than ponderosa pine, although little experimental evidence has been gathered to test this hypothesis.

Thorne (1977), Vogl (1976), Borchert and Hibberd (1984), and Küchler (1977) all agree that 2100 m

Table 5.11. *Absolute densities for trees in the midmontane mixed conifer forest in the easternmost Transverse Ranges*

Species	Density (trees ha^{-1})
Pinus ponderosa	75
Quercus kelloggii	75
Pinus jeffreyi	47
Abies concolor var. *concolor*	47
Calocedrus decurrens	17
Pinus lambertiana	11
Quercus chrysolepis	8

Note: Data are averages of six stands, 1500–2100 m elevation, in the San Bernardino Mountains.
Source: Gemmill (1980).

elevation in the southern California mountains is a transition point between dominance by ponderosa and Jeffrey pines. A lower zone, 1400–2100 m elevation , is thus defined as a "*Pinus ponderosa* forest," and a higher zone, 2100–2700 m, as a "*P. jeffreyi* forest," which are phases of a "yellow pine forest" (Thorne 1977). Only Minnich (1976), working in the San Bernardino Mountains, did not recognize a transition elevation. He described an "open western coniferous forest" at 1800–2700 m as containing *P. ponderosa, P, jeffreyi, P. lambertiana, Abies concolor* var. *concolor, Calocedrus decurrens, Quercus kelloggii,* and *Q. chrysolepis.* Note that in southern California the white fir is the Rocky Mountain variety (or subspecies), *A. concolor* var. *concolor*, which differs from *A. concolor* var. *lowiana* in needle tip shape and in stomatal arrangement (Hamrick and Libby 1972).

Mixed conifer forest is poorly represented in the Sierra Juárez, as the highest peaks are only 1400 m elevation. In the Sierra San Pedro Mártir, at 1400–2400 m elevation, is an open forest dominated by *Pinus quadrifolia* and *P. jeffreyi* (Wiggins 1980); *Pinus ponderosa* is present, but not common, and *P. lambertiana* is also present. *Abies concolor* (presumably var. *concolor*) and *Calocedrus decurrens* are restricted to the most mesic habitats within that elevational zone. Parry pinyon (*P. quadrifolia*) is abundant up to 2100 m. From that point, Jeffrey pine becomes more of a dominant, and it continues to the higher peaks at 3100 m, associated with lodgepole pine and quaking aspen. I have found no quantitative descriptions of these forests. From general descriptions by Wiggins (1980) and photographs by Nelson (1922), it seems that components of midmontane and uppermontane forests coexist between 2100 and 2400 m and that a depauperate upper montane forest exists above that, to the summit.

White fir phase: Sierra Nevada. At low elevations, *Abies concolor* var. *lowiana* is restricted to the most mesic sites, but with increasing elevation it becomes dominant over a broad spectrum of habitats. Many regional ecologists recognize the presence of a "white fir forest" (Thorne 1976; Rundel et al. 1977; Matyas and Parker 1980), even though the elevational zone may overlap that of "ponderosa pine mixed conifer forest" (Fig. 5.10). I think it is reasonable to view California white fir as a distinct phase of the midmontane conifer forest. Some elements of the upper montane forest penetrate locally into the white fir phase: *Abies magnifica* in the more mesic portion of the gradient, *Pinus contorta* var. *murrayana* and *P. jeffreyi* on the more xeric end. The tree canopy cover is generally greater than 75% (Vankat and Major 1978), but herb and shrub covers are lower than those of the mixed conifer forest.

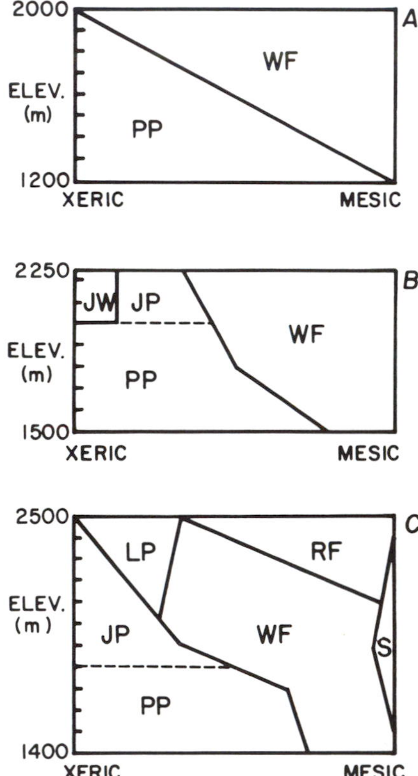

Figure 5.10. Diagrammatic relationships between several phases of the midmontane conifer forest in the Sierra Nevada. A: Central Sierra, Yosemite National Park (Parker 1982). B: Southern Sierra, Sequoia–Kings Canyon National Park (Vankat 1982). C: Same location, but with more detail (Rundel et al. 1977). PP = ponderosa pine mixed conifer phase; WF = white fir phase; JP = Jeffrey pine phase; S = Sequoiadendron phase; JW = Juniper woodland; LP = lodgepole pine; RF = red fir. The latter three are upper montane, but penetrate into the midmontane forest.

Table 5.12. *Mixed conifer and white fir phases of the midmontane forest, southern Sierra Nevada, 36° 30′ N latitude, generally 1500–2200 m elevation*

	Mixed conifer				White fir
Species	C	D	BA	IV	IV
Calocedrus decurrens	31	2764	10	89	73
Quercus kelloggii	46	1364	16	76	16
Pinus ponderosa	32	382	11	47	<1
Albies concolor var. *lowiana*	22	755	9	48	135
Pinus lambertiana	9	482	2	31	36
Quercus chrysolepis	1	45	<1	9	4
Cornus nuttallii	<1	<1	<1	<1	11
Pinus jeffreyi	—	—	—	—	1
Populus trichocarpa	—	—	—	—	<1
Corylus cornuta	—	—	—	—	7
Abies magnifica	—	—	—	—	14
Torreya californica	—	—	—	—	<1

Note: C = Cover (%); D = stems ha^{-1}; BA = basal area (m^2 ha^{-1}); IV = relative frequency + relative density + relative basal area. Averages of 27 stand surveys and 28 transects by Vankat and Major (1978).

Perhaps the most detailed comparison of white fir and mixed conifer stands in one area is that by Vankat and Major (1978) (Table 5.12). The importance value of *Abies concolor* was three times as great in the white fir phase as in the mixed conifer, whereas *Pinus ponderosa*'s IV fell to less than 1. *Calocedrus decurrens* remained unchanged in the two phases, much as it remains constant in Yosemite National Park stands studied by Parker (1982). *Abies magnifica*, from the upper montane forest, had a significant presence. A coefficient of community or similarity index between the two types in Table 5.12, weighted by IV, is 58, which is relatively high–as high as coastal and Sierran versions of mixed conifer (Table 5.9).

Farther north, white fir stands in Sierra County are common on mesic sites between 1500 and 2000 m (Conard and Radosevich 1982). These stands show an even more pronounced dominance by white fir, with an importance value of 187 out of 200. *Pinus ponderosa* and *Calocedrus decurrens* may be absent, but there is a consistent presence of *Abies magnifica*.

As summarized by Vankat (1970), there is evidence from many studies that tree density has increased and shrub cover decreased in the white fir phase over the past 100 yr. The change is usually ascribed to lack of ground fires, because young white firs are easily killed by such fires, whereas the pines are not. It may also be possible that the light requirements for adequate white fir sapling growth are lower than those for the pines (Lanini and Radosevich 1986), but I have seen no experimental data to support this hypothesis.

Sequoiadendron phase: Sierra Nevada. The Sierra "big tree," Sierra redwood, or giant sequoia (*Sequoiadendron giganteum*) is distributed in 75 groves from Placer County to the southern boundary of Tulare County (Rundel 1972a). Elevations range from 825 to 2680 m. The groves are situated in particularly mesic microenvironments, and grove boundaries appear to have been stable for the past 500 yr (Rundel 1971, 1972b). Because of the brittleness of the wood, logging was not extensive and had removed only 34% of big-tree acreage up to the time that most stands were placed in public ownership and protected (Hartesveldt et al. 1975). Their maximum age is estimated to be 3200 yr.

If all trees larger than 3 cm dbh are tallied in a giant sequoia forest, it is apparent from Table 5.13 that *Abies concolor* var. *lowiana* is the major tree species. *Pinus ponderosa* is insignificant, but *P. lambertiana* and *Calocedrus decurrens* contribute significantly to stand structure. Apart from the imposing presence of *Sequoiadendron* and a richer collection of herbs and shrubs, the floristic and physiognomic characteristics of the forest are not significantly different from those in the white fir phase.

Jeffrey pine phase. As previously mentioned, "yellow pine forest" is one synonym for the mixed conifer forest. It is a useful name, for it allows one to focus on both *Pinus ponderosa* and *P. jeffreyi* as potential dominants or at least as continuous threads always present that hold the various phases together. In southern California, the shift in dominance from *P. ponderosa* to *P. jeffreyi* occurs at about 2100 m elevation.

Table 5.13. *Absolute densities for trees in the* Sequoiadendron *phase of the midmontane conifer forest*

Species	Density (trees ha^{-1})	IV
Abies concolor var. *lowiana*	941	131
Pinus lambertiana	109	30
Calocedrus decurrens	117	42
Abies magnifica	67	<1
Sequoiadendron giganteum	49	98
Quercus kelloggii	33	<1
Q. chrysolepis	33	<1
Pinus ponderosa	17	<1
Cornus nuttallii	16	<1

Note: Summaries of 21 stands in several groves from Rundel (1971), Vankat and Major (1978), and Barbour et al. (1981). Importance values (relative density + relative basal area + relative frequency) are only for trees >12.8 cm dbh in three groves sampled by Rundel (1971).

Haller (1962) examined the traits of the two yellow pines and of putative hybrids. Hybrids are regularly found, but not in high frequencies. The hybrid frequency is kept low by seasonal differences in dates of pollen maturity, by the reduced viability of hybrid seed, and by failure of hybrids to become established. Haller found the elevational zone for hybrids to be relatively narrow, about ±275 m around a mean elevation of 1970 m.

The Jeffrey pine forest in the southern Sierra Nevada extends well into the upper montane, beyond 2500 m elevation (Rundel et al. 1977; Vankat 1982). Jeffrey pine appears to respond more to edaphic or microenvironmental factors than to macroenvironmental lapse rates of temperature and precipitation. It is often on more arid south- or west-facing slopes and on shallow soils. At its lower limits one may think of it as a phase of the mixed conifer forest, and at its upper limits as a xeric phase of the upper montane forest normally dominated by red fir (*Abies magnifica*) or lodgepole pine (*Pinus contorta* var. *murrayana*). Dominance is shared among *P. jeffreyi*, *Calocedrus decurrens*, and *Abies concolor* var. *lowiana*. The trees are large and somewhat scattered, the tree canopy cover being 40–60%. Shrub cover is variable, and herb cover is about 20% (Vankat and Major 1978).

Farther north, in Yosemite National Park, Parker (1982) placed all his stands with significant Jeffrey pine well within the upper montane red fir forest, and my own field experience in the central and northern Sierra Nevada supports his treatment. However, considering the entire range of Jeffrey pine, I am placing it as a phase of the midmontane forest.

Haller (1959, 1962), Wright (1968), Yeaton and associates (1980), and Yeaton (1978, 1981, 1983a, 1983b) have all examined the elevational replacement of montane pines in some detail. With respect to the low-elevation replacement of *P. sabiniana* by *P. ponderosa*, both Yeaton (1981) and Griffin (1965) implicated competition for soil moisture at the sapling stage. The reasoning is deductive, rather than based on experimental manipulation, but it is persuasive. Digger pine's habitat range becomes more and more restricted with increasing elevation to the most open habitats on serpentine. In mixed stands (700–800 m elevation for northern and central Sierra Nevada), saplings of ponderosa pine do have the faster growth rates, implying a more favorable carbon balance, but I have seen no data on photosynthetic rates. Both are diploxylon pines, a taxonomic group to which Yeaton (1981) ascribes the following traits: colonizing arid, rather open habitats; reproducing at a young age; reproducing frequently; producing small seeds; in general, possessing r-type traits. Consequently, the two diploxylon pines are more likely to compete for similar habitats than is a haploxylon-diploxylon combination.

At 1800 m elevation, *P. jeffreyi* begins to replace *P. ponderosa* (both are diploxylon pines), and *P. monticola* begins to replace *P. lambertiana* (both are haploxylon pines). Again, Yeaton (1981) hypothesized that competition for soil moisture is the driving mechanism. To test this hypothesis, we need comparative data on water-use efficiency, photosynethetic rates, and carbon allocation patterns. Axelrod (1976) ascribed the shift to decreasing warmth and length of the growing season.

Fire Ecology

Several extensive studies of fire scars in various phases of the midmontane conifer forest have been reviewed by Biswell (1967), Komarek (1967), Rundel and associates (1977), and Kilgore and Taylor (1979). The autecological and synecological roles of fire in this vegetation have been discussed by Kilgore (1973), Vogl (1973), Hartesveldt and associates (1975), Vankat and Major (1978), and Harvey and associates (1980). The summary paragraphs that follow come largely from those sources.

Fire frequency prior to 1875 averaged about one fire every 8 yr in pine-dominated sites, and about one per 16 yr in more mesic fir-dominated sites. Most of these fires appear to have been ground fires of low intensity and of limited areal extent. It is not possible to determine the percentage of fires that were ignited by Indians or by lightning; we can only conclude that both were significant in producing these frequencies. The seed release, germination, and seedling establishment phases of many woody species in the midmontane conifer forest are enhanced by the effects of ground fires. Furthermore, the open structure of the forest and, to some extent, the balance in relative abundances of the dominant tree species were maintained by ground fires. *Abies concolor* var. *lowiana* and, to a lesser extent, *Calocedrus decurrens* have increased in importance dramatically since 1900, when a policy of fire suppression was instituted. White fir seedling density and white fir sapling mortality are inversely related to fire frequency.

The policy of fire suppression has resulted in a large increase in dead and living ground fuel biomass, which increases the likelihood of ground fires becoming crown fires of greater areal extent. The USDA Forest Service, the National Park Service, the California Department of Forestry, and the California Department of Parks and Recreation have finally begun programs of controlled burning and have modified their responses to natural fires, moving to suppress fires only in certain localities, and allowing others to burn within limits.

A few species are nearly obligate in their requirement for fire in order to complete or continue their life cycles. The cones of *Sequoiadendron giganteum*, for example, generally remain closed and attached to parental trees until opened by hot updrafts from fires; seeds of *Ceanothus integerrimus* remain dormant until cracked by modestly hot temperatures; seedlings of *Pinus ponderosa* experience high mortality unless mineral soil is very close to the surface; and stands of the closed-cone *Pinus attenuata* do not regenerate in the absence of fire. Although some montane brush areas are edaphic climaxes, others are seral stages which follow forest crown fires. Conard and Radosevich (1982) have shown that such seral brush fields both encourage and suppress conifer saplings beneath their canopies for many decades, and Lanini and Radosevich (1986) have demonstrated that suppression of ponderosa pine, sugar pine,and white fir was largely due to competition for soil moisture with the shrubs.

Effects of Air Pollution

Air pollutants that affect montane vegetation in California are primarily those released in automobile exhausts and secondarily those that result from exhaust fumes interacting with oxygen in the presence of sunlight. Major phytotoxicants are ozone, nitrogen oxides, hydrocarbons such as PAN (peroxyacyl nitrates), and sulfur dioxide (NAPCA 1970). The collection of pollutants is commonly called "smog," and ozone appears to be the most important phytotoxicant in the complex. During summer months, temperature inversions are common, and the result is that smog may collect and accumulate along mountain slopes up to the inversion boundary layer, commonly 1700 m elevation.

Ozone is a powerful oxidant that externally causes visible mottling or necrosis on leaves and internally affects major metabolic pathways such as photosynthesis. Sensitive plants show ozone injury within 8 hr when exposed to ozone at a concentration of 0.03 ppm (NAPCA 1970). Current federal (EPA) standards say that ozone should not anywhere exceed a concentration of 0.12 ppm for more than 1 hr yr^{-1}. There is evidence now that mixed conifer forests in the Transverse, Peninsular, and Sierra Nevada ranges frequently experience ozone concentrations that exceed this limit (Miller 1973; McBride et al. 1975; Williams et al. 1977; Pronos and Vogler 1981; Williams 1983).

In a study of the San Bernardino Mountains (Kickert et al. 1976; Taylor 1980), ponderosa pine was shown to be the tree species most sensitive to ozone chlorotic mottle disease. The associated, less sensitive trees were *Calocedrus decurrens*, *Abies concolor* var. *concolor*, *Pinus lambertiana*, and *Quercus kelloggii*. Apart from mottling, the disease symptoms were smaller needles, 67% decline in needle life span, reduced height growth, reduced diameter growth, and increased susceptibility to root rot and bark beetle infestation. Miller (1973) hypothesized that a long-term continuation of these symptoms would result in succession to brush fields on xeric sites and to a white fir–incense cedar forest on mesic sites, and that such changes "may not enhance the recreational uses, wildlife habitat and watershed values of the area."

In a smaller study of forests between 1000 and 3000 m elevation in the Sequoia–Kings Canyon National Park area of the Sierra Nevada (Williams et al. 1977; Pronos and Vogler 1981; Williams 1983), the following species exhibited ozone chlorotic mottle: *Calocedrus decurrens*, *Pinus ponderosa*, *P. jeffreyi*, *P. contorta* var. *murrayana*, *Quercus kelloggii*, and *Sequoiadendron giganteum*. *Abies concolor* var. *lowiana* and *Pinus lambertiana* were much less affected. Again, leaf longevity was reduced, but no significant effect on needle length or annual ring width was found. Between 70% and 100% of all trees in plots at 1300–1900 m elevation showed chlorosis. Ponderosa pines were affected at altitudes as low as 1000 m, but the most intense damage occurred at 1200–1800 m, and damage declined above 2600 m.

Chlorotic Jeffrey and sugar pines are commonly seen in the Lake Tahoe basin of the central Sierra Nevada, at 1800 m elevation, and ozone damage on pines has been observed downwind from Lake Tahoe toward Luther Pass (CDFR 1979). Little attention has yet been paid to determining the extent of ozone damage in the northern Sierra Nevada.

Oxides of nitrogen and sulfur may also be pollutants in the form of acid rain. McColl (1981) recently reviewed reports on acid rain in California and concluded "that acid rain does occur in California and is quite widespread in its geographical distribution. . . . Most of the acid is HNO_3 derived from NO_x, and only about a third . . . is attributed to H_2SO_4 derived from SO_2." McColl experimented with germination and growth of Douglas fir and ponderosa pine (among other species) as affected by acid spray and modified by soil substrate. Germination, seedling mortality, growth, and needle longevity were all adversely affected by a spray at pH 2. Further, granitic soils with low degrees of base saturation and having shallow depths were most susceptible to acidification; these are typical soils at high elevations in the Sierra Nevada.

Bradford and associates (1981) agreed that Sierran soils are potentially sensitive, but their survey of 170 Sierran lakes showed essentially no change in acidity during the past 15 yr. They concluded that acid rain observed in coastal and interior valleys near population centers was diluted along its path to montane elevations to such an extent that it had no measurable effect.

UPPER MONTANE CONIFER FOREST

One species that provides a thread of continuity through the upper montane forest, from the northern Sierra Nevada to the Sierra San Pedro Mártir, is lodgepole pine (*Pinus contorta* var. *murrayana*). It would be inappropriate, however, to give this forest the overall name "lodgepole pine forest," because lodgepole pine also can extend, in open stands, into the subalpine zone, and it can be supplanted by red fir (*Abies magnifica*) within the upper montane zone.

This forest corresponds to the Canadian Life Zone of Merriam (1898), a term still widely used despite general criticisms of his system (Daubenmire 1938). Some southern California ecologists (Hanes 1976; Thorne 1977, 1982) refer to all lodgepole pine forests as being subalpine, whereas Sierran ecologists, such as Vankat (1982), typically place lodgepole pine forests in the upper montane zone. In a recent review of subalpine forests, MacMahon and Andersen (1982) seemed to define the subalpine zone of the Sierra Nevada as the habitat of *Pinus albicaulis* (in the north) and *P. flexilis* (in the south), implying that closed forests of lodgepole pine and red fir are upper montane. I adopt their view here.

Upper montane forests cover about 3% of the state's area (Barbour and Major 1977). Elevational limits are about 1800–2400 m in the northern Sierra Nevada, 2200–3000 m in the southern Sierra Nevada, 2400–2800 m in the Transverse and Peninsular ranges, and from 2400 m to the tops of peaks at 3100 m in the Sierra San Pedro Mártir in Baja California. This forest is also present from 1900 m to the tops of six peaks in the southern Yolla Bolly Mountains, the tallest of which is 2270 m.

Soils are shallow, rocky inceptisols or entisols. Weather data from 10 stations throughout the Sierra Nevada (USDC 1964; Major 1967, 1977; Smith 1978a, 1978b) indicate a mean annual precipitation of 820 mm. There is some drop with declining latitude, and northernmost sites commonly receive 1000–1600 mm. A central conclusion, however, is that there is little difference in total annual precipitation between midmontane conifer forest and upper montane conifer forest. Instead, major climatic differences between the two zones have to do with the form of precipitation and the mean annual temperature. About 70–90% of all precipitation falls as snow (compared with 33% in the midmontane), leading to snowpack depths of 2.5–4 m and snow duration of close to 200 d. The mean annual temperature is 5°C, well below the 11°C of the midmontane forest. Few climatic data are available for southern California and Baja California.

Lodgepole pine forests are moderately dense, with 55–80% cover, and are of modest stature, the trees typically are <300 yr old, <20 m tall, and <70 cm dbh (Fig. 5.11). Shrub cover is generally insignificant, but may in places reach >15%. Prominent genera include *Arctostaphylos, Ceanothus, Castanopsis*, and *Ribes*. Except where lodgepole encroaches on meadows, herb cover is also insignificant. Unlike the Rocky Mountain lodgepole pine forest, the Sierran equivalent is not a fire-type community (Parker 1986).

In the Sierra Nevada, the most mesic sites within this elevational zone are dominated by red fir (Oosting and Billings 1943). Lodgepole pine is here a consistent, but minor, element. *Abies magnifica* is among the largest species of the genus *Abies* on earth. The physiognomy of this forest contrasts with that of lodgepole pine forest. Mature overstory trees are commonly 30–45 m tall, 115 cm dbh, and over 300 yr in age. Stands are usually open and park-like, the impressive columnar boles seemingly evenly spaced (Fig. 5.12). Overstory cover averages 60%. In some stands there is a lower canopy layer of young trees 1–3 m tall, but these are scattered in dense patches. Herbs and shrubs are uncommon and contribute less than 5% ground cover.

Figure 5.11. Lodgepole pine forest, San Gabriel Mountains in the Transverse Ranges, 2440 m elevation. Openings occupied by Ceanothus cordulatus and Castanopsis sempervirens (courtesy of R.F. Thorne).

Figure 5.12. Understory view of red fir forest, central Sierra Nevada, Placer County, 2000 m elevation. Notice wolf lichen on trunks.

Lodgepole Pine Phase

For some taxonomists, *Pinus murrayana* is an acceptable synonym for *P. contorta* var. (or ssp.) *murrayana* (Critchfield 1957). In this latter scheme, it is closely related to the Rocky Mountain *P. contorta* var. (or ssp.) *latifolia*, differing in such traits as cone serotiny. More is known of stand structure, population dynamics, and autecology for the Rocky Mountain taxon than for the Pacific montane taxon.

An average lodgepole stand, based on nine sites in Sequoia National Park (Vankat and Major 1978) and six sites in Yosemite National Park (Parker 1986), exhibits 49% canopy cover, 2327 lodgepole trees per hectare, with 60 m^2 ha^{-1} basal area, and a lodgepole IV of >200 (out of 300 possible).

The bimodal habitat distribution of lodgepole pine in California is well known: It can occur on arid, windswept sites on shallow soils, and it dominates relatively wet sites at the edges of meadows or lakes that also receive cold-air drainage (Rundel et al. 1977). In these wetter sites, red fir often dominates the understory, suggesting a succession to red fir dominance. Stands of *Populus tremuloides* var. *aurea* or *P. trichocarpa* may alternate with lodgepole stands in similar habitats.

Vankat and Major (1978) reviewed the evidence for lodgepole invasion of meadows in the southern Sierra Nevada and concluded that the most recent wave of increase began about 1900. They ascribed the increase to elimination of sheep grazing, but added that there may be an overall cycle relating to an interaction between stand density and fire frequency that ultimately regulates lodgepole movement into and away from meadows. Many stands, however, do have an age structure that implies a climax state (Eyre 1980; Parker 1986). Other biologists theorize that lodgepole stand dynamics are driven by episodic infections of the needle miner *Coleotechnites milleri* (= *Recurvaria milleri*). Strubble (1973) has compiled a review of the biology and control of this interaction that includes Sierran as well as Rocky Mountain data.

I have found no published quantitative studies of lodgepole pine forest for southern and Baja California. Commonly associated trees are *Abies concolor* var. *concolor*, *Cercocarpus ledifolius*, *Juniperus occidentalis* var. *australis*, and *Pinus jeffreyi*. At higher elevations, *Pinus flexilis* is also present (Thorne 1982).

If one examines the total flora within the upper montane and subalpine zones, then there appears to be considerable diversity within the southern California region, and considerable difference between that region and the Sierra Nevada (Taylor 1977). The San Bernardino Mountains and Mount San Jacinto have a floristic similarity of only 40%, and as a group they show only a 30% similarity with the southern Cascades–Sierra Nevada flora.

Red Fir Phase

In pioneering work four decades ago, Oosting and Billings (1943) described the *Abietum magnificae* association. Their work was based on rather limited sampling data. Barbour and Woodward (1985) have recently completed a review of the red fir forest, and their summary is abstracted in Table 5.14. More than 80 plots, ranging from the southern Yolla Bolly Mountains in the North Coast Range to the southern Sierra Nevada, are represented. It is clear from the range of values that there is considerable variation from stand to stand, but overall there is an overwhelming dominance by *Abies magnifica*, with an average importance value of 251 (highest possible value 300). Total basal area is the highest for any forest type in montane California. Associated trees (in the stands in Table 5.14) are *Abies concolor* var. *lowiana*, *Pinus contorta* var. *murrayana*, *P. jeffreyi*, *P. lambertiana*, *P. monticola*, and *Tsuga mertensiana*.

The range of red fir is nearly endemic to California, extending for a short distance in the southern Cascades into Oregon. It is not genetically homogeneous throughout this range, but exists in three forms: the species itself, as Shasta red fir (*Abies magnifica* var. *shastensis*), and as a series of hybrids in the southern Cascades between *A. magnifica* and noble fir (*Abies procera*). Shasta red fir differs from red fir in terms of length of cone bracts, seed weight, cotyledon number, and terpene composition (Silen et al. 1965; Ustin 1976; Zavarin et al. 1978; Franklin et al. 1980). It has a disjunct distribution, primarily restricted to the southern part of the Sierra Nevada and to the southern Cascade–Klamath–northern Coast Range area (Hallin 1957). It appears to be completely sympatric and interfertile with red fir, and in terms of stand structure, Barbour and Woodward (1985) found no significant differences between stands dominated by either taxon. Shasta red fir has a fossil record extending back 21 million yr (D. I. Axelrod, pers. commun.).

Relatively little has been published on red fir demography, stand dynamics, or autecology. Consequently, prescriptions for optimum recovery following harvests are not yet in hand. Seedling density in undisturbed stands appears to be highly variable from year to year. If one combines seedlings and saplings (juveniles less than 3 cm dbh or younger than 45 yr) into one category, the density ranges from 100 to 10,000 ha^{-1} from stand to stand (Barbour and Woodward 1985). Trees appear to be distributed in age as one might expect of a climax population: steeply declining numbers with increasing age (Fig. 5.13). Ustin and associates (1984) and Selter (1983) showed that seedling distribution in undisturbed stands was closely correlated with low duration and intensity of sun flecks on the ground floor. Seedlings were largely absent from areas that received half-strength full sun, and the photosynthetic saturation point was seen to be well below that, about 500 $\mu E\ m^{-2}\ sec^{-1}$. It appears that natural regeneration of red fir should be inhibited by any system of logging and site preparation that fails to provide significant shade from slash, understory plants, or remaining overstory trees. To some extent recovery studies by Gordon (1970a, 1970b) substantiate this prediction. Saplings >10 yr of age, however, do respond positively to release from shade.

The ecotone between midmontane and upper montane zones in the Sierra Nevada is an important one for several species pairs. *Abies concolor* var. *lowiana* gives way to *A. magnifica*, *Pinus ponderosa* to *P. jeffreyi*, and *P. lambertiana* to *P. monticola*. My field

Table 5.14. *Red fir contribution to 86 red fir stands from the southern Yolla Bolly Mountains through the Sierra Nevada*

	Trees (ha^{-1})		BA (m^2 ha^{-1})		Cover (%)		
	Abs	Rel	Abs	Rel	Abs	Rel	IV
Range	70–1832	8–100	8–160	20–100	16–90	39–100	67–300
Mean	418	82	69	83	55	89	251

Note: Ranges are shown for absolute (Abs) and relative (Rel) values. IV = relative cover + relative density + relative basal area. Commonly associated trees included: *Abies concolor* var. *lowiana, Pinus monticola,* and *P. jeffreyi.*
Source: Barbour and Woodward (1985).

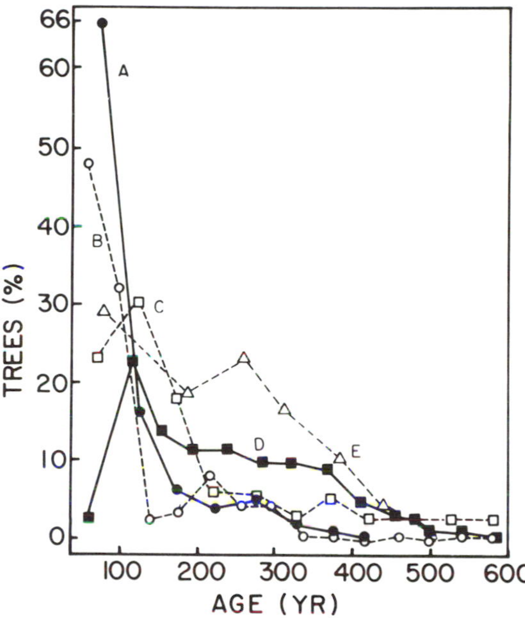

Figure 5.13. Age structure of red fir and Shasta red fir trees (>3 cm dbh) in stands from five different regions: A, mean of 9 central Sierran stands from Talley (1977b); B, mean of 3 southern Sierran stands from Pitcher (1981); C, mean of 3 central Sierran stands from Talley (1977a); D, mean of 16 Northern Coast Range stands from Barbour and Woodward (1985); E, mean of 14 southern Sierran stands from Barbour and Woodward (1985). Lines B, D, and E are for Shasta red fir populations; A and C are for red fir populations.

experience indicates that the elevational width of the ecotone is about 150 m. White and red fir seedlings are similar in their photosynthetic light curves (compensation point 25–50 μE m^{-2} sec^{-1}, 90% of maximum net photosynthesis at 500–600 μE m^{-2} sec^{-1}, optimum temperature at 20°C) (Conard and Radosevich 1981. Ustin et al. 1984; R. Sage, unpublished data), but it may be that their phenological responses to temperature and soil moisture differ. Leaf bud opening is earlier by 2–3 wk for white fir growing adjacent to red fir in the central Sierra Nevada, and leaf conductance for white fir is lower than for adjacent red fir (R. Sage, unpublished data).

Aspen Parkland Phase

Quaking aspen (*Populus tremuloides*) is the most widely distributed tree species in North America (Fowells 1965). It may be mainly represented in montane California by the variety *aurea*, which has been monographed by Barry (1971). In the Sierra Nevada, quaking aspen occurs from elevations as low as 1500 m in the north (40° N) to near timberline above 3000 m in the south (36° N). It is nearly absent from southern California, occurring in a stand at 2200 m in the San Gorgonio Wilderness Area of the San Bernardino Mountains (Minnich 1976; Thorne 1977) and in a stand west of Lake Arrowhead in the San Bernardino Mountains (D. I. Axelrod, pers. commun.). It reappears in the Sierra San Pedro Mártir above 2100 m (Goldman 1916; Wiggins 1980).

Quaking aspen stands can be spatially adjacent to montane forest, subalpine forest, dry *Wyethia* meadows, and wet *Carex-Salix* meadows. Along the east slope of the Sierra Nevada, they may also be adjacent to Jeffrey pine forest and sagebrush scrub. In his study of Murphy Meadows in Placer County, at 2450 m elevation, Barry (1971) found ecotones between aspen parkland and five other communities all within a 9-ha area. Ecotones are often abrupt and may relate to soil moisture, temperature, and pH.

Aspen parkland typically exists in discrete patches 1–10 ha in area, and the hundreds of trees included in a patch may represent only one to several clones. Genetic differences detected in Sierran clones include bole color (white vs. yellow-green), time of spring bud break, radial growth rate, fall leaf coloration pattern, and time of fall leaf drop. *Populus trichocarpa* stands are more narrowly riparian.

Figure 5.14. Aspen parkland, central Sierra Nevada, Placer County, 2400 m elevation.

I have found no published quantitative studies of Sierran aspen stands. A 50-yr record of publications from the Pacific Southwest Forest and Range Experiment Station (Aitro 1977) fails to include even one publication on aspen. It appears that mature overstory trees may commonly reach 65 cm dbh and 20 m in height; canopy cover may reach 100% (Fig. 5.14). *Abies concolor* var. *lowiana* and *A. magnifica* saplings can be common in the herbaceous understory [Barry (1971) reported densities of 2500 ha^{-1}], implying that some stands are seral. Aspen stands are not resistant to fire (Fowells 1965), but the roots remain alive after fire, and these are capable of sending up new shoots.

MIXED SUBALPINE WOODLAND

This vegetation class qualifies more as a woodland than a forest. Although exceptional individuals reach 25 m in height, and some stands in locally mesic or protected sites approach the density of upper montane forests, typical subalpine woodland has a canopy height of 10–15 m and consists of clusters of widely spaced individuals that contribute 5–40% cover (Fig. 5.15). At the lower limits of the subalpine zone the trees are erect, with single

Figure 5.15 Aspect of subalpine woodland, about 2800 m elevation, along the slopes of Slide Mountain in the Carson Range of the northeastern Sierra Nevada. Prominent trees include Pinus albicualis, P. contorta *var.* murrayana, P. monticola, *and* Tsuga mertensiana.

trunks, but with increasing elevation the trees of most species become dwarfed and multistemmed, finally reaching their upper limits as shrubs (krummholz) 1 m or less in height.

Dominant conifers include whitebark pine (*Pinus albicaulis*), limber pine (*Pinus flexilis*), mountain hemlock (*Tsuga mertensiana*), lodgepole pine (*Pinus contorta* var. *murrayana*), western white pine (*Pinus monticola*), and foxtail pine (*Pinus balfouriana* ssp. *austrina*). *Pinus monticola* does not form krummholz, but persists as trees to the treeline; foxtail pine rarely forms krummholz (Ryerson 1984). I have adopted the "mixed subalpine" name applied by the Society of American Foresters (Eyre 1980) (SAF no. 256) to emphasize the pattern of shared or shifting dominance exhibited by these six taxa.

Pure stands of different taxa may occur in adjacent patches, or several taxa may share dominance in a single stand. In addition, there are north-south replacements. Only *Pinus contorta* var. *murrayana* extends throughout the Sierra Nevada and the Transverse and Peninsular ranges. A northern trio extends through all or most of the Sierra Nevada, to south of 36° N, but not into southern California: *Pinus albicaulis, P. monticola*, and *Tsuga mertensiana*. Foxtail pine has a patchy, discontinuous distribution in northern and central California, and it does not extend into southern California. It is closely related to the Great Basin bristlecone pine (*Pinus aristata*), with which it is interfertile, but they differ in tree form, bark color, and minor chemical compositions (Bailey 1970; Critchfield 1977; Sigg 1983; Rourke 1986). Its northern and central California forms appear to be worthy of subspecific rank (Mastroguiseppe and Mastroguiseppe 1980). Limber pine extends from the central Sierra Nevada, at 38° 20′ N, through the rest of the Sierra Nevada and into southern California. Northern stands thus tend to be dominated by *Pinus albicaulis* and *Tsuga mertensiana*, and southern stands by *Pinus flexilis*.

Secondary species commonly associated with the six listed earlier include *Populus tremuloides* var. *aurea, Pinus jeffreyi, P. monophylla, Abies concolor* var. *lowiana, Abies magnifica*, and *Juniperus occidentalis* ssp. *australis*.

Patches of prostrate shrubs are characteristic of the subalpine woodland, but their cover is of minor importance. Common genera include *Arctostaphylos, Artemisia, Holodiscus, Phyllodoce, Salix, Ribes*, and *Vaccinium*.

Except where the woodland encroaches on watercourses or wet meadows, herb cover and diversity are also low (about 5% cover). Common genera include *Agropyron, Antennaria, Bromus, Carex, Eriogonum, Lupinus, Sitanion*, and *Stipa*.

Physical Factors

There are some discrepancies from author to author as to the elevational limits of the subalpine woodland, in part because some taxa (*Pinus contorta* var. *murrayana*) extend well into the upper montane, and in part because the upper limit is very dependent on microenvironment and thus fluctuates widely. The elevational limits cited later are averages of limits given by Smiley (1921), Major and Taylor (1977), Minnich (1976), Thorne (1977, 1982), Eyre (1980), and Vankat (1982).

In the northern Sierra Nevada, about 39° 30′ N, the subalpine woodland extends from 2300 to 2900 m elevation; in the southern Sierra Nevada, about 36° 30′ N, it extends from 3000 to 3400 m; in the Transverse and Peninsular ranges it extends from 2800 m to the summits of the tallest peaks at 3500 m; it is absent from peaks in Baja California, which reach only to 3100 m.

Annual precipitation on the west flank of the Sierra Nevada in this zone is 750–1400 mm, with 85% or more of it falling as snow (Major 1967, 1977; Taylor 1976; Rundel et al. 1977; Eyre 1980). Precipitation on the east flank and in southern California localities is only 350–750 mm (Major 1967; Lepper 1974). Snow depth in late March may average 2 m (Klikoff 1965), and trees may be absent where snow accumulation is less than 1 m (Taylor 1976). The growing season (mid-July to mid-September) is 2 mo long, but hard frosts may occur at any time during that period. Climatic data are few. On the basis of only four sites described by Lepper (1974), Major (1967, 1977), and Taylor (1976), the mean annual temperature appears to be less than 4°C. It is likely that the upper elevational limit corresponds to a mean temperature for the warmest month of less than 10°C, based on tundra-taiga boundaries elsewhere in North America.

Soils are shallow, rocky, coarse-textured Inceptisols, judging from limited studies by Klikoff (1965) and Lepper (1974). The maximum depth is 15–25 cm, with clays accounting for less than 8%; the pH is 5–6, and total nitrogen is less than 0.2%. These soils exceed or approach a water potential of −1.5 MPa by the end of the growing season. Conifer establishment appears to be positively correlated with the depth of winter snowpack, but negatively correlated with the duration of snowpack into the growing season. Successful establishment may require seed dispersal and burial by animals, such as Clark's nutcracker (Tomback 1986).

Tree life spans are typically 500–1000 yr, but annual growth is small, so that trunks greater than 1 m dbh are uncommon. Needle longevity may reach 30⁺ yr. We may presume that the modest surplus of net photosynthesis over respiration on a yearly basis characteristic of *Pinus aristata* in the White Mountains (Mooney et al. 1964) is representative of other subalpine conifers. *Pinus flexilis* has a maximum net photosynthesis rate of 1.3–2.8 mg CO_2 g^{-1} hr^{-1} at 15°C, which is not appreciably higher than the 1.2 rate at 20°C reported for bristlecone pine (Lepper 1974, 1980). Temperatures above 25°C depress net photosynthesis significantly in *Pinus flexilis*, as does soil moisture at −1.3 to −2.0 MPa; however, its drought tolerance appears to exceed that of lower-elevation pines (Lepper 1974). *Pinus albicaulis* and *Tsuga mertensiana* both appear to accumulate nonstructural carbohydrate in leaves over winter, but this could be an artifact of translocation, rather than evidence of photosynthesis (Wanner 1981).

Timberline

The physical factors thought to correlate with the timberline have been explored in general by Daubenmire (1954) and locally for one area in the central Sierra Nevada by Klikoff (1965). Daubenmire concluded that the most likely limiting factor is heat or warmth during the growing season. Klikoff showed that soil moisture (water potential) beneath *Pinus albicaulis* forest and krummholz is not significantly different from that beneath upland *Carex* meadow, and he concluded that snow depth and duration (as modified by exposure) are more important in determining the location of the treeline.

In the central Sierra Nevada, the elevational distance between the limit of tree growth form and the limit of krummholz is about 150 m. Based largely on field observations, Clausen (1965) developed a hypothesis that krummholz forms of trees in the central Sierra Nevada are ecotypes of erect forms—that is, that they were genetically dwarfed. To my knowledge, this has not been tested experimentally, and many ecologists ascribe the dwarfing to winter desiccation and ice blast above the snowpack.

Stand Structure

Some intensive sampling of subalpine woodland has been conducted in the south central part of the Sierra Nevada. A summary of 31 stands (Table 5.15) shows an overall strong dominance by *Pinus balfouriana* ssp. *austrina*, but a stand-by-stand analysis shows that individual stands are dominated by *P. flexilis*, *P. albicaulis*, *P. monticola*, or *P. contorta* var. *murrayana*, as well as by *P. balfouriana*. Average absolute basal area and tree density for all taxa in a stand were 50 m^2 ha^{-1} and 300 ha^{-1}, respectively, values that are about 70% of those for red fir forest in the upper montane zone. *Tsuga mertensiana*, which reaches its southern limit at about 37° N, was not represented in these stands. Samples from the northern Sierra Nevada would show shared dominance by *Tsuga mertensiana* and *Pinus albicaulis* and an absence of *P. balfouriana* ssp. *austrina*.

Table 5.15. *Summary of 31 subalpine woodland stands, south central Sierra Nevada*

Species	PR	BA	D	IV
Pinus balfouriana ssp. *austrina*	81	48	52	100
Pinus albicaulis	32	20	13	33
Pinus flexilis	35	11	13	24
Abies magnifica (var. *shastensis*?)	26	9	7	16
Pinus contorta var. *murrayana*	35	8	7	15
Pinus monticola	16	4	3	7
Abies concolor	13	1	3	4
Populus trichocarpa	3	<1	2	2
Juniperus occidentalis ssp. *australis*	6	<1	<1	<1
Pinus monophylla	3	<1	<1	<1
Pinus jeffreyi	3	<1	<1	<1

Note: PR = presence (%); BA = relative basal area; D = relative density; IV = relative basal area + relative density.
Source: Data from Ryerson (1983), Lepper (1974), Rundel et al. (1977), and Vankat and Major (1978).

Lepper (1974) sampled three subalpine stands in southern California, in the Santa Rosa, San Bernardino, and San Gabriel mountains. On the average, *Pinus flexilis* contributed 43% relative density, *Abies concolor* var. *concolor* contributed 29%, *P. contorta* var. *murrayana* 18%, *Juniperus occidentalis* ssp. *australis* 6%, *P. jeffreyi* 4%, and *P. monophylla* 1%. Lepper found her California stands to fit well within a regional western association that she called the *Pinus flexilis–Ribes cereum–Artemisia tridentata* community. Other commonly associated species were *Koeleria cristata*, *Populus tremuloides*, *Sitanion hystrix*, and *Symphoricarpos vaccinoides*.

Lepper also examined stand age structure and growth rates, as estimated from cores. The maximum tree age in her Sierran–southern California stands ranged from 85 to 1176 yr and averaged 337 yr. Age (dbh) distribution curves were generally very "flat," showing uniform, sporadic establishment over a long period of time (Fig. 5.16). Seedling density was quite low, about 15 ha^{-1}. Growth rates did not correlate positively with importance values—that is, *P. flexilis* grew fastest in sites where it had a low IV, and where establishment of young plants was also low.

Ryerson (1983, 1984) investigated 15 southern

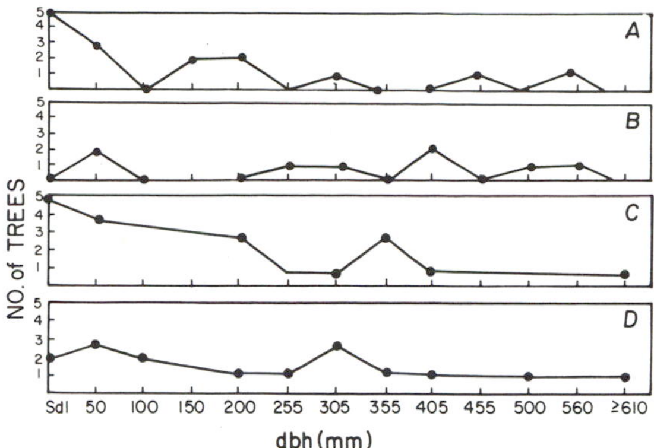

Figure 5.16. Age (dbh) distributions of Pinus flexilis *in 0.2 ha plots: A and B, Onion Valley, southern Sierra Nevada, 36° 46′ N; C, Santa Rosa Mountains, southern limit of limber pine; D, San Bernardino Mountains (from Lepper 1974).*

Sierran stands that contained *Pinus balfouriana*. The sites were at 2700–3600 m elevation, on both east and west slopes. She found diameter growth rates of foxtail pine to be similar to those of limber and whitebark pines (7–10 cm per century) and to be about half the growth rates of associated species from lower-elevation zones, such as red fir. Demographic profiles showed foxtail pine to be reproducing well enough over the past 500 yr (though sometimes with 200–300-yr gaps) to maintain constant populations. There is some evidence to show a downslope movement along the east side in the past several hundred years. Trees younger than 50 yr do not reproduce sexually. Senescence (associated with significant amounts of heart rot) appears to begin by 1000 yr of age, and the maximum tree age encountered was 3300 yr.

Some of her stand data for tree species have been incorporated into Table 5.15. Common associated shrubs were (in order of cover): *Chrysolepis sempervirens, Holodiscus microphyllus, Ribes* species, *Symphoricarpos parishii,* and *Cercocarpus ledifolius*. Associated herbaceous species were largely xerophytic. Foxtail pine appeared less tolerant of cool, short growing seasons than whitebark pine, and less tolerant of warm, arid growing seasons than limber pine.

Taylor (1976) sampled 26 subalpine woodland stands in the Carson Pass area of the Sierra Nevada (38° 40′N). These were dominated either by *Pinus albicaulis* or *Tsuga mertensiana* or both. Tree basal area and cover averaged 26 m^2 ha^{-1} and 39%, respectively. Lodgepole pine, western white pine, and red fir were common associates. *Pinus albicaulis* dominated the drier sites, as also indicated by summer pressure bomb readings 0.3–0.7 MPa more negative than those for hemlock.

EAST-SIDE SIERRAN FORESTS

The eastern, desert-facing slopes of the Sierra Nevada are physically quite different from the western slopes. The elevation changes precipitously, soils are more skeletal, and forest cover is less continuous than along the west face. In addition, the eastern escarpment is within a rainshadow, as prevailing winds bring precipitation from the west. Major (1977) showed temperature and precipitation lapse rates to be significantly steeper on the east slope than on the west slope.

Forest zonation is also different, being less marked into discrete communities or phases. Subalpine and upper montane tree species, for example, tend to associate in the same stands, and midmontane and lower montane zones are both thoroughly dominated by *Pinus jeffreyi*. Further, relatively few stands have been quantitatively described. Consequently, I shall divide east-side forests into only two phases: upper montane (which includes a sprinkling of subalpine conifers associated with *Abies magnifica, A. concolor* var. *lowiana,* and *P. contorta* var. *murrayana*); and lower montane (dominated by *P. jeffreyi*).

In the northern Sierra Nevada, the upper montane east-side zone generally extends from 2500 to 2900 m elevation; in the southern Sierra Nevada it is at 2800–3400 m (Rundel et al. 1977). *Abies magnifica* is a prominent species, but it does not often form pure stands of large individuals as it does on the west slope, because the area available is much smaller.

The lower montane east-side zone extends from 2000 to 2500 m elevation in the north and from 2600 to 2800 m elevation in the south. *Pinus jeffreyi* is the typical dominant. Forests are often open and parklike (Fig. 5.17), with fewer than 200 Jeffrey pines per hectare, and tree canopy cover less than 65%. On better sites, the largest overstory trees may reach 40 m in height and 120 cm dbh. Apart from occasional intrusions by patches of aspen (*Populus tremuloides* var. *aurea* or, along streams, *P. trichocarpa*), understory deciduous trees are rare. A summary of six Jeffrey pine stands appears in Table 5.16. At scattered locations in the northern Sierra Nevada, *P. washoensis* can also be associated. The type locality of this narrowly distributed tree is near Mt. Rose in Washoe County, Nevada (Billings 1954), and its relationship to eastern ponderosa pine (*Pinus ponderosa* var. *scopulorum*) has been summarized by Critchfield (1984).

Shrub cover is variable, but may average 20%, and the list of common species shows a cold desert flavor: *Arctostaphylos patula, Artemisia tridentata, Ceanothus prostratus, C. velutinus, Cercocarpus ledi-*

Figure 5.17. Jeffrey pine forest, east side of the central Sierra Nevada, Mono County, about 2400 m elevation.

Table 5.16. *Composition of east-side Jeffrey pine forest.*

Species	C	D	BA	IV	No. stands
Pinus jeffreyi	52	21	53	126	6
Calocedrus decurrens	22	33	28	83	6
Abies concolor var. *lowiana*	16	42	13	71	5
Populus trichocarpa	9	12	4	25	2
Pinus contorta var. *murrayana*	2	2	<1	4	1
Pinus ponderosa	<1	1	<1	1	1
Juniperus occidentalis ssp. *australis*	<1	<1	<1	<1	2
Quercus kelloggii	<1	<1	2	2	1
Pinus lambertiana	<1	<1	<1	<1	1
Totals:					
Absolute Cover (%)	58				
Absolute D (ha^{-1})	872				
Absolute BA (m^2 ha^{-1})	47				

Note: Data are means of six locations, from 36 to 40° N. C = relative cover; D = relative density; BA = relative basal area; IV = relative cover + relative density + relative basal area.
Source: Rundel et al. (1977).

folius, Chrysothamnus parryi, Haplopappus bloomeri, Leptodactylon pungens, Purshia tridentata.

Herb cover may exceed 10% cover at its seasonal peak. The most common species include *Bromus tectorum, Deschampsia elongata, Elymus glaucus, Sitanion hystrix, Stipa occidentalis,* and *Wyethia mollis* (Rundel et al. 1977).

Climatic data from Major (1967, 1977), Axelrod

(1981), and McDonald (1982) indicate that this east-side Jeffrey pine forest typically receives 500–1000 mm precipitation per year and experiences a mean annual temperature of 5–8°C; thus, it is a colder and drier habitat than that of west-side montane forest. Axelrod (1981) examined islands of mixed conifer forest within Jeffrey pine forest in the north-eastern Sierra Nevada and determined that the mean annual temperature was about 4°C cooler than for comparable forest on the west slope. Equability was also lower, about 45, in contrast to 53 for west-side forest. Most precipitation is in the form of snow, and the frost-free period is only 70–130 d.

AREAS FOR FUTURE RESEARCH

Significant contributions to our ecological understanding of Californian vegetation can be made at all levels: basic stand descriptions, autecological studies of key species, demographic studies of stand dynamics, and construction of predictive models. My list of specific topics that follows is biased, and it reflects the topics addressed in this chapter.

Comparative data on productivity, photosynthetic rate, carbon allocation patterns, and water relations of California oak species are few. Closely related deciduous and evergreen species and their hybrids could be compared to test several general hypotheses already in the literature. There is little quantified information on stand data for southern oak woodland phases, especially *Quercus agrifolia* woodland and walnut woodland. There is little information on natural fire frequency and stand response to fire in the various oak woodland phases of California.

Stand data on all the phases of mixed evergreen forest are few. Autecological studies of such important, wide-ranging dominants in mixed evergreen forest as *Pinus coulteri*, *Quercus chrysolepis*, and *Pseudotsuga macrocarpa* have not been conducted.

In the midmontane and upper montane conifer forests, comparative autecological studies have yet to be conducted for such important taxonomic pairs as: ponderosa pine–Jeffrey pine, white fir–red fir, the two varieties of white fir, east- and west-slope populations of Jeffrey pine, and sugar pine–western white pine. The closed-cone conifer *Pinus attenuata* has not been studied in northern California. The extent of ozone mottle disease in central and northern Sierra Nevada forests is not well known and needs clarification.

As the intensity of logging moves upslope, we need to understand the autecological requirements, demography, and stand dynamics of upper montane species. At present, some data on red fir exist, but most information on lodgepole pine comes from non-Californian varieties, and there seems to have been no attention paid to Sierran quaking aspen by the USDA Forest Service for the past 50 yr.

Climatic data need to be accumulated for the subalpine woodland zone. There is little information on stand structure for the mixed woodland in the central and northern Sierra Nevada, and autecological information exists for only one of the subalpine species: *Pinus flexilis*. The genetic status of krummholz forms remains untested.

Finally, there are essentially no quantitative descriptions of east-side Sierran forests or of Baja California forests.

REFERENCES

Aitro, V. P. (ed). 1977. Fifty years of forestry research, 1926–1975. USDA Forest Sevice general technical report PSW-23, Berkeley, Calif.

Axelrod, D. I. 1965. A method for determining the altitudes of Tertiary floras. Paleobotanist 14:144–171.

Axelrod, D. I. 1976. History of the coniferous forests, California and Nevada. Univ. Calif. Pub. Bot. 70: 1–62.

Axelrod, D. I. 1977. Outline history of California vegetation, pp. 139–192 in M. G. Barbour and J. Major (eds.), Terrestrial vegetation of California. Wiley, New York.

Axelrod, D. I. 1981. Holocene climatic changes in relation to vegetation disjunction and speciation. American Naturalist 117:847–870.

Bailey, D. K. 1970. Phytogeography and taxonomy of *Pinus* subsection Balfourianae. Ann. Miss. Bot. Gardens 57:210–249.

Baker, G. A., P. W. Rundel, and D. J. Parsons. 1981. Ecological relationships of *Quercus douglasii* (Fagaceae) in the foothill zone of Sequoia National Park, California. Madroño 28:1–12.

Barbour, M. G., and J. Major. 1977. Introduction, pp. 3–10 in M. G. Barbour and J. Major (eds.), Terrestrial vegetation of California. Wiley, New York.

Barbour, M. G., T. C. Wainwright, and C. Manansala. 1981. Preparation of a vegetation map of Calaveras Big Trees State Park, final report. California Department of Parks and Recreation, Sacramento.

Barbour, M. G., and R. A. Woodward 1985. The Shasta red fir forest of California. Can. J. Forestry 15:570–576.

Barry, W. J. 1971. The ecology of *Populus tremuloides*, a monographic approach. Ph.D. dissertation, University of California, Davis.

Billings, W. D. 1954. Nevada trees. Univ. Nev. Agric. Ext. Serv. Bull. 94:1–125.

Biswell, H. H. 1967. Forest fire in perspective, pp. 43–63 in Proceedings, California Tall Timbers Fire Ecology Conference, Tall Timbers Research Station, Tallahassee, Fla.

Borchert, M. 1985. Serotiny and cone-habit variation in populations of *Pinus coulteri* (Pinaceae) in the southern Coast Ranges of California. Madroño 32:29–48.

Borchert, M., and M. Hibberd. 1984. Gradient analysis of a north slope montane forest in the western Transverse Ranges of southern California. Madroño 31:129–139.
Bowerman, M. L. 1944. The flowering plants and ferns of Mount Diablo, California. Gillick Press, Berkeley.
Bradford, G. G., A. L. Page, and I. R. Straughan. 1981. Are Sierra lakes becoming acid? Calif. Agr. (May-June):6–7.
Brooks, W. H. 1971. A quantitative ecological study of the vegetation in selected stands of the grass-oak woodland in Sequoia National Park, California. Progress report to superintendent, Sequoia–Kings Canyon National Park, Three Rivers, Calif.
CDFR (California Department of Forestry). 1979. California's forest resources. Sacramento.
Cheatham, N. H., and J. R. Haller. 1973. An annotated list of California habitat types. Unpublished manuscript, Biological Sciences Department, University of California, Santa Barbara.
Chabot, B. F., and D. J. Hicks. 1982. The ecology of leaf life spans. Ann. Rev. Ecol. Syst. 13:229–259.
Clausen, J. 1965. Population studies of alpine and subalpine races of conifers and willows in the California high Sierra Nevada. Evolution 19:56–68.
Colwell, W. L., Jr. 1977. The status of vegetation mapping in California today, pp. 195–200 in M. G. Barbour and J. Major (eds.), Terrestrial vegetation of California. Wiley, New York.
Conard, S. G., and S. R. Radosevich. 1981. Photosynthesis, xylem pressure potential and leaf conductance of three montane chaparral species in California. Forest Sci. 27:627–639.
Conard, S. G., and S. R. Radosevich. 1982. Post-fire succession in white fir (*Abies concolor*) vegetation of the northern Sierra Nevada. Madroño 29:42–56.
Cooper, W. S. 1922. The broad-sclerophyll vegetation of California. Carnegie Institution of Washington publication 319.
Critchfield, W. B. 1957. Geographic variation in *Pinus contorta*. Maria Moors Cabot Foundation, publication 3, Harvard University, Cambridge, Mass.
Critchfield, W. B. 1977. Hybridization of foxtail and bristlecone pines. Madroño 24:193–212.
Critchfield, W. B. 1984. Crossability and relationships of Washoe pine. Madroño 31:144–170.
Daubenmire, R. F. 1938. Merriam's life zones of North America. Q. Rev. Biol. 13:327–332.
Daubenmire, R. F. 1954. Alpine timberlines in the Americas and their interpretation. Butler Univ. Bot. Studies 11:119–136.
Dunn, P. H. 1980. Nutrient-microbial considerations in oak management, pp. 148–160 in T. R. Plumb (ed.), Ecology, management and utilization of California oaks. USDA Forest Service, PSW Forest and Range Experiment Station, PSW-44.
Eyre, F. H. (ed.). 1980. Forest cover types of the United States and Canada. Society of American Foresters, Washington, D.C.
Fowells, H. A. (ed.). 1965. Silvics of forest trees of the United States. USDA Forest Service agricultural handbook 271, Washington, D.C.
Franklin, J. F., F. C. Sorensen, and R. K. Campbell. 1980. Summarization of the ecology and genetics of the noble and California red fir complex, pp. 133–139 in Proceedings IUFRO joint meeting of workshop parties, vol. 1. USDA Forest Service, Washington, D.C.
Gemmill, B. 1980. Radial growth of California black oak in the San Bernardino Mountains, pp. 128–135 in T. R. Plumb (ed.), Ecology, management and utilization of California oaks. USDA Forest Service, PSW Forest and Range Experiment Station, PSW-44.
Goldman, E. A. 1916. Plant records of an expedition to Lower California. Contr. U.S. Natl. Herbarium 16:309–371.
Gordon, D. T. 1970a. Natural regeneration of white and red fir . . . influence of several factors. USDA Forest Service research paper PSW-58, Berkeley, Calif.
Gordon, D. T. 1970b. Shade improves survival rate of outplanted 2-0 red fir seedlings. USDA Forest Service research note PSW-210, Berkeley, Calif.
Gray, J. T. 1978. The vegetation of two California mountain slopes. Madroño 25:177–185.
Griffin, J. R. 1965. Digger pine seedling response to serpentinite and non-serpentinite soil. Ecology 46:801–807.
Griffin, J. R. 1971. Oak regeneration in the upper Carmel Valley, California. Ecology 52:862–868.
Griffin, J. R. 1973. Xylem sap tension in three woodland oaks of central California. Ecology 54:152–159.
Griffin, J. R. 1976. Regeneration in *Quercus lobata* savannas, Santa Lucia Mountains, California. Amer. Midl. Nat. 95:422–435.
Griffin, J. R. 1977. Oak woodland, pp. 383–415 in M. G. Barbour and J. Major (eds.), Terrestrial vegetation of California. Wiley, New York.
Griffin, J. R. 1978. The marble-cone fire ten months later. Fremontia 6(2):8–14.
Griffin, J. R. 1980a. Animal damage to valley oak acorns and seedlings, Carmel Valley, California, pp. 242–245 in T. R. Plumb (ed.), Ecology, management and utilization of California oaks. USDA Forest Services, PSW Forest and Range Experiment Station PSW-44.
Griffin, J. R. 1980b. Sprouting in fire-damaged valley oaks, Chews Ridge, California, pp. 216–219 in T. R. Plumb (ed.), Ecology, management and utilization of California oaks. USDA Forest Service, PSW Forest and Range Experiment Station PSW-44.
Griffin, J. R., and W. B. Critchfield. 1972. The distribution of forest trees in California. USDA Forest Service research paper PSW-82, Berkeley, Calif.
Haller, J. R. 1959. Factors affecting the distribution of ponderosa and Jeffrey pines in California. Madroño 15:65–71.
Haller, J. R. 1962. Variation and hybridization in ponderosa and Jeffrey pines. Univ. Calif. Pub. Bot. 34:123–166.
Hallin, W. E. 1957. Silvical characteristics of California red fir and Shasta red fir. USDA Forest Service technical paper 16, Berkeley, Calif.
Hamrick, J. L., and W. J. Libby. 1972. Variation and selection in western U.S. montane species. I. White fir. Silvae Genet. 21:29–35.
Hanes, T. L. 1976. Vegetation types of the San Gabriel

Mountains, pp. 65–76 in J. Latting (ed.), Plant communities of southern California. Special publication 2, California Native Plant Society, Berkeley.

Hartesveldt, R. J., H. T. Harvey, H. S. Shellhammer, and R. E. Stecker. 1975. The giant sequoia of the Sierra Nevada. U.S. Department of the Interior, National Park Service, Washington, D.C.

Harvey, H. T., H. S. Shellhammer, and R. E. Stecker. 1980. Giant sequoia ecology. USDI, National Park Service, Scientific Monograph Series No. 12, Washington, D.C.

Hastings, J. R., and R. R. Humphrey (eds.). 1969. Climatological data and statistics for Baja California. University of Arizona Institute of Atmospheric Physics, technical report 18, Tempe.

Holland, R. F. 1986. Preliminary descriptions of the terrestrial natural communities of California. California Resources Agency, Department Fish and Game, Sacramento.

Holland, V. L. 1980. Effect of blue oak on rangeland forage production in central California, pp. 314–318 in T. R. Plumb (ed.), Ecology, management and utilization of California oaks. USDA Forest Service, PSW Forest and Range Experiment Station PSW-44.

Holland, V. L., and J. Morton. 1980. Effect of blue oak on nutritional quality of rangeland forage in central California, pp. 319–322 in T. R. Plumb (ed.), Ecology, management and utilization of California oaks. USDA Forest Services, PSW Forest and Range Experiment Station PSW-44.

Holstein, G. 1981. Special plant communities of California–working lists. The Nature Conservancy, San Francisco.

Kay, B. L., and O. A. Leonard. 1980. Effect of blue oak removal on herbaceous forage production in the north Sierra foothills, pp. 323–328 in T. R. Plumb (ed.), Ecology, management and utilization of California oaks. USDA Forest Service, PSW Forest and Range Experiment Station PSW-44.

Kickert, R. N., et al. 1976. Photochemical air pollutant effects on mixed conifer ecosystems: a progress report. U.S. Environmental Protection Agency, Office of Research and Development, CERL-026, Corvallis, Ore.

Kilgore, B. M. 1973. The ecological role of fire in Sierran conifer forests: its application to national park management. Quat. Res. 3:496–513.

Kilgore, B. M., and D. Taylor. 1979. Fire history of a sequoia–mixed conifer forest. Ecology 60:129–142.

Klikoff, L. 1965. Microenvironmental influence on vegetational pattern near timberline in the central Sierra Nevada. Ecol. Monogr. 35:187–211.

Komarek, E. C. 1967. The nature of lightning fires. Proc. Tall Timbers Fire Ecol. Conf. 7:5–41.

Küchler, A. W. 1964. Potential natural vegetation of the coterminous United States. American Geographical Society special publication 36, New York.

Küchler, A. W. 1977. The map of the natural vegetation of California, pp. 909–938 in M. G. Barbour and J. Major (eds.), Terrestrial vegetation of California. Wiley, New York.

Laacke, R. J. 1979. California forest soils. University of California, Division of Agricultural Science publication 4094, Berkeley.

Lanini, W. T., and S. R. Radosevich. 1986. Responses of three conifer species to site preparation and shrub control. Forest Sci. 32:61–77.

Larcher, W. 1980. Physiological plant ecology, 2nd ed. Springer-Verlag, New York.

Lepper, M. G. 1974. *Pinus flexilus* James, and its environmental relationships. Ph.D. dissertation, University of California, Davis.

Lepper, M. G. 1980. Carbon dioxide exchange in *Pinus flexilis* and *P. strobiformis* (Pinaceae). Madroño 27:17–24.

Littrell, E. E., and P. M. McDonald. 1974. Within-stand dynamics of bigcone Douglas-fir in southern California. Unpublished report, PSW Forest and Range Experiment Station, Berkeley, Calif.

McBride, J. R., V. P. Semion, and P. R. Miller. 1975. Impact of air pollution on the growth of ponderosa pine. Calif. Agr. 29(12):8–9.

McColl, J. G. 1981. Effects of acid rain on plants and soils in California, final report. California Air Resources Board, Contract A8-136-31, Sacramento.

McDonald, P. M. 1980. Growth of thinned and unthinned hardwood stands in the northern Sierra Nevada . . . preliminary findings, pp. 119–127 in T. R. Plumb (ed.), Ecology, management, and utilization of California oaks. USDA Forest Service, PSW Forest and Range Experiment Station PSW-44.

McDonald, P. M. 1982. Climate, history and vegetation of the eastside pine type in California, pp. 1–16 in Management of the eastside pine type in northeastern California. Cooperative Extension, University of California, Berkeley.

MacMahon, J. A., and D. C. Andersen. 1982. Subalpine forests: a world perspective with emphasis on western North America. Prog. Phy. Geogr. 6:368–425.

Major, J. 1967. Potential evapotranspiration and plant distribution in western states with emphasis on California, pp. 93–126 in R. H. Shaw (ed.), Ground level climatology. AAAS, Washington, D.C.

Major, J. 1977. California climate in relation to vegetation, pp. 11–74 in M. G. Barbour and J. Major (eds.), Terrestrial vegetation of California. Wiley, New York.

Major, J., and D. W. Taylor. 1977. Alpine, pp. 601–675 in M. G. Barbour and J. Major (eds.), Terrestrial vegetation of California. Wiley, New York.

Mason, H. L. 1970. The scenic, scientific, and educational values of the natural landscape of California. California Department of Parks and Recreation, Sacramento.

Mastroguiseppe, R. J., and J. D. Mastroguiseppe. 1980. A study of *Pinus balfouriana* Grev. and Balf. (Pinaceae). Syst. Bot. 5:86–104.

Matyas, W. J., and I. Parker. 1980. CALVEG, mosaic of existing vegetation of California. Regional Ecology Group, USDA Forest Service, San Francisco.

Merriam, C. H. 1898. Life zones and crop zones of the United States. USDA Biol. Surv. Div. Bull. 10:9–79.

Miller, P. L. 1973. Oxidant-induced community change

in a mixed conifer forest, pp. 101–117 in J. Naegle (ed.), Air pollution damage to vegetation. Advances in chemistry series 122, American Chemical Society, Washington, D.C.

Minnich, R. A. 1976. Vegetation of the San Bernardino Mountains, pp. 99–124 in J. Latting (ed.), Plant communities of southern California. Special publication 2, California Native Plant Society, Berkeley.

Minnich, R. A. 1980. Wildfire and the geographic relationships between canyon live oak, Coulter pine, and bigcone Douglas-fir forests, pp. 55–61 in T. R. Plumb (ed.), Ecology, management and utilization of California oaks. USDA Forest Service, PSW Forest and Range Experiment Station PSW-44.

Mooney, H. A. 1972. The carbon balance of plants. Ann. Rev. Ecol. Syst. 3:315–346.

Mooney, H.A., R. D. Wright, and B. R. Strain. 1964. Field measurements of the metabolic responses of bristlecone pine and big sagebrush in the White Mountains of California. Amer. Midl. Nat. 72:281–297.

Mueller-Dombois, D., and H. Ellenberg. 1974. Aims and methods of vegetation ecology. Wiley, New York.

Munz, P. A., and D. D. Keck. 1959. A California flora. University of California Press, Berkeley.

Myatt, R. G. 1980. Canyon live oak vegetation in the Sierra Nevada, pp. 86–91 in T. R. Plumb (ed.), Ecology, management, and utilization of California oaks. USDA Forest Service, PSW Forest and Range Experiment Station PSW-44.

NAPCA (National Air Pollution Control Administration). 1970. Air quality criteria for photochemical oxidants. U.S. DHEW, Washington, D.C.

Nelson, E. W. 1922. Lower California and its natural resources. National Academy of Science, XVI (first memoir). Washington, D.C.

Oosting, H. J., and W. D. Billings. 1943. Abietum magnificae: the red fir forest of the Sierra Nevada. Ecol. Monogr. 13:259–274.

Parker, A. J. 1982. Environmental and compositional ordinations of conifer forests in Yosemite National Park, California. Madroño 29:109–118.

Parker, A. J. 1986. Persistence of lodgepole pine forests in the central Sierra Nevada. Ecology 67: 1560–1567.

Pavlik, B. M. 1976. A natural history of southern California oaks. Report to the West Los Angeles County Resource Conservation District, Van Nuys, Calif.

Paysen, T. E., J. A. Derby, H. Balck, Jr., V. C. Bleich, and J. W. Mincks. 1980. A vegetation classification system applied to southern California. USDA Forest Service, PSW Forest and Range Experiment Station general technical report PSW-45.

Paysen, T. E., J. A. Derby, and C. E. Conrad. 1982. A vegetation classification system for use in California. USDA Forest Service, PSW Forest and Range Experiment Station general technical report PSW-63.

Pitcher, D. C. 1981. The ecological effects of fire on stand structure and field dynamics in red fir forests of Mineral King, Sequoia National Park, California. M.S. thesis, University of California, Berkeley.

Plumb, T. R. 1980. Response of oaks to fire, pp. 202–215 in T. R. Plumb (ed.), Ecology, management and utilization of California oaks. USDA Forest Service, PSW Forest and Range Experiment Station PSW-44.

Pronos, J., and D. R. Vogler. 1981. Assessment of ozone injury to pines in the southern Sierra Nevada, 1979/1980. USDA Forest Service, forest and pest management report 81-20. San Francisco.

Raven, P. H., and D. I. Axelrod. 1978. Origin and relationships of California flora. Univ. Calif. Pub. Bot. 72:1–134.

Rourke, M. D. 1986. A preliminary cladistic analysis of the Balfourianae pines based strictly on published character data, pp. 77–83 in C. A. Hall and D. J. Young (eds.), Natural history of the White-Inyo Range, Eastern California and Western Nevada. White Mountain Research Station, Bishop, California.

Rundel, P. W. 1971. Community structure and stability in the giant sequoia groves of the Sierra Nevada, California. Amer. Midl. Nat. 85:478–492.

Rundel, P. W. 1972a. An annotated check list of the groves of *Sequoiadendron giganteum* in the Sierra Nevada, California. Madroño 21:319–328.

Rundel, P. W. 1972b. Habitat restriction in giant sequoia: the environmental control of grove boundaries. Amer. Midl. Nat. 87:81–99.

Rundel, P. W. 1980. Adaptations of mediterranean-climate oaks to environmental stress, pp. 43–54 in T. R. Plumb (ed.), Ecology, management and utilization of California oaks. USDA Forest Service, PSW Forest and Range Experiment Station, PSW-44.

Rundel, P. W., D. J. Parsons, and D. T. Gordon. 1977. Montane and subalpine vegetation of the Sierra Nevada and Cascade Ranges, pp. 559–599 in M. G. Barbour and J. Major (eds.), Terrestrial vegetation of California. Wiley, New York.

Ryerson, A. D. 1983. Population structure of *Pinus balfouriano* Grev. & Balf. along the margins of its distribution area in the Sierran and Klamath regions of California. M.A. thesis, California State University, Sacramento.

Ryerson, A. D. 1984. Krummholz foxtail pines. Fremontia 11(4):30.

Sawyer, J. O., D. A. Thornburg, and J. R. Griffin. 1977. Mixed evergreen forest, pp. 359–415 in M. G. Barbour and J. Major (eds.), Terrestrial vegetation of California. Wiley, New York.

Schmidle, C. J. 1983. Mesa Oaks Reserve feasibility study. California Department of Parks and Recreation, Sacramento.

Selter, C. M. 1983. Site microenvironment characteristics and seedling survival of Shasta red fir (*Abies magnifica* var. *shastensis*). M.S. dissertation, San Jose State University.

Sigg, J. 1983. The foxtail pine of the Sierra. Fremontia 11(1):3–8.

Silen, R. R., W. B. Critchfield, and J. F. Franklin. 1965. Early verification of a hybrid between noble and California red firs. Forest Sci. 11:460–462.

Smiley, F. J. 1921. A report upon the boreal flora of the Sierra Nevada of California. Univ. Calif. Pub. Bot. 9:1–423.

Smith, J. L. 1978a. Snowpack characteristics and the simulated effects of weather modification upon them. Central Sierra Snow Laboratory, USDA Forest Service, PSW Forest and Range Experiment Station, Berkeley, Calif.

Smith, J. L. 1978b. Historical climatic regime and the projected impact of weather modification upon precipitation and temperature at the Central Sierra Snow Laboratory. USDA Forest Service, PSW Forest and Range Experiment Station, Berkeley, Calif.

Snow, G. E. 1980. The fire resistance of Engelmann and coast live oak seedlings, pp. 62–66 in T. R. Plumb (ed.), Ecology, management and utilization of California oaks. USDA Forest Service, PSW Forest and Range Experiment Station PSW-44.

Stebbins, G. L., and J. Major. 1965. Endemism and speciation in the California flora. Ecol. Monogr. 35:1–35.

Strubble, G. R. 1973. Biology, ecology, and control of the lodgepole needle miner. USDA technical bulletin 1458, Washington, D.C.

Talley, S. N. 1974. The ecology of Santa Lucia fir (*Abies bracteata*), a narrow endemic of California. Ph.D. dissertation, Duke University, Durham, N.C.

Talley, S. N. 1977a. An ecological survey of the Babbit Peak candidate research natural area on the Tahoe National Forest. Report on PO 1222-PSW-75, USDA Forest Service, Berkeley, Calif.

Talley, S. N. 1977b. An ecological survey of the Onion Creek candidate research natural area on the Tahoe National Forest, California. Report on PO 896-PSW-75, USDA Forest Service, Berkeley, Calif.

Talley, S. N., and J. R. Griffin, 1980. Fire ecology of a montane pine forest, Junipero Serra Peak, California. Madroño 27:49–60.

Taylor, D. W. 1976. Vegetation near timberline at Carson Pass, central Sierra Nevada, California. Ph.D. dissertation, University of California, Davis.

Taylor, D. W. 1977. Floristic relationships along the Cascade-Sierran axis. Amer. Midl. Nat. 97:333–349.

Taylor, O.C. 1980. Photochemical oxidant air pollution effects on a mixed conifer forest ecosystem. EPA-600/3-80-002, U.S. Environmental Protection Agency, Corvallis, Ore.

Thorne, R. F. 1976. The vascular plant communities of California, pp. 1–31 in J. Latting (ed.), Plant communities of southern California. Special publication 2, California Native Plant Society, Berkeley, Calif.

Thorne, R. F. 1977. Montane and subalpine forests of the Transverse and Peninsular Ranges, pp. 537–557 in M. G. Barbour and J. Major (eds.), Terrestrial vegetation of California. Wiley, New York.

Thorne, R. F. 1982. The desert and other transmontane plant communities of southern California. Aliso 102:219–257.

Tomback, D. F. 1986. Post-fire regeneration of krummholz whitebark pine. Madroño 33:100–110.

Tucker, J. M. 1980. Taxonomy of California oaks, pp. 19–29 in T. R Plumb (ed.), Ecology, management and utilization of California oaks. USDA Forest Service, PSW Forest and Range Experiment Station PSW-44.

Twisselman, E. D. 1967. A flora of Kern County, California. Wasmann J. Biol. 25:1–395.

UNESCO (United Nations Educational and Scientific Council). 1973. International classification and mapping of vegetation. Paris.

USDC (United States Department of Commerce). 1964. Climatic summary of the United States, supplement for 1951 through 1960, California. Washington, D.C.

Ustin, S. L. 1976. Geographic variation in relative cone bract length, cotyledon number, and monoterpene composition of *Abies magnifica* in the southern Sierra Nevada. M.S. thesis, California State University, Hayward.

Ustin, S. L., R. A. Woodward, M. G. Barbour, and J. L. Hatfield. 1984. Relationships between sunfleck dynamics and red fir seedling distribution. Ecology 65:1420–1428.

Vankat, J. L. 1970. Vegetation change in Sequoia National Park, California. Ph.D. diss., Univ. California, Davis, 198 pp.

Vankat, J. L. 1982. A gradient perspective on the vegetation of Sequoia National Park, California. Madroño 29:200–214.

Vankat, J. L., and J. Major. 1978. Vegetation changes in Sequoia National Park, California. J. Biogeography 5:377–402.

Vogl, R. J. 1973. Ecology of knobcone pine in the Santa Ana Mountains, California. Ecol. Mongr. 43:125–143.

Vogl, R. J. 1976. An introduction to the plant communities of the Santa Ana and San Jacinto Mountains, pp. 77–98 in J. Latting (ed.), Plant communities of southern California. Special publication 2, California Native Plant Society, Berkeley.

Vogl, R. J., and B. C. Miller. 1968. The vegetational composition of the south slope of Mt. Pinos, California. Madroño 19:225–234.

Wainwright, T. C., and M. G. Barbour. 1984. Characteristics of mixed evergreen forest in the Sonoma Mountains of California. Madroño 31:219–230.

Wanner, J. L. 1981. Winter carbohydrate allocation in *Pinus albicaulis* and *Tsuga mertensiana* at timberline in the Sierra Nevada. M.S. thesis, California State University, Sacramento.

White, K. L. 1966. Structure and composition of foothill woodland in central coastal California. Ecology 47:229–237.

Wiggins, I. L. 1980. Flora of Baja California. Stanford University Press.

Williams, W. T. 1983. Tree growth and smog disease in the forests of California: case history, ponderosa pine in the southern Sierra Nevada. Environ. Pollut. (Series A) 30:59–75.

Williams, W. T., M. Brady, and S. C. Willison. 1977. Air pollution damage to the forests of the Sierra Nevada Mountains of California. J. Air Pollut. Control Assoc. 27:230–234.

Wright, R. D. 1968. Lower elevational limits of montane trees. II. Environment-keyed responses of three conifer species. Bot. Gaz. 129:219–226.

Yeaton, R. I. 1978. Some ecological aspects of reproduction in the genus *Pinus* L. Bull. Torrey Botanical Club 105:306–311.

Yeaton, R. I. 1981. Seedling characteristics and elevational distributions of pine (Pinaceae) in the Sierra Nevada of central California; a hypothesis.

Madroño 28:67–77.
Yeaton, R. I. 1983a. The effect of predation on the elevational replacement of digger pine by ponderosa pine on the western slopes of the Sierra Nevada. Bull. Torrey Botanical Club 110:31–38.
Yeaton, R. I. 1983b. The successional replacement of ponderosa pine by sugar pine in the Sierra Nevada. Bull. Torrey Botanical Club 110:292–297.
Yeaton, R. I., R. W. Yeaton, and J. E. Horenstein. 1980. The altitudinal replacement of digger pine by ponderosa pine on the western slopes of the Sierra Nevada. Bull. Torrey Botanical Club 107:487–495.
Zavarin, E., W. B. Critchfield, and K. Snajberk. 1978. Geographic differentiation of monoterpenes from *Abies procera* and *Abies magnifica*. Biochem. Syst. Ecol. 6:267–278.

Note added in proof. A major descriptive work on the distribution of 22 forest and woodland tree species in Baja California was published recently. Though appearing too late to be utilized in this chapter, its importance can at least be acknowledged by citing it here: Minnich, R. A. 1987. The distribution of forest trees in Northern Baja California, Mexico. Madroño 34:98–127.

Chapter 6

Chaparral

JON E. KEELEY
STERLING C. KEELEY

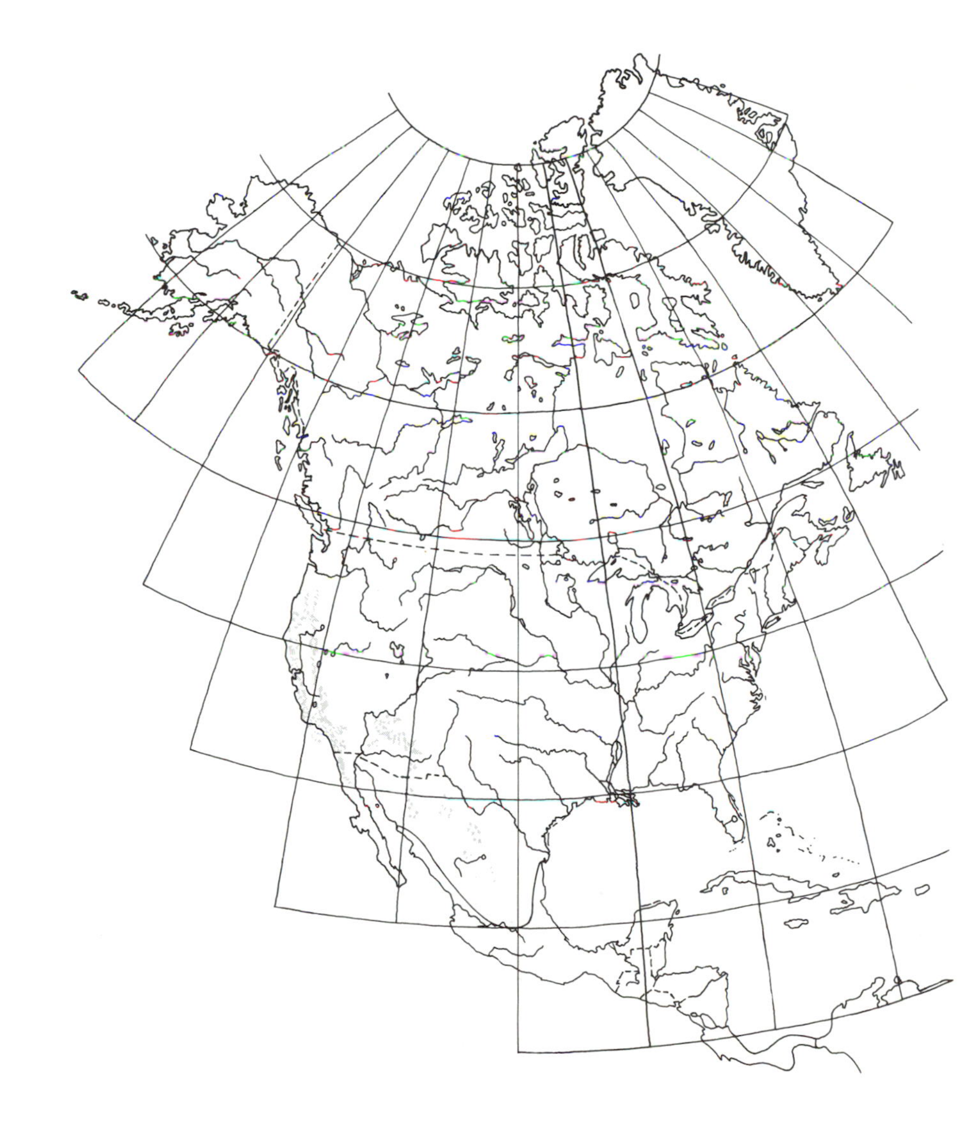

INTRODUCTION

Chaparral is the name applied to the sclerophyllous shrub vegetation of southwestern North America. This chapter concentrates on the evergreen chaparral centered in California, although related vegetation–including the interior chaparral of Arizona and northern Mexico, the winter-deciduous "petran" chaparral of the Rocky Mountains, and the West Coast summer-deciduous "soft chaparral" or coastal sage scrub–will also be discussed. For a complete bibliography of chaparral literature, see Keeley (1984a), and for reviews with a more historical perspective, see Mooney and Parsons (1973) and Hanes (1977). For in-depth comparisons of chaparral with other mediterranean-climate sclerophyllous shrub communities, see Mooney (1977b), Miller (1981), Shmida and Barbour (1982), and Kruger and associates (1983).

California chaparral dominates the foothills from the Sierra Nevada to the Pacific Ocean (Wieslander and Gleason 1954). The northern limits are the drier parts of the Rogue River watershed in Oregon (43° N latitude) (Detling 1961), and the southern limits are the San Pedro Mártir Mountains of Baja California (30° N) (Shreve 1936), and extending to 27° 30′ N in isolated patches (Axelrod 1973).

Throughout this region, chaparral characteristically forms a nearly continuous cover of closely spaced shrubs 1–4 m tall, with intertwining branches (Fig. 6.1). Herbaceous vegetation is generally lacking, except after fires, which are frequent throughout the range. Chaparral occurs from sea level to 2000 m on rocky, nutrient-poor soils and is best developed on steep slopes. Because of complex patterns of topographic, edaphic, and climatic variations, chaparral may form a mosaic pattern in which patches of oak woodland, grassland, or coniferous forest appear, often in sharp juxtaposition. The fire frequency and other environmental factors play important roles in determining distribution. Chaparral is replaced by grassland in frequently burned regions, especially at low elevations, and by oak woodland on mesic slopes where fires occur infrequently.

California chaparral is distributed in a region of mediterranean climate: cool, wet winters and hot, dry summers (Fig. 6.2). Rainfall is 200–1000 mm annually, two-thirds of which falls November to April in storms of several days' duration (Miller and Hajek 1981). Because of the episodic nature of the winter rains, there may be prolonged dry spells, even during the wet season. The annual rainfall variance is significantly greater than in other regions, and extreme droughts are not uncommon (Varney 1925; Major 1977). Significant summer

Figure 6.1. View of southern California mixed chaparral.

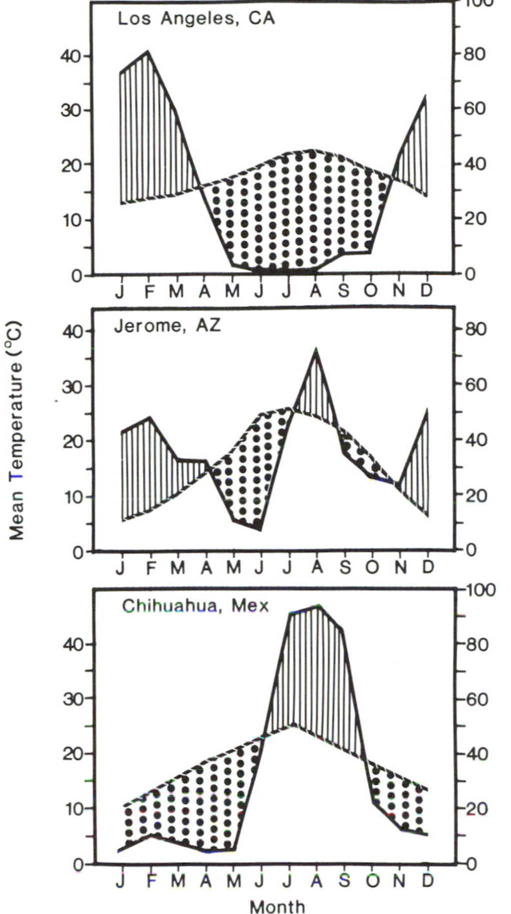

Figure 6.2. Climatic data for three chaparral sites (vertical lines, ppt > evap; dotted area, evap > ppt; solid line = precipitation, dashed line = mean temperature): Los Angeles, California (75 m; 34°05′, 118°15′), Jerome, Arizona (1600 m; 34°45′, 112°07′), and Chihuahua, Mexico (1350 m; 28°42′, 105°57′). In addition to differences in seasonal distribution of precipitation, the annual variance in precipitation is much greater in California than interior regions. For example, 40 yr of data for Los Angeles showed 5 mo with a coefficient of variation between 220% and 350%, whereas the Arizona site had no month with more than 120% (data from U.S. Department of Commerce, Climatic Summary of the United States).

precipitation is rare and arises from convectional storms in the higher elevations or tropical storms in the south. Mean winter temperatures range from less than 0°C at montane sites to greater than 10°C at lower elevations. Summer temperatures often exceed 40°C, but are more moderate along the coast and at the upper elevational limits.

The climate is dominated by the subtropical high-pressure cell that forms over the Pacific Ocean. During the summer, this air mass moves northward and blocks polar fronts from reaching land. During the winter, this high-pressure cell moves toward the equator and allows winter storms to pass onto land. The climate is wettest in the north, where the effect of the Pacific High is least, and becomes progressively drier to the south.

Another important factor is the orographic effect. Air cools adiabatically with increasing elevation, so that temperature decreases and precipitation increases with elevation. The interior sides of mountains lie within rainshadows where it is hotter and drier than in coastal exposures at comparable elevations.

A factor of local importance in southern California is the Santa Ana wind, which is the result of a high-pressure cell in the interior of the United States driving dry desert air toward the coast. These föhn-type winds may exceed 100 km hr^{-1} and bring high temperatures and low humidities. Santa Anas are most common in spring and fall, and some of the most catastrophic wildfires occur during these conditions.

COMMUNITY COMPOSITION

More than 100 evergreen shrub species occur in chaparral, but only a fraction of these are widespread (Table 6.1). There may be as few as one or more than 20 species at a given site, depending on available moisture, slope aspect, slope angle, distance from the coast, elevation, latitude, and fire history. Generalizations about species preferences toward edaphic and topographic features are difficult to make in that they often change with the region. Attempts to ascribe site preferences for most species have generally produced weak correlations (Gauss 1964; Wilson and Vogl 1965; Zenan 1967; Hanes 1971; Steward and Webber 1981).

Examples of species compositions and coverage values for a range of mature chaparral stands are shown in Table 6.2. Because of the overlapping of branches, areal coverage often exceeds 100% (ground-surface cover). Typically, bare ground will be much less than 10%, but on drier desertic sites (sites VII and VIII in Table 6.2) or on serpentine soils the percentage of bare ground may be much greater. It is not uncommon to encounter stands dominated by a single species. In the absence of disturbance, species diversity is said to decrease with age, but it is clear from Table 6.2 that site factors play a major role in determining community diversity. Detailed demographic patterns for three chaparral stands are shown in Table 6.3. Sometimes a species that represents a minor part of the cover may be numerically important, as is *Cercocarpus betuloides* at site 1. This particular species often spreads over short distances by rhizomes, but

Table 6.1. *Widespread dominant California evergreen chaparral shrub species*

Family and species	Common name	Distribution: California	Distribution: Interior	Distribution: Non-chaparral communities
Anacardiaceae				
Rhus integrifolia	Lemonadeberry	s, B		css
R. ovata	Sugar bush	s, B	Az	
Malosma (Rhus) laurina (Young 1974)	Laurel sumac	s, B		css
Ericaceae				
Arctostaphylos	Manzanita			
**A. canescens*		n	O	ypf
A. crustaceae		c-n		
A. glandulosa		s-n, B	O	
**A. glauca*		s-c, B		
**A. manzanita*		c-n		ypf
A. mewukka		c		ypf
**A. parryana*		s		ypf
A. patula (Wells 1968)		s	O, Ut	ypf
A. peninsularis (Wells 1972)		s?, B		
**A. pringlei*		s, B	Az	ypf
**A. pungens*		c-s, B	Az-Mx	owd, ypf
**A. stanfordiana*		n	O	
A. tomentosa		c-n		ccf
**A. viscida* (Wells 1968)		c-n	O	owd, ypf
Comarostaphylis diversifolia	Summer holly	s, B		
Xylococcus bicolor	Mission manzanita	s, B		
Fabaceae				
Pickeringia montana	Chaparral-pea	s-n, B		
Fagaceae				
Quercus dumosa	Scrub oak	s-n, B		
Q. durata		c-n		
Q. turbinella		s-c, B	Az-Tx	
Garryaceae				
Garrya buxifolia	Silk-tassel bush	n	O	ewd, ypf
G. congdoni		c-n		
G. elliptica		s-n	O	ewd
G. flavescens		s-c	Az	ypf, pjw
G. fremontii		s-n	O-Ws	ypf, ewd,
G. veatchii		s-c, B		owd
Papaveraceae				
Dendromecon rigida	Tree poppy	s-n, B		
Rhamnaceae				
Ceanothus	Buckbrush, California lilac			
[Section *Cerastes*]				
**C. crassifolius*		s, B		
**C. cuneatus*		s-n, B	O	
**C. greggii*		s-c,	Az-Mx	pjw
**C. jepsonii*		n		

Table 6.1. *(Cont.)*

Family and species	Common name	Distribution: California	Interior	Non-chaparral communities
*C. *jepsonii*		n		
*C. *megacarpus*		s-c		
*C. *pumilus*		n	O	
*C. *ramulosus*		c-n		
*C. *verrucosus*		s, B		
[Section *Euceanothus*]				
C. *cordulatus*		s-n, B	O, Nv	ypf
C. *dentatus*		c		ccf, ewd
C. *intergerrimus*		s-n	O-Ws. Az	ypf, ewd
C. *leucodermis*		s-n, B		
C. *oliganthus*		s-c		
C. *palmeri*		s-n, B		ypf
C. *papillosus*		s-c		ewd
C. *parryi*		n		
C. *sorediatus*		s-n		owd, ewd
C. *spinosus*		s-c, B		css
C. *thyrsiflorus*		s-n	O	ypf
C. *tomentosus*		s-c, B		
C. *velutinus*		c-n	O-BC-Co	ypf, ewd
Rhamnus crocea	Redberry	s-n, B	Az	owd, ypf
R. californica	Coffee berry	s-n, B	O, AZ-NM	css, ewd, owd, ypf
Rosaceae				
Adenostoma fasciculatum	Chamise	s-n, B		
A. sparsifolium	Red shanks	s-c, B		
Cercocarpus betuloides	Mountain mahogany	s-n, B	O, Az	owd
C. ledifolius		s-n, B	O-Ws-Co-A	pjw, sbs
Heteromeles arbutifolia	Chaparral holly	s-n, B		owd
Prunus ilicifolia	Chaparral cherry	s-n, B		owd
Stercullaceae				
Fremontia [Fremontodendron]	Flannel bush			
F. californica		s-n	Az	
F. mexicana		s, B		

Note: s = southern, c = central, n = northern, B = Baja California, O = Oregon, Az = Arizona, BC = British Columbia, Co = Colorado, Mx = Mexico, Nv = Nevada, NM = New Mexico, Tx = Texas, Ut = Utah, Ws = Washington. Non-chaparral communities: ccf = closed-cone forest, css = coastal sage scrub, ewd = evergreen woodland, owd = oak woodland, pjw = pinyon-juniper woodland, sbs = sagebrush scrub, ypf = yellow pine forest.

*Indicates obligate-seeding species (i.e., taxa without the ability to regenerate after tops are killed), with exceptions: *Ceanothus cuneatus* is a very weak sprouter in certain high-elevation Sierra Nevada sites; *C. greggii* is reported to have weak sprouting ability in Arizona; various *Arctostaphylos* taxa show subspecific variation in this regard (Wells 1968, 1972).

Source: Nomenclature according to Munz (1959) except where indicated otherwise.

Table 6.2. *Areal coverage values for 11 selected stands (I–XI) of mature chaparral throughout the range*

	Percentage ground surface cover										
Species	I	II	III	IV	V	VI	VII	VIII	IX	X	XI
Adenostoma fasciculatum	138	49	80	7		5			33	9	5
Arctostaphylos auriculata										64	
A. glandulosa						2			58		
A. glauca			33				17	19			
A. pajaroensis											50
Ceanothus cuneatus										10	
C. greggii		22					9		7		
C. megacarpus				79							
C. spinosus					136						
Cercocarpus betuloides				26	3	11	5				
Heteromeles arbutifolia					5	10					12
Quercus dumosa		8				93					
Other species	3		1	13	10	12	18	42	5		16
Bare ground (%)	—	—	7	7	3	5	>47	>37	—	12	16
Species diversity (H')	0.068	1.531	0.793	1.827	1.958	2.964	—	—	1.913	—	—
Years since last fire	35	22	95	55	55	>115	—	—	>45	—	—

Notes:

I. Parsons (1976): south-facing slope at 800 m in the southern Sierra Nevada.
II. Keeley and Johnson (1977): east-facing slope at 1000 m in the interior Peninsular Range.
III. J. Keeley (unpublished data): south-facing slope at 1100 m in the southern Sierra Nevada.
IV. J. Keeley (unpublished data): west-facing slope at 400 m in the coastal Transverse Ranges.
V. J. Keeley (unpubished data): north-facing slope at 300 m in the coastal Transverse Ranges.
VI. J. Keeley (unpublished data): north-facing slope at 1000 m in the coastal Peninsular Ranges.
VII. Vasek and Clovis (1976): east-facing slope at 1100 m in the interior Transverse Ranges.
VIII. Vasek and Clovis (1976): a level site at 1300 m in the interior Transverse Ranges.
IX. Schorr (1970): southeast-facing slope at 1000 m in the interior Peninsular Ranges.
X. Davis (1972): south-facing slope at 400 m in the northern Central Coast Ranges.
XI. Davis (1972): south-facing slope at 100 m in the northern Central Coast Ranges.

this is atypical for most species. Sprouting shrubs such as *Adenostoma fasciculatum* commonly produce three to four stems per plant, although on more open sites the number may be much higher. Nonsprouting species of *Ceanothus* and *Arctostaphylos* usually have a single stem per shrub. Species in the latter genus may reach tree-like proportions and dominate a site, despite being numerically less important than other species (site 2 in Table 6.3).

The most widely distributed chaparral shrub is *Adenostoma fasciculatum* (Fig. 6.3). This species is found from Baja to northern California in pure stands (chamise chaparral) or mixed stands with other shrub species. It often dominates at low elevations and on xeric south-facing slopes, with 60–90% cover. The short needle-like leaves produce a sparse foliage, and soil litter layers are poorly developed. Along its lower elevational limits, *A. fasciculatum* intergrades with subligneous coastal sage subshrubs, particularly *Salvia mellifera*, *S. apiana*, and *Eriogonum fasciculatum*.

Adenostoma fasciculatum is often codominant with one or more species of *Arctostaphylos* or *Ceanothus*. Two such mixed chaparral stands are shown in Table 6.3. Site 1 is a low-elevation coastal stand dominated by a species of *Ceanothus*; as is commonly the case in chaparral along the lower elevational border, subshrubs form a significant part of the cover. Site 2 is a south-facing site dominated by *Arctostaphylos glauca*. Species of *Ceanothus* and *Arctostaphylos* predominate in mixed chaparral from middle to high elevations and also can occur in pure stands (manzanita chaparral or *Ceanothus* chaparral). There are more than 60 species in each of these two genera. Some are highly restricted, whereas others are nearly as widespread as *Adenostoma*. Most species are endemic to the California chaparral and show specific suites of characters that make them well adapted to this community. Both genera contain species that respond to fire by resprouting, as well as species that have no capacity for vegetative regeneration but require fire scarification of seeds (Table 6.1). The nonsprouting (obligate-seedling) species tend to be more abundant on south-facing slopes, ridge tops, and desert exposures. Sprouting species are more important on mesic slopes and at higher elevations (Keeley 1977b).

Adenostoma, *Arctostaphylos*, and *Ceanothus* spe-

Figure 6.3. The needle-leaved Adenostoma fasciculatum *(chamise) is the most widely distributed of all California chaparral shrubs, but it is absent from Arizona chaparral.*

cies predominate in the drier areas of chaparral, but as conditions become more mesic, other broad-leaved sclerophyllous shrubs become important (site 3 in Table 6.3). This association, sometimes referred to as broad-sclerophyll chaparral, is more diverse and includes *Quercus dumosa* (Fig. 6.4), *Heteromeles arbutifolia, Prunus ilicifolia, Cercocarpus betuloides, Rhamnus* species, *Garrya* species, *Rhus* species, and *Malosma laurina*. Shrubs are generally taller in this chaparral, 3–6+ m, and because of overlapping canopies, areal coverage often exceeds 100%. Light levels below the canopy are often quite low, and soil temperatures seldom approach levels found in more open stands. These stands commonly have a well-developed soil litter layer. Some of the species in this association are long-lived and, if left undisturbed, are capable of becoming small trees; for example, *Heteromeles* and *Prunus* can reach 11 m or more (J. Keeley, pers. observ.), and *Malosma* was considered to be one of the dominant arboreal species of southern California (Hall 1903). Most of these species are common components of other communities such as oak woodland. In moist ravines, many of these broad-sclerophyll species coexist with small winter-deciduous trees such as *Sambucus* species and *Fraxinus dipetala*.

Montane chaparral at the upper elevational limits has a somewhat different physiognomy. The evergreen shrubs have a more rounded, compact shape, with foliage to the ground surface. Some sites may be covered by snow for many months. The association is dominated by vigorous sprouting species capable of dense coppice growth after fire, resulting in nearly impenetrable thickets with more than 100% cover (Wilson and Vogl 1965; Conard and Radosevich 1982). Often montane chaparral is dominated by species that are more typically found as understory or gap-phase coniferous forest shrubs (e.g., *Castanopsis sempervirens, Quercus vaccinifolia*, prostrate species of *Arctostaphylos* and *Ceanothus*, and winter-deciduous shrubs such as species of *Prunus, Ribes, Amelanchier*, and *Symphoricarpos*).

Regional Composition

Evergreen chaparral is best developed at middle elevations (300–1500 m) in southern California. This area, sometimes described as the South Coast Region (Sampson 1944), includes the Transverse and Peninsular ranges, which extend from Ventura County to northern Baja California. Here chamise chaparral forms a blanket-like cover over large areas from the coast to the mesas and foothills and into the mountains. Near the coast, chaparral commonly gives way to the summer-deciduous coastal sage scrub, although the evergreen sclerophylls *Rhus integrifolia* and *Malosma laurina* are often associated with coastal sage vegetation. In parts of this region, *Adenostoma fasciculatum* is replaced by *A. sparsifolium*. This latter shrub is distributed from the southern part of the south Coast Ranges through the coastal section of the Transverse Ranges and the interior parts of the Peninsular Ranges (Marion 1943); it appears to replace *A. fasciculatum* on more mesic and fertile sites (Beatty 1984). The species coexist on some sites, and Hanes (1965) suggested that phenological and physiological differences are important in promoting coexistence, although Beatty (1984) contended that even within the same stand of chaparral these congeners are distributed in different soil microhabitats. Epling and Lewis (1942) considered this South Coast Region to hold special significance because of the occurrence of numerous chaparral shrubs that are endemic or reach their most northern distribution here. Throughout this region, chaparral occurs on a variety of soils and substrates, including Jurassic, Upper Cretaceous, and Eocene sedimentary rocks and

Table 6.3. *Demographic structures of three older stands of chaparral in southern California*

Species	Post-fire regeneration		Density: Individuals ha^{-1}	Density: Stems ha^{-1}	m^2 basal stem per hectare
Site 1. West-facing slope (30–35°) in Santa Monica Mountains (400 m), 55 yr old					
Ceanothus megacarpus	(OS)[a]	Alive	2,530	2,970	34.9
		Dead	950	1,110	5.0
Adenostoma fasciculatum	(FR)	Alive	1,140	4,140	4.8
		Dead	1,170	5,420	2.7
Rhamnus crocea	(OR)	Alive	190	330	0.5
		Dead	110	280	0.6
Cercocarpus betuloides	(OR)	Alive	610	34,190	2.1
		Dead	500	5,920	0.6
Four other shrub species	(OR)	Alive	140	2,330	4.1
		Dead	80	830	1.0
Six subshrub species	(FR & OR)	Alive	3,170	7,000	10.1
		Dead	1,750	5,640	6.7
Three woody vine species	(OR)	Alive	250	2,250	<0.1
		Dead	110	220	<0.1
Site 2. Southeast-facing slope (25–30°) in San Gabriel Mountains (1000 m), 88 yr old					
Arctostaphylos glauca	(OS)	Alive	2,000	5,300	41.8
		Dead	1,700	5,400	8.1
Adenostoma fasciculatum	(FR)	Alive	5,200	17,800	21.9
		Dead	4,900	18,000	9.7
Ceanothus crassifolius	(OS)	Alive	1,400	2,400	17.3
		Dead	2,800	5,600	16.6
Quercus dumosa	(OR)	Alive	200	1,300	2.3
		Dead	30	30	<0.1
Two other shrub species & one subshrub species	(OR)	Alive	150	260	<0.1
		Dead	90	300	<0.1
Site 3. North-facing slope (35–40°) in San Gabriel Mountains (900 m), 65 yr old					
Adenostoma fasciculatum	(FR)	Alive	1,100	3,900	16.0
		Dead	1,100	5,300	21.2
Heteromeles arbutifolia	(OR)	Alive	300	7,400	12.3
		Dead	0	2,700	3.5
Prunus ilicifolia	(OR)	Alive	300	5,900	10.3
		Dead	30	3,000	5.7
		Seedlings	4,528	4,528	0.7
Quercus dumosa	(OR)	Alive	800	5,200	8.8
		Dead	0	900	1.2
		Seedlings	80	80	<0.1
Garrya veatchii	(OR)	Alive	400	1,900	7.4
		Dead	30	1,700	5.7
Ceanothus crassifolius	(OS)	Alive	200	250	0.8
		Dead	700	1,000	4.9
Arctostaphylos glauca	(OS)	Alive	200	200	0.6
		Dead	30	200	1.0
Four other shrub species	(OR)	Alive	310	6,590	5.5
		Dead	80	910	1.3
Two subshrub species	(FR & OR)	Alive	110	2,390	3.6
		Dead	0	410	1.3
One woody vine species	(OR)	Alive	60	170	<0.1
		Dead	0	0	

[a]OS = obligate-seeder (no capability for resprouting); FR = facultative-resprouter (resprouts and establishes seedlings); OR = obligate-resprouter (typically does not establish seedlings after fire).

Source: J. Keeley and T. Montygierd-Loyba (unpublished data).

Figure 6.4. Quercus dumosa *is a broad-sclerophyll species commonly forming nearly pure stands known as scrub oak chaparral. This species is replaced in Arizona by the very closely related (if not conspecific)* Q. turbinella, *which is one of the most widely distributed of the interior chaparral species.*

Tertiary volcanics in coastal ranges, and granitic substrates of Precenozoic metamorphic and metavolcanic rock on interior ranges (Minnich and Howard 1984).

In the central coastal regions of California, the chaparral is less continuous and is part of a mosaic with grassland, coastal sage scrub, and broadleaf and coniferous forest (Shreve 1927). Fire is believed to be the determining factor in this mosaic distribution because no consistent pattern of edaphic or topographic factors coincides with the distribution of chaparral (Wells 1962). A noteworthy feature of this region is the large number of endemic species of *Arctostaphylos* and *Ceanothus*, many of which are restricted to particular substrates (Wells 1962) or coastal areas under marine influences (Griffin 1978). Davis (1972) described the distribution of six species in the *Arctostaphylos andersonii* complex in the Santa Cruz Mountains. He suggested that the group had its origin on forested sites and radiated into chaparral, with each species occupying a habitat characterized by a distinct combination of soil conditions, including water-holding capacity, texture, pH, depth, root penetrability, and fertility. Both *Arctostaphylos* and *Ceanothus* have endemic species on the Channel Islands. Here the chaparral has a more open, woodland aspect (Bjorndalen 1978), apparently because of less frequent fires and more intensive grazing by feral animals (Minnich 1980a).

In the north coastal region, chaparral becomes a less important component of the landscape and is restricted to the driest sites (Clark 1937). *Adenostoma fasciculatum* is common in the drier interior valleys, whereas broad-sclerophyll species are more important on the coastal slopes, where some (such as *Prunus ilicifolia*) form small tree-like communities (Oberlander 1953). Throughout this region, localized outcrops of serpentine substrate produce a more open vegetation referred to as serpentine chaparral. Low levels of Ca and high levels of Mg, plus potentially toxic levels of Ni and Cr, in these soils (Koenigs et al. 1982) restrict the distribution of many species (Whittaker 1960; Kruckeberg 1969). Serpentine endemics include shrubs such as *Q. durata, Garrya congdoni, Ceanothus jepsonii, C. ferrisae*, various subspecific taxa of *Arctostaphylos*, and herbaceous species. There is good evidence that other more widespread species have evolved serpentine ecotypes. Mechanisms for tolerating serpentine soils vary with the species; for example, White (1971) found that *Arctostaphylos nevadensis* was able to selectively take up Ca over Mg, whereas *Ceanothus pumilus* was not able to do so, but it could regulate Mg, Ni, and Cr uptake. The exclusion of serpentine endemics from other sites has been attributed to competition (Kruckeberg 1954), although Tadros (1957) and Wicklow (1964) found that in the case of the fire-following annual *Emmenanthe rosea*, restriction to serpentine sites was due to an inability to establish on more fertile soils that supported greater microbial growth.

Chaparral is absent from the Central Valley of California; however, some claim that this is an artifact of human disturbance (Cooper 1922; Bauer 1930). Above 500 m in the Sierra Nevada foothills, grasslands or xeric woodlands of *Pinus sabiniana* and *Aesculus californica* intergrade into mixed chaparral or nearly pure stands of *Ceanothus cuneatus* or *Adenostoma fasciculatum* (Graves 1932; Rundel and Parsons 1979; Vankat 1982). Above 1000 m, montane chaparral occurs, eventually giving way to coniferous forest above 2000 m. This upper border is dynamic and strongly influenced by fire frequency (Wilken 1967).

On the interior side of the Sierra Nevada, montane chaparral forms a mosaic with coniferous forest, pinyon-juniper woodlands, or scrub vegetation with Great Basin affinities (Skau et al. 1970). On these sites montane chaparral may replace coniferous forest after wildfires and remain for 50+ yr (Townsend 1966). In the northern Sierra Nevada and Cascade ranges, extending as far north as Bend, Oregon (W. D. Billings, pers. commun.), montane chaparral becomes more restricted and forms a mosaic with ponderosa pine forest and a variety of other, more mesic vegetation. It often forms associations with winter-deciduous shrubs, especially on the eastern slopes of the Sierra Nevada and adjacent ranges.

Community Response to Wildfire

Wildfires are a dominant part of the environment. At present, the fire frequency averages once every two to three decades, but this may be more frequent than in the historical past (Byrne et al. 1977, unpublished data). Lightning strikes are the natural source of fire ignition, but today humans are responsible for most wildfires (Keeley 1977b). Lightning-ignited fires increase with elevation, latitude, and distance from the coast, whereas human-ignited fires show the opposite pattern and peak in different months (Parsons 1981; Keeley 1982). Throughout much of its range, chaparral forms a continuous cover over great distances, and as a result, huge wildfires that cover tens of thousands of hectares are not uncommon, particularly during Santa Ana wind conditions. Minnich (1983) suggested that fires of this size are an artifact due to modern-day fire suppression that results in unnaturally large accumulations of fuel. In support of this, he reported that large wildfires are relatively unknown from northern Baja California. Although Santa Ana winds are less common in this region, the difference is probably largely due to the fact that fire prevention is not encouraged, nor is fire suppression practiced, in Baja California. Consequently, fires are three times more common (largely ignited by humans), and this higher burning frequency keeps the vegetation more open and also produces a mosaic of different fuel conditions. Minnich (1983) argued that this represents the more "natural" situation for chaparral; however, this assumes that the fire-recurrence interval for a stand of chaparral is entirely a function of fuel load and, under primeval conditions, is not limited by sources of ignition. In the central Coast Ranges, Greenlee and Langenheim (1980) did a careful survey of the distribution of lightning-caused fires in conjunction with known patterns of fire behavior. They concluded that the "natural fire cycle" for the inland reaches of Santa Cruz County may have ranged upward to 100 yr and was probably far longer in the coastal and lower-elevation areas.

Fires typically kill all aboveground biomass, although much of the belowground shrub biomass on sprouting species remains alive. In the first year after fire, there is an abundant growth of herbaceous and suffrutescent vegetation (Fig. 6.5), the extent of which varies with site and year (Sampson 1944; Horton and Kraebel 1955; Sweeney 1956; Stocking 1966; Ammirati 1967; Keeley 1977c; Keeley et al. 1981). This "temporary" vegetation is relatively short-lived, and by the fourth year shrubs commonly dominate the site (Fig. 6.6). The rate of shrub recovery, however, varies with elevation, with slope aspect and inclination, and with coastal versus desert sites (Hanes 1971).

The herbaceous flora arising after fire is dominated by annuals, and species diversity is typically greatest the first year after fire. Sweeney (1956) studied 10 chaparral burns in northern California and reported 214 herbaceous species, two-thirds

Figure 6.5. This lush herbaceous growth the first spring after wildfire is in marked contrast to the depauperate herbaceous vegetation under the mature-chaparral shrub canopy.

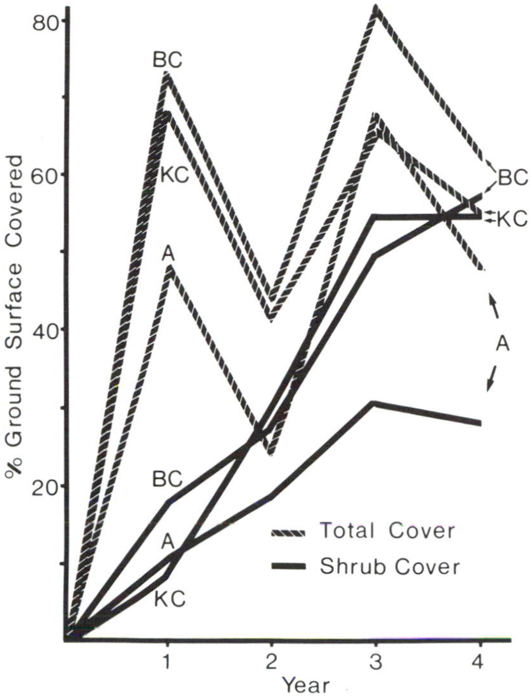

Figure 6.6. Changes in shrub cover and total cover (shrubs + "temporary" species) after fire at three sites at 560 m (A), 1000 m (BC), and 1670 m (KC) in southern California (from Keeley and Keeley 1981). The first-year increase total cover is largely due to herbaceous species, but by the fourth year after fire, shrubs account for nearly all of the cover except at the lowest elevation. The drop in total cover in the second post-fire year was apparently due to low rainfall (Keeley 1977c).

Figure 6.7. Resprout from lignotuber of Adenostoma fasciculatum *in the first spring after fire.*

of which were annual species. Few of these were widespread, and nearly one-third of the herb species were found on only a single burn. Annuals dominated most sites the first year after fire, but in subsequent years herbaceous perennials became a more prominent part of the herb flora. Ammirati (1967) noted that herbaceous perennials were more important than annuals on the more mesic coastal slopes of ranges in northern California. In southern California, Keeley and associates (1981) studied sites at three elevations burned in the same month (same sites as in Fig. 6.6) and found 99 herb and suffrutescent species, of which two-thirds were annuals. At all sites the number of herb species was greatest in the first year after fire. The number of herb species ranged from 19 to 38 on different slopes at the lowest-elevation site, and from 12 to 19 at the highest elevation. Annuals accounted for 87–97% of the herb coverage in the first year, but only 20–28% by the fourth year.

Recovery of shrub biomass is from basal resprouts (Fig. 6.7) and seedlings. After a spring or early summer burn, sprouts may arise within a few weeks, whereas after a fall burn, sprout production may be delayed until winter (Biswell 1974). Seed germination, on the other hand, is delayed until the following spring. After the first year, shrub seedling establishment is uncommon.

Fire-caused mortality of potentially resprouting shrubs is variable, depending on species, site, and fire characteristics. Some, such as *Quercus dumosa*, *Heteromeles arbutifolia*, and *Malosma laurina*, are seldom killed, whereas others, such as *Adenostoma fasciculatum* and various *Ceanothus* species, sometimes suffer extensive mortality (Keeley 1977b). Factors that may be involved include season of burn, elevation, soil moisture, plant size, and physiological condition (Laude et al. 1961; Plumb 1961; Keeley and Zedler 1978; Tratz 1978; Baker et al. 1982a; Stohlgren et al. 1984). Comparisons of pre-fire soil seed pools with post-fire seedling densities suggest that vast numbers of seeds are killed (Keeley 1977a; Bullock 1982; Davey 1982). Generalizations about the temperatures that shrub bases or seeds are exposed to during fires are difficult to make, because temperatures vary greatly with depth of burial, stand age and composition, weather conditions, and burning patterns (Sampson 1944; Bentley and Fenner 1958; DeBano et al. 1977; Anfuso 1982). For example, surface temperatures may remain higher than 500°C for more than 5 min during some fires, but not exceed 250°C in others. Temperatures at 2.5 cm depth are more commonly in the range of 50–200°C, but often persist for half an hour or longer.

Seedling mortality is generally high during the first year and is concentrated in the spring (Bullock 1982; Mills 1983). Seedlings are strikingly smaller than resprouts (Sampson 1944; Horton and Kraebel 1955; Keeley and Keeley 1981) and are most successful in gaps between the resprouting shrubs. Herbivory of seedlings and resprouts is important at this time (Davis 1967; Howe 1982; Mills 1983) and remains a factor in mature chaparral (Bartholomew 1970; Christensen and Muller 1975b; Schlising 1976).

Agencies concerned with managing chaparral lands often seed recently burned sites with nonnative herbs, *Lolium perenne* (ryegrass) in particular. Such artificial seeding has been carried out for two purposes: "Type-conversion" programs may seed in order to produce fuel loads sufficient for repeat burns in successive years, that will replace chaparral with grassland (Sampson 1944). More commonly, the justification for seeding is that species such as *L. perenne* are thought to establish a better plant cover and reduce soil erosion. There is evidence that this practice is having negative effects on the natural regeneration of chaparral (Schultz et al. 1955; Corbett and Green 1965; Gautier 1981; Keeley et al. 1981; Griffin 1982).

It has been hypothesized that in the absence of fire, chaparral would be replaced by other types of vegetation. Sampson (1944) suggested that in northern California, grassland would eventually replace chaparral. Hedrick (1951), however, found no evidence of this in San Benito County chaparral free from fire for more than 90 yr. He commented that "the most striking feature of this old chamise stand is the lack of evidence that it is dying out or being replace by herbaceous vegetation. It is true that old plants become defoliated and many stems die but these seem to be replaced by additional crown sprouts." This notion of resprouting shrubs rejuvenating themselves in the absence of fire is at odds with much published dogma on chaparral, but recent work suggests that it may be more important than previously thought, as discussed in the next section. In addition to Hedrick's study, other investigations have shown no evidence of old chaparral being replaced (sites III and VI in Table 6.2, sites 2 and 3 in Table 6.3) (Keeley and Zedler 1978), and it appears to be resilient to fire-recurrence intervals of 100+ yr (Keeley 1981, 1986). However, for most, but not all shrub species, mortality in old stands of chaparral is not accompanied by seedling establishment (Table 6.3). When there are unusually long intervals between fires, chaparral may be replaced by sclerophyllous woodland (Cooper 1922; Wells 1962). This will be a function of available seed sources, site factors, and particularly the life-history characteristics of the shrub species already established.

PLANT LIFE HISTORIES

Chaparral has a rich diversity of species with a variety of life histories closely linked to fire (Table 6.4).

Shrubs

Shrub species form a continuum in terms of mode of post-fire regeneration, ranging from species entirely dependent on seedling recruitment to sprouting species that rarely establish seedlings after fire.

Obligate-seeding species establish seedlings in the first year after fire, but seedling recruitment is almost nonexistent in subsequent years. Consequently, populations are usually even-aged (Schlesinger and Gill 1978). Post-fire seedlings of obligate seeders arise from a long-lived seed pool that lies dormant in the soil (Keeley 1977a; Davey 1982) until germination is cued either by intense heat (Quick 1935; Hadley 1961; Quick and Quick 1961) or by chemical stimulus from charred wood (Keeley 1987).

Sprouting species of *Arctostaphylos* and *Ceanothus* and *Adenostoma fasciculatum* often establish many seedlings after fire, although the proportion of resprouting shrubs to seedlings is variable with the species, site, and fire (Keeley 1977b). As with obligate-seeding species, seedling establishment by these facultative resprouters is confined to the first post-fire year, cued by heat or charred wood. *Adenostoma fasciculatum* is reported to produce two types of seeds: those that germinate readily at maturity, and a portion that require intense heat shock (Stone and Juhren 1951, 1953). Thus, the former seed type could germinate in the absence of fire. Although successful seedling establishment under the shrub canopy is in fact nonexistent, such seeds do contribute to colonization of other types of disturbance.

Obligate resprouters such as *Heteromeles arbutifolia*, *Quercus dumosa*, *Prunus ilicifolia*, *Cercocarpus betuloides*, and *Rhamnus* species seldom establish seedlings after fire. In mature chaparral, these shrubs do produce substantial seed crops that are

Table 6.4. *Life-history modes in chaparral, including only native higher vascular plants and not meant to be an exhaustive list*

Evergreen sclerophyllous shrubs (see Table 6.1)	
Obligate-seeders	
Sprouters with post-fire seedling recruitment	
Sprouters without significant post-fire seedling recruitment	
Evergreen sclerophyllous trees	
Cupressaceae	*Cupressus* spp.
Pinaceae	*Pinus* spp.
	Pseudotsuga macrocarpa
Ericaceae	*Arbutus menziesii*
Fagaceae	*Quercus* spp.
Lauraceae	*Umbellularia californica*
Semi-deciduous subshrubs	
Anacardiaceae	*Rhus trilobata*
Asteraceae	*Artemisia californica*
	Gutierrezia sarothrae
	Haplopappus squarrosus
Hydrophyllaceae	*Eriodictyon* spp.
Lamiaceae	*Salvia* spp.
	Lepechinia spp.
	Trichostema lanatum, T. parishii
Malvaceae	*Malacothamnus fasciculatus, M. fremontii*
Polygonaceae	*Eriogonum fasciculatum*
Saxifragaceae	*Ribes* spp.
Scrophulariaceae	*Mimulus aurantiacus, M. longiflorus, M. puniceus*
Suffrutescents	
Asteraceae	*Eriophyllum confertiflorum*
Cistaceae	*Helianthemum scoparium*
Fabaceae	*Lotus scoparius*
Fumariaceae	*Dicentra chrysantha, D. ochroleuca*
Hydrophyllaceae	*Turricula parryi*
Papaveraceae	*Romneya coulteri, R. trichocalyx*
Scropulariaceae	*Penstemon centranthifolius, P. heterophyllus, P. spectabilis*
Agavaceae	*Yucca whipplei*
Woody vines	
Anacardiaceae	*Toxicodendron (Rhus) diversiloba* (Munz 1974)
Caprifoliaceae	*Lonicera interrupta, L. subspicata*
Ranunculaceae	*Clematis lasiantha, C. ligusticifolia, C. pauciflora*
Rubiaceae	*Galium* spp.
Scrophulariaceae	*Keckiella (Penstemon) cordifolia, K. ternata* (Munz 1974)
Herbaceous vines	
Convolvulaceae	*Convolvulus* spp.
Cucurbitaceae	*Marah fabaceus, M. macrocarpus*
Cuscutaceae	*Cuscuta* spp.
Fabaceae	*Lathyrus laetiflorus, L. splendens, L. vestitus*
Perennial herbs	
Apiaceae	*Lomatium* spp.
	Sanicula spp.
Asteraceae	*Helianthus gracilentus*
	Perezia microcephala
Caryophyllaceae	*Silene californica, S. laciniata*
Lamiaceae	*Salvia sonomensis*
Paeoniaceae	*Paeonia californica*
Ranunculaceae	*Delphinium* spp.
Scrophulariaceae	*Scrophularia californica*
Solanaceae	*Solanum* spp.
Amaryllidaceae	*Allium* spp.
	Bloomeria crocea, B. clevelandii
	Brodiaea spp.

(Cont.)

Table 6.4. *(Cont.)*

Iridaceae	*Sisyrinchium bellum*
Liliaceae	*Calochortus* spp.
	Chlorogalum pomeridianum
Poaceae	*Elymus condensatus*
	Melica imperfecta
	Stipa spp.
	Zigadenus fremontii, Z. micranthus
Annuals	
Apiaceae	*Apiastrum angustifolium*
	Daucus pusillus
Asteraceae	*Chaenactis* spp.
	Filago arizonica, F. californica
	Gnaphalium spp. (biennials)
	Heterotheca grandiflora
	Malacothrix clevelandii
	Rafinesquia californica
	Stephanomeria virgata
Boraginaceae	*Cryptantha* spp.
	Plagiobothrys spp.
Brassicaceae	*Caulanthus* spp.
	Descurainia pinnata
	Lepidium nitidum
	Streptanthus spp.
Caryophyllaceae	*Silene multinervia*
Fabaceae	*Lotus salsuginosus, L. strigosus*
	Lupinus spp.
Hydrophyllaceae	*Eucrypta chrysanthemifolia*
	Phacelia spp.
Lamiaceae	*Salvia columbariae*
Onagraceae	*Camissonia* spp. (Munz 1974)
	Clarkia spp.
Papaveraceae	*Eschscholzia* spp.
	Papaver californicum
Polemoniaceae	*Allophyllum* spp.
	Gilia spp.
	Linanthus spp.
	Navarretia spp.
Polygonaceae	*Chorizanthe* spp.
	Pterostegia drymarioides
Portulacaceae	*Calandrinia breweri*
	Calyptridium monandrum
	Montia perfoliata
Scrophulariaceae	*Antirrhinum* spp.
	Collinsia parryi
	Cordylanthus spp.
	Mimulus spp.
Poaceae	*Festuca* spp.

Source: Nomenclature according to Munz (1959) except where indicated otherwise.

widely dispersed (Bullock 1978; Hom 1984); the seeds, however, are short-lived and germinate readily with adequate moisture, and thus a dormant seed pool does not build up in the soil (Keeley 1987). This, coupled with the observation that these seeds are easily killed by intense heat, accounts for the failure to establish seedlings after fire.

On recently burned sites, obligate-seeding species typically have very high seedling densities, but low coverage relative to sprouting species (Sampson 1944; Horton and Kraebel 1955; Vogl and Schorr 1972; Keeley and Zedler 1978; Keeley and Keeley 1981). These obligate-seeding species require 5–15 yr before substantial seed crops are produced, and thus fires at intervals more frequent than this can produce localized extinctions (Zedler et al. 1983). In the absence of fire, these species produce seed crops that are largely deposited in the soil around the parent plant and lie dormant until the next fire. Despite relatively large seed crops by

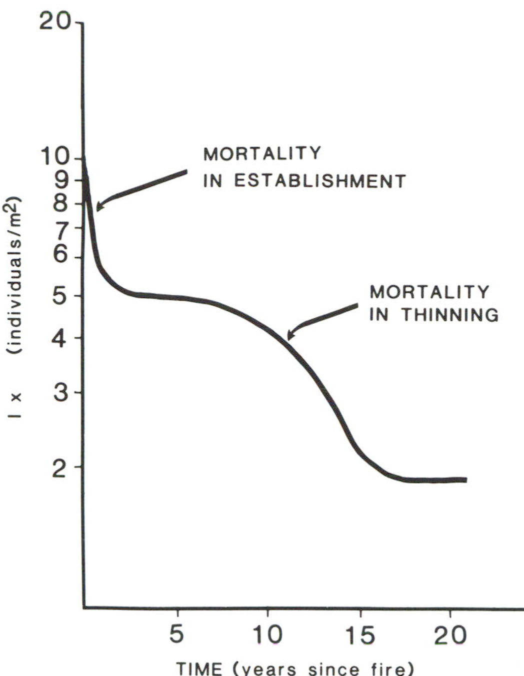

Figure 6.8. Generalized survivorship (l_x) curve for individuals in pure even-aged stands of Ceanothus megacarpus *in the south Coast Range (redrawn with permission from* Botanical Review *48: 86 1982, Schlesinger et al. and The New York Botanical Garden). Not shown are variance/mean ratios calculated for living stems per plot, which showed a drop with thinning, suggesting that shrubs are becoming less clumped as the stand ages.*

these species, accumulation in the soil appears to be relatively slow because a large portion of the seed crop is removed from the soil by predators (Keeley and Hays 1976; Davey 1982). In one study, two *Arctostaphylos* species were found to produce more seeds in a single year than were present in the soil seed pool (Keeley 1977a). A follow-up study 10 yr later showed no statistically significant changes in the sizes of the soil seed pools (J. Keeley, unpublished data).

As stands age and the canopy closes, there is intense competition and density-dependent thinning, resulting in high mortality (Fig. 6.8). In stands older than 50 yr, the fate of obligate-seeding species varies with the species and site. On mesic slopes, these shrubs are commonly outlived ("outcompeted"?) by sprouting species, as illustrated by the ratio live : dead plants on the north-facing slope (site 3) in Table 6.3. On drier sites (site 2 in Table 6.3), obligate-seeding shrubs are dominant for much longer, although certain species such as *Arctostaphylos glauca* are capable of great size (>9 m) and readily shade out *Ceanothus* species (site 2) (Keeley

Table 6.5. *North- versus south-facing slope comparison for* Ceanothus crassifolius *populations in the San Gabriel Mountains (800 m) of southern California*

Parameter	North	South	*p*
Total plant cover (% ground surface)	>100%	<80%	—
C. crassifolius			
Height (m)	2.6	2.6	n.s.
Areal coverage, individual (m^2)	2.6	5.6	<0.01
Mortality over 6 yr (% per yr)	2.9	0	<0.01
Fruit production over 6 yr (fruits per m^2 areal coverage)	254	481	<0.01

Note: Each slope aspect represents 15 shrubs on two different slopes ($N = 30$). Shrubs were 24 yr old at the end of the study.
Source: J. Keeley (unpublished data).

and Zedler 1978). On the most xeric sites, nonsprouting *Ceanothus* species may still dominate the stand even after 50 yr (e.g., *C. megacarpus*, site 1 in Table 6.3). In general, obligate-seeding species increase in abundance, diversity, and longevity with increasing aridity (Keeley 1975, 1986). Even within a site, obligate-seeding species often do much better on drier south-facing slopes than on adjacent north-facing slopes (Table 6.5). Not surprisingly, there is evidence that seedlings of these species are more tolerant to drought than seedlings of associated sprouters (Musick 1972; Jacks 1984).

Sprouting shrubs are distinctly different from obligate-seeding species in that the aboveground stems are not even-aged. All sprouting species continually produce new shoots from the root crown throughout their life spans (Fig. 6.9). Figure 6.10 illustrates that the demographic pattern of stem recruitment is not the same in all such species.

The timing of seedling recruitment is not well documented for those obligate-resprouting species that fail to establish seedlings after fire. In general, establishment appears to be restricted to older, more mesic stands of chaparral, although seldom are seedlings of these species very abundant. Successful reproduction does occur under some conditions, as illustrated by the age distributions of *Quercus dumosa* and *Rhamnus crocea* seedlings in one very old chaparral stand (Fig. 6.11); some individuals do survive from most cohorts, and thus recruitment into the adult population seems likely. Seedlings in mature chaparral have also been noted for *Cercocarpus betuloides*, *Prunus ilicifolia*, and *Heteromeles arbutifolia* (Gibbens and Schultz 1963; Patric and Hanes 1964; Zedler 1982).

Figure 6.9. Multistemmed Adenostoma fasciculatum *shrub with various-aged shoots arising from a common root crown. The distribution of stem ages for this individual is shown in Fig. 6.10.*

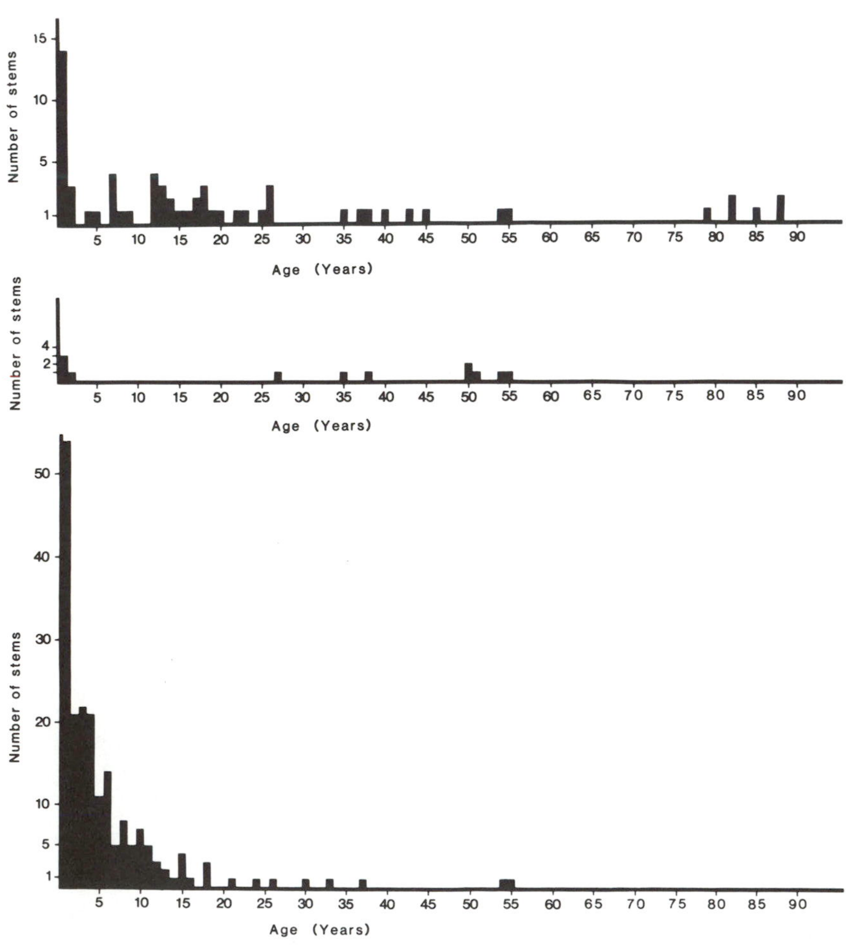

Figure 6.10. Age histograms for all stems on one individual each of Adenostoma fasciculatum *(top),* Ceanothus spinosus *(middle) and* Heteromeles arbutifolia *(bottom) from different sites in southern California (J. Keeley and T. Montygierd-Loyba, unpublished data).*

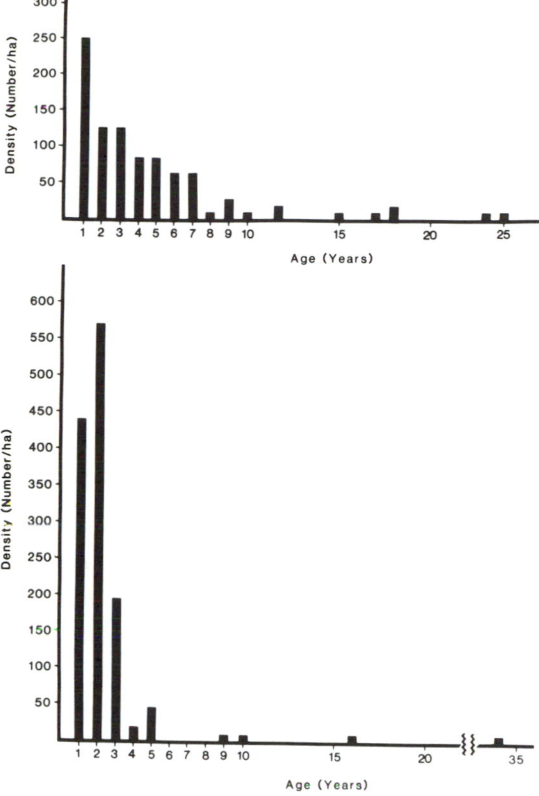

Figure 6.11. Age histograms for seedling populations of Rhamnus crocea *(top) and* Quercus dumosa *(bottom) in a north-facing broad-sclerophyll chaparral stand at 900 m in the Santa Ana Mountains of southern California unburned for more than 115 yr (site VI in Table 6.2). Mature shrub densities were 3420 ha^{-1} and 40 ha^{-1} for* Quercus *and* Rhamnus *,respectively (J. Keeley and T. Montygierd-Loyba, unpublished data).*

Trees

Evergreen coniferous trees, such as species of *Cupressus, Pinus attenuata,* and *P. muricata,* often form dense even-aged stands surrounded by a matrix of chaparral (Vogel et al. 1977; Zedler 1977, 1981), commonly on serpentine or other unusual substrates (McMillan 1956; Koenigs et al. 1982). These species do not resprout after fire, and seedling establishment is in the first post-fire year from a dormant seed pool. In this case, seeds are released after fire from serotinous cones held on the trees for many years prior to the fire. *Cupressus abramsii* in the Santa Cruz Mountains commonly exceeds 100 yr of age, and cones with viable seeds are known to survive intact for more than 40 yr (H. Kuhlmann, pers. commun.).

Pinus coulteri, P. sabiniana, and *P. torreyana* are, typically, nonserotinous species associated with chaparral in certain regions, but seldom in dense stands. Populations of *P. coulteri* are polymorphic with respect to cone serotiny, with this characteristic increasing toward the drier, chaparral-dominated end of its distribution (Borchert 1985). For all three species, seedling recruitment following fire is most abundant from seeds retained in previously opened cones, from the current crop of cones that commonly are unopened at the time of most fires, or from "seed trees" that survive the fire (Vale 1979; Minnich 1980b; McMaster and Zedler 1981). *Pseudotsuga macrocarpa* is the only conifer in chaparral capable of resprouting after fire. Sprouts arise from epicormic buds along the trunk and main branches. Seedling recruitment occurs under the chaparral or oak woodland canopy during fire-free periods (Bolton and Vogl 1969; McDonald and Litterell 1976), and thus populations of big-cone spruce are uneven-aged.

There is evidence that for most of these species, ranges have become more restricted in modern times because of increased fire frequency (Shantz 1947; Horton 1960; Gause 1966; Zedler 1977). In the case of *Cupressus forbesii,* Zedler (1981) has shown that reproductive success increases with the length of interval between fires. This obligate-seeding serotinous species starts producing cones at an early age, and cones accumulate on the trees. As a result, post-fire seedling establishment is an order of magnitude greater in stands over 50 yr of age at the time of burning than in stands 30 yr of age (Table 6.6). Long fire-free periods are also apparently necessary for successful seedling establishment by *Pseudotsuga macrocarpa* (Minnich 1980b).

Hardwood trees, in particular evergreen sclerophylls such as *Quercus agrifolia, Q. wizlizenii, Q. chrysolepis, Arbutus menziesii,* and *Umbellularia californica,* occasionally occur within chaparral vegetation. These species are best developed in ravines and mesic north-facing slopes. Because of the higher moisture content of the fuels, such sites often escape complete destruction by wildfires. Depending on the severity of the fire, these species may resprout from epicormic buds beneath the bark of stems or from the root crown. Seedling establishment after fire is rare because of the lack of a dormant seed pool stored in the soil or on the plant; the seeds are short-lived and germinate readily at maturity (Keeley 1987). These tree species more commonly form woodland or riparian communities, often in association with a variety of winter-deciduous tree species.

Subshrubs

These summer-deciduous species are dominants in the lower-elevation coastal sage scrub vegetation, and in chaparral they are most important along the xeric borders (Hanes 1971). They are readily shaded

Table 6.6. *Parameters related to recovery of burned* Cupressus forbesii *stands (Tecate cypress) in southern California*

Site	Age at time of fire (yr)	Pre-burn trees (m^{-2})	Cones per pre-burn trees	Seedlings per cone	Seedlings per Pre-burn trees	Seedlings (m^{-2})
Tecate A	10	0.05	0.8	0.33	0.28	0.02
Tecate B	32	8.90	3.7	0.25	0.07	0.60
Otay Mocogo	33	1.70	9.9	0.03	0.33	0.57
Tecate C	52	1.21	21.7	0.22	4.25	5.15
Tecate D	52	0.36	29.9	0.51	15.30	5.51
Tecate E	95	0.44	33.5	0.13	4.25	1.87

Source: Zedler (1981).

out by the evergreen chaparral species (McPherson and Muller 1967; Gray 1983) and thus occupy gaps in the chaparral canopy. Most have light, readily dispersed seeds (Wells 1962) and are capable of recruiting new individuals into gaps, as well as establishing after fire from a dormant seed pool. The relatively complex germination behavior of the subshrub *Salvia mellifera* (Table 6.7) may represent a response to this environmental pressure (seeds exposed to light germinate readily without other stimuli, increasing under a variable diurnal temperature regime such as might occur near the soil surface on open sites; buried seeds that are in the dark require the presence of charred wood for germination).

Suffrutescents

These species are abundant in the first year after fire, germinating from seeds in the soil. The germination cue may be heat for some and charred wood for others (Table 6.8). Depending on the site, one of these species may dominate a burned area in the second or third year after a fire, but as the shrub cover increases, suffrutescent species are eliminated or restricted to gaps (Horton and Kraebel 1955; Keeley et al. 1981).

On open rocky sites from central California southward, the rosette-forming *Yucca whipplei* is a conspicuous element. Seeds germinate readily with adequate moisture, and populations are various-aged (Keeley and Tufenkian 1984). In southern California chaparral, this species is monocarpic: The rosette grows for a decade or more before flowering and dying. Not all populations, however, are monocarpic (Haines 1941). On the desert slopes of the Transverse Ranges, plants produce up to 100 rosettes during the vegetative phase, and flowering may be spread over many decades; in the central Coast Ranges, *Y. whipplei* populations are capable of vegetative spread by rhizomes.

Vines and Perennial Herbs

Both woody and herbaceous vines are frequent in older stands of broad-sclerophyll chaparral. They

Table 6.7. *Seed germination for the subshrub* Salvia mellifera *collected from a chaparral site after heat treatment or placed with charred wood*

	Percentage germination					
Condition	Control	70°C (1 hr)	70°C (5 hr)	120°C (5 min)	Charred wood	*p*
Constant 23°C incubation						
Dark	1	2	5	0	24	<0.01
Light	23	24	17	25	25	>0.05
23°C/13°C diurnal incubation						
Dark	4	5	9	3	37	<0.01
Light	50	54	55	50	55	>0.05

Note: *N* = 3 dishes of 50 seeds each.
Source: Keeley (in press).

Table 6.8. *Seed germination for suffrutescent species after heat treatment or with charred wood or with potentially allelopathic leachate from* Adenostoma fasciculatum *foliage*

Condition	Percentage germination				p
	Control	80°C (2 hr)	120°C (5 min)	150° (5 min)	
Helianthemum scoparium					
Control	23	15	43	3	<0.01
Leachate	29	17	50	11	<0.01
Charred wood	23	16	37	4	<0.01
C + L	25	17	40	9	<0.01
p	>0.05	>0.05	>0.05	>0.05	
Eriophyllum confertiflorum					
Control	4	6	4	1	<0.05
Leachate	8	6	6	1	>0.05
Charred wood	52	58	66	2	<0.01
C + L	58	63	65	3	<0.01
p	<0.01	<0.01	<0.01	>0.05	

Note: C + L = charred wood + leachate; *N* = 8 dishes of 50 seeds each; similar variation is evident in other suffrutescent species; e.g., *Lotus scoparius* is stimulated by heat, whereas *Romneya* species and *Penstemon* species are stimulated by charred wood.
Source: Keeley et al. (1987); Keeley (1987).

grow into the shrub canopy and flower prolifically in wet years. Seeds of most of these species germinate readily, and seedlings are often produced under the shrub canopy. After fire, all of these species resprout from rootstocks, and seedling recruitment is uncommon at this time.

Perennial herbs are most conspicuous in the first spring after fire, and their presence is due to resprouting from bulbs or other buried parts. As the shrub canopy returns, these species persist, but they produce very depauperate growth in most years. Because of low light levels, they seldom flower under the canopy (Stone 1951). The timing of seedling recruitment is unknown, although it is clear that most of these species have seeds that germinate readily without treatment other than cold stratification (Sweeney 1956; Keeley et al. 1985; Keeley and Keeley in press).

Annuals

Annuals make up the most diverse component of the chaparral flora. They are most abundant in disturbed areas and produce spectacular floral displays in the first spring after fire. Some of these species, such as *Phacelia* species, *Emmenanthe penduliflora*, and *Papaver californica*, have been referred to as "fire annuals" or "pyrophyte endemics" because of the fact that they may dominate a site in the first year after fire and then disappear until the next fire. These species do not have well-developed dispersal characteristics; hence, the long-lived seeds (Went 1969) simply wait in the soil for disturbance to come to them. Germination is apparently not stimulated by heat (Sweeney 1956; Keeley et al. 1985) but rather is dependent on the presence of charred wood (Wicklow 1977; Jones and Schlesinger 1980; Keeley and Keeley 1982); germination is stimulated by a water-soluble organic compound leached from charred, but not ashed, wood (Keeley and Nitzberg 1984; Keeley et al. 1985). The compound is apparently a breakdown product of lignin and hemicellulose from any species of wood and is produced at temperatures greater than 175°C (Keeley and Pizzorno 1986).

Other native annuals are quite opportunistic in that they are most abundant on burned sites, but persist within gaps in the chaparral canopy. Some of these species have polymorphic seed pools in which a portion of the seeds germinate readily and another portion is refractory, requiring the stimulus of intense heat or charred wood for germination (Table 6.9). The mechanisms by which these factors stimulate germination probably are different. Species with heat-stimulated germination typically have seed coats with a water-impermeable cuticle that appears to crack under heat treatment (Keeley and Kelley in press). None of the species stimulated by charred wood has an obvious cuticle, and it is known, at least for *Emmenanthe penduliflora*, that the seed coat is freely permeable to water (Sweeney 1956).

Some native annuals are typically most abundant in gaps in mature chaparral, and their seeds

Table 6.9. *Seed germination for native annuals after heat treatment or with charred wood*

	Percentage germination			
	Control	120°C (5 min)	Charred wood	*p*
Antirrhinum coulterianum	2	3	42	<0.01
A. kelloggii	40	45	63	<0.01
A. nuttallianum	69	56	58	>0.05
Camissonia californica	3	6	48	<0.01
C. hirtella	30	66	26	<0.01
Chorizanthe fimbriata	37	43	45	>0.05
Clarkia purpurea	40	40	72	<0.01
Collinsia parryi	24	12	77	<0.01
Cordylanthus filifolius	57	27	62	<0.01
Cryptantha muricata	24	37	67	<0.01
Gilia australis	31	32	80	<0.01
G. capitata				
Burn	20	25	69	<0.01
Mature[a]	15	27	67	<0.01
Lotus salsuginosus	24	40	20	<0.01
Malacothrix clevelandii	9	10	35	<0.01
Rafinesquia californica	4	3	55	<0.01
Silene multinervia	36	43	83	<0.01

[a]Seeds from a recent burned site and adjacent mature chaparral; see also Grant (1949).
Source: Keeley and Keeley (in press).

germinate readily without special treatment (*Cordylanthus filifolius* in Table 6.9). Such species, as well as those with polymorphic seed pools, increase in abundance in more open communities, arising from repeated disturbances or along xeric margins. On such sites, very closely related species may have quite different microsite requirements. For example, Shimida and Whittaker (1981) reported that *Cryptantha muricata* and *Lotus strigosus* were restricted to the open, whereas the congeners *C. intermedia* and *L. salsuginosus* increased nearer to clumps of *Adenostoma fasciculatum*.

The striking contrast between the depauperate herb growth under mature chaparral and the flush of herbs after fire has been hypothesized to be due to allelopathic suppression of germination by the overstory shrubs. After field and laboratory studies, McPherson and Muller (1969) concluded that "nearly all seeds in the soil of mature *A. fasciculatum* stands are prevented from germinating by the toxin (leached from the shrub overstory) which is most abundantly present during the normal germination period." Fire consumes the shrubs and destroys the toxin, thus releasing the herb seeds from inhibition. Their theory is widely cited in textbooks and other sources, although as a note of caution we suggest that the following points should be considered: (1) The majority of their work focused on the effects of leached inhibitors on *growth* of non-native seedlings, not on germination of native herbs. (2) Leachate from *Adenostoma* foliage may inhibit the *germination* of some species, but apparently has no effect on many others (Table 6.8) (McPherson and Muller 1969; Christensen and Muller 1975a; Keeley et al. 1985). (3) Temperatures applied to the soils that resulted in enhanced germination were far lower than the temperatures needed to degrade the suspected toxins (Chou and Muller 1972). (4) The concentration of toxins McPherson and associates (1971) found necessary for inhibition were much higher than those in the soil (Kaminsky 1981). (5) Christensen and Muller (1975a) found that the concentrations of suspected allelopathic toxins were greatest in soils from recently burned sites. (6) Seeds in soil that had been heat-treated and then returned to beneath the shrub canopy (and exposure to the putatively allelopathic leachate) showed high germination (Christensen and Muller 1975b). (7) Seeds of many chaparral herbs fail to germinate even if they are never exposed to so-called allelopathic toxins.

Kaminsky (1981) hypothesized that toxins produced by soil microbes are responsible for inhibiting herb germination under chaparral, and he demonstrated such an inhibitory effect with lettuce seeds. Pack (1987), however, could not duplicate this effect with seeds of native species. Also, it appears that the potential for microbial inhibition of germination is greater in soils from burned sites (Keeley 1984b), which is consistent with the fact that fungal and bacterial populations in soils increase after fire (Christensen and Muller 1975a).

Field studies suggest that any seedlings that do establish under mature chaparral are consumed by small animals. In all studies in which animals were excluded from plots beneath chaparral, there were highly significant increases in herb densities, although they were not comparable in species composition or density to those observed after fire (McPherson and Muller 1969; Christensen and Muller 1975b; Quinn 1986).

In summary, it appears that in response to the poor conditions of low light, limited water, insufficient nutrients, and high predation under the shrub canopy, many species of annuals have evolved mechanisms that ensure seed dormancy until the canopy is removed. A germination requirement for intense heat or charred wood cues establishment to the post-fire environment. Opportunistic annuals colonize gaps in the chaparral canopy with a portion of their seed pool that is nonrefractory. Such a system of germination cues could be fine-tuned if these seeds were sensitive to "allelopathic" compounds, using such chemicals as cues to inhibit germination when under the canopy (Koller 1972); however, this remains to be demonstrated.

Non-native annual grasses and forbs are found throughout chaparral regions. Under a regime of frequent fires, they readily displace the native herb flora, and if fires are frequent enough, chaparral can be converted to grasslands dominated by these non-native species (Cooper 1922; Sampson 1944; Arnold et al. 1951; Hedrick 1951; Wells 1962). In the absence of fire, seeds of non-natives have a low residence time in the soil, and thus the presence of these species on burned sites is due to colonization after fire. Most, such as *Bromus* species, *Erodium* species, and *Centaurea melitensis*, disperse prior to the summer fire season and consequently are uncommon in first-year burns, but are common in the second and third years (Sampson 1944; Horton and Kraebel 1955; Keeley et al. 1981). Fall-fruiting species such as *Lactuca serriola* and *Conyza canadensis* are often present in first-year burns.

SHRUB MORPHOLOGY AND PHENOLOGY

Leaves

The dominant shrubs are evergreen, with small, sclerified, heavily cutinized leaves (Cooper 1922). The widespread *Adenostoma fasciculatum* has a linear-terete isofacial leaf less than 1 cm in length (0.06 cm^2) that is markedly smaller than leaves of other chaparral shrubs (Fishbeck and Kummerow 1977). These leaves are produced individually on new growth and in short-shoot fascicles on old growth (Jow et al. 1980). Leaves on seedlings and fire-induced basal resprouts (but not on basal sprouts from mature shrubs) are bifacial and deeply lobed (compare Figs. 6.3 and 6.7). Similar juvenile-type leaves occur on mature plants under abnormally mesic conditions. Other shrub species have broad-sclerophyll leaves remarkably convergent in size, shape, and anatomy. Most are simple (<5 cm in length and 1–5 cm^2), with average leaf thickness ~300 μ, plus 5–10 μ cuticles on upper and lower surfaces (Cooper 1922; Fishbeck and Kummerow 1977). Many species have sharply serrated leaves (Fig. 6.4). Most *Arctostaphylos* species and *Dendromecon rigida* have isofacial leaves with an upper and lower palisade and stomata on both surfaces. In *Arctostaphylos*, such leaf types are largely restricted to interior species, whereas coastal taxa have stomata restricted to the lower leaf surface, and some have bifacial leaves (Howell 1945). Species in other genera have stomata restricted to the lower leaf surface. Sunken stomata are not uncommon, ranging from slightly sunken stomata in *Heteromeles arbutifolia* to the extreme case in certain *Ceanothus* species with stomatal crypts that are invaginated to over half the width of the leaf. These stomatal crypts are characteristic of the subgenus *Cerastes* (except in the seedling stage) and are absent in *Euceanothus* (Nobs 1963). Species of *Cerastes* also have markedly thicker leaves, higher leaf specific weights, and thicker cuticles than *Euceanothus* species (Barnes 1979).

Leaf orientation is variable among shrub species and depends on environmental conditions. Nearly vertical leaves are prominent in *Dendromecon rigida* and *Arctostaphylos* species, and many other species have leaf angles greater than 50° (Kummerow et al. 1981). In *Arctostaphylos* species, the leaf angle increases with aridity (Shaver 1978), and in *H. arbutifolia*, sun leaves have significantly steeper angles that shade leaves (H. A. Mooney, unpublished data). Other leaf characteristics are also highly modifiable, depending on the microenvironment (Cooper 1922; Mortenson 1973; Krause and Kummerow 1977; Hochberg 1980; Ball et al. 1983).

Stems and Growth Forms

Chaparral shrubs show a remarkable degree of convergence in growth form. Across areas of similar topography, shrubs are of similar heights and give the impression of a smooth blanket of vegetation. Detailed studies of plant structure show similar degrees of convergence in distribution of leaves and stems among unrelated taxa (Mooney et al. 1977; Kummerow et al. 1981). Growth forms vary with environment, as on north- versus south-facing slopes (Table 6.5) or on coastal versus desert sites (Vasek and Clovis 1976). Growth forms may also change temporally, as in *Arctostaphylos* and *Ceanothus* species, which in older stands grow horizontally to "escape" shading by adjacent plants. Often such branches will root and spread vegetatively by layering, often forming large clones (James 1984).

Fires affect growth form in that many species capable of tree-like proportions resprout after fire, giving rise to a multistemmed shrubby growth form. Resprouting ability is found in all chaparral shrub species, with the obvious exception of obligate-seeding *Ceanothus* and *Arctostaphylos* taxa (Table 6.1). The mode of sprouting, however, is variable. *Adenostoma fasciculatum* and sprouting taxa of *Ceanothus* and *Arctostaphylos* initiate a basal lignotuber or burl as a normal part of seedling development (Wieslander and Schreiber 1939), although populations of seedlings with and without lignotubers are known for *A. fasciculatum* on mesas north of San Diego and for *Arctostaphylos rudis* on mesas east of Lompoc (J Keeley, unpublished data). Some species, such as *Heteromeles arbutifolia* and *Quercus dumosa*, may have large lignotuberous structures that are induced by repeated coppice growth after fires. Others, such as *Cercocarpus betuloides*, sprout from rhizomes a meter or more distal to the main shoot system (Site 1 in Table 6.3).

Lignotubers are uncommon in shrubs outside of mediterranean ecosystems (Keeley 1981). These "burls" are often large and commonly exceed the aboveground biomass (Kummerow and Mangan 1981). They differ anatomically among species (Anfuso 1982; Lopez 1983; James 1984), but in all cases they proliferate adventitious buds that are suppressed to various degrees by the dominant stems. After fire, new shoots are initiated from these burls, fed by carbohydrate stores in the burls (Lopez 1983) and roots (Jones and Laude 1960; Laude et al. 1961). Reserves appear sufficient to sustain the roots for more than 1 yr (Kummerow et al. 1977). Storage of inorganic nutrients may also be an important function of lignotubers. After fire, sprouts from burls are more robust because of a much larger pith (Watkins 1939), and often such sprouts will branch and proliferate multiple shoots at the ground level. Sprout production continues in the absence of fire, replacing stems that die (Figs. 6.9 and 6.10).

Stems of most species are ring-porous, with well-developed annual rings, but stems tend toward diffuse-porous, with poorly defined rings, in such others, as *Prunus ilicifolia* and *Malosma laurina* (Webber 1936; Watkins 1939; Young 1974; Carlquist 1980). Wood storage products such as tannins and calcium oxalate are abundant in many species. Older stems of some species (e.g., *Adenostoma fasciculatum*) tend to rot, whereas others (e.g., *Quercus dumosa*) remain resistant for much longer periods.

Stem development in species of *Arctostaphylos* and *Ceanothus* (*Cerastes*) follows a peculiar pattern (Fig. 6.12). In *Arctostaphylos*, large stripes of bark die, leaving behind only a ribbon of living tissue (Adams 1934). In *Ceanothus* species, living tissues grow around these dead stripes, producing a flanged appearance called "longitudinal fissioning" by Jepson (1928). These stripes of dead stem tissue are connected to shaded branches or roots in unfavorable microsites and are often produced during severe droughts (Parsons et al. 1981). This characteristic of allowing selected strands of vascular tissue to die may have evolved as a way of decreasing the amount of stem cortical surface needed to maintain productive parts of the canopy or root system, and this apparently increases the longevity of these nonsprouting species (Davis 1973; Keeley 1975).

Figure 6.12 Stems showing patterning described as bark striping in Arctostaphylos *(far left stem) or longitudinal fissioning in* Ceanothus *(far right stem).*

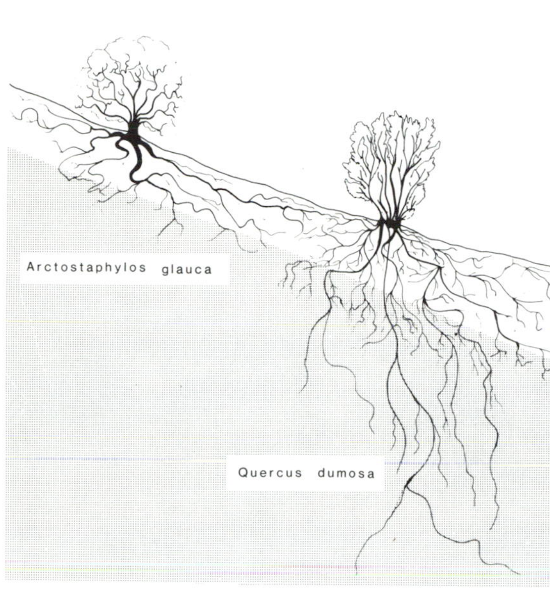

Figure 6.13. Root distribution of the nonsprouting Arctostaphylos glauca *and the sprouting* Quercus dumosa *(redrawn from Hellmers et al. 1955b by T. Montygierd-Loyba).*

Roots

Root systems vary with species and soil depth. Sprouting shrub species tend to have more deeply penetrating roots than nonsprouting species (Fig. 6.13) (Cooper 1922; Hellmers et al. 1955b). Soil depth limits root penetration, but in highly fractured substrates, roots may penetrate bedrock to 9 m or more (Hellmers et al. 1955b). Deep roots may not be a prerequisite for surviving summer drought. Kummerow and associates (1977) excavated a site in southern California on shallow soil overlying bedrock and found that *Adenostoma fasciculatum* roots penetrated to less than 60 cm, and over two-thirds of all root biomass was in the top 20 cm of the soil profile. At another site on deeper soil, roots were distributed deeper for sprouting species *A. fasciculatum* and *Quercus dumosa*, but not for the nonsprouting *Ceanothus greggii* (Kummerow and Mangan 1981). In northern California, Davis (1972) excavated nonsprouting *Arctostaphylos* species and found much of the root mass concentrated in the upper 20 cm of the soil profile. Popenoe (1974) found that soil depth was positively correlated with plant height for shrub species in mixed chaparral.

Fine roots (defined as <0.25 cm in diameter) tend to be concentrated below the canopy; however, there is a great deal of overlap in the distribution of these feeder roots between adjacent plants (Fig. 6.14), suggesting the potential for direct competition for water and nutrients. David (1972) reported that on sites with a well-developed litter layer, *Arctostaphylos* species proliferate feeder roots near the soil surface, and these roots penetrate the decomposing litter mat. In general, for most shrubs, the radial spread of roots is several times greater than the canopy, although root : shoot biomass ratios are less than 1 (Kummerow 1981). Despite the fact that sprouting species maintain their major roots between fires, root : shoot ratios are similar between sprouting and nonsprouting species.

Root nodules with symbiotic nitrogen-fixing actinomycetes are known for species of *Cercocarpus* (Vlamis et al. 1964) and *Ceanothus*, but their presence is dependent on various site factors (Vlamis et al. 1958; Furman 1959; Hellmers and Kelleher 1959; White 1967; Youngberg and Wollum 1976; Kummerow et al. 1978). Ectomycorrhizal associations are common with *Quercus dumosa* and *Arctostaphylos glauca*, and vesicular-arbuscular mycorrhizae with *Adenostoma fasciculatum*, *Ceanothus greggi*, and *Rhus ovata* (Kummerow 1981). Root grafting is apparently not common, although it has been observed in *Q. dumosa* (Hellmers et al. 1955b) and *Prunus ilicifolia* (Bullock 1981).

Vegetative Phenology

Rates of development and growth are controlled by the interaction of low temperatures and irradiance in winter, and low soil moisture and high temperatures in summer. After winter rains have replenished soil moisture, growth initiation is dependent on the onset of higher temperatures. In general, phenological events begin and end later on north- than on south-facing exposures and are delayed with increasing elevation and latitude. Even at the same site, phenological events are not synchronized. In southern California, mean dates of stem elongation range from March to June, depending on the species, and the growing season ranges from 2 mo for *Adenostoma fasciculatum* to 1 mo for *Rhus ovata* (Kummerow et al. 1981). Such patterns also vary spatially; Bedell and Heady (1959) noted a 3-mo growing season for *A. fasciculatum* in northern California. Some species, such as *Malosma laurina* (Watkins and DeForest 1941) and *Adenostoma sparsifolium* (Hanes 1965), may continue growth during the summer months. In years of severe drought, there may be no new leaf production or stem elongation (Harvey and Mooney 1964).

Leaf longevity averages 1.5 yr in *Ceanothus mega-*

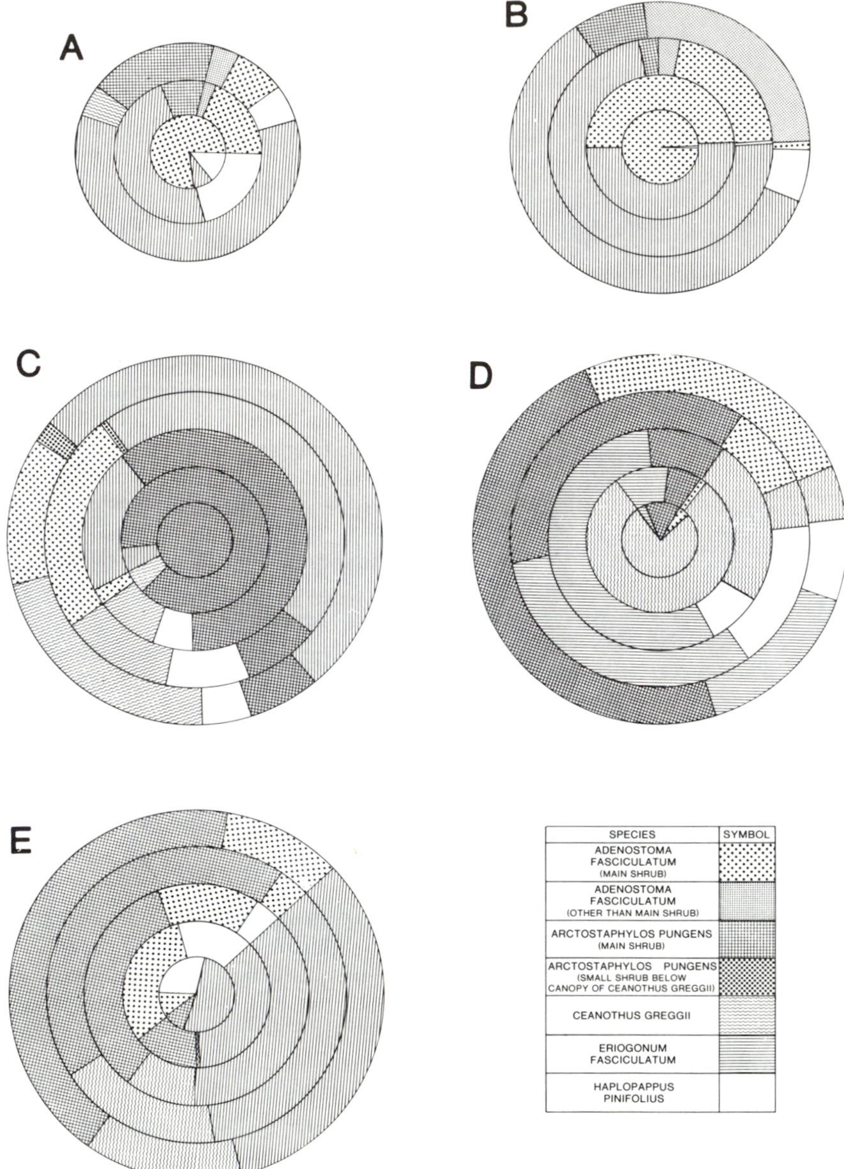

Figure 6.14. Graphic representation of fine-root distribution around the root crowns of five chaparral shrubs in southern California. Each ring represents a width of 1 m. The fine-root amount is expressed as a percentage of the total fine-root mass in the indicated area. The central core of each diagram represents one "soil unit" (= 0.6 m^3) containing the root crown of the respective shrub. The first ring represents 8, the second 16, the third 24, and the fifth 32 soil units. A: Adenostoma fasciculatum, *main shrub. B:* A. fasciculatum, *stump sprout. C:* Arctostaphylos pungens. *D:* Ceanothus greggii. *E:* Haplopappus pinifolius *(from Kummerow et al. 1977).*

carpus (Gray 1982) and 2 yr in *Adenostoma fasciculatum* (Jow et al. 1980). *Heteromeles arbutifolia* leaves average 2 yr (maximum 3 yr) in full sunlight and 3 yr (7 yr maximum) in shade (H. A. Mooney, unpublished data). Litter fall peaks following the completion of primary growth in most species (Mooney et al. 1977). It is almost entirely composed of leaves, as most shrubs have a marked tendency to retain dead twigs and branches.

Secondary stem growth begins earlier and extends later into the season than primary growth (Avila et al. 1975); in *Rhus* species and *Malosma laurina* it may occur year-round. The width of growth rings is significantly lower in drought years

(Gray 1982) and was shown to be sensitive to the level of late winter and spring precipitation (Guntle 1974).

Fine root growth follows a pattern of growth similar to aboveground growth, with peak biomass levels in midsummer and a massive die-off as soil moisture is depleted (Kummerow et al. 1978).

Reproductive Phenology

Flowers of most shrub species are small and are borne in large showy clusters, but few of the flowers produce mature fruits. Most species are largely self-incompatible (Raven 1973; Moldenke 1975). Fulton and Carpenter (1979) reported self-compatibility in *Arctostaphylos pringlei*, but Brum (1975) found this species to be entirely self-incompatible.

Most species are insect-pollinated, and this may have selected for the markedly asynchronous nature of flowering phenology in different species (Mooney 1977b). *Arctostaphylos* and *Garrya* species flower earliest in the season, prior to the initiation of vegetative growth, followed by *Ceanothus* and *Rhus* species. Early flowering in these four genera may be related to the fact that they flower on old growth, from floral buds initiated during the previous year's growing season (J. Keeley, unpublished data). *Adenostoma fasciculatum*, *Heteromeles arbutifolia*, and *Malosma laurina* flower later, on new growth after stem elongation is completed (Bauer 1936; Kummerow et al. 1981), and *Adenostoma sparsifolium*, with stem growth extending well into summer, is one of the latest flowering species, typically not beginning until August (Hanes 1965).

Annual flower and fruit production patterns are variable. Some species (e.g., *Heteromeles* and *Malosma*) tend to flower (and fruit) more or less annually, whereas others, such as *Arctostaphylos* and *Ceanothus* species, are typically biennial bearers.

PHYSIOLOGY AND PRODUCTIVITY

Water Relations

Chaparral shrubs vary in water-relations characteristics largely in accordance with species-specific differences in rooting habit (Poole et al. 1981). Shallow-rooted species are able to respond to elevated soil moisture levels early in the rainy season. During the summer, shallow-rooted nonsprouting species of *Arctostaphylos* and *Ceanothus* (section *Cerastes*) are exposed to extremely negative soil water potentials. At this time they commonly have pre-dawn stem xylem water potentials of −6.5 to −8 MPa versus −3 to −4 MPa during the summer drought for deeper-rooted shrubs such as *Rhus ovata*, *R. integrifolia*, *Malosma laurina*, *Heteromeles arbutifolia*, sprouting species of *Arctostaphylos* and *Ceanothus* (section *Euceanothus*), *Prunus ilicifolia*, and *Rhamnus californica* (Poole and Miller 1975, 1981; Dunn et al. 1976; Burk 1978; Barnes 1979; Miller and Poole 1979; Schlesinger and Gill 1980; Parsons et al. 1981).

Stem water potentials are more negative during summers following low-rainfall seasons (Poole and Miller 1981), although the monthly distribution of rainfall may be as influential as the seasonal total (Gill 1985). In a study by R. Macdonald (unpublished data) in northern California, water potentials in *Adenostoma fasciculatum* and *Ceanothus cuneatus* fell from −2 to −5 MPa during July to October of a normal-rainfall year (1974), whereas in a drought year (1979) these species fell an additional −1.5 to −3.5 MPa; species in other communities showed less reduction, with the magnitude decreasing in the order savanna > woodland > alluvial forest, and with riparian trees showing little change from 1974. During a severe drought in the Sierra Nevada foothills in 1977, the shallow-rooted *Arctostaphylos viscida* reached pre-dawn potentials of −7 MPa and showed very little diurnal variation, whereas the following summer, after a very wet winter, the lowest pre-dawn potentials were −4 MPa, and there was a large diurnal change (Parsons et al. 1981). For most species, water potentials are more negative at their lower elevational and southern latitudinal limits than in the center of their distribution (Poole and Miller 1975; Mooney et al. 1977).

New resprouts may have stem water potentials one-third as negative as mature shrubs, but any differences disappear by the second year after fire (Radosevich and Conard 1980; Oechel and Hastings 1983). Seedlings of *Adenostoma fasciculatum* and *Ceanothus greggii* in recent burns may have water potentials below −6 MPa during the summer (Barro 1982). Schlesinger and Gill (1980) showed that water potentials for the shallow-rooted *C. megacarpus* changed markedly as the stand developed after fire. Comparing stands 6, 13, and 22 yr old, they found that throughout the year, values were always more negative for plants in the youngest-aged stands. Within the 6-yr-old stand, smaller plants suffered more severe water stress during the summer drought than did larger plants; there was a significant correlation ($r = 0.68$) between stem diameter and xylem pressure potential, and one small stem had an amazing pre-dawn water potential of −12 MPa (Schlesinger et al. 1982). They contended that water stress is a major factor in stand thinning.

The relationship between xylem water potential and leaf conductance (Fig. 6.15) shows that most

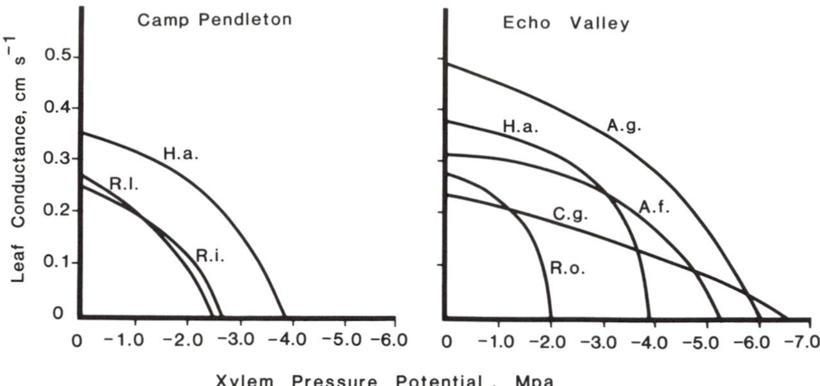

Figure 6.15. Relation between leaf conductance and xylem pressure potential in evergreen chaparral shrubs in southern California: Echo Valley, an interior site at 1000 m, and Camp Pendleton, a coastal site at 100 m. Species are Heteromeles arbutifolia *(Ha.),* Malosma (Rhus) laurina *(Rl.),* Rhus integrifolia *(Ri.),* Arctostaphylos glauca *(Ag.),* Adenostoma fasciculatum *(Af.),* Ceanothus greggii *(Cg.), and* Rhus ovata *(Ro.) (redrawn from Poole et al. 1981).*

deep-rooted species exhibit stomatal closure at far lower water stress than shallow-rooted species. A similar pattern was observed by Barnes (1979) for shallow-rooted nonsprouting *Ceanothus* species in section *Cerastes* versus deeper-rooted species in section *Euceanothus*. The coastal and inland populations of *Heteromeles arbutifolia* illustrated in Fig. 6.15, as well as comparisons of northern and southern California plants (Harrison 1971), show that the relationship between stomatal conductance and xylem water potential is relatively constant for a species.

Stomatal conductances during the winter remain high throughout the day for most shrubs, although the absolute levels are typically twice as high for shallow-rooted species as for deeper-rooted species (Poole et al. 1981). Stomatal conductances of 0.5 cm sec^{-1} observed for the shallow-rooted *Arctostaphylos glauca* and *Ceanothus greggii* are among the highest for chaparral evergreen shrubs, but they are only about half as high as conductances recorded for other associated-life-history types, such as subshrubs and herbs.

During the growing season, there is a midday depression in stomatal conductance that begins earlier in the day as the season progresses. Shallow-rooted *Arctostaphylos* and *Ceanothus* species are likely to maintain more active photosynthesis longer into the drought than many deep-rooted shrubs. On the other hand, by late summer, these shallow-rooted species may show complete stomatal closure for a month or more, whereas deeper-rooted species commonly have a brief period of stomatal conductance each day. There is some evidence that these patterns may be the result of seasonal and diurnal osmotic adjustments (Roberts 1982; Calkin and Pearcy 1984).

These water-relations characteristics produce distinctive species-specific patterns of annual transpiration that vary with elevation and slope exposure and predict different patterns of productivity for these species. Poole and Miller (1981) hypothesized that chaparral communities on different sites should converge at maturity in terms of transpiration per unit of leaf area, and they put that figure at 150−200 mm yr^{-1}. Parker (1984) found that maximum transpiration rates are consistently higher for shallow-rooted obligate-seeding species of *Arctostaphylos* and *Ceanothus* than for deeper-rooted sprouting shrub species. He hypothesized that this characteristic results in more rapid seedling growth rates and hence a better potential for establishment in comparison with seedlings of resprouting species.

Nutrients

Enhanced vegetative and reproductive growth after fertilizer application suggests that chaparral shrubs are nutrient-limited (Hellmers et al. 1955a; Vlamis et al. 1958; Christensen and Muller 1975a; McMaster et al. 1982; Gray and Schlesinger 1983). *Adenostoma fasciculatum* is clearly nitrogen-limited, whereas others such as *Ceanothus megacarpus* and *C. greggii* apparently are not, although the latter species does respond to phosphorus addition. *Ceanothus* species are nitrogen-fixers, and there is some evidence that such species are capable of enhancing the nitrogen status in soils surrounding them (Quick 1944; Hellmers and Kelleher 1959; Vlamis et al. 1964; Kummerow et al. 1978). Asymbiotic nitrogen fixation may also be important (Ellis 1982; Poth 1982).

Foliar leaching of nitrate with the first fall rains may result in a pulse of nitrogen input to the soil

(Christensen 1973). Schlesinger and Hasey (1980) found that this was largely from atmospheric deposition, and this, plus foliar-leached ammonium, could exceed the input by symbiotic nitrogen fixation. Quantitatively, litter fall is the most important means of returning nutrients to the soil (Gray and Schlesinger 1981). Litter fall is concentrated in summer, and decomposition is relatively rapid (Schlesinger and Hasey 1981). The highest concentrations of soil nutrients tend to be in the upper soil layers (Christensen and Muller 1975a), thus, shallow-rooted shrub species may have a competitive advantage. In *Adenostoma fasciculatum*, most of the nitrogen and phosphorus uptake occurs in winter, prior to growth, and Mooney and Rundel (1979) suggested that this may reduce leaching losses from the soil. These winter uptake patterns, however, are not typical of all species (Shaver 1981; Gray 1983).

Fire has a marked effect on the nutrient status of chaparral soils. By recycling nutrients tied up in plant matter, soil levels of most nutrients increase after fire (Sampson 1944; Christensen and Muller 1975a; Gray and Schlesinger 1981). Fires, however, result in substantial ecosystem losses of K and N through volatilization and runoff (DeBano and Conrad 1978) that may require 60–100 yr to replace (Schlesinger and Gray 1982). The first year after fire, the foliage concentrations of important nutrients are very high, although by the second or third year, nutrient levels may be comparable to levels observed for mature vegetation (Sampson 1944; Rundel and Parsons 1980, 1984). On some sites, the post-fire proliferation of suffrutescent and annual legumes may add nitrogen through their symbioses with nitrogen-fixing *Rhizobium* bacteria (Poth 1982).

Productivity

Chaparral shrubs are all C_3 plants (Mooney et al. 1974b) and have the capacity to fix carbon year-round. Maximum photosynthetic rates range from 5 to 15mg CO_2 dm^{-2} hr^{-1} (Mooney 1981). These relatively low rates result from internal limitations inherent in the sclerophyllous leaf structure (Dunn 1975) and strategies of nitrogen use efficiency (Field et al. 1983). For most species there is a broad temperature optimum range for photosynthesis between 10°C and 30°C, and there is apparently little capacity for temperature acclimation (Oechel et al. 1981). Some species are light-limited at less than one-third full sunlight (e.g., *Heteromeles arbutifolia*) (Harrison 1971), whereas others, such as *Ceanothus greggii*, are not saturated at two-thirds full sunlight (Oechel 1982). In general, sprouting species such as *H. arbutifolia*, *Quercus dumosa*, *Prunus ilicifolia*, and *Rhus* species are reasonably shade-tolerant, but nonsprouting species seldom survive under the canopy of adjacent plants. Part of the explanation may lie in the observation that although leaves of *Ceanothus megacarpus* in low-light environments show a net carbon gain, they have lower water use efficiency than leaves under higher irradiances (Mahall and Schlesinger 1982). Field and associates (1983) reported that for five species of evergreens in northern California, photosynthetic capacity was highly correlated with stomatal conductance, and water use efficiency was highest in species commonly found in the driest habitats and lowest in the species common in the wettest habitats. However, species from the wetter habitats had the highest photosynthesis per unit of leaf nitrogen, and this is consistent with a similar finding by Rundel (1982).

The thermal insensitivity of photosynthesis suggests that low temperatures are likely to play a minor role in limiting wintertime carbon gain. Mooney and associates (1975) calculated photosynthetic rates for one evergreen shrub under typical seasonal limitations of light, temperature, and water. Their simulation suggests that wintertime depression of photosynthesis is largely a result of limited irradiance. Peak photosynthetic rates are typically observed only during the spring growing season, and as the season progresses, water availability becomes the major limiting factor for photosynthesis. Oechel (1982) suggested that xylem water potentials of −4 to −4.5 MPa mark the point of zero daily carbon uptake for *Adenostoma fasciculatum* and *Arctostaphylos glauca*. Because tissue water stress, through its effect on stomatal conductance, limits daily CO_2 uptake, carbon-gain patterns vary between species and years. Dunn and associates (1976) showed that the carbon balance for a summer-fall season following a wet winter was several times greater than for a summer following a dry winter for *Heteromeles arbutifolia*, *Rhus ovata*, and *Prunus ilicifolia* (but *A. fasciculatum* was little affected). Similar results have been observed between natural and irrigated plants (Gigon 1979).

After fire, there is a rapid increase in aboveground shrub biomass that continues for several decades (Fig. 6.16). The estimated annual biomass production ranges from 840 to 1750 kg ha^{-1}, with the lower values being from southern California (Ehleringer and Mooney 1983). After the first 20 yr, primary production slows, although the living biomass remains stable for 60 yr or more (Fig. 6.16). These older stands may be nutrient-limited, but this idea has been questioned (Schlesinger et al. 1982). Old stands of chaparral are frequently described as "decadent," "senescent," or "senile,"

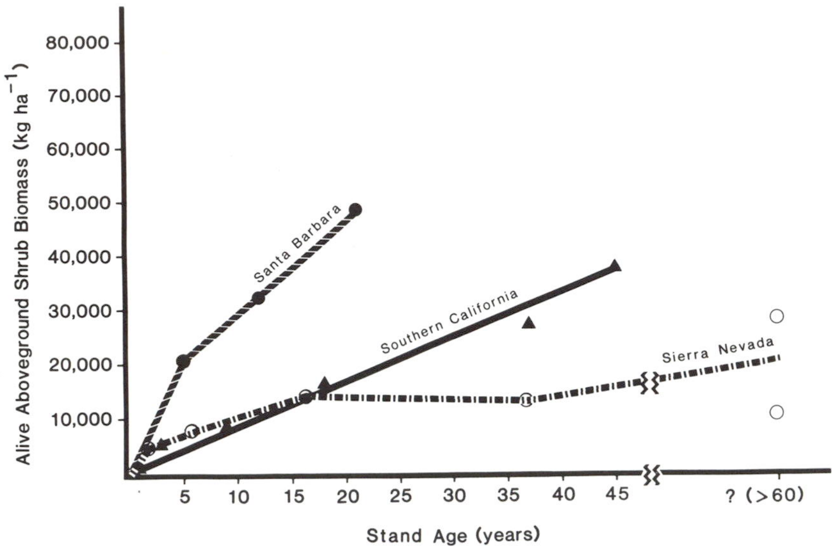

Figure 6.16. Standing living biomass in chaparral stands as a function of age since last fire: southern California mixed chaparral at San Dimas Forest from Specht (1969) (last datum point from Conrad and DeBano 1974); Santa Barbara Ceanothus megacarpus *chaparral from Schlesinger and Gill (1980); Sierra Nevada foothills chamise chaparral from Rundel and Parsons (1979), Stohlgren et al. (1984), and Stohlgren (pers. commun.).*

terms that lack clear definition and are based on little more than anecdotal observations. These terms apparently derive from the fact that as chaparral matures, there is a natural thinning of shrub density (Fig. 6.8), and dead stems accumulate and give the impression that the stands are "trashy" (Hanes 1971). Dead stems, however, are continually replaced by new basal sprouts in sprouting shrub species (Figs. 6.9 and 6.10), and surviving shrubs often remain quite healthy. For example, Rundel and Parsons (1979 and unpublished erratum) found that *Adenostoma fasciculatum* shrubs older than 60 yr had more live biomass than 40-yr-old shrubs. Nonsprouting species clearly are unable to replace older stems from basal sprouts, and this lack may account for the evolution of peculiar stem morphologies that appear to increase longevity (Fig. 6.12). Vigorous populations of nonsprouting *Arctostaphylos* and *Ceanothus* unburned for half a century or more are not uncommon (Table 6.3).

Chaparral stands older than 60 yr have been described as unproductive, with little annual growth (Hanes 1977), an idea apparently derived from measurements of browse production for wildlife (Biswell et al. 1952; Hiehle 1961; Gibbens and Schultz 1963). These studies showed that older chaparral produces very little deer browse. However, those were not valid measures of productivity, because production above 1.5 m, which is normally unavailable to deer, was not included; most new growth in older stands occurs higher than 2 m. There is evidence, in fact, that older chaparral shrubs are capable of considerable annual growth. In a 1-yr study, *Arctostaphylos glauca* and *A. glandulosa* shrubs unburned for 90 yr produced as much or more terminal shoot growth and fruits (per unit of areal coverage) as individuals of the same species in a younger 23-yr-old chaparral stand (Keeley and Keeley 1977). A similar pattern has been recorded for *Adenostoma fasciculatum* in the same stands (J. Keeley and S. Keeley, unpublished data).

RELATED PLANT COMMUNITIES

California Coastal Sage Scrub

Often called "soft chaparral," this summer-deciduous vegetation tolerates more xeric conditions than evergreen chaparral (Mooney 1977a). The name derives from its Coastal Range distribution, where it occurs on both coastal and interior slopes. Coastal sage is most common at lower elevations below chaparral, but patches occur at higher elevations on outcroppings of shallow or fine-textured soils or on excessively disturbed sites. The dominants include all of the subshrubs and suffrutescent species that often are associated with chaparral, in gaps or after fires (Table 6.4), plus more restricted species such as *Encelia californica*, *Baccharis pilularis*, *Viguera laciniata*, and *Lepidospar-*

tum squamatum. Coastal sage scrub is lower (<1.5 m) and more open than chaparral and frequently has some herbaceous understory (Westman 1981a). Various associations have been delineated on the basis of latitudinal changes in species composition from Baja to northern California (Axelrod 1978; Kirkpatrick and Hutchinson 1980; Westman 1983) and show alliances with the northern coastal scrub (Heady et al. 1977). Succulents in the Crassulaceae and Cactaceae are important components near the coast and at the southern limits (Mooney and Harrison 1972). Oftentimes, evergreen sclerophyllous shrubs, such as *Malosma laurina* and *Rhus integrifolia*, will be distributed singly at various intervals throughout coastal sage stands, apparently exploiting infrequent favorable soil microsites.

Most dominants have nonsclerified malacophyllous leaves (3–6 cm^2) that abscise and are replaced during the drought by a few smaller leaves surrounding terminal buds (Harrison et al. 1971; Gray and Schlesinger 1981; Westman 1981b), and these two leaf types may differ physiologically (Gigon 1979; Gulmon 1983). In *Salvia* species, a portion of the leaves may curl up during the summer drought, but then expand during the following growing season, making these plants technically evergreen (Mahall 1985). During the spring growing season, maximum leaf conductances, transpiration rates, and photosynthetic rates may be more than double those observed for sclerophyllous shrubs (Harrison et al. 1971; Oechel et al. 1981; Poole et al. 1981). These shallow-rooted subshrubs avoid drought by losing their foliage; however, they respond to summer or fall precipitation very rapidly and can initiate meristematic activity within a week of such events. Vigorous growth begins in early winter and extends until soil moisture, temperature, and photoperiod induce leaf abscission (Nilsen and Muller 1981; Gray 1982). Flowering is on new growth and thus is delayed until summer or fall (Mooney 1977a).

Volatilization of aromatic compounds from leaves is a notable feature of the coastal sage dominants *Artemisia californica*, *Salvia mellifera*, and *S. leucophylla*. Muller and associates (1964) showed that these compounds were potentially allelopathic to herb growth and suggested that this accounted for the typical "bare zone" of a meter or more that forms between coastal sage and grasslands. Such bare zones also occur between grassland and nonaromatic vegetation such as chamise and scrub oak chaparral. Exclosure experiments in both vegetation types, however, have shown small mammals are a major factor in bare-zone formation because of their propensity to forage on grassland species as close to the protective shrub canopy as possible (Bartholomew 1970; Halligan 1974; Bradford 1976). Allelopathy still may play some role. The facts that coastal sage aromatics represent a substantial carbon drain on the plant (Tyson et al. 1974) and are potentially toxic (Muller and del Moral 1966; Muller and Hague 1967; Halligan 1975) argue strongly for an adaptive role; however, antitranspirant or antiherbivore functions have not been explored.

Coastal sage scrub is intermediate between grassland and chaparral in its resilience following frequent fires (Wells 1962; Kirkpatrick and Hutchinson 1980; Keeley 1981). Fire-recurrence intervals of 5–10 yr may result in chaparral being replaced with coastal sage scrub. More frequent fires, however, will eliminate sage, leaving such sites dominated by non-native grasses. The mode of post-fire recovery of coastal sage scrub is distinctly different from that of the dominant chaparral shrubs. Most sage scrub dominants establish few seedlings in the first year after fire; however, all are capable of resprouting and flowering vigorously in the first year (Westman et al. 1981; Malanson and O'Leary 1982; Keeley and Keeley 1984). These seed crops result in a huge pulse of seedling establishment in the second year after fire, because the seeds of most species (*Artemisia californica*, *Encelia californica*, *Haplopappus squarrosus*, and *Eriogonum fasciculatum*) are nonrefractory and germinate readily (Keeley in press). Coastal sage scrub communities have a post-fire burst of fire annuals and other herbs composed of the same species as in chaparral (Keeley and Keeley 1984). There is some evidence that resprouting of the dominant coastal sage species is greatly reduced on inland sites, and thus recovery is much slower (Westman et al. 1981; Westman 1982). In the absence of fire, coastal sage species are capable of regenerating their canopy from basal sprouts, similar to the pattern observed for sprouting chaparral shrubs (Figs. 6.9 and 6.10).

Interior Chaparral

Interior regions of western North America have vegetation showing various degrees of similarity to California chaparral. Of these, the Arizona chaparral is the most closely related, despite being separated by more than 200 km of desert. The majority of the shrub species are shared with California chaparral (Table 6.1), and others are closely related if not conspecific with California taxa (e.g. *Quercus turbinella*) (Tucker 1953). The nearly ubiquitous Californian *Adenostoma fasciculatum* is noticeably absent, and *Q. turbinella* is to Arizona chaparral what *A. fasciculatum* is to California chaparral: It occurs throughout the chaparral region and dominates most sites (Carmichael et al. 1978). Vast stretches of chaparral, such as in southern California, are uncommon in Arizona; rather, chaparral is distributed in widely disjunct patches (1000–2000 m), and it intergrades with desert scrub or grassland at the lower margin and yellow pine forest or pinyon-juniper woodland at the higher elevation.

Across this range, precipitation is 400–650 mm yr^{-1}; however, in contrast to the situation for the Californian mediterranean climate (Fig. 6.2) 35% of the annual rainfall comes as high-intensity summer thunderstorms (Carmichael et al. 1978). Because summer rains are of short duration and occur at a time of high evaporative loss which has been preceded by a several-month drought, this precipitation may not always recharge the soil volume occupied by the deep-rooted shrubs. Thus, the climate could effectively be mediterranean for these shrubs. If this were true, then the lower elevational limit would be about 260 mm yr^{-1} "available" precipitation, comparable to that for California chaparral (Mooney and Miller 1985). This border occurs at a higher elevation in Arizona, and the colder winter temperatures at this level may account for the absence of the xerophyllous *Adenostoma fasciculatum*. Recent studies by J. Vankat (pers. commun.) show that water stress in Arizona chaparral shrubs is lessened by summer precipitation. However, the extent to which these plants remain photosynthetically active through the summer is unknown, although Swank (1958) noted that most growth is in the spring, and summer growth is unpredictable. Lessened availability of summer precipitation is suggested by the fact that for a specific level of average annual rainfall, Arizona chaparral sites consistently produce sparser and more open communities than California chaparral (Cable 1957). Areal coverage is typically from 35% to 80% ground surface in Arizona chaparral communities (Cable 1957; Carmichael et al. 1978) (cf. Table 6.2). The similarity of flowering behavior in *Arctostaphylos* and *Ceanothus* in Arizona and California–a pattern quite distinct from that recorded from northern Mexico, with *only* summer rain (Fig. 6.17)–further suggests that summer precipitation may not be reliable.

Because of the sparser cover, wildfires are apparently infrequent, occurring every 50–100 yr (Cable 1957), although the responses to fire of the dominant shrubs are similar to Californian plants. Obligate-seeding species such as *Ceanothus greggii*, *Arctostaphylos pringlei*, and *A. pungens* establish seedlings from soil-stored seeds only after fire (Pase 1965; Pase and Lindenmuth 1971). Other species sprout prolifically after fire. Perhaps the most striking difference is the lack of a distinct "fire-type" temporary flora. Herbs and suffrutescent species are more abundant after fire, but they are different from those of California chaparral. Total herb growth may be two orders of magnitude lower, and the herb flora is composed of only a few herbaceous perennial species (Pase and Pond 1964). None of the subshrubs, suffrutescents, or annuals with seed germination cued to post-fire conditions are found in Arizona chaparral.

On the west slopes of the Sierra Madre Oriental of northeastern Mexico are isolated patches of vegetation with strong similarities to California chaparral (Muller 1939, 1947; Shreve 1942). Muller (1939) suggested that the slightly lower precipitation and greater diurnal and annual temperature fluctuations in this region accounted for the replacement of thorn scrub by evergreen sclerophyllous vegetation. The strictly summer-rain climate of this region is markedly unlike that of other chaparral regions (Fig. 6.2). This Mexican chaparral is restricted to limestone or shallow rocky soils at 2000–3000 m, with desert scrub below and evergreen forest above. It is distinguished from the surrounding vegetation types by its predominance of shrubs, the importance of evergreen species, greater density of plant cover, and paucity of herbs. Important genera include *Quercus*, *Garrya*, *Cercocarpus*, *Rhus*, *Ceanothus*, and *Arctostaphylos*, including two California obligate-seeding species, *C. greggii* and *A. pungens*. Rzedowski (1978) considered this zone to have a high fire frequency, and he noted that the several scrub oak species were vigorous sprouters. A similar chaparral-type formation, often with greater numbers of deciduous elements, has been described from other isolated montane sites to the south of the Sierra Madre (Rzedowski 1978).

Petran "chaparral" is a largely winter-deciduous shrub vegetation at 2000–3000 m in the central Rocky Mountains (Vestal 1917; Daubenmire 1943). Despite its high elevation, annual precipitation is 380–535 mm yr^{-1}, well within the range of values for other chaparral regions (Pase and Brown 1982). Winter temperatures are below freezing, and the summer growing season may be <100 d long. The vegetation has an overall physiognomy similar to evergreen chaparral in its height and thicket-like aspect (Hayward 1948). *Quercus gambelii* is the dominant throughout the range, and it, like the majority of species, is winter-deciduous. Evergreen species, including a few from California chaparral (Table 6.1), are minor components, although some, such as *Cercocarpus* species, are sometimes locally abundant (Brooks 1962; Davis 1976). A well-developed herb flora is characteristic of the mature vegetation (Christensen 1949; Allman 1952). Fires occur, and *Q. gambelii* responds like other scrub oaks by sprouting vigorously from the rootstock

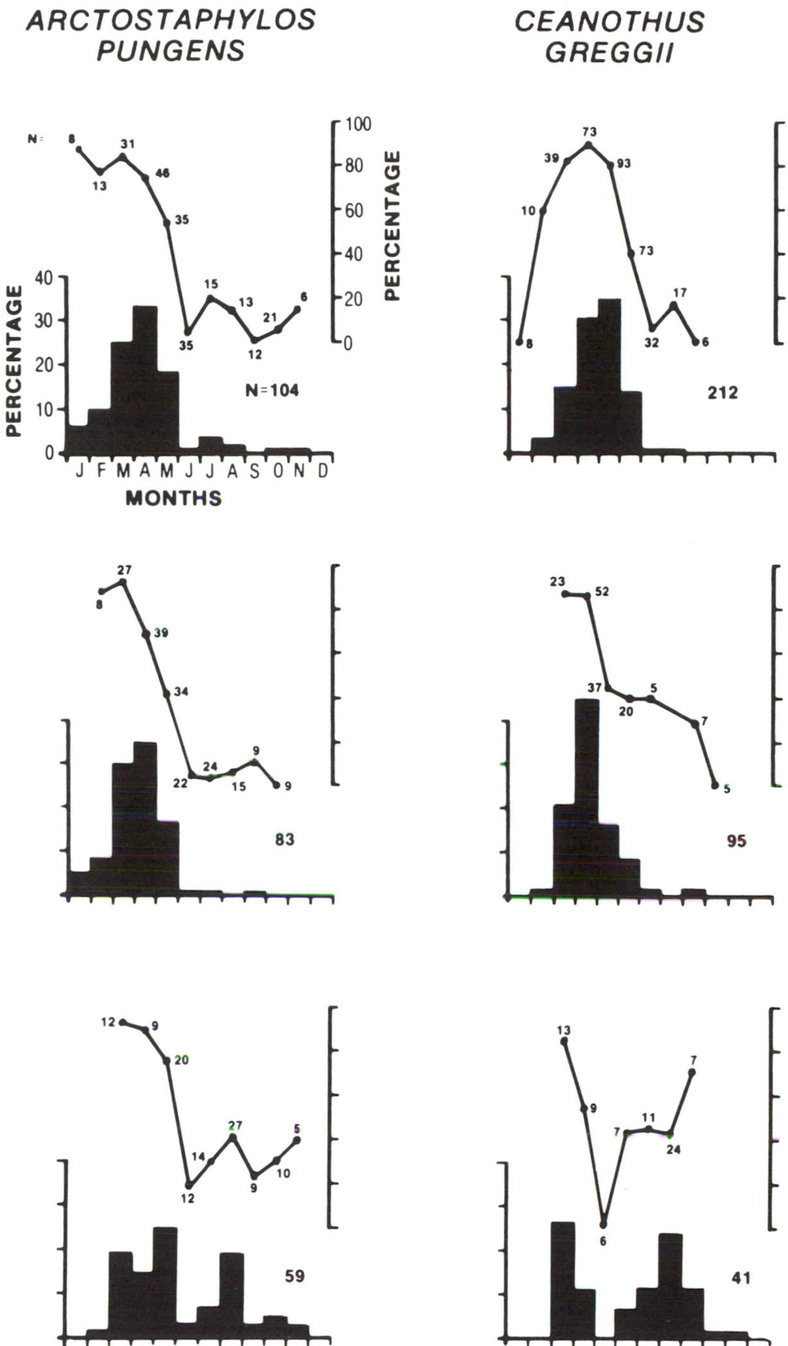

Figure 6.17. Season of flowering for Arctostaphylos pungens *and* Ceanothus greggii *in California, Arizona, and northeastern Mexico. Bars indicate percentage of all herbarium sheets in flower, by month, and lines show percentage of each month's sheets in flower; based on samples from 20 herbaria, at least five collections each (J. Vankat, unpublished data).*

(Brown 1958; Kunzler and Harper 1980). All other common shrub species also sprout vigorously after fire, and obligate-seeding species are infrequent (McKell 1950). Most species maintain themselves in the absence of fire through additional sprouts and seedling recruitment, although disturbance-free periods favor some species over others (Allman 1952; Eastmond 1968).

EVOLUTION

Evergreen Cretaceous vegetation responded to the increasing aridity of the Tertiary period by the evolution of drought-deciduous and evergreen sclerophyllous taxa. California chaparral sclerophylls owe their origins to physiognomically similar taxa that appeared early in the Tertiary under conditions quite unlike the present mediterranean climate (Axelrod 1973). In light of the ample summer rains of that period, it is most reasonable that these species evolved in outcroppings of unusually stressful substrates. By analogy with modern vegetation, we can infer the critical factors to have been soil moisture and nutrient stress. Low-nutrient, coarse soils, with high infiltration rates, produce severe surface soil drought, but they also retain deeper water during droughts. Deep-rooted, woody evergreen sclerophylls can exploit these sites by their enhanced nutrient use efficiency and ability to remain metabolically active further into the drought. Very severe droughts, fine-textured soils, or high-nutrient conditions favor drought-deciduous types.

By the middle of the Miocene, evergreen sclerophylls were widespread across the Southwest, forming various broad-leaved sclerophyllous woodlands and shrublands associated with subtropical species no longer found in this region (Axelrod 1975). Closed-cone pine forests dominated the more mesic coastal regions (Raven and Axelrod 1978), suggesting that fires (and consequently droughts at some time of the year) must have been a predictable feature of the Miocene environment.

By the close of the Tertiary, the climate of California was taking on a mediterranean flavor, possibly a bimodal precipitation regime similar to that in Arizona today (Fig. 6.2), with a greater range of sites exposed to periodic droughts. This, coupled with increased tectonic activity and uplift of mountain ranges, increased the extent of well-drained shallow rocky soils and thus enhanced the spread of evergreen sclerophylls.

The Pleistocene marked the firm establishment of a mediterranean climate in California, accompanied by greater temperature extremes, elimination of summer-rain-dependent taxa, and widespread distribution of modern chaparral species within southern and central California.

The present spatial pattern of chaparral distribution, from the summer-rain region of northern Mexico, through Arizona with bimodal rainfall, to mediterranean California, may represent a useful model of the temporal development of California chaparral. It suggests a pattern of evolution of chaparral taxa on islands of poor soils and seasonal drought, in an otherwise mesic landscape. As the climate changed, these drought-prone "islands" coalesced into larger patches, with consequent elimination of other vegetation. Despite the fact that chaparral taxa originated under a summer-rain climate, there is evidence that evolutionary changes have occurred in response to a mediterranean climate. For example, the flowering phenology of *Arctostaphylos pungens* and *Ceanothus greggii* is bimodally distributed in the summer-rain climate of Mexico (Fig. 6.17). Unimodal flowering in California is the result of flower-bud dormancy through the summer and fall and may have been selected for by the mediterranean climate. Lignotubers, as ontogenetic traits, such as in species of *Adenostoma*, *Arctostaphylos*, and *Ceanothus* (Keeley 1981), or fused endocarp segments in *Arctostaphylos* fruits (Wells 1972), are traits presently absent from Arizona chaparral species, and they possibly arose in response to a greater predictability of fire in a mediterranean environment.

The present distribution patterns of most species date to the xerothermic of recent times (8000–3000 BP) (Raven and Axelrod 1978), a time of most severe soil drought conditions in California that initiated a major expansion of chaparral vegetation. Evidence of this is seen in the elimination of more mesic forests throughout coastal southern and central California (Axelrod 1973; Warter 1976; Heusser 1978). It is quite likely that these climatic changes were exacerbated by humans in a scenario not unlike the "Pleistocene overkill" model (Martin 1973). Wells (1962) suggested that early Holocene humans in California played a major role in shaping present vegetation patterns through widespread use of fire. Lightning-ignited fires have played a role in chaparral evolution from the beginning, but humans have accelerated this process.

The evolutionary history is better known for some chaparral taxa than others. The ubiquitous *Adenostoma fasciculatum* is unknown from the fossil record (Axelrod 1973). It likely evolved on the most severe sites, and this in itself may account for it not being recorded in the fossil record. Most of the broad-sclerophyll species, in more or less present form, date back to middle Miocene times, although as components of various other vegetation types (Axelrod 1975). Many herbaceous genera and *Arctostaphylos* and *Ceanothus* underwent rapid speciation during the Pleistocene, in part because of the creation of new habitats during extensive mountain building (Raven and Axelrod 1978).

Hybridization has played a major role in the evolution of the latter two genera. All *Ceanothus* species are $2n = 24$, and although crosses between

Cerastes and *Euceanothus* are rare, hybrids within the subgenera are common (McMinn 1944; Nobs 1963; Phillips 1966). In *Arctostaphylos*, there are diploid ($2n = 13$) and tetraploid taxa. In this genus, foliaceous floral bracts are considered ancestral, and these are restricted to tetraploid lignotuberous taxa (Wells 1969). These are taxonomically difficult species, because the resprouting mode allows for perpetuation of many forms; across their range, numerous subspecies or varieties are recognized, and even within a population there may be a number of very different forms. Obligate-seeding taxa all have the (apparently) derived characteristic of reduced bracts and are mostly diploid, although tetraploid populations are known for some (Wells 1968; Roof 1978). *Arctostaphylos rudis* and *A. patula* are diploids, with reduced bracts, and they are typically burl-forming resprouters; both species, however, are variable in this respect. Hybridization has been implicated in explanations of variation for over 30 taxa in the genus (Keeley 1976), most commonly between diploid species, although diploid × tetraploid crosses are thought to account for the origin of many taxa (Wells 1968, 1972; Keeley 1974). The most thoroughly documented case for hybridization in *Arctostaphylos* involves the low-elevation obligate-seeding *A. viscida* ssp. *mariposa* and the montane-sprouting *A. patula*. At 1000 m, near Mather in the Sierra Nevada, sympatric populations occur, along with forms intermediate to these species and shrubs with unique combinations of traits (Dobzhansky 1953); isozymic evidence indicates hybridization and introgression (N. Elstrand, unpublished data). Ball and associates (1983) have documented the microhabitat distribution and physiological characteristics of the hybrids and parents.

Wells (1969) suggested that speciation in *Arctostaphylos* and *Ceanothus* is a result of the obligate-seeding reproductive mode that increased the number of sexual generations, relative to resprouting species. In *Ceanothus*, all species in the subgenus *Cerastes* have the obligate-seeding mode. In *Euceanothus*, some species are easily killed by fire, but "true"obligate seeders–taxa lacking adventitous buds in the root crown–are uncommon. The obligate-seeding mode in *Arctostaphylos* has clearly been polyphyletic, because many species have both obligate-seeding and burl-forming sprouting subspecies. In both genera, the obligate-seeding mode is undoubtedly a derived condition, because sprouting from adventitious buds in the root crown is a nearly ubiquitous trait in woody dicots and has undoubtedly been selectively maintained because of its value in recovering from many types of disturbances. Raven (1973) suggested that loss of the crown-sprouting trait allowed for a more rapid fine-tuning of adaptation to the relatively recent mediterranean climate. Others have suggested that evolution of the obligate-seeding mode was tied to conditions that created large gaps for seedling establishment after fire, and thus it was favored along arid borders or in places of infrequent, intense fires (Keeley and Zedler 1978). Under such conditions, allocation of energy to seeds, as opposed to lignotubers, would be adaptive (Keeley and Keeley 1977). However, the energetic cost of lignotubers may not be high, and elimination of that structure does not preclude maintaining the ability to lay down adventitious buds in root and stem material.

A comparison of the structural and functional characteristics of obligate-seeders and sprouters in both genera suggests that they represent quite different syndromes (Keeley 1986). Sprouting shrubs are typically deep-rooted and require access to deep reserves of soil moisture during the summer drought. Once established, they preempt space for considerable periods of time, but as aridity increases, the number of safe sites for establishment goes down. Safe sites for shallow-rooted obligate-seeders are essentially open sites between resprouting shrubs. As the number of resprouting shrubs that can be maintained on a site decreases, the number of safe sites for seeders goes up. This strategy, however, depends on an ability to withstand severe water stress. Physiological data suggest a marked divergence between sprouters and obligate-seeders in this regard, and evolution of this divergence may have been enhanced by the greater number of sexual generations resulting from loss of the resprouting ability.

AREAS FOR FUTURE RESEARCH

Long-term ecosystem studies are needed in order to evaluate management practices. Possible problems include the following:

1. The negative effect of artifical seeding after wildfires on reestablishment of the native flora.
2. The present fire frequency of every two to three decades may be producing significant alterations in nutrient pools, as well as eliminating certain obligate-seeding species.
3. The detrimental ecological and esthetic effects of type-conversion projects, biomass programs, and the present fuel-break system.

Much of the early research in chaparral focused on the community response to wildfire. There has been much speculation about community-level changes in the absence of fire, and although much dogma has been published, basically no studies have investigated the long-term changes in productivity and species diversity.

Demographic studies have the potential for explaining a great deal about the patterns observed in chaparral. Considering the wealth of species, relatively little is known of the life-history characteristics for species of all growth forms. Specific areas in need of study include the following:

1. Resilience of species to short as well as long fire-free intervals.
2. The timing and extent of seedling establishment for many sprouting shrubs and all herbaceous perennial species.
3. Seed-germination biology of most species, particularly the mechanism of charred-wood-stimulated germination, the distribution and control of germination polymorphisms, the degree of allelopathic control of germination, and longevity of seeds in the soil.

Physiological studies of chaparral shrubs have produced very detailed pictures of how a few species function. Broader comparisons with many more species and different growth forms are needed. Recent studies show that much is yet to be learned about the trade-offs involved in maximizing water use and nutrient use efficiency.

Detailed biosystematic studies in large woody genera such as *Arctostaphylos* and large herbaceous genera such as *Phacelia* are needed for a clearer picture of the evolutionary changes involved in adapting to the mediterranean climate.

ACKNOWLEDGMENTS

We thank Jochen Kummerow and John Vankat for helpful comments on the manuscript, Hal Mooney and many SDSU faculty, particularly Paul Zedler and Al Johnson, for introducing us to chaparral and providing many useful insights, Norm Elstrand, H. Kuhlmann, J. Kummerow, H. Mooney, Walt Oechel, Rod Macdonald, and J. Vankat for providing data and published figures, and Teresa Montygierd-Łoyba, Cheryl Swift, and Amy Gonzalez for the orginal figures.

REFERENCES

Adams, J. E. 1934. Some observations on two species of *Arctostaphylos*. Madroño 2:147–152.

Allman, V. P. 1952. A preliminary study of the vegetation in an exclosure in the chaparral of the Wasatch Mountains, Utah. M.S. thesis, Brigham Young University, Provo, Utah.

Ammirati, J. F. 1967. The occurrence of annual perennial plants on chaparral burns. M.S. thesis, San Francisco State University.

Anfuso, R. F. 1982. Fire temperature relationships of *Adenostoma fasciculatum*. M.S. thesis, California State University, Los Angeles.

Arnold, K., L. T. Burcham, R. L. Fenner, and R. F. Grah. 1951. Use of fire in land clearing. Calif. Agr. 5(3):9–11; 5(4):4–5, 13, 15; 5(5):11–12; 5(6):13–15; 5(7):6, 15.

Avila, G., M. Lajaro, S. Araya, G. Montenegro, and J. Kummerow. 1975. The seasonal cambium activity of Chilean and Californian shrubs. Amer. J. Bot. 62:473–478.

Axelrod, D. I. 1973. History of the mediterranean ecosystem in California, pp. 225–277 in F. de Castri and H. A. Mooney (eds.), Mediterranean ecosystems. Origin and structure. Springer, New York.

Axelrod, D. I. 1975. Evolution and biogeography of Madrean-Tethyan sclerophyll vegetation. Ann. Missouri Bot. Gardens 62:280–334.

Axelrod, D. I. 1978. The origin of coastal sage vegetation, Alta and Baja California. Amer. J. Bot. 65:117–131.

Baker, G. A., P. W. Rundel, and D. J. Parsons. 1982a. Postfire recovery of chamise chaparral in Sequoia National Park, California, p. 584 in C. E. Conrad and W. C. Oechel (eds.), Proceedings of the symposium on dynamics and management of mediterranean-type ecosystems. USDA Forest Service, Pacific Southwest Forest and Range Experiment Station general technical report PSW-58.

Baker, G. A., P. W. Rundel, and D. J. Parsons. 1982b. Comparative phenology and growth in three chaparral shrubs. Bot. Gaz. 143:94–100.

Ball, C. T., J. Keeley, H. Mooney, J. Seaman, and W. Winner. 1983. Relationship between form, function, and distribution of two *Arctostaphylos* species (Ericaceae) and their putative hybrids. Oecologia Plantarum 4:153–164.

Barnes, F. S. 1979. Water relations of four species of *Ceanothus*. M.A. thesis, San Jose State University, San Jose, Calif.

Barro, S. C. 1982. Water relations of *Ceanothus greggii* and *Adenostoma fasciculatum* seedlings in California chaparral. M.A. thesis, California State University, Fullerton.

Bartholomew, B. 1970. Bare zone between California shrub and grassland communities: the role of animals. Science 170:1210–1212.

Bauer, H. L. 1930. On the flora of the Tehachapi Mountains, California. Bull. South. Calif. Acad. Sci. 29:96–99.

Bauer, H. L. 1936. Moisture relations in the chaparral of the Santa Monica Mountains, California. Ecol. Monogr. 6:409–454.

Beatty, S. W. 1984. Vegetation and soil patterns in southern California chaparral communities, pp. 4–5 in B. Dell (ed.), Medecos IV: proceedings 4th international conference on mediterranean ecosystems. Botany Department, University of Western Australia, Nedlands.

Bedell, T. E., and H. F. Heady. 1959. Rate of twig elongation of chamise. J. Range Management 12:116–121.

Bentley, J. R., and R. L. Fenner. 1958. Soil temperatures during fires on California foothills: how to recognize post-fire seedsheds. J. Forestry 56:738.

Biswell, H. H. 1974. Effects of fire on chaparral, pp. 321–364 in T. T. Kozlowski and C. E. Ahlgren (eds.), Fire and ecosystems. Academic, New York.

Biswell, H. H., R. D. Taber, W. W. Hedrick, and A. M. Schultz. 1952. Management of chamise brushlands for game in the north coast region of California. Calif. Fish Game 38:453–484.

Bjorndalen, J. E. 1978. The chaparral vegetation of Santa Cruz Island, California. Norwegian J. Bot. 25:255–269.

Bolton, R. B., and R. J. Vogl. 1969. Ecological requirements of *Pseudotsuga macrocarpa* in the Santa Ana Mountains, California. J. Forestry 69:112–119.

Borchert, M. 1985. Serotiny and cone-habit variation in populations of *Pinus coulteri* (Pinaceae) in the southern Coast Ranges of Calfironia. Madroño 32:29–48.

Bradford, D. F. 1976. Space utilization by rodents in *Adenostoma* chaparral. J. Mammology 57:576–579.

Brooks, A. C. 1962. An ecological study of *Cercocarpus montanus* and adjacent communities in part of the Laramie basin. M.S. thesis, University of Wyoming, Laramie.

Brown, H. W. 1958. Gambel oak in west-central Colorado. Ecology 39:317–327.

Brum, G. D. 1975. Floral biology and pollination strategies of *Arctostaphylos glauca* and *A. pringlei* var. *drupaceae* (Ericaceae). Ph.D. dissertation, University of California, Riverside.

Bullock, S. H. 1978. Plant abundance and distribution in relation to types of seed dispersal in chaparral. Madroño 25:104–105.

Bullock, S. H. 1981. Aggregation of *Prunus ilicifolia* (Rosaceae) during dispersal and its effect on survival and growth. Madroño 28:94–95.

Bullock, S. 1982. Reproductive ecology of *Ceanothus cordulatus*. M.A. thesis, California State University, Fresno.

Burk, J. H. 1978. Seasonal and diurnal water potentials in selected chaparral shrubs. Amer. Midl. Nat. 99:244–248.

Byrne, R., J. Michaelsen, and A. Soutar. 1977. Fossil charcoal as measure of wildfire frequency in southern California: a preliminary analysis, pp. 361–367 in H. A. Mooney and C. E. Conrad (eds.), Proceedings of the symposium on environmental consequences of fire and fuel management in mediterranean ecosystems. USDA Forest Service, general technical report WO-3.

Cable, D. R. 1957. Recovery of chaparral following burning and seeding in central Arizona. USDA Forest Service, Rocky Mountain Forest and Range Experiment Station research note RM-28.

Calkin, H. W., and R. W. Pearcy. 1984. Seasonal progressions of tissue and cell water relations parameters in evergreen and deciduous perennials. Plant Cell Environ. 7:347–352.

Carlquist, S. 1980. Further concepts in ecological wood anatomy, with comments on recent work in wood anatomy and evolution. Aliso 9:499–553.

Carmichael, R. S., O. D. Knipe, C. P. Pase, and W. W. Brady. 1978. Arizona chaparral: plant associations and ecology. USDA Forest Service, Rocky Mountain Forest and Range Experiment Station research paper RM-202.

Chou, C.-H., and C. H. Muller. 1972. Allelopathic mechanisms of *Arctostaphylos glandulosa* var. *zacaensis*. Amer. Midl. Nat. 88:324–347.

Christensen, E. M. 1949. The ecology and geographic distribution of oak (*Quercus gambelii*) in Utah. M.S. thesis, University of Utah, Salt Lake City.

Christensen, N. L. 1973. Fire and the nitrogen cycle in California chaparral. Science 181:66–68.

Christensen, N. L., and C. H. Muller. 1975a. Effects of fire on factors controlling plant growth in *Adenostoma* chaparral. Ecol. Monogr. 45:29–55.

Christensen, N. L., and C. H. Muller. 1975b. Relative importance of factors controlling germination and seedling survival in *Adenostoma* chaparral. Amer. Midl. Nat. 93:71–78.

Clark, H. W. 1937. Association types in the north Coast Ranges of California. Ecology 18:214–230.

Conard, S. G., and S. R. Radosevich. 1981. Photosynthesis, xylem pressure potential, and leaf conductance of three montane chaparral species in California. Forest Sci. 27:627–639.

Conard, S. G., and S. R. Radosevich. 1982. Post-fire succession in white fir (*Abies concolor*) vegetation of the northern Sierra Nevada. Madroño 29:42–56.

Conard, C. E., and L. F. DeBano. 1974. Recovery of southern California chaparral. American Society of Civil Engineers national meeting on water resources engineering, meeting reprint 2167.

Cooper, W. S. 1922. The broad-sclerophyll vegetation of California. An ecological study of the chaparral and its related communities. Carnegie Institution of Washington publication 319.

Corbett, E. S., and L. R. Green. 1965. Emergency revegetation to rehabilitate burned watersheds in southern California. USDA Forest Service, Pacific Southwest Forest and Range Experiment Station research paper PSW-22.

Daubenmire, R. F. 1943. Vegetational zonation in the Rocky Mountains. Bot. Rev. 9:325–393.

Davey, J. R. 1982. Stand replacement in *Ceanothus crassifolius*. M.S. thesis, California State Polytechnic University, Pomona.

Davis, C. B. 1972. Comparative ecology of six members of the *Arctostaphylos andersonii* complex. Ph.D. dissertation, University of California, Davis.

Davis, C. B. 1973. "Bark striping" in *Arctostaphylos* (Ericaceae). Madroño 22:145–149.

Davis, J. 1967. Some effects of deer browsing on chamise sprouts after fire. Amer. Midl. Nat. 77:234–238.

Davis, J. N. 1976. Ecological investigation in *Cercocarpus ledifolius* Nutt. communities of Utah. M.S. thesis, Brigham Young University, Provo, Utah.

DeBano, L. F., and C. E. Conrad. 1978. The effect of fire on nutrients in a chaparral ecosystem. Ecology 59:489–497.

DeBano, L. F., P. H. Dunn, and C. E. Conrad. 1977. Fire's effect on physical and chemical properties of chaparral soils, pp. 65–74 in H. A. Mooney and C. E. Conrad (eds.), Proceedings of the symposium on environmental consequences of fire and fuel management in mediterranean ecosystems. USDA Forest Service, general technical report WO-3.

Detling, L. E. 1961. The chaparral formation of southeastern Oregon with consideration of its postglacial history. Ecology 42:348–357.

Dobzhansky, T. 1953. Natural hybrids of two species

of *Arctostaphylos* in the Yosemite region of California. Heredity 7:73–79.

Dunn, E. L. 1975. Environmental stresses and inherent limitations affecting CO_2 exchange in evergreen sclerophylls in mediterranean climates, pp. 159–181 in D. M. Gates and R. B. Schmeri (eds.), Perspectives in biophysical ecology. Springer, New York.

Dunn, E. L., F. M. Shropshire, L. C. Song, and H. A. Mooney. 1976. The water factor and convergent evolution in mediterranean-type vegetation, pp. 492–505 in O. L. Lange, L. Kappen, and E. D. Schulz (eds.), Water and plant life. Springer, New York.

Eastmond, R. J. 1968. Vegetational changes in a mountain brush community of Utah during 18 years. M.S. thesis, Brigham Young University, Provo, Utah.

Ehleringer, J., and H. A. Mooney. 1983. Productivity of desert and mediterranean-climate plants, pp. 205–231 in O. L. Lange, P. S. Nobel, C. B. Osmond, and H. Zeigler (eds.), Physiological plant ecology, IV. Springer, New York.

Ellis, B. A. 1982. Asymbiotic N_2 fixation and nitrogen content of bulk precipitation in southern California chaparral, p. 595 in C. E. Conrad and W. C. Oechel (eds.), Proceedings of the symposium on dynamics and management of mediterranean-type ecosystems. USDA Forest Service, Pacific Southwest Forest and Range Experiment Station general technical report PSW-58.

Epling, C., and H. Lewis. 1942. The centers of distribution of the chaparral and coastal sage. Amer. Midl. Nat. 27:445–462.

Field, C., J. Merino, and H. A. Mooney. 1983. Compromises between water-use efficiency and nitrogen-use efficiency in five species of California evergreens. Oecologia 60:384–389.

Fishbeck, K., and J. Kummerow. 1977. Comparative wood and leaf anatomy, pp. 148–161 in N. J. W. Thrower and D. E. Bradbury (eds.), Chile-California mediterranean scrub atlas: a comparative analysis. Dowden, Hutchinson and Ross, Stroudsburg, Pa.

Fulton, R. E., and F. L. Carpenter. 1979. Pollination, reproduction, and fire in *Arctostaphylos*. Oecologia 38:147–157.

Furman, T. E. 1959. The structure of the root nodules of *Ceanothus sanguineus* and *Ceanothus velutinus*, with special reference to the endophyte. Amer. J. Bot. 46:698–703.

Gause, G. W. 1966. Silvical characteristics of bigcone Douglas-fir (*Pseudotsuga macrocarpa*) [Vasey] Mayr). USDA Forest Service, Pacific Southwest Forest and Range Experiment Station research paper PSW-39.

Gauss, N. M. 1964. Distribution of selected plant species in a portion of the Santa Monica Mountains, California on the basis of site. M.A. thesis, University of California, Los Angeles.

Gautier, C. R. 1981. The effect of rye grass on erosion and native vegetation recovery in a burned southern California watershed. M.S. thesis, San Diego State University.

Gibbens, R. P., and A. M. Schultz. 1963. Brush manipulation on a deer winter range. Calif. Fish Game 49:95–118.

Gigon, A. 1979. CO_2-gas exchange, water relations and convergence of mediterranean shrub-types from California and Chile. Oecologia Plantarum 14:129–150.

Gill, D. S. 1985. A quantitative description of the phenology of an evergreen and a deciduous shrub species with reference to temperature and water relations in the Santa Ynez Mountains, Santa Barbara County, California. M.A. thesis, University of California, Santa Barbara.

Grant, V. 1949. Seed germination in *Gila capitata* and its relatives. Madroño 10:87–93.

Graves, G. W. 1932. Ecological relationships of *Pinus sabiniana*. Bot. Gaz. 94:106–133.

Gray, J. T. 1982. Community structure and productivity in *Ceanothus* chaparral and coastal sage scrub of southern California. Ecol. Monogr. 52:415–435.

Gray, J. T. 1983. Nutrient use by evergreen and deciduous shrubs in southern California. I. Community nutrient cycling and nutrient-use efficiency. J. Ecol. 71:21–41.

Gray, J. T., and W. H. Schlesinger. 1981. Nutrient cycling in mediterranean type ecosystems, pp. 259–285 in P. C. Miller (ed.), Resource use by chaparral and matorral. Springer, New York.

Gray, J. T., and W. H. Schlesinger. 1983. Nutrient use by evergreen and deciduous shrubs in southern California. II. Experimental investigations of the relationship between growth, nitrogen uptake and nitrogen availability. J. Ecol. 71:43–56.

Greenlee, J. M., and J. H. Langenheim. 1980. The history of wildfires in the region of Monterey Bay. Unpublished report, California State Department of Parks and Recreation.

Griffin, J. R. 1978. Maritime chaparral and endemic shrubs of the Monterey Bay Region, California. Madroño 25:65–81.

Griffin, J. R. 1982. Pine seedlings, native ground cover, and *Lolium multiflorum* on the marble-cone burn, Santa Lucia Range, California. Madroño 29:177–188.

Gulmon, S. L. 1983. Carbon and nitrogen economy of *Diplacus aurantiacus* a Californian mediterranean climate drought-deciduous shrub, pp. 167–176 in F. Kruger, D. T. Mitchell, and J. Jarvis (eds.), Mediterranean-type ecosystems. The role of nutrients. Springer, New York.

Guntle, G. R. 1974. Correlation of annual growth in *Ceanothus crassifolius* Torr. and *Arctostaphylos glauca* Lindl. to annual precipitation in the San Gabriel Mountains. M.A. thesis, California State Polytechnic University, Pomona.

Hadley, E. B. 1961. Influence of temperature and other factors on *Ceanothus megacarpus* seed germination. Madroño 16:132–138.

Haines, .L. 1941. Variation in *Yucca whipplei*. Madroño 6:33–45.

Hall, H. M. 1903. Botanical survey of San Jacinto Mountains. Univ. Calif. Pub. Bot. 1:1–140.

Halligan, J. 1974. Relationship between animal activity and bare areas associated with California sagebrush in annual grassland. J. Range Management 27:358–363.

Halligan, J. P. 1975. Toxic terpenes from *Artemisia californica*. Ecology 56:999–1003.

Hanes, T. L. 1965. Ecological studies on two closely

related chaparral shrubs in southern California. Ecol. Monogr. 35:213–235.

Hanes, T. L. 1971. Succession after fire in the chaparral of southern California. Ecol. Monogr. 41:27–52.

Hanes, T. L. 1977. California chaparral, pp. 417–470 in M. G. Barbour and J. Major (eds.), Terrestrial vegetation of California. Wiley, New York.

Harrison, A. T. 1971. Temperature related effects on photosynthesis in *Heteromeles arbutifolia* M. Roem. Ph.D. dissertation, Stanford University.

Harrison, A. T., E. Small, and H. A. Mooney. 1971. Drought relationships and distribution of two mediterranean-climate California plant communities. Ecology 52:869–875.

Harvey, R. A., and H. A. Mooney. 1964. Extended dormancy of chaparral shrubs during severe drought. Madroño 17:161–163.

Hayward, C. L. 1948. Biotic communities of the Wasatch chaparral, Utah. Ecol. Monogr. 18:473–506.

Heady, H. F., T. C. Foin, M. M. Hektner, D. W. Taylor, M. G. Barbour, and W. J. Barry. 1977. Coastal prairie and northern coastal scrub, pp. 733–757 in M. G. Barbour and J. Major (eds.), Terrestrial vegetation of California. Wiley, New York.

Hedrick, D. W. 1951. Studies on the succession and manipulation of chamise brushlands in California. Ph.D. dissertation, Texas A&M College, College Station.

Hellmers, H., J. F. Bonner, and J. M. Kelleher. 1955a. Soil fertility: a watershed management problem in the San Gabriel Mountains of southern California. Soil Sci. 80:189–197.

Hellmers, H., J. S. Horton, G. Juhren, and J. O'Keefe. 1955b. Root systems of some chaparral plants in southern California. Ecology 36:667–678.

Hellmers, H., and J. M. Kelleher. 1959. *Ceanothus leucodermis* and soil nitrogen in southern California mountains. Forest Sci. 5:275–278.

Heusser, L. 1978. Pollen in the Santa Barbara Basin, California: a 12,000-yr record. Geol. Soc. Amer. Bull. 89:673–678.

Hiehle, J. L. 1961. Measurement of browse growth and utilization. Calif. Fish Game 50:148–151.

Hochberg, M. C. 1980. Factors affecting leaf size of the chaparral on the California islands, pp. 189–206 in D. M. Power (ed.), The California Islands: proceedings of a multidisciplinary symposium. Santa Barbara Museum of Natural History.

Hom, S. 1984. Bird dispersal of toyon (*Heteromeles arbutifolia*). M.S. thesis, California State University, Hayward.

Horton, J. S. 1960. Vegetation types of the San Bernardino Mountains. USDA Forest Service, Pacific Southwest Forest and Range Experiment Station technical paper 44.

Horton, J. S., and C. J. Kraebel. 1955. Development of vegetation after fire in the chamise chaparral of southern California. Ecology 36:244–262.

Howe, C. F. 1982. Death of chamise (*Adenostoma fasciculatum*) shrubs after fire as a result of herbivore browsing. Bull. South. Calif. Acad. Sci. 80:138–143.

Howell, J. T. 1945. Concerning stomata on leaves in *Arctostaphylos*. Wasmann Collector 6:57–65.

Jacks, P. M. 1984. The drought tolerance of *Adenostoma fasciculatum* and *Ceanothus crassifolius* seedlings and vegetation change in the San Gabriel chaparral. M.S. thesis, San Diego State University.

James, S. M. 1984. Lignotubers and burls–their structure, function and ecological significance in mediterranean ecosystems. Bot. Rev. 50:225–266.

Jepson, W. L. 1928. Biological peculiarities of California flowering plants, part I. Madroño 1:190–192.

Jones, C. S., and W. H. Schlesinger. 1980. *Emmenanthe penduliflora* (Hydrophyllaceae): further consideration of germination response. Madroño 27:122–125.

Jones, M. D., and H. M. Laude. 1960. Relationships between sprouting in chamise and physiological condition of the plant. J. Range Management 13:210–214.

Jow, W., S. H. Bullock, and J. Kummerow. 1980. Leaf turnover rates of *Adenostoma fasciculatum* (Rosaceae). Amer. J. Bot. 67:256–261.

Kaminsky, R. 1981. The microbial origin of the allelopathic potential of *Adenostoma fasciculatum* H & A. Ecol. Monogr. 51:365–382.

Keeley, J. E. 1974. Notes on *Arctostaphylos glauca* Lindl. var. *puberula* J. T. Howell. Madroño 22:403.

Keeley, J. E. 1975. The longevity of nonsprouting *Ceanothus*. Amer. Midl. Nat. 93:504–507.

Keeley, J. E. 1976. Morphological evidence of hybridization between *Arctostaphylos glauca* and *A. pungens* (Ericaceae). Madroño 23:427–434.

Keeley, J. E. 1977a. Seed production, seed populations in soil, and seedling production after fire for two congeneric pairs of sprouting and non-sprouting chaparral shrubs. Ecology 58:820–829.

Keeley, J. E. 1977b. Fire dependent reproductive strategies in *Arctostaphylos* and *Ceanothus*, pp. 371–376 in H. A. Mooney and C. E. Conrad (eds.), Proceedings of the symposium on environmental consequences of fire and fuel management in mediterranean ecosystems. USDA Forest Service, general technical report WO-3.

Keeley, J. E. 1981. Reproductive cycles and fire regimes, pp. 231–277 in H. A. Mooney, T. M. Bonnicksen, N. L. Christensen, J. E. Lotan, and W. A. Reiners (eds.), Proceedings of the conference fire regimes and ecosystem properties. USDA Forest Service, general technical report WO-26.

Keeley, J. E. 1982. Distribution of lightning and man-caused wildfires in California, pp. 431–437 in C. E. Conrad and W. C. Oechel (eds.), Proceedings of the symposium on dynamics and management of mediterranean-type ecosystems. USDA Forest Service, Pacific Southwest Forest and Range Experiment Station general technical report PSW-58.

Keeley, J. E. 1984a. Bibliographies on chaparral and the fire ecology of other mediterranean systems. Report 38, California Water Resources Center, University of California, Davis.

Keeley, J. E. 1984b. Factors affecting germination of chaparral seeds. Bull. South. Calif. Acad. Sci. 83:113–120.

Keeley, S. 1984c. Stimulation of post-fire herb germination in the California chaparral by burned shrub stems and heated wood components, pp. 79–80 in B. Dell (ed.), Medecos IV: proceedings 4th international conference on mediterranean ecosystems. Botany Department, University of Western Australia, Nedlands.

Keeley, J. E. 1986. Resilience of mediterranean shrub communities to fire, pp. 95–112 in B. Dell, A. J. M. Hopkins, and B. B. Lamont (eds.), Resilience in mediterranean-type ecosystems. Dr. W. Junk, Dordrecht.

Keeley, J. E. 1987. Role of fire in the seed germination of woody taxa in California chaparral. Ecology.

Keeley, J. E., and R. L. Hays. 1976. Differential seed predation on two species of *Arctostaphylos* (Ericaceae). Oecologia 24:71–81.

Keeley, J. E., and S. C. Keeley. 1977. Energy allocation patterns of sprouting and non-sprouting species of *Arctostaphylos* in the California chaparral. Amer. Midl. Nat. 98:1–10.

Keeley, J. E., and S. C. Keeley. 1981. Postfire regeneration of California chaparral. Amer. J. Bot. 68:524–530.

Keeley, J. E., and S. C. Keeley. 1984. Postfire recovery of California coastal sage scrub. Amer. Midl.. Nat. 111:105–117.

Keeley, J. E., and S. C. Keeley in press. Role of fire in the germination of chaparral herbs and suffrutescents. Madroño

Keeley, J. E., B. A. Morton, A. Pedrosa, and P. Trotter. 1985. The role of allelopathy, heat and charred wood in the germination of chaparral herbs and suffrutescents. J. Ecol. 73:445–458.

Keeley, J. E., and M. Nitzberg. 1984. The role of charred wood in the germination of the chaparral herbs *Emmenanthe penduliflora* and *Eriophyllum penduliflora*. Madroño 31:208–218.

Keeley, J. E., and D. A. Tufenkian. 1984. Garden comparison of germination and seedling growth of *Yucca whipplei* subspecies (Agavaceae). Madroño 31:24–29.

Keeley, J. E., and P. H. Zedler. 1978. Reproduction of chaparral shrubs after fire: a comparison of sprouting and seeding strategies. Amer. Midl. Nat. 99:142–161.

Keeley, S. C. 1977c. The relationship of precipitation to post-fire succession in the southern California chaparral, pp. 387–390 in H. A. Mooney and C. E. Conrad (eds.), Proceedings of the symposium on environmental consequences of fire and fuel management in mediterranean ecosystems. USDA Forest Service, general technical report WO-3.

Keeley, S. C., and A. W. Johnson. 1977. A comparison of the pattern of herb and shrub growth in comparable sites in Chile and California. Amer. Midl. Nat. 97:120–132.

Keeley, S. C., and J. E. Keeley. 1982. The role of allelopathy, heat and charred wood in the germination of chaparral herbs, pp. 128–134 in C. E. Conrad and W. C. Oechel (eds.), Proceedings of the symposium on dynamics and management of mediterranean-type ecosystems. USDA Forest Service, Pacific Southwest Forest and Range Experiment Station general technical report PSW-58.

Keeley, S. C., J. E. Keeley, S. M. Hutchinson, and A. W. Johnson. 1981. Postfire succession of the herbaceous flora in southern California chaparral. Ecology 62:1608–1621.

Keeley, S. C., and M. Pizzorno. 1986. Charred wood stimulated germination of two fire-following herbs of the California chaparral and the role of hemicellulose. Amer. J. Bot. 73:1289–1297.

Kirkpatrick, J. B., and C. F. Hutchinson. 1980. The environmental relationships of California coastal sage scrub and some of its component communities and species. J. Biogeography 7:23–28.

Koenigs, R. L., W. A. Williams, and M. B. Jones. 1982. Factors affecting vegetation on a serpentine soil. I. Principal components analysis of vegetation data. Hilgardia 50(4):1–14.

Koller, D. 1972. Environmental control of seed germination, pp. 1–107 in T. T. Kozlowski (ed.), Seed biology, vol. II. Academic, New York.

Krause, D., and J. Kummerow. 1977. Xeromorphic structure and soil moisture in the chaparral. Oecologia Plantarium 12:133–148.

Kruckeberg, A. R. 1954. The ecology of serpentine soils. III. Plant species in relation to serpentine soils. Ecology 35:267–274.

Kruckeberg, A. R. 1969. Soil diversity and the distribution of plants, with examples from western North America. Madroño 20:129–154.

Kruger, F. J., D. T. Michell, and J. U. M. Jarvis (eds.). 1983. Mediterranean-type ecosystems. The role of nutrients. Springer, New York.

Kummerow, J. 1981. Structure of roots and root systems, pp. 269–288 in F. di Castri, D. W. Goodall, and R. L. Specht (eds.), Ecosystems of the world. II. Mediterranean-type shrublands. Elsevier Scientific, New York.

Kummerow, J., J. V. Alexander, J. W. Neel, and K. Fishbeck. 1978. Symbiotic nitrogen fixation in *Ceanothus* roots. Amer. J. Bot. 65:63–69.

Kummerow, J., D. Krause, and W. Jow. 1977. Root systems of chaparral shrubs. Oecologia 29:163–177.

Kummerow, J., and R. Mangan. 1981. Root systems in *Quercus dumosa* dominated chaparral in southern California. Oecologia Plantarium 2:177–188.

Kummerow, J., G. Montenegro, and D. Krause. 1981. Biomass, phenology and growth, pp. 69–96 in P. C. Miller (ed.), Resource use of chaparral and matorral. Springer, New York.

Kunzler, L. M., and K. T. Harper. 1980. Recovery of gambel oak after fire in central Utah. Great Basin Naturalist 40:127–130.

Laude, H. M., M. B. Jones, and W. F. Moon. 1961. Annual variability in indicators of sprouting potential in chamise. J. Range Management 14:323–326.

Lopez, E. N. 1983. Contribution of stored nutrients to post-fire regeneration of *Quercus dumosa*. M. S. thesis, California State University, Los Angeles.

McDonald, P. M., and E. E. Litterell. 1976. The bigcone Douglas-fir–canyon live oak community in southern California. Madroño 23:310–320.

McKell, C. M. 1950. A study of plant succession in the oak brush (*Quercus gambelii*) zone after fire. M.S. thesis, University of Utah, Salt Lake City.

McMaster, G. S., W. Jow, and J. Kummerow. 1982. Response of *Adenostoma fasciculatum* and *Ceanothus greggii* chaparral to nutrient additions. J. Ecol. 70:745–756.

McMaster, G. S., and P. H. Zedler. 1981. Delayed seed dispersal in *Pinus torreyana* (torrey pine).

Oecologia 51:62–66.
McMillan, C. 1956. Edaphic restriction of *Cupressus* and *Pinus* in the Coast Ranges of central California. Ecol. Monogr. 26:177–212.
McMinn, H. E. 1944. The importance of field hybrids in determining species in the genus *Ceanothus*. Proc. Calif. Acad. Sci. 25:323–356.
McPherson, J. K., C. H. Chou, and C. H. Miller. 1971. Allelopathic constituents of the chaparral shrub *Adenostoma fasciculatum*. Phytochemistry 10:2925–2933.
McPherson, J. K., and C. H. Muller. 1967. Light competition between *Ceanothus* and *Salvia* shrubs. Bull. Torrey Botanical Club 94:41–55.
McPherson, J. K., and C. H. Muller. 1969. Allelopathic effects of *Adenostoma fasciculatum*, "chamise," in the California chaparral. Ecol. Monogr. 39:177–198.
Mahall, B. E. 1985. Chaparral shrubs and their endurance of extreme drought. Chaparral ecosystem research, program & abstracts, May 16–17. University of California, Santa Barbara.
Mahall, B. E., and W. H. Schlesinger. 1982. Effects of irradiance on growth, photosynthesis and water use efficiency of seedlings of the chaparral shrub *Ceanothus megacarpus*. Oecologia 54:291–299.
Major, J. 1977. California climate in relation to vegetation, pp. 11–74 in M. G. Barbour and J. Major (eds.), Terrestrial vegetation of California. Wiley, New York.
Malanson, G. P., and J. F. O'Leary. 1982. Post-fire regeneration strategies of California coastal sage shrubs. Oecologia 53:355–358.
Marion, L. H. 1943. The distribution of *Adenostoma sparsifolium*. Amer. Midl. Nat. 29:106–116.
Martin, P. S. 1973. The discovery of America. Science 179:969–974.
Miller, P. C. (ed.). 1981. Resource use by chaparral and matorral. Springer, New York.
Miller, P. C., and E. Hajek. 1981. Resource availability and environmental characteristics of mediterranean type ecosystems, pp. 17–41 in P. C. Miller (ed.), Resource use by chaparral and matorral. Springer, New York.
Miller, P. C., and D. K. Poole. 1979. Patterns of water use by shrubs in southern California. Forest Sci. 25:84–98.
Mills, J. N. 1983. Herbivory and seedling establishment in post-fire southern California chaparral. Oecologia 60:267–270.
Minnich, R. A. 1980a. Vegetation of Santa Cruz and Santa Catalina Island, pp. 123–127 in D. M. Power (ed.), The California Islands – proceedings of a multidisciplinary symposium. Santa Barbara Botanic Garden.
Minnich, R. A. 1980b. Wildfire and the geographic relationships between canyon live oak, Coulter pine, and bigcone Douglas-fir forests, pp. 55–61 in T. R. Plumb (ed.), Proceedings of the symposium on ecology, management and utilization of California oaks. USDA Forest Service, Pacific Southwest Forest and Range Experiment Station general technical report PSW-44.
Minnich, R. A. 1983. Fire mosaics in southern California and north Baja California. Science 219:1287–1294.
Minnich, R. A., and C. Howard. 1984. Biogeography and prehistory of shrublands, pp. 8–24 in J. J. DeVries (ed.), Shrublands in California: literature review and research needed for management. Contribution 191, Water Resources Center, University of California, Davis.
Moldenke, A. R. 1975. Niche specialization and species diversity along an altitudinal transect in California. Oecologia 21:219–242.
Mooney, H. A. 1977a. Southern coastal scrub, pp. 471–478 in M. G. Barbour and J. Major (eds.), Terrestrial vegetation of California. Wiley, New York.
Mooney, H. A. (ed.). 1977b. Convergent evolution of Chile and California – mediterranean climate ecosystems. Dowden, Hutchinson and Ross, Stroudsburg, Pa.
Mooney, H. A. 1981. Primary production in mediterranean-climate regions, pp. 249–255 in F. di Castri, D. W. Goodall, and R. L. Specht (eds.), Ecosystems of the world. II. Mediterranean-type shrublands. Elsevier Scientific, New York.
Mooney, H. A., S. L. Gulmon, D. J. Parsons, and A. T. Harrison. 1974a. Morphological changes within the chaparral vegetation type as related to elevational gradients. Madroño 22:281–285.
Mooney, H. A., and A. T. Harrison. 1972. The vegetational gradient on the lower slopes of the Sierra San Pedro Mártir in northwest Baja California. Madroño 21:439–445.
Mooney, H. A., A. T. Harrison, and P. A. Morrow. 1975. Environmental limitation of photosynthesis on a California evergreen shrub. Oecologia 19:293–301.
Mooney, H. A., J. Kummerow, A. W. Johnson, D. J. Parsons, S. Keeley, A. Hoffman, R. J. Hays, J. Giliberto, and C. Chu. 1977. The producers – their resources and adaptive response, pp. 85–143 in H. A. Mooney (ed.), Convergent evolution of Chile and California – mediterranean climate ecosystems. Dowden, Hutchinson and Ross, Stroudsburg, Pa.
Mooney, H. A., and P. C. Miller. 1985. Chaparral, pp. 213–231 in B. F. Chabot and H. A. Mooney (eds.), Physiological ecology of North American plant communities. Chapman & Hall, New York.
Mooney, H. A., and D. J. Parsons. 1973. Structure and function of the California chaparral an example from San Dimas, pp. 83–112 in F. di Castri and H. A. Mooney (eds.), Mediterranean ecosystems: origin and structure. Springer, New York.
Mooney, H. A., and P. W. Rundel. 1979. Nutrient relations of the evergreen shrub, *Adenostoma fasciculatum*, in the California chaparral. Bot. Gaz. 140:109–113.
Mooney, H. A., J. Troughton, and J. Berry. 1974b. Arid climates and photosynthetic systems. Carnegie Institution Yearbook 73:793–805.
Mortenson, T. H. 1973. Ecological variation in the leaf anatomy of selected species of *Cercocarpus*. Aliso 8:19–48.
Muller, C. N. 1939. Relations of the vegetation and climate types in Nuevo Leon, Mexico. Amer. Midl. Nat. 21:687–729.
Muller, C. N. 1947. Vegetation and climate of Coahuila, Mexico. Madroño 9:33–57.
Muller, C. N., and R. del Moral. 1966. Soil toxicity induced by terpenes from *Salvia leucophylla*. Bull. Torrey Botanical Club 93:130–137.
Muller, C. N., and P. Hague. 1967. Volatile growth in-

hibitors produced by *Salvia leucophylla*: effect on seedling anatomy. Bull. Torrey Botanical Club 94:182–191.

Muller, C. N., W. H. Muller, and B. L. Haines. 1964. Volatile growth inhibitors produced by aromatic shrubs. Science 143:471–473.

Munz, P. A. 1959. A California flora. University of California Press, Berkeley.

Munz, P. A. 1974. A flora of southern California. University of California Press, Berkeley.

Musick, H. B., Jr. 1972. Post-fire seedling ecology of two *Ceanothus* species in relation to slope exposure. M.A. thesis, University of California, Santa Barbara.

Nilsen, E. T., and W. H. Muller. 1981. Phenology of the drought deciduous shrub *Lotus scoparius*. Oecologia 53:79–83.

Nobs, M. A. 1963. Experimental studies on species relationships in *Ceanothus*. Carnegie Institution of Washington Publication 623.

Oberlander, G. T. 1953. The taxonomy and ecology of the flora of the San Francisco watershed reserve. Ph.D. dissertation, Stanford University.

Oechel, W. C. 1982. Carbon balance studies in chaparral shrubs: implications for biomass production, pp 158–166 in C. E. Conrad and W. C. Oechel (eds.), Proceedings of the symposium on dynamics and management of mediterranean-type ecosystems. USDA Forest Service, Pacific Southwest Forest and Range Experiment Station general technical report PSW-58.

Oechel, W. C., and S. J. Hastings. 1983. The effects of fire on photosynthesis in chaparral resprouts, pp. 274–285 in F. J. Kruger, D. T. Mitchell, and J. U. M. Jarvis (eds.), Mediterranean-type ecosystems. The role of nutrients. Springer, New York.

Oechel, W. C., W. T. Lawrence, J. Mustafa, and J. Martinez. 1981. Energy and carbon acquisition, pp. 151–183 in P. C. Miller (ed.), Resource use by chaparral and matorral. Springer, New York.

Pack, P. 1987. Germination of chaparral herb seeds. M.A. thesis, Occidental College, Los Angeles.

Parker, V. T. 1984. Correlations of physiological divergence with reproductive mode in chaparral shrubs. Madroño 31:231–242.

Parsons, D. J. 1976. The role of fire in natural communities: an example from the southern Sierra Nevada, California. Environ. Conserv. 3:41–99.

Parsons, D. J. 1981. The historical role of fire in the foothill communities of Sequoia National Park. Madroño 28:111–120.

Parsons, D. J., P. W. Rundel, R. Hedlund, and G. A. Baker. 1981. Survival of severe drought by a non-sprouting chaparral shrub. Amer. J. Bot. 68:215–220.

Pase, C. P. 1965. Shrub seedling regeneration after controlled burning and herbicidal treatment of dense Pringle manzanita chaparral. USDA Forest Service, Rocky Mountain Forest and Range Experiment Station research note RM-56.

Pase, C. P. 1969. Survival of *Quercus turbinella* and *Q. emoryi* seedlings in an Arizona chaparral community. Southwestern Naturalist 14:149–155.

Pase, C. P., and D. E. Brown. 1982. Interior chaparral. Desert Plants 4:95–99.

Pase, C. P., and A. W. Lindenmuth, Jr. 1971. Effects of prescribed fire on vegetation and sediment in oak-mountain mahogany chaparral. J. Forestry 69:800–805.

Pase, C. P., and F. W. Pond. 1964. Vegetation changes following the Mingus Mountain burn. USDA Forest Service, Rocky Mountain Forest and Range Experiment Station research note RM-18.

Patric, J. H., and T. L. Hanes. 1964. Chaparral succession in a San Gabriel Mountain area of California. Ecology 45:353–360.

Phillips, P. W. 1966. Variation and hybridization in *Ceanothus cuneatus* and *Ceanothus megacarpus*. M.A. thesis, University of California, Santa Barbara.

Plumb, T. R. 1961. Sprouting of chaparral by December after a wildfire in July. USDA Forest Service, Pacific Southwest Forest and Range Experiment Station technical paper 57.

Poole, D. K., and P. C. Miller. 1975. Water relations of selected species of chaparral and coastal sage communities. Ecology 56:1118–1128.

Poole, D. K., and P. C. Miller. 1981. The distribution of plant water stress and vegetation characteristics in southern California chaparral. Amer. Midl. Nat. 105:32–43.

Poole, D. K., S. W. Roberts, and P. C. Miller. 1981. Water utilization, pp. 123–149 in P. C. Miller (ed.), Resource use by chaparral and matorral. Springer, New York.

Popenoe, J. H. 1974. Vegetation patterns on Otay Mountain, California. M.S. thesis, San Diego State University.

Poth, M. 1982. Biological dinitrogen fixation in chaparral, pp. 285–290 in C. E. Conrad and W. C. Oechel (eds.), Proceedings of the symposium on dynamics and management of mediterranean-type ecosystems. USDA Forest Service, Pacific Southwest Forest and Range Experiment Station general technical report PSW-58.

Quick, C. R. 1935. Notes on the germination of *Ceanothus* seeds. Madroño 3:135–140.

Quick, C. R. 1944. Effects of snowbrush on the growth of sierra gooseberry. J. Foresty 32:827–932.

Quick, C. R., and A. S. Quick. 1961. Germination of *Ceanothus* seeds. Madroño 16:23–30.

Quinn, R. D. 1986. Mammalian herbivory and resilience in mediterranean-climate ecosystems, pp. 113–128 in B. Dell, A. J. M. Hopkins, and B. B. Lamot (eds.), Resilience in mediterranean-type ecosystems. Dr. W. Junk, Dordrecht.

Radosevich, S. R., and S. G. Conard. 1980. Physiological control of chamise shoot growth after fire. Amer. J. Bot. 67:1442–1447.

Raven, P. H. 1973. The evolution of mediterranean floras, pp. 213–223 in F. di Castri and H. A. Mooney (eds.), Mediterranean ecosystems: origin and structure. Springer, New York.

Raven, P. H., and D. I. Axelrod. 1978. Origin and relationships of the California flora. Univ. Calif. Publ. Bot. 72:1–134.

Roberts, S. W. 1982. Some recent aspects and problems of chaparral plant water relations, pp. 351–357 in C. E. Conrad and W. C. Oechel (eds.), Proceedings of the symposium on dynamics and management of mediterranean-type ecosystems. USDA Forest Service, Pacific Southwest Forest and Range Experiment Station general technical report PSW-58.

Roof, J. B. 1978. Studies in *Arctostaphylos* (Ericaceae). Four Seasons 5(4):2–24.

Rowntree, L. 1939. Flowering shrubs of California. Stanford University Press.

Rundel, P. W. 1982. Nitrogen use efficiency in mediterranean-climate shrubs of California and Chile. Oecologia 55:409–413.

Rundel, P. W., and D. J. Parsons. 1979. Structural changes in chamise (*Adenostoma fasciculatum*) along a fire induced age-gradient. J. Range Management 32:462–466 (and unpublished erratum).

Rundel, P. W., and D. J. Parsons. 1980. Nutrient changes in two chaparral shrubs along a fire-induced age gradient. Amer. J. Bot.. 67:51–58.

Rundel, P. W., and D. J. Parsons. 1984. Post-fire uptake of nutrients by diverse ephemeral herbs in chamise chaparral. Oecologia 61:285–288.

Rzedowski, J. 1978. Vegetacion de Mexico. Editorial Limusa, Mexico City.

Sampson, A. W. 1944. Plant succession on burned chaparral lands in northern California. Agricultural Experiment Station Bulletin 685, University of California, Berkeley.

Schlesinger, W. H., and D. S. Gill. 1978. Demographic studies of the chaparral shrub, *Ceanothus megacarpus*, in the Santa Ynez Mountains, California. Ecology 59:1256–1263.

Schlesinger, W. H., and D. S. Gill. 1980. Biomass, production, and changes in the availability of light, water, and nutrients during development of pure stands of the chaparral shrubs, *Ceanothus megacarpus*, after fire. Ecology 61:781–789.

Schlesinger, W. H., and J. T. Gray. 1982. Atmospheric precipitation as a source of nutrients in chaparral ecosystems, pp. 279–284 in C. E. Conrad and W. C. Oechel (eds.), Proceedings of the symposium on dynamics and management of mediterranean-type ecosystems. USDA Forest Service, Pacific Southwest Forest and Range Experiment Station general technical report PSW-58.

Schlesinger, W. H., J. T. Gray, D. S. Gill, and B. E. Mahall. 1982. *Ceanothus megacarpus* chaparral: a synthesis of ecosystem properties during development and annual growth. Bot. Rev. 48:71–117.

Schlesinger, W. H., and M. M. Hasey. 1980. The nutrient content of precipitation, dry fallout, and intercepted aerosols in the chaparral of southern California. Amer. Midl. Nat. 103:114–122.

Schlesinger, W. H., and M. M. Hasey. 1981. Decomposition of chaparral shrub foliage: losses of organic and inorganic constituents from deciduous and evergreen leaves. Ecology 62:762–774.

Schlising, R. A. 1976. Reproductive proficiency in *Paeonia californica* (Paeoniaceae). Amer. J. Bot. 63:1095–1103.

Schorr, P. K. 1970. The effects of fire on manzanita chaparral in the San Jacinto Mountains of southern California. M.A. thesis, California State University, Los Angeles.

Schultz, A. M., J. L. Baunchbauch, and H. H. Biswell. 1955. Relationship between grass density and brush seedling survival. Ecology 36:226–238.

Shantz, H. L. 1947. The use of fire as a tool in the management of brush ranges of California. State of California, Department of Natural Resources, Division of Forests.

Shaver, G. R. 1978. Leaf angle and light absorbance of *Arctostaphylos* species (Ericaceae) along environmental gradients. Madroño 25:133–138.

Shaver, G. R. 1981. Mineral nutrient and nonstructural carbon utilization, pp. 237–257 in P. C. Miller (ed.), Resource use by chaparral and matorral. Springer, New York.

Shimida, A., and M. Barbour. 1982. A comparison of two types of mediterranean scrub in Israel and California, pp. 100–106 in C. E. Conrad and W. C. Oechel (eds.), Proceedings of the symposium on dynamics and management of mediterranean-type ecosystems. USDA Forest Service, Pacific Southwest Forest and Range Experiment Station general technical report PSW-58.

Shimida, A., and R. H. Whittaker. 1981. Pattern and biological microsite effects in two shrub communities, southern California. Ecology 62:234–251.

Shreve, F. 1927. The vegetation of a coastal mountain range. Ecology 8:37–40.

Shreve, F. 1936. The transition from desert to chaparral in Baja California. Madroño 3:257–264.

Shreve, F. 1942. Grassland and related vegetation in northern Mexico. Madroño 6:190–198.

Skau, C. M., R. O. Meeuwig, and T. W. Townsend. 1970. Ecology of eastside chaparral – a literature review. Agricultural Experimental Station, University of Nevada, Reno.

Specht, T. L. 1969. A comparison of the sclerophyllous vegetation characteristics of mediterranean type climates in France, California, and southern Australia. I. Structure, morphology and succession. Australian J. Bot. 17:277–292.

Steward, D., and P. J. Webber. 1981. The plant communities and their environments, pp. 43–68 in P. C. Miller (ed.), Resource use by chaparral and matorral. Springer, New York.

Stocking, S. K. 1966. Influence of fire and sodium-calcium borate on chaparral vegetation. Madroño 18:193–203.

Stohlgren, T. J., D. J. Parsons, and P. W. Rundel. 1984. Population structure of *Adenostoma fasciculatum* in mature stands of chamise chaparral in the southern Sierra Nevada, California. Oecologia 64:87–91.

Stone, E. C. 1951. The stimulative effect of fire on the flowering of the golden brodiae (*Broadiae ixiodes* Wats. var. *lugens* Jeps.). Ecology 32:534–537.

Stone, E. C., and G. Juhren. 1951. The effect of fire on the germination of the seed of *Rhus ovata* Wats. Amer. J. Bot. 38:368–372.

Stone, E. C., and G. Juhren. 1953. Fire stimulated germination. Calif. Agr. 7(9):13–14.

Swank, W. G. 1958. The mule deer in Arizona chaparral and an analysis of other important deer herds. Arizona Game and Fish Department, wildlife bulletin 3.

Sweeney, J. R. 1956. Responses of vegetation to fire. A study of the herbaceous vegetation following chaparral fires. Univ. Calif. Pub. Bot. 28:143–216.

Tadros, T. M. 1957. Evidence of the presence of an edapho-biotic factor in the problem of serpentine tolerance. Ecology 38:14–23.

Townsend, T. W. 1966. Plant characteristics relating to the desirability of rehabilitating the *Arctostaphylos patula–Ceanothus veluntinus–Ceanothus prostratus* association on the east slope of the Sierra Nevada. M.S. thesis, University of Nevada, Reno.

Tratz, W. M. 1978. Postfire vegetational recovery, productivity, and herbivore utilization of a chaparral-desert ecotone. M.S. thesis, California State University, Los Angeles.

Tucker, J. M. 1953. The relationship between *Quercus dumosa* and *Quercus turbinella*. Madroño 12:49–60.

Tyson, B. J., W. A. Dement, and H. A. Mooney. 1974. Volatilization of terpenes from *Salvia mellifera*. Nature 252:119–120.

Vale, T. R. 1979. *Pinus coulteri* and wildfire on Mount Diablo, California. Madroño 26:135–140.

Vankat, J. L. 1982. A gradient perspective on the vegetation of Sequoia National Park, California. Madroño 29:200–214.

Varney, B. M. 1925. Seasonal precipitation in California and its variability. Monthly Weather Review 53:208–218.

Vasek, F. C., and J. F. Clovis. 1976. Growth forms in *Arctostaphylos glauca*. Amer. J. Bot. 63:189–195.

Vestal, A. G. 1917. Foothills vegetation in the Colorado Front Range. Bot. Gaz. 64:353–385.

Vlamis, J., A. M. Schultz, and H. H. Biswell. 1958. Nitrogen-fixation by deerbrush. Calif. Agr. 12(1):11, 15.

Vlamis, J., A. M. Schultz, and H. H. Biswell. 1964. Nitrogen fixation by root nodules of western mountain mahogany. J. Range Management 17:73–74.

Vogl, R. J., W. P. Armstrong, K. L. White, and K. L. Cole. 1977. The closed-cone pines and cypresses, pp. 295–358 in M. G. Barbour and J. Major (eds.), Terrestrial vegetation of California. Wiley, New York.

Vogl, R. J., and P. K. Schorr. 1972. Fire and manzanita chaparral in the San Jacinto Mountains, California. Ecology 53:1179–1188.

Warter, J. K. 1976. Late Pleistocene plant communities – evidence from the Rancho La Brea tar pits, pp. 32–39 in J. Latting (ed.), Proceedings of the symposium on plant communities of southern California. Special publication 2, California Nature Plant Society, Berkeley.

Watkins, K. S. 1939. Comparative stem anatomy of dominant chaparral plants of southern California. M.A. thesis, University of California, Los Angeles.

Watkins, V. M., and H. DeForest. 1941. Growth in some chaparral shrubs of California. Ecology 22:79–83.

Webber, I. E. 1936. The woods of sclerophyllous and desert shrubs of California. Amer. J. Bot. 33:181–188.

Wells, P. V. 1962. Vegetation in relation to geological substratum and fire in the San Luis Obispo Quadrangle, California. Ecol. Monogr. 32:79–103.

Wells, P. V. 1968. New taxa, combinations and chromosome numbers in *Arctostaphylos*. Madroño 19:193–210.

Wells, P. V. 1969. The relation between mode of reproduction and extent of speciation in woody genera of the California chaparral. Evolution 23:264–267.

Wells, P. V. 1972. The manzanitas of Baja California, including a new species of *Arctostaphylos*. Madroño 21:268–273.

Went, F. W. 1969. A long term test of seed longevity. II. Aliso 7:1–12.

Westman, W. E. 1981a. Diversity relations and succession in California coastal sage scrub. Ecology 62:170–184.

Westman, W. E. 1981b. Seasonal dimorphism of foliage in California coastal sage scrub. Oecologia 51:385–388.

Westman, W. E. 1982. Coastal sage scrub succession, pp. 91–99 in C. E. Conrad and W. E. Oechel (eds.), Proceedings of the symposium on dynamics and management of mediterranean-type ecosystems. USDA Forest Service, Pacific Southwest Forest and Range Experiment Station general technical report PSW-58.

Westman, W. E. 1983. Xeric mediterranean-type shrubland association of Alta and Baja California and the community/continium debate. Vegetatio 52:3–19.

Westman, W. E., J. F. O'Leary, and G. P. Malanson. 1981. The effects of fire intensity, aspect and substrate on post-fire growth of Californian coastal sage scrub, pp. 151–179 in N. S. Margaris and H. A. Mooney (eds.), Components of productivity of mediterranean regions – basic and applied aspects. Dr. W. Junk, The Hague.

White, C. D. 1967. Absence of nodule formation on *Ceanothus cuneatus* in serpentine soils. Nature 215:875.

White, C. D. 1971. Vegetation-soil chemistry correlations in serpentine ecosystems. Ph.D. dissertation, University of Oregon, Eugene.

Whittaker, R. H. 1960. Vegetation of the Siskiyou Mountains, Oregon and California. Ecol. Monogr. 30:279–338.

Wicklow, D. T. 1964. A biotic factor in serpentine endemism. M.A. thesis, San Francisco State University.

Wicklow, D. T. 1977. Germination response in *Emmenanthe penduliflora* (Hydrophyllaceae). Ecology 58:201–205.

Wieslander, A. E., and C. H. Gleason. 1954. Major brushland areas of the Coastal Ranges and Sierra Cascades Foothills in California. USDA Forest Service, California Forest and Range Experiment Station miscellaneous paper 15.

Wieslander, A. E., and B. O. Schreiber. 1939. Notes on the genus *Arctostaphylos*. Madroño 58:38–47.

Wilken, C. C. 1967. History and fire record of a timberland brush field in the Seirra Nevada of California. Ecology 48:302–304.

Wilson, R. C., and R. J. Vogl. 1965. Manzanita chaparral in the Santa Ana Mountains, California. Madroño 18:47–62.

Young, D. A. 1974. Comparative wood anatomy of *Malosma* and related genera (Anacardiaceae). Aliso 8:133–146.

Youngberg, C. T., and A. G. Wollum II. 1976. Nitrogen accretion in developing *Ceanothus velutinus* stands. Soil Sci. Soc. Amer. J. 40:109–112.

Zedler, P. H. 1977. Life history attributes of plants and the fire cycle: a case study in chaparral dominated by *Cupressus forbesii*, pp. 451–458 in H. A. Mooney and C. E. Conrad (eds.), Proceedings of the symposium on environmental consequences of fire and fuel management in mediterranean ecosystems. USDA Forest Service, general technical report WO-3.

Zedler, P. H. 1981. Vegetation change in chaparral

and desert communities in San Diego County, California, pp. 406–430 in D. C. West, H. H. Shugart, and D. Botkin (eds.), Forest succession. Concepts and applications. Springer, New York.

Zedler, P. H. 1982. Demography and chaparral management in southern California, pp. 123–127 in C. E. Conrad and W. C. Oechel (eds.), Proceedings of the symposium on dynamics and management of mediterranean-type ecosystems. USDA Forest Service, Pacific Southwest Forest and Range Experiment Station general technical report PSW-58.

Zedler, P. H., C. R. Gautier, and G. S. McMaster. 1983. Vegetation change in response to extreme events. The effect of a short interval between fires in California chaparral and coastal scrub. Ecology 64:809–818.

Zenan, A. J. 1967. Site differences and the microdistributions of chaparral species. M.A. thesis, University of California, Los Angeles.

Chapter 7

Intermountain Deserts, Shrub Steppes, and Woodlands

NEIL E. WEST

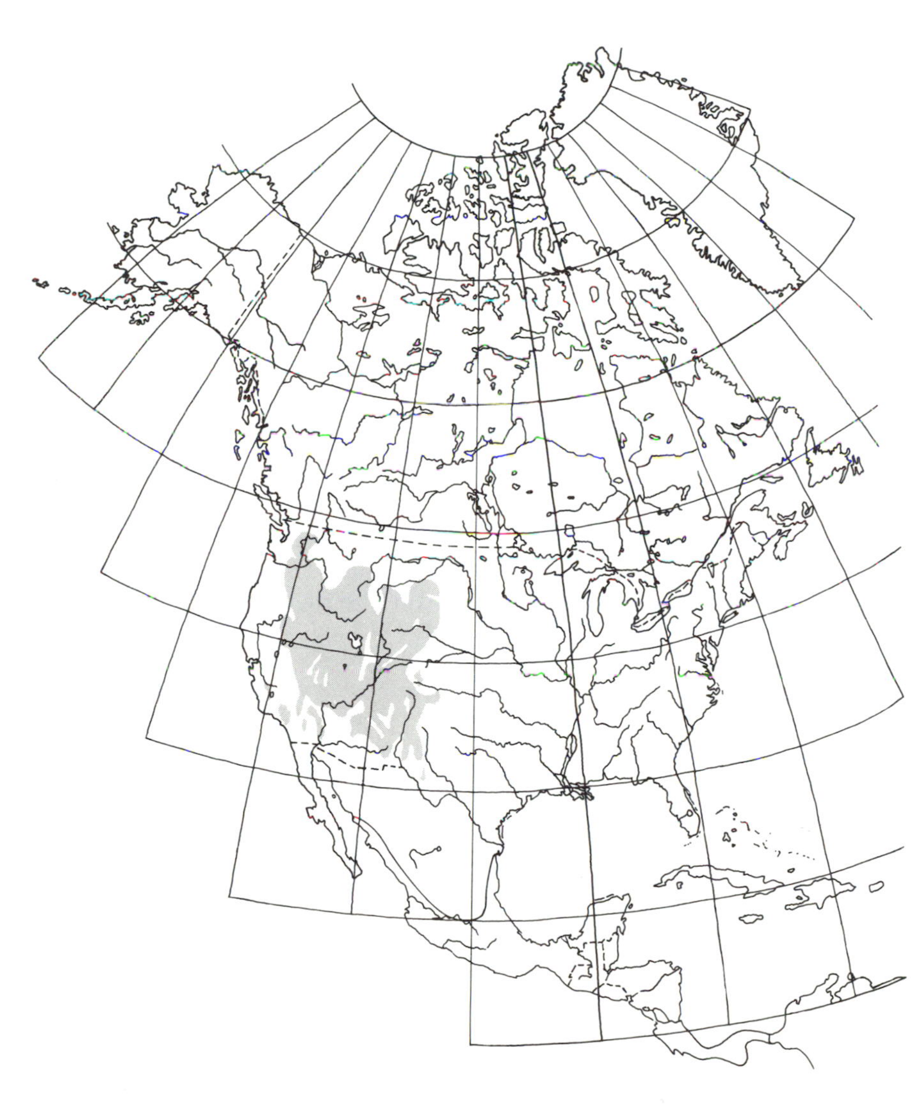

INTRODUCTION

Of concern here is the vegetation that occupies the comparatively lower elevations and more xeric portions of the basins, valleys, lower plateaus, and mountain slopes of the intermountain region of western North America. Küchler's map (1970) of potential natural vegetation types is used for convenience. The Küchler vegetation types and reference numbers included here are as follows:

Sagebrush steppe (*Artemisia-Agropyron*), type 49
Great Basin sagebrush (*Artemisia*), type 32
Saltbush-greasewood (*Atriplex-Sarcobatus*), type 34
Blackbrush (*Coleogyne*), type 33
Galleta–three-awn shrub steppe (*Hilaria-Aristida*), type 51
Juniper-pinyon woodland (*Juniperus-Pinus*), type 21
Mountain mahogany–oak scrub (*Cercocarpus-Quercus*), type 31

Forest types at higher elevations in the Rocky Mountain region are covered in Chapter 3. Adjacent coniferous forests in the Pacific Northwest are discussed in Chapter 4. Coniferous forests to the west, in California, are considered in Chapter 5. The grasslands of the Palouse region of eastern Washington, northern Idaho, northeastern Oregon, and southern British Columbia, the four-corners region of New Mexico, Arizona, Colorado, and Utah, and the Great Plains are described in Chapter 9. Finally, the "warm" desert areas to the south are discussed in Chapter 8.

Tule marshes (Küchler's type 42), plus riparian forests and wet meadows, have been omitted because they occupy collectively small areas on sites with permanent water tables above or very near the soil surface. The marshes are characterized by rather open stands of *Typha* species, *Scirpus* species, and/or *Phragmites*. The riparian forests are dominated by *Populus* species and *Salix* species where they have not been displaced by *Tamarix*.

My focus is on what has often been called the "cold" desert biome (Shelford 1963). I dislike that term, however, because this region is cold only in winter and is not to be confused with the permafrost- and glacial-ice-dominated environments of the polar deserts discussed in Chapter 1. Furthermore, only small areas of the intermountain lowlands are deserts in the international sense. Most of the area to be discussed is fairly well vegetated and is more properly termed a semidesert scrub or shrub-steppe landscape.

These areas, although "lowlands" in the relative sense, are nearly all above 1000 m in elevation. This, plus the latitudinal position of the region, creates temperate climates, as contrasted with the subtropical climates of the warm deserts to the south (see Chapter 8). Thus, I titled a major recent review of the ecosystems in the region *Temperate Deserts and Semi-deserts* (West 1983a).

The pygmy conifer (juniper-pinyon) woodlands and the mountain mahogany–oak scrub (Wasatch and Arizona chaparral types), positioned elevationally directly above the semideserts, are also found in semiarid environments. Thus, they are more easily understood by ecologists familiar with semidesert shrublands than by those with experience only in forests.

I shall start with a brief general discussion of the landforms, geology, climates, and soils of the region. One also needs to consider the paleoecological forces that have influenced what we find today. These sections on environmental features are followed by sections on each of the vegetation types, where more detailed information has been brought together.

LANDFORMS AND GEOLOGY

The region under consideration is bounded by high mountain chains on both the east and west. The Rocky Mountain system, positioned along the east, has over 60 peaks with elevations exceeding 4266m (14,000 ft). The only major gap in these mountains is in central Wyoming, where there is a broad transition from shrub steppe to shortgrass prairie on high, rolling plain and basin topography.

Westward from the Rocky Mountains, one encounters differing types of topography depending on the geologic history of the area. In eastern Washington, eastern Oregon, and southern Idaho there are various depths of loess and volcanic ash deposited on basalt plains of largely Miocene-Pliocene age. The kinds and depths of the surface materials depend mainly on distance and direction from sites where geologic catastrophes struck. For instance, Pleistocene ice plugs formed temporary dams. In other places, lakes went over earth barriers. In both cases, dramatic floods resulted. Some of these areas have been scoured down to basaltic bedrock, where only scattered shrubs can now grow. Elsewhere, alluvial and loessial deposits of fines promote the success of native bunchgrass or, when tilled, of agronomic crops. Volcanic deposits have also affected present-day soils to a considerable degree; the recent (1980) eruption of Mount St. Helens is an example.

The nearly flat to rolling topography of the Columbia–Snake River Plateau is punctured by a few higher mountains, particularly in northeastern Oregon, but most of the land surface here is elevationally lower than in other intermountain areas;

that is, about 70% of the area lies lower than 1500 m (Hunt 1974).

Moving west from the Rocky Mountains in western Colorado requires one to traverse two other distinctive physiographic provinces: the Colorado Plateau and the Great Basin.

The Colorado Plateau is a vast epeirogenic upwarp of largely horizontal Mesozoic sedimentary strata that has been deeply dissected by the Colorado River and its tributaries. The elevation of the plateau varies from about 1200 m to 3000 m, with 45% below 1850 m (West 1969). The soils vary from badlands on marine shales to skeletal residual profiles on mesa tops. Small areas of colluvium next to cliffs, sand dunes on the tablelands, alluvial fill along the rivers, and washes also occur.

West of the high, forested plateaus running north-south through central Utah is the Great Basin. This is a vast tract of broad valleys situated on a high plateau (all above 1200 m). The Great Basin contains more than 200 fault-block mountain ranges created largely by expansional faulting. A few of these mountains exceed 4000 m in elevation. The mountain masses above 1850 m (the approximate average lower boundary of present juniper-pinyon woodlands) occupy about 40% of the area (Billings 1978). The other 60% of the landscape has been coated with gently sloping loessal, alluvial, or lacustrine fill. The Great Basin has no drainage to the sea, and therefore many lakes existed in its valleys during the Pleistocene. This explains much of the fine-textured, halomorphic soil in the valleys at the present time.

PRESENT CLIMATE

The major reason why semideserts prevail in the intermountain lowlands is the orographic "rain shadow" created by the Sierra Nevada and Cascade mountains on the western boundary of the region. These mountains intercept the moist winter air masses brought by the prevailing westerly winds. The result is a pronounced decline in precipitation to the east (Fig. 7.1).

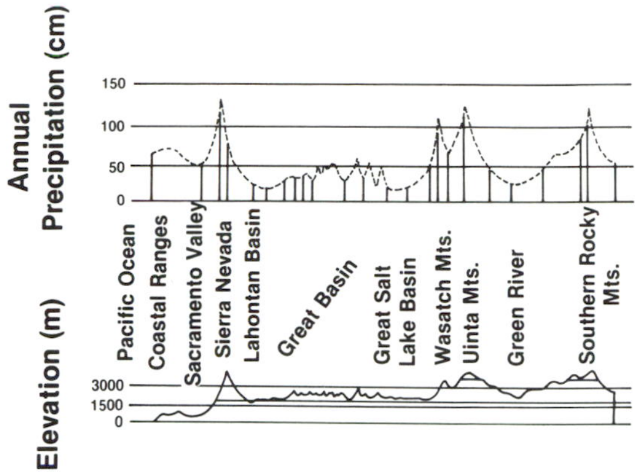

Figure 7.1. Profiles of average elevation (solid line) and mean annual precipitation (dotted line) along 40° N latitude in the western United States (after U.S.D.A. Yearbook *1941).*

The Rocky Mountains also block some of the weather fronts developing in the Great Plains. Some "monsoonal" storms, however, develop to the south, and precipitation coming from sources other than the westerlies increases in the southeastern portion of the region, especially during the summer.

Whereas the snow-dominated precipitation of winter usually melts slowly, summer storms are high-intensity, short-duration rainfall events. Infiltration of water from summer storms into fine-textured soils is usually minimal and of marginal value to most plants. Gradual snowmelt from winter precipitation, however, usually results in deeper infiltration. The dominance of winter precipitation, combined with either fine-textured or rocky soils, is the main reason for shrub dominance in the intermountain region. Where the rarer deep loams or sands prevail, perennial grasses thrive, if not excessively grazed.

The relatively high elevations of the intermountain lowlands produce relatively cool average temperatures. Even though generally less precipitation falls on this region than in the desert region to the south, it comes largely during winter and spring, when evaporation and transpiration are minimal. Thus, about half of the precipitation enters the soil profile on most intermountain lowlands. The deserts to the south have much higher average temperatures, evaporation rates, and runoff rates. Therefore, they have much less capacity to store moisture in the soil per unit of precipitation (see Chapter 8).

Paleoecological Influences

The present is always at least partially determined by conditions of the past. This is especially so in the intermountain lowlands. Some dramatic changes in environment have occurred over comparatively short time spans.

Although present-day geologic conditions and soils have been influenced by the conditions that prevailed in earlier times, when the continental block was farther south and east, the evolution of the present flora is linked most strongly to conditions during the Cenozoic era. During the early Tertiary period, the western half of North America was largely a level plain occupied by forests (Wolfe 1978; Axelrod 1979; Barnosky 1984). There was an uplift of the region, beginning in the Oligocene, and involving especially the higher portions of the Sierras and Cascades in the Pliocene, that created the drier, cooler environments favoring the evolution of semidesert plants and extinction of many precursors with more mesic preferences. There also was an immigration of some of the presently dominant genera from Eurasia (Shmida and Whittaker 1979).

Quaternary influences center on the indirect effects of glaciations during the Pleistocene. No part of the intermountain lowlands was glaciated. The decrease in temperature was linked to an increase in precipitation. Lower summer temperatures meant less evaporation from the approximately 90 lake surfaces in the Great Basin. Conifer forests became established on the lower mountain slopes and possibly, in some places, in those parts of the valleys not covered by lakes (Wells 1983). Death of the trees and expansion of grasslands and shrublands likely took place during the warm interglacials. Holocene climatic changes caused some less pronounced movements of vegetation boundaries (LaMarche 1974; Van Devender and Spaulding 1977; Betancourt 1984; Davis 1984).

Although data are ambivalent, the prevailing view is that there was a relatively drier and warmer (xerothermic) interval from about 7000 to 4000 BP (Mehringer 1977). Postxerothermic cooling may have caused elevational and latitudinal depression of vegetational zones. The most recent evidence (Davis 1984) indicates that these effects differed between vegetation types, depending on the climatic variables controlling growth of their dominants.

The migration of primitive humans into the region at least by 13,000 BP (Jennings 1978) may have influenced vegetation, mainly by reducing the numbers of browsers (Mehringer 1977). The switch from hunter-gatherers to sedentary life styles brought locally heavy impacts, at least in the four-corners region of the Colorado Plateau in the last millennium (Samuels and Betancourt 1982). Europeans, in about 150 yr of occupation, have brought about more profound changes than all those of the previous 13,000 yr. Some of these will be mentioned in the following sections on vegetation.

MAJOR VEGETATION TYPES

Sagebrush Steppe

Woody species of *Artemisia* (sagebrush) probably are the most characteristic and widespread vegetation dominants in the intermountain lowlands. This is a group that has undergone rapid evolutionary radiation into about 12 species (McArthur 1983). The major species, *Artemisia tridentata* (big sagebrush) has four subspecies or forms that should be recognized if one is to fully interpret site differences (Winward 1983).

The sizes and degrees of dominance by plants of the several species of *Artemisia* vary greatly with site and disturbance history (Fig. 7.2). Sagebrush density is generally greater, but stature is lower, on more xeric sites. Sagebrush also increases in abundance with excessive livestock grazing combined with lowered fire frequency.

One can find a continuum of situations, from almost pure, stunted sagebrush on the most xeric sites to increasingly more robust stands of brush with increasing amounts of perennial herbaceous species, on more mesic sites. However, all variations in sagebrush community structure cannot be dealt with here. Accordingly, the lead of Küchler (1970) has been followed. He recognized two potential natural vegetation types where sagebrush is a dominant: the sagebrush steppe and the Great Basin sagebrush types.

The sagebrush steppe occurs in the northern portion of the region (Fig. 7.3). This is where there was a more or less codominance of *Artemisia* with perennial bunchgrasses under presettlement and somewhat pristine conditions (Fig. 7.4). The sagebrush steppe is the largest of the North American semidesert vegetation types, occupying 44.8×10^6 ha (West 1983b). Some of this area is now farmland, and some of it has been so degraded by excessive livestock grazing and burning that its relationship to its origins is no longer easily recognizable. After repeated fires, sagebrush steppe has, in many places, been replaced by European annual grasses such as *Bromus tectorum* and *Taeniatherum caput-medusae*.

The floristic diversity of the sagebrush steppe is moderate (West 1983b). On relict sites in central Washington, Daubenmire (1975) found an average of 20 vascular plant species on 1000-m^2 plots. Tisdale and associates (1965) found 13 to 24 higher plant species in three ungrazed stands in southern Idaho. Zamora and Tueller (1973) found a total of 54 vascular plant species in a set of 39 high-condition stands in the mountains of northern Nevada.

The vertical and horizontal structures are remarkably similar in all relatively undisturbed examples of this vegetation type. The shrub layer reaches approximately 0.5–1.0 m in height (Fig. 7.5). The shrubs have a cover of about 10%–80%, depending on site and successional status. The grass-and-forb stratum reaches to about 30–40 cm during the growing season. The herbaceous cover varies widely, depending on site and successional status. On relict sites, the sum of cover values, species by spe-

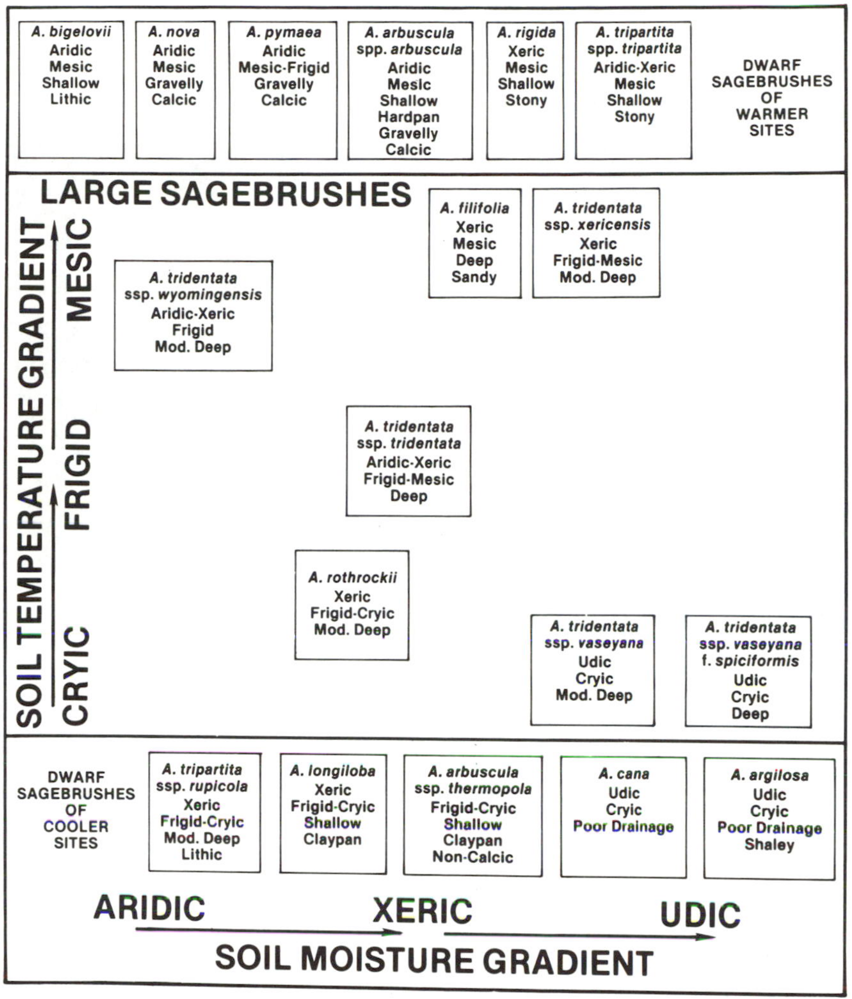

Figure 7.2. Ordination of major sagebrush taxa against gradients of soil temperature and soil moisture (adapted from Hironaka 1979, with additions by the author; from Robertson et al. 1966; McArthur 1983). For explanation of terms relating to soil temperature and moisture regimes, see Soil Survey (1975).

cies, usually exceeds 80% and can approach 200% on the most mesic sites (Daubenmire 1970).

The herbaceous life form most prevalent on relict sites is the hemicryptophyte (Daubenmire 1975), although the proportion of therophytes has increased markedly with disturbance (West 1983b). The proportion of geophytes is around 20%. A microphytic crust dominated by mosses, lichens, and algae is commonly found in the interspaces between the perennials.

The dominance of *Artemisia* is due to many factors, not the least of which is its seasonal dimorphism of leaves (McDonough et al. 1975). That is, there are large, ephemeral leaves that develop in the spring and remain on the plant until soil moisture stress develops in the summer, and there are also smaller, persistent, overwintering leaves that develop in late spring, but last through the winter, even carrying on photosynthesis then (Caldwell 1979).

Artemisia tridentata has both a fibrous root system that can draw water and nutrients near the surface and a taproot that can feed from deep in the soil profile (Sturges 1977). However, if flooding creates anaerobic conditions, it does not survive (Caldwell 1979). Other *Artemisia* species, namely *A. arbuscula* and *A. cana*, may owe their dominance to occasionally supersaturated soils (Passey et al. 1982). None of the major sagebrushes has the capacity to resprout after being burned. As we shall see, this is an important feature in explaining successional patterns.

Perennial grasses associated with *Artemisia* vary greatly throughout the vegetation type (West

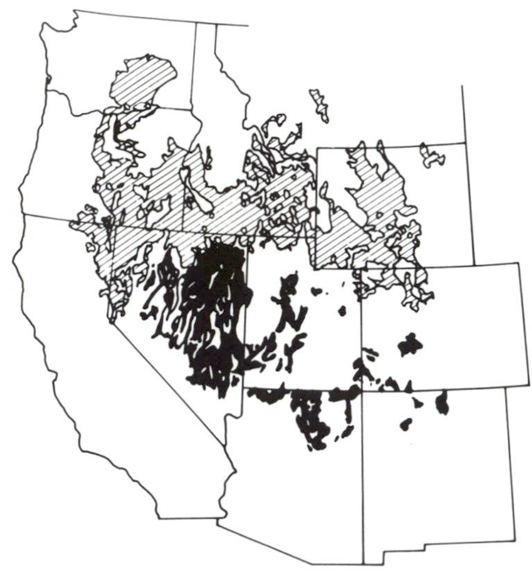

Figure 7.3. Map of the sagebrush steppe and the Great Basin sagebrush types (adapted from Küchler 1979). Some sagebrush vegetation in California is not shown. ▨ =Sagebrush-Steppe; ■ =Great Basin Sagebrush.

Figure 7.4. Pristine example of sagebrush steppe on the Carey kipuka, Craters of the Moon National Monument, Idaho (from Tisdale et al. 1975; photo courtesy M. Hironaka).

1983b). *Festuca idahoensis* and congeners are common in the northwestern part of the vegetation type and at higher elevations and latitudes elsewhere. *Pseudoroegneria spicata* (*Agropyron spicatum*) is probably the most widespread and important herbaceous component of this vegetation type. Sod-forming Triticeae [*Pascopyrum smithii* (*Agropyron smithii*), *Elymus lanceolatus* (*Agropyron dasystachum*) etc.] become more important in the eastern portion of the type because greater growing-season precipitation there encourages these warm-season grasses. *Stipa* species are important codominants along the southwestern boundary of the sagebrush steppe. *Stipa thurberiana*, *S. arida*, and *S. speciosa* sort out along a complex gradient related to decreasing elevation (Young et al. 1977).

The perennial grasses must build most of their above-ground tissues during the narrow "window" of favorable temperature and adequate soil moisture in the late spring and early summer (Fig. 7.6). Perennial forbs are subject to the same constraints, but can store more carbohydrate reserves in their fleshier roots.

Phenological progression is more rapid for forbs than for grasses or shrubs (Fig. 7.7). This could be related to greater leaf size, requiring greater transpiration for cooling of forbs. Shrub roots go deeper and thus draw on soil moisture that is recharged over the winter. Summer precipitation usually is light and is ineffective for recharging the soil moisture profile to depths where it contributes to plant growth. Instead, most of the moisture from summer precipitation is simply lost through evaporation, because the active roots of these species are deep within the soil profile.

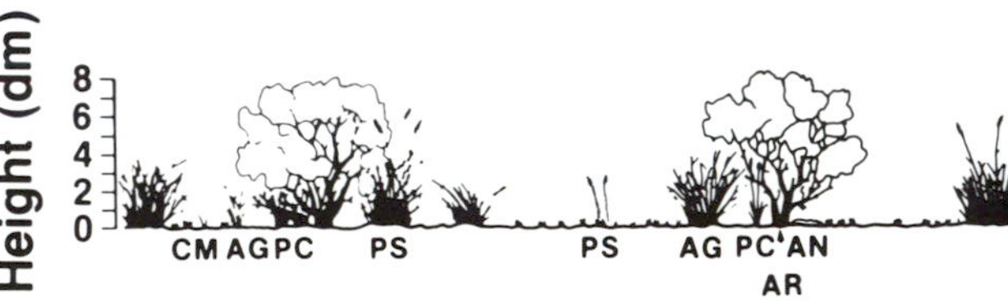

Figure 7.5. A bisect of relict Artemisia tridentata–Agropyron spicatum stand, Yakima County, Washington (from Daubenmire (1970). Drawings to scale, including all vascular plants with any basal area impinging on a transect 2 cm wide by 330 cm long. Height (0-8 dm) is on the vertical axis. Key to species symbols: AG, Agropyron spicatum (now Pseudoroegneria spicata); An, Antennaria dimorpha; AR, Artemisia tridentata, CM, Calochortus macrocarpos; PC, Poa cusickii; PS, Poa secunda.

Artemisia plants are long-lived once they make it past the seedling stage. West and associates (1979) found that individuals of *A. tripartita* in southeastern Idaho lived an average life span of about 4 yr once they survived the first year. The maximum longevity for this species at the site studied exceeded 40 yr. Plants of *A. tridentata* can live to be more than 100 yr of age (Ferguson 1964).

The pristine sagebrush steppe evolved with large

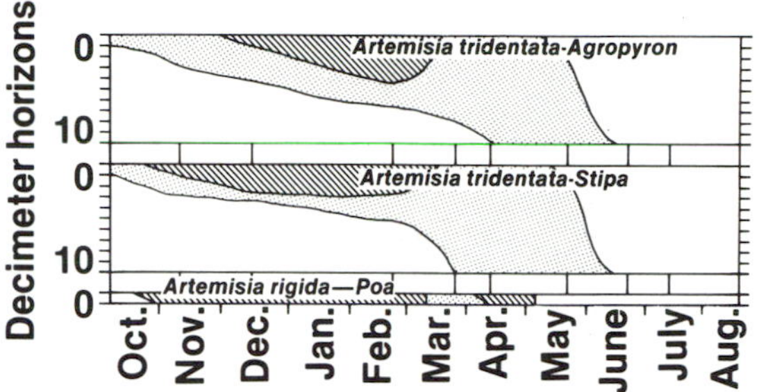

Figure 7.6. Soil moisture status for three climax sagebrush steppe communities in eastern Washington between 2 October 1962 to 17 August 1963. Limits of decimeter horizons shown on ordinate. Shaded areas indicate water content in excess of field capacity; stippled areas indicate water content between field capacity and wilting coefficient; unshaded areas lack growth water. Vertical lines between horizontal panels show actual dates of sampling (from "Annual cycles of soil moisture and temperature as related to grass development in the steppe region of eastern Washington" by R. F. Daubenmire, Ecology, 1972, 53, 419–424. Copyright 1972 Duke University Press, Durham, N.C., reprinted by permission).

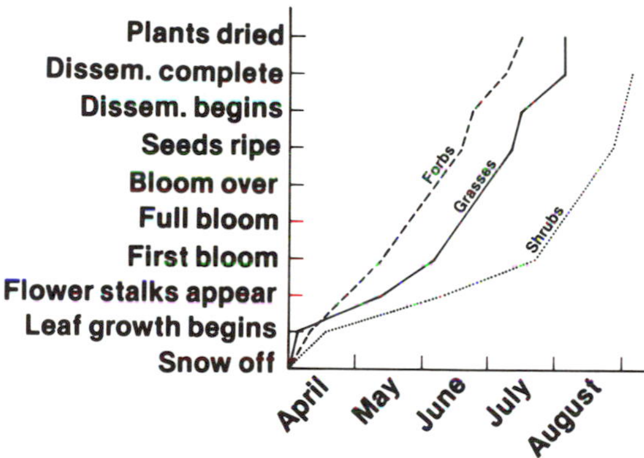

Figure 7.7. Average phenological progression for grasses, forbs, and shrubs at the U.S. Sheep Experimental Station, Dubois, Idaho, 1941–7 (from Blaisdell 1958).

browsers, but most of them had disappeared by about 12,000 BP (Martin 1970). Graminivore populations were also low (Mack and Thompson 1982). The small populations of aboriginal hunter-gatherers had little direct impact on the vegetation. It took European colonization to drastically change the native vegetation. Some of the land was and still is being converted to intensive agriculture. The most widespread early influence, however, was that of livestock, whose populations built up rapidly in the late nineteenth century. Griffiths (1902) judged that the grazing capacity of these ranges had definitely been exceeded by the year 1900. Hull (1976) examined historical documents and concluded that the loss of native perennial grass and expansion of shrubs took only 10–15 yr.

The primeval vegetation was only weakly stable, because of the ecophysiological and reproductive disadvantages of the native perennial grasses and the great competitive advantage of the shrubs. Herbaceous species have to rebuild all of their aboveground tissues each growing season. This requires upward translocation of a considerable amount of food reserves, water, and nutrients during the short growing season (late spring and early summer). To maintain their dominance, shrubs have to rebuild only enough tissue to replace the comparatively smaller amount that was shed as litter during the previous years.

Brush foliage, furthermore, has chemical defenses against herbivory, whereas grass is extremely palatable when it is green. The native bunchgrasses sustain high mortality when grazed heavily in the spring (Stoddart 1946). In addition, they rarely produce good seed crops (Young et al. 1977).

The only time that the grasses and forbs have an advantage over brush is when this vegetation is burned. Herbs resprout easily, especially if fires occur during the summer. The major species of *Artemisia* have to reestablish from seed.

Europeans introduced another force: aggressive annual weeds. Foremost among these is *Bromus tectorum* (cheatgrass). Although introduced to the Pacific Coast in the 1870s (Mack 1981), it was not until around 1928 that it reached its present distribution, and not until the 1940s and 1950s did it become dominant. This grass outcompetes the native grasses, mainly by beginning growth in the fall and growing root systems throughout the winter (Harris 1977). Cheatgrass completes its life cycle and produces seeds by late June or early July. The fine, continuous fuel load it creates makes the region susceptible to earlier and more frequent fires than occurred in the past, and many of the native herbs and shrubs are destroyed by these fires (Wright and Bailey 1982), with the result that the organic matter and nutrient pools in the soil decline. If soils are without cover during summer convectional storms, soil erosion usually is severe. The result, on much of this area, has been a downward spiral of degradation (Fig. 7.8).

Other groups of species that have increased because of excessive livestock grazing and an accelerated fire frequency are members of the shrub genera *Chrysothamnus* (rabbit brush), *Ephedra* (Mormon tea), and *Tetradymia* (horse brush). This is because they can resprout after fire (Tisdale and Hironaka 1981). On heavier soils, the exotic annual grass *Taeniatherum caput-medusae* (medusa head) has even replaced *Bromus tectorum*. Other problem taxa that

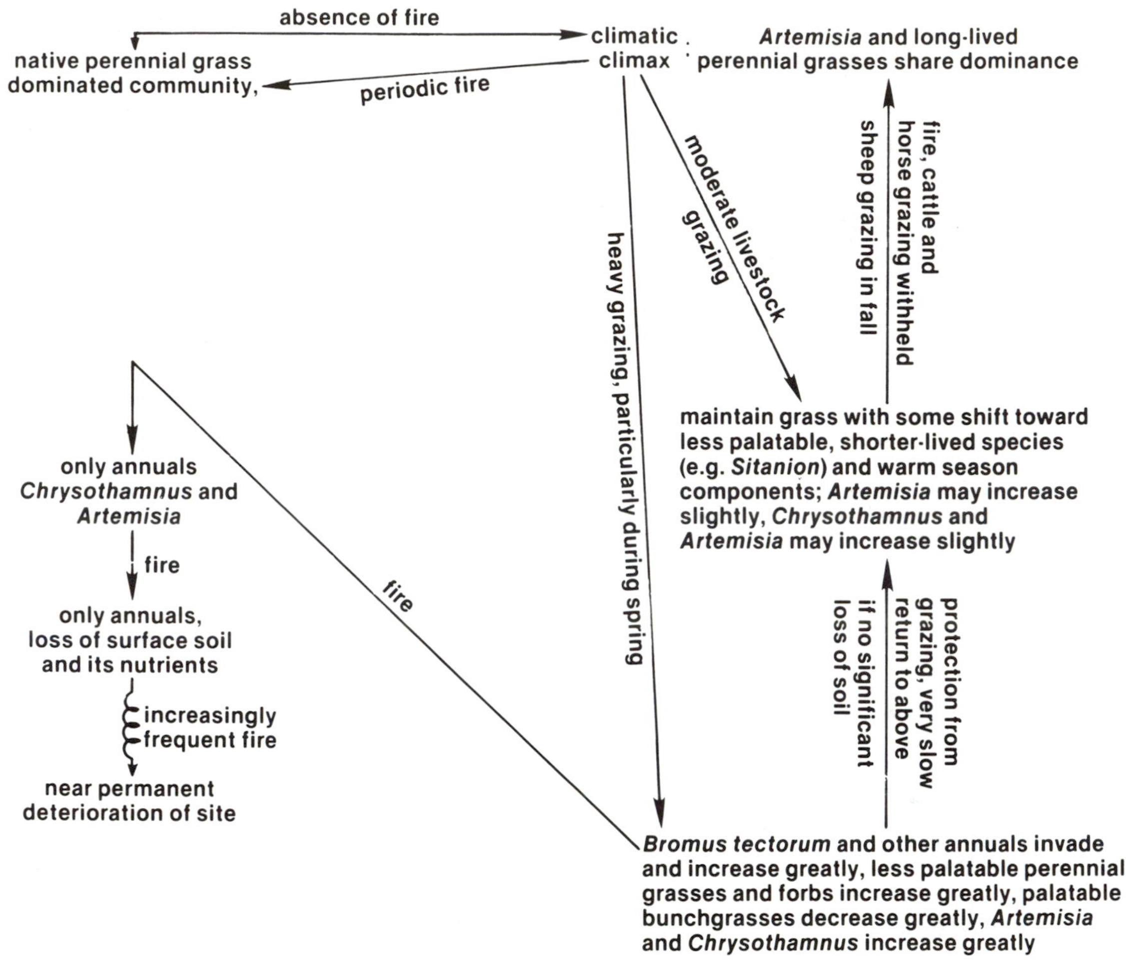

Figure 7.8. Major pathways of progressive and retrogressive succession in lower-elevation sagebrush steppe and Great Basin sagebrush vegetation (from West 1979).

currently are localized, but are expanding, are *Aegilops cylindrica* (goat grass), *Salvia aethiopis* (Mediterranean sage), *Isatis tinctoria* (dyer's woad), and several species of *Centaurea* (knapweed). There has thus been a vast replacement of long-lived perennials by shorter-lived taxa, especially annual grasses. Controlling fires entirely is an impossibility, and reduction of livestock will not result in a rapid return to vegetation similar to that in the pristine condition (Anderson and Holte 1981). The new annual-dominated vegetation appears to have created its own, new equilibrium (Hanley 1979). Our major means of obtaining greater dependability of forage production, while at the same time reducing the chance of fire, is to plant introduced Triticeae (Keller 1979). However, this can be done easily only on relatively level sites with deep soils; the remaining, rougher topography awaits more imaginative management that perhaps will be aided by basic research in plant ecology.

Great Basin Sagebrush

The second vegetation type in which *Artemisia* dominates is found to the south of the sagebrush steppe (Fig. 7.3). Although centered in the Great Basin, there is considerable acreage on the Colorado Plateau and in adjacent physiographic provinces, for a total of 17.9×10^6 ha (West 1983c). The sagebrush steppe and Great Basin sagebrush vegetation types, because of similarities in gross vegetation structure, have often been considered together in reviews (College of Natural Resources 1979; Tisdale and Hironaka 1981; Blaisdell et al. 1982). I consider this to be a mistake, however, because floristic diversity, production, and responses to perturbations are significantly lower and slower in the Great Basin sagebrush type.

The areas of the Great Basin sagebrush type are much more arid and thus akin to deserts, whereas the sagebrush steppe is similar to a semiarid grassland. In the Great Basin sagebrush type, *Artemisia* prevails, without much grass, even in pristine or late seral conditions (Fig. 7.9).

The shrub layer in this vegetation type seldom reaches over 1 m in height. The shrubs usually are less dense than in the sagebrush steppe, and the considerable interspace is composed mainly of vesicular (Fig. 7.10) and/or microphytic crusts, whereas herbaceous vascular plants are located among or near the shrubs. The shrubs are commonly located on hummocks of elevated microrelief because of differential erosion and deposition. These serve as "islands of fertility," providing an ameliorated microclimate within a more hostile matrix (West 1983c).

Artemisia commonly makes up more than 70% of the relative vegetational cover and more than 90% of the phytomass, regardless of successional status. This high degree of dominance by low shrubs lends a characteristic grayish green color to the landscape. An ordination of major *Artemisia* taxa in this vegetation type against elevation and soil moisture gradients is shown in Figure 7.11.

Major associated plant species on relatively undisturbed sites are perennial bunchgrasses and perennial forbs. There is a trend from cool-season grasses in the Great Basin to an increasing component of warm-season sod grasses on the Colorado Plateau (West 1979). Temperature and moisture patterns generally are more variable than those illustrated in Fig. 7.6.

Figure 7.9. Relict area with sagebrush-dominated vegetation, Cedar Mesa, Canyonlands National Park, Utah, 20 September 1983. See Van Pelt (1978) for description of this site.

This vegetation type appears deceptively monotonous and simple at first glance. A low floristic diversity and concentration of dominance in so few species lead to loose species packing. This is apparently compensated for by great intraspecific variation in major taxa such as *Artemisia* (McArthur 1983), *Chrysothamnus* (Anderson 1975), and *Sitanion hystrix* (Clary 1975). Taxa with only slight morphological variations actually have considerable genetic and ecological variations, with functionally different communities resulting.

The same ecophysiological and autecological attributes of *Artemisia*, the native perennial bunchgrasses, and the introduced annuals previously discussed apply to successional pathways here also (Fig. 7.8). Original stability probably was even more precarious than in the sagebrush steppe. Modern soil erosion has been more severe because there has been less plant and litter cover. Plants and litter are concentrated around the mounds of microrelief (Fig. 7.10), where soil organic matter, microbial activity, and nutrients are concentrated. Anything that alters this pattern diminishes productivity on a landscape basis, because these desert-like environments have too little moisture, nutrient input, and storage to carry on a high level of production across the entire landscape. Removal of livestock does not result in the sagebrush releasing its dominance and allowing perennial grass to increase (Potter and Krenetsky 1967; Rice and Westoby 1978; West et al. 1984). Because of a lower site potential, the economic feasibility of vegetation type conversion is much less here than in the sagebrush steppe (West 1983c).

Undesirable plants have increased in abundance. In addition to those already mentioned for sagebrush steppe, one must contend with *Halogeton glomeratus* and *Ceratocephalus* (*Ranunculus*) *testiculatus*, both of which are poisonous to livestock. The conversion to annual dominants seems to be a widespread and permanent trend (Rogers 1982).

Saltbush-Greasewood

Vegetation dominated by perennial chenopod shrubs and half-shrubs makes up another considerable portion of the intermountain lowlands. Because such communities usually are associated with halomorphic soils, the descriptor "salt-desert

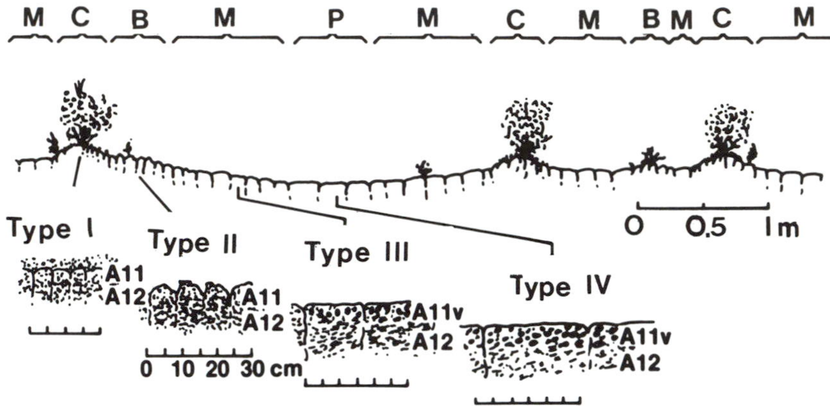

*Figure 7.10. Schematic cross-sectional diagrams of the microtopographic positions and associated surface soil morphological types of gently sloping, shallowly loess-mantled Argids of the Humboldt loess belt with plant communities dominated by Wyoming big sagebrush (*Artemisia tridentata ssp. wyomingensis*). Microtopographic positions: C, coppice; B, coppice bench; M, intercoppice microplains; P, playette. Vertical scale somewhat exaggerated; intercoppice microplains and playettes may be much wider than shown here; several coppices may be linked together. Vertical lines under soil indicate sides of crust polygons (A11v) that continue downward as sides of prisms in the compoundly weak prismatic and moderately platy A12 horizon. Type 1 is covered by litter. Circles indicate vesicles in crust (A11v). Only types III and IV are significantly crusted (from Eckert et al. 1978).*

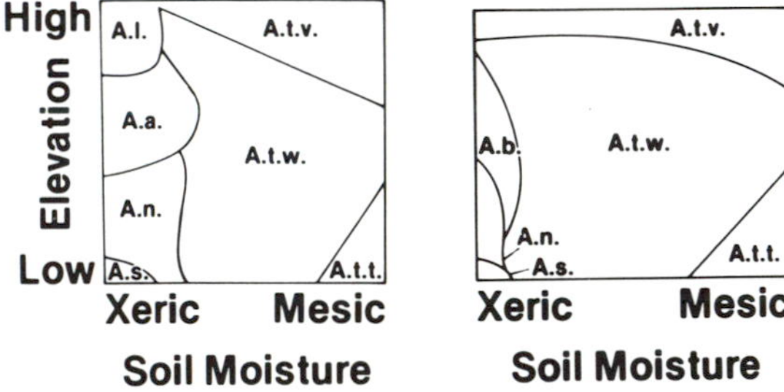

*Figure 7.11. Ordinations of major sagebrush taxa found in the Great Basin (left) and Colorado Plateau (right) against gradients of elevation and effective soil moisture. A.b = *Artemisia bigelovii*, and so forth, as derived from the first letters of the genus, species, and subspecies names from Fig. 7.2 (adapted from West 1979).*

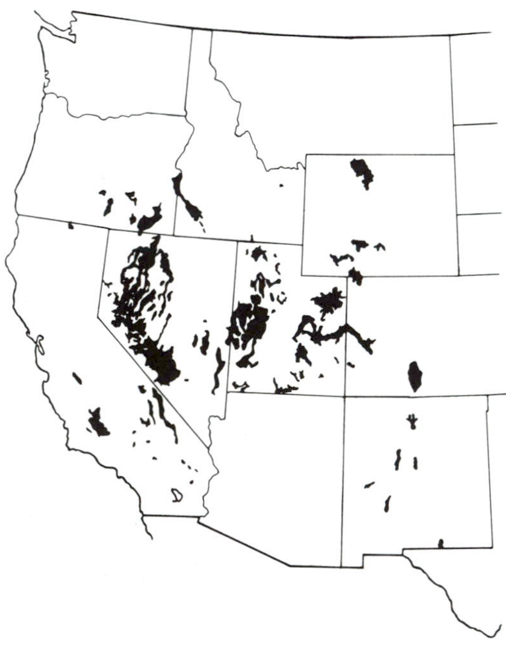

Figure 7.12. Map of the saltbush-greasewood vegetation type (adapted from Küchler 1970).

shrub" has gained wide usage (West 1983d). This is synonymous with Küchler's (1970) saltbush-greasewood vegetation type, which occupies a total area of 16.9×10^6 ha, scattered over all four of the regional deserts of North America, plus special locations in the Central Valley of California, the upper Rio Grande drainage, and the Great Plains (Fig. 7.12).

Although the correlation of this vegetation with halomorphic soils is strong, it is not universal. Included are some chenopod-dominated areas on dry, desert-pavement-covered but nonsaline soils in the extreme rain shadow of the Sierra Nevada in western Nevada and adjacent California (Billings 1949). Glycophytes, such as most *Artemisia* species, cannot thrive there. Where extreme aridity, occasional flooding, and/or salinity or alkalinity are found, the type approaches the international definition of desert—that is, it is devoid of vegetation altogether

Figure 7.13. Surface of the Pine Valley saltpan, southwestern Utah, summer 1968.

Figure 7.14. Atriplex confertifolia–Ceratoides lanata *stand in Curlew Valley, northwestern Utah, July 1964.*

(Fig. 7.13). Where aridity, salinity, and/or alkalinity are more moderate, the landscape has plant communities dominated by more closely spaced shrubs (Fig. 7.14).

There are three major types of habitats involved: uplands dominated by xerohalophytes, where the permanent water table is well below 1 m; lowlands dominated by hydrohalophytes, where the water table remains within about 1 m of the soil surface; marshlands and meadows dominated by grasses and grass-like plants, where the permanent free water table is above the soil surface for much of the year. Figure 7.15 shows the positioning of plant associations in relation to those gradients in a valley in western Nevada. The marshlands and meadows will not be considered further here; see Bolen (1964) for details.

The major salt-desert shrub species and their habitat preferences are listed in Table 7.1. Halomorphic soils develop from either flat, deep lacustrine deposits (Fig. 7.13) or from badlands of marine shales that are excessively drained, as well as being halomorphic. Variations in vegetation composition and productivity are so intimately related to soil characteristics that considerable study of these interrelationships has been done in this vegetation type (West 1983d).

Most of the vascular plants found in salt-desert shrub areas are members of the Chenopodiaceae. Only occasional representatives of the Asteraceae, Brassicaceae, Fabaceae, and Poaceae are found, and none can be called dominants. About 20% of the vegetation type can be described as essentially mosaics of monocultures of single-perennial-species dominance. This is because few plants can tolerate such dry and salty habitats, and topographic gradi-

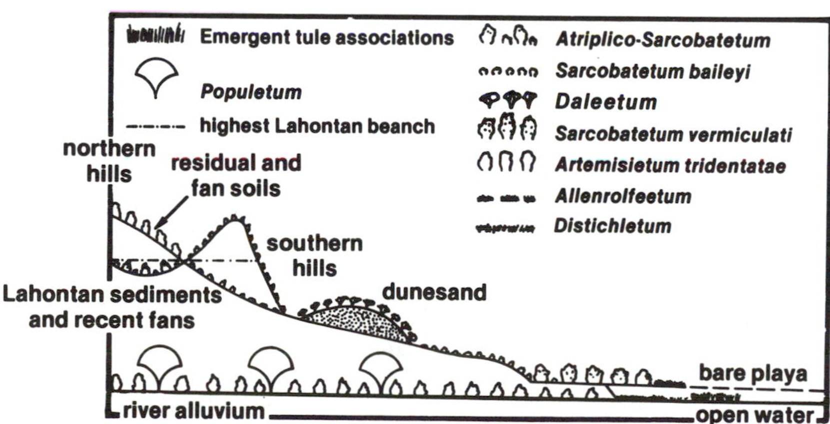

Figure 7.15. Diagrammatic representation of the topographic and geologic positions of the principal plant associations in the Carson Desert region of western Nevada (from Billings 1945).

Table 7.1. *Major vascular species of the salt-desert shrub type according to growth form and preference in regard to subsurface moisture*

	Growth forms		
Habitat–habit	Shrubs	Half-shrubs	Herbs
Lowland–hygrohalophytes: free water table at least occasionally present at the surface and usually remaining within about 1 m	*Sarcobatus vermiculatus*	*Allenrolfea occidentalis* *Salicornia utahensis* *Suaeda torreyana* *Suaeda fruticosa*	*Distichlis stricta* *Sporobolus airoides* *Suaeda depressa* *Suaeda diffusa*
Upland–xerohalophytes: water table well below 1 m	*Artemisia spinescens* *Atriplex confertifolia* *A. (Grayia) spinosa*	*Atriplex corrugata* *A. cuneata* *A. falcata* *A. gardneri* *A. tridentata* *Ceratoides (Eurotia) lanata* *Kochia americana*	*Bromus tectorum* *Elymus cinereus* *Halogeton glomeratus* *Lepidium perfoliatum* *Oryzopsis hymenoides* *Salsola kali* *Sitanion hystrix*

ents are very gentle, perhaps allowing competitive sorting to become evident (West 1983d). Because the positioning of species may follow different sequences in different valleys, at least some of the variation may be due to ecotypic variation. Rapid evolution of these Chenopodiaceae is strongly suspected (Stutz 1978).

Total cover of higher plants varies from zero on the most saline shales or saltpans up to about 25% on some upland sites. Soil textures in the uplands greatly influence water infiltration and evaporative losses, with gravelly soils having the greatest cover and production because of the inverse-texture principle (Noy-Meir 1973).

Variation in cover on sites with seasonal or permanent water tables depends on the chemical characteristics of the soils and waters there. Species apparently sort out in relation to seasonal patterns of matric and osmotic potentials (Detling and Klikoff 1973). Physiological drought is not a problem for the native species, however, because either they have means to exclude uptake of salts or they take them up and anatomically isolate and/or excrete them.

Shrubs are widely spaced and usually occur in clusters (West and Goodall 1986). Regeneration of new plants usually occurs where old plants exist or have existed, rather than in the interspaces (West 1983d). This apparently results from a more favorable situation involving a number of factors (soil moisture, soil organic matter, temperature, nutrients, etc.) in these spots. These advantages ostensibly outweigh any competitive interactions. The interspaces usually are covered with microphytic crusts, whose surfaces will be soft and rugose if the soil has not been compacted by the feet of animals or wheels of vehicles. These crusts apparently reduce soil erosion and contribute to nitrogen input (West and Skujins 1977).

The maximum height for most upland salt-desert shrubs is usually less than 50 cm. Greater heights and densities are found in the phreatophytic *Sarcobatus* (greasewood) stands.

Upland xerohalophytes experience a surge in growth in the spring (West and Gasto 1978), because they can draw in only vadose soil moisture. Several of the dominants feature leaf dimorphism; that is, they have a set of larger, spring leaves that are lost as soil drought develops and a second set of much smaller, overwintering leaves that can carry on some photosynthesis over the winter. The hydrohalophytes of high-water-table sites are more deciduous, developing new leaves in midsummer. Most of them are C_4 species (but not *Sarcobatus*) and are subirrigated.

The harsh environments of salt deserts slow down community dynamics, but because the same species or species similar in appearance and stature often succeed each other after disturbances, "autosuccession" probably best describes what occurs over the shorter run (West 1982). On a longer time scale, we can consider primary succession that occurs with waxing and waning of saline lakes. The primary driving factor is either flooding, which destroys previous vegetation, or a deepening water table during drying periods. Figure 7.16 outlines the usual sequence of native higher-plant species located along this latter gradient. Species change largely in response to variations in salinity. For instance, Rickard (1964) described how *Sarcobatus* can cause *Artemisia* to decline as the microtopography is eroded downward.

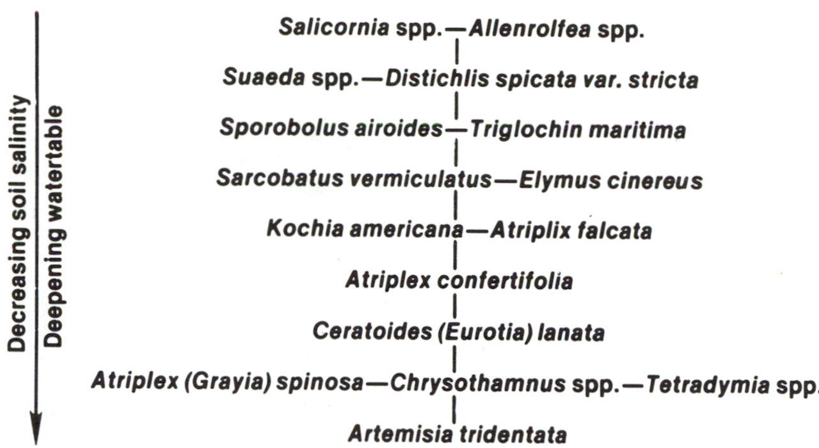

Figure 7.16. Probable halosere around retreating saline lakes of the Great Basin (adapted from Flowers and Evans 1966).

As with the sagebrush-dominated vegetation types, livestock grazing over approximately the last 100 yr has been the major instrument for change in these communities. We have little concrete evidence of changes, however, because the topography of the area does not lend itself to preservation of any relict areas. Productivity was so low, forage so coarse, and water so scarce that overwintering sheep were the main animals allowed to graze there. This meant that there were few fences to create the unintentional experiments that fence-line contrasts provide. Control over livestock numbers did not occur until after 1934, when the Taylor Grazing Act became law.

Not much research was done on this vegetation type until after the scare provided by the *Halogeton* (a poisonous annual from Eurasia) invasion beginning in the 1940s. Thus, our perspective is short, and the data are inadequate to decipher retrogression. We do know that the most palatable shrubs–*Artemisia spinescens, Ceratoides lanata,* and *Kochia americana*–declined substantially, especially when grazing use extended into the spring (Blaisdell and Holmgren 1984). Unfortunately, these shrubs also had the least reproductive capacity. The less palatable species–*Atriplex confertifolia, A. gardneri, A. falcata, A. tridentata, A. cuneata, A. corrugata*–have come back more rapidly after control of livestock grazing. These trends, however, are difficult to distinguish from annual fluctuations and the effects of longer-term climatic influences (Norton 1978).

Population explosions involving insects such as round-headed borers and cutworms, as well as species of wild mammals such as jackrabbits (*Lepus californicus*), can greatly change a plant community even in the absence of livestock (West 1982). Interactions between animal population irruptions and drought or excessive soil moisture are likely, but cannot yet be verified. Extensive die-off of shrubs has been observed during the wet trend of the past 5 yr.

Because higher-plant cover is generally so sparse, this has been one of the few extensive vegetation types in which we have not had to worry about wildfire. In 1983, this pattern was broken, and fires occurred even in salt-desert shrub communities of the Great Basin. These fires were possible only because of profusion of (mostly exotic) annuals after a sequence of years in which precipitation exceeded the average precipitation by a factor of 2 or more. We shall now learn how the various species can respond to these new kinds of impacts.

Unfortunately, the expansion of annuals does more than bring fire to these communities. *Halogeton*, for instance, may permanently change the soil surface by means of salt pumping (Eckert and Kinsinger 1960), which impedes moisture infiltration and enhances evaporation, contributing to further xerification of these areas. Fortunately, *Halogeton* and the other major invaders–*Bromus tectorum, B. rubens, Salsola kali, Lepidium perfoliatum, Malcolmia africana, Ceratocephalus* (*Ranunculus*) *testiculatus*, and several annual *Atriplex* and *Chenopodium* species–are not so aggressive that they invade closed communities in middle to late seral status. This is especially important because there is currently no proven technology available to consistently revegetate such areas at reasonable cost. Nevertheless, the spread of annuals has been noted even under moderate livestock grazing (West 1983d). The prognosis for maintaining a diverse, stable, and economically valuable plant cover in some of these areas is not good.

Blackbrush

In the lower but nonsaline parts of the Colorado Plateau, and also where the Mohave and Great Basin deserts merge, there is another type of shrub-dominated vegetation termed the blackbrush (*Coleogyne ramosissima*) semidesert (West 1983e). This species of shrub also occurs at higher elevations, dominating inclusions within grasslands or dominating the woodland understory wherever soil depth is restricted. Blackbrush is one of the few kinds of plants in our region that seems to prefer old pediment slopes and bajadas with petrocalcic (caliche) horizons. Thatcher and associates (1976) have noted a gradient of declining brush and increasing amounts of grasses with increasing depth to the petrocalcic or other types of restrictive layers, even on areas that have never been grazed by livestock (Fig. 7.17).

Figure 7.17. Coleogyne ramosissima*-dominated vegetation on pristine site never grazed by livestock; isolated mesa in the western portion of Grand Canyon National Park, Arizona, 11 June 1981.*

Not many species thrive in these difficult environments transitional between the "cold" and "warm" deserts. Only in the spring is there much of a showing of annuals. *Coleogyne* dominates in terms of both height and cover (Fig. 7.17). *Coleogyne* is a round shrub seldom reaching a height greater than 0.5 m. The round shape is the result of terminal twigs dying back and forming spines; all subsequent twig growth is from subterminal buds (Bowns and West 1976). Also, intercalary cork is formed between stem segments during drought, creating multiple stems. Other kinds of shrubs are rare where *Coleogyne* occurs. Only *Prunus fasciculata, Atriplex spinosa, Thamnosa montana, Ephedra* species, and *Gutierrezia* species are worthy of mention. Some occasional *Yucca, Opuntia,* and *Agave* plants can be found interspersed in some variants of the blackbrush vegetation type. Perennial herbaceous plants occurring here often are limited to *Hilaria rigida, H. jamesii, Oryzopsis hymenoides, Aristida* species, *Stipa* species, *Bouteloua eriopoda,* and *Muhlenbergia porteri.* The amounts of these are ordinarily low, and very little of the vegetation type can be described as shrub steppe.

Despite the simple floristics, the total higher-plant cover can be surprising (37%–51%) (Beatley 1975), especially after wet antecedent conditions. This, plus the resinous nature of the brush, leads to great susceptibility to fires. There is typically a well-developed microphytic crust on the soil surface between shrubs, particularly where livestock grazing and human traffic have not been excessive.

Winter recharging of the soil moisture from light snows apparently drives development of the shrubs and winter annuals. Their growth begins quite early (mid-March). Summer rainfall activates the microphytic crusts and warm-season C_4 grasses (*Hilaria, Bouteloua,* and *Aristida*). *Coleogyne* is protected against heavy ungulate herbivory by the combination of its woodiness, its low nutrient quality, and its heavy load of secondary compounds (Provenza et al. 1983). Despite its generally low and quite variable primary production, this vegetation has been greatly influenced by livestock grazing and fire. Those interested in livestock production have regarded fire as desirable (Bates and Menke 1984). *Coleogyne* does not ordinarily resprout after fire, and it reseeds itself with difficulty. Various stem sprouters and annuals (rarely perennial grasses) occur after fire, foremost among them being *Gutierrezia* species, *Bromus tectorum,* and *Bromus rubens.* Once these gain dominance, the recurrence rate for fires increases. Although forage production is higher during wet years, it is almost nil during drought. Soil erosion apparently has been accelerated by widespread prescribed burning and uncontrolled wildfires (West 1983e). Feral horses and burros have also had an undesirable impact. We currently do not have either promising replacement species or the technology to consistently effect revegetation in this community type.

Galleta-Three-Awn Shrub Steppe

This is a small (0.5×10^6 ha) and little-known vegetation type centered in southeastern Utah (Küchler 1970). It occupies relatively deep and undeveloped

Figure 7.18. Hilaria-Aristida shrub steppe near Looking Glass Rock, San Juan County, Utah, July 1967. Shrubs are Ceratoides lanata *in foreground and* Atriplex canescens *in background.*

Figure 7.19. Microphytic crusts in Virginia Park, Canyonlands National Park, Utah, October 1967. Hat in center of photo provides scale.

sandy soils of the Canyonlands section of the Colorado Plateau (Fig. 7.18). Comparing relict areas to those that have been left open to livestock use leads one to believe that during pre-Columbian times there was a more abundant perennial grass cover, and less shrubby and weedy annual cover, than is now found in most of this area (West 1983f). There is some affinity between this community type and the grama-galleta steppes to the south, as discussed in Chapter 9.

The flora is a mixture of about a dozen bunchgrasses and sod-forming grasses and palatable half-shrubs. *Hilaria jamesii* (galleta) is a major long-lived sod-forming grass. *Aristida* species (three-awn) are short-lived and often weedy bunchgrasses. Both are known to increase following grazing disturbance (West 1983f). The shrubs and half-shrubs are quite scattered and also generally increase with disturbance. Thus, had more been known about the area prior to erection of the Küchler classification, this probably would have been labeled a type of semidesert grassland, with too many codominants to have listed only *Hilaria* and *Aristida*.

The major reason that grasses prevail here is that they are most efficient in quickly utilizing moisture that rapidly infiltrates the deep, sandy soils. The concentration of precipitation in high-intensity late-summer storms means that there is relatively little soil moisture recharge in winter to favor shrub growth.

The total vascular plant cover found here is high for a semidesert environment (25%−60%), and the remaining space usually is covered by microphytic crust (Fig. 7.19) if the rate of disturbance has been low. In ungrazed areas, the plants of each species occupy the microenvironment for which each is best suited, creating a mosaic of nearly monospecific patches (Kleiner and Harper 1972). Livestock grazing breaks the crusts and creates a more spatially mixed community. The palatable shrubs and half-shrubs decline, and less desirable species invade, especially short-lived, less palatable grasses (*Aristida* species and *Muhlenbergia torreyana*). If the cover is broken too much, microdunes start to form; less palatable shrubs such as *Coleogyne*, *Ephedra*, and *Quercus* catch the sand. The height of the sward due to sod-formers is usually less than 30 cm, whereas bunchgrasses, especially *Stipa*, are taller.

Like other semideserts, these grasslands apparently were only weakly stable communities even prior to the introduction of livestock. Secondary succession has been slow (Loope 1975). Whether recovery will lead to an approximation of pre-Columbian communities or to different kinds of quasi-stable communities is unknown. Fortunately, the fuel loads rarely are great enough to carry fires. Few exotic weed species have taken over in great abundance. Nevertheless, we have no proven methods of artificially regenerating the vegetation of such areas.

Juniper-Pinyon Woodlands

At elevations slightly above all those for the previous vegetation types we have considered, one usually passes into a woodland dominated by scattered *Juniperus* and cembroid pines. The collective area of such woodlands is huge, at least 17×10^6 ha, and far-flung (Fig. 7.20). Although all of this vegetation type is on semiarid sites, with 25−50 cm total an-

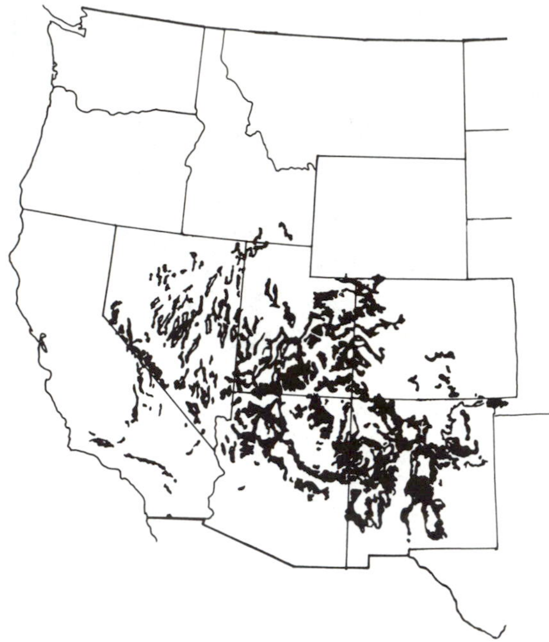

Figure 7.20. Map of the juniper-pinyon woodland vegetation type; (adapted from Küchler 1970).

Table 7.2. *Distribution of principal tree species in pinyon-juniper woodland in United States*

Colorado Plateau (eastern Utah, western Colorado, northern Arizona, northwestern New Mexico)	
	Utah juniper (*Juniperus osteosperma*)
	Colorado pinyon (*Pinus edulis*)
	Single-seed juniper (*Juniperus monosperma*)
Great Basin (Nevada, western Utah, California east of the Sierra Nevada)	
	Utah juniper (*Juniperus osteosperma*)
	Single-needle pinyon (*Pinus monophylla*)
Mohave border (southern California)	
	Single-needle pinyon (*Pinus monophylla*)
	California juniper (*Juniperus californica*)
	Sierra Juarez pinyon [*Pinus juarezensis* (*P.* × *quadrifolia*)]
Pacific Northwest (eastern Oregon, southwestern Idaho, northeastern California)	
	Western juniper (*Juniperus occidentalis*)
Northern Rockies (Wyoming, northern Colorado Front Range, Montana, eastern Idaho)	
	Utah juniper (*Juniperus osteosperma*)
	Rocky Mountain juniper (*Juniperus scopulorum*)
Southern Rockies (southern Colorado, northern New Mexico)	
	Single-seeded juniper (*Juniperus monosperma*)
	Colorado pinyon (*Pinus edulis*)
Mogollon rim (central Arizona)	
	Utah juniper (*Juniperus osteosperma*)
	Single-needle pinyon (*Juniperus monosperma*)
	Alligator-bark juniper (*Juniperus deppeana*)
	Arizona cypress (*Cupressus arizonica*)
Sonoran Desert border (southern Arizona)	
	Mexican pinyon (*Pinus cembroides* var. *bicolor*)
	Alligator-bark juniper (*Juniperus deppeana*)
	Colorado pinyon (*Pinus edulis*)
	Redberry juniper (*Juniperus erythrocarpa*)
Edwards Plateau (central Texas)	
	Pinchot's juniper (*Juniperus pinchotii*)
	Ashe's juniper (*Juniperus ashei*)
	Texas pinyon pine (*Pinus cembroides* var. *remota*)
Trans-Pecos-Chihuahuan Desert border (western Texas and southeastern New Mexico)	
	Texas pinyon pine (*Pinus cembroides* var. *remota*)
	Redberry juniper (*Juniperus erythrocarpa*)
	Pinchot's juniper (*Juniperus pinchotii*)

nual precipitation, the seasonality and effectiveness of this precipitation vary greatly. Consequently, the species of *Juniperus* and *Pinus* vary, and the understory varies even more.

The major tree species in various portions of this vegetation type are listed in Table 7.2. To the uninitiated, the different juniper-pinyon woodlands look much alike. The junipers and pines are similar in height: 10–15 m at maturity. Junipers are more widespread geographically and elevationally, going into drier and colder habitats (Fig. 7.21), but in the central parts of the vegetation type and in the woodland belt, usually one juniper species and one pinyon species form the tree guild; hence the name for these woodlands.

The understory varies so greatly over the vegetation type (West et al. 1975) that it is easier to say that it is similar to adjacent grasslands and shrub steppes; that is, cool-season bunchgrasses and sagebrushes prevail in the northern and western Great Basin, whereas there are more warm-season sod grasses and fewer shrubs in the woodlands of the Colorado Plateau and upper Rio Grande basin, where "monsoonal" rainfall patterns prevail.

The tree crowns rarely touch in these open woodlands. Tree root systems extent out two to three times as far as crown diameters. Tierney and Foxx (1982) have reported that *Juniperus monosperma* and *Pinus edulis* have rooted to depths of 6.4 m in cracks in tuff near Los Alamos, New Mexico. No other data on rooting depths have been located. Tree height and density increase with site favorableness, which usually is tied to elevation (West 1984). Total higher-plant cover varies from 40% to 80%. Most of the lower-stature vegetation is in the interspaces rather than under the tree crowns, where litter accumulates, often in mounds centered on the boles.

Juniper-pinyon woodlands have changed enormously because of human uses, and these impacts have by no means been limited to Europeans. There is mounting evidence that primitive humans

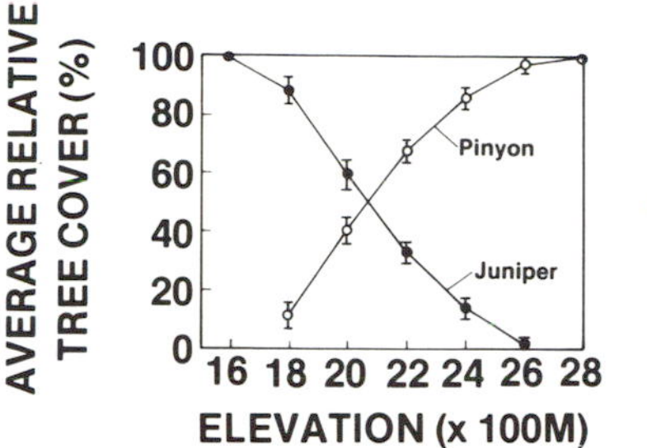

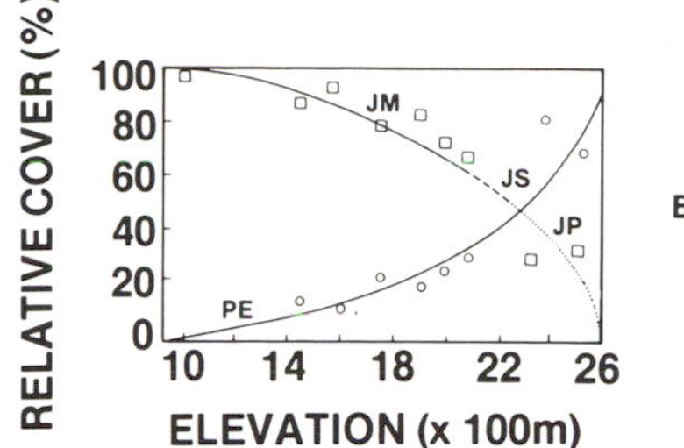

Figure 7.21. Elevational sorting of pinyon and juniper. A: In Great Basin (from Tueller et al. 1979). The juniper is J. osteosperma*; most of the pinyon is* P. monophylla*, except in the eastern extremities, where* P. edulis *occurs. Vertical lines are 95% confidence intervals. B: In southern Rocky Mountains in New Mexico, from "Juniper-pinyon east of the Continental Divide analysed by the line strip method" by H. E. Woodin and A. A. Lindsey,* Ecology*, 1954, 35, 473–489. Copyright 1954 by the Ecological Society of America, reprinted by permission). The circular points represent relative cover of* P. edulis*; the solid squares represent relative cover of* Juniperus*;* J. monosperma *is associated with the solid line,* J. scopulorum *is associated with the dashed line, and* J. pachyphloea *(now* J. deppeana*) with the dotted line.*

brought about some localized changes because of fuel harvesting (Samuels and Betancourt 1982). However, the more pervasive influences have been due to the livestock of European colonists.

Examinations of relict areas, tree age-class structures, fire scars, and historical documents (West 1984) lead us to believe that much of the juniper-pinyon woodland was once more savannah-like. Fires were frequent enough to keep the oldest trees restricted to steep, rock, and/or dissected topography. The ability of vegetation to carry fire on gentler topography was due to the abundance of fine fuel, mainly grasses. Livestock found the grasses virtually the only valuable component of this community. When grazing was excessive, fire could no longer carry and perform its natural thinning function. Shrubs and then trees increased in abundance, with shrubs often serving as nurse plants for tree seedlings (West et al. 1975).

Because of greater average tissue longevity, trees and shrubs are able to build more phytomass per unit of soil moisture and nutrients. Their roots go farther, both vertically and horizontally, to obtain these resources. They also outcompete herbaceous species through the casting of shade and litter with allelochemic properties. Humans also have aided this lignification of the community by conscious fire control. This has been countered in some places by harvesting of wood for fence posts, mine props, firewood, and charcoal. The net effect, however, has been an increase in density of trees where originally there was savannah, as well as expansion both upslope and downslope into grasslands and shrub steppes that were being simultaneously degraded.

In addition to the loss of forage for livestock and wild ungulates through this retrogressive succession, there may have been an increase in soil erosion rates: Tree root systems extend laterally near soil surfaces for great distances from the boles. Other vegetation is thus competitively restricted from interspaces. We can now observe much of the woodland composed of nothing but trees on mounded microtopography, with rills or gullies forming in the interspaces. This could mean that the system is degrading to a new level of lower potential. Because the individual trees can live for at least hundreds of years, it will take a long time to decipher fully the trajectories of retrogression. Carrara and Carroll (1979) have used these characteristics of great-tree longevity and erosional exposure of roots to demonstrate that soil erosion rates in at least one part of the juniper-pinyon woodland have increased over 400% during the past century, as compared with the previous three centuries. This rate change coincides with the introduction of livestock about 100 yr ago in the vicinity of this northwestern Colorado area. Recent reductions in livestock and increases in wood harvest have had no appreciable impact on recovery of understory vegetation, apparently because seed banks and sources have been lost (Koniak and Everett 1982).

Land managers, recognizing the foregoing successional changes, began in the 1950s and 1960s to alter this vegetation on relatively level sites by mechanical means. A battleship anchor chain or cable would be drawn between two crawler tractors, pulling over the trees. This may have been preceded by aerial seeding of grasses, or the debris might be ricked into piles, and then drills used to implant seeds. The result was an enormous increase in grasses on the more thoroughly treated sites. Most such treatments were carried out before petroleum prices soared and environmental-impact statements were required. Both economic and environmental considerations have recently reduced such type conversions.

Successional trajectories set over 100 yr ago are still leading to tree dominance. The accumulating wood still is not valuable enough to manage, except in northern New Mexico, where wood harvests are exceeding annual increments (Gray et al. 1982). Prescribed burning usually cannot be done under reasonable conditions because the fine fuels are sparse to entirely lacking. Crown fires will eventually open up the more mesic part of the vegetation type. If we are ready for that contingency, we can artificially seed more desirable species on such areas (Koniak 1983; West 1984). If we do not, annuals and closely recurrent fires will occur and result in accelerated erosion.

Mountain Mahogany–Oak Scrub

The transition zone from montane coniferous forests to treeless plains and plateaus at margins of the Rocky Mountains usually is occupied by broad-leaved scrub. Chaparral-like vegetation can also be found above or intermingled with juniper-pinyon woodlands in the Great Basin. This type is most widespread and best developed in the southern Rockies and along the south side of the Mogollon rim (Fig. 7.22). The belt narrows and becomes discontinuous farther north. This vegetation is discussed in Chapter 6.

The most widespread dominant is *Cercocarpus ledifolius* (curl-leaf mountain mahogany). *Cercocarpus montanus* is restricted to the central Rockies, where it joins with deciduous oaks (*Quercus gambelii, Q. gunnisoni, Q. undulata, Q. fendleri,* and others). Beginning in southern Utah, the evergreen oaks (*Q. turbinella, Q. emoryi, Q. dumosa, Q. chrysolepis*) and *Cercocarpus breviflorus* and *C. betuloides* begin their appearance and become dominant farther south (Carmichael et al. 1978; Pase and Brown 1982).

Other shrubby associates include *Artemisia tridentata* ssp. *vaseyana, Acer grandidentatum, Purshia tridentata, Rhus trilobata, Rhamnus crocea, Fallugia paradoxa, Cowania mexicana, Amelanchier* species, *Symphoricarpos* species, *Berberis* species, *Arctostaphylos pungens, A. pringlei, Ceanothus greggii, Garrya flavescens, G. wrightii,* and *Eriodictyon angustifolium*, any one of which may assume local dominance. For instance, many of the upper elevations on medium-size mountain ranges in Nevada lack, because of paleoecological influences, montane forests above the juniper-pinyon belts. The tops of these mountains are covered by *Artemisia-Symphoricarpos*-dominated vegetation, and *Cercocarpus ledifolius* diminishes to small patches.

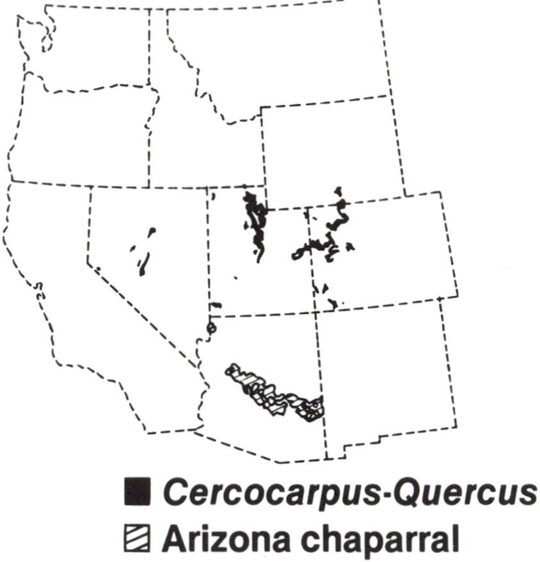

Figure 7.22. Map of mountain mahogany–oak scrub and Arizona chaparral vegetation types (adapted from Küchler 1970).

The taller shrubs rarely exhibit continuous cover, but occur as dense clumps ("mottes"), separated by areas of grassland or low shrub steppe. The height of the shrub cover is 1–5 m, depending on species, site, and recent fire history. Most of the shrubs resprout readily after burning and have their seed germination stimulated by fires (e.g., *Ceanothus, Arctostaphylos*) (See Chapter 6).

As in the case of the other vegetation types we have discussed, there has been a marked increase in woody species as herbaceous components have been reduced by livestock grazing over the past 100–130 yr (Pase and Brown 1982). Loss of fine fuels and human control over fire have also influenced the increased height of brush and its expansion into the interspaces between the older mottes (Rogers 1982). Excessive deer and elk browsing has "high-lined" stands of some species, especially *Cercocarpus ledifolius*, and prevented adequate regeneration. Occasional late spring frosts can set back deciduous oak growth (Nielsen and Wullstein 1983) and provide more fuel for wildfire.

Where tall, dense oaks dominate, livestock and big-game interests have attempted to control the brush component, followed by seeding of herbaceous species. The resprouting characteristics of the shrubs have thwarted all but the most vigorous efforts to maintain dominance by herbaceous species. There are many introduced species that can be seeded successfully (Plummer et al. 1968), because soils on relatively level sites usually are quite rich in nutrients and have high water-holding capacity.

In the last decade there has been increased human activity in these communities because of firewood gathering. This is likely to continue as these possibilities become more restricted in adjacent coniferous forests.

AREAS FOR FUTURE RESEARCH

I hope the reader now can appreciate that there is great variation in the vegetation across the lower elevations of the intermountain West. These are vast areas that are sparsely populated between the few urban oases at the feet of the higher mountains where perennial streams emerge. The low cover and productivity of these desert-to-woodland communities have not encouraged much applied research as compared with that done in more mesic parts of the continent. In fact, those residing in the more populous parts of the country often have regarded the region as an unused wasteland roamed only by a few cowboys, miners, and shepherds, and where "nuisance" activities such as military research and training, electricity-generating stations, and nuclear wastes might be placed. The striking geology showing through this low, sparse vegetation has attracted mineral and recreational exploration. The geologic, topographic, and climatic complexities have also led to evolution of some remarkable plants. There is much more that we need to know about them and the vegetation they compose.

In addition, this region has experienced remarkable vegetation changes in the short span of a century, mainly because of human-modified fire and grazing regimes. Future research directions should include development of economical techniques for revegetation of sagebrush steppe, saltbush-greasewood scrubs, blackbrush scrub, galleta-three-awn shrub steppe, and juniper-pinyon woodlands. Whether the objective is to reconstitute pristine-like vegetation or to replace degraded types with something new, productive, and stable, we need significant studies in basic plant autecology; we also need basic descriptive information on the pristine nature of saltbush-greasewood scrub and galleta-three-awn shrub steppe. These latter types have been so completely modified in short spans of time that their original structures are largely conjectural.

REFERENCES

Anderson, J. E., and K. E. Holte. 1981. Vegetation development over 25 years without grazing on sagebrush dominated rangeland in southeastern Idaho. J. Range Management 34:25–29.

Anderson, L. C. 1975. Modes of adaptation to desert conditions in *Chrysothamnus*. p. 141 in H. C. Stutz (ed.), Procedings of symposium and workshop on wildland shrubs. USDA, Forest Service, Shrub Sciences Laboratory, Provo, Utah.

Axelrod, D. I. 1979. Desert vegetation, its age and origin. pp. 1–72 in J. R. Goodin and D. K. Northington (eds.), Arid land plant resources. International Center for Arid and Semi-arid Land Studies, Texas Technological University, Lubbock.

Barnosky, C. W. 1984. Late Miocene vegetational and climatic variations inferred from a pollen record in northwest Wyoming. Science 223:49–51.

Bates, P. A., and J. W. Menke. 1984. Role of fire in blackbrush succession. In Abstracts, annual meeting of the Society of Range Management, Rapid City, S.D. no. 149.

Beatley, J. C. 1975. Climates and vegetation patterns across the Mojave/Great Basin transition of southern Nevada. Amer. Midl. Nat. 93:53–70.

Betancourt, J. L. 1984. Late Quaternary plant zonation and climate in southeastern Utah. Great Basin Naturalist 44:1–35.

Billings, W. D. 1945. The plant associations of the Carson Desert region, western Nevada. Butler Univ. Bot. Studies 8:89–123.

Billings, W. D. 1949. The shadscale vegetation zone of Nevada and eastern California in relation to climate and soils. Amer. Midl. Nat. 42:87–109.

Billings, W. D. 1978. Alpine phytogeography across the Great Basin. Great Basin Naturalist Memoirs 2:105–118.

Blaisdell, J. P. 1958. Seasonal development and yield of native plants in the upper Snake River Plains and their relation to certain climatic factors. USDA technical bulletin 1190.

Blaisdell, J. P., and R. C. Holmgren. 1984. Managing intermountain rangelands–salt desert shrub ranges. USDA Forest Service general technical report INT-163, Intermountain Forest and Range Experiment Station. Ogden, Utah.

Blaisdell, J. P., R. B. Murray, and E. D. McArthur. 1982. Managing intermountain rangeland–sagebrush-grass ranges. USDA Forest Service general technical report INT-134, Intermountain Forest and Range Experiment Station, Ogden, Utah.

Bolen, E. G. 1964. Plant ecology of spring-fed salt marshes in western Utah. Ecol. Monogr. 34:143–166.

Bowns, J. E., and N. E. West. 1976. Blackbrush (*Coleogyne ramosissima* Torr.) on southwestern Utah rangelands. Utah State University Agricultural Experiment Station research report 27.

Caldwell, M. M. 1979. Physiology of sagebrush. pp. 74–85 in The sagebrush ecosystem: a symposium. College of Natural Resources, Utah State University, Logan.

Carmichael, R. S., O. D. Knipe, C. P. Pase, and W. W. Brady. 1978. Arizona chaparral: plant associations and ecology. USDA Forest Service research paper RM-202. Rocky Mountain Forest and Range Experiment Station, Ft. Collins, Colo.

Carrara, P. E., and T. R. Carroll. 1979. The determination of erosion rates from exposed tree roots in the Piceance Basin, Colorado. Earth Surface Processes 4:307–317.

Clary, W. P. 1975. Ecotypic variation in *Sitanion hystrix*. Ecology 56:1407–1415.

College of Natural Resources (CNR). 1979. The sagebrush ecosystem: a symposium. CNR, Utah State University, Logan.

Daubenmire, R. 1970. Steppe vegetation of Washington. Washington State University Agricultural Experiment Station technical bulletin 62.

Daubenmire, R. F. 1972. Annual cycles of soil moisture and temperature as related to grass development in the steppe of eastern Washington. Ecology 53:419–424.

Daubenmire, R. 1975. An analysis of structural and functional characters along a steppe-forest catena. Northwest Sci. 49:120–140.

Davis, O. K. 1984. Multiple thermal maxima during the Holocene. Science 225:617–619.

Detling, J. K., and L. G. Klikoff. 1973. Physiological response to moisture stress as a factor in halophyte distribution. Amer. Midl. Nat. 90:307–318.

Eckert, R. E., Jr., and F. E. Kinsinger. 1960. Effects of *Halogeton glomeratus* leachate on chemical and physical characteristics of soils. Ecology 41:764–772.

Eckert, R. E., Jr., M. K. Wood, W. H. Blackburn, F. F. Peterson, J. L. Stephens, and M. S. Meurisse. 1978. Effects of surface-soil morphology on improvement and management of some arid and semi-arid rangelands. pp. 299–302. in D. N. Hyder (ed.), Proceedings of the First International Rangeland Congress. Society for Range Management, Denver.

Ferguson, C. W. 1964. Annual rings in big sagebrush. University of Arizona Press, Tucson.

Flowers, S., F. R. Evans. 1966. The flora and fauna of the Great Salt Lake region, Utah. pp. 367–393 in H. Boyko (ed.), Salinity and aridity, Monographics biologicae 16. W. Junk, The Hague.

Gray, J. R., J. F. Fowler, and M. A. Bray. 1982. Free-use fuelwood in New Mexico: inventory, exhaustion and energy equations. J. Forestry 80:23–26.

Griffiths, D. 1902. Forage conditions on the northern border of the Great Basin. USDA Bureau of Plant Industry bulletin 15.

Hanley, T. A. 1979. Application of an herbivore-plant model to rest-rotation grazing management on shrub-steppe ranges. J. Range Management 32:115–118.

Harris, G. A. 1977. Root phenology as a factor of competition among grass seedlings. J. Range Management 30:172–176.

Hironaka, M. 1979. Basic synecological relationships of the Columbia River sagebrush type. pp. 27–30 in The sagebrush ecosystem: a symposium. College of Natural Resources, Utah State University, Logan.

Hull, A. C., Jr. 1976. Rangeland use and management in the Mormon West. In Symposium on agriculture, food and man–a century of progress. Brigham Young University, Provo, Utah.

Hunt, C. B. 1974. Natural regions of the United States and Canada. W. H. Freeman, San Francisco.

Jennings, J. D. 1978. Prehistory of Utah and the eastern Great Basin. University of Utah Anthropological Papers 98.

Keller, W. 1979. Species and methods for seeding in the sagebrush ecosystem. pp. 129–136 in The sagebrush ecosystem: a symposium. College of Natural Resources, Utah State University, Logan.

Kleiner, E. F., and K. L. Harper. 1972. Environmental and community organization in grasslands of Canyonlands National Park. Ecology 53:299–309.

Koniak, S. 1983. Broadcast seeding success in eight pinyon-juniper stands after wildlife. USDA Forest Service research note INT-334, Intermountain Forest and Range Experiment Station, Ogden, Utah.

Koniak, S., and R. L. Everett. 1982. Seed reserves in soils of successional stages of pinyon woodlands. Amer. Midl. Nat. 108:295–303.

Küchler, A. W. 1970. Potential natural vegetation (map at scale 1 : 7,500,000). pp. 90–91 in The national atlas of the U.S.A. U.S. Government Printing Office, Washington, D.C.

LaMarche, V. C., Jr. 1974. Paleoclimatic inferences from long tree-ring records. Science 198:1043–1048.

Loope, W. L. 1975. Vegetation in relation to environment of Canyonlands National Park, Utah. Ph.D. dissertation Utah State University, Logan.

McArthur, E. D. 1983. Taxonomy, origin and distribution of big sagebrush (*Artemisia tridentata*) and allies (subgenus Tridentatae). pp. 3–11 in R. L. Johnson (ed.), Proceedings of first Utah shrub ecology workshop. College of Natural Resources, Utah State University, Logan.

McDonough, W. T., R. O. Harniss, and R. B. Campbell. 1975. Morphology of perennial and persistent leaves of three subspecies of big sagebrush grown in a uniform environment. Great Basin Naturalist 35:325–326.

Mack, R. N. 1981. Invasion of *Bromus tectorum* L. into western North America: an ecological chronicle. Agro-ecosystems 7:145–165.

Mack, R. N., and J. N. Thompson. 1982. Evolution in steppe with few large, hooved animals. American Naturalist 119:757–773.

Martin, P. S. 1970. Pleistocene niches for alien animals. BioScience 20:218–221.

Mehringer, P. J., Jr. 1977. Great Basin Late Quaternary environments and chronology. pp. 113–167 in D. D. Fowler (ed.), Models and Great Basin prehistory. Desert Research Institute Publications in Social Sciences 12. University of Nevada, Reno.

Nielsen, R. P., and C. H. Wullstein. 1983. Biogeography of two southwest American oaks in relation to atmospheric dynamics. J. Biogeography 10:275–298.

Norton, B. E. 1978. The impact of sheep grazing on long-term successional trends in salt desert shrub vegetation of southwestern Utah. pp. 610–613 in D. N. Hyder (ed.), Proceedings of first international rangeland congress. Society for Range Management, Denver.

Noy-Meir, I. 1973. Desert ecosystems: environment and producers. Ann. Rev. Ecol. Syst. 4:25–51.

Pase, C. P., and D. E. Brown. 1982. Interior chaparral. pp. 95–99 in D. E. Brown (ed.), Biotic communities of the American Southwest–United States and Mexico. Boyce Thompson Institute of Plant Biology, Superior, Ariz.

Passey, H. B., V. K. Hugie, and E. W. Williams. 1982. Relationships between soil, plant community, and climate on rangelands of the intermountain West. USDA Soil Conservation Service technical bulletin 1669.

Plummer, A. P., D. R. Christensen, and S. B. Monsen. 1968. Restoring big-game range in Utah. Utah Division of Fish and Game publication 68–3, State of Utah, Department of Natural Resources, Salt Lake City, Utah.

Potter, L. D., and J. C. Krenetsky. 1967. Plant succession with released grazing on New Mexico rangelands. J. Range Management 20:145–151.

Provenza, F. D., J. E. Bowns, P. J. Urness, and J. E. Butcher. 1983. Biological manipulation of blackbrush by goat browsing. J. Range Management 36:513–518.

Rice, B., and M. Westoby. 1978. Vegetation responses of some Great Basin shrub communities protected against jackrabbits or domestic stock. J. Range Management 31:28–33.

Rickard, W. H. 1964. Demise of sagebrush through soil changes. BioScience 14:43–44.

Robertson, D. R., J. L. Nielsen, and N. H. Bare. 1966. Vegetation and soils of alkali sagebrush and adjacent big sagebrush ranges in North Park, Colorado. J. Range Management 19:17–20.

Rogers, G. G. 1982. Then and now: a photographic history of vegetation change in the central Great Basin desert. University of Utah Press, Salt Lake City.

Samuels, M. L., and J. L. Betancourt, 1982. Modeling the long-term effects of fuel wood harvests on pinyon-juniper woodlands. Environ. Management 6:505–515.

Shelford, V. E. 1963. The ecology of North America. University of Illinois Press, Urbana.

Shmida, A., and R. H. Whittaker. 1979. Convergent evolution of arid regions in the New and Old Worlds. pp. 437–450 in R. Tuxen (ed.), Vegetation and history. International Vereing Vegetationskunde, Rinteln.

Soil Survey Staff. 1975. Soil taxonomy: a basic system of soil classification for making and interpreting soil surveys. USDA agricultural handbook 436. U.S. Government Printing Office, Washington, D.C.

Stoddart, L. A. 1946. Some physical and chemical responses of *Agropyron spicatum* to herbage removal at various seasons. Utah State University Agricultural Experiment Station bulletin 324.

Sturges, D. L. 1977. Soil water withdrawal and root characteristics of big sagebrush. Amer. Midl. Nat. 98:257–274.

Stutz, H. C. 1978. Explosive evolution of perennial *Atriplex* in western North America. Great Basin Naturalist Memoirs 2:161–168.

Thatcher, A. P., J. W. Doughty, and D. L. Richmond. 1976. Amount of blackbrush in the pristine plant communities controlled by edaphic conditions. p. 12 in Abstracts of annual meeting of the Society of Range Management, Omaha, Neb. vol. 27.

Tierney, G. D., and T. S. Foxx. 1982. Floristic composition and plant succession on near-surface radioactive waste disposal facilities in the Los Alamos National Laboratory. Los Alamos National Laboratory report LA-9212-MS.

Tisdale, E. W., and M. Hironaka. 1981. The sagebrush-grass region: a review of the ecological literature. Forest, Wildlife and Range Experiment Station bulletin 33, University of Idaho, Moscow.

Tisdale, E. W., M. Hironaka, and F. A. Fosberg. 1965. An area of pristine vegetation in Craters of the Moon National Monument, Idaho. Ecology 46:349–352.

Tueller, P. T., C. D. Beeson, R. J. Tausch, N. E. West, and K. H. Rea. 1979. Pinyon-juniper woodlands of the Great Basin: distribution, flora, vegetal cover. USDA Forest Service research paper INT-229, Intermountain Forest and Range Experiment Station, Ogden, Utah.

U.S. Department of Agriculture. 1941. Climate and man. Agricultural yearbook. USDA, Washington, D.C.

Van Devender, T. R., and W. G. Spaulding. 1977. Development of vegetation and climate in the southwestern United States. Science 204:701–710.

Van Pelt, N. 1978. Woodland parks of southeastern Utah. M.S. thesis, University of Salt Lake City.

Wells, P. V. 1983. Paleobiogeography of montane islands in the Great Basin since the last glaciopluvial. Ecol. Monogr. 53:341–382.

West, N. E. 1969. Soil-vegetation relationships in arid southeastern Utah. International conference on arid lands in a changing world. University of Arizona, Tucson.

West, N. E. 1979. Basic synecological relationships of sagebrush-dominated lands in the Great Basin and the Colorado Plateau pp. 33–41 in The sagebrush ecosystem: a symposium. College of Natural Resources, Utah State University, Logan.

West, N. E. 1982. Dynamics of plant communities dominated by chenopod shrubs. Int. J. Ecol. Environ. Sci. 8:73–84.

West, N. E. 1983a. Overview of North American temperate deserts and semi-deserts. pp. 321–330 in N. E. West (ed.), Temperate deserts and semi-deserts, vol. 5, Ecosystems of the world. Elsevier, Amsterdam.

West, N.E. 1983b. Western intermountain sagebrush steppe. pp. 351–374 in N. E. West (ed.), Temperate deserts and semi-deserts, vol. 5, Ecosystems of the world. Elsevier, Amsterdam.

West, N. E. 1983c. Great Basin–Colorado Plateau sagebrush semi-desert, pp. 331–349 in N. E. West (ed.), Temperate deserts and semi-deserts, vol. 5, Ecosystems of the world. Elsevier, Amsterdam.

West, N.E. 1983d. Intermountain salt-desert shrubland. pp. 375–397 in N. E. West (ed.), Temperate deserts and semi-deserts, vol. 5, Ecosystems of the world. Elsevier, Amsterdam.

West, N. E. 1983e. Colorado Plateau–Mohavian blackbrush semi-desert. pp. 399–411 in N. E. West (ed.), Temperate deserts and semi-deserts, vol. 5, Ecosystems of the world. Elsevier, Amsterdam.

West, N. E. 1983f. Southeastern Utah galleta-threeawn shrub steppe. pp. 413–421 in N. E. West (ed.), Temperate deserts and semi-deserts, vol. 5, Ecosystems of the world. Elsevier, Amsterdam.

West, N. E. 1984. Successional patterns and productivity potentials of pinyon-juniper ecosystems. pp. 1301–1332. in Developing strategies for range management. Westview Press, Boulder, Colo.

West, N. E., and J. Gasto. 1978. Phenology of the aerial portions of shadscale and winterfat in Curlew Valley, Utah. J. Range Management 31:43–45.

West, N. E., and D. W. Goodall, 1986. Dispersion patterns in relation to successional status of salt desert shrub vegetation. Abstracta Botanica 10:87-201.

West, N. E., F. D. Provenza, P. S. Johnson, and M. K. Owens. 1984. Vegetation change after 13 years of livestock grazing exclusion on sagebrush semidesert in west central Utah. J. Range Management 37:262–264.

West, N. E., K. H. Rea, and R. O. Harniss. 1979. Plant demographic studies in sagebrush-grass communities of southeastern Idaho. Ecology 60:376–388.

West, N. E., K. H. Rea, and R. J. Tausch. 1975. Basic synecological relationships in pinyon-juniper woodlands. pp. 41–58 in G. F. Gifford and F. E. Busby (eds.), The pinyon-juniper ecosystem: a symposium. Utah State University Agricultural Experiment Station, Logan.

West, N. E., and J. Skujins. 1977. The nitrogen cycle in North American cold-winter semi-desert ecosystems. Oecologia Plantarium 12:45–53.

Winward, A. H. 1983. Using sagebrush ecology in wildland management. pp. 15–19 in K. L. Johnson (ed.), First Utah shrub ecology workshop. College of Natural Resources, Utah State University, Logan.

Wolfe, J. A. 1978. A paleobotanical interpretation of Tertiary climates in the northern hemisphere. Amer. Sci. 66:694–703.

Woodin, H. E., and A. A. Lindsey. 1954. Juniper-pinyon east of the continental divide analyzed by the line-strip method. Ecology 35:473–489.

Wright, H. A., and A. W. Bailey. 1982. Fire ecology: United States and southern Canada. Wiley, New York.

Young, J. A., R. A. Evans, and J. Major. 1977. Sagebrush steppe. pp. 763–796 in M. G. Barbour and J. Major (eds.), Terrestrial vegetation of California. New York, Wiley.

Zamora, B., and P. T. Tueller. 1973. *Artemisia arbuscula, A. longliloba* and *A. nova* habitat types in northern Nevada. Great Basin Naturalist 33:225–242.

Chapter 8

Warm Deserts

JAMES A. MacMAHON

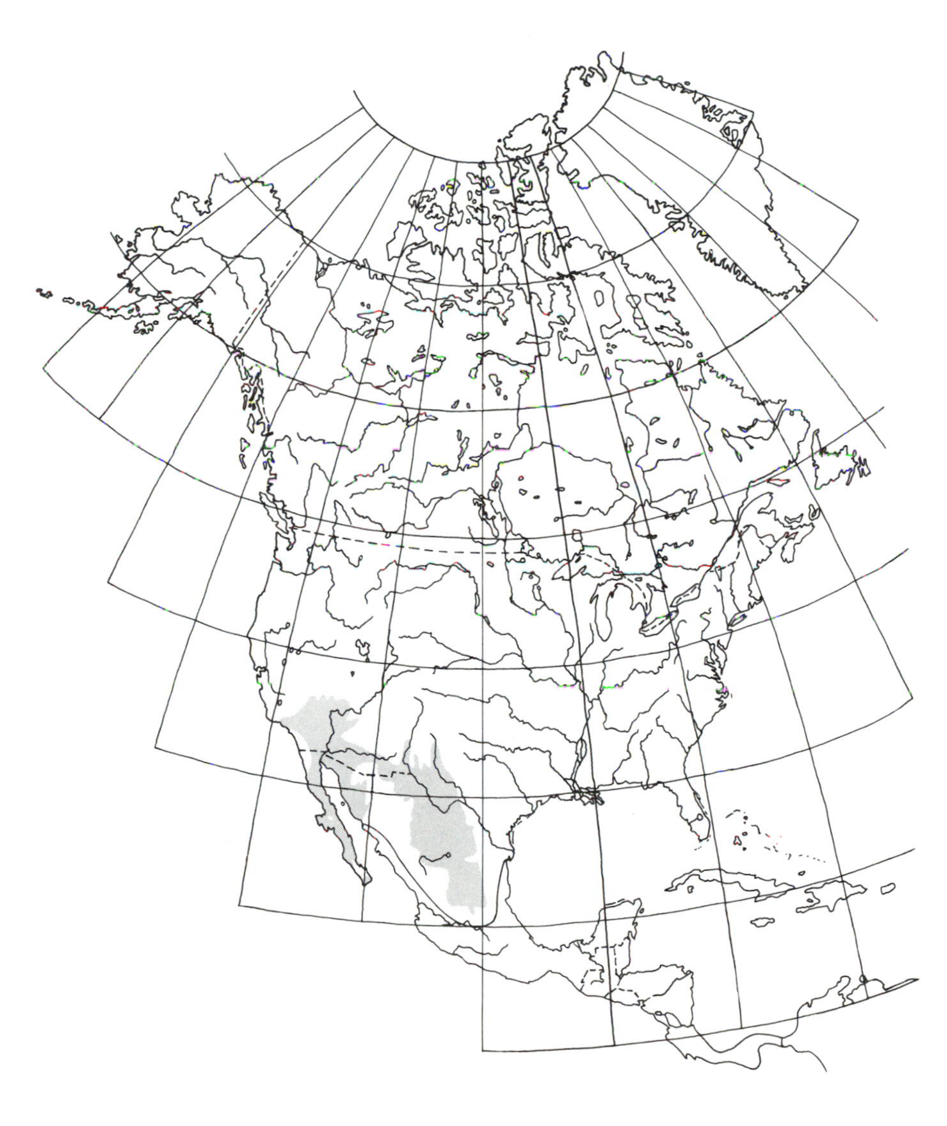

INTRODUCTION

This chapter treats the Mojave, Sonoran, and Chihuahuan deserts, the "warm deserts" of North America (Fig. 8.1). These three quite different vegetation units occupy 68% of the area indicated as desert by MacMahon (1979) (Table 8.1). In general, these deserts are the low-elevation, xeric sites of the southwestern portion of the United States and the northern quarter of Mexico. Based on climatic considerations, the Mojave and Chihuahuan deserts are termed warm-temperate deserts, whereas the Sonoran Desert is subtropical in nature. The three are lumped under the term "warm deserts" not so much because of their average annual temperatures but more because their precipitation is in the form of rain, even if it occurs in the winter. This contrasts with the Great Basin Desert (see Chapter 7), generally called a cold desert, that usually has over 60% of its precipitation in the form of snow.

Because of their wide elevational span (−86 m in Death Valley in the Mojave to 1525 m in southern sections of the Chihuahuan Desert in Mexico) and latitudinal span (36½–22° N latitude), the warm deserts form transitions with a wide variety of vegetation types: the Great Basin Desert in the north (see Chapter 7), grasslands in the east and at higher elevations (see Chapter 9), and subtropical thorn forests to the south (see Chapter 12).

In the last 8 yr there has been a great deal written about North American deserts. This coverage has included consideration of a range of topics: all North American deserts (MacMahon 1979; Bender 1982); all or a significant part of the warm deserts (Barbour and Major 1977; Wauer and Riskind 1977; Rzedowski 1978; McGinnies 1981; Brown 1982a; MacMahon and Wagner 1985); comparisons between our deserts and other warm deserts (Orians and Solbrig 1977); water (MacMahon and Schimpf 1981) and nitrogen (West and Skujins 1978) as they relate to desert vegetation; the biology of major plants such as mesquite (Simpson 1977), creosote bush (Mabry et al. 1977), cacti (Gibson and Nobel 1986), and agaves (Gentry 1978, 1982; Pinkava and Gentry 1985).

Additionally, identification of warm-desert plants has been made easier by the appearance of a number of books that, when used with the various state floras, cover many plants in detail and provide a wealth of illustrations. Such works cover cacti (Benson 1982), woody perennials (Benson and Darrow 1981), wildflowers (Niehaus and Ripper 1976; Niehaus et al. 1984), and a variety of aspects of desert biology (MacMahon 1985).

Because of the intensity of recent coverage of warm deserts, as indicated by the references cited, this chapter emphasizes work published in the last 10 yr; the references cited earlier give the reader entrée to thousands of older studies. This chapter briefly considers the physiography, climate, and soils of our warm deserts, followed by an analysis of the vegetation for each of the three warm deserts and their subdivisions, where appropriate. Finally, those vegetation types common to all three of the deserts, here termed azonal vegetation, are considered in a separate section.

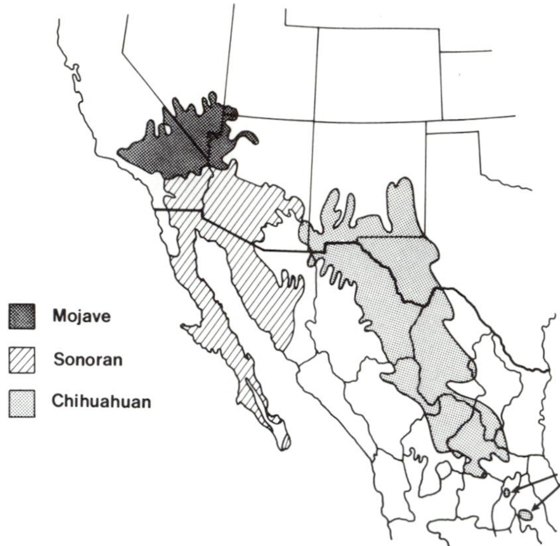

Figure 8.1. Distributions of Mojave, Sonoran, and Chihuahuan warm deserts as adopted in this chapter. Details of this delimitation can be found in MacMahon (1979) and MacMahon and Wagner (1985).

Table 8.1. *Approximate areas of North American deserts based on boundaries presented by MacMahon (1979)*

Unit	Area (km^2)	Portion of North American desert area (%)
Great Basin Desert	409,000	32.0
Sonoran Desert	275,000	21.5
Mojave Desert	140,000	11.0
Chihuahuan Desert	453,000	35.5
Total desert	1,277,000	100
Warm deserts	868,000	68.0
Cold deserts	409,000	32.0
Basin and Range Province	1,717,000	—

PHYSIOGRAPHY

The physiographic area that is referred to as the Basin and Range Province of North America essen-

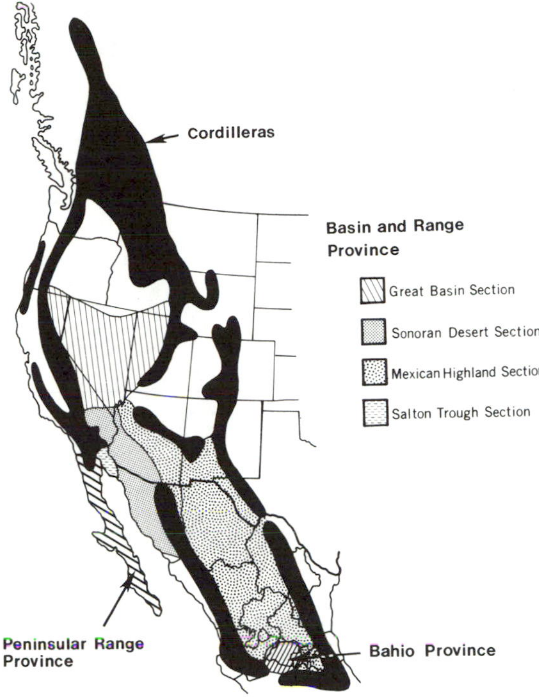

Figure 8.2. North American physiographic provinces and their sections as they apply to desert areas. Data from MacMahon (1979) and other sources.

tially contains all of our deserts, hot and cold. The main exceptions are (1) a piece of the Great Basin desert that laps onto the Columbia Plateau Province and (2) Baja California, generally considered to be a province by itself (West 1964), the Peninsular Range Province. The Basin and Range Province extends over 30° of latitude and 12° of longitude, encompassing an area of 1,717,000 km^2 (Table 8.1). The province is essentially an area surrounded by the main western North American mountain masses, the Rockies and the Sierra Nevada in the United States and the Sierra Madre Occidental and Sierra Madre Oriental in Mexico (Fig. 8.2).

The name of the province derives from its general appearance from the air. The entire landscape is composed of rather large basins dotted with much smaller, generally north-south-trending mountain ranges. Usually, more than 50% of the land surface is covered by the basins, though the figure may exceed 80%, especially in the Sonoran Desert. For example, at Organ Pipe Cactus National Monument, the basins occupy 82.6% of the area. The mountain ranges number more than 300. One worker, seeing them on a physiographic map, said they looked like "an army of caterpillars crawling northward out of Mexico" (King 1959). Elevations of the basins may be lower than −85 m in Death Valley and up to about 1525 m in Utah. The surrounding ranges may exceed 3950 m.

The basins of the province are dotted by lakes or their remnants, depending on the season and year. These lakes are dry for at least part of the year and often remain dry for years at a time. Over 200 examples of such lakes, generally referred to as playas, occur in the province. Playas form when the runoff of precipitation from the mountains accumulates in the depressions of the surrounding basins. The standing water evaporates, leaving deposits high in various salts of calcium or sodium. Playas usually are small (less than 100 km^2), and though they often look similar, the 50,000 or so that occur worldwide often are geomorphically quite distinct (Neal 1969, 1975).

Another structural feature of the desert landscape is the alluvial fan. These are cone-like deposits that originate from canyons in the mountains and fan out from the mouth of a canyon into a valley. Where several alluvial fans from adjacent canyons coalesce, a bajada is formed. Bajadas (Spanish for "slope") have smooth surfaces and may cover up to 75% of a valley; they have slopes of 6° to 9° where they meet mountain ranges, lessening to about 1° where they flow onto the plains or the playas of valley floors. Upper bajadas have coarse-textured soils, whereas their lower ends and the associated playas may have very fine soils (e.g., silts). The importance of bajadas will be emphasized often in our discussion of the distribution of plants.

The Basin and Range Province is subdivided into a number of units called sections, whose boundaries are somewhat arbitrary (Fig. 8.2), and they do not coincide with the biotically defined desert boundaries (Fig. 8.3). The Great Basin Section occupies about one-third of the province, including most of Nevada, the western half of Utah, and minute portions of Idaho, Oregon, and California. This section has few drainage outlets other than an area emptying into the Snake River in the northeast, a portion of the northwest that drains via the Pit River to the Sacramento River, and a part of the southeast corner that empties into the Colorado River via the Virgin River. Part of the northern Mojave Desert of Nevada and southwestern Utah is contained in this section. The Death Valley region falls into this section (Fiero 1986).

The Sonoran Desert Section includes the southwestern quarter of Arizona, the adjacent desert areas of California, and about one-third of Sonora, Mexico. Most of the Mojave and Sonoran deserts are contained in this section. Common rock substrates of the section include Precambrian granites and gneisses. Lakes were common in the Pleisto-

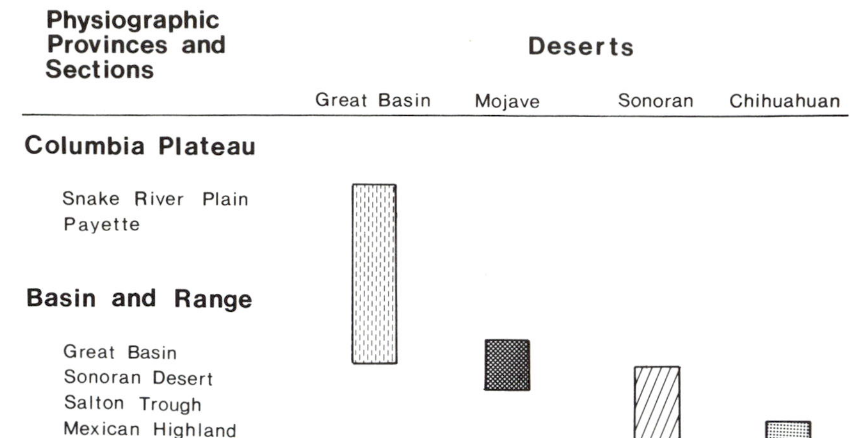

Figure 8.3 Differences between the extent of the biologically based desert subdivisions of North America and the physiographic provinces and sections of the geographers, as depicted in Fig. 8.2. The spans of sections in which the deserts occur are indicated by bars. Clearly, the biological deserts overlap some physiographic sections (i.e., one physiographic boundary may contain more than one desert type).

cene. Because slopes often are composed of metamorphic rocks, they may have rises of up to 20%. The elevations of the valleys are low, seldom exceeding 650 m.

The Salton Sea Trough Section is an extension, onto the land, of the 1600-km trough occupied by the Gulf of California. In the United States, the section includes a very small area centered around Brawley and El Centro, California. It occurs just south of the mountains forming the southern boundary of Joshua Tree National Monument. A main feature of the trough is the Salton Sea. At one time, perhaps only a few hundred years ago, the trough was occupied by Lake Cahuilla.

The Mexican Highland Section occurs as a diagonal band across the middle third of Arizona, the southwest quarter of New Mexico, and southward into Mexico, where it abuts the Bahio Province near the Transverse Volcanic Belt. Excluding the mountainous portions, the Chihuahuan Desert and the Chihuahuan-Sonoran transitions are essentially contained within this section. The area is characterized by high valleys (up to 1525 m). Generally, the lowest elevations of the section are along the Rio Grande, rising both to the north and to the south. Unlike other portions of the Basin and Range Province, this section contains well-developed drainage systems, though internal drainages and their associated playas do occur.

The Basin and Range Province is so vast and its geologic history so varied that it is difficult to generalize about it. The landscape throughout is composed of a mosaic of sedimentary rocks high in calcium carbonate patched with acidic volcanic materials. These substrates have created complex soil mosaics that influence the patterns of vegetation. Add to this the major changes in surface geomorphology caused by the arroyo cutting in the last 100 yr (Cooke and Reeves 1976), and one can see that vegetation analysis must be conducted in a circumspect manner. General treatments of desert geomorphic features are readily available (Cooke and Warren 1973; Goudie and Wilkinson 1977; Mabbutt 1977).

SOILS

Generally, arid-zone soils are low in organic matter, have slightly acidic to alkaline surface soils, and develop calcium carbonate accumulations in the upper 2 m of soil. They generally do not show excellent profile development (Dregne 1976, 1979; Hendricks 1985). Additionally, they may have long periods of low biological activity. Soils having all of these characteristics are generally termed Aridisols and they form under strong influences from wind, low but often torrential supplies of water, and high temperatures.

Some soil horizons in desert areas form hardpans because of cementing action by calcium carbonate, silica, or even iron compounds. Calcium carbonate cementation may form a water-impervious petrocalcic layer termed a caliche (Schlesinger 1985). Caliches can be quite thick (90 m), and they can be at the surface or buried to great depths (230 m) in some valleys (Shreve and Mallery 1933). It should be noted that carbonate layers, because of complex solution dynamics, can occur in desert soils that are not especially high in carbonate (Gile et al. 1966).

The presence of a caliche layer greatly influences the distribution of plant species. Creosote bush (*Larrea tridentata*), a dominant warm-desert shrub, requires high-calcium, gravelly soils (Hallmark and Allen 1975).

The carbon in calcium carbonate may represent a storage pool important for the global balance of carbon. In most of the world's soils, the main carbon fraction is in the form of organic matter from decomposition of plants and animals. In deserts, the ratio of carbonate to organic carbon may exceed 10 : 1, and for desert areas the size of Arizona this can mean carbonate (c) stores of 396 × 10^7 Mt (Schlesinger 1982).

Many North American desert landscapes, especially those of southeastern California, adjacent Mexico, and scattered localities in the Chihuahuan Desert, are dominated by dunes (Smith 1982). The soils in these areas are composed of sand-size particles of either silica or gypsum. Both types of dune systems support characteristic floras. Dune soils (Regosols) allow for rapid percolation and subsequent storage of soil water (Bowers 1982), and thus these are among the most mesic sites in deserts.

One important soil catena in deserts is represented by the particle size and salinity gradients seen on bajadas. Upper bajadas have a more diverse vegetation, a higher proportion of large soil particle sizes, and lower salinity than the lower portions of bajadas (Solbrig et al. 1977; Phillips and MacMahon 1978). The presence of arroyos or even shallow rills on a bajada can modify vegetation and soil patterns.

Two desert soil surface phenomena should be mentioned. In many areas, the surface of the soil is covered with stones or pebbles of relatively uniform size spaced so closely together that the soil surface is obscured. These areas, termed "desert pavement," may cover a thousand hectares in some areas. Their exact origin isn't clear, but it probably involves the effects of wind removing fine particles on these generally flat surfaces (Fuller 1975).

The stones forming the pavement often are covered with a dark patina opposed to 70% clay, but containing oxides of iron and manganese that give them their dark color. It has been argued that these varnishes have completely abiotic origins (Moore and Elvidge 1982); however, Dorn and Oberlander (1981) argue for the involvement of microorganisms, and there is no reason to believe that these two mechanisms are mutually exclusive.

The second soil surface phenomenon of interest is the presence of hardened soil crusts that are inhabited and generated by cyanobacteria, algae, and lichens. These cryptogamic crusts form most often on clay or silty soils. The crusts stabilize the soil surface and may fix atmospheric nitrogen in low quantities, but significant quantities from the perspective of desert ecosystem nitrogen cycling (Skujins 1984).

Saline soils are common in warm deserts on the flats and valley bottoms, especially around playas. Their significant interactions with vegetation are discussed in a later section.

Soils of desert areas are especially susceptible to damage by human use. Improper irrigation can cause salinity; hiking and the use of off-road vehicles can cause compaction (Eckert et al. 1979; Webb and Wilshire 1983), as can overgrazing (Webb and Stielstra 1979). All of these effects can be reversed only very slowly (Webb and Wilshire 1980), and thus may have long-range effects on the native vegetation and potential land uses.

CLIMATE

By definition, deserts are warm areas with low rainfall and high rates of evapotranspiration. There have been various schemes to classify the world into arid, semiarid, humid zones, and so forth. Often the North American deserts are thought to be semideserts, rather than true deserts, because of their relatively lush vegetation. Figure 8.4 clearly suggests, however, that although many sites in the Chihuahuan Desert border on being semiarid, there are sites within each of our warm deserts that are equal in aridity to those elsewhere in the world. Different results might be obtained if one used a different aridity index, such as one based on water-balance parameters (Oberlander 1979).

Precipitation is low in amount and highly variable from year to year in deserts (Fig. 8.5). For example, Furnace Creek, in Death Valley, California (Mojave Desert), has a long-term mean rainfall of 4.2 cm yr^{-1}. However, between 1912 and 1962, the averages for approximately 10-yr periods showed wet episodes in which rainfall was 6.7 cm yr^{-1} and dry periods with one-third of that (2.1 cm yr^{-1}). In two of those years, 1929 and 1954, there was no rainfall over 12-mo periods (Hunt 1975).

Seasonal rainfall patterns vary significantly among our warm deserts. North American deserts lie in a zone of the earth's surface in which strong seasonal shifts occur in storm tracks because of seasonal heating of the earth's surface and global patterns of winds. In the winter, storms originating in the Pacific Ocean move inland and are pushed across the mountain chains of the Coast Ranges, Sierra Nevada, and Sierra Madre Occidental, causing adiabatic cooling, condensation, and rain. Thus, the areas near the coast of western North America

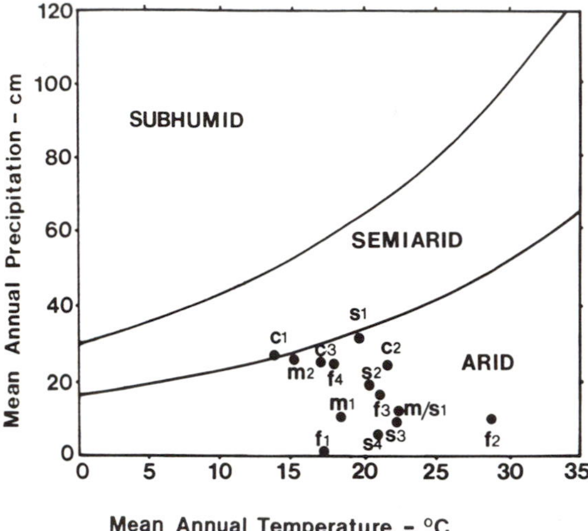

Figure 8.4. Depiction of arid and semiarid moisture-province boundaries as presented by Bailey (1979), with data from some desert localities superimposed. Mojave desert = m1 (Las Vegas, Nevada); m2 (St. George, Utah); m/s1 (Needles, California, a Mojave–Sonoran transition site): Sonoran desert = s1 (Tucson, Arizona); s2 (Phoenix, Arizona), s3 (Yuma, Arizona); s4 (Brawley, California); Chihuahuan desert = c1 (Socorro, New Mexico), c2 (Ojinaga, Chihuahua), c3 (El Paso, Texas); sites on other continents = f1 (Atacama Desert, Antofagasta, Chile), f2 (southern Sahara Desert, Tessalit, Mali), f3 (northern Sahara Desert, Biskra, Algeria), f4 (Karroo Desert, Beaufort West, South Africa). Climagrams for the North American sites are given in Fig. 8.5 or by MacMahon and Wagner (1985). Data for other sites are from Walter (1971). Note that despite the common assertions that North American deserts are not really deserts, their aridity is of a magnitude similar to that for many "true" deserts throughout the world.

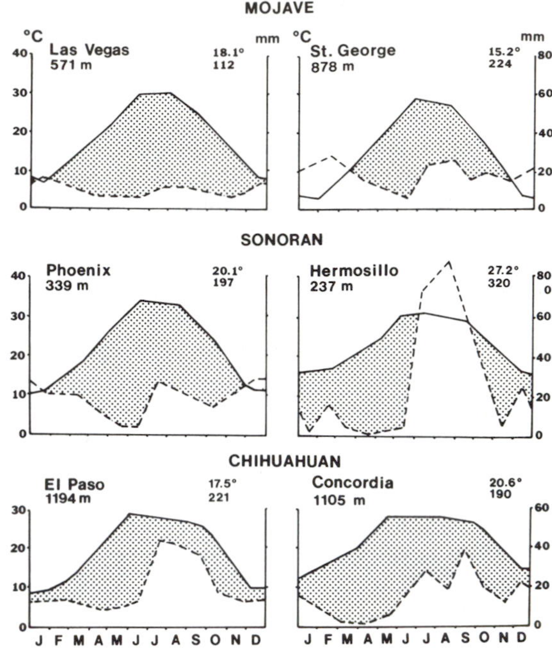

Figure 8.5. Simplified climate diagrams for some weather stations within the North American warm deserts (adapted from Walter and Lieth 1967). The number in the upper right of each station's graph is mean annual temperature (°C); the number below that is the mean annual precipitation (mm). The dashed curve depicts monthly rainfall from January to December. The solid curve depicts mean monthly temperature. Shaded area = months when potential evaporation exceeds precipitation.

receive fall to spring rains (winter precipitation). Storm tracks responsible for winter precipitation move north of the United States in the spring, decreasing the flow of air from the Pacific Ocean. At this time, stronger storm cells originate from the Gulf of Mexico, moving westward and northwestward. This movement causes spring-to-fall rains (summer rainfall). Each of these patterns becomes less pronounced as one moves away from the places where the storms originate. The result is that the Mojave Desert, being close to the Pacific Ocean, gets winter rain, whereas the Chihuahuan Desert farther east is more directly in the path of the summer-rainfall track. The Sonoran Desert is intermediate between the two storm systems, and so it receives biseasonal rainfall whose exact proportions depend on location along the Mojave-Chihuahuan axis (Fig. 8.6).

Summer and winter rains differ not only in seasonality but also in other attributes: Winter rains

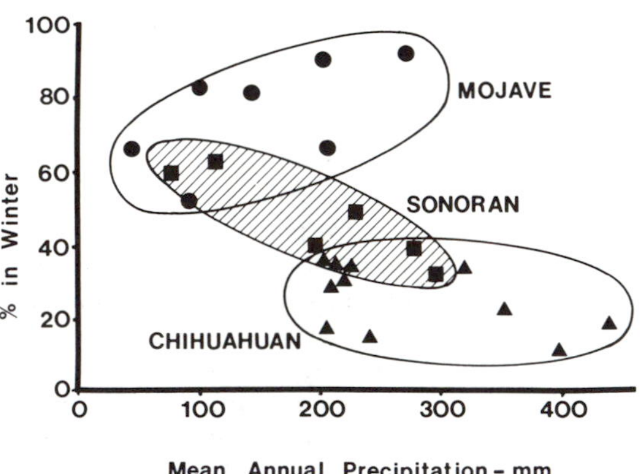

Figure 8.6. Percentage of winter rainfall plotted against mean annual precipitation for a variety of warm-desert sites. Site assignment to a desert type was by knowledge of vegetation at each site (adapted from MacMahon and Wagner 1985).

are of long duration and low intensity, and they cover large areas at a time; summer rains are cyclonic thunderstorms of short duration (minutes to hours) and high intensity, and they are limited in areal extent.

High-intensity, short-duration rainfall can cause surface runoff of water, and so the rain that falls on a given site may not be a good measure of the water percolating to a plant's rooting zone at that site (Schlesinger and Jones 1984), with topography being an important determining factor (Osborn et al. 1980). This lack of correlation between rainfall and the amount of water available for plant growth makes it difficult to use certain indices of plant-growth–rainfall relationships, such as the rain-use efficiency metric (Le Houerou 1984).

Whereas average temperatures in our warm deserts are high (Fig. 8.5), there are differences among the deserts. The Sonoran Desert is generally the warmest. In part, this is because of its low elevation (less than 600 m) as compared with the Mojave Desert (three-quarters of the area is at 600–1200 m) or the Chihuahuan Desert (more than half above 1200 m). In contrast to this generalization, Death Valley in the Mojave Desert has the highest absolute temperatures and the lowest elevations of our warm deserts.

Of equal importance to the high temperatures are the periods in which freezing temperatures persist in excess of 36 hr. These bouts cause significant mortality among many desert species (Hastings and Turner 1965), the results of which are important in some areas, such as the Sonoran Desert (Lowe and Steenbergh 1981). Thus, the composition of vegetation in some areas may be due to unusual occurrences of prolonged freezes, rather than the result of tens of years of "normal" temperature and precipitation.

Evaporation and wind are important to the vegetation in desert areas. Interestingly, water loss from a plot containing *Larrea tridentata* was found to be virtually the same as from a plot of bare soil (Sammis and Gay 1979). Transpiration accounted for only 7% of the total evapotranspiration.

Fog and dew are seldom considered as important climatic variables in North American deserts. It is clear, however, that many desert plants, including mosses, lichens, and vascular plants, can use these two forms of water for hydration. Although conditions necessary for fog or dew development are rare in inland portions of warm deserts, coastal areas, such as some of the Sonoran Desert portions of Baja California, experience both fog and dew, and these can be locally significant in determining the nature of plant communities (Nash et al. 1979).

MOJAVE DESERT

Many scientists have suggested that the Mojave Desert does not represent a discrete vegetation type but rather is a zone of transition or an ecotone. This position emphasizes the Great Basin-like character of the vegetation in the north and the Sonoran Desert-like aspect in the south. Overall, about one-fourth of the plant species of the Mojave Desert are endemics; for annuals, nearly 80% of the 250 or so species are endemics. Because of this large number of Mojave Desert endemics, it seems to me that the vegetation is sufficiently circumscribed to be discussed as an entity in its own right.

In the north, the Mojave Desert is limited by the higher elevations of the Great Basin. If one starts in Las Vegas, Nevada (670 m), in typical Mojave Desert vegetation, and drives northward, one is in the Mojave–Great Basin transition at Beatty (1170 m), and in the Great Basin at Tonopah (1840 m). The transitional vegetation has been well studied in the area of the Nevada Test Site, at Mercury, Nevada (Beatley 1976). Here, *Larrea*-dominated communities give way to those dominated by *Artemisia tridentata* (big sagebrush) and *Atriplex confertifolia* (shadscale). Also typical of these sites is *Coleogyne ramosissima* (blackbrush) (see Chapter 7), *Lycium* species (wolfberries), and *Atriplex (Grayia) spinosa* (hopsage).

This transitional vegetation is intermingled in complex ways with typical Mojave Desert plants. For example, two plots 90 m apart, each having 12% cover and separated in elevation by only 1.5 m, differed in that one had *Larrea* as 2.7% cover and *Lycium shockleyi* at 4.5%, whereas the other had zero *Larrea* cover and 10.1% *Lycium* cover (Beatley 1974). Typical Mojave Desert sites differ from our other warm deserts in that they are dominated by low-growing, often widely spaced perennial shrubs, representing relatively few species, and although cacti are present, they are generally of low stature. Yuccas can be locally common.

Subdivisions of the Mojave Desert vegetation have been proposed in several recent studies. Vasek and Barbour (1977) proposed, and Turner (1982) accepted, a classification including five general vegetation types, termed series: creosote bush, shadscale, saltbush, blackbrush, and Joshua tree. Rowlands and associates (1982) referred to 11 vegetation types. The difference between the two systems is the further subdivision of the vegetation of saline areas by Rowlands. For our purposes, the five-unit system will suffice. The shadscale series was discussed by West in Chapter 7, as was the blackbrush series. The saltbush series, and saline-

*Figure 8.7. General view of a typical Mojave Desert creosote bush flat. The taller shrub is creosote bush (*Larrea tridentata*), and the shorter, lighter bush is white bursage (*Ambrosia dumosa*). Site is on the Arizona-Utah border, Washington County, Utah.*

soil vegetation in general, will be covered in a later section of this chapter. Thus, we are left with two broad vegetation types: the creosote bush and the Joshua tree series.

Creosote Bush Series

The most common association of plants is dominated by *Larrea tridentata* and *Ambrosia dumosa* (white bursage) (Fig. 8.7). Perhaps 70% of the Mojave Desert is covered with these two species as codominants, especially on lower portions of bajadas and valley floors.

Creosote bush is virtually synonymous with the warm deserts, because its distribution coincides closely with their distribution. Even the approximate boundaries between our various warm deserts are roughly indicated by the distribution of chromosome numbers within creosote bush (Mojave, $N=39$; Sonoran, $N = 26$; Chihuahuan, $N = 13$) (Yang 1970). Creosote bush occurs from 73 m below sea level in Death Valley to 1585 m on southern exposures (Hunt 1966). Although it occurs on soils ranging from sand dunes to quite rocky soils, creosote bush is limited to areas that are fairly well aerated, that have low salinity, that receive less than 18 cm of rainfall, and that experience freezing temperatures for no more than 6 d consecutively (Beatley 1976). Because of its wide ecological and geographical distribution, creosote bush is associated with a variety of other species, especially on bajadas and nonsaline flats.

Typical bajadas in parts of Death Valley National Monument show significant vegetation changes from bottom to top (Hunt 1975). On the fine soils of the bajada bottom, one of two saltbush species often forms nearly pure stands. Soils with high percentages of carbonate rocks are dominated by *Atriplex hymenelytra* (desert holly), which may average densities of 120 individuals per hectare, but may reach 550 individuals per hectare. Where there is less carbonate, *Atriplex polycarpa* (cattle spinach), a large species, predominates at about the same density. Farther up the bajada, creosote bush appears and mixes in some places with cattle spinach or desert holly.

Creosote bush densities vary considerably, ranging from 100 or so per hectare to over 1000 per hectare. It is difficult to count individual creosote bushes because they reproduce vegetatively; clones produced in this manner may persist for thousands of years (Vasek et al. 1975b; Vasek 1980).

The middle portions of bajadas are typical *Larrea–Ambrosia dumosa* communities, though varia-

*Figure 8.8. General aspects of three examples of Mojave Desert sites: (a) Joshua tree (*Yucca brevifolia*) visually dominates an upper-elevation site near Wikieup, Mohave County, Arizona; (b) a creosote bush site with some* Yucca schidigera *north of Las Vegas, Clark County, Nevada; (c) a site partially up a bajada, ascending the Spring Mountains, Nye County, Nevada. This site contains 13 shrub species, including members of* Thamnosma, Salazaria, Ceratoides, Ephedra, *and others.*

tions do occur. In some sections of the Mojave Desert, including southern Death Valley, *Encelia farinosa* (brittlebush) may be abundant.

Other shrubs that are locally abundant, associated with the creosote bush and characteristic of the Mojave Desert bajadas, include *Menodora spinescens* (spiny menodora), *Lycium pallidum* or *L. andersonii* (wolfberries), *Ephedra* (Mormon tea), *Krameria parvifolia* (ratany), *Acamptopappus schockleyi* (goldenhead), *Dalea fremontii* (Fremont dalea), and *Psilostrophe cooperi* (yellow paper daisy) (Fig. 8.8c).

Three yuccas commonly occur on middle portions of bajadas: *Yucca brevifolia* (Joshua tree) (Fig. 8.8a), *Y. schidigera* (Mojave yucca) (Fig. 8.8b), and *Y. baccata* (banana yucca). A fourth species, *Y. whipplei* (desert Spanish bayonet), occurs along the western edge of the Mojave Desert.

Many cacti occur on the bajadas, and some species or varieties are Mojave Desert endemics. Benson (1982) listed 23 cacti occurring primarily in the Mojave Desert. Conspicuous species include *Opuntia basilaris* (beavertail), *O. echinocarpa*, and *O. acanthocarpa* (chollas), the many-stemmed *Echinocactus polycephalus*, and *Ferocactus acanthodes* (a barrel).

Where there are rills or small channels, *Hymenoclea salsola* (cheesebush) may dominate, in association with *Cassia armata*, *Ambrosia eriocentra*, *Brickellia incana*, and *Acacia greggii* (catclaw).

On highly calcareous soils, frequently with well-developed pavements, creosote bush often associates with *Atriplex confertifolia*. Such sites often develop petrocalcic layers (caliche) that can inhibit deep penetration of plant roots. These kinds of sites recur over major portions of the Mojave Desert. Locally, the codominant shrubs vary considerably. However, common associates include *Krameria*, *Ephedra*, *Lycium*, and one or another *Yucca*. Beatley (1976) pointed out that *Opuntia ramosissima* is known in the Mojave Desert only from this type of vegetation. She also listed some 40 species of herbs found in this association.

Higher-elevation sites or northern sites, with poorly developed pavements and essentially no caliche layer, support stands of creosote bush mixed with *Lycium andersonii* and *Atriplex (Grayia) spinosa*. These sites often contain Joshua tree (*Yucca brevifolia*) and thus represent a transition to the next series.

Table 8.2. *Density (D,ha^{-1}) and cover (C,%) of perennials at five southern California localities dominated by creosote bush*

Species	Baker (79)[a] (D/C)	Inyokern (86) (D/C)	Barstow (96) D/C)	Lucerne Valley (106) (D/C)	Mojave (126) (D/C)
Acamptopappus sphaerocephalus		348/0.7			
Ambrosia dumosa	404/0.2	2876/8.6	2284/0.5	568/0.5	524/0.9
Atriplex polycarpa	148/0.2			44/0.6	
Atriplex spinifera			932/1.5		
Atriplex (Grayia) spinosa		40/0.2		4/<0.1	
Cassia armata				48/<0.1	
Chrysothamnus paniculatus					272/0.3
Dalea fremontii		308/1.7			
Ephedra nevadensis				24/0.1	
Haplopappus linearifolius				124/0.1	8/<0.1
Larrea tridentata	196/1.3	400/7.8	144/0.9	392/3.0	400/7.8
Lepidium fremontii				4/<0.1	
Lycium andersonii		12/0.1			
Lycium pallidum			36/0.2		
Opuntia echinocarpa				4/<0.1	
Total					
Species	3	6	4	9	4
Density	748	3984	3396	1212	1204
Cover	1.7	19.1	3.1	4.3	9.0

Note: [a] Annual ppt (mm)
Source: Based on Phillips and MacMahon (1981)

Representative stand data for some *Larrea*-dominated sites in the Mojave Desert are presented in Table 8.2.

Joshua Tree Series

As the series name implies, this vegetation type is usually dominated by *Yucca brevifolia*, the tallest nonriparian plant of the Mojave Desert (Fig. 8.8a). This conspicuous species is, in many people's minds, the most characteristic species of the Mojave Desert. Although its mapped distribution (Fig. 8.9) approximates the Mojave Desert boundary, the Joshua tree is actually confined to higher elevations or slightly cooler and moister sites, rather than occurring over the low-elevation, drier sites that predominate in the Mojave Desert. At the upper elevations of this vegetation type, a transition occurs with pinyon-juniper woodland.

The Joshua tree occurred at much lower elevations during the cool periods of the Pleistocene. For

Figure 8.9. *Distribution of Joshua tree, a species whose distribution essentially outlines the Mojave Desert, but it is elevationally restricted to higher sites (adapted from Benson and Darrow (1981).*

example, in Death Valley, where the Joshua tree currently occurs only above about 1700 m, it was present near the valley floor around 19,550 BP (Wells and Woodcock 1985).

Despite its visual dominance on some sites, Joshua tree densities may be as low as 104–125 per hectare, with a cover of 0.20% in a vegetation containing 13 species of perennial shrubs with a total coverage of 10.15% and a density of 7250 plants per hectare (Vasek and Barbour 1977).

The perennials associated with the Joshua tree vary considerably, depending on whether one samples sites at its lower-elevation limit, where it may mix with *Larrea* and *Ambrosia dumosa* (e.g., southern Utah), or sites at middle to upper elevations, where *Ephedra nevadensis, Ceratoides lanata, Lycium, Salazaria mexicana, Thamnosma montana,* or *Coleogyne ramosissima* may be numerical dominants. Additionally, several cacti, both opuntias and barrels (*Ferocactus*), and several perennial grasses (*Hilaria rigida, Muhlenbergia porteri,* and *Stipa* species) occur.

Where the Sonoran and Mojave deserts abut (northeastern Arizona), saguaros, junipers, Joshua trees, and paloverdes may co-occur in stands of

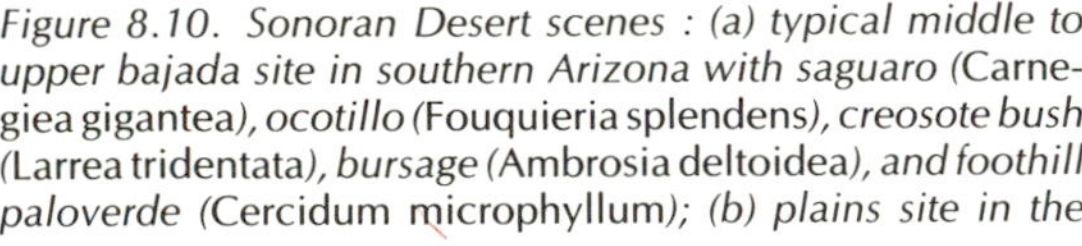

*Figure 8.10. Sonoran Desert scenes : (a) typical middle to upper bajada site in southern Arizona with saguaro (*Carnegiea gigantea*), ocotillo (*Fouquieria splendens*), creosote bush (*Larrea tridentata*), bursage (*Ambrosia deltoidea*), and foothill paloverde (*Cercidum microphyllum*); (b) plains site in the Lower Colorado Valley subdivision, southern Arizona, dominated by creosote bush; (c) an upper-elevation site with a special abundance of saguaros in the Tucson Mountains, Arizona.*

great complexity and beauty. Also characteristic of these transitional sites is crucifixion thorn (*Canotia holacantha*).

SONORAN DESERT

The Sonoran Desert, with its biseasonal rainfall and tropical affinities, contains a complex biota. Numerous plant species representing a variety of life forms (Crosswhite and Crosswhite 1984) often coexist, creating architecturally varied environments (Fig. 8.10a and c). On some soils, however, the mix of species and their physiognomy is virtually the same as in the Mojave Desert (Fig. 8.10b).

This great diversity of associations was subdivided by Shreve (1951) into seven divisions. Brown and Lowe (1980) redefined subdivision boundaries, relegated some areas to a thornscrub series, and settled on six major subdivisions that were accepted by Turner and Brown (1982) and that I follow here (Fig. 8.11). Turner and Brown used data from 162 weather stations to show that major changes in mean annual temperature and precipitation, as well as continentality (the difference between average temperatures of summer and winter), coincide, generally, with vegetation boundaries determined by physiognomy and species composition.

Many Sonoran Desert genera also occur in the Monte Desert of Argentina (Solbrig 1972), and these two deserts are therefore structurally and functionally similar (Orians and Solbrig 1977).

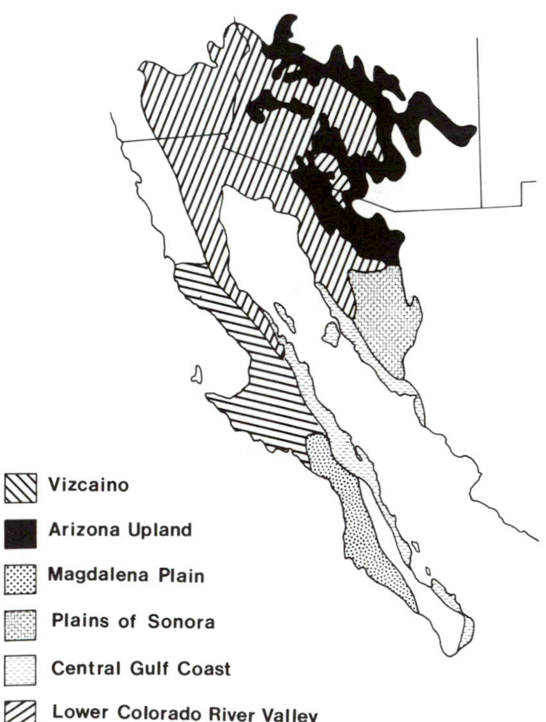

Figure 8.11 Subdivisions of the Sonoran Desert based originally on ideas of Forrest Shreve, but modified most recently by Turner and Brown (1982).

Lower Colorado Valley Subdivision

Because of its relatively high winter and summer temperatures and low annual precipitation, this largest Sonoran Desert subdivision is generally the most xeric. Adding to the stress produced by its low rainfall amounts is the fact that the rainfall frequency is low and its pattern highly irregular, a situation that contrasts with that in the adjacent Arizona Upland subdivision (Ezcurra and Rodrigues 1986). The vegetation in many areas reflects these environmental extremes by its simple species composition and lack of structural complexity. Large areas are dominated by *Ambrosia dumosa* and *Larrea*, with few other perennial shrubs present (Fig. 8.10b, Table 8.3). Throughout the Sonoran Desert, and others also, *Larrea* is strongly associated with calcareous soils that have a caliche layer. Some areas, especially those covered with pavements, support virtually no shrubs unless the landscape is traversed by rills or runnels (Turner and Brown 1982).

Many of North America's most extensive sand-dune systems occur in this subdivision. Adjacent flats with conspicuously sandy soils support stands of perennial grasses (e.g., *Hilaria rigida*) and shrubs such as *Ephedra trifurca, Dalea schottii, D. emoryi, Eriogonum deserticola*, and *Petalonyx thurberi* (Felger 1980).

In areas with significant development of washes, *Cercidium floridum, Dalea spinosa, Sapium biloculare, Condalia lycioides*, and *Prosopis glandulosa*, or any one of some 20 other species, may dominate locally (Crosswhite and Crosswhite 1982).

In some areas of the subdivision, brittlebush (*Encelia farinosa*) is abundant and especially conspicuous because its light gray leaves and showy yellow flowers stand out against the dark volcanic soils it frequently inhabits. Also on volcanic substrates one encounters *Agave deserti, Trixis californica, Hyptis emoryi*, and *Peucephyllum schottii* (Zabriskie 1979).

On bajadas in the subdivision, the simple flora of the plains gives way to a complex that includes *Fouquieria splendens, Cercidium microphyllum, Acacia greggii*, and saguaro (*Carnegiea gigantea*), a mixture also encountered in the Arizona Upland.

The vegetation of bajadas throughout the Sonoran Desert increases in physiognomic and species

Table 8.3. *Density and cover of perennials at four Sonoran Desert sites*

Species	Lower Colorado Valley (D/C)	Arizona Upland (D/C)	Central Gulf Coast (D/C)	Plains of Sonora (D/C)
Shrubs				
Ambrosia ambrosioides		1.01/—		
Ambrosia deltoidea		87.8/0.5		
Ambrosia dumosa	84/0.1	549.7/2.7		
Caesalpinia pumila			104.0/2.5	
Calliandra eriophylla			45.6/0.2	
Coursetia glandulosa			5.3/0.3	
Croton sonorae			104.0/0.4	
Encelia farinosa		316.7/2.0	1067.4/4.9	2882/1.6
Hymenoclea monogyra		151.3/1.1		
Jatropha cuneata			40.3/2.9	
Krameria parvifolia			58.4/0.3	
Larrea tridentata	448/5.5	437.7/6.3	1.1/—	
Lycium andersonii		88.8/1.1	1.1/—	
L. berlandieri			9.5/0.6	
Mimosa laxiflora			37.1/0.4	
Trixis californica		3.0/—		
Other shrubs			54.1/0.7	
Subtrees				
Acacia constricta			82.8/4.6	
Bursera laxiflora			33.0/2.3	1/0.3
B. microphylla			17.0/2.2	
Cericidium microphyllum		26.2/6.7	19.2/3.7	24/7
Eysenhardtia orthocarpa			4.2/0.1	
Fouquieria macdougalii			4.2/0.3	17/0.7
Forchammeria watsoni			1.1/—	
Guaiacum coulteri				5/0.1
Olneya tesota		1.0/0.1	17.0/2.4	65/9
Pithecollobium mexicanum			6.8/0.2	
Prosopis glandulosa				
Large cacti				
Carnegiea gigantea		11.0/—		
Lemaireocereus thurberi			0.1/—	
Lophocereus schottii				8/0.5
Pachycereus pringlei				1/0.1
Small cacti				
Opuntia acanthocarpa		27.2/0.2		
Opuntia fulgida			20.2/—	
Opuntia ramosissma	144/0.5			
Totals				
Species	3	12	23	8
Density	676	1701	1734	3003
Cover	6.1	20.7	29.0	19.3

Note: Density (D) = number per hectare; cover (C) = percentage. A dash indicates cover less than 0.1%. Sites: Lower Colorado Valley, near Gila Band, Arizona; Arizona Upland, near Maricopa, Arizona; Central Gulf Coast, near Guaymas, Sonora, Mexico; Plains of Sonora, Mexico, 25 km north of Hermosillo, Sonora, Mexico. MacMahon, unpub. data.

complexity as one ascends them (Fig. 8.12). These plant associations are composed of nonrandom sets of species with regard to form (Bowers and Lowe 1986). Patterns of compositional change correlate well with soil physical properties and the related soil water relations (Phillips and MacMahon 1978).

In flat areas associated with major rivers, where periodic floods historically have taken place, soils

Figure 8.12. A stylized cross section of vegetation change along a Sonoran Desert bajada in the Arizona Upland subdivision. From right to left (ascending the bajada), species from Larrea-Ambrosia *flats become mixed with subtrees, usually* Cercidium, Olneya, *and* Prosopis, *with ocotillo* (Fouquieria), *and with a variety of prickly pear and cholla cacti* (Opuntia) *and columnar cacti such as saguaro* (Carnegiea gigantea). *Middle to upper bajada sites, with subtrees and columnar cacti, typify the Sonoran Desert. Lower bajadas look, physiognomically, like vast areas of the Mojave Desert or Chihuahuan Desert.*

are often slightly to moderately saline. Where this obtains, *Atriplex polycarpa* may be an obvious dominant, often forming nearly pure stands. Other species that do well in such sites include one or another species of *Isocoma*, a *Lycium* (often *L. fremontii*), and another *Atriplex* (usually either *A. canescens* or *A. lentiformis*). In some areas, especially on sandy soils, screwbean mesquite (*Prosopis pubescens*) is abundant, and it may occur with *Tessaria sericea*, as along the Gila River of Arizona (Rea 1983). Many of these areas, where salinity is not extreme, currently are used for agricultural purposes and are quite productive, such as California's Coachella Valley.

Arizona Upland Subdivision

The Arizona Upland subdivision is remarkable for its diversity of species and life forms. Many areas are visually dominated by the presence of subtrees and tall cacti such as saguaros. In fact, the saguaro-dominated landscape is synonymous in the minds of most North Americans with the word "desert." The biology of saguaro has been extensively documented (Steenbergh and Lowe 1976, 1977, 1983).

The rainfall totals between 100 and 300 mm and has a biseasonal distribution of about equal amounts. This pattern supports an annual flora in each period of mesic conditions and adds to the species richness of sites in the subdivision.

Valleys and lower portions of bajadas are dominated by creosote bush, where the density in pure stands may reach 3700 per hectare and represent 29% cover. Although white bursage codominates in many areas, *Ambrosia deltoidea* is a frequent dominant on coarse soils (Niering and Lowe 1984). Acacias and several chollas, such as *Opuntia fulgida* (Fig. 8.13a), are locally common.

On the slopes, saguaros are common (Fig. 8.10c), especially where larger shrubs and subtrees occur. Young saguaros (and other cacti, large and small) frequently establish beneath the canopy of a "nurse" plant that perhaps affords some advantage in avoiding heat stress and the effects of frost, predators, and so forth (Fig. 8.13d) (Nobel 1980; Vandermeer 1980). As the saguaro grows, it may impede its nurse and cause an increase in stem dieback, at least for one common nurse, *Cercidium microphyllum* (McAuliffe 1984). This case of facilitation for the cactus, followed by competition, is but one example of a complex of biotic interactions that have been documented within this subdivision. Yeaton and associates (1977) detailed a complex competitive milieu involving five species in Organ Pipe Cactus National Monument. In their study, saguaro did not compete with any of the other species (*Larrea, Ambrosia deltoidea, Opuntia fulgida, Fouquieria splendens*), whereas *Larrea* competed with all species except saguaro.

As one travels up the slopes of this subdivision, subtrees increase in abundance. Species include *Cercidium microphyllum, Fouquieria splendens, Olneya tesota, Celtis pallida*, and *Condalia warnockii*. In the northern localities of this subdivision, a subtree characteristic of transitional sites, *Canotia holacantha*, is abundant. In southern sites, *Bursera microphylla* and two large cacti, *Lophocereus schottii* and *Lemaireocereus thurberi* (Fig. 8.13c), are locally conspicuous.

Shrubs of the middle to upper bajadas include *Calliandra eriophylla, Zinnia grandiflora, Psilostrophe cooperi, Jatropha cardiophylla, Simmondsia chinensis, Encelia farinosa, Krameria parvifolia*, and a host of others.

A variety of small to medium-size cacti also respond to the soil gradient of bajadas (Yeaton and Cody 1979).

It should be noted that three subtrees (*Cercidium floridum, C. microphyllum*, and *Olneya tesota*) and the saguaro (*Carnegiea gigantea*) that so typify this subdivision also occur prominently in several other subdivisions of the Sonoran Desert (Fig. 8.14). Especially widespread are *C. microphyllum* and *O. tesota* (Fig. 8. 14c, and d) two species that occur in nearly every subdivision except on the Pacific Ocean side of Baja California.

Finally, a long-term study (72 yr) within perma-

*Figure 8.13. Some conspicuous plant forms found in the Sonoran Desert: (a) cholla (*Opuntia fulgida*); (b) ocotillo (*Fouquieria splendens*); (c) columnar cactus, such as organ pipe cactus (*Lemaireocereus thurberi*); (d) subtrees, such as foothill paloverde (*Cercidium microphyllum*). Note that in part d, young saguaros are using a paloverde as a nurse plant.*

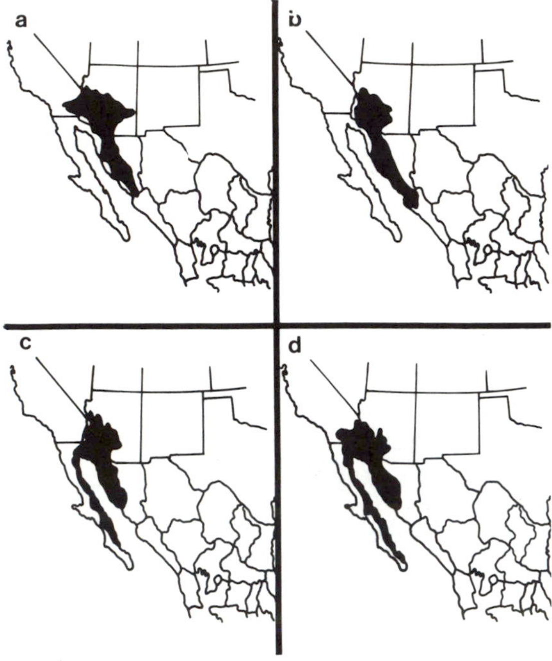

*Figure 8.14. Distributions of four widely ranging Sonoran Desert dominants that impact this desert's unique physiognomy: (a) blue paloverde (*Cercidium floridum*); (b) saguaro (*Carnegiea gigantea*); (c) foothil paloverde (*Cercidium microphyllum*); (d) ironwood (*Olneya tesota*). Maps are based on data from Benson and Darrow (1981) and Hastings and associates (1972).*

nent plots in this subdivision (Tumamock Hill, near Tucson, Arizona) showed no consistent directional changes in vegetation composition despite large fluctuations in cover and density of species. The relative covers by most dominants remained about the same, albeit absolutely varying with rainfall regimes (Goldberg and Turner 1986). Other sites, however, especially disturbed sites, may change directionally (Hastings and Turner 1965).

Central Gulf Coast Subdivision

This subdivision is split into two portions on either side of the Gulf of California. The mainland portion in Sonora consists of a narrow band of vegetation ranging from 27° N to 30° N latitude. A similar band exists in Baja California, extending nearly as far north, but reaching the tip of Baja at 23° N.

The climate patterns, especially the seasonal distribution and reliability of rainfall, vary greatly along this rather large north-south gradient (Humphrey 1974; Turner and Brown 1982).

The vegetation of the area is dominated by subtrees and tall cacti, with some medium-size shrubs. Conspicuously absent is the layer of small shrubs (e.g., one of the *Ambrosia* species) that is so obvious in the Sonoran Desert in the United States. The result is that the vegetation has a sparse, coarse appearance. Soils are often rocky.

The composition of the vegetation on the two sides of the gulf differ in that the mainland portion does not contain all of the conspicuous dominants found on the peninsula side. Some species that do occur on the mainland have very limited distribution. The highly photogenic boojum (*Fouquieria columnaris*) is such a species. In Baja, boojum spans a north-south distance of 400 km, whereas in Sonora it is confined to a 50-km section of the Sierra Bacha, and there it is confined to within a few kilometers of the coast (Humphrey 1974), perhaps because of humidity (Humphrey and Marx 1980).

Shreve and Wiggins (1964) referred to this subdivision as the *Bursera-Jatropha* region. This designation indicates the dominance by *Bursera microphylla, B. hindsiana, Jatropha cuneata,* and *J. cinerea,* in association with the more widely distributed *Olneya tesota, Cercidium floridum, Fouquieria splendens,* and teddy bear cholla (*Opuntia bigelovii*). Creosote bush is much less prominent than in the Arizona Upland subdivision. Close to shores, *Frankenia palmeri* mixes with *Atriplex, Lycium, Suaeda, Encelia,* and *Ambrosia* to form an association with a physiognomy of very low form that extends north into the Lower Colorado Valley subdivision and into the Vizcaino subdivision in Baja.

On bajadas and inland, the *Bursera-Jatropha* admixture is punctuated by a variety of species, including *Opuntia cholla, Lysiloma candida, Fouquieria columnaris, Ficus palmeri,* one of several *Ambrosia* species, *Viscainoa geniculata, Solanum hindsianum, Hyptis emoryi,* or *Justica californica* (Humphrey 1974: Bratz, 1976; Felger and Lowe 1976; Stromberg and Krischan 1983).

Turner and Brown (1982) recognized three "series" within this subdivision. These include a *Bursera-Pachycereus pringlei* grouping on deep granite soils and an ocotillo–*Jatropha*–creosote bush series on bajadas. Both of these occur on both sides of the Gulf of California. A third series, cactus-mesquite-saltbush, occurs on the coastal plain between Empalme and Potam, Sonora. This grouping contains five columnar cacti (saguaro, organ pipe, senita, *Pachycereus pecten-aboriginum,* and *Lemaireocereus alamosensis*) in association with burseras, jatrophas, mesquites, or paloverdes.

Plains of Sonora Subdivision

The Plains of Sonora subdivision is the smallest and least diversified region of the Sonoran Desert.

"Over most of the area the impression given by the vegetation is that of a very open forest of small, low-branching trees, with irregularly placed colonies of shrubs which are not tall enough to impair the view, and with large but very widely spaced columnar cacti" (Shreve and Wiggins 1964).

Ironwood (*Olneya tesota*) is a very prominent member of this association, as are paloverdes (*Cercidium floridum, C. microphyllum, C. praecox*), *Parkinsonia aculeata, Atamisquea emarginata, Fouquieria splendens*, and mesquite (*Prosopis*). Other subtrees reach their northern terminus in this association, especially along washes. Included in this mixture are *Forchammeria watsoni, Jatropha cordata, Bursera laxiflora, Guaiacum coulteri, Fouquieria macdougalii, Piscidia mollus, Ipomoea arborescens*, and *Ceiba acuminata* (Table 8.3).

Brittlebush (*Encelia*) is widespread and often common in some places, as are several columnar cacti. Several chollas (*Opuntia fulgida, O. arbuscula, O. leptocaulis*) are locally abundant.

Common shrubs include *Caesalpinia pumila, Coursetia glandulosa, Calliandra, Eysenhardtia*, and *Mimosa laxiflora*, among others.

Locally, *Cordia parvifolia, Croton sonorae, Jacobinia ovata, Tecoma stans*, and *Zizyphus obtusifolia* are conspicuous.

Magdalena Plain Subdivision

The Magdalena Plain of southern Baja California is the southernmost section of the Sonoran Desert, reaching only north to 27° N latitude. Rainfall is low (generally less than 200 mm). Turner and Brown (1982) emphasized the affinity between some Magdalenan sites and thornscrub or San Lucan deciduous shrubs. The presence of *Larrea* as the dominant in many sites causes me to include it as a Sonoran Desert complex, albeit somewhat transient in its nature at some sites.

Several conspicuous cacti, including *Pachycereus pringlei, Lophocereus schottii, Lemaireocereus thurberi*, and *Machaerocereus gummosus*, occur commonly.

Paloverdes (*Cercidium praecox, C. microphyllum*, or hybrids) occur in combination with *Jatropha cuneata, J. cinerea, Bursera microphylla, Fouquieria peninsularis, Lycium brevipes, Fagonia californica, Encelia farinosa*, and *Krameria parvifolia*.

All of these and *Lysiloma candida* give this area its characteristic appearance of scattered subtrees within a sparse shrub matrix. This rather stark visage contrasts with that of the adjacent Vizcaino, with its architecture accented by *Fouquieria columnaris* and *Pachycormus discolor*. Also absent are most yuccas and agaves, with the stark exception of *Yucca valida*, especially in the south.

A cholla, *Opuntia cholla*, occurs throughout the subdivision, and it is sometimes associated with only one or two other perennial species on fine soils of volcanic origin.

Vizcaino Subdivision

The Vizcaino subdivison lies to the west of the Lower Colorado Valley subdivision in the north and the Central Gulf Coast subdivision to the south. Although it extends inland, it is generally open toward the Pacific Ocean side and thus receives the cooling effect of westerly breezes and the low-vapor-pressure deficits of moist air and fog, despite a generally low mean annual rainfall of about 100 mm (Turner and Brown 1982).

These coastal–inland gradients for temperature and humidity influence one of the visually conspicuous aspects of the Vizcaino: the lichen flora. Going from the coast to 70 km inland (at four sites), lichen species density dropped (25, 21, 9, and 2 species per site), as did cover (15.2%, 6.3%, 0.9%, and 0.1%). This was the reverse of the pattern for the higher-plant species richness (12, 9, 17, and 16 species per site) and cover (38.9%, 17.9%, 45.4%, and 45.1%), which increased inland. In general, fruticose species dominated the coasts, whereas foliose and crustose species dominated inland areas (Nash et al. 1979).

Similar patterns for richness and abundance are observed for all the warm deserts of North America as one moves from the ocean, inland (Nash and Moser 1982). Differences in lichen biomass between ocean and inland sites may reach 5000-fold.

The higher plants of the Vizcaino subdivision present a very complex pattern of species richness and physiognomy. Boojum dominates many areas visually because of its height. Locally it may reach coverages of up to 2% and densities close to 200 per hectare. Its relative role and the general pattern of Vizcaino vegetation can be seen by summarizing data from two "series" presented by Turner and Brown (1982) (Table 8.4).

Shreve and Wiggins (1964) termed this subdivision the *Agave-Franseria* (*Ambrosia*) region, in deference to the dominance by these two genera. On gentle slopes and loams, *Ambrosia chenopodifolia* predominates, but is replaced on clays by *Ambrosia camphorata*, though the two species co-occur. Maguey (*Agave shawii*) can be quite abundant, especially in the north and along the coasts. Also, in inland areas, one finds *Viguiera laciniata, Simmondsia, Eriogonum fasciculatum, Opuntia prolifera, Yucca valida*, and some *Larrea, Agave deserti*, and several large cacti.

In areas of the north, a stark vegetation of *Am-*

Table 8.4. *Density and cover of perennials at two Vizcaino subdivison sites.*

Species	Catavina (D/C)	Punta Prieta (D/C)
Agave cerulata	156/0.3	1580/5.7
Ambrosia chenopodifolia	972/3.0	182/0.7
Ambrosia magdalenae	—	382/3.6
Atriplex polycarpa	6/0.1	233/1.2
Encelia californica	204/0.7	—
Encelia farinosa	—	449/0.5
Eriogonum fasciculatum	2864/0.9	—
Euphorbia misera	—	151/1.4
Fagonia californica	228/0.1	1806/0.7
Fouquieria columnaris	154/1.3	63/1.1
Fouquieria splendens	215/—	13/0.5
Krameria grayi	196/0.8	—
Larrea tridentata	108/1.1	45/1.0
Pachycormus discolor	22/0.7	—
Pedilanthus macrocarpus	—	158/0.6
Prosopis juliflora	7/0.4	383/1.8
Simmondsia chinensis	107/0.5	7/0.1
Stenocereus (Machaerocereus) gummosus	—	366/3.3
Viguiera laciniata	2447/9.0	184/1.8
Totals		
Species	36	37
Density	9594	7268
Cover	21.6	27.8

Note: Density (D) = number per hectare; cover (C) = percentage. A dash indicates cover less than 0.1%. Sites are a ragged-leaf goldeneye–boojum series near Catavina and an agave-boojum series near Punta Prieta, both in Baja California del Norte, Mexico. Totals of columns do not equal summary totals because species of low coverage or density were not included in this table.
Source: Data from Turner and Brown (1982).

brosia and *Opuntia* persists, whereas nearby areas, on granite, support complex groups such as the ragged-leaf goldeneye–boojum admixture described by Table 8.4.

South of Catavina, valleys contain creosote bush and *Atriplex polycarpa*, and the slopes include *Lycium californicum*, *Fouquieria splendens*, *Yucca whipplei*, and *Ephedra*. Chollas are noticeably uncommon, as are columnar cacti, but barrels (*Ferocactus*) are prominent.

Boojum and *Pachycormus* increase their importance southward in the subdivision. On some sites, 75% of the individuals are *Pachycormus*, associated with ocotillo and elephant trees (*Bursera*).

The Vizcaino plain, below 100 m, contains many of the species in Table 8.3, but also one encounters *Fouquieria peninsularis*, *Opuntia calmalliana*, and *Triteliopsis palmeri*.

CHIHUAHUAN DESERT

The general vegetation of the Chihuahuan Desert (Fig. 8.15) has been studied less well than that of any other North American desert. In stark contrast to this dearth of published broad-scale information is a series of detailed studies of the functioning of desert communities, plant and animal components, that have been conducted by Robert Chew on the Arizona–New Mexico border near Portal, Arizona, and by a large group of workers associated with Walter Whitford working on the Jornada Experimental Range, near Las Cruces, New Mexico. When the long-awaited book on the flora of the Chihuahuan Desert area is completed, we may expect to see an increase in general vegetation surveys, especially in the Mexican portion.

The Chihuahuan Desert is elevationally high, with many sites in the Mexican basin being above 1000 m, and up to 2000 m in the south. Most sites are between 1100 and 1500 m. Its lowest elevations, those along the Rio Grande, are near 400 m. Much of the landscape is dominated by limestones, though gypsum and igneous rocks are the parent materials in some areas. Gypsic "sand" dunes, a characteristic feature of the Chihuahuan Desert, contain a host of endemic plant species.

The Chihuahuan Desert is cooler, partly because of high elevations, and has more rainfall than our other warm deserts: The yearly mean temperature for a suite of stations across the whole desert is 18.6°C (14–23°C), and the average precipitation is 235 mm (150–400 mm) (Schmidt 1986). The cool, moist aspect of the desert, coupled with the limestone substrates, probably is responsible for a significant grass component in this desert as compared with other warm deserts.

Interestingly, though the area is relatively unstudied, generally similar boundaries for the desert have been recognized whether based on climate (Schmidt 1979), reptile and amphibian distributions (Morafka 1977), or vegetation (Henrickson and Straw 1976; Brown 1982b). Morafka (1977) subdivided the Chihuahuan Desert into three regions. The northernmost region, the Trans-Pecos region, encompasses about 40% of the desert and includes all of the sections in the United States and more than half of the desert areas of Chihuahua. The middle region, the Mapimian, includes part of eastern Chihuahua, Coahuila, and part of Durango. This area is dominated by basin and limestone range topography and contains many playas. The third and most southern region, the Saladan, includes Zacatecas and San Luis Potosi and small parts of other states. Elevational ranges here are

*Figure 8.15. Chihuahuan Desert scene in the Big Bend National Park, Texas. Leaf succulents, especially lechuguilla (*Agave lechuguilla*) and a yucca (*Yucca torreyi*) are apparent. Creosote bush is the dominant shrub. In the far background one can see* Dasylirion leiophyllum*. Grasses include fluff grass (*Erioneuron pulchellum*) and* Muhlenbergia porteri*, among others.*

extreme, with valley floors near 500 m and mountain peaks, both limestone and igneous, exceeding 3000 m.

Henrickson and Johnston (1986) accepted these three divisions based on the flora and outlined a system of plant community delineation. Their system includes eight primary subdivisions of desert scrub and woodlands: Chihuahuan Desert Scrub, Lechuguilla Scrub, *Yucca* Woodland, *Prosopis-Atriplex* Scrub, Alkali Scrub, Gypsophilous Scrub, Cactus Scrub, and Riparian Woodland. Each of these subdivisions is a conspicuous floristic unit. On a more local scale of classification, the Chihuahuan Desert Scrub is subdivided into five phases: *Larrea* Scrub, Mixed Desert Scrub, Sandy Arroyo Scrub, Canyon Scrub, and Sand Dune Scrub. Obviously, some of these community types and phases are part of what I have termed azonal vegetation (e.g., the Riparian Woodland community type and the Sand Dune Scrub phase of the Chihuahuan Desert Scrub). However, the general classification is useful and will be followed here to some extent.

The five phases of the Chihuahuan Desert Scrub cover about 70% of the entire area of the Chihuahuan Desert. The *Larrea* Scrub phase covers 40% of the area. This phase varies in its composition, but over vast areas, 40%–80% of the plant cover can be composed of *Larrea* or *Flourensia cernua* or a mixture of the two (Fig. 8.16). Table 8.5 shows three sites of this type. Often mixed in with the two dominants are *Parthenium incanum, Jatropha dioica, Koeberlinia spinosa*, a *Lycium* species, perhaps an acacia, often *A. neovernicosa, Mortonia scabrella*, a *Dalea* (e.g., *D. formosa*), a *Krameria*, and an *Ephedra*.

In the Saladan region, a *Yucca filifera* woodland may develop. Some individuals of this species may reach nearly 15 m in height. In the north, some sites contain smaller species of yucca, such as *Y. torreyi* or *Y. elata* (soaptree, Fig. 8.17a, p.252). The latter is most abundant in desert grasslands, but occurs commonly and conspicuously in "true" desert on mesic sites. It tends to be clumped in its dispersion because of its vegetative reproduction and because of influences of cattle and fire (Smith and Ludwig 1978).

Ascending the slopes of bajadas, one encounters an enrichment of shrub species and a greater sharing of dominance. On those sites which become the Mixed Desert Scrub phase, cacti (Fig. 8.17b), ocotillo (*Fouquieria splendens*), and lechu-

Figure 8.16. Rolling limestone hills of the Big Bend area in Texas: (a) Flats are dominated by Larrea *and* Flourensia. *Upper slopes contain more dominants. (b) Monotony of a creosote bush flat with some lechuguilla.*

guilla (*Agave lechu guilla*) (Fig. 8.17c) become common. Ocotillo and lechuguilla occur over most of the Chihuahuan Desert and are among the best Chihuahuan Desert "indicators" (Fig. 8.18, p. 252). Also locally common are allthorn (*Koeberlinia spinosa*) and one or another zinnia (*Zinnia*). Sites where the Mixed Desert Scrub phase might be expected are often dominated by dense stands of

Table 8.5. *Density and cover of perennials for three sites in the United States portion of the northern Chihuahua Desert*

Species	*Larrea* (D/C)	*Flourensia* (D/C)	Jornada (D/C)
Acacia greggii		230/2.0	
Flourensia cernua	550/1.1	1460/3.8	196/1.0
Larrea tridentata	4460/17.4		4844/23.7
Opuntia phaeacantha	130/—	600/1.3	86/0.1
Opuntia spinosior		110/—	
Parthenium incanum	660/1.6	300/—	32/0.2
Prosopis glandulosa			27/1.1
Xanthocephalum microcephalum		1250/—	
Xanthocephalum sarothrae			261/0.7
Yucca baccata			33/0.2
Yucca elata			123/0.4
Others	340/0.6	270/1.6	72/0.4
Totals (without grasses)			
Species	4+	6+	8+
Density	6140	4220	5674
Cover	20.7	8.7+	27.8

Note: Density (D) = number per hectare; cover (C) = percentage. A dash indicates cover less than 0.1%. Sites are one *Larrea*- and one *Flourensia*-dominated site from near Portal, Arizona (Chew and Chew 1965) and the well-studied Jornada Bajada, near Las Cruces, New Mexico (Whitford 1973).

lechuguilla and are differentiated by Henrickson and Johnston as the Lechuguilla Scrub. This community type contains species from surrounding communities and is recognized only by the relative importance of its members.

The patchwork of limestone and igneous rocks obviously influences local vegetation. On slopes, desert scrub communities prevail on limestone at elevations that support grassland on igneous rocks (Wentworth 1981; Aide and Van Auken 1985). Similar relationships have been described for Sonoran Desert mountains (Whittaker and Niering 1968).

On limestone, *Agave lechuguilla* and *Fouquieria* are common, as are *Hechtia scariosa* and *Leucophyllum* species (e.g., *L. frutescens*). *Euphorbia antisyphylitica*, a species that is economically important for its wax, is common on, but not confined to, limestone. Also occurring with these species are dogweeds (*Dyssodia* species), *Condalia* (one of four or so species), and *Viguiera stenoloba*. At ground level, the holly-like leaves of *Perezia nana* are often abundant (Rzedowski 1966, 1978). One species of particular interest on the limestone slopes, and less commonly on the plains, is guayule (*Parthenium argentatum*), a locally common, widespread species whose rubber is nearly equal in quality to that of *Hevea brasiliensis* (Miller 1986).

Sites with igneous substrates, in Mexico contain many *Opuntia* species, large cacti such as *Myrtillocactus geometrizans*, and *Yucca carnerosana*. In their best development, such sites form the arborescent Cactus Scrub subdivision of Henrickson and Johnston. Two areas of this vegetation type occur, both in the Saladan region. In the southern Saladan, forests of these species occur. Also present are *Acacia schaffneri*, *Agave filifera*, *Opuntia robusta*, *O. leucotricha*, and *O. streptacantha*. Often these species form mosaics of vegetated areas separated from each other by open spaces. Within these mosaics there are three concentric microhabitats occupied by different plants. In the middle is a legume (e.g., *Acacia schaffneri* or *Mimosa biuncifera*), then a zone containing an erect platyopuntia (*Opuntia cantabrigiensis* or *O. streptacantha*), and an outer zone of decumbent platyoptuntias (*O. rastrera* or *O. robusta*) near the boundary. Also associated with these areas, but not in such a patterned manner, are *Agave salmiana*, *Jatropha dioica*, and *Mimosa biuncifera* (Yeaton and Manzanares 1986).

Grasses are obvious in most areas of the Chihuahuan Desert. Large swale areas can be covered by tobosa (*Hilaria mutica*). Genera such as *Sporobolus*, *Muhlenbergia*, and *Bouteloua* are often common. Bush muhly (*Muhlenbergia porteri*) frequently occurs growing among the branches of creosote bush (Welsh and Beck 1976). In sandy areas, *Oryzopsis hymenoides* is common and conspicuous. Throughout many areas, even in creosote bush flats, *Erioneuron pulchellum* occurs, sometimes with species such as *Bouteloua eriopoda* or *Sporobolus airoides*.

One community type that, in my mind, could be

*Figure 8.17. Three conspicuous plants of the Chihuahuan Desert: (a) soaptree (*Yucca elata*), a species of grassland-desert transitions or on mesic sites in the desert; (b)* Opuntia phaeacantha, *a widespread, common prickly pear cactus; (c) lechuguilla (*Agave lechuguilla*), a Chihuahuan Desert "indicator" species.*

considered a transition to grasslands is the *Yucca* Woodland. In Mexico, some areas just above the Chihuahuan Desert Scrub are dominated by large yuccas and sotols (*Dasylirion*). Especially prominent are *Yucca carnerosana, Y. faxoniana, Y. torreyi, Y. filifera*, and either *Dasylirion leiophyllum* or *D. texanum*. In the northern portions of the desert, the same genera, or even species (e.g., *D. leiophyllum*), form a transition; however, it usually does not take on a dense, woodland appearance.

Plains areas with fine soils are dominated by one of three community types. On gypsum, a unique Gypsophilous Scrub develops. This type will be treated in the later section on azonal vegetation. On alluvium with low salinity, vast stands of *Prosopis glandulosa* or, in the Saladan region, *P. laevigata* dominate. These are joined by *Atriplex canescens, Lycium* species, and *Ziziphus*, often *Z. obtusifolia*. In areas of southeastern New Mexico, this habitat type also includes *Microrhamnus ericoides* and *Xanthocephalum sarothrae* (Secor et al. 1983). As salinity increases, an Alkali Scrub community develops. *Atriplex* species (*A. acanthocarpa, A. canescens, A. obovata*) mix with *Allenrolfea occidentalis, Suaeda* species, *Sesuvium verrucosum*, and salt grass (*Distichlis spicata*).

In the Chihuahuan Desert, many species of cacti prevail. *Opuntia phaeacantha* can occur in dense stands over large areas, as can *Echinocactus horizonthalonius*. Many small to medium-size species occur in complex mixes and represent such a great variety

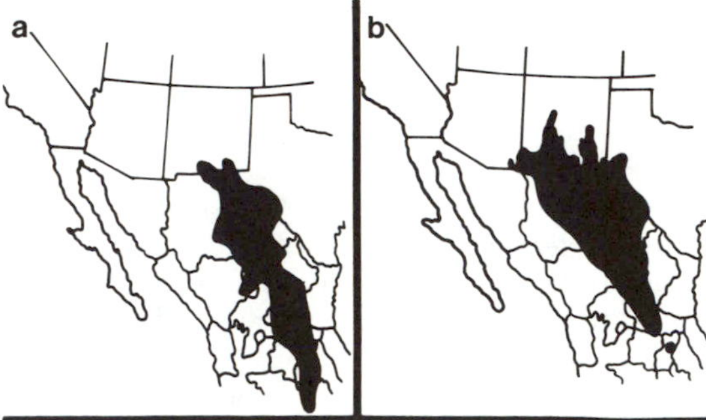

*Figure 8.18. Distributions of two Chihuahuan Desert indicator species: (a) lechuguilla (*Agave lechuguilla*) (adapted from Gentry 1982); (b) tarbush (*Flourensia cernua*) (adapted from Dillon 1984).*

of forms that the Chihuahuan Desert is a cactophile's paradise. Some species (e.g., *Opuntia leptocaulis*) have interesting cyclical replacement relationships with shrubs such as creosote bush (Yeaton 1978).

Despite the profusion of cacti in the Chihuahuan Desert, the leaf succulents dominate the visual landscape and give the Chihuahuan Desert its special appearance, in contrast to the Sonoran Desert subtree–columnar cactus profile or the low, scattered aspect of the Mojave Desert.

AZONAL VEGETATION

Some habitats occur on similar sites across all of our deserts; that is, they are not confined to only one geographical zone. The plant species occupying these sites, regardless of the desert involved, are often similar or even the same species. Because of this similarity, I treat some of these vegetation types separately here.

Sand Dunes

Sand dunes occur in all of our deserts, but are especially apparent in southern California and southwestern Arizona (Bowers 1984). Despite their commonness, the conditions under which dunes form and the factors that control their size and spacing are not well understood (Lancaster 1984). Sand movement characterizes most dunes. The degree of this movement can alter the vegetation significantly (compare Fig. 8.19a and c). When sand moves, plants can be covered in zones of accumulation, or uncovered in zones of deflation. Many plants that grow in dunes show rapid growth as an adaptation to sand movement (Bowers 1982).

Dune vegetation is often formed of a mix of widely distributed species, species occurring in a limited geographical area and those endemic to a particular dune system. Widely distributed species include sandy-soil specialists such as sand sage (*Artemisia filifolia*), a species occurring from the Mojave Desert through the Sonoran Desert and Chihuahuan Desert as far south as Coahuila, Mexico, and species of more catholic substrate affinities, such as *Larrea* and *Prosopis glandulosa* (Fig. 8.19b), both species of wide occurrence. Species of more constrained geographical occurrence, such as those limited to one desert type, are exemplified by species that are essentially restricted to parts of the Sonoran Desert, such as *Ammobroma sonorae, Tiquilia palmeri, Eriogonum deserticola*, and *Hesperocallis undulata*. At a finer level of geographical differentiation, various dunes have their share of endemics, usually ranging from 5% to 15% of the species in that system (Bowers 1982). A good example is the grass *Swallenia alexandrae*, which occurs only in the Eureka Valley dunes in California (Pavlik 1980).

Dune floras seem to break into eastern and western groups, with little overlap of species. These two floras correlate, roughly, with winter-versus summer-rainfall regimes (Bowers 1984).

One specialized "dune" system is formed where *Prosopis glandulosa* occurs in sandy areas. The mesquite takes on a prostrate, coppice form that traps sand, whereas deflation occurs in the poorly vegetated interdunal areas (Fig. 8.19b). Soils of the dunes and the interdunes are virtually identical (Hennessy et al. 1985). These areas probably developed in the last 100 yr in response to overgrazing, through the cause–effect relationships for the dune origins are complex (Wright 1982).

All dune systems support large numbers of showy, annual herbs during rainy periods. Especially conspicuous are genera such as *Euphorbia, Pectis, Boerhavia, Allionia*, and *Oenothera*.

Gypsum Soils

Some dune areas of the warm deserts, especially in the Chihuahuan Desert, are composed of gypsum (hydrous calcium sulfate) (Fig. 8.20a) rather than silica sands. These areas contain several widely distributed species such as *Yucca elata*, Indian rice grass (*Oryzopsis hymenoides*), *Atriplex canescens, Hilaria mutica, Rhus trilobata*, and even an occasional cottonwood (*Populus*) (Shields 1953). There is a significant degree of endemism among members of gypsum-soil plant communities. Some 70 species are confined to Chihuahuan Desert gypsums (Henrickson and Johnston 1986). These include some endemic genera (*Dicranocarpus, Marshalljohnstonia, Strotheria*) and others mainly associated with gypsum (*Selinocarpus, Nerisyrenia, Sartwellia*, and *Pseudoclappia*). Additionally, many widespread genera contain species more or less endemic to gypsum soils (Fig. 8.20a and b).

Most gypsum communities are depauperate compared with adjacent areas. For example, at White Sands National Monument, New Mexico (Fig. 8.20c), 28 species common to the surrounding areas are absent from the gypsum dunes. Most gypsum species seem to be herbs or dwarf shrubs; annuals are uncommon (Parsons 1976; Powell and Turner 1977).

Some genera of gypsum-soil plants show significant regional differentiation of species within the Chihuahuan Desert subregions. In the Trans-Pecos region, *Nerisyrenia linearifolia* occurs, but it is re-

Figure 8.19. Azonal sandy desert sites: (a) sand dunes dominated by sand sage (Artemisia filifolia) (light color), a species that occurs on loose sands in all North American warm deserts, and some creosote bush (Mojave Desert, near St. George, Utah); (b) the prostrate shrub form of mesquite (Prosopis glandulosa) that dominates vast areas of loose sands in North American deserts (Chihuahuan Desert, east of El Paso, Texas); (c) stabilized sand area, 100 m from site in part a. Note increase in importance of creosote bush and the presence of other shrubs, including Krameria, Ambrosia dumosa, and Hymenoclea.

placed by *N. castillonii* in the Mapimian region and *N. gracilis* in the Saladan region. Similar situations exist for *Sartwellia* and *Nama*. Gypsum soils in our other deserts contain species in many of the genera mentioned earlier.

Riparian and Wash Vegetation

Water channels ranging from small rills to major river courses influence desert vegetation. In some creosote bush stands, on what appear to be flat plains, one can find cheesebrush (*Hymenoclea salsola*), often arrayed in lines along almost imperceptible rills.

In contrast to this situation, large river courses can be surrounded by extensive, dense forests (bosques) of willows (*Salix*, e.g., *S. gooddingii*), mesquites (*Prosopis glandulosa* or *P. pubescens*, and the aggressive introduced tamarisks or salt cedars (*Tamarix*, e.g., *T. chinensis* or *T. ramosissima*). Often the banks of the river will be dominated by large cottonwoods (*Populus fremontii*). Various ashes (*Fraxinus*) and desert willow (*Chilopsis linearis*) are also common. In some areas, thickets of the native

Figure 8.20. (a) Gypsum "sand" area on east side of Guadalupe Mountains National Park, Texas. Obvious species include Yucca elata, Ephedra torreyana, *and several grasses, including the gypsum specialist* Bouteloua breviseta; *(b) Indian blanket* (Gaillardia multiceps), *a species confined mainly to gypsum soils; site is the same as in part a; (c) White Sands National Monument, New Mexico, an extensive shifting gypsum dune system.*

arrow weed (*Pluchea sericea*) or the introduced *Arundo donax* border rivers.

Along washes, several of the foregoing species are common, as are hackberries (*Celtis* species), various species of *Rhus*, several *Dalea* and *Lycium* species, and Apache plume (*Fallugia paradoxa*). Several acacias occur in similar sites. In the Mojave Desert and Sonoran Desert, catclaw (*Acacia greggii*) is abundant, whereas *A. constricta* is often the Chihuahuan Desert dominant.

In a study of Chihuahuan Desert washes, Gardner (1951) listed the plants occurring in 93 New Mexico washes. Nine species were common, and an additional 22 species were recorded. The numbers of washes in which some species occurred were *Prosopis glandulosa* 54, *Hymenoclea monogyra* 45, *Fallugia paradoxa* 44, *Rhus microphylla* 40, *Larrea* 36, *Brickellia laciniata* 33, and *Chilopsis linearis* 12. In the Mojave Desert, Beatley (1976) listed 22 arroyo shrub species and 30 herbs. A "typical" riparian zone is pictured in Fig. 8.21.

Saline Areas

In saline sites (more than 2% salt), the number of species of plants is rather limited. In the entire

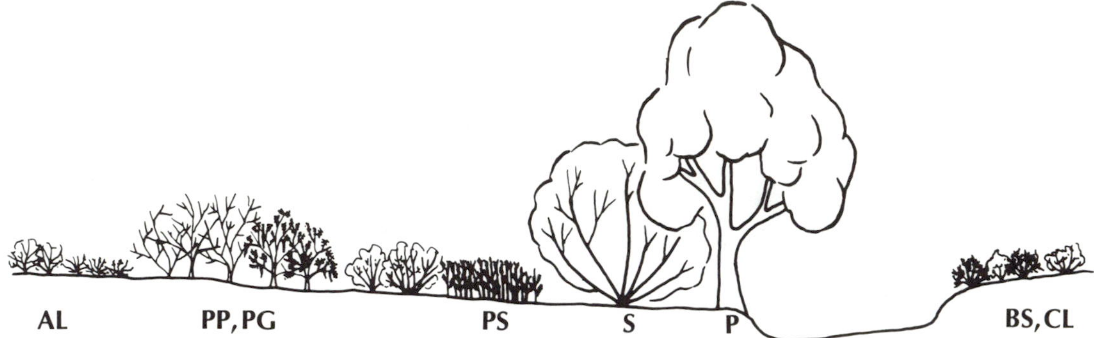

*Figure 8.21. Stylized transect from a Sonoran Desert river channel upslope (right to left) to show riparian vegetation. Plants include cottonwood (*Populus*) and willows (*Salix*), often followed by a band of arrow weed (*Pluchea (Tessaria) sericea*) and in turn by mesquites (*Prosopis pubescens *or* P. glandulosa*) and perhaps quail bush (*Atriplex lentiformis*). Minor channel species (right of riverbed) include seep willow (*Baccharis sarothroides*) and desert willow (*Chilopsis linearis*), among others.*

Chihuahuan Desert, Henrickson (1977) found only 40 halophytic vascular plant taxa. Interestingly, 25 of these were Chihuahuan endemics, including three genera: *Meiomeria*, *Reederochloa*, and *Pseudoclappia*. Despite this high degree of endemism, some plants occur in saline sites in all of our deserts, as well as in nondesert habitats. These include *Allenrolfea occidentalis*, *Atriplex canescens*, *Distichlis spicata*, and *Sporobolus airoides*. The genus *Atriplex* has many salt-tolerant species, though the presence of the genus cannot be used as a sign of high soil salinity. Similarly, *Tidestromia*, *Suaeda*, *Juncus*, *Salicornia*, and *Nitrophila* all contain common salt-tolerant species. In the Mojave Desert, and somewhat into the Sonoran Desert, greasewood (*Sarcobatus vermiculatus*) occupies saline sites that have standing water or are periodically inundated.

Mesquite often occurs in quite saline areas and thus is frequently a conspicuous plant around the edges of playas, where moderate salinity occurs.

Transitions

At the upper elevations or eastern limits of our deserts, grasslands usually prevail. Often elevation transitions include xeric-adapted trees such as oaks (*Quercus*) or junipers (*Juniperus*) (Fig. 8.22a), whereas the eastern boundaries are more dominantly grasslands (Fig. 8.22b) (Burgess and Northington 1977; Marroquin 1977). To the south, both the Chihuahuan and Sonoran deserts blend into subtropical thorn scrub (Fig. 8.22c), a vegetation type that shares many genera with the adjacent deserts. To the north, the Sonoran Desert blends with the Mojave Desert, and the Mojave with the Great Basin Desert. Much of the northern Chihuahuan Desert grades to grasslands. All of these vegetation types are covered extensively elsewhere in this volume.

POTPOURRI OF VEGETATION TOPICS

This chapter has emphasized the floristic and vegetational aspects of warm deserts. In using this approach, several important topics have been omitted. These include consideration of dispersion patterns, primary productivity, patterns of photosynthetic pathways, and annuals. Each of these topics will be considered briefly in order to provide an introduction to an extensive literature related to these topics, a literature beyond the scope of this book.

Finally, because deserts often are considered wastelands, I want to at least mention some potentially important economic factors related to deserts and their use for and by humans.

Productivity

Primary productivity is generally thought to be low in deserts. A range of 30–300 g dry weight per square meter per year has generally been accepted as encompassing North American deserts, although low values may fall to 2.6 (a sand dune) and high values may reach 816 (Chihuahuan Desert) (Szarek 1979; Hadley and Szarek 1981). An especially productive site was studied by Sharifi and associates (1982): a *Prosopis glandulosa* stand (90% of the plant cover was mesquite) in the Sonoran Desert of southern California. Stand biomass ranged from 23,000 kg ha^{-1} near a wash to 3500 kg ha^{-1} on the fringe of the stand. Mean dry-weight produc-

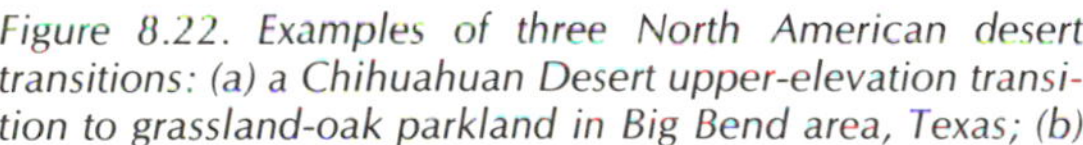

Figure 8.22. Examples of three North American desert transitions: (a) a Chihuahuan Desert upper-elevation transition to grassland-oak parkland in Big Bend area, Texas; (b) grassland-desert transition in the Sonoran Desert on the Santa Rita Experimental Range, near Tucson, Arizona ; (c) Sonoran Desert–thorn scrub transition, southern Sonora, Mexico.

tion in 1980 was 3650 kg ha^{-1}; 51% of that was allocated to wood, and 33.6% to leaves.

Although desert primary production is, in general, related to abiotic variables (Lane et al. 1984), the specific stand composition is also very important (Webb et al. 1983).

Dispersion

The pattern of distribution of plants with respect to one another is termed "dispersion." Three general dispersion patterns are usually recognized: random, regular, and clumped. Desert shrubs were at one time alleged to show almost classic regularity in their distributions. This was supposed to have been the result of competition for water or nutrients and/or a response to water-soluble toxins exuded through the plant's roots. Barbour (1973) questioned the generality of a regular dispersion pattern. Subsequently it has become clear that not all desert shrubs show regular patterns (Fonteyn and Mahall 1981; Phillips and MacMahon 1981), even though competition for water may occur under stress conditions (Fonteyn and Mahall 1978, 1981). Probably, as is the case for most natural phenomena, the answer to the question whether or not desert shrubs are regularly dispersed is, "It depends." For example, it depends on the species involved, their sizes, the values for the abiotic variables of the plant's environment, and a host of other factors. Clearly, many plants are clumped at one time in their life cycles (e.g., the nurse-tree–cactus interaction). Just as clearly, this pattern may change.

Recently, pattern regularity has been inferred to be due, in part, to the methodological problems of discerning when a clone-forming plant like creosote bush is an individual as opposed to being part of a clonal clump (Ebert and McMaster 1981; King and Woodell 1984). More experimentation is needed to elucidate the factors involved in an interaction as complex as the development of a community pattern, and it is highly unlikely that one factor will explain dispersion, or that one dispersion pattern will exist to the exclusion of all others.

Metabolic Pathways in Relation to Vegetation

Plants vary with regard to the first chemical products associated with the process of photosynthesis.

In many plants, a 3-carbon acid (phosphoglyceric acid) is the first, whereas in others there is a 4-carbon acid. Additionally, there are two different pathways that use a 4-carbon acid. In one case, the assimilation proceeds in light; in the other, CO_2 is changed to malate at night and stored until day. These three pathways are termed C_3, C_4, and CAM (Crassulacean Acid Metaobolism), respectively. C_3 plants are represented by the vast majority of common plants in a host of families. C_4 plants have a special sheath surrounding the vascular traces in their leaves and include families such as Asteraceae, Chenopodiaceae, Euphorbiaceae, Nyctaginaceae, Poaceae, and Zygophyllaceae, to mention a few. CAM plants are usually succulents, either leaf or stem forms, and include the Agavaceae, Cactaceae, and Crassulaceae. C_4 plants demonstrate more favorable high-temperature physiological performance and water use efficiency than C_3 forms. CAM plants also have high water use efficiencies. Clearly, we would anticipate that C_4 and CAM plants should be better suited to arid conditions, and a number of workers have used transect data to demonstrate this point. It is possible that at any site, various communities can be formed of any combination of these three types, ranging from pure stands of only one type to mixtures of all pairwise combinations and even to combinations of all three.

In Johnson's analysis (1976) of about 1000 desert species in California, 85% were C_3, 11% were C_4, and 4% were CAM. More than half of the C_4 species were grasses. Johnson found that communities on alkali soils contained mostly C_4 plants, that many desert sites contained high proportions of C_3 species, and that mixes of all three strategies occurred throughout the Mojave Desert. Sites dominated only by CAM-species communities were rare.

A study of plants along an aridity gradient that included the Chihuahuan Desert in the Big Bend area of Texas showed that—for 88 nonherbaceous species—as aridity decreased, the dominant plants changed in order from CAM to C_4 to C_3 species, as we would infer (Eickmeier 1978) from what was said previously. Interestingly, the CAM species had somewhat of a bimodal distribution (Eickmeier 1979).

Looking only at evergreen rosette plants (Liliaceae) of the Chihuahuan Desert, a group one would expect to be C_3 species, Kemp and Gardetto (1982) found that *Yucca baccata* and *Y. torreyi* were CAM species, whereas *Yucca elata, Y. campestris, Nolina microcarpa,* and *Dasylirion wheeleri* were C_3. Despite the fact that some species are known to be able to shift between C_3 and CAM, none of these species did. The important point is that although C_4 and CAM appear to be the species best adapted to deserts, many desert species are C_3, and one needs to look carefully at the specific microhabitat when interpreting relationships between patterns of distribution and photosynthetic pathways. Mesic microsites or times of the year permit C_3 species to flourish in areas that appear, in general, to be very arid.

The importance of seasonally mesic conditions was well demonstrated by Kemp (1983). He studied the phenological patterns of Chihuahuan Desert plants in relation to the timing of water availability. The perennials on his study site consisted of C_3 and C_4 forbs. C_4 grasses, C_3 shrubs, and CAM shrubs. Remembering that rainfall occurs predominantly in the summer in the Chihuahuan Desert, it is interesting that C_3 forbs showed greatest activity in spring or fall, whereas C_4 grasses were active in the summer and fall. The C_3 and CAM shrubs were active at various times and thus were not as dependent on soil moisture. These results suggest that the differences in responses between C_3 and C_4 species were greater for variations in temperature regime than for variations in available moisture. This last point is important. Many general studies of aridity gradients have suggested that the proportions of the metabolic pathway that involve the various photosynthetic types should segregate along the gradient. And whereas in some cases this generally occurs, often there is considerable variability in the relationships. "Aridity" gradients often involve parallel changes in moisture availability and temperature; these two factors are not independent of one another. A plant establishes and lives at a point in space and time, not on an "average" landscape. Thus, our crude correlations of C_3, C_4, or CAM strategies to yearly or monthly measurements of average values of abiotic variables across a whole landscape may obscure the biologically most important variables. In a sense, we measure the abiotic factors on a broad integrated scale, but the plant is responding to a highly localized series of variables acting where the plant happens to occur.

Annuals

I have ignored annual plants in this treatment because in many years they are virtually nonexistent as adults, although in other years they dominate the landscape. This variability makes characterization of deserts based on annuals difficult. This omission is not meant to demean their importance. In terms of number of species on a site, annuals often dominate. For a Mojave Desert site at Rock

Valley, Nevada, there are 10 common perennials, but 41 annuals; at a Chihuahuan Desert site near Tucson, the ratio is 8 to 48. The productivity of annuals, in favorable rainfall years, can exceed that of perennials. For example, at Rock Valley, Nevada, annuals produced a dry-weight biomass of 616 kg ha^{-1} yr^{-1}, compared with less than 600 for perennials (MacMahon and Wagner 1985). One specific site in the Mojave Desert makes the importance of annuals quite clear: Beatley (1976) studied sites that, in 1000 m^2, contained 30 species, 30% cover, 975 plants per square meter, and a standing crop of 616 kg ha^{-1}.

Annuals are generally C_3 or C_4 species. In the Mojave Desert, some sites contain virtually all winter annuals with C_3 metabolism, whereas other sites contain some summer annuals that are either C_3 or C_4 (Mulroy and Rundel 1977). This is generally the case across all of our deserts: The majority of winter annuals are C_3; summer annuals are C_4 or a mix of C_3 and C_4.

Winter annuals tend to have long lives and high mortality rates and occur both under shrubs and in the interspaces. Summer annuals are short-lived and occur mainly in the interspaces. All deserts have mixtures of winter and summer annuals, but usually one group predominates. For example, because the west coast of Baja California receives abundant winter precipitation, it supports 114 winter-annual species out of a total of 400–500 annuals. The Plains of Sonora subdivision shares about 100 annual species with southern Arizona; of the 84 species listed by Shreve and Wiggins (1964) as common, 56 species (66.7%) also occur in Texas; they are predominantly summer annuals.

Numerous genera contain annual species that can be seen in all three deserts. Some conspicuous examples are *Astragalus, Dyssodia, Eriogonum, Plantago, Eriophyllum, Phacelia*, and *Cryptantha*.

Some species of annuals occur in all of our warm deserts. Winter forms include *Monolepis nuttalliana, Lepidium lasiocarpum, Descurainia pinnata*, and *Calycoseris wrightii*. Summer forms include *Bouteloua barbata, Cenchrus echinatus, Tidestromia lanuginosa*, several *Boerhavia* species, *Euphorbia micromera, Baileya multiradiata* (one of the most conspicuous plants of desert roadsides), *Bahia absinthifolia*, and *Pectis papposa*.

Early workers suggested that annuals did not grow under the canopies of many shrub and subtree species. Although this may be true in some cases, it is more generally true that annuals do occur under a wide variety of shrubs. In fact, in relatively dry years, annuals may be more abundant under shrubs than in the adjacent interspaces. The density of annuals even varies under the shrub, depending on compass position (Patten 1978).

Human Uses of Deserts and Their Plants

There are myriad potential uses of deserts and their plants. As human populations expand, old cities will expand, and new ones will form in the "sun belt" of the Americas. Expansion of human populations requires construction of power-transmission lines (Vasek et al. 1975a), pipelines (Vasek et al. 1975b), roadways, and reservoirs (Rea 1983). All of these activities influence desert vegetation, often in ways that have long-term deleterious effects.

We cannot afford to alter desert vegetation in an unthinking manner. For one thing, many species of desert plants contain highly desirable chemicals; for example, an oil in jojoba (*Simmondsia chinensis*) can replace whale oil, and a rubber substitute can be produced from guayule (*Parthenium argentatum*). Many other species can be used as food, forage, and sources of medicines and industrial goods (McKell 1985). Clearly, desert species represent a vast, relatively untapped source of products for human use.

Currently, desert landscapes are used as sites for recreation, especially by off-road-vehicle enthusiasts, and as places to raise cattle. Both of these activities can have damaging effects. For example, grazing can reduce the diversity of annuals (Waser and Price 1981) or alter the composition of the community so that undesirable species predominate (Chew 1982). Uncontrolled use of off-road vehicles can alter desert soil in a way that accelerates erosion, causes compaction, and ultimately destroys native vegetation (Luckenbach and Bury 1983; Webb and Wilshire 1983).

The examples cited mandate that we use deserts in a prudent fashion. There are many positive benefits that we can derive from deserts if we temper our activities with knowledge of the potential influences of our actions. For many of us, the best use of the desert is as an environment in which to find solitude and contemplate life. I have never been to a place that permits clear thinking and introspection better than does a desert.

AREAS FOR FUTURE RESEARCH

Clearly, from a vegetation perspective, warm deserts are both well known and unknown. They are well known in a floristic sense; that is, the distributions and co-occurrences of species are reasonably well documented. They are unknown in the sense that quantitative data on composition of vegetation at the level of species, life forms, or functional groups are generally lacking.

Correlation, in a cause–effect sense, between vegetation attributes and abiotic factors has been attempted for a few species, as well as for numerous species in a very local area or over a broad area, but it has been based on very superficial measurements. To truly understand the causes of the structure and functioning of our deserts, we must accumulate more data to permit correlations between vegetation characteristics and physical factors over a greater variety of habitat types across the full span of our deserts.

Additionally, we know that the biotic environment is extremely important to species and consequently to communities. More careful studies of species-species interactions are needed to further our understanding of community origins and organization.

Finally, because deserts show extreme variation, both biotically and abiotically, over time spans of 1 yr or less, more long-term ecological research of even a simple monitoring nature must be conducted. The responses of vegetation to episodic extreme conditions may give us better clues as to what drives these communities than do studies conducted in "average" years.

The foregoing comments are constrained to consideration of studies that might explain community composition. Community functioning is beyond the scope of this treatment, although it is of great importance to understanding these ecosystems.

REFERENCES

Aide, M., and O. W. Van Auken. 1985. Chihuahuan desert vegetation of limestone and basalt slopes in west Texas. Southwestern Naturalist 30:533–542.

Bailey, H. P. 1979. Semi-arid climates: their definition and distribution. pp. 73–97 in A. E. Hall, G. H. Cannell, and H. W. Lawton (eds.), Agriculture in semi-arid environments. Springer-verlag, Berlin.

Barbour, M. G. 1973. Desert dogma reexamined: root/shoot productivity and plant spacing. Amer. Midl. Nat. 89:41–57.

Barbour, M. G., and J. Major (eds.). 1977. Terrestrial vegetation of California. Wiley, New York.

Beatley, J. C. 1974. Effects of rainfall and temperature on the distribution and behavior of *Larrea tridentata* (creosote-bush) in the Mojave Desert of Nevada. Ecology 55:245–261.

Beatley, J. C. 1976. Vascular plants of the Nevada test site and central-southern Nevada: ecologic and geographic distributions, TID-26881. Energy Research and Development Administration, NTIS, Springfield, Va.

Bender, G. L. (ed.). 1982. Reference handbook on the deserts of North America. Greenwood Press, Westport, Conn.

Benson, L. 1982. The cacti of the United States and Canada. Stanford University Press.

Benson, L., and R. A. Darrow. 1981. Trees and shrubs of the southwestern deserts, 3rd ed. University of Arizona Press, Tucson.

Bowers, J. E. 1982. The plant ecology of inland dunes in western North America. J. Arid Environ. 5:199–220.

Bowers, J. E. 1984. Plant geography of southwestern sand dunes. Desert Plants 6:31–42, 51–54.

Bowers, M. A., and C. H. Lowe. 1986. Plant-form gradients on Sonoran Desert bajadas. Oikos 46:284–291.

Bratz, R. D. 1976. The central desert of Baja California. J. Idaho Acad. Sci. 12:58–72.

Brown, D. E. (ed.). 1982a. Biotic communities of the American Southwest–United States and Mexico. Desert Plants 4:1–342.

Brown, D. E. 1982b. Chihuahuan desertscrub. Desert Plants 4:169–179.

Brown, D. E., and C. H. Lowe. (eds.). 1980. Biotic communities of the Southwest. USDA Forest Service general technical report RM-78. Rocky Mountain Forest and Range Experiment Station, Ft. Collins, Colo.

Burgess, T. L., and D. K. Northington. 1977. Desert vegetation in the Guadalupe Mountains region. pp. 229–242 in R. H. Wauer and D. H. Riskind (eds.), Transactions of the symposium on the biological resources of the Chihuahuan Desert region, United States and Mexico. USDI, National Park Service, Transactions and Proceedings Series, no. 3. U. S. Government Printing Office, Washington, D.C.

Chew, R. M. 1982. Changes in herbaceous and suffrutescent perennials in grazed and ungrazed desertified grassland in southeastern Arizona, 1958–1978. Amer. Midl. Nat. 108:159–169.

Chew, R. M., and A. E. Chew. 1965. The primary productivity of a desert shrub (*Larrea tridentata*) community. Ecol. Monogr. 35:355–375.

Cooke, R. U., and R. W. Reeves. 1976. Arroyos and environmental change in the American South-West. Clarendon Press, Oxford.

Cooke, R. U., and A. Warren. 1973. Geomorphology in deserts. University of California Press, Berkeley.

Crosswhite, F. S., and C. D. Crosswhite. 1982. The Sonoran Desert. pp. 163–319 in G. L. Bender (ed.), Reference handbook on the deserts of North America. Greenwood Press, Westport, Conn.

Crosswhite, F. S., and C. D. Crosswhite. 1984. A classification of life forms of the Sonoran Desert, with emphasis on the seed plants and their survival strategies. Desert Plants 5:131–161.

Dillon, M. O. 1984. A systematic study of Flourensia (Asteraceae, Heliantheae). Fieldiana publication 1357, botany, new series, no. 16. Field Museum of Natural History, Chicago.

Dorn, R. I., and T. M. Oberlander. 1981. Microbial origin of desert varnish. Science 213:1245–1247.

Dregne, H. E. 1976. Soils of arid regions. Elsevier, Amsterdam.

Dregne, H. E. 1979. Desert soils. pp. 73–81 in J. R. Goodin and D. K. Northington (eds.), Arid land plant resources. International Center for Arid and Semi-arid Land Studies, Texas Technological University, Lubbock.

Ebert, T. A., and G. S. McMaster. 1981. Regular

pattern of desert shrubs: a sampling artefact? J. Ecol. 69:559–564.

Eckert, R. E., Jr., M. K. Wood, W. H. Blackburn, and F. F. Peterson. 1979. Impacts of off-road vehicles on infiltration and sediment production of two desert soils. J. Range Management 32:394–397.

Eickmeier, W. G. 1978. Photosynthetic pathway distributions along an aridity gradient in Big Bend National Park, and implications for enhanced resource partitioning. Photosynthetica 12:290–297.

Eickmeier, W. G. 1979. Eco-physiology differences between high and low elevation CAM species in Big Bend National Park, Texas. Amer. Midl. Nat. 101:118–126.

Ezcurra, E., and V. Rodrigues. 1986. Rainfall patterns in the Gran Desierto, Sonora, Mexico. J. Arid Environ. 10:13–28.

Felger, R. S. 1980. Vegetation and flora of the Gran Desierto, Sonora, Mexico. Desert Plants 2:87–114.

Felger, R. S., and C. H. Lowe. 1976. The island and coastal vegetation and flora of the northern part of the Gulf of California. Contributions in Science 285. Natural History Museum of Los Angeles County.

Fiero, B. 1986. Geology of the Great Basin. University of Nevada Press, Reno.

Fonteyn, P. J., and B. E. Mahall. 1978. Competition among desert perennials. Nature (London) 275:544–545.

Fonteyn, P. J., and B. E. Mahall. 1981. An experimental analysis of structure in a desert plant community. J. Ecol. 69:883–896.

Fuller, W. H. 1975. Soils of the desert Southwest. University of Arizona Press, Tucson.

Gardner, J. L. 1951. Vegetation of the creosotebush area of the Rio Grande Valley in New Mexico. Ecol. Monogr. 21:379–403.

Gentry, H. S. 1978. The agaves of Baja California. Occasional Papers of the California Academy of Science, no. 130.

Gentry, H. S. 1982. Agaves of continental North America. University of Arizona Press, Tucson.

Gibson, A. C., and P. S. Nobel. (eds.), 1986. The cactus primer. Harvard University Press, Cambridge, Mass.

Gile, L. H., F. F. Peterson, and R. B. Grossman. 1966. Morphological and genetic sequences of carbonate accumulation in desert soils. Soil Sci. 101:347–360.

Goldberg, D. E., and R. M. Turner. 1986. Vegetation change and plant demography in permanent plots in the Sonoran Desert. Ecology 67:695–712.

Goudie, A., and J. Wilkinson. 1977. The warm desert environment. Cambridge University Press.

Hadley, N. F., and S. R. Szarek. 1981. Productivity of desert ecosystems. BioScience 31:747–753.

Hallmark, C. T., and B. L. Allen. 1975. The distribution of creosotebush in west Texas and eastern New Mexico as affected by selected soil properties. Soil Sci. Soc. Amer. Proc. 39:120–124.

Hastings, J. R., and R. M. Turner. 1965. The changing mile: an ecological study of vegetation change with time in the lower mile of an arid and semiarid region. University of Arizona Press, Tucson.

Hastings, J. R., R. M. Turner, and D. K. Warren. 1972. An atlas of some plant distributions in the Sonoran Desert. Technical Reports on the Meteorology and Climatology of Arid Regions, no. 21. University of Arizona Institute of Atmospheric Physics, Tucson.

Hendricks, D. M. 1985. Arizona soils. College of Agriculture, University of Arizona, Tucson.

Hennessy, J. T., R. P. Gibbens, J. M. Tromble, and M. Cardenas. 1985. Mesquite (*Prosopis glandulosa* Torr.) dunes and interdunes in southern New Mexico: a study of soil properties and soil water relations. J. Arid Environ. 9:27–38.

Henrickson, J. 1977. Saline habitats and halophytic vegetation of the Chihuahuan Desert region. pp. 289–314 in R. H. Wauer and D. H. Riskind (eds.), Transactions of the symposium on the biological resources of the Chihuahuan Desert Region, United States and Mexico. USDI, National Park Service, Transactions and Proceedings Series, no. 3. U. S. Government Printing Office, Washington, D.C.

Henrickson, J., and M. C. Johnston. 1986. Vegetation and community types of the Chihuahuan Desert. pp. 20–39 in J. C. Barlow, A. M. Powell, and B. N. Timmermann (eds.), Chihuahuan Desert–U.S. and Mexico, vol. 11. Chihuahuan Desert Research Institute, Sul Ross State University, Alpine, Tex.

Henrickson, J., and R. M. Straw. 1976. A gazetteer of the Chihuahuan Desert region. A supplement to the Chihuahuan Desert flora. California State University, Los Angeles.

Humphrey, R. R. 1974. The boojum and its home. University of Arizona Press, Tucson.

Humphrey, R. R., and D. B. Marx. 1980. Distribution of the boojum tree (*Idria columnaris*) on the coast of Sonora, Mexico as influenced by climate. Desert Plants 2:183–196.

Hunt, C. B. 1966. Plant ecology of Death Valley, California. USDI, geological survey professional paper 509. U. S. Government Printing Office, Washington, D.C.

Hunt, C. B. 1975. Death Valley. Geology, ecology, archaeology. University of California Press, Berkeley.

Johnson, H. B. 1976. Vegetation and plant communities of southern California deserts–a functional view. pp. 125–164 in J. Latting (ed.), Plant communities of Southern California. Special publication no. 2. California Native Plant Society, Berkeley, Calif.

Kemp, P. R. 1983. Phenological patterns of Chihuahuan desert plants in relation to the timing of water availability. J. Ecol. 71:427–436.

Kemp, P. R., and P. E. Gardetto. 1982. Photosynthetic pathway types of evergreen rosette plants (Liliaceae) of the Chihuahuan desert. Oecologia 55:149–156.

King, P. B. 1959. The evolution of North America. Princeton University Press, Princeton, N.J.

King, T. J., and S. R. J. Woodell. 1984. Are regular patterns in desert shrubs artefacts of sampling? J. Ecol. 72:295–298.

Lancaster, N. 1984. Aeolian sediments, processes and land forms. J. Arid Environ. 7:249–254.

Lane, L. J., E. M. Romney, and T. E. Hakonson. 1984. Water balance calculations and net production of perennial vegetation in the northern Mojave Desert. J. Range Management 37:12–18.

LeHouerou, H. N. 1984. Rain use efficiency: a unifying concept in arid-land ecology. J. Arid Environ. 7:1–35.

Lowe, C. H., and W. F. Steenbergh. 1981. On the Cenozoic ecology and evolution of the Sahuaro. Desert Plants 3:83–86.

Luckenbach, R. A., and R. B. Bury. 1983. Effects of off-road vehicles on the biota of the Algodones Dunes, Imperial County, California. J. Appl. Ecol. 20:265–286.

Mabbutt, J. A. 1977. Desert landforms, vol. 2. MIT Press, Cambridge, Mass.

Mabry, T. J., J. H. Hunziker, and D. R. DiFeo. (eds.). 1977. Creosote bush. Biology and chemistry of *Larrea* in New World deserts. US/IBP 6. Dowden, Hutchinson & Ross, Inc., Stroudsburg, Pa.

McAuliffe, J. R. 1984. Sahuaro–nurse tree associations in the Sonoran Desert: competitive effects of Sahuaros. Oecologia 64:319–321.

McGinnies, W. G. 1981. Discovering the desert. University of Arizona Press, Tucson.

McKell, C. M. 1985. North America. pp. 187–232 in J. R. Goodin and D. K. Northington (eds.), Plant resources of arid and semiarid lands, a global perspective. Academic, Orlando.

MacMahon, J. A. 1979. North American deserts: their floral and faunal components. pp. 21–82 in D. W. Goodall and R. A. Perry (eds.), Arid-land ecosystems: structure, functioning and management, vol. 1. IBP 16. Cambridge University Press.

MacMahon, J. A. 1985. The Audubon Society field guide to North American desert habitats. Chanticleer Press, New York.

MacMahon, J. A., and D. J. Schimpf. 1981. Water as a factor in the biology of North American desert plants. pp. 119–171 in D. D. Evans and J. L. Thames (eds.), Water in desert ecosystems. US/IBP 12. Dowden, Hutchinson and Ross, Inc., Stroudsburg, Pa.

MacMahon, J. A., and F. H. Wagner. 1985. The Mojave, Sonoran and Chihuahuan deserts of North America. pp. 105–202 in M. Evenari et al. (eds.), Hot deserts and arid shrublands. Elsevier, Amsterdam.

Marroquin, J. S. 1977. A physiognomic analysis of the types of transitional vegetation in the eastern parts of the Chihuahuan Desert in Coahuila, Mexico. pp. 249–272 in R. H. Wauer and D. H. Riskind (eds.), Transactions of the symposium on the biological resources of the Chihuahuan Desert region, United States and Mexico. USDI, National Park Service, Transactions and Proceedings Series, no. 3. U. S. Government Printing Office, Washington, D.C.

Miller, J. M. 1986. Phytogeography and potential economic use of the guayule rubber plant on Chihuahuan Desert limestone geologic formation. J. Arid Environ. 10:153–162.

Moore, C. B., and C. Elvidge. 1982. Desert varnish. pp. 527–536 in G. L. Bender (ed.), Reference handbook on the deserts of North America. Greenwood Press, Westport, Conn.

Morafka, D. J. 1977. A biograpical analysis of the Chihuahuan Desert through its herpetofauna. Biogeographic 9. Junk, The Hague.

Mulroy, T. W., and P. W. Rundel. 1977. Annual plants: adaptations to desert environments. BioScience 27:109–114.

Nash, T. H., and T. J. Moser. 1982. Vegetational and physiological patterns of lichens in North American deserts. J. Hattori Bot. Lab. 53:331–336.

Nash, T. H., III, G. T. Nebeker, T. J. Moser, and T. Reeves. 1979. Lichen vegetational gradients in relation to the Pacific Coast of Baja California: the maritime influence. Madroño 26:149–163.

Neal, J. T. 1969. Playa variation. pp. 15–44 in W. G. McGinnies and B. J. Goldman (eds.), Arid lands in perspective. University of Arizona Press, Tucson.

Neal, J. T. (ed.). 1975. Playas and dried lakes. Occurrence and development. Benchmark Papers in Geology 120. Dowden, Hutchinson & Ross, Inc., Stroudsburg, Pa.

Niehaus, T. F., and C. L. Ripper. 1976. A field guide to Pacific states wildflowers. Houghton Mifflin, Boston.

Niehaus, T. F., C. L. Ripper, and V. Savage. 1984. A field guide to southwestern and Texas wildflowers. Houghton Mifflin, Boston.

Niering, W. A., and C. H. Lowe, 1984. Vegetation of the Santa Catalina Mountains: community types and dynamics. Vegetatio 58:3–28.

Nobel, P. S. 1980. Morphology nurse plants, and minimal apical temperatures for young *Carnegiea gigantea*. Bot. Gaz. 141:188–191.

Oberlander, T. M. 1979. Characterization of arid climates according to combined water balance parameters. J. Arid Environ. 2:219–241.

Orians, G. H., and O. T. Solbrig. (eds.). 1977. Convergent evolution in warm deserts. US/IBP 3. Dowden, Hutchinson & Ross, Inc., Stroudsburg, Pa.

Osborn, H. B., E. D. Shirley, D. R. Davis, and R. B. Koehler. 1980. Model of time and space distribution of rainfall in Arizona and New Mexico. Agricultural Reviews and Manuals, ARM-W-14. USDA, Science and Education Administration, Oakland, Calif.

Parsons, R. F. 1976. Gypsophily in plants–a review. Amer. Midl. Nat. 96:1–20.

Patten, D. T. 1978. Productivity and production efficiency of an upper Sonoran Desert ephemeral community. Amer. J. Bot. 65:891–895.

Pavlik, B. M. 1980. Patterns of water potential and photosynthesis of desert sand dune plants, Eureka Valley, California. Oecologia 46:147–154.

Phillips, D. L., and J. A. MacMahon. 1978. Gradient analysis of a Sonoran Desert bajada. Southwestern Naturalist 23:669–680.

Phillips, D. L., and J. A. MacMahon. 1981. Competition and spacing patterns in desert shrubs. J. Ecol. 69:97–115.

Pinkava, D. J., and H. S. Gentry (eds.). 1985. Symposium on the genus *Agave*, Desert Botanical Garden, Phoenix, March 7–9. 1985. Desert Plants 7:1.

Powell, A. M., and B. L. Turner. 1977. Aspects of the plant biology of the gypsum outcrops of the Chihuahuan Desert. pp. 315–325 in R. H. Wauer and D. H. Riskind (eds.), Transactions of the symposium on the biological resources of the Chihuahuan Desert region, United States and Mexico. USDI, National Park Service, Transactions and Proceedings Series, no. 3. U.S. Government Printing Office, Washington, D.C.

Rea, A. M. 1983. Once a river. Bird life and habitat changes on the middle Gila. University of Arizona Press, Tucson.

Rowlands, P., H. Johnson, E. Riter, and A. Endo. 1982. The Mojave Desert. pp. 103–145 in G. L. Bender (ed.), Reference handbook on the deserts of North America. Greenwood Press, Westport, Conn.

Rzedowski, J. 1966. Vegetacion del Estado de San Luis Potosi. Acta Cientifica Potosina 5:5–291.

Rzedowski, J. 1978. Vegetacion de Mexico. Editorial Limusa, Mexico City.

Sammis, T. W., and L. W. Gay. 1979. Evapotranspiration from an arid zone plant community. J. Arid Environ. 2:313–321.

Schlesinger, W. H. 1982. Carbon storage in the caliche of arid soils: a case study from Arizona. Soil Sci. 133:247–255.

Schlesinger, W. H. 1985. The formation of caliche in soils of the Mojave Desert, California. Geochim. Cosmochim. Acta 49:57–66.

Schlesinger, W. H., and C. S. Jones. 1984. The comparative importance of overland runoff and mean annual rainfall to shrub communities of the Mojave Desert. Bot. Gaz. 145:116–124.

Schmidt, R. H., Jr. 1979. A climatic delineation of the "real" Chihuahuan Desert. J. Arid Environ. 2:243–250.

Schmidt, R. H., Jr. 1986. Chihuahuan climate. pp. 40–63 in J. C. Barlow, A. M. Powell, and B. N. Timmermann (eds.), Chihuahuan Desert–U.S. and Mexico, vol. II. Chihuahuan Desert Research Institute, Sul Ross State University, Alpine, Tex.

Secor, J. B., S. Shamash, D. Smeal, and A. L. Gennaro. 1983. Soil characteristics of two desert plant community types that occur in the Los Medanos area of southeastern New Mexico. Soil Sci. 136:133–144.

Sharifi, M. R., E. T. Nilsen, and P. W. Rundel. 1982. Biomass and net primary production of *Prosopis glandulosa* (Fabaceae) in the Sonoran Desert of California. Amer. J. Bot. 69:760–767.

Shields, L. M. 1953. Gross modifications in certain plant species tolerant of calcium sulfate dunes. Amer. Midl. Nat. 50:224–237.

Shreve, F. 1951. Vegetation of the Sonoran Desert. Carnegie Institution of Washington publication 591.

Shreve, F., and T. D. Mallery. 1933. The relation of caliche to desert plants. Soil Sci. 35:99–113.

Shreve, F. and I. L. Wiggins. 1964. Vegetation and flora of the Sonoran Desert, vol. I. Stanford University Press, Stanford, Calif.

Simpson, B. B. (ed.). 1977. Mesquite. Its biology in two desert scrub ecosystems. US/IBP 4. Dowden, Hutchinson & Ross, Inc., Stroudsburg, Pa.

Skujins, J. 1984. Microbial ecology of desert soils. pp. 49–91 in C. C. Marshall (ed.), Advances in microbial ecology, vol. 7. Plenum, New York.

Smith, R. S. U. 1982. Sand dunes in the North American deserts. pp. 481–524 in G. L. Bender (ed.), Reference handbook on the deserts of North America. Greenwood Press, Westport, Conn.

Smith, S. D., and J. A. Ludwig. 1978. The distribution and phytosociology of *Yucca elata* in southern New Mexico. Amer. Midl. Nat. 100:202–212.

Solbrig, O. T. 1972. The floristic disjunctions between the "Monte" in Argentina and the "Sonoran Desert" in Mexico and the United States. Ann. Missouri Bot. Garden 59:218–223.

Solbrig, O. T., M. G. Barbour, J. Cross, G. Goldstein, C. H. Lowe, J. Morello, and T. W. Yang. 1977. The strategies and community patterns of desert plants. pp. 67–106 in G. H. Orians and O. T. Solbrig (eds.), Convergent evolution in warm deserts. US/IBP 3. Dowden, Hutchinson & Ross, Inc. Stroudsburg, Pa.

Steenbergh, W. F., and C. H. Lowe. 1976. Ecology of the saguaro. I. The role of freezing weather on a warm-desert plant population. Research in the Parks, National Park Service Symposium Ser. 1:49–92.

Steenbergh, W. F., and C. H. Lowe. 1977. Ecology of the saguaro. II. Reproduction, germination, establishment, growth and survival of the young plant. National Park Service, Scientific Monograph Series, no 8. U.S. Government Printing Office, Washington, D.C.

Steenbergh, W. F., and C. H. Lowe. 1983. Ecology of the saguaro. III. Growth and demography. National Park Service, Scientific Monograph Series, no 17. U.S. Government Printing Office, Washington, D.C.

Stromberg, J. C., and T. M. Krischan. 1983. Vegetation structure at Punta Cirio, Sonora, Mexico. Southwestern Naturalist 28:211–214.

Szarek, S. R. 1979. Primary production in four North American deserts: indices of efficiency. J. Arid Environ. 2:187–209.

Turner, R. M. 1982. Mohave desertscrub. Desert Plants 4:157–168.

Turner, R. M., and D. E. Brown. 1982. Sonoran desertscrub. Desert Plants 4:181–222.

Vandermeer, J. 1980. Saguaros and nurse trees: a new hypothesis to account for the population fluctuations. Southwestern Naturalist 25:357–360.

Vasek, F. C. 1980. Creosote bush: long-lived clones in the Mojave Desert. Amer. J. Bot. 67:246–255.

Vasek, F. C., and M. G. Barbour. 1977. Mojave desert scrub vegetation. pp. 835–867 in M. G. Barbour and J. Major (eds.), Terrestrial vegetation of California. Wiley, New York.

Vasek, F. C., H. B. Johnson, and G. D. Brum. 1975a. Effects of power transmission lines on vegetation of the Mojave Desert. Madroño 23:114–130.

Vasek, F. C., H. B. Johnson, and D. H. Eslinger. 1975b. Effects of pipeline construction on creosote bush scrub vegetation of the Mojave Desert. Madroño 23:1–13.

Walter, H. 1971. Ecology of tropical and subtropical vegetation. Oliver & Boyd, Edinburgh.

Walter, H., and H. Lieth. 1967. Klimadiagramm-Weltatlas. Gustav Fischer-Verlag, Jena.

Waser, N. M., and M. V. Price. 1981. Effects of grazing on diversity of annual plants in the Sonoran Desert. Oecologia 50:407–411.

Wauer, R. H., and D. H. Riskind. (eds.). 1977. Transactions of the symposium on the biological resources of the Chihuahuan Desert Region, United States and Mexico. USDI, National Park Service Transactions and Proceedings Series, no.

3. U.S. Government Printing Office, Washington, D.C.

Webb, R. H., and S. S. Stielstra. 1979. Sheep grazing effects on Mojave Desert vegetation and soils. Environ. Management 3:517–529.

Webb, R. H., and H. G. Wilshire. 1980. Recovery of soils and vegetation in a Mojave Desert ghost town, Nevada, U.S A. J. Arid Environ. 3:291–303.

Webb, R. H., and H. G. Wilshire. (eds.), 1983. Environmental effects of off-road vehicles. Springer-Verlag, New York.

Webb, W. L., W. K. Lauenroth, S. R. Szarek, and R. S. Kinerson. 1983. Primary production and abiotic controls in forests, grasslands, and desert ecosystems in the United States. Ecology 64:134–151.

Wells, P. V., and D. Woodcock. 1985. Full-glacial vegetation of Death Valley, California: juniper woodland opening to yucca semidesert. Madroño 32:11–23.

Welsh, R. G., and R. F. Beck. 1976. Some ecological relationships between creosotebush and bush muhly. J. Range Management 29:472–475.

Wentworth, T. R. 1981. Vegetation on limestone and granite in the Mule Mountains, Arizona. Ecology 62:469–482.

West, N. E., and J. J. Skujins, (eds.). 1978. Nitrogen in desert ecosystems. US/IBP 9. Dowden, Hutchinson & Ross, Inc., Stroudsburg, Pa.

West, R. C. 1964. Surface configuration and associated geology of Middle America. pp. 33–83 in R. C. West (ed.), Natural environment and early cultures. University of Texas Press, Austin.

Whitford, W. G. 1973. Jornada validation site report. US/IBP, desert biome research memorandum 73-4. Utah State University, Logan.

Whittaker, R. H., and W. A. Niering. 1968. Vegetation of the Santa Catalina Mountains, Arizona. IV. Limestone and acid soils. J. Ecol. 56:523–544.

Wright, R. A. 1982. Aspects of desertification in *Prosopis* dunelands of southern New Mexico, U.S.A. J. Arid Environ. 5:277–284.

Yang, T. W. 1970. Major chromosome races of *Larrea divaricata* in North America. J. Arizona Acad. Sci. 6:41–45.

Yeaton, R. I. 1978. A cyclical relationship between *Larrea tridentata* and *Opuntia leptocaulis* in the northern Chihuahuan desert. J. Ecol. 66:651–656.

Yeaton, R. I., and M. L. Cody. 1979. The distribution of cacti along environmental gradients in the Sonora and Mohave deserts. J. Ecol. 67:529–541.

Yeaton, R. I., and A. R. Manzanares. 1986. Organization of vegetation mosiacs in the *Acacia schaffneri–Opuntia streptacantha* association, southern Chihuahuan desert, Mexico. J. Ecol. 74:211–217.

Yeaton, R. I., J. Travis, and E. Gilinsky. 1977. Competition and spacing in plant communities: the Arizona upland association. J. Ecol. 65:587–595.

Zabriskie, J. G. 1979. Plants of Deep Canyon and the Central Coachella Valley, California. University of California, Riverside.

Chapter 9

Grasslands

PHILLIP L. SIMS

INTRODUCTION

North American grasslands stretch from southeastern Alberta, central Saskatchewan, and southeastern Manitoba to the highlands of central Mexico and from eastern Indiana to California. They range from desert grasslands in the lower plateaus of the arid Southwest to mountain grasslands in the Rocky Mountains. Küchler's map (1975) of potential natural vegetation types and the maps of Rzedowski (1978) and Bailey (1976) were used to delimit the plant communities of concern in this chapter (Table 9.1). The primary foci of this chapter are the humid, temperate-to-arid grasslands, with special emphasis on the plains, California, Palouse, and desert grasslands. The plains grasslands are discussed by grouping Küchler's potential plant communities into those conventionally called tall-grass, mixed-grass, and shortgrass prairies.

Table 9.1. Extents of the major grassland types of the contiguous United States

Grassland type	Area (ha)	Percentage
Tall-grass prairie	57,351,100	22
Mixed-grass prairie	56,617,400	21
Shortgrass prairie	61,522,300	23
California grassland	5,268,800	2
Palouse prairie	64,471,600	24
Desert grassland	20,756,500	8
Total	265,987,700	100

Source: Data from Kücher (1964), as planimetered by Risser et al. (1981).

Grasslands, the largest of the four major natural vegetation formations covering the earth's surface (Gould 1968), occur on most continents and are accompanied by certain degrees of consistency in climate, flora, fauna, growth form, and physiognomy (Carpenter 1940). Grasslands, including steppes (Russia), velds (South Africa), pampas (South America), puszta (Hungary), and prairies (North America), account for 24% of the earth's vegetation, covering more than 4.6 billion ha (Shantz 1954). Grasslands have immense scenic, aesthetic, and watershed values and provide forage and habitat for large numbers of domestic and wild animals. Grasses have the widest range of adaptation of any group of plants (Archer and Bunch 1953). Around the world, they have contributed the germ plasm for the principal human food crops, and they contain a priceless array of germ plasm for future study and incorporation into domesticated foods.

Grasslands are the largest of the North American vegetation formations, originally covering almost 300 million of the 770 million ha in the United States (Küchler 1964) and about 50 and 20 million ha in Canada and Mexico, respectively, as estimated from the data of Rowe (1972) and Rzedowski (1978). Even today, grasslands remain the largest of the natural biomes in the United States, covering more than 125 million ha (U.S. Forest Service 1980). Most of the productive, arable lands in North America were once grasslands.

Originally, grasslands dominated central North America and occurred as important and sometimes extensive islands of vegetation throughout the western United States, Canada, and Mexico. Major North American grasslands include the tall-grass, mixed-grass, and shortgrass prairies of the central plains, the desert grasslands of the southwestern United States and Mexico, the California grasslands, and the Palouse prairie in the intermountain region of the northwestern United States and British Columbia, Canada (Fig. 9.1). The annual and desert grasslands account for about 2% and 8%, respectively, of all grasslands, with the remaining 90% rather uniformly distributed as shortgrass, mixed-grass, tall-grass, and Palouse prairies (Table 9.1). Small pockets of mountain grasslands occur within the western coniferous forest. The mountain grasslands are not extensive in contiguous distribution, but are important components of valuable, scenic watersheds in western North America. Risser and associates (1981) estimated that about 27 million ha of grassland occur in association with *Pinus ponderosa* in the Rocky Mountains. Extensive grassland-forest combinations are also found in the transition zone to the east of the Great Plains.

Grasslands are dominated primarily by grasses (Gramineae) and grass-like plants (mostly Cyperaceae). Three or four species usually characterize a grassland site and produce a majority of the biomass (Coupland 1974; Sims et al. 1978). Floristically, graminoids often compose no more than 20% of the total number of species in a typical grassland community. Forbs (non-grass-like herbs) are seasonally important and, along with some dwarf shrubs, may dominate the grassland aspect. Compositae species are the most numerous herbs, followed by the Leguminosae (Coupland 1979).

The numbers of plant species that occur in grasslands increase as the growing-season environment becomes more mesic, where topographic variations increase, and where humans have had the least impact (Coupland 1979). Coupland and associates (1973) recorded 50 vascular species in a temperate, semiarid, nearly level grassland, compared with more than 200 species in a subhumid prairie (Steiger 1930). Over 330 plant species contributed significantly to the biomass structure of an array of

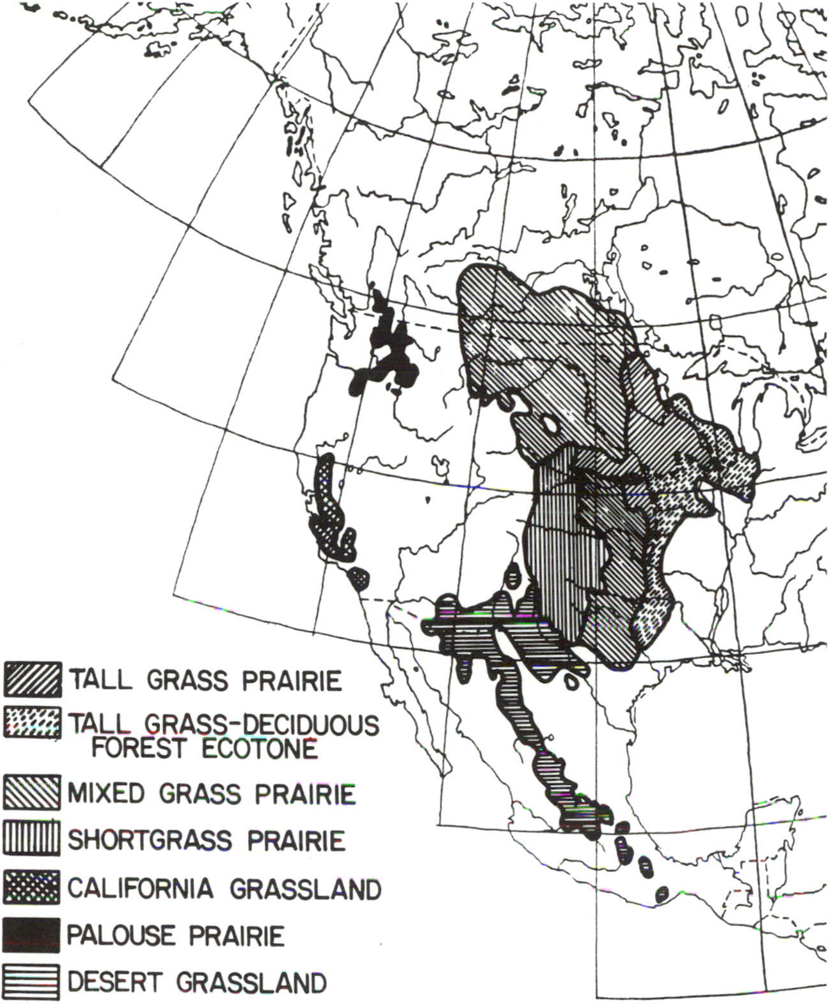

Figure 9.1. Distribution of the major grasslands types of North America (adapted from Rae 1972, Küchler 1975, Wright and Bailey 1982, and Rzedowski 1978).

10 western North American grasslands (Sims et al. 1978) studied by the U.S. International Biome Program Grassland Biome Project (Van Dyne 1971; Breymeyer and Van Dyne 1980). These biome studies were conducted in the tall-grass prairie (Osage site), mixed-grass prairie (Dickinson, North Dakota, Cottonwood, South Dakota, and Hays, Kansas), shortgrass prairie (Pawnee, Colorado, and Pantex, Texas), Palouse prairie (ALE, Richland, Washington) and mountain grassland (Bison and Bridger, Montana). Generally, forbs outnumbered grasses by threefold to fourfold, and shrubs, half-shrubs, succulents, and trees were minor components that provided important differentiating features to the grasslands.

The central grasslands of North America have few unique taxa. Wells (1970) reported no plant families and few genera and species (and these were primarily forbs) endemic to the central plains. The same is true for insects and birds (Axelrod 1985). In contrast, numerous endemic plant and animal species occur in the forests bordering grasslands and the deserts to the south. The lack of endemism was attributed primarily to the youthfulness of these grasslands – they attained their present extent only in the post-glacial period (Sears 1935, 1948; Schmidt 1938).

Although the extent and nature of the prehistoric animal populations are largely unknown, there is little doubt that large numbers of herbivores were present in the North American grasslands at the time of settlement by Europeans (Sims et al. 1982). Large herds of bison, antelope, elk, and deer and vast "towns" of prairie dogs were present and significantly impacted local areas at times (Nelson 1925; Seton 1929; Larson 1940; England and DeVos 1969; Vankat 1979; McDonald 1981). All but the latter were migratory in nature, and their transitory grazing patterns allowed the vegetation to recover from cyclic grazing events (Curtis 1962). The large herbivores migrated in search of green forage and in response to patterns of precipitation, drought, and fire. Heady (1975) speculated that the migratory nature of pristine herbivore grazing was beneficial to plants because the impact was intermittent, allowing rest between grazing events. For example, bison used a particular area for a short period of time, then moved to a new area – a pattern that became more or less repetitive from year to year. This relatively fixed pattern of migration exerted seasonal grazing pressures of the vegetation, which became adapted to these pressures through natural selection.

Climate

North American grasslands are characterized by a number of different climates, with some similarities in seasonal rainfall and temperature distributions. Grassland climates generally have distinct wet and dry seasons and are noted for temperature and precipitation extremes. Grassland vegetation in the central plains is favored over the deciduous forests to the east by a deficiency of rainfall late in the growing season.

Grassland climates range from continental in the central grasslands to mediterranean in the California and Palouse grasslands and to dry subtropical in the desert grassland. The Great Plains climate has been characterized as having severe, windy, dry winters, and with little snow accumulation, relatively moist springs in most years, and summers that are often droughty and punctuated by thunderstorms (Borchert 1950). Except in the California grasslands and the Palouse prairie, approximately two-thirds of the precipitation in the North American grasslands occurs during the growing season (Sims et al. 1978).

The seasonal dynamics and movements of air masses (maritime polar, maritime tropical, continental polar, and continental tropical) over long distances and physiographic features (Thornthwaite 1933, Willet 1949; Brunnschweiler 1952; Harlan 1956) result in a climatic range for grasslands greater than that for any other North American biome (Collins 1969).

The distribution of the grasslands roughly parallels the prevailing north-south precipitation zones (Collins 1969). The historical distribution and present distribution of this youthful biome, however, are functions of complex factors, such as precipitation-evaporation ratios, seasonal precipitation-temperature regimes, precipitation-soil interactions (McMillan 1959a; Risser et al. 1981; McNaughton et al. 1982), fires of natural and human origin (Sauer 1950; Wright and Bailey 1982; Axelrod 1985), and the impact of the large prehistoric browsers (Axelrod 1985) and the more recent grazers (Heady 1975; Anderson 1982).

Solar radiation across the North American grasslands ranges from 5×10^9 J m^{-2} yr^{-1} in the northern mixed-grass prairie to 8×10^9 J m^{-2} in the desert grassland (Sims et al. 1978). Mean annual air temperatures vary from about 3°C in the mountain grassland to 4–8°C in the northern mixed-grass prairie to 15°C in the tall-grass and southern shortgrass prairies and the desert grassland. Average annual precipitation is greatest in the tall-grass prairie (100 cm) and decreases westward across the mixed-grass prairie (50 cm) and shortgrass prairies (30 cm) to a low in the southwestern desert grassland (20 cm). Precipitation equals or exceeds potential evaporation only in the eastern tall-grass prairie and in the mountain grassland; elsewhere, evaporation exceeds precipitation. In the desert grassland, evaporation is fourfold greater than precipitation. The relatively high wind velocities across the plains are major contributors to evaporative stress.

The length of the potential growing season (the number of consecutive days with a 15-day-running mean annual temperature ≥4.4°C) ranges from about 100 d in mountain grassland to 335 d in desert grassland (Sims et al. 1978). The growing season begins as early as mid-January in desert grassland and as late as June in mountain grassland. In the mixed-grass prairie of the Great Plains, the growing season varies from 168 d at Dickinson, North Dakota, to 200 d at Cottonwood, South Dakota, to 226 d at Hays, Kansas. The growing season in the central Great Plains shortgrass prairie averages 193 d, compared with 269 d in a southern Great Plains shortgrass prairie. The tall-grass prairie site in northeastern Oklahoma has a growing season of 270 d.

Paleobotanical evidence indicates a gradual shift in vegetation in the central plains, between the middle Miocene and the early Holocene, from a largely semiopen forest and woodland, with scattered grassy areas, to an open grassland, with trees and woodlands along breaks and escarpments (Axelrod 1985). The shift to grasslands occurred as aridity increased and drought became more frequent west of the 100th meridian. Natural and aborigine-caused fires, associated with ample flammable plant material during dry periods following a few years of adequate rainfall, moved uninterrupted across the relatively level plains at sufficient frequency to restrict the occurrence of trees and shrubs. Thus, the grasslands as the European settlers first saw them and as we know them today may be as much the result of recurrent fires (Gleason 1913, 1923; Sauer 1950; Curtis 1962; Axelrod 1985) as of climate (Clements 1916; Thornthwaite 1933; Borchert 1950; Coupland 1979). Anderson (1982) suggested that the confusion regarding the roles of fire and climate in grassland maintenance and distribution can be lessened by understanding that fire and climate do not have separate and unrelated effects on grasslands and that the responses of grasslands to fires can be quite different in different climatic regimes.

Physiography

North American grasslands are found in several physiographic regions in the western three-fifths of the continent. The central grasslands occur in three physiographic regions: Great Plains, Coastal Plains, and Central Lowlands. These compose a vast plain that slopes gently downward to the east, at a rate of about 1.2 m km^{-1}, from the base of the Rocky Mountains on the west to the Central Lowland on the banks of the Mississippi River in the upper Midwest (Hunt 1972; Lewis and Engle 1980). These plains extend from the Mackenzie River delta in Canada to the Balcones Fault in south Texas (Hunt 1972).

The pristine grasslands of California occupied the central valleys of the Pacific Border physiographic province between the Coast Range to the west and the Sierra Nevada on the east. They also occupied a narrow coastal strip in central to northern California and a broader coastal zone in southern California now covered by the Los Angeles–San Diego urban area (Barry 1972). Central Valley alluvial-floodplain elevations vary from sea level to 520 m, with the majority of the area being less than 150 m. The region is composed of a network of gently sloping valleys bordered by sloping terraces and fans and steep uplands (Austin 1965).

The Palouse prairie, in eastern Washington and Oregon and western Idaho, lies principally in the lava-rich Walla Walla section of the Columbia Plateau, between the Rocky and Cascade mountains. This inland basin, in an orographic rain shadow of the Cascades, slopes gradually downward from east to west (1070–120 m) and was formed by glacial abrasion and deposition, followed by erosion and sedimentation. Loess and ash deposited over the basalt plain are moderately dissected by deep canyons along the major streams, with resulting hilly areas and steep slopes. Elevations range from 180 m along streams to 600–1200 m over most of the plain (Austin 1965).

The desert grasslands are in the Colorado Plateau and the southern and eastern extensions of the Basin and Range physiographic regions. They lie between the Colorado River in Arizona and the Pecos River in New Mexico and extend from the Rocky Mountains on the north, southward into Mexico. The Colorado Pleateau is a large structural depression composed of the deeper San Juan Basin in New Mexico and a shallower basin underlying Black Mesa to the west in Arizona (Hunt 1972). It consists of extremely dissected tablelands, sloping to the west and south of the Rocky Mountains (WLGU 1964). In Mexico, the desert grasslands are found in northern Sonora and along the east slope of the Sierra Madre Occidental in Chihuahua, Durango, and Zacatecas (Rzedowski 1978).

Soils

The compositions of grassland soils are functions of regional physiography and interacting climate and vegetation. Nearly all of the wetter and cooler grasslands are associated with Mollisols, soils developed through processes unique to grasslands. Aridisols are the common soils of the arid grasslands. Entisols and Inceptisols, which are soils with little or limited profile development, are often found in grasslands in association with Mollisols and Aridisols.

The dominant soil-forming process for Mollisols is melanization: darkening of the soil profile by addition of organic matter (Buol et al. 1980). A mollic epipedon (dark surface horizon) is formed, and it extends to varying depths, depending on the amount of rainfall and temperature. Melanization results from root penetration into the developing soil and a partial decaying of this root material, leaving dark, relatively stable compounds (Hole and Nielsen 1968). Soil animals, such as rodents, ants, earthworms, moles, and cicada nymphs, rework the soil and organic matter and form characteristic dark soil–organic-matter complexes, krotovinas, and mounds. Through further eluviation and illuviation of organic and mineral colloids, the surfaces become coated with dark cutans. Decay-resistant lignoprotein residues give rise to the dark colors that remain in many grassland soils even after long periods of disturbance.

The processes for Aridisol development are similar to those for Mollisols, except that the reactions are less intense. Aridisols are dry most of the year, even when temperatures are adequate for plant growth (Buol et al. 1980). Aridisols formed where potential evapotranspiration greatly exceeded precipitation during most of the year and very little water percolated through the soil (Buol et al. 1980). The leaching that may be evident in aridisols probably was caused by erratic rainfall patterns in periods of humid paleoclimates (Smith 1965).

Grassland soils differ markedly from those found under a forest canopy because the processes of soil formation and the influences of organisms, temperature, and moisture are unique to each vegetation type (Coupland 1979). In grasslands, leaching is usually restricted because of scarcity of water. Consequently, grassland soils are more basic than forest soils. The acidity of forest soils reduces the rate of organic-matter decomposition and increases mineralization and translocation of the solution below the rooting zone. The organic debris tends to accumulate at the surface in forest soils, which have very little humus within their profiles. Generally, there is organic matter throughout the grassland soil profile. Grassland soil colors range from black in wetter communities to brown in semiarid areas. Forest soils are brown to gray, depending on the degree of leaching, which removes iron while the silica remains in the profile.

The 195 soil mapping units identified in the Great Plains by Aandahl (1982) illustrate the general influences of temperature and moisture on the morphogenesis of soils derived from diverse parent materials. The topography of the region of mollisols generally ranges from level to undulating, with local differences being less than 10 m. More diverse topography occurs in the Badlands, the bluffs of the Missouri River, the Missouri Cotean, the Prairie Coteau of the Dakotas, the Nebraska sandhills, the Kansas and Oklahoma flint hills, and the Edwards Plateau of Texas.

Glacial deposits and associated outwash sands and gravels are the principal parent materials north of the Missouri River in the northern Great Plains. Sandstone- and shale-derived soils are found south and west of the Missouri River in Montana and the Dakotas and in northeastern Wyoming and northwestern Nebraska. Extensive loess and eolian sand deposits are found in Nebraska, Kansas, eastern Colorado, southeastern Wyoming, the southern high plains, and along major streams. Areas just east of the Rocky Mountains have soils that are deep loamy sediments of loess, eolian sand, alluvium, and mountain-outwash origin. Large areas of fine-textured soils are found in the Texas panhandle and adjacent areas.

Soils of the southern plains vary from fine sands to heavy clay loams. A strip of relatively fertile, medium brown Alfisols stretches from southeastern Kansas southward across east central Oklahoma into central Texas, generally following the Cross-timbers vegetation type, as mapped by Küchler (1975). Alfisols are also found in much of the southern high plains shortgrass prairie in west Texas and eastern New Mexico. Mollisols dominate the remainder of the southern plains. Mollisols are also the primary soils of the Palouse prairie, with Aridisols being an important secondary soil group. California grassland soils are principally Entisols, whereas the desert grassland soils are Aridisols.

MAJOR GRASSLAND TYPES

North American grasslands contain approximately 7500 plant species from about 600 genera of grasses, plus numerous grass-like plants, forbs, and woody plants (Hartley 1950, 1964; Risser 1985). The graminoids rank third in number of genera, fifth in number of species, and first in geographical distribution, and comprise the greatest percentage of the total world vegetation (Gould 1968).

The nature of the biotic components of grasslands is controlled by climate. Plants of the cool, temperate element function best at the lower temperatures and moderate evaporative stresses found at the higher elevations and northern latitudes (Fig. 9.2). Taxa of the desert grassland and the shortgrass and mixed-grass prairies primarily have southwestern affinities, whereas the more mesic eastern tall-grass prairie has eastern and southeastern forest-border species (Transeau 1935; Harlan 1956; Axelrod 1985). Important North American grassland genera representing these two elements are shown in Table 9.2.

Grasslands are represented by several major associations and societies interrelated by the common occurrence of similar or allied species of several genera (Axelrod 1985). For convenience, grassland vegetation has been abstracted into discrete com-

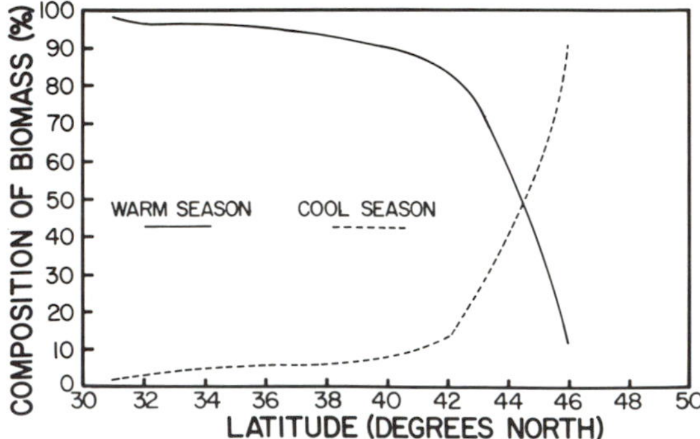

Figure 9.2. Changes in the relative compositions of warm-season and cool-season plants across the North American grasslands from the southern to the northern latitudes based on data from Sims, Singh, and Lauenroth 1978).

Table 9.2. *Genera of cool-temperate and warm-temperate origins that are important in the grasslands of North America*

Cool-temperate	Warm-temperate
Agropyron	*Andropogon*
Agrostis	*Bouteloua*
Bromus	*Buchloë*
Dactylis	*Chloris*
Danthonia	*Eragrostis*
Deschampsia	*Muhlenbergia*
Elymus	*Panicum*
Festuca	*Paspalum*
Hordeum	*Schizachyrium*
Poa	*Setaria*
Secale	*Sorghastrum*
Stipa	*Sporobolus*
Triticum	

munities such as tall-grass, mixed-grass, and short-grass prairies, and so forth. Such generalization should be used carefully, because two grassland communities containing the same species may still differ ecologically because of genetically variable ecotypes (McMillan 1959b). The east-west precipitation gradient, on top of a north-south temperature gradient, gives rise to diversity in soils (Jenny 1930) that supports floristically and functionally complex plant communities. The abiotic gradients that exist across the landscape are inhabited by genetically specialized plants that match the constraints of the environment (Beetle 1947).

Figure 9.3. A tall-grass prairie site in the Flint Hills of Kansas showing the grassland vegetation interspersed with trees and shrubs along the breaks.

Plains Grasslands: Tall-Grass Prairie

The tall-grass prairie is the most mesic of the grasslands of the central plains. According to Risser and associates (1981), the tall-grass prairie receives the most rainfall and has the greatest north-south diversity and the largest number of dominant species of any association within the grassland formation. The tall-grass prairie, like the other central grasslands, has plant species originating from several geographical sources. These species have been subjected to a wide range of climates over the long term and consequently have developed rather broad ecological amplitudes and are distributed over relatively large geographical ranges (Risser et al. 1981). The vegetation is primarily of long-lived perennials that are highly developed and integrated. Although the physiognomy of the tall-grass prairie is relatively homogeneous, it varies across the landscape with changing climate and soils (Weaver 1954 (Fig. 9.3).

Three grassland associations are commonly grouped into the tall-grass prairie (Küchler 1975). The bluestem prairie, also known as the "true" prairie, extending from the southern tip of Manitoba through eastern North Dakota and western Minnesota southward to eastern Oklahoma, is dominated by *Andropogon gerardii, Panicum virgatum,* and *Sorghastrum nutans*. The second community, dominated by *Agropyron, Andropogon*, and *Stipa* species, originally occurred from south central Canada down through east central North and South Dakota and Nebraska to north central Kansas. The third area is the Nebraska Sandhills, with an *Andropogon, Calamovilfa*, and *Stipa* community. Bailey ((1976, 1980) combined the two latter communities to form a wheatgrass, bluestem, and needlegrass prairie. He also included the bluestem-grama prairie of western Kansas and Oklahoma in the tall-grass prairie. I shall consider the bluestem-grama prairie as a part of the mixed-grass prairie.

To the east of the tall-grass prairie lies the grassland-forest ecotone. Küchler (1975) identified two communities that border the tall-grass prairie: (1) the upper Midwest *Quercus-Andropogon* type in North Dakota and around the Great Lakes and (2) the *Juniperus-Quercus-Sporobolus-Andropogon* type of western Tennessee, Alabama, Missouri, and Arkansas. See also Chapter 10.

The vegetation of the tall-grass prairie is composed of bunchgrasses and sod-forming grasses. The dominant grasses are *Andropogon gerardii* and *Schizachyrium scoparium*, with *Sorghastrum nutans* and *Panicum virgatum* contributing significantly to aspect and biomass. *Sporobolus asper* is an important intermediate-height grass, especially in grazed

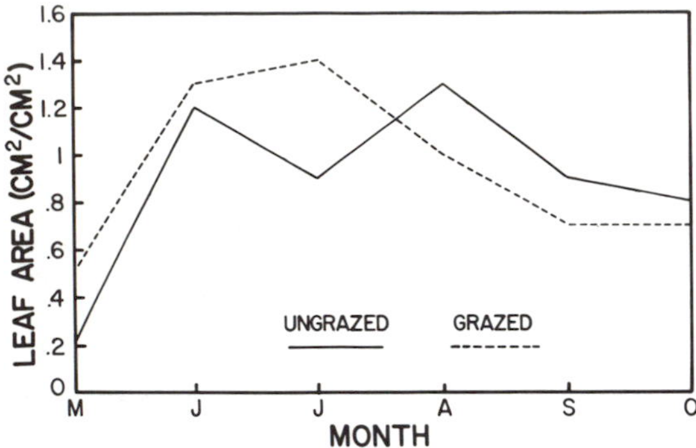

Figure 9.4. Seasonal progression of the actual leaf area during the growing season for ungrazed and grazed tall-grass prairie (adapted from Cnant and Risser 1974).

areas. Although grasses produce 80%–90% of the biomass, forb species exceed graminoid species by threefold to fourfold (Barbour et al. 1980).

Andropogon gerardii-dominated communities are the most prevalent in the bluestem prairie, particularly on the lowlands and other wetter sites (Weaver 1954). *Schizachyrium scoparium* dominates the uplands, especially on the more shallow slopes, whereas *Stipa spartea* and *Sporobolus heterolepis* are important on upland associations where soils are more shallow, rocky, or sandy.

Most of the canopy is less than 1 m tall. The average canopy height is 25 cm at the beginning of the growing season and increases to 50 cm by late summer. Leaf area is uniformly distributed up to average canopy height, and above that it rapidly decreases. The leaf area for live tissue peaks at about 1.3 $cm^2\ cm^{-2}$ in the middle of the growing season (Fig. 9.4). There is a relatively large amount of standing dead material within the canopy at all times, and Conant and Risser (1974) found as much as 1.8 $cm^2\ cm^{-2}$ in the tall-grass prairie of northeastern Oklahoma.

Most of the tall-grass prairie is now in cultivation. The true prairie remains as important grasslands in the Osage and Flint Hills of Oklahoma and Kansas, in the Nebraska Sandhills, and in isolated locations throughout the Central Lowlands geographical region. Livestock grazing is a primary use, and many ranchers take pride in maintaining the grassland in good condition. There are isolated tracts in the Great Plains and a few areas in the northern part of the Great Plains. The tall-grass prairie that remains does so usually because of topography or rockiness that makes farming untenable.

The effects of grazing on the tall-grass prairie at three widely spaced locations in Texas, Oklahoma, and eastern Nebraska have been studied (Dyksterhuis 1949; Voight and Weaver 1951; Sims and

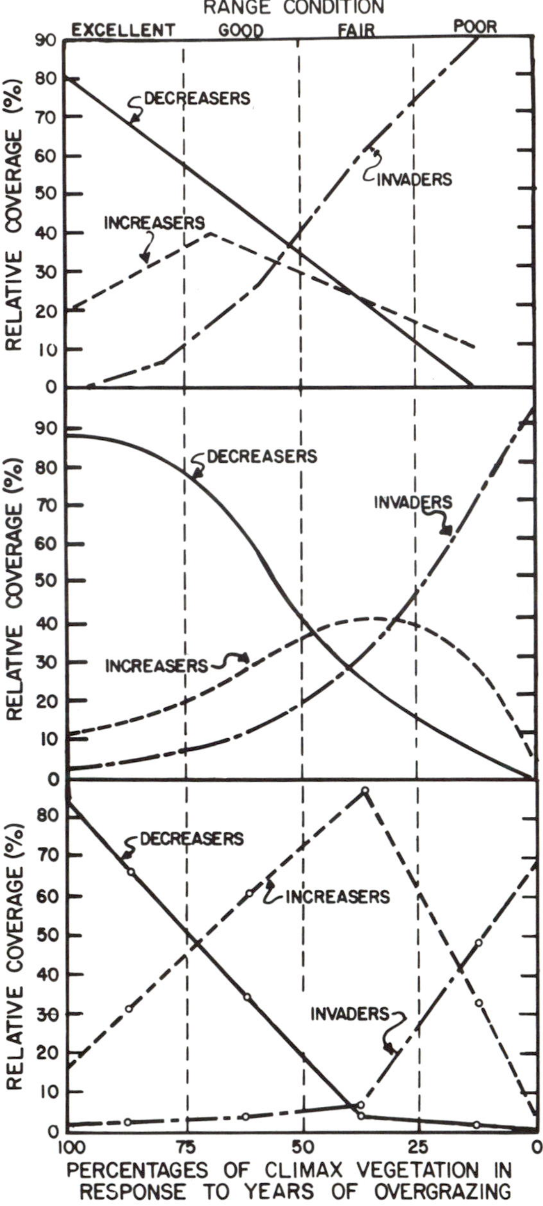

Figure 9.5. Relative changes in species compositions for (top to bottom) southern (Fort Worth Prairie) (Dyksterhuis 1949), central (north central Oklahoma) (Sims and Dwyer 1965), and northern (eastern Nebraska) (Voight and Weaver 1951) tall-grass prairie sites.

Dwyer 1965) (Fig. 9.5). In the Fort Worth prairie, the decreaser species–the most palatable and productive plants–declined from 67% on excellent-condition rangeland to less than 2% on poor-condition range. Increaser species–the less palatable species present in climax grasslands that replace the decreaser species–increased from 3% on excellent-condition prairie to 88% on fair-condition sites. Invader plants—those that were not present originally, but invade the site as a plant community is opened up—increased from 2% on excellent-condition rangeland to 47% on poor-condition rangeland.

Schizachyrium scoparium, Andropogon gerardii, Sorghastrum nutans, and *Panicum virgatum* were the decreaser species found in the tall-grass prairie in north central Oklahoma (Sims and Dwyer 1965) (Fig. 9.5). These species composed 80%–93% of the vegetation in the excellent-condition grasslands studied. As overgrazing occurred, the species composition of these palatable plants declined totally or to an absolute minimum. *Panicum virgatum* and *Sorghastrum nutans* were components of the best-condition grassland and were the first of the tall grasses to disappear. *Schizachyrium scoparium*, dominant in excellent-condition grasslands, declined with heavier grazing.

Andropogon gerardii was the second most abundant species in excellent-condition grasslands and persisted even in the most overgrazed sites, probably because of its short, strong rhizomes. It did, however, decline with mowing pressures more than from heavy grazing. In contrast to *Andropogon gerardii, Sorghastrum nutans* was favored by mowing and reduced by grazing.

Responses of forb populations to grazing pressure are somewhat more erratic and fluctuate with current moisture conditions more than do the responses of perennial grasses. The mulch cover declines with grazing intensity, and this decline results in less infiltration and more runoff. Consequently, with equal amounts of rainfall, soil moisture conditions are lower on poor-condition tall-grass prairie sites than where the prairie is in a more pristine state.

Bouteloua curtipendula, B. gracilis, and other increaser plants become more abundant as the tall grasses decline, up to a point. The primary invader was found to be *Bouchloë dactyloides,* which increased from zero on excellent-condition pastures to 70% relative cover on poor-condition rangeland. Because this plant is a sod-forming species, the basal cover increased from 12.5% to more than 23% as range condition declined.

Degeneration of the tall-grass prairie in the Nebraska Sandhills area has been less marked than in other places. The known fragility of the soils may have cautioned ranchers to avoid destructive grazing practices. Changes that have occurred include decreases in *Schizachyrium scoparium, Stipa comata,* and *Andropogon hallii* and increases in *Bouteloua hirsuta, Calamovilfa longifolia,* and *Sporobolos cryptandrus.*

Historically, fire played a beneficial role in preserving the tall-grass prairie. Long-term protection of the tall-grass prairie from fire has led to an increase in woody vegetation (Kucera 1960; Penfound 1964). Fire will supress the encroachment of trees and shrubs, reduce competition from cool-season invaders such as *Poa pratensis* and *Bromus inermis,* and improve the palatability and nutritional value of the grazable forage (Wright and Bailey 1980). Late spring burning generally benefits the major tall grasses. Forbs may be reduced in growth, but their diversity is unaffected (Anderson 1965; McMurphy and Anderson 1965; Kucera 1970).

The Crosstimbers (Küchler 1975), often associated with the tall-grass prairie, lies in a band 10–180 km wide from the southern edge of the bluestem prairie in Kansas southward across east central Oklahoma down to the Trinity River in east Texas. The overstory vegetation consists primarily of *Quercus stellata* and *Q. marilandica* in a 2 : 1 ratio.

Impacts caused by humans through grazing animals, erosion, and reduction of fire have drastically changed the vegetation of Crosstimbers. Dyksterhuis (1948) found that the understory cover in undisturbed Crosstimbers areas was largely *Schizachyrium scoparium* at 65%, *Sorghastrum nutans* at 6%, and 2%–3% each for *Andropogon gerardii, Bouteloua hirsuta, B. curtipendula,* and *Sporobolus asper.* Dyksterhuis (1948) sampled 269 areas and determined that about 84% had burned within the last 10 yr. Basal cover increased 4%–7% following grazing and burning, with an invasion of annual grasses and forbs (35% relative cover) and warm-season perennial grasses (30% relative cover) such as *Buchloë dactyloides, Aristida* species, *Paspalum ciliatifolium, Stipa leucotricha, Bothriochloa saccharoides,* and *Cynodon dactylon.* The original tall grasses have been virtually eliminated, whereas *Bouteloua hirsuta* has increased somewhat, and *B. curtipendula* has remained about the same as in the original undisturbed stands.

The Blackland Prairie is intermingled with the Crosstimbers at its southern end and is often associated with the tall-grass prairie; however, soils of the Blackland Prairie are Vertisols rather than Mollisols. The immature profiles lie over soft limestone parent material. This vegetation has been largely converted to farmland. A few areas remain in grassland used for ranching.

Gould (1962) indicated that the Blackland Prairie should be classed as true prairie, with *Schizachyium scoparium* as the climax dominant. Important associated species are *Andropogon gerardii, Sorghastrum nutans, Panicum virgatum, Bouteloua curtipendula, B. hirsuta, Sporobolus asper, Bothriochloa saccharoides,* and *Stipa leucotricha.* Heavy grazing causes *Stipa leucotricha* to increase and *Buchloë dactyloides* and *B. rigidiseta* to invade. Annual grasses such as *Hordeum*

Figure 9.6. A mixed-grass prairie site near Antelope, South Dakota, where the overstory of cool-season grasses hides the shorter warm-season grasses.

pusillum, Bromus japonicus, B. catharticus, and *Vulpia octoflora* invade much of the same as in Crosstimbers.

Plains Grassland: Mixed-Grass Prairie

The mixed-grass prairie, recognized as a distinct plant association by Clements (1920), is a blend of the vegetation of the tall-grass and shortgrass prairies. The boundaries of this grassland are less well defined, because it serves as an ecotone between the tall-grass and shortgrass prairies. In the northern Great Plains, the mixed-grass prairie lies to the west of the tall-grass prairie and is found in the western Dakotas, northeastern Wyoming, eastern Montana, and then on into the southern parts of the central Canadian provinces (Fig. 9.6). The mixed-grass prairie in Wyoming and Montana is similar in appearance to the shortgrass prairie. South of the Nebraska Sandhills, a broad belt of mixed-grass prairie extends from southwest Nebraska through western and central Kansas and Oklahoma,and on into central Texas.

In the mixed-grass prairie, an array of intermediate-and short-stature grasses is sometimes complemented by tall grasses. These grasses, along with a large number of forbs, a few suffrutescents, and a scattering of low-growing shrubs, result in the richest floristic complexity of all the grasslands (Barbour et al. 1980). Typical dominants of this grassland are species of *Schyzachrium, Stipa, Agropyron, Calamovilfa, Bouteloua,* and *Sporobolus*. These grasses combine with the tall grasses in the wetter areas and a number of short grasses in the drier sites in the mixed-grass prairie. Primary short grasses are species of *Bouteloua, Buchloë, Muhlenbergia,* and *Aristida*. Sedges (*Carex* species) are also important in the mixed-grass prairie.

Whitman (1941) conducted intensive studies of the mixed-grass prairie in North Dakota. He listed the dominant species as *Bouteloua gracilis, Schizachyrium scoparium, Stipa comata, Agropyron smithii, Carex filifolia, Koelaria cristata,* and *Poa secunda*. Redmann (1975) made a detailed study of 10 excellent-condition mixed-prairie stands representing certain features of soils, topography, and aspect that effected a moisture gradient. Redmann was able to group the stands into three communities corresponding to soil types. Stands on the rolling upland with fine-textured soils were dominated by *Agropyron smithii, Carex pennsylvania,* and *Stipa comata* or *S. viridula*; stands in the lower topographic positions on medium-textured soils were dominated by *Sporobolus heterolepis*; sites at the higher topographic position on coarse-textured soils were dominated by *Schizachyrium scoparium*. Additional stands were identified that were more or less slope- and aspect-related. A stand on a steep north-facing slope was dominated by *Stipa spartea* and numerous forbs such as *Helianthus rigida, Aster ericoides,* and *Brauneria angustifolia*. In contrast, a stand on a south-facing slope was dominated by *Schizachyrium scoparium* and *Andropogon gerardii*.

Albertson (1937) described a mixed-grass prairie in west central Kansas, and he separated the vegetation into three distinct communities, with *Bouteloua*

Table 9.3. *Comparison of the relative contributions (%) of warm- and cool-season grasses to season-long biomass production in ungrazed (U) and grazed (G) areas for three mixed-grass prairies in the central grasslands of North America*

	Grasses				Others[a]	
	Warm		Cool			
Prairie site	U	G	U	G	U	G
Dickinson, N.D.	17	27	58	45	25	28
Cottonwood, S.D.	19	74	78	18	3	8
Hays, Kns.	85	80	<1	4	15	16

[a]Primarily warm-season forbs.
Source: Sims et al. (1978)

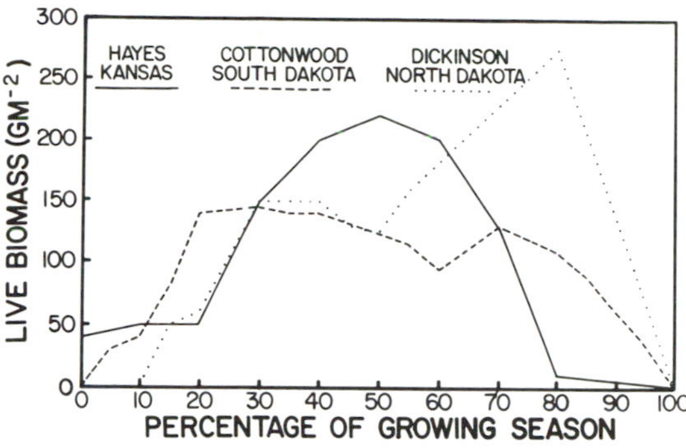

Figure 9.7. Distribution of seasonal live biomass at three locations of the mixed-grass prairie in the central and northern Great Plains (based on data from Sims and Singh 1978a).

gracilis–Buchloë dactyloides, Schizachyrium scoparium, or *Andropogon gerardii* dominating as the habitat ranged from a dry upland site to more mesic slopes and down to a well-watered lowland site.

The seasonal distribution of live biomass has been found to vary depending on the mix of cool- and warm-season plants present in the mixed-grass prairie community (Table 9.3) (Sims et al. 1978). A single peak in live biomass occurred in the mixed-grass prairie in central Kansas that was dominated by warm-season plants (Fig. 9.7), but at the northern sites in South and North Dakota, a bimodal pattern emerged as a result of production of biomass from first the cool-season plants and then the warm-season plants (Sims and Singh 1978b).

Ode and associates (1980) found 27 C_4 and 278 C_3 species for a total of 305 species in the mixed "Ordway" prairie in north central South Dakota. The production activity of the C_3 and C_4 plants, as measured by delta ^{13}C values (Tieszen et al. 1979a, 1979b), indicated a seasonal displacement in production activity for C_3 and C_4 grasses. The relative contributions of $C_3 : C_4$ plants early, midway, and late in the growing season for an upland site were about 90 : 10, 55 : 45, and 85 : 15, respectively. Corresponding figures for a lowland site were 100 : 0, 85 : 15, and 95 : 5.

The mixed-grass prairie is the most dynamic of the central grasslands because of the climatic extremes across this layered vegetation, with its flora of diverse origins. The mixed-grass prairie is, in some respects, a tension zone between the tall-grass and shortgrass prairies, with the vegetation fluctuating with climate, suppression of fire, and the degree and frequency of grazing by domestic livestock and wildlife. As the climatic cycle becomes drier and grazing further increases the aridity of the environment, the shorter and more drought-tolerant species become more important in the plant community.

The general topography of this grassland is rolling, with an intermingling of uplands and low-lying areas at the bases of slopes of varying lengths. The uplands are codominated by the sod-forming *Bouteloua gracilis* and *Buchloë dactyloides* during more normal climates (Table 9.4). Associated secondary taller grasses, such as *Agropyron smithii, Schizachyrium scoparium,* and *Bouteloua curtipendula,* form a distinct upper layer, but are rarely abundant.

Albertson (1937) found that drought cycles had reduced the density and cover of the short grasses by as much as 70%–80% and those for the bunch-grasses, along with many of the native forbs, almost to zero. When normal precipitation returned, the stoloniferous *Buchloë dactyloides* was the most aggressive and assumed a more dominant position than the slow-to-recover *Bouteloua gracilis. Sphaeralcea coccinea* is an important component of the short-grass habitat in the mixed-grass community. Albertson (1937) fund that this forb tended to withstand dry cycles and remained in the stand, but most of the native forbs were readily replaced during dry cycles by an array of weedy and invasive forbs.

Schizachyrium scoparium occurred in distinct bunches and dominated the slopes, accompanied by scattered clumps of *Andropogon gerardii* and isolated plants of *Panicum virgatum* and *Sorghastrum nutans.* Small tufts of *B. gracilis* and *B. hirsuta* were present. During excessively dry years, *S. scoparium* declined significantly, whereas the deeper-rooted and strongly rhizomatous *A. gerardii* increased (Albertson 1937).

Weaver and Albertson (1956) found that the *A. gerardii* habitat type in the mixed-grass prairie was closely allied to the tall-grass prairie. It occurred at the bases of the slopes and in ravines and valleys

Table 9.4. *Dominant, principal, and secondary species for three habitat types in the mixed-grass prairie in west central Kansas*

Short grasses	Little bluestem	Big bluestem
Dominants		
Bouteloua gracilis	*Schizachyrium scoparium*	*Andropogon gerardii*
Buchloë dactyloides		*Agropyron smithii*
		Bouteloua curtipendula
		Sporobolus drummundii
Principal grasses and sedges		
Agropyron smithii	*Andropogon gerardii*	*Andropogon torreyanus*
Schizachyrium scoparium	*Bouteloua curtipendula*	*Carex gravida*
Aristida purpurea	*Bouteloua gracilis*	*Elymus canadensis*
Bouteloua cutipendula	*Bouteloua hirsuta*	*Elymus virginicus*
Carex praegracilis	*Panicum virgatun*	*Panicum virgatum*
Sitanion elymoides	*Sorghastrum nutans*	*Poa arida*
		Sorghastrum nutans
Grasses of secondary importance		
Alopecurus carolinianus	*Buchloë dactyloides*	*Bouteloua gracilis*
Andropogon gerardii	*Eatonia obtusata*	*Buchloë dactyloides*
Distichlis stricta	*Vulpia octoflora*	*Sporobolus asper*
Festuca octòflora	*Koeleria cristata*	*Sporobolus cryptandrus*
Hordeum pusillum	*Sporobolus asper*	
Munroa squarrosa	*Sprobolus cryptandrus*	
Sporobolus asper	*Triodia acuminata*	
Sporobolus cryptandrus		
Principal forbs		
Ambrosia psilostachya	*Ambrosia psilostachya*	*Amorpha canescens*
Anemone caroliniana	*Amorpha canascens*	*Aster multiflorus*
Antennaria campestris	*Echinacea pallida*	*Erigeron ramosus*
Aster multiflorus	*Lacinaria punctata*	*Psoralea tenuiflora*
Astragulus spp.	*Meriolix serrulata*	*Salvia pitcheri*
Cirsium undulatum	*Psoralea tenuiflora*	*Verbena stricta*
Gaura coccinea	*Tetraneuris stenophylla*	*Veronia baldwini*

Source: Albertson (1937).

where soil moisture conditions were enhanced by runoff from the uplands and slopes, where snow accumulated, and where moisture losses from soil and plants were reduced because of decreased wind (Weaver and Albertson 1956). *A. gerardii* composed 50% – 90% of the vegetation, with *S. nutans* a common but not abundant associate. *P. virgatum, Elymus canadensis*, and *E. virginicus*, along with the tall sedge *Carex gravida*, were often intermixed, but were not abundant except on the wetter expressions of this habitat type. Many rather large forbs occurred in this type (Table 9.4). Insufficient light penetrated the dense overstory of the tall species to allow the short grasses to exist, even though the basal cover of the tall-grass prairie was only 8% – 13%.

The major changes in vegetation in the mixed-grass and shortgrass prairies have primarily been associated with both grazing pressure and drought. Fire was only of secondary importance in this regard. During the drought of the 1930s, basal cover of grasses on even moderately grazed grasslands, but especially on heavily grazed grasslands, declined from 80% or more to less than 10% in 3 – 5 yr. The perennial grasses declined, and annuals such as *Salsola kali* increased and dominated certain grasslands.

E. H. McIlvain and M. C. Shoop (unpublished data) summarized 20 yr of grazing studies in the mixed grassland of northwest Oklahoma (Fig. 9.8). They found that after the drought of the 1930s, the basal cover of tall grasses, *Bouteloua gracilis*, and *Paspalum stramineum* expanded at the expense of *Sporobolus cryptandrus*, an aggressive drought-tolerant plant. Similarly, during the drought of the 1950s, *S. cryptandrus* increased, while *B. gracilis* and the tall grasses declined. The drought in the early 1950s caused a marked decline in the tall grasses and in *B. gracilis* and *P. stramineum*, with a slight increase in *S. cryptandrus*. As the drought ended and more nearly average precipitation returned, tall grasses expanded, along with *B. gracilis* and *P. stramineum*, while *S. cryptandrus* declined.

The same vegetation trends occurred under grazing and protection from grazing (Fig. 9.8).

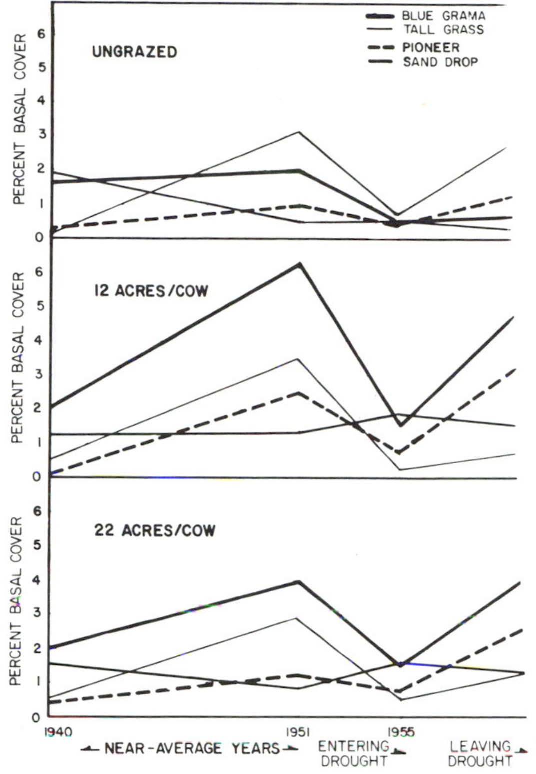

Figure 9.8. Changes in an Oklahoma mixed-grass prairie in relation to changing climatic conditions over time and grazing intensity (McIlvain and Shoop, unpublished data).

However, as the grazing intensity increased to 12 acres per cow (a somewhat heavy grazing intensity), the changes became more dynamic. For example, *Bouteloua gracilis* fluctuated from about 2% basal cover in ungrazed areas to about 6% with heavy grazing, compared with somewhat less than 4% for light grazing. During the drought of the 1950s, which was a very severe drought, perhaps even worse than the drought of the 1930s, blue grama in ungrazed grasslands declined to much less than 1% of basal area. In contrast, under heavy grazing, blue grama expanded to 6%, but then declined during the severe drought of the early 1950s to less than 2% (still more than was present in the ungrazed condition). Following the drought of the 1950s, blue grama increased at all three levels of grazing. The dynamics of species changes in the mixed-grass prairie are functions of climate, but the magnitudes of these changes are greatly influenced by the intensity of grazing.

McMillan (1956a, 1956b, 1957, 1959a) found that the mechanism controlling grassland distribution involved the intricate relationships between vegetation genetics and habitat gradients. Climate was a dominant gradient factor, although not the only factor. Clones, collected from Canada to Mexico, of *Andropogon gerardii, Bouteloua curtipendula, B. gracilis, Elymus canadensis, Koeleria cristata, Oryzopsis hymenoides, Panicum virgatum, Schizachyrium scoparium, Sorghastrum nutans, Sporobolus heterolepis, Stipa comata*, and *S. spartea* have been evaluated by McMillan. Other work on the photoperiodic responses of major grass species includes studies of *Schizachyrium scoparium* (Larsen 1947), *Bouteloua curtipendula* (Olmsted 1944), *B. gracilis* (Riegel 1940), and *B. curtipendula, Sorghastrum nutans, Panicum virgatum, Schizachyrium scoparium*, and *Andropogon gerardii* (Rice 1950).

McMillan (1959a) defined three general patterns of flowering behavior from transplant garden studies at Lincoln, Nebraska. The most prevalent pattern was one of earlier flowering of northern clones, followed by progressively later flowering of other clones within the same species that were collected farther south and east. Major central grassland species with this flowering pattern were *Panicum virgatum, Schizachyrium scoparium*, the *Andropogon gerardii-hallii* complex, *Sorghastrum nutans, Sporobolus heterolepis, Koeleria cristata, Bouteloua gracilis*, and *B. curtipendula*. *Elymus canadensis* essentially had a reverse flowering pattern. *Stipa spartea, S. comata*, and *Oryzopsis hymenoides* displayed no geographically oriented flowering trends. In these studies, McMillan observed the point of initiation of spring activity, the time of floral initiation, flowering patterns, height correlations, caryopsis production, and the time of senescence.

Fire has been used to increase forage production and the palatability of coarse grasses in the southern plains mixed-grass prairie (Wright and Bailey 1980). Most grasses, except *Bouteloua curtipendula* and the cool-season bluegrasses, appear to be relatively tolerant of prescribed fires (Wright 1974). Planned fires have been used to eliminate undesirable annual grasses and forbs and to suppress *Prosopis glandulosa* var. *glandulosa, Juniperus virginiana* and *Opuntia* species (Wright and Bailey 1980).

Limited studies in the northern mixed-grass prairie indicate that early spring burns may enhance the stand of native grasses, whereas fires later in the growing season reduce both the warm- and cool-season grasses (Schripsema 1977). The bluestem-forest ecotone sites in the Black Hills of South Dakota may be more tolerant of fire than other vegetation communities in the northern plains. Gartner and Thompson (1972) found that big and little bluestem and forbs increased following fire, whereas the frequency of cool-season grasses declined. Winter burns had the opposite effect of increasing the cool-season *Agropyron smithii* and *Stipa* species at the expense of the warm-season grasses.

Plains Grassland: Shortgrass Prairie

The shortgrass prairie is a large contiguous expanse of vegetation that lies to the east of the Rocky Mountains from the Nebraska panhandle and southeastern Wyoming through eastern Colorado and western Kansas southward through the high plains of Oklahoma, Texas, and New Mexico. Similar grasslands are found in northern Arizona (Stoddard et al. 1975). The intermediate-height grasses of the mixed-grass prairie are secondary in importance, and only a sprinkling of tall grasses can be found in this grassland on wetter sites.

The shortgrass prairie was called a grazing disclimax by Weaver and Clements (1938) and McComb and Loomis (1944), who considered it a part of the original mixed-grass prairie, based on climatic factors. Larson (1940), however, called the shortgrass prairie a true climax, based on historical records of conditions at the time of the arrival of Europeans in this region and by equating the impact of domestic animal grazing to that of the buffalo of earlier days. Sampson (1950) indicated that grazing may have had an impact on the extent of the shortgrass prairie at its more mesic distribution, but the drought-enduring short grasses were climax in the vast, arid areas of the Great Plains.

Bouteloua gracilis and *Buchloë dactyloides* are the dominant sod-forming grasses of the shortgrass prairie (Fig. 9.9). Other grasses that characterize this prairie are *Agropyron smithii, Sporobolus cryptandus, Muhlenbergia torreyi, Stipa comata, Koeleria cristata*, and *Hilaria jamesii*.

Figure 9.9. A shortgrass prairie in eastern Colorado with Opuntia species interspersed with Buchloë dactyloides and Bouteloua gracilis.

In eastern Colorado and in Montana and Wyoming, *Artemisia frigida, Carex filifolia*, and *Koeleria cristata* are more important to the shortgrass community (Sampson 1950). Stoddard and associates (1975) stated that as the shortgrass prairie extends into Canada it takes on an aspect typical of the mixed-grass prairie characterized by Weaver and Albertson (1956). Because of the vast expanse of the shortgrass prairie, the impacts of two environmental gradients are quite apparent: decreasing temperature from south to north and increasing precipitation from west to east. Consequently, cool-season species have more importance toward the north, short grasses dominate in the western extremes, and mid-grasses and tall grasses play larger roles at the eastward edge of the shortgrass prairie.

When the shortgrass prairie was first grazed by domestic livestock, the original grasses persisted, probably because of their low stature and natural resistance to grazing pressure. As abuse occurred and the grasses declined, weedy perennial species of *Opuntia, Gutierrezia*, and *Yucca* increased. Invader annual species have come from the *Bromus, Salsola, Hordeum*, and *Festuca* genera. Stoddart and associates (1975) noted that dryland cultivation had been attempted unsuccessfully on large areas of the shortgrass prairie. Where natural precipitation was augmented by irrigation, high levels of crop production were attained.

The infamous dust bowl of the 1930s originated in southeastern Colorado, southwestern Kansas, and the panhandles of Texas and Oklahoma, where the shortgrass prairie was plowed for dryland farming. Many "old fields" remain today, 50 yr after cultivation attempts had failed and fields had been left to revegetate naturally. The persistence of *Aristida* species may be the result of changes in the soil following plowing that require long periods of time for restoration; a reduction in soil phosphorus may leave the site more suitable for this species than for the climax plants that are so slow to reestablish.

In the Texas high plains—a high, relatively level plateau to the southwest of the Caprock Escarpment—much of the shortgrass prairie either has been converted to irrigated farming or has been invaded by *Prosopis glandulosa* var. *glandulosa* to form a shrubland or savannah with an understory of the shortgrass prairie species. *Prosopis glandulosa* stands on upland sites with finer-textured soils are often associated with other brush and suffrutescent species: *Opuntia imbricata* var. *imbricata, O.*

polycantha, O. leptocaulis, Zizyphus obtusifolia, and *Berberis trifoliolata* (Scifres 1980). *Quercus harvardii* is the dominant brush species on sandy soils of the high plains.

Generally, fire has been found to be detrimental to shortgrass prairie plants (Wright and Bailey 1980, 1982), particularly through reduction of forage yields (Launchbaugh 1972). Lowered yields have been attributed to reduced numbers of tillers and shorter growth of plants caused by more runoff and less infiltration of soil moisture (Launchbaugh 1964). Much of the information on fire effects has been gathered from wildfires; knowledge of the effects of prescribed fires, following precise conditions, on many plant communities is limited.

California Grasslands

The original dominants of the Pacific prairie of California, which covered more than 5 million ha in California, were cool-season, perennial bunchgrasses such as *Stipa pulchra, S. cernua, Elymus* species, *Poa scabrella, Aristida* species, *Koeleria cristata, Melica imperfecta,* and *Muhlenbergia rigins* (Weaver and Clements 1938; Heady 1977; Heady et al. 1977). These perennial bunchgrasses were complemented by an array of annual and perennial grasses and forbs that flowered in the late spring.

The origin of the annual grassland vegetation is varied. Some of the species are native to the region, and others came from the north. The remaining species either originated in the south and migrated into California during interglacial periods or were introduced from other continents (Beetle 1947). The Pacific prairie shares an ecotone with the Palouse prairie in northern California and exhibits some similarity in species relationships. Weaver and Clements (1938) contended that this association also had an affinity with the central mixed-grass prairie, because relicts of *Bouteloua, Andropogon, Hilaria,* and *Aristida* species could be found in southern California, and also because *Stipa lepida* and *S. speciosa* from the Pacific grassland were also found in Arizona and New Mexico.

Currently, the Pacific prairie is dominated by "weedy" annuals of *Avena, Bromus, Lolium,* and *Erodium* species, plus a large number of other annual forbs. These invaders are extremely persistent and resist reestablishment of the original bunchgrasses. Heady (1977) called them "new natives" to emphasize their permanent nature.

Ranching began in California with the Spanish colonists in 1769 (Burham 1951). Prior to that, the grasslands were grazed occasionally by Pronghorn antelope (*Antilocapra americana*), tule elk (*Cervus elaphus nannodes*), and deer (*Odocoileus hemionus*) (Barry 1972). Antelope may have been abundant in the Central Valley, but evidently the grazing pressure was in balance with the vegetation. Livestock numbers grew gradually until the gold rush of 1849, when demand for meat brought rapid expansion. Cattle numbers peaked at about 3.5 million head in about 1860, as farming began to expand (Branson 1985). Large areas of the Central Valley have been used for cultivation, large urban centers, and industrial expansion.

European settlers introduced numerous plants to California. Species from similar environments (southern Europe, Chile, South Australia, and South Africa) have adapted well to this grassland environment (Barry 1972). The native grasses apparently lacked the ability to withstand heavy grazing and to compete with the exotic introductions. These exotics quickly spread and formed stable communities.

Vegetation changes in the California grasslands have been more drastic and perhaps more rapid than in any other North American grassland. The original perennial vegetation had largely been replaced by mediterranean annuals by 1860. Extensive heavy grazing began with the large Spanish land grants of 1824 (Barry 1972). The grassland remaining a century later–primarily in the foothills surrounding the Central Valley and along the coastal mountains–was dominated by exotic annuals (Biswell 1956). By 1918, *Avena fatua* had invaded most of the Central Valley and was dominant throughout (Beetle 1947); less than 5% of the original area remained as original perennial vegetation (Sampson 1950).

Pristine California grasslands were dominated by *Stipa cernua* (Stoddart et al. 1975), *S. pulchra* (Burcham 1957), and an assortment of other perennials, including *Danthonia californica, Elymus triticoides, Melica californica, M. imperfecta, Poa scabrella, Sitanion hystrix,* and *Stipa pulchra*. Mediterranean annuals that replaced these perennials were *Avena fatua, A. barbata, Lolium multiflorum, Bromus hordeaceous, B. mollis, B. rubens, B. rigidus, B. tectorum, Taeniatherum asperum, Erodium ciccutarium,* among many others.

The impact of fire on the original perennials is largely unknown, although *Stipa pulchra* probably was fire-tolerant (Wright and Bailey 1982). This perennial persisted along railroads, where it was protected from grazing but frequently burned. Fire shifts the annual grassland species from grasses to legumes and other forbs. Yields of annual grasses may be reduced by about 25% the first year following fire, whereas yields of *Medicago hispida, Erodium cicutarium,* and *E. botrys* have increased more than fivefold (Hervey 1949).

Palouse Prairie

This intermountain bunchgrass vegetation type originally extended throughout southwest Canada, eastern Washington and Oregon, and southwestern Idaho and even into western Montana (Stoddart et al. 1975). The Palouse prairie grassland was dominated originally by *Agropyron spicatum, Festuca idahoensis,* and *Elymus condensatus* and the associated species *Poa scabrella, Koeleria cristata, Elymus sitanion, Stipa comata,* and *Agropyron smithii* from the annual grasslands of California and the mixed-grass prairie of the central grasslands (Weaver and Clements 1938). Although it has a climate similar to that of the California grassland, its aspect more nearly resembles the mixed-grass vegetation, except that there are no short grasses present. Cool-season grasses dominate and compose more than 80% of the flora, with the remainder being predominantly C_4 grasses (Barbour et al. 1980).

Excessive grazing in the past resulted in the demise of many of the perennial grasses and led to an abundance of *Artemisia* and *Bromus* species. Intense fires that occurred during the summer caused considerable damage to the perennial grasses (Daubenmire 1970). Annual forbs (*Amsinckia, Helianthus, Lactuca, Leptilon,* and *Sisymbrium*) rapidly invaded the community following such fires. Even cool-burning fires, though not so damaging to the perennial grasses, favored the annual *Bromus* species (Wright and Bailey 1982) that became widespread in the Palouse grassland. Fire also reduced *Descurania pinnata, Sisymbrium altissimum,* and *Opuntia* species, while favoring *Lithophragma bulbifera* and *Plantago patagonia.*

Today the Palouse prairie is a shrub-steppe grassland, with dominance shared by perennial-grass remnants and *Artemisia* and *Bromus* species. The species of *Bromus* that invaded this plant community changed these grasslands from a useful year-long source of forage to seasonal rangelands, because the ephemerals introduced have a short growth and high-nutrition period.

Massive changes in the Palouse prairie vegetation have been caused by cultivation, grazing, and plant introductions (Franklin and Dyrness 1969; Young et al. 1977; Mack 1981). The major dominants, *Festuca idahoensis, F. scabrella,* and *Poa secunda,* were replaced by *Bromus tectorum* and *Poa pratensis.* Fire and grazing favored annuals over perennials, and there is evidence to show that annuals were better competitors for soil moisture (Harris 1967). Broadleaf herbs of the *Balsamorhiza, Lupinus, Achillea, Wyethia,*and *Helianthella* genera were locally abundant (Stoddart et al. 1975). Branson (1985) suggested that the original perennials were not resistant to grazing and that the soil moisture patterns favored the cool-season annuals over the perennials.

In the more moist, cooler, northern areas, overgrazing decreased *Festuca* species and *Agropyron spicatum,* but increased *Bromus tectorum, Stipa columbiana, Poa pratensis, Antennaria parviflora,* and *Poa secunda* (Tisdale 1947; Branson 1985). Farther south, in the warmer and drier zones, *B. tectorum,* primarily, along with *Poa secunda,* replaced *Festuca idahoensis.* Where *Artemisia* species originally occurred, they and *B. tectorum* have increased at the expense of the perennial grasses.

Desert Grassland

The desert grasslands extend from the shortgrass area in Texas and New Mexico through the southwest into northern Mexico. Originally, *Bouteloua eriopoda, Hilaria belangeri, Oryzopsis hymenoides,* and *Muhlenbergia porteri* were the important species because of their forage value and/or coverage of vast areas of land (Stoddart et al. 1975). Those plateau grasslands that occur at the edge of the Sonoran and Chihuahuan deserts above 1000 m elevation originally had a short-bunchgrass appearance, with the desert scrub restricted to ravines, knolls, and sites where soils were particularly poor, thus limiting the more productive grasses (Barbour et al. 1980).

Changes in the vegetation of the desert grassland have been documented in several places across the southwestern United States (Humphrey 1953, 1958, 1962; Humphrey and Mehrhoff 1958; Buffington and Herbel 1965; Branson 1985). Grasses have gradually been replaced by desert shrubs such as *Larrea tridentata, Flourensia cernua,* and *Prosopis juliflora.* Overgrazing by domestic livestock and subsequent erosion of the thin topsoil, fire suppression, and, perhaps, a gradual warming of the climate of the Southwest are correlated with these marked vegetation changes.

Three varieties of *Prosopis* (mesquite) are distributed across the Southwest (Parker and Martin 1952): *P. juliflora* var. *glandulosa* is an abundant tree or shrub in Texas; *P. juliflora* var. *torreyana* is a common shrub in west Texas, southern New Mexico, and northern Mexico; and *P. juliflora* var. *velutina* is common in Arizona. *P. juliflora* var. *glandulosa* has spread into *Bouteloua eriopoda* grasslands of New Mexico and displaced numerous forbs and shrubs (Hennessy et al. 1983). The grasses have disappeared on both grazed and ungrazed areas, while *Sporobolus flexuosus, Erioneuron pulchellum,* and *Xanthocephalum sarothrae* have increased with the mesquite. Mesquite has increased on all soil types in the desert grassland, but more so on sandy soils. In certain areas the increase in this shrub was associated with the development of a sand dune aspect (Gibbens et al. 1983). Buffington and Herbel (1965) suggested that a generally drier climate and increased seed dispersal achieved by grazing animals, particularly during droughts, hastened the invasion and spread of *P. juliflora* var. *glandulosa* across the Jornada Experimental Range between 1858 and 1963.

Many areas of the desert grassland have been modified for so long that no standard for comparison exists (Martin 1975). Generally, shifts from grassland to shrubland have occurred since the time domestic livestock were first introduced, which probably occurred earlier in this area than in most of the other North American grasslands. Although overgrazing is generally implicated as the cause, Malin (1953) believes that *P. juliflora* has been widely distributed across Texas since 1800 in a savannah form, but not as the present-day thicket.

In Arizona, shrubs have invaded grazed grasslands as well as grasslands from which both large and small herbivores have been excluded. As a result, Brown (1950) concluded that the desert grassland is a subclimax to the desert shrub climax in southern Arizona.

AREAS FOR FUTURE RESEARCH

Resource Allocation and Abiotic Stress

Understanding how grasslands were originally distributed across the North American continent and how they respond to current perturbations requires knowledge of the mechanisms controlling the temporal and spatial responses of individual plants to variations in site conditions (McMillan 1959a). Additional research is needed to further define the mechanisms that support the geographical repetition of an apparently uniform vegetation array across a number of diverse, widespread habitats.

The adaptability of grassland plants to a myriad of environments is related to their tremendous genetic amplitude, as expressed through their physiological and morphological response mechanisms that operate within the more complex ecological processes. Risser (1985) reviewed the literature on the mechanisms of grassland responses to water and nutrient stresses, solar radiation, temperature fluctuations, and fire and grazing impacts. Although he found no generalized morphological-physiological model that fully characterized a particular plant species or grassland community, he described several adaptive strategies important to grassland communities. Each of the strategies proposed by Risser (1985) as listed below, requires investigation to determine its floristic extent and competitive importance.

1. Grassland plants heavily invest carbohydrates in structural development early in the growing season when moisture conditions are generally adequate. Thus, plants are better able to cope with the stresses that commonly occur during the annual cycle in grasslands. During drought or following heavy grazing, grassland plants can utilize stored labile carbohydrates, decrease dark respiration to conserve substrate, and maintain gas-exchange processes under water potentials of -2 to -4 MPa.

2. Grassland plants physically respond to drought stress by closing stomata and by curling leaves to reduce water losses. Pubescence and paraheliotropism also contribute to water use efficiency. The C_4 plants, generally favored in drier climates, tend to be more water-use-efficient than the C_3 plants.

3. The ability of seed or other propagules to become dormant or to germinate in relatively dry soil characterizes many grassland species. The primary and adventitious roots of grasses undergo rapid expansion when conditions permit, increasing their ability to cope with ensuing stress. Physically, the roots of grassland plants have the strength to withstand shrink-swell characteristics of clay soils.

4. Nutrient uptake is rapid when moisture is available and plants are growing. At that time, forage quality is highest, and the plants are at their most palatable and are most tolerant of grazing. As plants mature, the relative proportion of coarse stems increases, and they become less palatable. Removal of a portion of the photosynthetic tissue results in an increase in the rate of photosynthesis by the remaining tissue. The intercalary meristems of grasses permit growth following grazing.

The responses of grassland plants to temperature stress are related to carbon-fixation pathways and these may affect the ratio of $C_3 : C_4$ species (Risser 1985). Slack and associates (1974) found that C_4 plants were intolerant of low temperatures during growth. Teeri and Stowe (1976) found a strong correlation between the distribution of C_4 grasses and relatively high growing-season temperatures. C_3 plants, which have greater quantum yields at lower leaf temperatures (Ehleringer and Björkman 1977), may have the advantage in grasslands with low to moderate growing-season temperatures

(Ehleringer 1978). The relative advantages of each photosynthetic syndrome still need clarification.

Physiological studies using delta ^{13}C techniques (Ode et al. 1980), as proposed by Tieszen and associates (1979a, 1979b), and carbon-assimilation and -allocation investigations (Caldwell et al. 1977a, 1977b,) may provide valuable insight toward a better understanding of the functional relationships of plant community structures to herbivore grazing (Caswell et al. 1973; Sims and Singh 1978b) and grassland fire responses (Ode et al. 1980).

Both temperature and precipitation have been shown to explain some of the functions of the grassland ecosystem (Sims and Singh 1978a, 1978b, 1978c). Grasslands with a mean annual temperature of 10°C or less are dominated by cool-season grasses. Grasslands in the southern latitudes, with mean annual temperatures greater than 10°C, are dominated by warm-season grasses. Cool-season forbs and shrubs compose 15%–40% of the biomass in the cooler grasslands. Warm-season forbs are similarly important in the southern grasslands, especially in the desert grassland of New Mexico. The major abiotic factors controlling biomass (B) for cool-season (CS) and warm-season (WS) plants are thought to be long-term mean annual temperature (T), growing-season precipitation (P_g), annual usable solar radiation (R_a) or growing-season usable solar radiation (R_g) and annual actual evapotranspiration (E_a) or growing-season actual evapotranspiration (E_g) (Sims et al. 1978). The percentages of cool-season and warm-season plant biomass are related by simple or multiple regression to these abiotic variables by the following formulas:

$$B_{CS} = 1.13 - 0.07T \qquad (r^2 = 0.51,\ p = 0.001)$$

or,

$$B_{CS} = 2.20 - 0.00023P_g - 0.0000012R_a + 0.00018E_a \qquad (r^2 = 0.81,\ p = 0.001)$$

$$B_{WS} = 0.06 + 0.06T \qquad (r^2 = 0.42,\ p = 0.01)$$

or,

$$B_{WS} = 0.84 + 0.0022P_g + 0.0000008R_g - 0.0016E_g \qquad (r^2 = 0.70,\ p = 0.001)$$

In the northern shortgrass and mixed-grass prairies, grazing by domestic livestock shifted the species composition from cool-season to warm-season species. Ungrazed shortgrass and mixed-grass prairies had both cool-season and warm-season plants, and grazing resulted in a drier, warmer habitat suitable for the plants of more southerly origin (Sims et al. 1978). It was found that in excellent-condition northern mixed-grass prairie of South Dakota, *Agropyron smithii* dominated, and there was an understory of the warm-season short grasses *Buchloë dactyloides* and *Bouteloua gracilis* (Smolik and Lewis 1982). As the grazing pressure was increased and the condition of the grassland declined, the short grasses dominated.

Further research is needed to understand the potential of the grassland vegetation and the mechanisms that produce harmony between vegetation and habitat. Perpetuation of grasslands will require more than protection of a few grassland preserves. Detailed and comprehensive studies of grasslands to develop new insight into the ecology of the largest and most extensive of the North American vegetation formations will provide a basis for enhanced yields of food, fiber, and contentment in the future.

Role of Fire

The history of fire in the grasslands has been reviewed by Wright and Bailey (1982) and Axelrod (1985). Fire has been almost a constant event at one place or another in the grasslands, and some of these fires have been severe and enormous in size. Fires once were caused by lightning and by Indians; with the development of the West, they came to be caused by cowboys, cooks, and steam engines. The intensities of fires and their frequency increased with the degree of drought. In years following a period of average or better precipitation that left an abundance of continuous fuel, wildfires covered vast distances, particularly when winds and air temperatures were high and relative humidity low. Grassland fires are limited today by the firebreaks provided by cultivated land, highways, and roads, and by the wide availability of fire-fighting personnel and equipment.

The earlier frequency of fire in the Great Plains grasslands cannot be determined directly. Wright and Bailey (1982) extrapolated from data on forest with grassland understories and concluded that the natural fire frequency on level to rolling prairie grassland may have been every 5–10 yr. In grasslands dissected by breaks and streams, the fire frequency may have been every 20–30 yr.

If fire has played a significant role in the development and maintenance of grasslands, we need to know much more about the impact of fire on individual plants and on plant communities. Managing our grasslands and at the same time reducing our use of herbicides will require further research on the importance of controlled burning at particular times or at particular intensities.

Grazing History

A critical factor limiting the development of better grassland management strategies is lack of understanding of the history of the vegetation and the animals (Barnard and Frankel 1966). Platou and Tueller (1985) concluded that a better understanding of natural pristine grazing systems could improve the design and implementation of contemporary grazing management strategies, help design innovative grazing systems, and lead to more efficient utilization of watersheds, grazing lands, and recreational resources.

REFERENCES

Aandahl, A. R. 1982. Soils of the Great Plains–land use, crops and grasses. University of Nebraska Press, Lincoln.

Albertson, F. W. 1937. Ecology of mixed prairie in west central Kansas. Ecol. Monogr. 7:481–547.

Anderson, K. L. 1965. Fire ecology–some Kansas prairie forbs. Proc. Tall Timbers Fire Ecol. Conf. 4:153–160.

Anderson, R. C. 1982. An evolutionary model summarizing the roles of fire, climate, and grazing animals in the origin and maintenance of grasslands. pp. 297–308 in J. R. Estes, R. J. Tyrl, and J. N. Brunken (eds.), Grasses and grasslands, systematics and ecology. University of Oklahoma Press, Norman.

Archer, S. G., and C. E. Bunch. 1953. The American grass book. A manual of pasture and range practices. University of Oklahoma Press, Norman.

Austin, M. E. 1965. Land resource regions and major land resource areas of the United States (exclusive of Alaska and Hawaii) USDA, Soil Conservation Service, agricultural handbook 296.

Axelrod, D. I. 1985. Rise of the grassland biome, central North America. Bio. Rev. 51:163–201.

Bailey, R. G. 1976. Ecoregions of the United States [Map]. USDA Forest Service, Ogden, Utah.

Bailey, R. G. 1980. Description of the ecoregions of the United States. USDA Forest Service, Ogden, Utah.

Barbour, M. G., J. H. Burk, and W. D. Pitts. 1980. Terrestrial plant ecology. Benjamin/Cummings Publishing, Menlo Park, Calif.

Barnard, C., and O. H. Frankel, 1966. Grass, grazing animals and man, historic perspective of grasses and grasslands. pp. 1–12 in C. Barnard (ed.), Grasses and grasslands. Macmillan, New York.

Barry, W. J. 1972. The central valley prairie, vol. 1, California Prairie Ecosystem. California Department of Parks and Recreation, Sacramento.

Beetle, A. A. 1947. Distribution of the native grasses of California. Hilgardia 17:309–357.

Biswell, H. H. 1956. Ecology of California grasslands. J. Range Management 9:19–24.

Borchert, J. R. 1950. The climate of the central North American grassland. Ann. Assoc. Amer. Geogr. 40:1–39.

Branson, F. A. 1985. Vegetation changes on western rangelands. Society for Range Management, Denver, Colo.

Breymeyer, A. I., and G. M. Van Dyne (eds.). 1980. Grasslands, systems analysis and man. Cambridge University Press.

Brown, A. L. 1950. Shrub invasion of southern Arizona desert grassland. J. Range Management 3:172–177.

Brunnschweiler, D. H. 1952. The geographic distribution of air masses in North America. Vierteljahrsschr. Naturforsch. Ges. Zür. 97:42–49.

Buffington, L. C., and C. H. Herbel. 1965. Vegetational changes of a semidesert grassland range from 1858 to 1963. Ecol. Monogr. 35:139–164.

Buol, S. W., F. D. Hole, and R. J. McCracken. 1980. Soil genesis and classification, 2nd ed. Iowa State University Press, Ames.

Burcham, L. T. 1951. Cattle and range forage in California: 1770–1880. Argic. Hist. 35:140–149.

Burcham, L. T. 1957. California rangeland. California Division of Forestry, Sacramento.

Caldwell, M. M., C. B. Osmond, and D. L. Nott. 1977a. C_4 pathway photosynthesis at low temperature in cold tolerant *Atriplex* species. Plant Physiol. 60:157–164.

Caldwell, M. M., R. S. White, R. T. Moore, and L. B. Camp. 1977b. Carbon balance, productivity, and water use of cold-winder desert shrub communities dominated by C_3 and C_4 species. Oecologia 29:275–300.

Carpenter, J. R. 1940. The grassland biome. Ecol. Monogr. 10:617–684.

Caswell, H., F. Reed, S. N. Stephenson, and P. A. Werner. 1973. Photosynethetic pathways and selective herbivory: a hypothesis. American Naturalist 107:465–481.

Clements, F. E. 1916. Plant succession: an analysis of the development of vegetation. Carnegie Institution of Washingon publication 242.

Clements, F. E. 1920. Plant indicators. Carnegie Institution of Washington publication 290.

Collins, D. D. 1969. Macroclimate and the grassland ecosystem. pp. 29–39 in R. L. Dix (ed.), The grassland ecosystem, a preliminary synthesis. Range Science Department Science Series no. 2, Colorado State University, Fort Collins.

Conant, S., and P. G. Risser. 1974. Canopy structure of a tall-grass prairie. J. Range Management 27:313–318.

Coupland, R. T. 1974. Grasslands. pp. 280–294 in The New Encyclopaedia Britannica, vol. 8.

Coupland, R. T. 1979. Grassland ecosystems of the world: analysis of grassland and their uses. Cambridge University Press.

Coupland, R. T., E. A. Ripley, and P. C. Robbins. 1973. Description of site, I, Floristic composition and canopy architecture of the vegetative cover. Canadian Committee for IBP, Matador Project report 11, University of Saskatchewan, Saskatoon.

Curtis, J. T. 1962. The modification of mid-latitude grasslands and forests by man. pp. 721–736 in W. L. Thomas, Jr. (ed.), Man's role in changing the face of the earth. University of Chicago Press.

Daubenmire, R. 1970. Steppe vegetation of Washington. Technical Bulletin no. 62, Washington Agricultural Experiment Station, Pullman.

Dyksterhuis, E. J. 1948. The vegetation of the western cross timbers. Ecol. Monogr. 18:327–376.

Dyksterhuis, E. J. 1949. Condition and management of range land based on quantitative ecology. J. Range Management 2:104–115.

Ehleringer. J. R. 1978. Implications of quantum yield differences on the distributions of C_3 and C_4 grasses. Oecologia 31:255–267.

Ehleringer, J. R., and O. Björkman. 1977. Quantum yields for CO_2 uptake in C_3 and C_4 plants: dependence on temperature, CO_2 and O_2 concentration. Plant Physiol. 59:86–90.

England, R. C., and A. DeVos. 1969. Influence of animals on the pristine conditions on the Canadian grasslands. J. Range Management 22:87–94.

Franklin, J. F., and C. T. Dyrness. 1969. Vegetation of Oregon and Washington. USDA Forest Service research paper PNW-80.

Gartner, F. G., and W. W. Thompson. 1972. Fire in the Black Hills forest-grass ecotone. Proc. Tall Timbers Fire Ecol. Conf. 12:37–68.

Gibbens, R. P., J. M. Tromble, J. T. Hennessy, and M. Cardenas. 1983. Soil movement in mesquite dunelands and former grasslands of southern New Mexico from 1933 to 1980. J. Range Management 36:145–148.

Gleason, H. A. 1913. The relation of forest distribution and prairie fires in the middle west. Torreya 13:173–181.

Gleason, H. A. 1923. The vegetational history of the Middle West. Ann. Assoc. Amer. Geogr. 12:39–85.

Gould, F. W. 1962. Texas plants—a checklist and ecological summary. Texas Agricultural Experiment Station bulletin MP-585.

Gould, F. W. 1968. Grass systematics. McGraw-Hill, New York.

Harlan, J. R. 1956. Theory and dynamics of grassland agriculture. D. Van Nostrand, New York.

Harris, G. A. 1967. Some competitive relationships between *Agropyron spicatum* and *Bromus tectorum*. Ecol. Monogr. 37:89–111.

Hartley, W. 1950. The global distribution of tribes of Gramineae in relation to historical and environmental factors. Aust. J. Agric. Res. 1:355–373.

Hartley, W. 1964. The distribution of grasses. pp. 29–64 in C. Barnard (ed.), Grasses and grasslands. Macmillan, New York.

Heady, H. F. 1975. Rangeland management. McGraw-Hill, New York.

Heady, H. F. 1977. Valley grassland. pp. 491–514. in M. G. Barbour and J. Major (eds.), Terrestrial vegetation of California. Wiley, New York.

Heady, H. F., T. C. Foin, J. J. Kektner, D. W. Taylor, M. G. Barbour, and W. J. Berry. 1977. Coastal prairie and northern coastal scrub. pp. 733–760 in M. G. Barbour and J. Major (eds.), Terrestrial vegetation of California. Wiley, New York.

Hennessy, J. T., R. P. Gibbens, J. M. Tromble, and M. Cardenas. 1983. Vegetation changes from 1935 to 1980 in mesquite dunelands and former grasslands of southern New Mexico. J. Range Management 36:370–374.

Hervey, D. F. 1949. Reaction of a California annual-plant community to fire. J. Range Management 2:116–121.

Hole, F. D., and G. A. Nielsen. 1968. Some processes of soil genesis under prairie. pp. 28–34 in P. Schramm (ed.), Proceedings of a symposium on prairie and prairie restoration. Knox College, Galesburg, Ill.

Humphrey, R. R. 1953. The desert grassland, past and present. J. Range Management 6:159–164.

Humphrey, R. R. 1958. The desert grassland. Bot. Rev. 24:193–252.

Humphrey, R. R. 1962. Range ecology. Ronald Press, New York.

Humphrey, R. R., and L. A. Mehrhoff. 1958. Vegetation changes on a southern Arizona grassland range. Ecology 39:720–726.

Hunt, C. B. 1972. Physiography of the United States. W. H. Freeman, San Francisco.

Jenny, H. 1930. A study of the influence of climate upon the nitrogen and organic matter content of soil. University of Missouri Agricultural Experiment Station bulletin 152.

Kucera, C. L. 1960. Forest encroachment in native prairie. Iowa State J. Sci. 34:635–639.

Kucera, C. L. 1970. Ecological effects of fire on tall grass prairie. Ecology 43:334–336.

Küchler, A. W. 1964. Potential natural vegetation of the conterminous United States. American Geographical Society special publication 36.

Küchler, A. W. 1975. Potential natural vegetation of the conterminous United States [Map]. American Geographical Society, New York.

Larsen, E. C. 1947. Photo-periodic responses of geographical strains of *Andropogon scoparius*. Bot. Gaz. 109:132–149.

Larson, F. 1940. The role of bison in maintaining the short grass plains. Ecology 21:113–121.

Launchbaugh, J. L. 1964. Effects of early spring burning on yields of native vegetation. J. Range Management 17:5–6.

Launchbaugh, J. L. 1972. Effect of fire on shortgrass and mixed prairie species. Proc. Tall Timbers Fire Ecol. Conf. 12:129–151.

Lewis, J. K., and D. M. Engle. 1980. Impacts of technologies of productivity and quality of rangelands in the Great Plains. Report to the Office of Technology Assessment, United States Congress.

McComb, A. L., and W. E. Loomis. 1944. Subclimax prairie. Bull. Torrey Botanical Club 71:46–76.

McDonald, J. N. 1981. North American bison; their classification and evolution. University of California Press, Berkeley.

Mack, R. N. 1981. The invasion of *Bromus tectorum* L. into western North America: an ecological chronicle. Agro-ecosystems 7:145–165.

McMillan, C. 1956a. Nature of the plant community, I, Uniform garden and light period studies of five grass taxa in Nebraska. Ecology 37:330–340.

McMillan, C. 1956b. Nature of the plant community, II, Variation in flowering behavior within populations of *Andropogon scoparius*. Amer. J. Bot. 43:429–436.

McMillan, C. 1957. Nature of the plant community, III, Flowering behavior within two grassland communities under reciprocal transplanting. Amer. J. Bot. 44:144–153.

McMillan, C. 1959a. Nature of the plant community, V, Variation within the true prairie community-type. Amer. J. Bot. 46:418–424.

McMillan, C. 1959b. The role of ecotypic variation in the distribution of the central grassland of North America. Ecol. Monogr. 29:287–308.

McMurphy, W. E., and K. L. Anderson. 1965. Burning Flint Hills range. J. Range Management 18:265–269.

McNaughton, S. J., M. B. Coughenhour, and L. L. Wallace. 1982. Interactive processes in grassland ecosystems. pp. 167–193 in J. R. Estes, R. J. Tyrl, and J. N. Brunken (eds.), Grasses and grasslands, systematics and ecology. University of Oklahoma Press, Norman.

Malin, J. C. 1953. Soil, animal, and plant relations of the grassland, historically reconsidered. Sci. Monthly 76:207–220.

Martin, S. C. 1975. Ecology and management of southwestern semi-desert grass-shrub ranges: the status of our knowledge. USDA Forest Service research paper RM-156.

Müller-Beck, H. 1966. Paleohunters in America: origins and diffusion. Science 152:1191–1210.

Nelson, E. W. 1925. Status of the pronghorned antelope, 1922–1924. USDA bulletin 1346.

Ode, D. J., L. L. Tieszen, and J. C. Lerman. 1980. The seasonal contribution of C_3 and C_4 plant species to primary production in a mixed prairie. Ecology 61:1304–1311.

Olmsted, C. E. 1944. Growth and development in range grasses, IV, Photoperiodic responses in twelve geographic strains of side-oats grama. Bot. Gaz. 106:46–74.

Parker, K. W., and S. C. Martin. 1952. The mesquite problem on southern Arizona ranges. USDA circular 908.

Penfound, W. T. 1964. The relation to grazing to plant succession in the tall grass prairie. J. Range Management 17:256–260.

Platou, K. A., and P. T. Tueller. 1985. Evolutionary implications for grazing management systems. Rangelands 7:57–61.

Redmann, R. E. 1975. Production ecology of grassland plant communities in western North Dakota. Ecol. Monogr. 45:83–106.

Rice, E. L. 1950. Growth and floral development of five species of range grasses in central Oklahoma. Bot. Gaz. 3:361–377.

Riegel, A. 1940. A study of the variations in the growth of blue grama grass from seed produced in various sections of the Great Plains Region. Kan. Acad. Sci. Trans. 43:155–171.

Risser, P. G. 1985. Grasslands. pp. 232–256 in B. F. Chabot and H. A. Mooney (eds.), Physiological ecology of North American plant communities. Chapman and Hall, New York.

Risser, P. G., E. C. Birney, H. D. Blocker, S. W. May, W. J. Parton, and J. A. Wiens. 1981. The true prairie ecosystem. Hutchinson Ross Publishing, Stroudsburg, Pa.

Rowe, J. S. 1972. Forest regions of Canada. Department of Environment, Canadian Forest Service, publication 1300.

Rzedowski, J. 1978. Vegetacion de Mexico. Editorial Limusa, Mexico City.

Sampson, A. W. 1950. Application of ecologic principles in determining condition of range lands. pp. 509–514 in United Nations conference on conservation and utiliization of resources, Lake Success, N.Y., vol. 6, Land resources. Department of Economic Affairs, United Nations, New York.

Saure, C. O. 1950. Grassland, climax, fire, and man. J. Range Management 3:16–22.

Schmidt, K. P. 1938. Post glacial steppes in North America. Ecology 19:396–407.

Schripsema, J. R. 1977. Ecological changes on pine-grassland burned in the spring, late spring, and winter. M.A. thesis, Biology Department, South Dakota State University, Brookings.

Scifres, C. J. 1980. Brush management, principles and practices for Texas and the Southwest. Texas A&M University Press, College Station.

Sears, P. B. 1935. Glacial and postglacial vegetation. Bot. Rev. 1:37–51.

Sears, P. B. 1948. Forest sequence and climatic change in northeastern North America since early Wisconsin time. Ecology 29:326–333.

Seton, E. T. 1929. Lives of game animals, vol. III, Part II. Doubleday, Doran, New York.

Shantz, H. L. 1954. The place of grasslands in the earth's cover of vegetation. Ecology 35:143–145.

Sims, P. L., and D. D. Dwyer. 1965. Pattern of retrogression of native vegetation in north-central Oklahoma. J. Range Management 18:20–25.

Sims, P. L., and J. S. Singh. 1978a. The structure and function of ten western North American grasslands, II, Intra-seasonal dynamics in primary producer compartments. J. Ecol. 66:547–572.

Sims, P. L., and J. S. Singh. 1978b. The structure and function of ten western North American grasslands, III, Net primary production, turnover, and efficiences of energy capture and water use. J. Ecol. 66:573–597.

Sims, P. L., and J. S. Singh. 1978c. The structure and function of ten western North American grasslands, IV, Compartmental transfers and energy flow within the ecosystem. J. Ecol. 66:983–1009.

Sims, P. L., J. S. Singh, and W. K. Lauenroth. 1978. The structure and function of ten western North American grasslands, I, Abiotic and vegetational characteristics. J. Ecol. 66:251–285.

Sims, P. L., R. E. Sosebee, and D. M. Engle. 1982. Plant responses to grazing management. pp. 4–31 in D. D. Briske and M. M. Kothmann (eds.), Proceedings of a national conference on grazing management technology, Texas A&M University, College Station.

Slack, C. R., P. G. Roughan, and H. C. M. Basset. 1974. Selective inhibition of mesophyll chloroplast development in some C_4 pathway species by low night temperature. Planta 118:57–73.

Smith, G. D. 1965. Lectures on soil classification. Pedologie special issue 4. Belgium Soil Science Society Rozier 6, Gent, Belgium.

Smolik, J. D., and J. K. Lewis. 1982. Effect of range condition on density and biomass of nematodes in a mixed prairie ecosystem. J. Range Management 35:657–663.

Steiger, T. L. 1930. Structure of prairie vegetation. Ecology 11:170–217.

Stoddart, L. A., A. D. Smith, and T. W. Box. 1975. Range management. McGraw-Hill, New York.

Teeri, J. A., and L. G. Stowe. 1976. Climatic patterns and the distribution of C_4 grasses in North America. Oecologia 23:1–2.

Thornthwaite, C. W. 1933. The climates of the earth. Geogr. Rev. 23:433–440.

Tieszen, L. L., D. Hein, S. A. Qvortrup, J. H. Troughton, and S. K. Imbamba. 1979a. Use of delta 13 C values to determine vegetation selectivity in East African herbivores. Oecologia 37:351–359.

Tieszen, L. L., M. M. Senyimba, S. K. Imbamba, and J. H. Troughton. 1979b. The distribution of C_3 and C_4 grasses and carbon isotope discrimination along an altitudinal and moisture gradient in Kenya. Oecologia 37:337–350.

Tisdale, E. W. 1947. The grasslands of southern British Columbia. Ecology 28:346–382.

Transeau, E. N. 1935. The prairie penninsula. Ecology 16:423–437.

U.S. Forest Service. 1980. A assessment of the forest

and range land situation in the United States. USDA publication FS-345.

Van Dyne, G. M. 1971. The U.S. IBP grassland biome study–an overview. pp. 1–9 in N. R. French (ed.), Preliminary analysis of structure and function in grasslands. Range Science Department Science Series 10, Colorado State University, Fort Collins.

Vankat, J. L. 1979. The natural vegetation of North America, an introduction. Wiley, New York.

Voight, J. W., and J. E. Weaver. 1951. Range condition classes of native midwestern pasture: an ecological analysis. Ecol. Monogr. 21:39–60.

Weaver, J. E. 1954. North American prairie. Johnsen Publishing, Lincoln, Neb.

Weaver, J. E., and F. W. Albertson. 1956. Grasslands of the Great Plains: their nature and use. Johnsen Publishing, Lincoln, Neb.

Weaver, J. E., and F. E. Clements. 1938. Plant ecology, 2nd ed. McGraw-Hill New York.

Wells, P. V. 1970. Historical factors controlling vegetation patterns and floristic distributions in the Central Plains region of North America. pp. 211–221 in W. Dort, Jr., and J. K. Jones. (eds.), Pleistocene and recent environments of the central Great Plains. University of Kansas, Department of Geology special publication 3, University of Kansas Press, Lawrence.

WLGU (Western Land Grant Universities and Colleges and Soil Conservation Service). 1964. Soils of the western United States. Regional publications, Washington State University, Pullman.

Whitman, W. C. 1941. The native grassland. pp. 5–7 in Grass. North Dakota Agricultural Experiment Station research bulletin 300.

Willet, H. C. 1949. Long period fluctuations in the general circulation of the atmosphere. J. Meteorol. 6:1.

Wright, H. A. 1974. Effect of fire on southern mixed prairie grasses. J. Range Management 27:417–419.

Wright, H. A., and A. W. Bailey. 1980. Fire ecology and prescribed burning in the Great Plains–a research review. USDA Forest Service general technical report INT-77, Intermountain Forest and Range Experiment Station, Ogden, Utah.

Wright, H.A., and A. W. Bailey. 1982. Fire ecology, United States and southern Canada. Wiley, New York.

Young, J. A., R. A. Evans, and J. Major. 1977. Alien plants in the Great Basin. J. Range Management 25:194–201.

Chapter 10

Deciduous Forest

ANDREW M. GRELLER

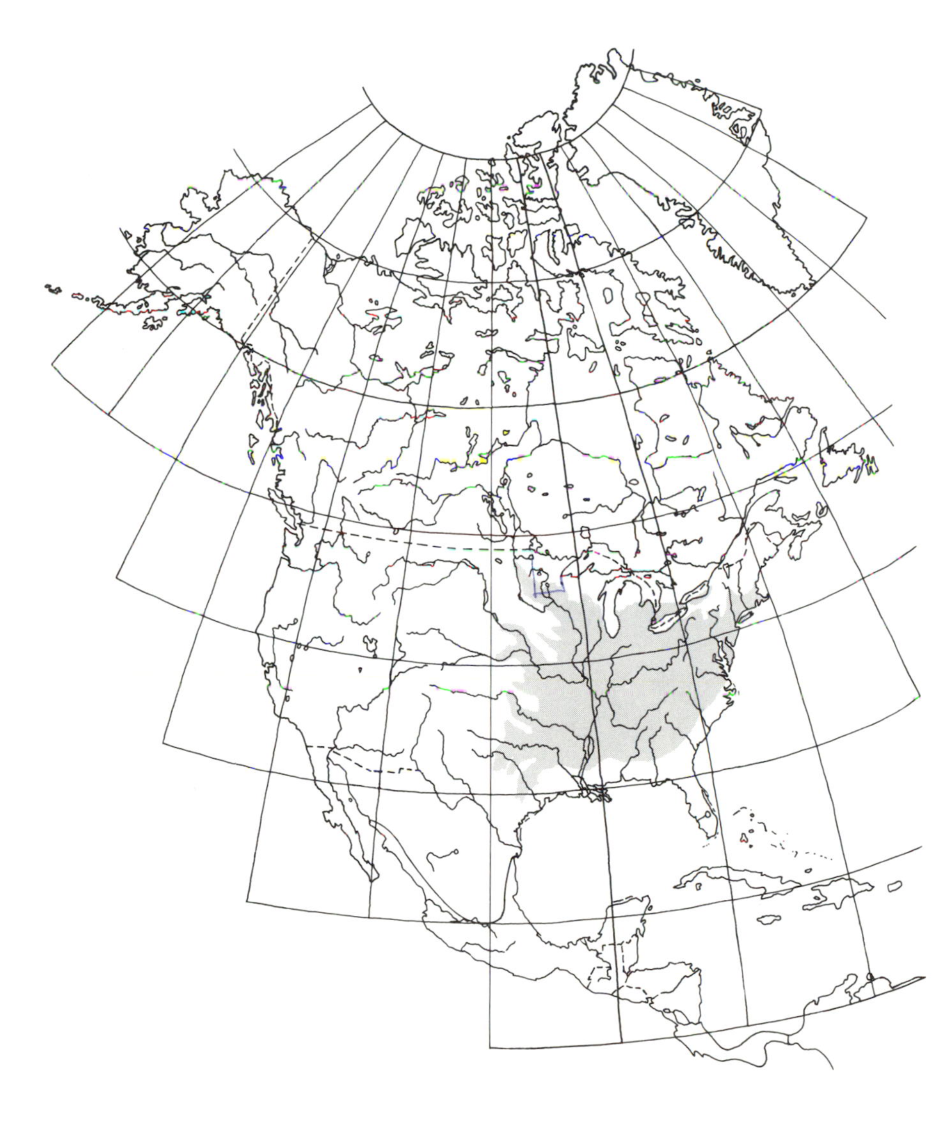

INTRODUCTION

Following Braun (1950), I characterize the deciduous forest of eastern North America as a tree-dominated vegetation type in which most of the woody taxa are winter-deciduous. Typically, the mature forest canopy reaches 30 m and is closed. In the central region of the deciduous forest, a distinct subcanopy is present, as well as one or two shrub layers and a well-developed herb layer; bryoids may be patchy, scattered, or absent. Lichens and mosses grow as epiphytes on the main trunks. Evergreen dicotyledonous taxa occur as trees, shrubs, and herbs. Evergreen dicotyledonous trees and shrubs occur as understory individuals at the southern transition of the deciduous forest, where warmth is great and frost is relatively infrequent. The general forest limits are 32–45° N latitude and 70–98° W longitude, and these lie almost wholly within the United States.

Evergreen dicotyledonous trees in overstory canopies occur at the southern edge of the deciduous forest (see Chapter 11), and evergreen conifers come to dominate at the northern edge (see Chapter 2). To the west, the deciduous forest becomes shorter in stature and poorer floristically; it may develop a more open canopy (woodland/savanna) and may include areas of graminoid-dominated vegetation (prairie) of variable extent. At its western extreme, the deciduous forest is confined to river floodplains (Weaver 1960). The deciduous forest shows a variety of transitions to the prairie region of central North America (see Chapter 9).

Gleason and Cronquist (1964) gave percentages for the deciduous-forest flora that exhibit each category of Raunkaier's life forms for Indiana, a state wholly within the deciduous-forest range. These percentages are (normal or world percentages in parentheses): phanerophytes, 14 (47); chamaephytes, 2 (9); hemicryptophytes, 49 (27); cryptophytes, 18 (4); therophytes, 17 (13). They ascribed the greater percentage of hemicryptophytes (herbs with buds at the ground surface) and the lesser percentage of phanerophytes (shrubs and trees with exposed buds) to harsh winter weather in the deciduous-forest region.

Leaves in the deciduous forest tend to be large and simple. My analysis of West Virginia, a state predominantly covered by deciduous forest, shows that few life forms exhibit compound leaves (Table 10.1). Dolph and Dilcher (1979) calculated the percentages of large leaves (mesophyllous, >20.25 cm^2) for woody taxa in the flora of North Carolina as being 60–70%, and 94% for Cedar Bluffs, Indiana.

Braun (1950) stated that the unity of the deciduous forest is manifested by the overlapping ranges of large numbers of genera and species. Brockman (1968) listed 67 trees and shrubs that have ranges centered on or restricted to the deciduous forest; Figure 10.1 shows the distributions of three tree taxa that correspond to the deciduous forest and the evergreen transitions to the north and south. Gleason and Cronquist (1964) listed 15 herb taxa that generally conform with the deciduous-forest region mapped here. Nomenclature in this chapter follows Gleason and Cronquist (1963).

Physiography, Geology, and Soils

Two major types of landform determine topography in the deciduous forests: platform areas that are flat or show only minor relief, and foldbelts that show pronounced relief (0–1600 m in elevation) and great geomorphological variations. These landform types are further subdividied as follows: (1) Appalachian Highlands, (2) Atlantic Rolling Plain, (3) Atlantic Coastal Flats (extreme northeastern section), (4) Lower New England, (5) North-Central Lake-Swamp-Moraine Plains, (6) Southwest Wisconsin Hills, (7) East-Central Drift and Lake-bed Flats, (8) Middle Western Upland Plain, (9) West-Central Rolling Hills, (10) Mid-Continent Plains and Escarpments, (11) Ozark-Ouchita Highlands, and (12) Eastern Interior Uplands and Basins (U.S. Geological Survey 1970). Topographic and geologic variations on a local scale, however, are more important in determining forest composition.

The geological substrata in the region of the deciduous forest are extremely varied. There are unconsolidated Quaternary and Tertiary sediments in the east and southwest. In the Appalachian Mountains there are wide age ranges for sedimentary rocks, which include extensively distributed metamorphic and igneous rocks, and also large sections of purely intrusive rocks, mainly granites. In the north and west, the rocks are mainly Paleozoic sedimentary types.

Glacially derived materials and associated landforms overlie the sedimentary and intrusive rocks in the northern half of the range of the deciduous forest (U.S. Geological Survey 1970).

Limestone-dolomite terrain is locally well developed throughout the deciduous-forest region and dominates tens of thousands of square kilometers in central Kentucky and adjacent states, in Missouri and Arkansas, and throughout the Appalachian foldbelt. Calcareous rock outcrops often harbor plants rarely found elsewhere (Steyermark 1940).

All of the soil orders in the region of the deciduous forest are classified as "warm" (mean soil tem-

Table 10.1. *Floristic analysis of leaf types of the woody flora of West Virginia, numbers of species (and percentages)*

Leaf type	Canopy trees	Sub-canopy	Shrubs	Vines
Simple, lanceolate, pinnate	10 (56)	14 (58)	21 (75)	1 (5)
Simple, ovate, palmate	4 (22)	5 (21)	3 (11)	10 (59)
Compound, pinnate	3 (17)	5 (21)	3 (11)	2 (12)
Compound, palmate	1 (5)	0 (0)	1 (3)	4 (24)
Total	18 (100)	24 (100)	28 (100)	17 (100)

Source: Based on data from Core (1966).

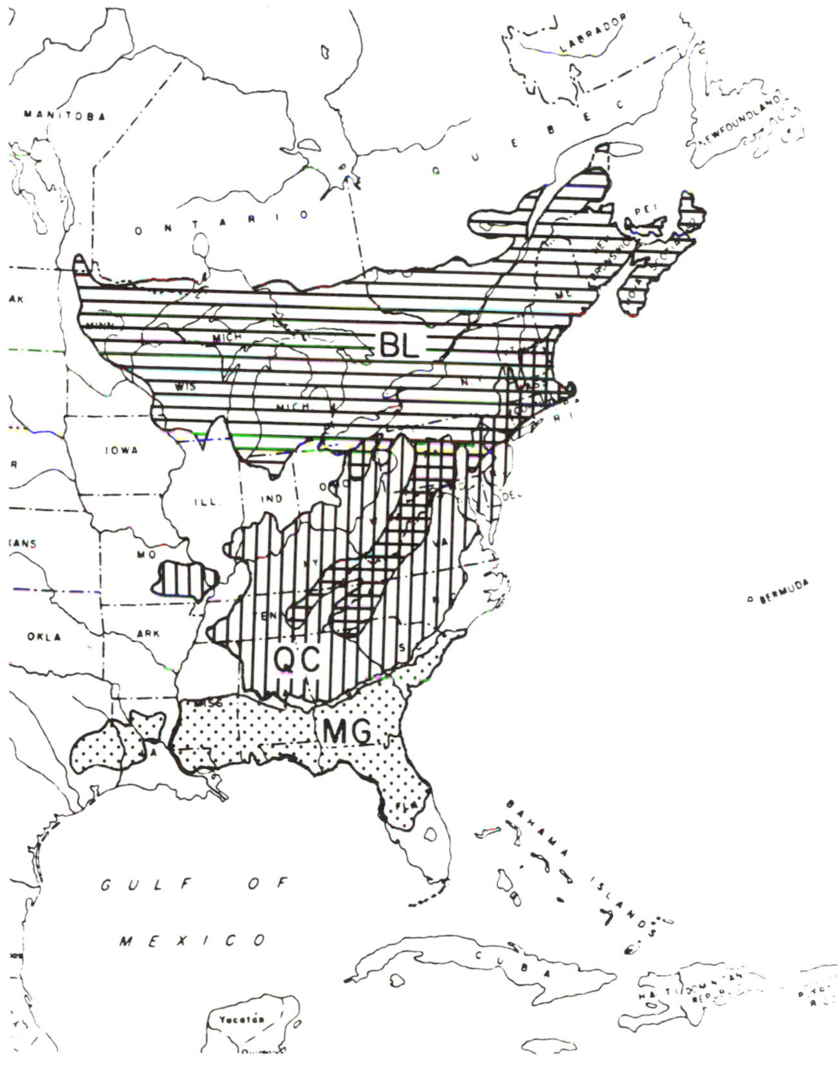

Figure 10.1. Ranges of three tree taxa in eastern North America (redrawn from Little 1971): Quercus coccinea *(QC), confined to the deciduous forest region;* Magnolia grandiflora *(MG), centered on the deciduous dicot, evergreen dicot, conifer forest region;* Betula lutea (=B. alleghenienisis) *(BL), centered on conifer, deciduous dicot forest region.*

perature higher than about 8.3°C). The distribution of soil orders appears to be correlated, generally, with associations (subdivisions) of the deciduous forest. The mixed mesophytic and oak-chestnut forests occur mainly on Inceptisols and Ultisols, the western mesophytic and oak-hickory forests on Ultisols and Mollisols, and the beech-maple and maple-basswood forests on Alfisols (U.S. Geological Survey 1970).

Climate

The deciduous forest region receives 300−400 langleys of solar radiation each year. Average January temperatures range from less than −6.7°C in the northwest portion to 4.4°C near the southern limit. Average January minimum temperatures range most often between −12.2°C and −1.1°C, but values for the southeastern portions are slightly higher. In the northern part of the deciduous forest there are 150 d with temperatures of 0°C or higher, but in the southwestern part there are as few as 60 d with such temperatures. The mean number of days between the last 0°C temperature in spring and the first 0°C temperature in autumn ranges from slightly fewer than 150 d in the north to approximately 210 d in the south. Average July temperatures range from 21.1°C to 26.7°C. There may be as many as 60 d yr^{-1} with temperatures greater than 32°C (southwestern portion), or as few as 5 d/yr^{-1} (northern part).

The mean annual precipitation ranges from 81 cm (northwestern portion) to 203 cm (Great Smoky Mountains). Most of the deciduous forest receives 81−122 cm. Precipitation is equally distributed throughout the year, although some southern and western stations receive peak precipitation in summer. The mean annual snowfall varies from 20 cm (southwest) to nearly 244 cm (northeast). The relative humidity is 60%−80% throughout the year throughout the deciduous forest.

When 14 stations at the extremes of the deciduous forest range are plotted on Bailey's nomogram (1960) (Fig. 10.2), the climatic limits of the deciduous forest can be expressed as warmth (W) = 12.5−15.2°C and temperateness (M) = 54−38, and between 2% and >20% of the hours in the year have temperatures of 0°C or less. Decreasing summer warmth appears to be correlated with the northern limits of the deciduous forest, whereas a combination of high summer temperatures (28°C) and infrequent frosts (<2% of the hours per year) appears to be correlated with the southern limits. In Bailey's nomenclature, the climate of the deciduous forest region is mild, subtemperate, and humid.

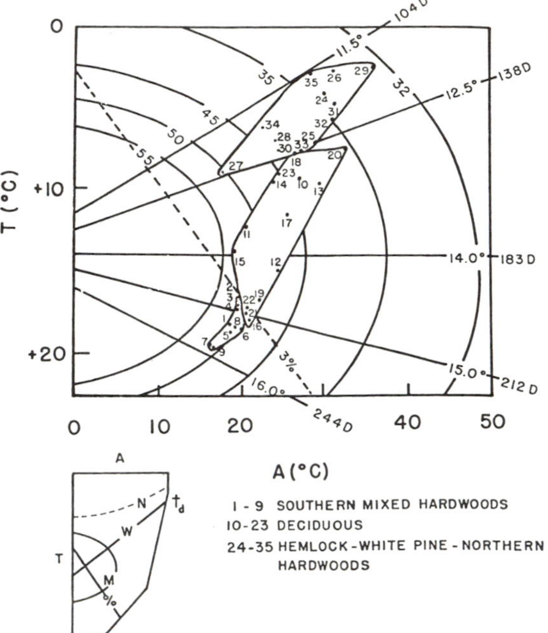

Figure 10.2. Nomograph (after Bailey 1960) of climatic stations well within, and at the extremes of, the geographical range of the deciduous forest. Stations 24−35 follow the boundaries of Braun (1950) for central hardwoods and hemlock−white pine−northern hardwood regions. Stations 1−9 follow the boundaries of Quarterman and Keever (1962) for southern mixed hardwoods. Additional information for North Carolina from Cooper (1979). T and A as on p. 138.

HISTORY

Fossil Record

No macrofossil assemblages (leaves, fruit, wood) have been found that would permit us to confirm the existence of a deciduous forest, as it presently exists, before historical times. Forests characteristic of the southeastern coastal plain, such as *Taxodium* swamp, *Pinus-Quercus* woodlands, and rich temperate evergreen forests, can be recognized in the Tertiary (Berry 1937; Dilcher 1971). Numerous reports on fossil pollen deposits suggest that genera now characteristic of the deciduous forest (*Betula, Carya, Castanea, Fagus, Juglans, Liquidambar, Liriodendron, Ostrya/Carpinus, Tilia,* and others) were present in the Eocene, in or near areas otherwise dominated by evergreen and other austral dicotyledonous taxa (Gray 1960; Dilcher 1971). A deciduous forest with evergreen dicotyledons, conifers, and tree ferns can be recognized from the Miocene of the New Jersey coastal plain (Rachele 1976). The closest modern counterpart of this forest is at higher elevations (1372 m) in eastern Mexico

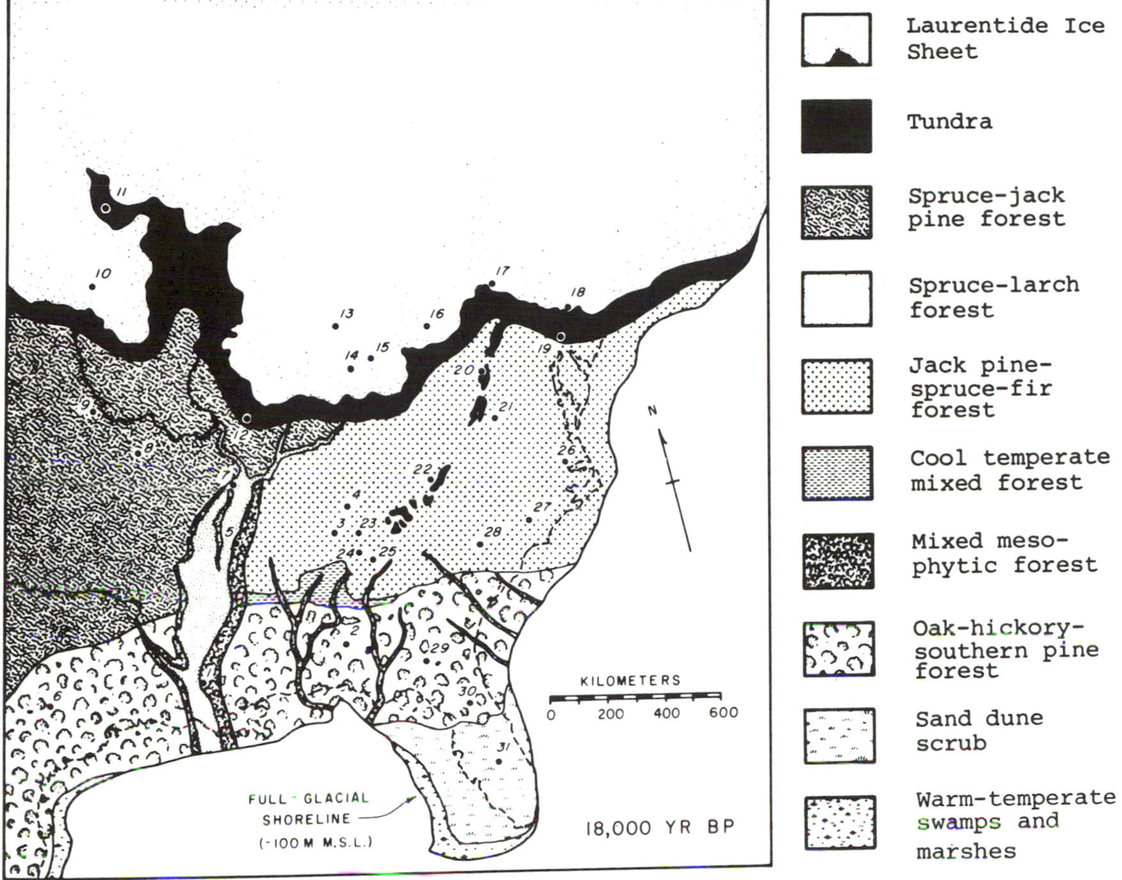

Figure 10.3. Map of full-glacial vegetation (18,000 BP) in eastern North America (Delcourt and Delcourt 1979). Numbered dots refer to fossil flora locations.

(Miranda and Sharp 1950; Greller and Rachele 1983).

Most of the 2 million yr of Quaternary time had a climate colder then at present, with 16 periods of extensive glaciation (Davis 1981). The maximum extent of the ice sheets during the last glaciation, the Wisconsin, was reached 18,000–20,000 BP, and retreat began about 16,500 BP (Delcourt and Delcourt 1979). Maximum temperatures for the present interglacial (the Holocene) most probably occurred 5000–8000 BP.

Delcourt and Delcourt (1981) recognized a deciduous forest along major southern rivers (mixed hardwoods phase) from at least 40,000 BP on the basis of comparisons of fossil pollen deposits with modern pollen assemblages. They found that by 25,000 BP there was an oak-hickory phase in the southeastern uplands, in addition to the mixed hardwoods in the river valleys. For 18,000 BP, the height of the Wisconsin glaciation, they mapped mixed deciduous hardwoods as extending north along the bluffs of the glacial-age Mississippi River to approximately southern Illinois, as well as along other major southern rivers (Fig. 10.3).

They showed a major advance of the deciduous forest into east central North America, following retreat of the glaciers at 10,000 BP, with oak-hickory confined to the western part of that region. For the period 8000–4000 BP, they argued that various forest zones were displaced to the north of their present ranges and that prairie penetrated eastward into what is now deciduous forest. The present distribution of forest zones in eastern North America has been developing since 5000 BP as a result of cooling and increased precipitation.

In the north, Davis's reconstructions (1981) of Holocene migration patterns show that elements of the modern spruce-fir forest, including *Picea, Abies, Larix,* and *Pinus* species, dominated the Great Lakes region as early as 11,000 BP. She also concluded that a mixed conifer-deciduous dicotyledonous forest that included *Ulmus, Acer, Fagus,*

Quercus, Tsuga canadensis, and *Pinus strobus* occurred in the region of eastern Virginia–South Carolina as early as 10,000 BP. She also stated that "the forests with which we are familiar seldom maintained a constant species composition for more than 2,000 or 3,000 years at a time." Her opinion is based on the changing compositions of fossil pollen floras at each given site, analyzed species by species.

These data suggest that modern forest types (mixed conifer-hardwood, boreal conifer, deciduous, etc.) could have been present in various species combinations at least by the beginning of the Holocene, if not earlier. Where highly equable conditions occurred during the Holocene, as postulated for the Tunica Hills by Delcourt and Delcourt (1979), unique species combinations, unknown in eastern North America today, could also have been present.

Historical Times

Bakeless (1950) summarized early (1500–1700 AD) accounts of the forests of North America, especially in the eastern, deciduous-forest region. In general, the distribution of tree species in the east is unchanged from that of pre-colonial times: white pine, hemlock, spruce, and deciduous dicots in the north; oaks, hickories, pines, and mixed deciduous dicots in the central region; live oak, cypress, cedar, magnolias, and so forth, in the south. Tree stature, however, was larger in the past. An eyewitness account of primeval deciduous forest in Ohio, quoted by Trautman (1977), mentions "red and white oaks . . . six feet and more in diameter . . . up to fifty or sixty feet without a limb . . . [tall] shellbark hickories and . . . maples . . . ash trees . . . over four feet through; elms and beeches, the great black walnuts and . . . sycamores, huge in limb and body, along the creek bottoms." There were 480–720 trees per hectare. The grapevines that often were present reached the tops of those trees. The forests occupied 95% of the land and were interrupted only by rivers and streams, natural prairies, hurricane-cleared patches, and Indian settlements.

Along the eastern coast, in addition to noting the nearly ubiquitous deciduous forests, early travelers observed that coastal sands and coarse tills were poorly vegetated. Natural grasslands and extensive coastal white cedar swamps were also present. Travelers commonly remarked on the sweet scent of the eastern forests, especially when many insect-pollinated trees were in bloom. Accounts of the forests also stressed the great beauty of the autumn coloration in the extensively forested landscape, as well as the edible fruits and nuts.

Native American Indians survived in heavily forested regions in small populations. They cleared and farmed small parcels (8–80 ha) of land, using fire and primitive implements. Their crops were corn, beans, squash, and tobacco. Soil fertility quickly decreased, so that after 8–10 yr, sites were abandoned. After 10–20 yr of abandonment, the sites were farmed again. Particularly productive sites were occupied repeatedly (Salomon 1984).

There are numerous records of the Indians using annual or semiannual fires to clear the forest understory to facilitate travel, to encourage the growth of forage plants for wild game, and to increase the production of edible fruits, wild nuts, and other plants. Fires probably were localized around settlements, with occasional, more extensive burns occurring when a fire got out of control. These fires appear to have had little lasting effect on forest structure or composition, except at the settlements, where frequent clearing, burning, and compaction of soil resulted in the persistence of meadows or savannahs (Russell 1983).

Until about 1800, settlement and clearing of the forests were undirected; and development after 1800 was directed by the federal government. After the War of 1812, when the threat from the British and their Indian allies was eliminated, the United States began removal of Indians from lands west of the central Appalachian Mountains to reservations farther west. After 1843, only a few scattered Indians remained in Ohio, for example. In 1790, there were 3000 non-Indians in Ohio, "45,000 by 1800, 585,000 by 1820, 2 million by 1850, 3.2 million by 1880, and 4.1 million by 1900" (Trautman 1977).

Potzger and associates (1956) and others have used land-survey records to reconstruct vegetation of that time, mainly in the Midwest. Section and quarter-section corners, as well as a number of intermediate points, were marked by slashing the bark of trees; thus, we have the locations, names, and diameters of thousands of "witness trees."

Chestnut blight. Chestnut (*Castanea dentata*), once an important canopy tree throughout the deciduous forest, and dominant in the oak-chestnut association region, has been relegated to the forest understory by the chestnut blight caused by *Endothia parasitica*, an ascomycete. Chestnut blight was first discovered affecting trees in New York City in 1904; its spread was completed by 1950 (Braun 1950). *Castanea dentata* survives today as a small understory tree, often with smaller secondary stems originating at the root collar. Future survival of the species, by slow expansion and division of the root crowns, seems likely (Paillet 1984). Former chestnut-oak forests were soon dominated by *Quercus* species, especially *Q. borealis* var. *maxima, Q. prinus,* and *Q. alba;* oaks are expected to remain dominants (Keever 1953). Gaps in the canopy left by chestnut usually are occupied by rapidly seeding, fast-growing forest associates. Some of the other tree taxa that have exhibited increased biomass since the decline of chestnut are *Carya* species, (e.g., *C. ovalis,* Johnson and Ware 1982), *Acer rubrum* (Nelson 1955), and *Betula lenta* (Woods and Shanks 1959; Good 1968; Johnson and Ware 1982).

Fulbright and associates (1983) noted that some chestnut trees planted in Michigan, 240–480 km northwest of their natural range, have become infected with a hypovirulent strain of chestnut blight. When the hypovirulent strain is present, the disease goes into remission, and the tree can recover. Many infected trees in the Michigan groves are showing recovery.

Acid precipitation. Smith (1981) defined acid precipitation as rain or snow with $pH < 5.6$. The pH of precipitation presently falling in the northeastern United States is in the range 3.0–5.5. Most of the acidity is attributable to sulfuric and nitric acids, whose precursors are presumed to be gaseous sulfur and nitrogen compounds. Anthropogenic sources (smelters, factories, etc.) are judged to be the main contributors. Smith considered the alteration of nutrient cycling to be the most important aspect of low-pH precipitation, specifically acceleration of leaching of nutrients from leaves and soil.

Odum (1983) has summarized the well-known effects of sulfuric acid fumes on the landscape of Copper Basin, Tennessee, which was originally covered with deciduous forest. Even with reduced emission of fumes, vegetation has failed to return in the most severely eroded areas. Partial recovery has followed heavy fertilization and artificial reforestation.

Johnson and associates (1984) documented "clear, widespread, and sustained decreases in tree growth rate" in the deciduous forest region over the last 20–30 yr. *Pinus rigida* and *Pinus echinata* are affected in southern New Jersey, *Picea rubens* in Vermont, and *Pinus strobus, P. ridiga,* and *Quercus prinus* in the Shawangunk Mountains of New York. *Picea rubens* (red spruce) is exhibiting widespread dieback. In the Green Mountains of Vermont, red spruce dieback has amounted to a 50% decline in density and basal area between 1965 and 1980, and the decline is seen over a wide range of elevations. Siccama and associates (1982) have made a detailed phytosociological analysis of the decline of red spruce in Vermont. Johnson and associates (1984) have implicated acid precipitation as the cause of declining forest productivity, but there is little direct evidence.

CLASSIFICATION OF ASSOCIATIONS

My classification of the deciduous forest and its regional parts is shown in Table 10.2. I follow the terminology of Walter (1979), who recognized major latitudinal climate belts over the earth's surface as "zonobiomes" (ZB). Figure 10.4 shows a map of the seven associations listed. In general, it follows an earlier map of Braun (1950).

Table 10.2. *Classification of deciduous-forest zonobiome, following terminology of Walter (1979)*

Zonobiome (ZB) or zonoecotone (ZE)	Eastern North American representative region
ZB VIII Cold-temperate (boreal)	(Boreal) conifer forest zone
ZE VI–VIII	(Boreal) conifer, (Boreal) deciduous forest zone 1. Hemlock, white pine, northern hardwoods 2. Spruce, northern hardwoods
ZB VI Typical temperate with a short period of frost (nemoral)	Deciduous forest zone Mixed mesophytic subzone 1. Mixed mesophytic association 2. Western mesophytic association 3. Oak-chestnut (oak) association Sugar maple–beech subzone 4. Beech-maple association 5. Maple-basswood association 6. Maple-beech-birch-buckeye [proposed association, p. 300]

(Cont.)

Table 10.2. *(Cont.)*

Zonobiome (ZB) or zonoecotone (ZE)	Eastern North American representative region
(and ZE VI–VII)	Oak-hickory subzone 7. Oak-hickory (and northern oak–hickory woodland/savanna) association
(and ZE VI–VII)	Oak, pine-hickory subzone 8. Oak-pine-hickory (and southern oak–hickory woodland/savanna) association
ZE VI–V	Deciduous dicot, evergreen dicot, (austral) conifer forest zone 1. Southern mixed hardwoods 2. Beech, magnolia
ZB V Warm-temperate maritime	Evergreen dicot, monocot (temperate broadleaved evergreen) forest zone

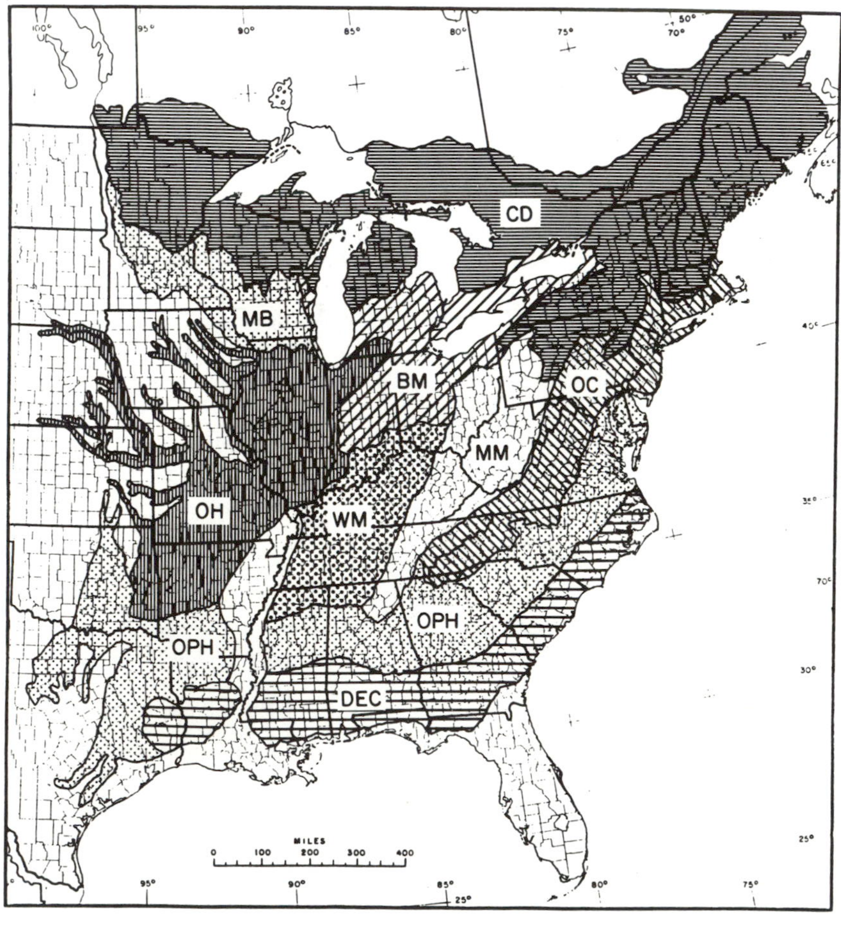

Figure 10.4. Map of the associations of the deciduous forest, based on Braun (1950), with my own modifications and terminology. Abbreviations for the nine associations: CD, conifer, deciduous; MB, maple, basswood; OH, oak, hickory; BM, beech, maple; WM, western mesophytic; MM, mixed mesophytic; OC, oak, chestnut; OPH, oak, pine, hickory; DEC, deciduous and evergreen dicots, conifer.

Figure 10.5. Cove forest in the southern Appalachian Mountains. The large trees are mainly Liriodendron tulipifera *(USDA Forest Service photo).*

Each of the seven geographical areas is named after the forest community type that is most widespread on well-drained, loamy uplands in the region. I use the term "association" to designate the entire vegetational complex co-occurring with that upland forest community over its geographical range. In most of the seven association regions, forest types typical of other associations are found in one or more areas. They usually occupy special, local, or edaphic sites in regions other than that of their best development. Transitions occur between association regions and are characterized by stand compositions that include dominants of both regions, or by a mosaic of stands typical of the two associations.

Mixed Mesophytic

I recognize a mixed mesophytic sub-zonobiome, following Knapp (1965). It comprises three associations: mixed mesophytic (MM), western mesophytic (WM), and oak-chestnut (OC). In the regions occupied by this sub-zonobiome, the most favorable sites (coves; well-drained, deep loam soils; sandy loams where water tables are suitably high) are dominated by a large number of aborescent taxa: *Fagus, Liriodendron, Acer saccharum, Tilia* species, *Aesculus, Castanea, Betula, Quercus* species, *Fraxinus*, and others (Fig. 10.5). On upper slopes and well-drained soils, *Quercus* species dominate. On ridges, rock outcrops, and sandy plains, a combination of *Quercus* species and *Pinus* species dominate in low, open stands. *Pinus* may dominate locally.

The MM association occupies the geographical center of the deciduous-forest region, primarily on the unglaciated Appalachian Plateau. According to Braun (1950), it is best developed in the Cumberland Mountains, where a mixture of trees dominate throughout the landscape. In the region of the OC association, stands of mixed deciduous trees dominate the coves or other extremely favorable sites. These contain oaks (*Quercus velutina, Q. borealis* var. *maxima, Q. alba*), with other genera as canopy subdominants (*Liriodendron, Fagus, Acer saccharum, Betula lenta, Carya* species). Formerly, chestnut (*Castanea dentata*) was a dominant or a codominant with oaks on slopes. In the WM geographical region there occurs a mosaic of types that reflects a transition from forests dominated by a mixture of trees in the east to forests dominated by *Quercus* species and *Carya* species in the west (Braun 1950). WM forest stands have almost as many canopy species as MM forest (Lindsey and Escobar 1976).

Figure 10.6. Beech-maple forest in Indiana. Summer view of Pioneer Mothers Woods. Foreground: two large Fagus grandifolia *trees. Center, rear:* Juglans nigra *(photograph courtesy of A. A. Lindsey originally published in "Natural Areas in Indiana," p. 198. Copyrighted 1969, copyright Indiana Natural Areas Survey, reprinted with permission).*

Sugar Maple

The sugar maple sub-zonobiome comprises at least two associations: beech-maple (BM) and maple-basswood (MB). This paragraph follows material presented by Lindsey and Escobar (1976). Sugar maple (*Acer saccharum*) is a dominant throughout the region. In combination with beech (*Fagus grandifolia*), it composes 80% of the canopy dominance in the BM region (Fig. 10.6). To the west of the BM region, beech reaches its western limit, and basswood (*Tilia americana*) gradually replaces it (Ward 1958). The relative dominance of sugar maple varies throughout the maple sub-zonobiome, depending on regional climate, microclimate, and habitat. Other forest types, dominated by oaks (*Quercus* species) and hickories (*Carya* species), may occur on drier sites, whereas mixed swamp forests, especially *Ulmus*, *Fraxinus*, and *Acer rubrum*, occupy the wettest soils. Sugar maple and its associates form a tightly closed canopy and produce a thick layer of humus and leaf litter that encourages the growth of spring perennial herbs and discourages bryophytes. The shrub layer is apparently better developed in the BM than in the MB association.

Oak-Hickory

The oak-hickory sub-zonobiome contains only the oak-hickory (OH) association of Braun (1950). Within the OH, a transition occurs from dense, closed, relatively tall deciduous forest to woodland and woodland/savanna. Clearly, decreasing precipitation is directly correlated with these transitions (Dyksterhuis 1957). At the western limit, OH forests are confined to ravine slopes and valleys, where the uplands are occupied by prairie (Braun 1950; Weaver 1960). Braun considered the best development of the OH type to be in the Ozark and Ouachita mountains (Fig. 10.7). It occupies extensive areas in the plains, rolling-hills, and drift-flats physiographic regions. Oaks are often accompanied by hickories (*Carya* species), which rarely dominate. Unusually favorable sites on north-facing slopes and deep valleys in the mountainous part of this region also have in the canopy *Acer* species, *Tilia americana*, *Magnolia* species, *Frauxinus*, and *Juglans nigra* (Braun 1950).

Oak-Pine-Hickory

The oak-pine-hickory sub-zonobiome (OPH) contains only the oak-pine association of Braun (1950), slightly modified following Bruner (1931) and Küchler (1974). It includes woodland and savannah dominated by *Quercus margaretta*, *Q. marilandica*, *Juniperus virginiana*, and *Carya texana* in eastern Oklahoma and adjacent Kansas. OPH occupies most of the Piedmont Plateau from Virginia southward and extends east-west from Georgia to Texas (Fig. 10.8). It can be recognized north of Virginia on the coastal plain to southern New Jersey (Bernard and Bernard 1971) and on the plains in eastern Oklahoma and adjacent southern Kansas as "oak-hickory savannah" (Bruner 1931) and "Cross-timbers, *Quercus-Andropogon*" (Küchler 1974). Throughout the major part of its range on the Piedmont, it is characterized by a rich mixture of oaks and hickories. Pines (*Pinus taeda* and *P. echinata*) dominate secondary forests, and post oak (*Quercus stellata*) and blackjack oak (*Q. marilandica*) dominate on poorer, thinner soils throughout the region (Braun 1950).

Figure 10.7. Oak-hickory forest in Missouri showing suboptimal development (USDA Forest Service photo).

Figure 10.8. Oak-hickory-pine forest in Norfolk, Virginia, dominated by pine (left and rear), tulip tree (right center), and oaks (extreme right rear).

Examples of Topographic-Edaphic Complexes

Southern Appalachian Mountains. Great variations in elevation, relief, geology, and physiography exist in the deciduous forest, and these permit wide variations in forest composition. It is useful to recognize forests of low, middle, and high elevations. At any given elevation, a variety of habitats may exist that will strongly influence the forest composition: rivers or streams; concave slopes ("coves"), convex slopes ("leads"), upper slopes, ridge tops, hill "crowns," and so forth; sandstone, limestone, granite, or serpentinite bedrock.

Whittaker's classic gradient analysis (1956) has described the mosaic of communities in the Great Smoky Mountains very well (Figs. 10.9 and 10.10). Coves and moist bottomlands below 900 m, throughout the MM sub-zonobiome, support a combination of approximately 25 species, of which 10–15 are often dominants (Table 10.3). The common dominants include *Fagus grandifolia, Liriodendron tulipifera, Tilia heterophylla* and other *Tilia* species,

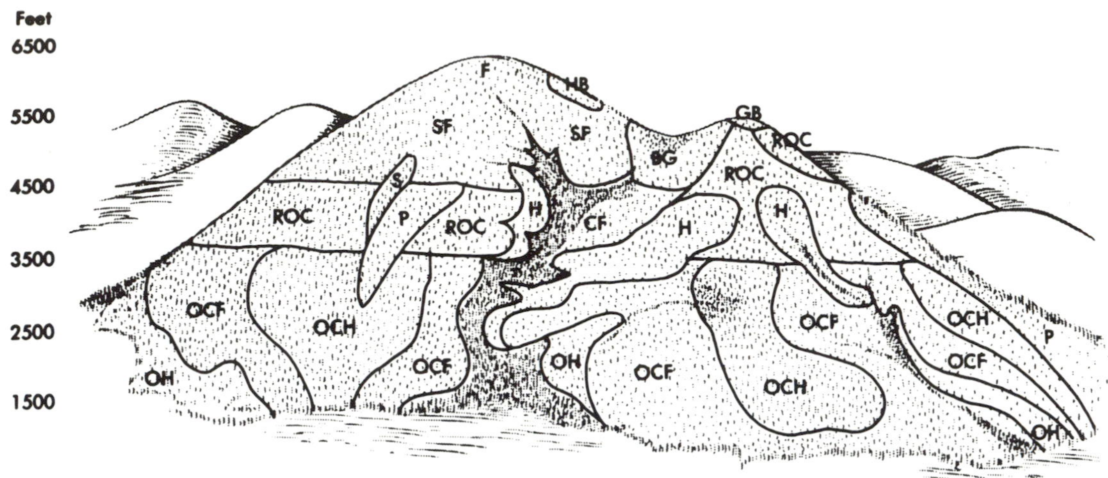

Figure 10.9. Topographic distribution of vegetation types on an idealized west-facing mountain and valley in the Great Smoky Mountains. Vegetation types: BG, beech gap; CF, cove forest (darker shading indicates steep slopes); F, Fraser fir forest; GB, grassy bald; H, hemlock forest; HB, heath bald; OCF, chestnut oak–chestnut forest; OCH, chestnut oak–chestnut heath; OH, oak-hickory forest; P, pine forest and pine heath; ROC, red oak–chestnut oak; S, spruce forest, SF, spruce-fir forest; WOC, white oak–chestnut forest. (from "Vegetation of the Great Smoky Mountains," by R. H. Whittaker, 1956, Ecological Monographs 26, 1-80. Copyright by ecological Society of America, reprinted by permission.)

Acer saccharum, *Castanea dentata* (formerly), *Aesculus octandra*, *Quercus borealis* var. *maxima*, *Quercus alba*, and *Tsuga canadensis*. The tulip tree (*Liriodendron*) is the giant of the forest. One individual was estimated by Core (1966) to contain 17,500 board feet of timber; it had a diameter of 2.64 m and reached 21.6 m to the first limb.

Ware (1982) performed a polar-ordination analysis of 34 stands of lowland forest sampled by Braun and concluded that certain groups of dominants tend to co-occur: *Aesculus octandra* and *Tilia heterophylla; Quercus alba* and *Castanea dentata*. Taxa such as *Fagus grandifolia* and *Tsuga canadensis* appear to be independent of other trees, whereas others, such as *Liriodendron* and *Acer saccharum*, are codominants with a large number of other taxa. Ware concluded that his analysis supported the unity of the MM community type of Braun.

Common subcanopy trees are *Cornus florida*, *Magnolia tripetala*, *M. fraseri*, *Oxydendrum arboreum*, *Acer pennsylvanicum*, *Cercis canadensis*, *Carpinus caroliniana*, *Ilex opaca* (evergreen), and *Amelanchier arborea*. Core (1966) gave an extensive list of taxa that are characteristic of MM subcanopy synusia.

Braun (1950) described transitions from lower-elevation, rich MM stands through simpler mixed deciduous stands at middle elevations to spruce forests at the highest elevations in the southern Appalachian Mountains (MM and OC coves). Braun referred to some of the middle-elevational

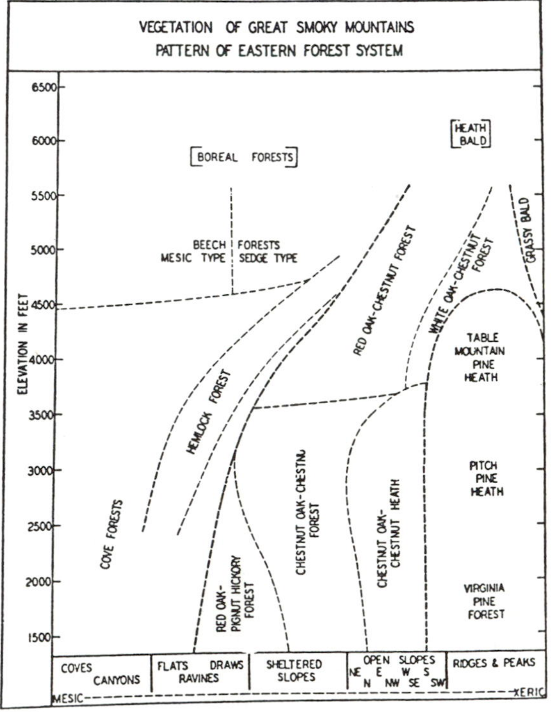

Figure 10.10. Vegetation of Great Smoky Mountains, below the subalpine conifer forest system, with respect to gradients of elevation and topography. (from Whittaker 1956, copyright Ecological Society of American, reprinted by permission).

Table 10.3. *Composition and percentage-importance values (≥ 5%) for species in a mixed mesophytic forest, Savage Gulf, Tennessee*

Species	Overstory	Understory
North-facing slope		
Tilia heterophylla	20.4	8.5
Tsuga canadensis	18.1	8.8
Acer saccharum	13.3	20.9
Liriodendron tulipifera	12.5	—
Aesculus octandra	7.7	7.2
Carya tomentosa	6.9	1.2
Fagus grandifolia	5.7	16.0
Cornus florida	—	17.6
Others (12)	15.4	19.8
South-facing slope		
Quercus prinus	29.0	4.8
Carya tomentosa	18.9	17.1
Quercus borealis var. *maxima*	18.9	5.5
Liriodendron tulipifera	7.8	—
Quercus alba	7.7	—
Carya ovata	5.9	4.1
Cornus florida	—	24.9
Cercis canadensis	—	17.9
Sassafras albidum	—	6.0
Nyssa sylvatica	—	6.0
Others (9)	11.8	13.7

Source: Quarterman et al. (1972).

stands as "maple-basswood-birch," and as "northern hardwoods." A rich herbaceous layer occurs throughout the elevational range of these predominantly deciduous, mixed forests. Mesic and mesoxeric oak forests extend to high elevations, above 1372 m, in the Great Smoky Mountains, according to Whittaker (1956). On the driest sites, pines (*Pinus virginiana, P. rigida,* and *P. pungens*) are associated with the oaks (*Quercus prinus, Q. coccinea*), or the pines form conifer stands (Racine 1966; Zobel 1969).

On lower, south-facing slopes, upper slopes, ridges, and hilltops below 914 m, a transition can be recognized from oak and hickory forests and oak and chestnut (formerly) forests to oak, oak and pine, and pine forests. With increasingly xeric conditions, oak dominants shift from *Quercus borealis* var. *maxima* (greatest moisture) to *Q. prinus* and *Q. coccinea* (driest) (Whittaker 1956; Racine 1971; Quarterman et al. 1972; Johnson and Ware 1982).

Hickories are associates of the oaks on poor sites highest in moisture: *Carya tomentosa, C. ovata, C. ovalis* (Whittaker 1956; Quarterman et al. 1972; Johnson and Ware 1982). On drier sites, chestnut (*Castanea dentata*) formerly was a dominant with *Quercus prinus* and some other *Quercus* species, (Whittaker 1956; Johnson and Ware 1982).

Russell (1953) noted that pure stands of *Fagus* ("beech gaps") occur above 1524 m on south-facing slopes of ridges, as deciduous "islands" in a zone of conifer forest that is dominated by *Picea rubens* and *Abies fraseri*, in the mountains of North Carolina and Tennessee. Beech gaps often occur at the high ends of coves in which mixed deciduous trees or *Tsuga-Fagus* form continuous forests from low elevations. Shanks (1956) found that beech gaps are always a few degrees warmer in soil temperature than spruce and fir forests at the same elevation. The variety of beech is the "Gray beech" (Camp 1951). The trees are gnarled, small, and broken. The understory consists of 88 species of herbaceous plants, but none is restricted to this vegetation type. There is a thin cover of deciduous leaf mold on the soil, and the soil pH is 4.5–6.0. In contrast, adjacent conifer sites have a thick litter of conifer needles and a pH of 3.6–3.8 (Russell 1953).

Crandall (1958) recognized a spruce-hardwoods forest (*Picea, Betula, Acer, Fagus, Tsuga*) in the Great Smoky Mountains, with three types of understories: (1) *Viburnum alnifolium* shrub, (2) evergreen *Rhododendron* shrub, and (3) a herbaceous type dominated by *Aster*, with many other forbs, some pteridophytes, and graminoids.

Balds. Brown (1941) described three types of "bald" communities on Roan Mountain in the southern Appalachians: grassy balds, *Rhododendron* (heath) balds, and alder balds. Mark (1958) defined "balds" as naturally occurring treeless vegetation located on well-drained sites below the climatic treeline, at 1524–1829 m elevation, in a predominently forested region. The forests surrounding balds are mixtures of deciduous dicots and conifers (*Picea, Abies*) or are purely deciduous dicots at their upper limits of range (*Quercus borealis* var. *maxima, Fagus grandifolia, Aesculus octandra*).

Heath balds or heath barrens have been described for mountaintops and other exposed sites with thin soils above 1067 m elevation. Dominant taxa are shrubs of *Rhododendron* species, *Kalmia latifolia, Gaylussacia* species, and *Vaccinium* species (Whittaker 1963; Core 1966). On Brushy Mountain, North Carolina, Whittaker (1963) recorded 4270 stems per 0.1 ha, with a basal area of $35 m^2 ha^{-1}$, and a volume estimated at $60.1 m^3 ha^{-1}$. The dominants are *Rhododendron catawbiense* and *Kalmia latifolia*. On Roan Mountain, the best development of *Rhododendron* balds occurs on flat summits between 1829 and 1875 m, according to Brown (1941).

Alder balds occur between 1753 and 1798 m. This is a three-layer community dominated by *Alnus alnobetula*, 0.5–2 m tall, a herb layer of *Dryopteris dilatata, Carex, Danthonia, Agrostis hiemalis,* and *Ru-*

Figure 10.11. Appalachian grass bald. Gregory Bald in Great Smoky Mountains National Park, 1934 (USDA Forest Service photo by C. C. Campbell).

mex acetosella, 15–46 cm tall, and a bryoid layer 2 cm tall dominated by *Polytrichum* (Brown 1941).

The grass balds are dominated by *Danthonia spicata*, *Viola* species, *Aster surculosa*, and other species (Fig. 10.11). Shrubs are occasional and are represented mainly by *Amelanchier laevis*. Mark (1958) reported that grass balds are in a state of dynamic equilibrium in high mountain ecotones, maintained by natural forest destruction, grazing by domesticated animals, and forest clearing by humans. They are being invaded by *Picea* and *Abies* trees, where seed sources are available. Mark further believed that the grass balds were initiated on forest soils as a result of recent climatic fluctuations that eliminated *Picea* and *Abies* forests from many Appalachian summits during the xerothermic. Spruce-fir forests are discussed at the end of this chapter.

Central Appalachian Mountains. Core (1966) discussed habitat preferences of major arborescent taxa in the hill region of western West Virginia. On the most mesic sites, oaks and a mixture of deciduous dicots dominate: *Quercus alba*, *Q. muhlenbergii*, *Carya ovata*, *Nyssa sylvatica*, *Juglans nigra*, *Prunus serotina*, *Acer saccharum*, and *Morus rubra* (Core 1966). On "loamy, well-drained soils," presumably drier than the foregoing habitat, a combination of oaks dominates, with deciduous dicots as subdominants: *Quercus alba*, *Q. velutina*, *Q. borealis* var. *maxima*, *Ulmus rubra*, *Acer rubrum*, *Juglans nigra*, *Tilia heterophylla*, *Robinia pseudoacacia*, *Fagus grandifolia*, and *Nyssa sylvatica*. At lower elevations, on southern exposures, and in areas of greater warmth or equability, southern trees such as *Castanea pumila*, *Liquidambar styraciflua*, and *Ilex opaca* are found as associates. Oaks 44.2 m tall and 2.7–4.0 m in diameter (dbh) have been recorded.

Core also described middle-elevation forests dominated by *Quercus borealis* var. *maxima*, *Tilia americana*, and *Fraxinus americana* in which subdominants included *Acer saccharum*, *Aesculus octandra*, *Betula lenta*, *Betula lutea*, *Castanea dentata* (formerly), and *Prunus serotina* (Table 10.4).

Reinhardt and Ware (1984), working in Virginia's Balsam Mountains, found a complex of forests dominated by broad-leaved deciduous tree taxa at elevations of 1200–1600 m, below the conifer and conifer-hardwood forests of the ridges and tops. They listed these in five types: beech-maple, dwarf beech, mesophytic-oak, mesophytic, and northern red oak. *Fagus grandifolia* is a major dominant in four of these types, *Acer saccharum* in three, *Betula lutea* and *Quercus borealis* in two, and *Aesculus octandra*, *Acer rubrum*, *Fraxinus americana*, *Tilia heterophylla*, and *Quercus prinus* in one each. If we remove from consideration the widely distributed *Quercus borealis*-dominated forest with its associated *Acer rubrum* and *Quercus prinus*, we are left with a complex of forests overwhelmingly dominated by *Fagus grandifolia* and *Acer saccharum*. These forests differ from the low-elevation MM stands in their fewer number of major dominants and (apparently) lower richness and diversity.

I propose that these high-elevation mesophytic forests be recognized as distinct from the MM type. They are related, instead, to the low-elevation northern BM and MB types. I suggest that a modification of Braun's name be adopted: maple-beech-birch-buckeye type. It is classified in Table 10.2 as (proposed) number 6.

Pine and oak barrens. On soils unfavorable to forest growth on the coastal plain of New Jersey and Long Island, near the northern limit of the OC, the following shrubland and grassland types have been described: "Oak Barrens" (*Quercus ilicifolia*), "Pine plains" (*Pinus rigida*) (Andresen 1959; Olsvig et al. 1979), and grass "Plains" (*Andropogon* species, *Sorghastrum*) (Cain et al. 1937). On sites with soils too shallow to support tree growth in the mountains of West Virginia, *Quercus ilicifolia* (bear or scrub oak) forms a "dense, impenetrable shrubland" in association with other taxa of the xeric forests (Core 1966).

Pitch pine (*Pinus rigida*) and pitch pine–oak forests are a prominent feature of the landscape in central New Jersey, central Long Island, the Albany vicinity in upstate New York, and Cape Cod, Massachusetts. These forests are widely recognized as protecting the last major underground reservoirs of pure water in one of the most densely populated

Table 10.4 Percentage abundance of canopy trees in stands of forest in the central Appalachian Mountains

Taxa	Allegheny Mountain, W. Va. (≤914 m)			Western Maryland (< 914 m)					Alleghany Mountain, W. Va. (> 1000 m) 1000–1220 m				Alleghany Mountain, W. Va. (> 1000 m) 1280–1340 m		
	1	2	3	4	5	6	7	8	9	10	11	12	13	14	15
Acer rubrum	5.0	14.0	—	—	2.3	5.3	0.6	10.4	12.9	16.2	6.6	13.6	1.1	—	13.2
Acer saccharum	5.0	7.8	22.1	20.5	—	—	2.5	24.0	20.1	5.0	13.2	9.1	31.2	5.8	—
Betula lenta	1.5	3.1	1.3	3.3	0.1	3.9	0.2	7.1	11.9	21.2	9.9	11.4	3.5	9.6	9.5
Betula lutea	—	—	—	4.0	—	—	6.6	—	5.2	5.0	4.4	6.8	4.0	13.4	12.5
Carya species	6.5	0.8	—	—	—	—	—	—	1.4	—	—	—	—	—	—
Castanea dentata	18.0	46.1	37.6	36.1	10.8	47.3	0.1	5.2	1.4	4.0	1.1	2.3	—	—	—
Fagus grandifolia	—	0.8	13.0	2.8	—	1.0	1.6	31.2	24.9	7.1	27.5	22.7	33.5	28.8	8.8
Fraxinus americana	—	3.9	1.3	—	—	—	—	5.2	6.2	5.0	14.3	6.8	1.7	—	—
Magnolia acuminata	6.5	6.2	2.6	—	—	—	—	1.9	4.3	3.0	9.9	18.2	1.1	—	—
Magnolia fraseri	—	—	1.3	—	—	—	—	—	0.5	—	—	—	—	—	—
Picea rubens	—	—	—	—	—	—	—	—	—	—	—	—	8.7	38.5	55.9
Prunus serotina	—	0.8	16.9	—	—	—	—	2.6	—	—	—	—	11.6	3.8	—
Quercus alba	11.5	2.3	—	2.5	81.1	10.7	0.3	—	—	—	—	—	—	—	—
Quercus borealis var. *maxima*	23.0	11.7	3.9	9.2	4.0	20.0	0.1	2.6	1.0	13.1	3.3	9.1	—	—	—
Quercus prinus	13.1	—	—	0.2	1.1	6.4	—	—	—	—	—	—	—	—	—
Tilia americana	—	—	—	—	—	—	—	9.7	10.0	20.2	9.9	—	3.5	—	—
Tilia species	10.0	1.6	—	8.7	—	0.2	1.2	—	—	—	—	—	—	—	—
Tsuga canadensis	—	—	—	—	—	0.3	86.0	—	—	—	—	—	—	—	—
Others	—	0.8	—	12.6	0.5	5.6	0.9	—	—	—	—	—	—	—	—

Source: Brown (1950).

Figure 10.12 Pitch pine woodland with dense understory of bear oak on High Knoll, Lebanon State Forest, New Jersey (photo by Silas Little, courtesy of David Gibson, Division of Pinelands Research, Center for Coastal and Environmental Studies, Rutgers–The State University, New Brunswick, New Jersey).

regions of North America. In central New Jersey, *Pinus rigida* dominates lowland and upland sites, accompanied, locally, by *Pinus echinata* and *Pinus virginiana* (Fig. 10.12). The pine stands are 4.6–7.6 m in height and have an understory dominated by shrub oaks and heaths: *Quercus ilicifolia, Q. prinoides, Q. marilandica, Q. stellata* (4.6 m tall), *Kalmia* species, *Gaylussacia* species, and *Vaccinium* species (to 1 m tall). Barden and Woods (1976) reported that "severe crown fires which removed more than 85 percent of the basal area and canopy coverage, encouraged re-establishment of pines." In the absence of these severe fires, the pines tend to be replaced by oaks (Little 1973). In many parts of these "Pine Barrens," *Quercus coccinea, Q. alba, Q. velutina*, and *Q. prinus* share dominance with the pines, or form dry oak communities (Harper 1908; Andresen 1959; McCormick 1970; Forman 1979). More details of pine and oak barrens appear in Chapter 11.

Rock outcrops. Throughout the deciduous-forest region, local geologic and physiographic conditions prevent normal soil processes from developing or from reaching maturity. Under these conditions, drought and excessive heat prevent vegetation from reaching its regional potential. Also, concentrations of certain minerals may occur in these shallow soils that are sufficient to be deleterious or even toxic to many taxa. Vegetation responses to edaphic and microclimatic extremes show broad similarities across different substrata. Plants of the shallowest soils are often summer or winter annuals, in contrast to the summer-active perennials of the deeper soils (Baskin and Baskin 1978). Floristic differences among the generally similar vegetal complexes appear to be due to differences in soil chemistry and macroclimate, as, for example, between stations at northern and southern extremes of the deciduous-forest region.

The generally similar environments of rock out-

Figure 10.13. Rock-crop vegetation showing several zones of vegetation: (0, community on rock dominated by lichens and mosses; 1, Sedum smallii *community on very shallow soil; 2,* Minuartia uniflora *community on slightly deeper soil; 3, perennial community,* Viguiera porteri *prominent, on still deeper soil (from Sharitz and McCormick 1973, copyright Ecological Society of America, reprinted by permission).*

crops result in some or all of the following habitats: a gradient of soil depth from bare rock to deep soils on which sit the surrounding mature vegetation, crevices where soil accumulates on the bare rock surfaces, temporary pools (granite), "dripways," overhangs, and caves.

On a gradient of soil depth on limestone in Kentucky, Baskin and Baskin (1978) showed an array of vegetation, the simplest of which was lichen-covered rock. With increasing soil depth one usually sees bryoids (soil 2 cm deep), then succulent forbs (2–5 cm deep), then grasslands (5–25 cm), then shrublands, and, finally, woodland on the deepest soils (Fig. 10.13).

Parmelia conspersa is often mentioned as a lichen of the bare rocks. In the bryoid zone are *Grimmia*, *Anomodon*, and other mosses, *Selaginella rupestris*, and *Cladonia* lichens. In the succulent forb zone are *Sedum* species, *Talinum* species, *Oputnia humifusa*, and *Agave virginica* on sandstone, and *Portulaca* on granite.

Many genera and species common in the low forb communities of shallow soils are rare elsewhere or even endemic to these sites, including many succulents and *Phlox* species, *Crotonopsis*, *Croton* species, *Viguiera*, *Allium oxyphilum* (on shale), *Allium allegheniense* (on limestone), *Clematis*, *Leavenworthia*, *Palafoxia callosa*, and *Arenaria* species.

On limestone, in Kentucky, on soils 5–10 cm deep, a simple grassland occurs, dominated by *Sporobolus vaginiflorus*. On soils 10–20 cm deep, *Sporobolus* is codominant with a number of forbs such as *Ruellia*, *Agave*, *Hypericum*, *Heliotropium*, and *Isanthus*. On soils 15–25 cm deep, *Andropogon scoparius* dominates, with *Andropogon gerardii* and *Sorghastrum nutans* as associates. In the grasslands, woody plants occur in deep, moist soils of cracks: *Celtis*, *Forestiera*, *Fraxinus*, *Symphoricarpos*, *Rhamnus*, *Ulmus*, *Juniperus*, and *Cercis*.

In southern Pennsylvania, on deeper serpentine soils adjacent to herbaceous communities, Harshberger (1903) described a "thicket" community dominated by *Smilax glauca* and *Smilax rotundifolia*, with *Juniperus virginiana* and *Nyssa sylvatica* as occasional emergents, and *Spiraea*, *Rosa*, *Rubus*, and *Kalmia* as more local dominants. *Juniperus virginiana*, it should be noted, forms conspicuous and extensive "cedar glade" woodlands on limestone in Missouri, Tennessee, and Kentucky. A xeric, oak-dominated community occurs on still deeper soils, with *Quercus stellata*, *Q. marilandica*, *Q. alba*, *Acer rubrum*, *Juniperus virginiana*, *Castanea dentata* (formerly), *Sassafras albidum*, and *Cornus florida*. There is an understory of mixed shrubs and woody vines, as well as a floristically poor herb layer, with *Hieracium*, *Pteridium*, *Antennaria*, *Baptisia*, *Rubus*, *Potentilla*, *Rumex*, *Veronica*, *Hypoxis*, and

Lysimachia. Knox (1984) recorded *Pinus virginiana* as an important tree in forests on serpentine in adjacent Maryland.

For further discussions of the flora and vegetation on different rock types in various forest regions, see Quarterman (1950), Kucera and Martin (1957), and Core (1966) for limestone, Platt (1952) and Core (1966) for shale, McVaugh (1943) and Larson and Batson (1978) for granite, Voight and Mohlenbrock (1964) for sandstone, and Pennell (1910) for serpentine.

ECOLOGY OF THE HERB LAYER

Distribution Patterns

As is the case for tree taxa, the southern Appalachian Mountains are the center of floristic richness for the herb layer. Rogers (1982) documented a gradient of decreasing richness, decreasing herb cover, and decreasing importance of the strictly vernal growth forms in a broad transect from central Ohio to southeast Minnesota and northwest Michigan (Table 10.5).

Table 10.5. *Mean percentage cover of herbaceous taxa, by region*

Species	Ohio, Indiana, southern Michigan	Southeastern Wisconsin	Eastern Minnesota	Combined deciduous-forest region	Combined conifer-hardwood forest region
Spring ephemeroids and annuals					
Dentaria laciniata	2.2	1.6	0.6	2.1	1.9
Dicentra cucullaria	3.2	0.7	3.1	3.1	2.1
Allium tricoccum	1.1	0.3	p[a]	1.0	0.9
Erythronium americanum	0.7	2.3	—	0.8	1.6
Dicentra canadensis	5.0	—	—	5.0	0.5
Galium aparine	0.8	0.2	0.4	0.7	p
Claytonia virginica	—	0.2	0.1	0.7	p
Erigenia bulbosa	0.4	—	—	0.4	—
Floerkia proserpinacoides	1.9	—	—	1.9	—
Cardamine douglasii	0.2	—	—	0.2	—
Anemonella thalictroides	0.1	p	—	0.1	—
Isopyrum biternatum	3.7	p	5.5	3.1	—
Erythronium albidum	—	0.7	11.5	5.0	—
Claytonia caroliniana	—	0.2	0.1	0.7	p
Herbs with persistent shoots					
Viola species	0.3	0.2	0.1	0.3	1.3
Osmorhiza claytoni	0.3	0.1	0.3	0.3	p
Sanguinaria canadensis	0.1	p	0.3	0.2	0.2
Trillium grandiflorum	0.2	0.1	0.3	0.2	0.3
Uvularia grandiflora	p	p	p	p	p
Anemone quinquefolia	0.2	p	0.2	p	0.2
Phlox divaricata	0.2	p	0.1	0.2	—
Geranium maculatum	0.4	p	0.3	—	—
Asarum canadense	0.5	p	0.4	0.5	—
Hydrophyllum virginianum	2.7	0.5	2.8	2.2	1.7
Hepatica acutiloba	0.3	0.1	0.3	0.2	p
Thalictrum dioicum	—	p	0.1	0.1	—
Amplectrum hyemale	p	p	p	p	p
Trillium gleasoni	—	—	p	p	—
Podophyllum peltatum	0.2	0.3	—	0.2	p
Trillium recurvatum	0.3	—	—	0.3	—
Hydrophyllum appendiculatum	0.4	—	—	0.4	—
Osmorhiza longistylis	2.4	—	—	2.4	—
Stylophorum diphyllum	p	—	—	p	0.6
Carex species	0.2	—	—	0.2	—
Geum canadense	p	—	—	p	—
Panax trifolius	0.1	—	—	0.1	0.1
Dentaria diphylla	p	—	—	p	p
Mitella diphylla	p	—	—	p	p
Trillium cernuum	—	—	—	—	0.2

[a]Present, but cover less than 0.1%.
Source: Rogers (1982).

Within a geographical region, herb richness and cover increase with nutrient content of the soil. This is especially true of the vernal flora, dominated by annuals and ephemeral perennials. Vernal flora is best developed in the leaf litter and humus under mixed deciduous dicot trees, as occur in cove forests and throughout the *Acer saccharum* subzonobiome (Braun 1950; Lindsey and Escobar 1976). *Quercus*-dominated forests lack a well-developed vernal flora (Greller 1977). On nutrient-poor soils, summer-green, semi-evergreen, and evergreen herb taxa increase in importance: for example, *Epigaea*, *Gaultheria*, *Kalmia*, and *Chimaphila* in the Ericaceae, and *Mitchella*, *Pyrola*, *Geum*, and *Galax*.

Within stands, microtopography and soil drainage are the most important factors influencing herbs (Bratton 1976; Rogers 1982). Bratton also noted as important the structure of the canopy, which influences the positions and sizes of openings, throughfall of rain, pattern of light, and distance from trees. Bratton noted the following features of microtopography as influencing the distribution of herbs: the distance from the base of a tree, from a log, or from rock and the presence of mounds from windthrown tree roots, vernal pools, and other disturbance-related phenomena. Using principal-components analysis, Bratton recognized groupings of herb species in a rich cove forest that could be related to these microtopographic factors.

Light and Temperature Relationships

Seasonal changes in temperature and rainfall are, of course, inherent factors in the nature of herb communities in the deciduous-forest region. Phenological differences are pronounced in these rich forests, and a temporal sequence is superimposed on microtopographic gradients (Fig. 10.14).

Sparling (1967) analyzed the light requirements of vernal, transitional, and aestival groups of herb taxa. Vernal herbs are shade-intolerant. Their leaves emerge early, before the canopy leaves, then expand with increasing warmth. Photosynthesis reaches a maximum when leaves are fully mature, temperature is greatest, and light most intense (for example, photosynthesis is saturated at one-half to one-third full sun). Herbs in this group are *Dicentra canadensis*, *D. cucullaria*, *Claytonia virginica*, *Caulophyllum thalictroides*, *Erythronium americanum*, and *Allium tricoccum*. When the canopy fills, these herbs exhibit yellowing and decay of their leaves.

A group of mesic herbs was recognized by Sparling as semi-shade-tolerant. Their leaves expand as the canopy fills. They show great variability in their light-compensation points, saturation light intensities (between $\frac{1}{40}$ and $\frac{1}{10}$ full sun), and photosynthetic rates at saturation light intensity. Listed in this group are *Asarum canadense*, *Fragaria virginiana*, *Podophyllum peltatum*, *Trillium erectum*, *Viola pubescens*, *V. sororia*, *Dentaria diphylla*, *D. laciniata*, and *Sanguinaria canadensis*.

A third large group of taxa is listed as shade-tolerant or "shade-obligate," whose leaves develop after the canopy expands. Photosynthesis is saturated usually below $\frac{1}{40}$ full sun, and compensation points often are below $\frac{1}{400}$ full sun. Shade-tolerant herbs are *Actaea pachypoda*, *Dryopteris marginalis*, *D. spinulosa*, *Epigaea repens*, *Hepatica* species, *Maianthemum canadense*, *Oxalis monana*, *Polygonatum pubescens*, *Polystichum acrostichoides*, *Smilacina racemosa*, and *Trillium grandiflorum*.

Hicks and Chabot (1985) summarized Sparling's data and included additional information (Table 10.6, p. 307). Spring ephemerals have a higher light-compensation point and saturation point than summer-green taxa. Maximum and minimum photosynthesis rates are about equally variable in both types. As might be expected, spring ephemerals have a leaf life span about half that for summer-green taxa. Evergreen herbs are variable in compensation point, saturation point, and maximum photosynthetic rates, overlapping the two other groups.

Bratton (1976) gave data on dominance (cover) for herbs in these tolerance categories for a cove forest of eastern Tennessee. Of the vernal herbs, *Phacelia fimbriata*, *Dicentra* species, and *Dentaria* species showed the greatest average percentage cover. Dominant late vernal herbs were *Stellaria pubera*, *Caulophyllum thalictroides*, and *Osmorhiza longistylis*. Aestival herb dominants were *Laportea canadensis* and *Cimicifuga racemosa*. Schemske and associates (1978) studied flowering and pollination relationships of vernal herbs in an *Acer saccharum–Quercus rubra* forest in central Illinois. They noted that most vernal herbs have white or whitish UV-light-absorbing flowers. They concluded that flowering of vernal herbs often is ill-timed to pollinator activity, making early spring flowering a "high-risk option."

Germination patterns among deciduous-forest herbs have received relatively little attention. Baskin and Baskin (1983, 1984) recently summarized that literature and described several patterns. In the pattern typified by *Osmorhiza longistylis*, a late vernal species, seeds require a long period of high temperatures, followed by a long period of low temperatures, in order to germinate. They germinate only in spring.

Figure 10.14. Phenological change of leaf cover at one site in a deciduous forest. Top: April 1985. Bottom: June 1985. Red-oak-dominated community in oak-chestnut region; Cunningham Park, Queens County, Long Island, New York.

BIOMASS, PRODUCTIVITY, AND NUTRIENT CYCLING

Table 10.7 (p. 308) has data on biomass and productivity for stands of vegetation in eastern North America, with emphasis on forests of the deciduous forest zonobiome. Undisturbed forests of mixed dominants exhibit the greatest dry-matter biomass, 61 kg m^{-2}. Forests on drier sites, dominated by *Quercus* and *Pinus*, exhibit smaller biomasses, 10.2 kg m^{-2}. With increasing elevation and decreasing warmth in the southern Appalachian Mountains, the biomass decreases, as shown by *Abies* forests at the highest elevations (20 kg m^{-2}). Productivity for a wide range of undisturbed lowland, broad-leaved forest types appears to be roughly 900–1200 g m^{-2} yr^{-1}. Productivity for disturbed forests is about twice that for undisturbed forests of the same type (Whittaker 1966).

Figure 10.15 (p. 308) shows an organic-matter budget for a mesic secondary forest in the southern Appalachian Mountains (Oak Ridge, Tennessee)

Table 10.6. *Photosynthetic characteristics of deciduous-forest herbaceous species*

Species	Compensation point[a]	Saturation point[a]	Maximum photosynthetic rate[b]	Leaf life span (wk)
Spring ephemerals				
Erythronium americanum	1.0	25.0	20[c]	11
	0.7	30.0	85.2	11
	1.0	~50.0	125	11
Allium tricoccum	0.5	20.0	82	10
	—	—	13.1	11
Dicentra canadensis	1.8	30.0	28.4	9
Claytonia caroliniana	1.8	15.0	26.5	11
Dentaria diphylla	0.4	3.8	36.6	10
Summer-greens				
Solidago flexicaulis	0.5	12.0	11[c]	28
Trillium grandiflorum	0.5	14.0	10[c]	22
	0.2	23.0	30.3	11
T. erectum	0.5	2.5	—	20
Podophyllum peltatum	0.5	2.5	27.1	18
	—	—	11.5[c]	11
Parthenocissus quinquefolia	—	—	4.7[c]	25
Caulophyllum thalictroides	1.0	30	49.2	20
Sanguinaria canadensis	0.2	5.0	70	20
Dryopteris marginalis	0.2	3.0	19.6	30
Maianthemum canadense	0.1	2.5	31	22
	0.3	3.0	82.7	11
Polygonatum pubescens	0.1	2.5	12	~22
Hydrophyllum appendiculatum	1.0	21.0	5.7[c]	
Evergreens				
Mitchella repens	0.4	7.0	43	120
Dryopteris spinulosa	0.2	2.5	20.8	54
Pyrola elliptica	0.4	7.0	55.5	54
Hexastylis arifolia	0.2–1.5	24.0	11[c]	50
Fragaria vesca	0.6–2.5	20–30	71(3.9)[c]	14–20

[a]Percentage of full sunlight.
[b]μmol Co_2 g^{-1} sec^{-1}.
[c]μmol Co_2 m^{-2} sec^{-1}.
Source: Hicks and Chabot (1985).

dominated by *Liriodendron*, with some *Quercus* species. The largest compartments for organic matter are the trees and the soil (including litter and decomposers).

Allocation of biomass between shoots and roots varies directly with overall community weight and structural complexity. For example, a cove forest with a biomass of 58.6 kg m^{-2} has 13.5% in the root compartment, whereas the Brookhaven *Quercus-Pinus* forest, with a biomass of 10.2 kg m^{-2}, has 34.2% in the root compartment, and an old field, with a maximum biomass of 2.9 kg m^{-2}, has 87% in the root compartment.

Southern swamp forest types, some near their northern limit, appear to be roughly comparable in biomass to moderately complex upland forests of similar age in the deciduous-forest region. Their productivity, however, may be roughly half the productivity of mature upland types, and perhaps a third or less of that for disturbed types (Table 10.7). Productivity for a southern old field is equal to that for a cove forest, although the biomass for the former is perhaps 5% of that for the latter.

Bormann and Likens (1979) studied inputs and outputs of some nutrients in a growing or aggrading secondary "northern hardwoods" forest at Hubbard Brook, New Hampshire. For nongaseous elements, inputs were recorded at rain-gauging stations, and outputs at weirs on small streams. Data for elements with a gaseous phase were less complete. Gross dissolved substance losses were 13 Mt km^{-2} yr^{-1}; particulate-matter losses were 2.5 Mt km^{-2} yr^{-1}. Precipitation is acidic, with sulfate and hydrogen ions dominating; nitrates have doubled since 1955. After precipitation passes through the forest ecosystem, its mineral

Table 10.7. *Some biomass and productivity estimates for forests of easternNorth America (aboveground and belowground unless otherwise stated)*

Forest region and type	Biomass (kg m^{-2})	Productivity (g m^{-2} y^{-1})	Author
Deciduous dicot, evergreen dicot conifer ZE (transition)			
Taxodium distichum swamp[a]	34.53		Dabel & Day (1977)
T. distichum swamp, small roots	1.53		Montague & Day (1980)
T. distichum swamp, leaf litter		528	Gomez & Day (1982)
Chamaecyparis thyoides swamp[a]	22.05		Dabel & Day (1977)
C. thyoides swamp, small roots	1.80		Montague & Day (1980)
C. thyoides swamp, leaf litter		506	Gomez & Day (1982)
Acer-Nyssa swamp[a]	19.57		Dabel & Day (1977)
Acer-Nyssa swamp, small roots	1.22		Montague & Day (1980)
Acer-Nyssa swamp, leaf litter		536	Gomez & Day (1982)
Mixed hardwood swamp[a]	19.46		Dabel & Day (1977)
Mixed hardwood swamp, small roots	3.10		Montague & Day (1980)
Mixed hardwood swamp, leaf litter		455	Gomez & Day (1982)
Deciduous-forest ZB			
Oak-hickory[a]	15.7	600	Monk et al. (1970)
Old field	2.9 (max.)	1210[b]	Golley (1960)
Oak-chestnut region			
Undisturbed cove forests	50−61	1000−1200	Whittaker (1966)
Mixed mesophytic (east Tenn.)	42	1510	Skeen (1973)
Disturbed cove forest	22	2408	Whittaker (1966)
Pinus-Quercus types	6−10	420−990	Whittaker (1966)
Quercus-Pinus, Brookhaven	10.2	1195	Whittaker & Woodwell (1969), Whittaker (1975)
Ilex opaca , sunken forest	17.1	1080	Art (1976)
Gray beech (*Fagus*),south-facing	17	906	Whittaker (1966)
Tsuga-Rhododendron	49	1022	Whittaker (1966)
Picea-Rhododendron	32.1	812	Whittaker (1966)
Picea-Abies, north-facing	30−35	940−1024	Whittaker (1966)
Abies	20	566−653	Whittaker (1966)

[a]Aboveground only.
[b]Estimated using 1 g = 4.08 kcal.

Table 10.8. *Nutrient budgets (kg ha^{-1} yr^{-1}) for forests in eastern United States*

Location (dominants)	Calcium			Magnesium			Sodium		
	Input[a]	Output[b]	Flux	Input	Output	Flux	Input	Output	Flux
Coshocton,Ohio (*Robinia, Pinus*)									
Coweeta, N.C. (*Quercus, Carya, Acer*)	6.2	6.9	−0.7	1.3	3.1	−1.8	5.4	9.7	−4.3
Hubbard Brook, N.H. (*Acer, Fagus, Betula*)	2.2	13.9	−11.7	0.6	3.3	−2.7	1.6	7.5	−5.9
Long Island, N.Y. (*Quercus, Pinus*)	3.3	9.6	−6.3	2.1	7.3	−5.2	17	23	−6.0
Taughannock Creek, N.Y. (*Acer, Tilia, Tsuga*)	14.9	182	−167	1.4	34.8	−33.4	2.3	18.9	−16.6
Walker Branch, Tenn. (*Quercus, Carya*)	14.3	148	−134	2.1	77.1	−75.0	3.9	4.5	−0.6

[a]Precipitation.
[b]Stream water.
Source: Likens et al. (1977).

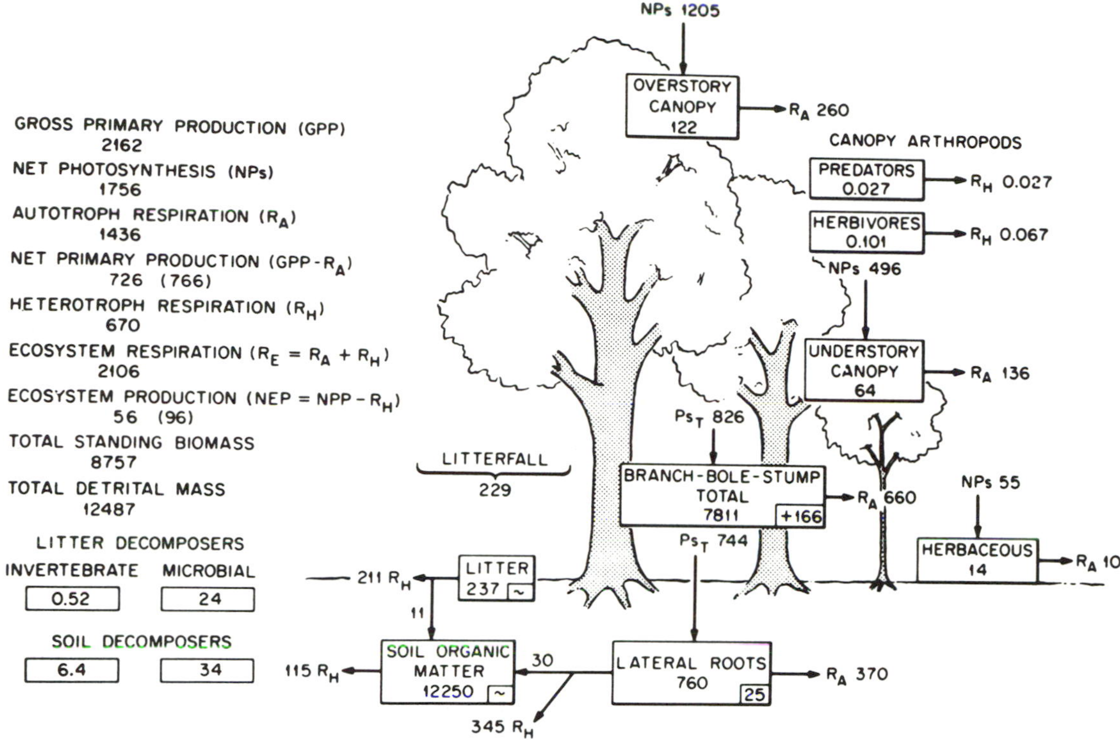

Figure 10.15. Abstract representation of a deciduous-forest ecosystem illustrating interrelationships of organic matter among the various biological and soil components (from Harris et al. 1975). Units are g G m^{-2} and g C m^{-2} yr^{-1}.

Potassium			Nitrogen			Phosphorus			Sulfate-sulfur		
Input	Output	Flux	Input	Output	Flux	Input	Output	Flux	Input	Output	Flux
			20.	2.5	+17.5	0.18	0.05	+0.13			
3.2	5.2	−2.0									
0.9	2.4	−1.5	20.7	4.0	+16.7	0.036	0.019	+0.017	18.8	17.6	+1.2
2.4	3.9	−1.5									
0.8	5.6	−4.8	9.7	5.6	+4.1	0.07	0.20	−0.13	21	38	−17
3.1	6.8	−3.7	8.7	1.8	+6.9	0.54	0.02	+0.52	18.8	11.3	+7.5

composition changes. Calcium ions increase 10-fold, ammonium ions decrease by sixfold, and the following ions are variously increased: Mg^{2+}, K^+, Na^+, Al^{3+}, SO_4^{2-}, NO_3^-, Cl^-. Likens and associates (1977) gave comparative nutrient budgets for five other eastern United States streams (Table 10.8). All showed ecosystem net losses of Ca^{2+}, Mg^{2+}, Na^+, and K^+, and net increases of nitrogen and phosphates. These authors concluded that these forested ecosystems act as filters, retaining and accumulating atmospheric contaminants (such as hydrogen ions, nitrogen, sulfur, phosphorus, and some heavy metals). They are not presently in equilibrium.

SUCCESSION

Old Fields

Keever (1983) has recently summarized information on the sequence of flora and vegetation on abandoned farmland. A pioneer vegetation of scattered forbs or grasses gradually develops into a dense meadow. This vegetation persists for greatly varying periods of time; it is eventually replaced by woody vegetation, which includes sapling trees, shrubs, and vines. Eventually, a closed canopy develops that excludes light-requiring plants and permits a shade-tolerant flora to enter. Eventually, a "climax forest," representative of the regional vegetation, will occupy the site (Fig. 10.16).

Flora involved in the various "stages" of old-field succession varies regionally with the macroclimate and locally with soil moisture, soil chemistry, and seasonal or logistical availability of seeds. The latter was strongly emphasized by Keever, who found conspicuously distinct seral communities in spring-, summer-, and fall-plowed fields. Egler (1954) argued that such variation in the initial floristic composition of the abandoned field was an important determinant of future composition.

Keever stated that, although the dominant species vary from place to place, a number of taxa occur widely in successional sequences. One of these is ragweed (*Ambrosia artemisiifolia*), common in first-year fields. Asters (*Aster pilosus* or *A. ericoides*) and goldenrods (*Solidago* species) often dominate by the third year and may persist for 20 yr (Bard 1952). Millets (*Setaria* species) are often characteristic plants in young fields.

Broomgrass (*Andropogon virginicus* or *A. scoparius*) may replace the early forbs, form a thick-sodded grassland, and persist for 45 yr (Bard 1952), or the asters and goldenrods may be replaced in as few as 4 yr by *Lonicera japonica, Rubus* species and *Rhus typhina* (Keever 1979).

Pines are important in old fields in the OPH region and may replace *Andropogon* completely in 10–15 yr. After one generation, overstory pines are replaced by slower-growing hardwoods emerging from the understory (Keever 1983). Peet and Christensen (1980) showed that the large, heavy seeds of oaks and hickories, which often are climax dominants, usually require a longer time to enter the area and rarely arrive initially in numbers sufficient to dominate the field.

The duration of any one stage of succession may be due to such microenvironmental changes as nutrient content of the soil, temperature and light conditions beneath the canopy, competition stresses, and allelopathy. Keever, however, does not believe that allelopathy plays an important role in old-field succession in the deciduous-forest region.

Gap Dynamics

Runkle (1982) calculated that canopy openings, or gaps, accounted for 9.5% of the total land area in seven areas of old-growth forest throughout the eastern United States. He concluded that new gaps are formed at a rate of 1% of the land each year, and old gaps are closed at a similar rate by sapling height growth.

The larger the gap, the richer its flora and the greater the total number of woody stems. In small gaps (1500 m^2) in the southern Appalachian Mountains, *Acer saccharum, Fagus grandifolia*, and *Tsuga canadensis* exhibited the greatest densities (Runkle 1982). In gaps of larger size, the small trees *Halesia carolina* and *Lindera benzoin* became very dense and (with *Acer saccharum*) were the leading species. The maximum density of woody stems developed in 10 yr in small gaps.

In gaps of the conifer–deciduous dicot zonoecotone region at Hubbard Brook Forest, New Hampshire, Forcier (1975) has observed population dynamics of the three dominant trees *Fagus grandifolia, Acer saccharum*, and *Betula allegheniensis* (*B. lutea*). *Betula* has the lightest seeds, which enter gaps early, germinate, and grow rapidly. *Fagus* has the heaviest seeds, but saplings usually originate asexually as root sprouts in the shade of a parent tree. Saplings grow rapidly up to the canopy after the death and fall of the parent tree. *Acer saccharum* is intermediate in seed mobility and sapling shade tolerance. His demographic data were used to postulate a model of cyclic change on small sites for the three species within a climax forest.

Marks (1975) noted that in large gaps and cleared areas, shade-intolerant trees like *Populus* species and *Prunus pennsylvanica* grew throughout most of

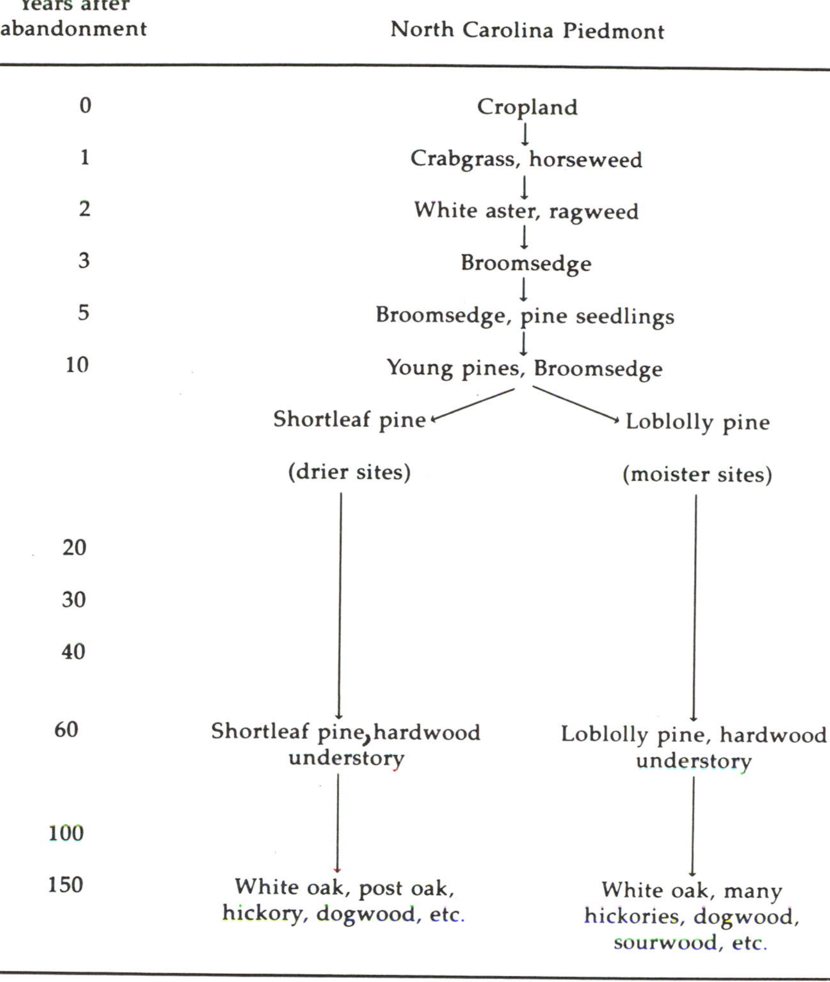

Figure 10.16. Diagram of old-field succession patterns in the Piedmont area of the OPH region of the deciduous forest.

the frost-free season. They exhibited indeterminate extension growth and therefore had the potential to grow more than 100 cm in a growing season. Shade-tolerant trees like *Acer saccharum* and *Fagus grandifolia*, characteristic of mature forests, showed extension growth for only 30 d. Their growth was determinate and did not exceed 25–40 cm.

Oosting and Humphreys (1940) investigated the seed composition in the soil under a mature, deciduous forest in the OPH region. They found that in addition to seeds from about two dozen herbaceous and woody forest taxa, seeds from old-field taxa were present in forest soils. If a large gap appears in the canopy, taxa typical of old fields are expected to be the first occupants of the newly disturbed soil. Livingston and Allessio (1968) obtained similar results from an investigation of seeds in the soil under an 80-yr-old white pine–hardwood forest in the conifer-hardwood transition region of central Massachusetts.

SPRUCE-FIR FORESTS OF THE SOUTHERN APPALACHIAN MOUNTAINS

A dominantly coniferous forest that is variously labeled "sub-alpine" or "spruce-fir" occurs above 1524 m in the Great Smoky Mountains of Tennessee and North Carolina (Fig. 10.17). It also occurs on the high plateaus of the Catskill Mountains (above 1280 m), above 762 m in the Green Mountains of Vermont and White Mountains of New Hampshire, and above 150 m in Maine (Oostring 1956; Davis 1966).

Spruce (*Picea rubens*) is the dominant of lower elevations in this forest type. The proportion of fir

Figure 10.17. Virgin red spruce forest in the southern Appalachian Mountains near Durbin, West Virginia, 1939 (USDA Forest Service photo).

(*Abies fraseri* in western North Carolina, eastern Tennessee, and southwest Virginia; *A. balsamea* in West Virginia and farther north) increases with elevation. Crandall (1958) recognized spruce forests in the elevational belt at 1555–1829 m, spruce-fir forests at 1676–1951 m, and fir forests at 1890–2012 m in the Great Smoky Mountains.

Oosting and Billings (1951) compared environments and compositions of virgin spruce-fir forests of the Great Smoky Mountains and the White Mountains of New Hampshire. The composition was fairly uniform from south to north (Table 10.9): *Picea rubens* is widespread, accompanied by *Abies fraseri* and *Betula lutea* in the south, and by *Abies balsamea* and *Betula papyrifera* in the north. *Abies* has higher densities, but *Picea* tends to have greater diameters, as well as greater average heights and ages (average 200–250 yr for *Picea* against 150 yr for *Abies*). Epiphytes of lichens and bryophytes are common (Crandall 1958; Core 1966).

Brown (1941) described the spruce-fir forest on Roan Mountain as having an overstory canopy (12–27 m tall) of *Picea rubens* and *Abies fraseri* and a subcanopy of *Picea*, *Abies*, and *Betula lutea* (4–12 m tall). Basal-area percentages varied widely from stand to stand, with either *Picea rubens* or *Abies fraseri* being the leading dominant. Crandall (1958)

Table 10.9. *Canopy composition in spruce-fir forests from the Great Smoky Mountains to the White Mountains: percentage basal area (and percentage presence)*

Species	Great Smoky Mountains[a]	Balsam Mountains, Va.[b]	Allegheny Mountains, W. Va.[c]	Catskill Mountains[d]	White Mountains, N.H.[a]
Abies balsamea			(0)[e]	(100)	(100)
Abies fraseri	25 (100)	60	(0)		
Acer pennsylvanicum			<1 (87)	(75)	(75)
Acer rubrum			19	(25)	
Acer saccharum			<1		
Acer spicatum	<1 (89)		<1 (25)	(50)	(100)
Aesculus octandra		1			
Amelanchier sp.	<1		<1	(50)	
Betula cordifolia				(75)	
Betula lenta			2		
Betula lutea	11 (100)	15	16 (87)	(75)	(75)
Betula papyrifera			(0)		(100)
Fagus grandifolia	(11)		5 (87)	(25)	
Picea rubens	64 (100)	19	33 (100)	(100)	(100)
Prunus pennsylvanica	(45)		(0)	(50)	
Prunus serotina			2 (75)		(25)
Sorbus americana	1 (100)		1 (50)	(75)	(100)
Tsuga canadensis			20 (37)	(25)	(25)
Others		5	3		

[a]Oosting and Billings (1951), McIntosh and Hurley (1964, Table VI), Crandall (1958, Table VI).
[b]Reinhardt and Ware (1984, Table I).
[c]Stephenson and Clovis (1983, Table I).
[d]McIntosh and Hurley (1964, Table VI).
[e](0) = present but at <1%.

noted that a five-layer understory can be recognized throughout this undisturbed conifer forest: (1) moss, (2) *Oxalis*, (3) fern, (4) low shrub, and (5) high shrub. Shrubs are 0.9–4 m tall, herbs are 0.2–0.9 m tall, and bryoids are less than 15 cm (Brown 1941). Crandall also listed nine different understory types in the high-elevation conifer forests of the Great Smoky Mountains. These can be aggregated into four major types: (1) moss-dominated (*Hylocomium-Vaccinium*); (2) herb-dominated (*Oxalis-Hylocomium, Oxalis-Dryopteris, Senecio*); (3) deciduous-shrub-dominated (*Viburnum-Vaccinium-Dryopteris, Viburnum-Vaccinium-Senecio, Viburnum-Vaccinium-Lycopodium*); and (4) evergreen-shrub-dominated (*Rhododendron, Rhododendron-Vibrunum*). Regardless of the understory type, bryoids (*Hylocomium splendens, Bazzania trilobata, Hypnum* species, *Dicranum scoparium, D. fuscescens*, and *Polytrichum* species) rarely account for less than 25% of total cover, and commonly more than 90%.

AREAS FOR FUTURE RESEARCH

Despite the enormous number of publications on the deciduous forest, much remains to be understood. The greatest gap in knowledge is historical and encompasses geologic time from the pre-Wisconsin to the Miocene. The whole southeastern United States is underlain by unconsolidated sediments that provide a more or less continuous record since the Cretaceous. Extensive urbanization in the area between Richmond, Virginia, and Boston, Massachusetts, would seem to provide a stimulus to explore for new microfossil and megafossil deposits.

Forest classification has lagged far behind ordination. This is especially true at the local and sectional (*sensu* Braun) levels. Studies of understory composition likely will result in finer resolution of differences among stands. To this end, compositional studies of understory guilds will be worthwhile. Reproduction of forest herbs deserves continued study; so does the ecology of rare and endangered herbs. Productivity studies, although they have been extensive enough to indicate the general parameters for vegetation types, need to be refined and systematized region by region. Nutrient cycling needs similar attention. Productivity and nutrient cycling must be evaluated in the context of dispersion of waste products from countless processes involved in mineral extraction and refining and industrial production. Both local and worldwide effects need to be considered and distinguished from, or related to, long-term macroclimatic phenomena.

REFERENCES

Andresen, J. W. 1959. A study of pseudo-nanism in *Pinus rigida* Mill. Ecol. Monogr. 29:309–332.

Art, H. W. 1976. Ecological studies of the Sunken Forest, Fire Island National Seashore, New York. National Park Service scientific monograph series, no. 7.

Bailey, H. P. 1960. A method of determining the warmth and temperateness of climate. Geografiska Annaler 40:196–215.

Bakeless, J. 1950. The eyes of discovery. Lippincott, New York.

Barbour, M. G., J. H. Burk, and W. D. Pitts. 1980. Terrestrial plant ecology. Benjamin/Cummings, Menlo Park, Calif.

Bard, G. E. 1952. Secondary succession on the piedmont of New Jersey. Ecol. Monogr. 22:195–215.

Barden, L. S., and F. W. Woods. 1976. Effects of fire on pine and pine-hardwood forests in the southern Appalachians. Forest Sci. 22:398–403.

Baskin, J. M., and C. C. Baskin. 1978. Plant ecology of Cedar Glades in the Big Barren region of Kentucky. Rhodora 80:545–557.

Baskin, J. M., and C. C. Baskin. 1983. Germination ecophysiology of eastern deciduous forest herbs: *Hydrophyllum macrophyllum*. Amer. Midl. Nat. 109:63–71.

Baskin, J. M., and C. C. Baskin. 1984. Germination ecophysiology of the woodland herb *Osmorhiza longistylis* (Umbelliferae). Amer. J. Bot. 71:687–692.

Bernard, J. M., and A. Bernard. 1971. Mature upland forests of Cape May County, New Jersey. Bull. Torrey Botanical Club 98:167–171.

Berry, E. W. 1937. Tertiary floras of eastern North America. Bot. Rev. 3:31–46.

Bormann, F. H., and G. E. Likens, 1979. Pattern and process in a forested ecosystem. Springer-Verlag, Berlin.

Bratton, S. P. 1976. Resource division in an understory herb community: responses to temporal and microtopographic gradients. American Naturalist 110:679–693.

Braun, E. L. 1950. Deciduous forests of eastern North America. Blakiston, Philadelphia.

Brockman, C. F. 1968. Trees of North America. Golden Press, New York.

Brown, D. M. 1941. The vegetation of Roan Mountain: phytosociological and successional study. Ecol. Monogr. 11:61–97.

Bruner, W. E. 1931. The vegetation of Oklahoma. Ecol. Monogr. 1:99–188.

Cain, S. A., M. Nelson, and W. McLean. 1937. *Andropogonetum hempsteadi:* a Long Island grassland vegetation type. Amer. Midl. Nat. 18:334–350.

Camp, W. H. 1951. A biogeographical and paragenetic analysis of the American beech (*Fagus*). Amer. Philos. Soc. Yearbook 1950:166–169.

Cooper, A. W. 1979. The natural vegetation of North Carolina. pp. 70–78 in H. Lieth and E. Landolt (eds.), Contributions to the knowledge of flora and vegetation in the Carolinas, vol. 1. Stiftung Rübel, Zürich.

Core, E. L. 1966. Vegetation of West Virginia. McClain Printing, Parsons, W. Va.

Crandall, D. L. 1958. Ground vegetation patterns of the spruce-fir area of the Great Smoky Mountains National Park. Ecol. Monogr. 28:337–360.

Dabel, C. V., and F. P. Day, Jr. 1977. Structural comparisons of four plant communities in the Great Dismal Swamp, Virginia. Bull. Torrey Botanical Club 104:352–360.

Davis, M. B. 1981. Quaternary history and the stability of forest communities. pp. 132–153 in D. C. West, H. H. Shugart, and D. B. Botkin (eds.), Forest succession, concepts and application. Springer-Verlag, Berlin.

Davis, R. B. 1966. Spruce-fir forests of the coast of Maine. Ecol. Monogr. 36:79–94.

Delcourt, P. A., and H. R. Delcourt. 1979. Late Pleistocene and Holocene distributional history of the deciduous forest in the southeastern United States. pp. 79–107. in H. Lieth and E. Landolt (eds.), Contributions to the knowledge of flora and vegetation in the Carolinas, vol. 1. Stiftung Rübel, Zürich.

Delcourt, P. A., and H. R. Delcourt. 1981. Vegetation maps for eastern North America; 40,000 yr B.P. to the present. pp. 123–165 in R. C. Romans (ed.), Geobotany, vol. II. Plenum Press, New York.

Dilcher, D. L. 1971. A revision of the Eocene flora of southeastern North America. Palaeobotanist 20:7–18.

Dolph, G. E. 1978. Variation in leaf size and margin type with respect to climate. Cour. Forsch. Inst. Senckenberg 30:153–158.

Dolph, G. E., and D. L. Dilcher. 1979. Foliar physiognomy as an aid in determining paleoclimate. Palaeontographica [Abt. B] 170:151–172.

Dyksterhuis, E. J. 1957. The savannah concept and its use. Ecology 38:435–442.

Egler, F. E. 1954. Vegetation science concepts. Initial floristic composition–a factor in old-field vegetation development. Vegetatio 4:412–417.

Forcier, L. K. 1975. Reproductive strategies and the cooccurrence of climax tree species. Science 189: 808–810.

Forman, R. T. T. 1979. Pine barrens: ecosystem and landscape. Academic Press, New York.

Fulbright, D. W., W. H. Weidlich, K. Z. Haufler, C. S. Thomas, and C. P. Paul. 1983. Chestnut blight and recovering American chestnut trees in Michigan. Can. J. Bot. 61:3164–3171.

Gleason, H. A., and A. Cronquist. 1963. Manual of vascular plants of northeastern United States and adjacent Canada. Van Nostrand Reinhold, New York.

Gleason, H. A., and A. Cronquist. 1964. The natural geography of plants. Columbia University Press, New York.

Golley, F. B. 1960. Energy dynamics of a food chain of an old-field community. Ecol. Monogr. 30: 187–206.

Gomez, M. M., and F. P. Day, Jr. 1982. Litter nutrient content and production in the Great Dismal Swamp. Amer. J. Bot. 69:1314–1321.

Good, N. F. 1968. A study of natural replacement of chestnut in six stands in the highlands of New Jersey. Bull. Torrey Botanical Club 95:240–253.

Gray, J. 1960. Temperate pollen genera in the Eocene (Claiborne) flora, Alabama. Science 132:808–810.

Greller, A. M. 1977. A vascular flora of the forested portion of Cunningham Park, Queens County, New York, with notes on the vegetation. Bull. Torrey Botanical Club 104:170–176.

Greller, A. M. 1980. Correlation of some climate statistics with distribution of broadleaved forest zones in Florida, U.S.A. Bull.Torrey Botanical Club 107:189–219.

Greller, A. M., and L. D. Rachele. 1983. Climatic limits of exotic genera in the Legler palynoflora, Miocene, New Jersey, U.S.A. Rev. Palaeobot. Palynol. 40:149–163.

Harper, R. M. 1908. The pine barrens of Babylon and Islip, Long Island. Torreya 8:1–9.

Harris, W. F., P. Sollins, N.T. Edwards, B. E. Dinger, and H. H. Shugart. 1975. Analysis of carbon flow and productivity in a temperate deciduous forest ecosystem. pp. 116–122 in Productivity of world ecosystems. National Academy of Science, Washington, D.C.

Harshberger, J. W. 1903. The flora of the serpentine barrens of southeast Pennsylvania. Science (n.s.) 18:339–343.

Hicks, D. J., and B. F. Chabot, 1985. Deciduous forest. pp. 257–277 in B. F. Chabot and H. A. Mooney (eds.), Physiological ecology of North American plant communities. Chapman & Hall, London.

Johnson, A. H., T. G. Siccama, R. S. Turner, and D. G. Lord. 1984. Assessing the possibility of a link between acid precipitation and decreased growth rates of trees in northeastern United States. pp. 81–109 in R. A. Linthurst (ed.), Direct and indirect effects of acidic deposition on vegetation. Butterworth, London.

Johnson, G. C., and S. Ware. 1982. Post-chestnut forests in the central Blue Ridge of Virginia. Castanea 47:329–343.

Keever, C. 1953. Present composition of some stands of the former oak-chestnut forest in the southern Blue Ridge Mountains. Ecology 34:44–54.

Keever, C. 1979. Mechanisms of plant succession on old fields of Lancaster County, Pennsylvania. Bull. Torrey Botanical Club 106:299–308.

Keever, C. 1983. A retrospective view of old-field succession after 35 years. Amer. Midl. Nat. 110:397–404.

Knapp, R. 1965. Die Vegetation von Nord- und Mittelamerika. Gustav Fischer Verlag, Stuttgart.

Knox, R. G. 1984. Age structure of forests on Soldier's Delight, a Maryland serpentine area. Bull. Torrey Botanical Club 111:498–501.

Kucera, C. L., and S. C. Martin. 1957. Vegetation and soil relationships in the glade region of the southwestern Missouri Ozarks. Ecology 38: 285–291.

Küchler, A. W. 1974. A new vegetation map of Kansas. Ecology 55:586–604.

Larson, S. S., and W. T. Batson. 1978. The vegetation of vertical rock faces in Pickens and Greenville Counties, South Carolina. Castanea 43:255–260.

Likens, G. E., F. H. Borman, R. S. Pierce, J. S. Eaton, and N. M. Johnson. 1977. Biogeochemistry of a forested ecosystem. Springer-Verlag, Berlin.

Lindsey, A. A., D. V. Schmelz, and S. A. Nichols. 1969. Natural areas in Indiana. Indiana Natural Areas Survey, Department of Biological Sciences, Purdue University, Lafayette, Ind.

Lindsey, A. A., and L. K. Escobar. 1976. Eastern Deciduous Forest, vol. 2, Beech-maple region. Inventory of Natural Areas and Sites Recommended as Potential Natural Landmarks. U. S. Department of the Interior, National Park Ser-

vice, publication 1481, Natural history theme studies, no. 3.
Little, E. L., Jr. 1971. Atlas of United States trees, vol.1, Conifers and important hardwoods. USDA miscellaneous publication 1146.
Little, S. 1973. Eighteen-year changes in the composition of a stand of *Pinus echinata* and *P. rigida* in southern New Jersey. Bull. Torrey Botanical Club 100:94–102.
Livingston, R. B., and M. L. Allessio. 1968. Buried viable seed in successional field and forest stands, Harvard Forest, Massachusetts. Bull. Torrey Botanical Club 95:58–69.
McCormick, J. 1970. The pine barrens: A preliminary ecological inventory. New Jersey State Museum, report 2. Trenton.
McIntosh, R. P., and R. T. Hurley. 1964. The spruce-fir forests of the Catskill Mountains. Ecology 45:314–326.
McVaugh, R. 1943. The vegetation of the granitic flatrocks of the southeastern United States. Ecol. Monogr. 13:119–166.
Mark, A. F. 1958. The ecology of the southern Appalacian grass balds. Ecol. Monogr. 28:293.
Marks, P. L. 1975. On the relation between extension growth and successional status of deciduous trees of the northeastern United States. Bull. Torrey Botanical Club 102:172–177.
Miranda, E., and A. J. Sharp. 1950. Characteristics of the vegetation in certain temperate regions of eastern Mexico. Ecology 31:313–333.
Monk, C. D., G. I. Child, and S. A. Nicholson. 1970. Biomass, litter and leaf surface area estimates of an oak-hickory forest. Oikos 21:138–141.
Montague, K. A., and F. P. Day, Jr. 1980. Belowground biomass of four plant communities of the Great Dismal Swamp, Virginia. Amer. Midl. Nat. 103:83–87.
Nelson, T. C. 1955. Chestnut replacement in the southern highlands. Ecology 36:352–353.
Odum, E. P. 1983. Basic ecology. W. B. Saunders, Philadelphia.
Olsvig, L. S., J. F. Cryan, and R. H. Whittaker. 1979. Vegetational gradients of the pine plains and barrens of Long Island, New York. pp. 265–282 in R. T. T. Forman (ed.), Pine barrens: ecosystem and landscape. Academic Press, New York.
Oosting, H. J. 1956. The study of plant communities, 2nd ed. W. H. Freeman, San Francisco.
Oosting, H. J., and W. D. Billings, 1951. A comparison of virgin spruce-fir forest in the northern and southern Appalachian system. Ecology 32:84–103.
Oosting, H. J., and M. E. Humphreys. 1940. Buried viable seeds in a successional series of old field and forest soils. Bull. Torrey Botanical Club 67:253–273.
Paillet, F. L. 1984. Growth-form and ecology of American chestnut sprout clones in northeastern Massachusetts. Bull. Torrey Botanical Club 111:316–328.
Peet, R. K., and N. L. Christensen. 1980. Succession: a population process. Vegetatio 43:131–140.
Pennell, F. W. 1910. Flora of the Conowingo Barrens of southeastern Pennsylvania. Acad. Nat. Sci. Phila. 62:541–584.
Platt, R. B. 1951. An ecological study of the mid-Appalachian shale barrens and of the plants endemic to them. Ecol. Monogr. 21:269–300.
Potzger, J. E., M. E. Potzger, and J. McCormick. 1956. The forest primeval of Indiana as recorded in the original U.S. land surveys and an evaluation of previous interpretations of Indiana vegetation. Butler Univ. Bot. Studies 13:95–111.
Quarterman, E. 1950. Major plant communities of Tennessee cedar glades. Ecology 31:234–254.
Quarterman, E., and C. Keever. 1962 Southern mixed hardwood forest: climax in the southeastern coastal plain, U.S.A. Ecol. Monogr. 32:167–185.
Quarterman, E., B. H. Turner, and T. E. Hemmerly. 1972. Analysis of virgin mixed mesophytic forests in Savage Gulf, Tennessee. Bull. Torrey Botanical Club 99:228–232.
Rachele, L. D. 1976. Palynology of the Legler lignite: a deposit in the Tertiary Cohansey formation of New Jersey, U.S.A. Rev. Palaeobot. Palynol. 22:225–252.
Racine, C. H. 1966. Pine communities and their site characteristics in the Blue Ridge escarpment. J. Elisha Mitchell Sci. Soc. 82:172–181.
Racine, C. H. 1971. Reproduction of three species of oak in relation to vegetational and environmental gradients in the southern Blue Ridge. Bull. Torrey Botanical Club 98:297–310.
Reinhardt, R. D., and S. A. Ware. 1984. Vegetation of the Balsam Mountains of southwest Virginia: a phytosociological study. Bull. Torrey Botanical Club 111:297–300.
Rogers, R. S. 1982. Early spring herb communities in mesophytic forests of the Great Lakes region. Ecology 63:1050–1063.
Runkle, J. R. 1982. Patterns of disturbance in some old-growth mesic forests of eastern North America. Ecology 63:1533–1546.
Russell, E. W. B. 1983. Indian-set fires in the forests of the northeastern United States. Ecology 64:78–88.
Russell, N. H. 1953. The beech gaps of the Great Smoky Mountains. Ecology 34:366–374.
Salomon, J. H. 1984. Indians that set the woods on fire. Conservationist (N.Y.S.). 38:34–39.
Schemske, D. W., M. F. Willson, M. N. Melampy, L. J. Miller, L. Verner, K. M. Schemske, and L. B. Best. 1978. Flowering ecology of some spring woodland herbs. Ecology 59:351–366.
Shanks, R. E. 1956. Altitudinal and microclimatic relationships of soil temperature under natural vegetation. Ecology 37:1–7.
Sharitz, R. R., and J. Frank McCormick. 1973. Population dynamics of two competing annual plant species. Ecology 54:723–739.
Siccama, T. G., M. Bliss, and H. W. Vogelmann. 1982. Decline of red spruce in the Green Mountains of Vermont. Bull. Torrey Botanical Club 109:162–168.
Skeen, J. N. 1973. Biomass and productivity estimates for a temperate mesic slope forest. J. Tennessee Acad. Sci. 48:103–105.
Smith, W. H. 1981. Air pollution and forests. Springer-Verlag. Berlin.
Sparling, J. H. 1967. Assimilation rates of some woodland herbs in Ontario. Bot. Gaz. 128:160–168.
Stephenson, S. L., and J. F. Clovis. 1983. Spruce forests of the Allegheny Mountains in central West Virginia. Castanea 48:1–12.
Steyermark, J. A. 1940. Studies of the vegetation of

Missouri, I, Natural plant associations and succession in the Ozarks of Missouri. Field Museum of Natural History Botanical Series 9:349–475.

Trautman, M. B. 1977. The Ohio country from 1750 to 1977–a naturalist's view. Ohio Biological Survey, biological notes no. 10.

U.S. Geological Survey. 1970. The national atlas of the United States of America. Department of the Interior, Washington, D.C.

Voight, J. W., and R. H. Mohlenbrock. 1964. Plant communities of Southern Illinois. Southern Illinois University Press, Carbondale.

Walter, H. 1979. Vegetation of the earth and ecological systems of the geo-biosphere, 2nd ed. Heidelberg Science Library, Springer-Verlag, Berlin.

Ward, R. T. 1958. The beech forests of Wisconsin–their phytosociology and relationships to forests of the state without beech. Ecology 39:444–457.

Ware, S. 1982. Polar ordination of Braun's Mixed Mesophytic Forest. Castanea 47:403–407.

Weaver, J. E. 1960. Floodplain vegetation of the central Missouri Valley and contacts to woodland with prairie. Ecol. Monog. 30:37–64.

Whittaker, R. H. 1956. Vegetation of the Great Smoky Mountains. Ecol. Monogr. 26:1–80.

Whittaker, R. H. 1963. Net production of heath balds and forest heaths in the Great Smoky Mountains. Ecology 44:176–182.

Whittaker, R. H. 1966. Forest dimensions and production in the Great Smoky Mountains. Ecology 47:103–121.

Whittaker, R. H. 1975. Communities and ecosystems, 2nd ed. Macmillan, New York.

Whittaker, R. H., and G. M. Woodwell. 1969. Structure, production, and diversity of the oak-pine forest at Brookhaven, New York. J. Ecol. 57:155–174.

Woods, F. W., and R. E. Shanks. 1959. Natural replacement of chestnut by other species in the Great Smoky Mountains National Park. Ecology 40:349–361.

Zobel, D. B. 1969. Factors affecting the distribution of *Pinus pungens*, an Appalachian endemic. Ecol. Monogr. 39:303–333.

Chapter 11

Vegetation of the Southeastern Coastal Plain

NORMAN L. CHRISTENSEN

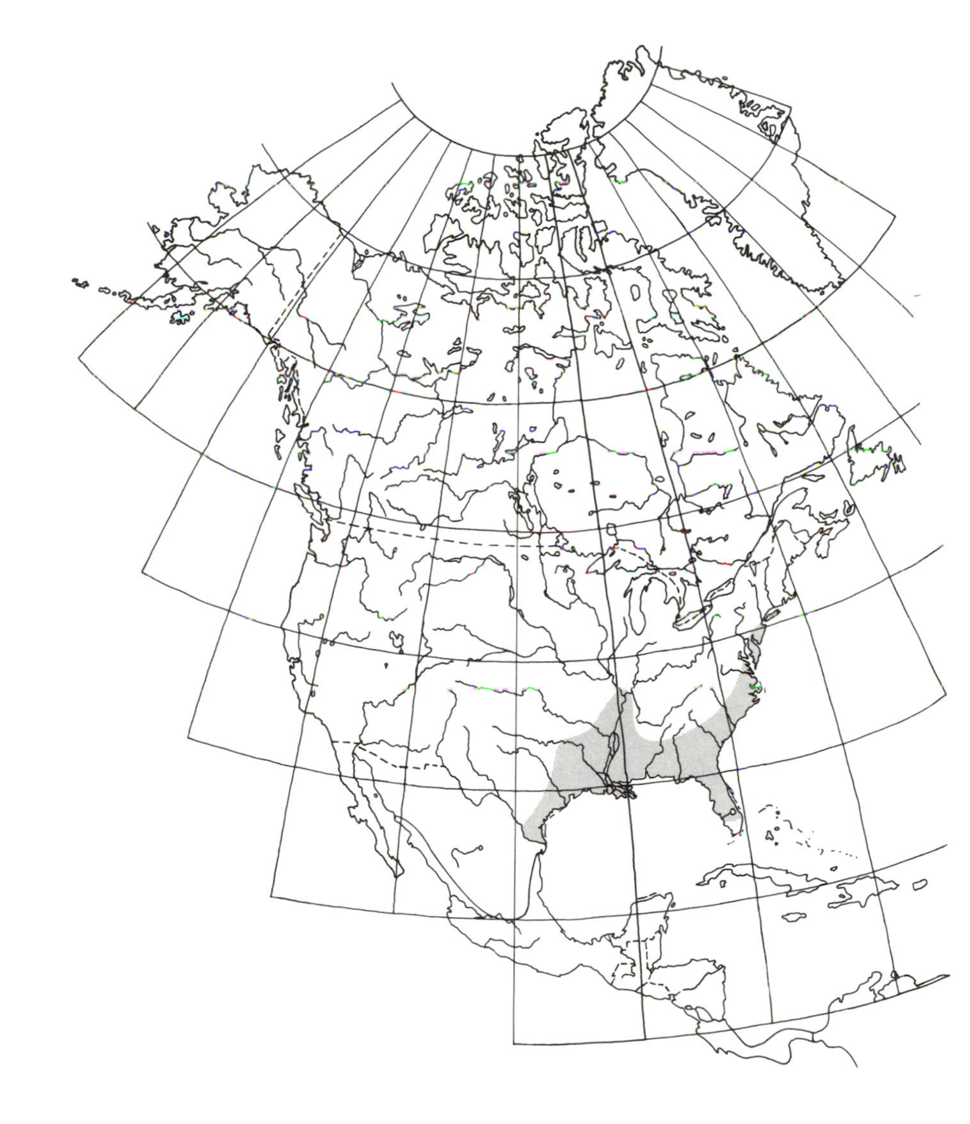

INTRODUCTION

The climax vegetation over most of the eastern United States is generally considered to be broad-leaved deciduous forest (Küchler 1964). The Coastal Plain of the Atlantic and Gulf states is the most striking and extensive exception to this statement. Community physiognomy varies across this landscape from grasslands and savannas to shrublands, to needle- and broad-leaved sclerophyllous woodlands, to rich mesophytic forest. These differences can be observed over a distance of only a few hundred meters and an elevational gradient of only 10 m. In addition, the southeastern Coastal Plain has the most diverse assemblage of freshwater wetland communities in North America, and its lengthy and complex shoreline is populated by a rich array of maritime ecosystems. Much of this variation is a consequence of dramatic gradients in physical and chemical characteristics of soils and hydrology. However, as Wells (1946) suggested, "succession would simplify the mosaic just outlined considerably were it not for fire," and where fire is not important, as in maritime and alluvial communities, other forms of chronic disturbances complicate community structure and function. Thus, succession and patch dynamics will be integral parts of any discussion of the vegetation of this province.

The southeastern Coastal Plain was an important laboratory for many prominent ecologists in the first half of this century, including J. W. Harshberger, R. M. Harper, B. W. Wells, and H. J. Oosting. The last major review of the ecology of this region was published by B. W. Wells in 1946. In the intervening 40 yr, North American community ecologists have come to accept the gradient nature of community variation, to deal with the complexity of successional pathways and mechanisms, and to recognize the role of natural disturbances within communities and across landscapes. All of these lessons were obvious to those early Coastal Plain ecologists.

VEGETATION HISTORY

The Coastal Plain of the southeastern United States had its beginnings during the Triassic period (~200 million BP) as North America separated from North Africa, creating the Atlantic Ocean. During the previous 400 million yr, the southern Appalachians had grown to a mountain range rivaling the present-day western Cordillera in majesty. Mountain building has virtually ceased in eastern North America since the Triassic, and the Coastal Plain Geologic Province has been built from the sediments eroded from that great massif. It is the distribution of these sediments that defines the extent of this province.

The oldest fossils of terrestrial vegetation on the Coastal Plain are from the Paleocene and Eocene. Macrofossils from the Mississippi embayment reveal a distinctly tropical flora in this region during that time (Dilcher 1973a, 1973b), and they suggest a mean annual temperature of about 27°C (Wolfe 1978). The general cooling trend that culminated in the Pleistocene ice ages appears to have begun during the Oligocene. Little is known of middle Tertiary vegetation in the Southeast; however, shallow marine deposition of phosphatic carbonates along the southeastern coast during the Miocene suggests considerable upwelling of cold, nutrient-rich water. Data from Miocene floras elsewhere in North America support the hypothesis that the southeastern climate was tropical to subtropical during this period (Axelrod and Bailey 1969).

Palynological analysis of Coastal Plain lake sediments has given us a detailed picture of the late Pleistocene history of this region (see Chapter 10 for a discussion of the relevance of these data to the entire Eastern Deciduous Forest Province). During the Altonian sub-age of the Wisconsinan (~40,000 BP), most of the Coastal Plain north of central South Carolina was dominated by a mixture of jack pine (*Pinus banksiana*) and spruce (*Picea* species), with perhaps a zone of mixed conifer–northern hardwoods extending across southern South Carolina (Watts 1980a, 1980b; Delcourt and Delcourt 1981). Communities similar to the Coastal Plain communities of today occupied a broad band extending from the coast of Georgia across the Mississippi embayment and into east Texas (Delcourt et al. 1980; Delcourt and Delcourt 1981). Peninsular Florida, greatly enlarged by the lowering of the sea level, was dominated by xeric sand-dune communities that appear to have no modern analogue (Watts 1975). Through the remainder of the Wisconsinan ice age, these vegetation zones waxed and waned in relation to shifts in the ice mass to the north. Although debate continues as to whether or not they constituted the sole refuge for the mixed mesophytic forest during full-glacial periods (Braun 1955; Delcourt and Delcourt 1979; Davis 1983), there is little doubt that most of the vascular plant constituents of this forest association could be found in river bluff habitats throughout the Southeast (Delcourt and Delcourt 1977a, 1980).

The northern extent of mixed mesophytic forest species along the Atlantic limb of the Coastal Plain is difficult to determine. During the full glacial, the sea level was more than 150 m lower, and the coastline in the mid-Atlantic region was 50–150 km east of its present location. Perhaps the southern forest elements extended farther north near the coast, much as they do today, but evidence to test this hypothesis is now submerged and perhaps obliterated by coastal processes.

During retreat of the continental glaciers 12,000–14,000 BP, the Southeast saw a marked warming trend (the hypsithermal or xerothermic period) (Wright 1976), with general extinction of boreal elements. Mixed deciduous forest dominated most of the northern Coastal Plain (Watts 1980a, 1980b; Delcourt and Delcourt 1981). Increased rainfall over the Florida peninsula favored widespread oak savanna in this area.

The sea level returned to near its modern position by 5000 BP, resulting in decreased drainage over many areas of the lower Coastal Plain coincident with the initiation of a general cooling trend (Watts 1971, 1980b) that probably lowered regional evapotranspiration rates. These two factors are likely responsible for the initiation of a period of paludification and bog formation in several localities along the Atlantic Coastal Plain (Whitehead 1972, 1981; Cohen 1973, 1974; Daniels et al. 1977; Cohen et al. 1984). Abundant charcoal in sediments indicates that fires were common across the Coastal Plain (Buell 1939, 1946; Delcourt 1980; Cohen et al. 1984).

Human habitation of the Coastal Plain began no less than 12,000 BP. At that time, hunters of the Paleoindian culture moved into the Southeast in pursuit of large mammals driven eastward by the hypsithermal drying of the Great Plains (Hudson 1976). The Archaic Indian culture was well established by 8,000–10,000 BP, coincident with the extinction of the megafauna. These people were hunter-gatherers whose food sources included deer, small mammals, fish, shellfish, and wild vegetables (Hudson 1976; Cowdrey 1983). Archaic Indian populations were largest near the coast and, as paludification proceeded, near swamp complexes such as the Great Dismal Swamp, the Green Swamp, and the Okefenokee Swamp (Hudson 1976; Wright 1984). These swamp complexes offered a wide variety of vegetable and wildlife resources (Wright 1984). Although they were seminomadic, Archaic Indians intensively managed the Coastal Plain landscape, particularly with the use of fire (Pyne 1982). Buell (1946) suggested that an abrupt increase in the charcoal content of strata in Jerome Bog, a peat-filled bay lake in North Carolina, might coincide either with the advent of the Indian or with climatic change. Undoubtedly, fire frequency increased as a consequence of both anthropogenic and climatic influences.

The Woodland culture, an Indian tradition incorporating a mixture of hunting, gathering, and primitive agriculture (including corn and squash), took shape along the Mississippi River about 3000 BP. Woodland Indians apparently relied most heavily on natural resources, using agriculture to subsidize their needs and perpetuating the patterns of land use of their Archaic predecessors (Hudson 1976; Cowdrey 1983).

The Mississippian tradition, a culture that was primarily sedentary and agricultural, arose on the floodplains of the Mississipppi embayment approximately 1300 BP. This was the most highly developed culture on the Coastal Plain prior to the arrival of Europeans; it was noted for its large walled towns and elegant tools. These people apparently never discovered the use of fertilizer, and the sterile nature of the Coastal Plain soils confined their agricultural activities to the nutrient-subsidized floodplains of large rivers (Cowdrey 1983). Up until the European colonization, the Archaic, Woodland, and Mississippian cultures coexisted in different habitats and regions of the southeastern Coastal Plain.

The Coastal Plain was the site where the first European colonists in temperate North America settled. To a people coming from a land where two millennia of intensive land use had removed every vestige of pristine forest and where most major industries were limited by the availability of wood, especially tall spars for shipbuilding, this landscape must have been a magnificent sight. Virtually every account by early explorers of the province emphasized the extensive and majestic forests and abundant natural resources. Just as frequently, early travelers in the region commented on the frequency of fires, both natural and those caused by humans (Catesby 1654; Lawson 1714; Byrd 1728; Bartram 1791). The impact of the subsequent exploitation of Coastal Plain natural resources on specific plant communities will be discussed later and have been outlined in detail by Cowdry (1983).

The effects of colonization on the various fire regimes of the Coastal Plain varied with location and the cultural tradition of the colonists. For example, settlers of Scottish and Irish descent were accustomed to using fire in the management of heaths and agricultural fields; so in many locations they perpetuated Indian burning practices. However, immigrants from central Europe were more apt to exclude fire from lands they managed (Pyne 1982). In general, however, most Coastal Plain colonists quickly learned the value of fire as a tool to

Figure 11.1. Boundaries and major physiographic features of the southeastern Coastal Plain.

manipulate the landscape, and from this region the accepted twentieth-century fire management practices later developed (Pyne 1982).

COASTAL PLAIN ENVIRONMENT

Physiography and Geology

The boundaries of the Coastal Plain are indicated in Fig. 11.1. On the Atlantic seaboard, the Coastal Plain merges with the Piedmont Plateau along the fall line or zone, a ragged line running roughly parallel to the trends of the mobile belts composing the Piedmont and Appalachian Highlands. In addition to rather abrupt changes in soils and their parent rocks, this zone is often marked by an abrupt change in topography from rolling hills to rather flat plain. The Coastal Plain border turns westward across Georgia and Alabama, then northward to the head of the Mississippi embayment in southern Illinois. The western margin of the Coastal Plain is not as clearly defined, either geologically or botanically, and the border indicated here is that suggested by Murray (1961).

The Atlantic Coastal Geologic Province extends from the fall line to the edge of the continental shelf as a geosynclinal wedge of alluvial and marine sediments. These strata rest on a basement of Paleozoic and Precambrian rocks that is actively subsiding in many places under the weight of this growing wedge. Much of the sediment, particularly at the surface, is siliceous alluvium; however, considerable carbonaceous sediments, often rich in phosphates, were deposited during a Miocene episode of coastal upwelling. These carbonates outcrop in many localities, but are most prominent throughout peninsular Florida and across the Gulf states, where they give rise to karst topography and weather to comparatively fertile soils.

Several physiographic sections are recognized (Fig. 11.1), although they are not necessarily separated by distinct boundaries. The Chesapeake-Delaware Downwarp lies north of central North Carolina. In this region, basement crystalline rocks are subsiding, and major river valleys have been "drowned" or embayed, forming extensive bays and sounds. Toward the south, these basement rocks warp upward to form the Cape Fear Arch, which reaches its peak along the border of North Carolina and South Carolina. Uplift in this region has resulted in erosion of Cenozoic sediments, exposing sandy Cretaceous sediments beneath. Crystalline rocks appear to be subsiding beneath the Sea Island Downwarp section, and the rising sea level in this region has created extensive tidal marshes and swamps. The Peninsular Florida section is a broad sandy plain resting on the Ocala Arch. The Mississippi Embayment and East Texas Embayment are characterized by broad deltaic plains along the coast. Inland is a "staircase" of Pleistocene depositional terrace surfaces and an inner region of belted topography developed on differentially eroded sediments (Murray 1961).

The transition from terrestrial to marine ecosystems along the thousands of kilometers of southeastern coastline is gradual and complex. In addition to having abundant river deltas, the coast is bordered by long ribbons of barrier or sea islands that protect extensive estuaries, lagoons, and sounds. These coastal features are in a constant state of change owing to erosion by wind and water and changes in the influx of sediment.

The surface of the Coastal Plain has been reworked considerably by coastal and fluvial processes during the last 2–3 million yr. The areal extent of the Coastal Plain has varied with changes in global climate and sea level. When the sea level was at its highest, Florida was nearly submerged; when the sea level was at its lowest, the peninsula was nearly 600 km across. Seven terraces and scarps have been recognized on the Atlantic segment of the Coastal Plain (Cooke 1931). These have been considered to be "high-water marks" formed by shifts in sea level during the Pleistocene. The elevation of the highest of these terraces is approximately 85 m above sea level. The origin and locations of these terraces, particularly those above 25 m, are still debated (Doering 1958; Murray 1961; Cronin et al. 1981). The fluvatile terraces along major river systems (especially the Mississippi) have been correlated with the coastal terraces and with alleged interglacial periods (Murray 1961). Coastal features such as dune fields, swales, and remains of mud flats are obvious on lower terraces and greatly influence vegetation patterns (Colquhoun 1969).

Hundreds of thousands of elliptical depressions, the so-called Carolina bays, occur on the Coastal Plain from southern Virginia to northern Florida. The axes of the ellipses are all oriented within a few degrees of a line trending northwest to southeast, and rims of coarse sterile sand often border them. The origin of these bays has been a matter of considerable debate (Cooke 1933; Johnson 1942; Prouty 1952; Wells and Boyce 1953; Savage 1982, Sharitz and Gibbons 1982), and the most widely accepted theories involve combinations of wind orientation and deflation, karst subsidence, and steamlining by groundwater. Recently, an old theory that they were created by a meteor shower has received renewed attention (Murray 1961; Savage 1982). These features have prominent effects on vegetation where they occur: The depressions vary from peat-filled bogs to open lakes or (where sediment is clayey) to savannas and "flatwoods," and the wind-worked sands of the bay rims are among the most sterile and dry Coastal Plain habitats.

Soils

The large-scale distribution of major soil orders across the Coastal Plain has been described in detail in Buol (1973). Entisols (soils with virtually no profile development) are common on the very well drained sands throughout the region; they represent the low extreme in water retention and have very low mineral adsorptive capacity, and hence are quite infertile. Inceptisols (soils with weakly developed horizons) are most common on alluvial plains; they are highly variable in texture and drainage and are generally infertile (although nutrient subsidization from river floods may increase their fertility). Typical cool-climate Spodosols are found only in the northern Coastal Plain (Tedrow 1979), but Aquods (groundwater podsols) are common throughout the region. Aquod soils form as soluble organic compounds, iron, and aluminum are leached to shallow water tables. Alfisols (soils with light-colored surface layers and a definite clay B horizon) border the alluvial plain of the Mississippi River and occur at other locations in Alabama, Florida, and South Carolina. Ultisols (highly weathered soils with B horizons containing appreciable amounts of translocated clay) occur widely over the Coastal Plain. These soils vary in fertility in relation to parent material: Soils from siliceous lagoonal sediments are nutrient-poor, whereas soils from carbonaceous sediment are fertile. Because Alfisols are less weathered than Ultisols, they retain considerable quantities of calcium and magnesium and are comparatively fertile. Drainage on Alfisols and Ultisols is highly variable; clay horizons may impede percolation and result in poor soil aeration. Histosols (organic soils) are associated with paludal wetlands throughout the Coastal Plain. Saprist or mucky peats form from herb wetlands such as marshes, whereas hemist and fibrist peats are characteristic of woody vegetation. Many of these soil orders may be found within a distance of a few hundred meters along catenas related to topography or hydrology (Daniels et al. 1984).

Climate

According to Köppen's classification (Trewartha 1968), the climate of the Coastal Plain is humid subtropical. Mean daily temperatures are between 0°C and 18°C in the coldest month and greater than 22°C in the warmest month; rainfall is distributed evenly throughout the year. Winter rainfall in this region is primarily a consequence of frontal cyclonic storms, whereas summer rain is usually associated with convectional thunderstorms.

Although there is considerable uniformity over this rather extensive area (Fig. 11.2), several trends should be noted. Seasonal variations in temperature increase away from the coast, and the length of the frost-free season tends to increase toward the coast and to the south. Rainfall is highest in the southeastern section, where the seasonal pattern is decidedly tropical (i.e., winter drought and summer peak). Rainfall tends to decrease away from the coast. Potential evapotranspiration (based on temperature) (Thornthwaite et al. 1958) varies from 700 to 1300 mm along a north-south gradient. Although annual precipitation exceeds potential evapotranspiration by 50–400 mm throughout most of the Coastal Plain, during much of the summer actual evapotranspiration is often less than potential owing to depleted soil reserves (Thornthwaite et al. 1958). In general, water deficits (i.e., potential minus actual evapotranspiration) increase to the south and west; topographic and soil factors make the variability in water deficit within any local area as great as that over the entire region.

Violent weather associated with hurricanes and convectional storms contributes to the disturbance mosaic in many Coastal Plain forests. Much tree windthrow is associated with such storms. Furthermore, this region has the highest frequency of lightning strikes of any region in North America (Komarek 1968), providing an ignition source for frequent fires.

UPLAND PINE FOREST VEGETATION

Northern Pine Barrens

The pine forests of the Atlantic Coast Plain north of Delaware Bay are quite distinct from their counterparts to the south. The dominant pine species is pitch pine (*Pinus rigida*),[1] although shortleaf pine (*Pinus echinata*) may co-occur in some localities. These forests have strong floristic affinities to the ridge-top pitch pine forests of the southern Appalachians (Whittaker 1979) (see Chapter 10). For example, *Comptonia peregrina, Kalmia latifolia, Gaylussacia baccata, Quercus ilicifolia*, and *Q. prinoides*–species characteristic of the shallow soils in the Appalachians–are common. Nevertheless, southern Atlantic Coastal Plain species such as *Ilex glabra, Gaylussacia frondosa, G. dumosa*, and *Clethra alnifolia* are also abundant. In fact, 109 species with centers of abundance on the southeastern Coastal Plain reach their northernmost limit in this region (Little 1979).

The unique features of the pine barrens flora were recognized by several early authors (Stone 1911, Harshberger 1916). Braun (1950) considered that the unique character was a consequence of colder climes, shorter growing season, and the podsolic nature of the soils in this region. The vegetation and ecology of pine barrens ecosystems have been reviewed in a set of excellent studies edited by Forman (1979). Buchholz and Good (1982a) have compiled a complete annotated bibliography of pine barrens literature.

Composition and structure. Lutz (1934) divided the pine-dominated ecosystems of this region into three types differentiated by stature and composition: the pine-oak or pine barrens forests, the pine–shrub oak forests (transitional forests), and the dwarf pine plains. This classification has been adopted by most later ecologists (Olsvig et al. 1979; Whittaker 1979; Forman and Boerner 1981).

Pine-oak forests, which correspond to the pine barrens community of Lutz (1934), have a well-

[1]The vascular plant taxonomy is according to Radford and associates (1968), except for central Florida, where I follow Wunderlin (1982). Bryophytes are according to Crum and Anderson (1981).

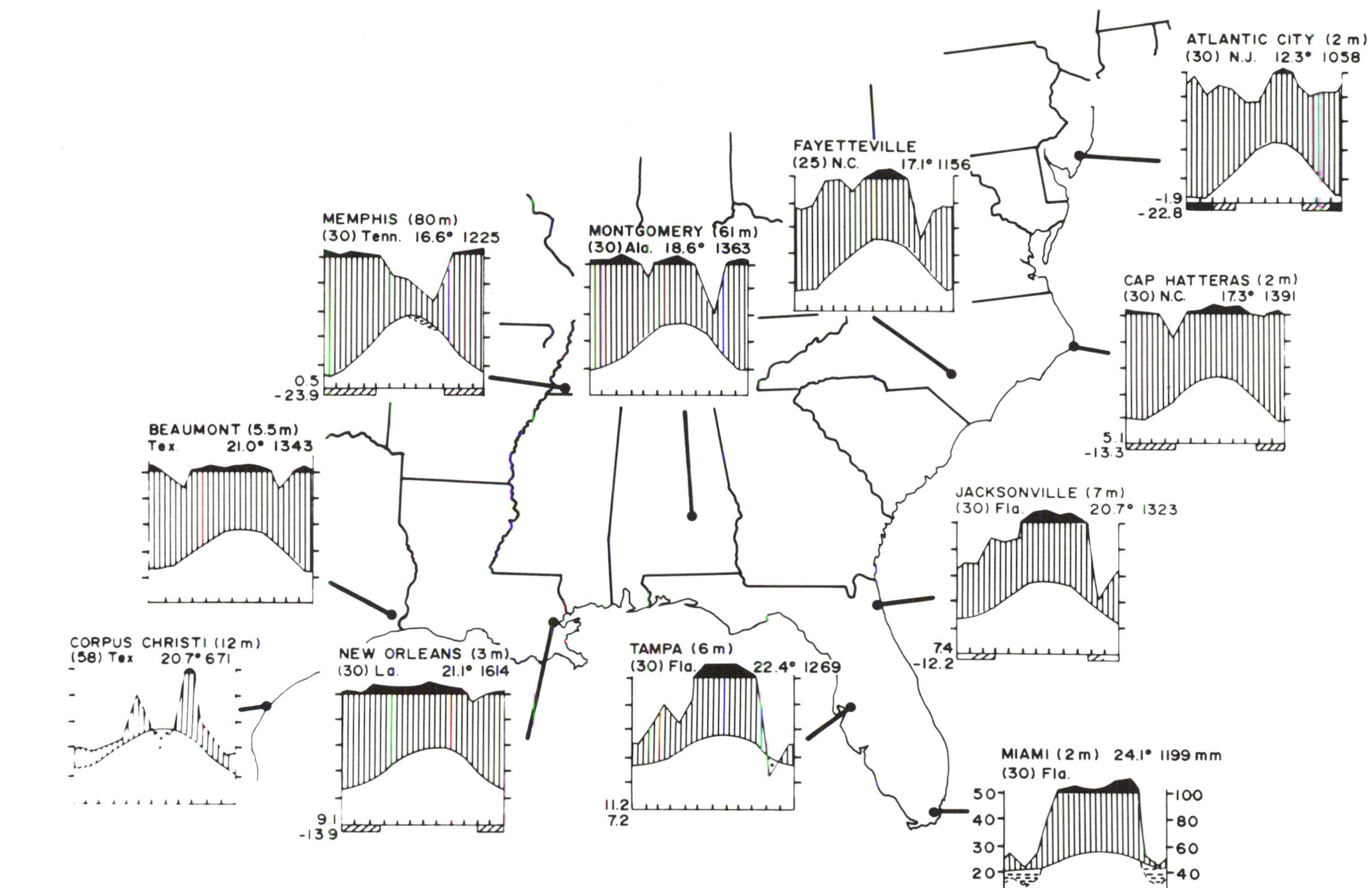

Figure 11.2 Climate diagrams for representative Coastal Plain stations. Graphs were prepared from Walter and Lieth (1967) using conventions outlined by Walter (1979). See the graph of Miami for axis labels. Data following place name include elevation (m), years of record, mean annual temperature (°C), and annual precipitation in mm. Left axis: mean monthly temperature; right axis: monthly precipitation. Vertical lines show months with precipitation > evaporation. Black area shows months with precipitation >100 mm.

Table 11.1. *Density, basal area (BA), and percentage cover for trees >1.2 cm dbh in pine-oak and oak-pine communities in the New Jersey pine barrens*

	Pine-oak			Oak-pine		
Species	Density (ha^{-1})	BA (m^2 ha^{-1})	Cover (%)	Density (ha^{-1})	BA (m^{2} $^{-1}$)	Cover (%)
Pinus rigida	388	11.3	29.8	204	3.79	16.6
Pinus echinata	24	0.89	2.9			2.0
Quercus marilandica	44	0.04	0.8	12	0.03	
Quercus stellata	44	0.74	1.6	28	0.11	1.4
Quercus velutina	8	0.78	6.1	212	5.16	39.9
Quercus prinus				476	6.31	43.0
Quercus alba			0.2			0.9
Sassafras albidum	4	0.02	0.4	4	0.001	
Totals	513	13.8	45.5	936	15.4	103.6

Source: Data from Buell and Cantlon (1950).

developed tree stratum, with scattered individuals of *Quercus stellata* and *Q. marilandica* (Table 11.1). There is normally an extensive shrub layer dominated by *Quercus ilicifolia, Gaylussacia baccata*, and *Vaccinium* species (Lutz 1934; Buell and Cantlon 1950; Olsvig et al. 1979). On finer-textured soils (Olsvig et al. 1979), or with decreased fire frequency (Forman and Boerner 1981), these forests grade into oak-pine forests. In more mesic sites, *Acer rubrum* and *Nyssa sylvatica* may occur (Whittaker 1979).

Pine–shrub oak forests are transitional between the pine plains and pine-oak forests. Tall-stature pitch pines (>10 m) are dominant, whereas tree oaks (*Quercus stellata* and *Q. marilandica*) are stunted or absent (Olsvig et al. 1979). The dominant shrubs are the same as those found in the plains.

Dwarf pine plains are dominated by a scrub ecotype of *Pinus rigida* and a dwarfed form of *Quercus marilandica*, both multistemmed from large irregularly shaped stools that are considerably older than the current stems (McCormick and Buell 1968; Little 1979). *Quercus ilicifolia* and other shrub species are also common. The shrub canopy is usually 0.5–1.5 m high and only rarely exceeds 3 m (Fig. 11.3). Although somewhat less conspicuous, the pyxie moss (*Pyxidanthera barbulata*, Diapensiaceae) is characteristic of these ecosystems (Good et al. 1979). Olsson (1979) distinguished two communities. In open denuded sites created by severe fire or other extreme disturbance, *Corema conradii* and *Arctostaphylos uva-ursi* are common; such areas are similar in composition to the Long Island heaths described by Olsvig and associates (1979). On more typical sites, the shrub canopy is closed, and the understory is characterized by the lichens *Cladonia caroliniana* and *C. strepsilis*.

Lutz (1934) listed 51 species of vascular plants from dwarf pine plains, 47 from the transition forests, and 39 from the pine-oak forests. Lichens (especially species of *Cladonia*) and mosses were also more abundant in the plains communities. Archard and Buell (1954), in a study of life-form spectra along this vegetational gradient, found that cryptophytes and hemicryptophytes were larger components of the flora in plains shrublands than in pine-oak forests.

Community dynamics and succession. Considerable effort has been directed toward understanding the factors responsible for the gradient from pine plains to pine-oak forests (Lutz 1934; Good et al. 1979; Olsvig et al. 1979). Factors that have been proposed include nutrient limitations, impervious subsoil, aluminum toxicity, and water deficits, but these factors are neither unique to the pine plains nor sufficient to explain the observed pattern. Good and associates (1979) concluded that the observed pattern is a consequence of variation in fire history. Lutz (1934) noted that fire return intervals in pine plains communities averaged 8 yr, compared with 16–26 yr in other pine barrens communities, and that the pine plains fires are intense and crown-killing, whereas pine-oak fires are lighter and more variable (Forman and Boerner 1981).

Good and associates (1979) proposed that severe, frequent fires have selected for a pine plains ecotype of *Pinus rigida* and that its shrubby growth form has a genetic basis (Andresen 1959). Common garden experiments have shown that pine plains genotypes show reduced apical dominance compared with those of larger-statured forests (Good and Good 1975) and that plains genotypes produce more intensely serotinous cones compared with those of other barrens genotypes (Frasco and Good 1976). Little (1979) noted that plains oaks begin producing acorns as soon as 3–4 yr following a fire,

Figure 11.3. View of pine plains vegetation near Lebanon, New Jersey. Shrubs and pines in the foreground are approximately 2 m tall (photo by W. D. Billings).

whereas tree oaks require 20 yr to begin seed production. It is not known whether or not other shrubby forms of tree species, such as *Quercus marilandica*, are ecotypes, but there is little doubt that frequent fire is responsible for their coppiced physiognomy in the plains.

Woodwell (1979) proposed that frequent fires increase nutrient loss from plains ecosystems, thus diminishing their nutrient capital. The nitrogen content of foliage and litter is lower in the plains than in other barrens communities (Lutz 1934). Lower litter nutrient content may reduce rates of decomposition (Vitousek 1982), resulting in the accumulation of flammable fuel and increased fire frequency (Christensen 1985). Given that at least some pine plains plant populations are genetically distinct from those of other barrens communities, it is unlikely that plains communities would rapidly succeed to pine-oak or oak-pine forests in the absence of fire (Good et al. 1979).

The environmental effects and vegetational responses to fire in pine barrens have been reviewed by Little (1979) and Boerner (1981, 1983; Boerner and Forman 1982). These responses are correlated with fire intensity, which in turn varies with climatic and fuel conditions. Following intense fires, oak reproduction is primarily from sprouts, because acorns usually are heat-killed (Little 1979; Gallagher and Good 1985). In the plains, abundant pitch pine seeds released from serotinous cones may result in large post-fire seedling populations. Pitch pine seeds germinate best on a mineral seedbed (Ledig and Little 1979), characteristic of intensely burned sites, whereas the oaks, with their large nutrient reserves, are quite successful in relatively thick mats of leaves (Little 1979). Shrub cover is reduced in the first few growing seasons following fire, but recovery is most rapid following low-intensity fires (Bierner 1981). In general, herb diversity is increased by fire, but the success of particular species appears to depend on the fire intensity. For example, Boerner (1981) found that growth of *Pteridium aquilinum* was greatest in pine-oak stands burned by intense wildfire, as compared with stands that were prescribed-burned. Olsson (1979) found that the floristic composition had shifted in the nearly 50 yr since Lutz's study (1934) and that herb diversity had declined in plains communities; he attributed this shift to decreased fire frequency.

Although surface fires in pine-oak stands may increase nutrient availability, losses of minerals via volatilization and leaching may be high (Wang 1984). A minimum of 8 yr is required for nutrient inputs to compensate for these losses. Intense wildfires result in greater nutrient loss than light prescribed fires (Boerner and Forman 1982; Boerner 1983), but in either case rapid vegetation regrowth minimizes these losses, and, as suggested by Woodwell (1979), these systems become less leaky with successional age (Boerner and Forman 1982).

Urbanization of the pine barrens region has resulted in a general lengthening of fire return intervals (Forman and Boerner 1981). Regional fire frequency (number of fires per year) actually increased from 1900 to 1940 and has since remained about the same; however, the average fire size decreased considerably. As a consequence, average fire return intervals for pine-oak forests have increased from 20 yr to about 65 yr in the period from 1900 to the present, and Forman and Boerner suggest that longer return intervals will result in changes in community structure in plains and pine-oak communities toward greater dominance of oaks and other less-fire-tolerant or -dependent species.

Xeric Sand Communities

Coarse sandy soils are abundant across the Coastal Plain, and they exhibit frequent water deficits owing to poor water retention and nutrient limitations resulting from low mineral adsorptive capacity. Extensive areas of sand are found in the fall-line sandhills, a more or less continuous formation of rolling hills extending from southern North Carolina through Georgia and parts of Alabama. These highly weathered sands are the residual product of underlying Cretaceous sediment. In addition, eolian and alluvial processes deposited sands in various locations across the middle and lower Coastal Plain during the Pleistocene; on the lower, more recently exposed terraces, relict dune fields and other coastal features provide abundant sandy habitats. These areas often form a so-called ridge-and-swale topography, with xeric conditions on the ridges and poor fens in the swales (Woodwell 1956; Christensen 1981; Ash et al. 1983). Two physiognomically and compositionally distinct ecosystems occur on these sands: the sandhill pine forests and the sand pine scrub.

Sandhill pine forests. Although two species, longleaf pine (*Pinus palustris*) and wiregrass (*Aristida stricta*), are characteristic throughout sandhill pine forests, I shall divide this community type into three subgroups: the pine–turkey oak sandridge, the fall-line sandhill, and the Florida sandhill associations.

Pine–turkey oak sandridge forests are found in the most austere habitats (e.g., ridge tops, the sand rims of the Carolina bays, and relict dune ridges of lower coastal terraces), where pine density is very low (50 ha^{-1}) and trees are often stunted and misshapen compared with those in more mesic habitats (Fig. 11.4). Turkey oak dominates the understory tree layer, with occasional individuals of black gum (*Nyssa sylvatica* var. *sylvatica*) and persimmon

Figure 11.4. Pine–turkey oak sandridge near Jones Lake, North Carolina. Note the large areas of bare sand and abundant lichens and mosses. A turpentine scar is obvious on the longleaf pine at the center of this photo.

(*Diospyros virginiana*) (Table 11.2). Low shrubs, including staggerbush (*Lyonia mariana*) and dwarf huckleberry (*Gaylussacia dumosa*), occur as scattered clumps (2% cover). Although not particularly species-rich, these communities have a unique herb layer. In open areas, foliose lichens (many of which are endemic to these habitats, such as species of *Cladonia* and *Cladina*) form a brittle carpet along with the sand-binding lichen *Lecidea uliginosa*. Beneath turkey oaks, the broom moss *Dicranum spurium* dominates. Among the most common and indicative vascular herb species are *Arenaria caroliniana*, *Cnidoscolus stimulosus*, *Selaginella arenicola* and *Stipulicida setacea*, which are usually scattered and display xeromorphic characteristics such as microphylly, glaucous pubescence, and succulence. Wells and Shunk (1931) suggested that these features were selected as a consequence of soil water deficits and high irradiance owing to the high albedo of the white sand. Indeed, the importance of vertical leaf orientation in seedlings of turkey oak has been clearly demonstrated to be related

Table 11.2. *Percentage cover of plants from Singletary Lake sand ridge*

Species	Cover (%)
Quercus laevis	15.5
Cladina species	5.6
Lecidia uliginosa	5.1
Arenaria caroliniana	1.2
Selaginella arenicola	1.2
Dicranum condensatum	0.9
Polygonella polygama	0.6
Areolaria pedicularia	0.5
Rhynchospora megalocarpa	0.2
Stipulicida setacea	0.1
Lyonia mariana	<0.1
Gaylussacia dumosa	<0.1
Aristida stricta	<0.1
Magnolia virginiana	<0.1
Total	30.9

Source: Unpublished data of S. McAlister for a 0.4-ha sample.

to high irradiance (Wells and Shunk 1931; Raff 1954).

The composition of the fall-line sandhill association varies with topography (Table 11.3) (Wells and Shunk 1931; Weaver 1969). On ridge tops, in the absence of clay horizons, turkey oak is the dominant understory tree. Downslope, clay horizons become better developed and occur nearer the soil surface (Daniels et al. 1984), and here blackjack oak (*Quercus marilandica*), sandhill post oak (*Quercus margaretta*), and bluejack (*Quercus incana*) share dominance. Other subcanopy trees include *Nyssa sylvatica* var. *sylvatica*, *Diospyros virginiana*, and sweet gum (*Liquidambar styraciflua*). Weaver (1969) speculated that vegetation changes along this catena are due to a combination of changes in drainage and soil fertility.

Various herb species and shrubs, such as *Gaylussacia dumosa* and *Vaccinium* species, are scattered throughout the dominant wiregrass matrix. Duke (1961) noted a high degree of endemism and habitat specificity among sandhill herb species; mosses and lichens are abundant in areas where growth of vascular plants is sparse. Where clay horizons are near the surface or moisture is more abundant, bracken fern (*Pteridium aquilinum*) may be abundant. Where the clay horizon meets the surface, seeps or springs occur, and shrub species typical of bog habitats, such as *Clethra alnifolia*, *Cyrilla racemiflora*, and *Lyonia lucida*, are often found (Wells and Shunk 1931).

Monk (1968) recognized three distinguishable "phases" of the Florida sandhill association dominated by turkey oak, bluejack oak, and southern red oak. Longleaf pine forms a broken canopy in each of these communities, but may share dominance with slash pine (*Pinus elliottii*). The turkey oak phase occurs on driest sites and is structurally similar to the pine–turkey oak sandridge vegetation already described. Scattered individuals of *Quercus incana* and *Diopyros virginiana* may also occur in this phase. *Diospyros* is most common on sites that have not been recently burned (Laessle 1942). The herb layer is dominated by two wiregrasses, *Aristida stricta* and *Sporobolus gracilis*. Shrubs are scarce in this community, except for the gopher apple, *Chrysobalanus oblongifolius*. The bluejack oak phase is characteristic on finer-textured, somewhat more fertile soils (Laessle 1942). Live oak (*Quercus virginiana*) is common in the overstory, and *Aristida stricta* forms a dense cover (Veno 1976). The southern red oak phase is more typical of calcareous soils and grades into southern mixed hardwood forest (Monk 1960, 1968).

The southern ridge sandhill community described by Abrahamson and associates (1984) is characteristic of sandy habitats in south central Florida. It is intermediate in composition and structure between the sand pine scrub and the pine-wiregrass association, with a distinctly three-layered open canopy. The overstory is 5–10 m in height (Fig. 11.5). The dominant canopy tree is the southern Florida slash pine (*Pinus elliottii* var. *densa*), which may co-occur with scattered individuals of longleaf pine and sand pine. Turkey oak (*Quercus laevis*) and scrub hickory (*Carya floridana*) are also important canopy dominants. Abrahamson and associates (1984) divided this community into two phases based on the relative prevalence of these two species.

Sandhill pine forests have also been described for the Big Thicket region of eastern Texas (Marks and Harcombe 1981). As in other sandhill forests, the pine density is low, the herb cover is relatively sparse, and there is considerable exposed sand. This area is several hundred miles west of the range of turkey oak, and the dominant oaks are *Quercus incana* and post oak (*Quercus stellata*). Here longleaf pine shares dominance with shortleaf pine (*Pinus echinata*) and loblolly pine (*Pinus taeda*). Yaupon (*Ilex vomitoria*) and flowering dogwood (*Cornus florida*) are common understory trees.

Sand pine scrub. The dominant overstory tree in this community is sand pine (*Pinus clausa*). The understory is dominated by a dense and rather diverse assemblage of evergreen schlerophyllous shrubs (Fig. 11.6). Abrahamson and associates (1984) differentiated two phases of the sand pine scrub community. The oak understory phase is a

Table 11.3. *Percentage cover and basal area (BA, trees) for plants in ridge-top and mesic-slope fall-line sandhill forests*

Species	Ridge top		Mesic slope	
	Cover (%)	BA (m^2 ha^{-1})	Cover (%)	BA (m^{2} $^{-1}$)
Trees				
Quercus laevis	28.8	4.8	2.2	0.9
Quercus marilandica	1.5	—	31.3	4.3
Quercus margaretta	—	—	11.4	1.5
Quercus incana	1.8	—	3.1	0.5
Carya species	—	—	3.3	0.4
Pinus palustris	8.6	3.0	8.3	2.0
Pinus taeda	—	—	3.5	1.1
Shrubs and herbs				
Aristida stricta	15.0		13.3	
Cryptogams	3.0		1.8	
Gaylussacia dumosa	3.3		2.4	
Tephrosia virginiana	1.0		2.7	
Andropogon species	1.0		2.5	
Clethra alnifolia	—		2.1	
Lyonia mariana	—		0.4	
Rhus radicans	—		0.8	
Solidago species	—		0.4	
Carphephorus bellidifolius	0.8		—	
Totals	64.8	7.8	89.3	10.7

Source: Data from Weaver (1969) for North Carolina.

three-layer community, with a lower shrub layer of saw palmetto (*Serenoa repens*) and scrub palmetto (*Sabal etonia*), an upper shrub layer dominated by scrub live oaks (*Quercus geminata*, *Q. myrtifolia*, and *Q. virgininia*), *Carya floridana*, and rusty lyonia (*Lyonia ferruginea*), and an overstory of sand pine. Herbs are particularly scarce in this phase. The rosemary phase of this community is a somewhat more open community dominated by even-aged stands of rosemary (*Ceratiola ericoides*). Another scrub oak, *Quercus inopina*, may share dominance with the rosemary. The pine canopy is much more broken in this phase, and herbs are more abundant and diverse. The rosemary phase appears to be characteristic of drier ridges and knolls. Although not usually abundant in either phase, *Rhynchospora dodecandra* and *Andropogon floridanus* are especially indicative of scrub communities (Laessle 1958). These communities are floristically similar to pine "flatwoods" and maritime scrub forests of Florida.

Two genetic races of sand pine are recognized. Choctawatchee sand pine (*Pinus clausa* var. *immuginata*) grows naturally in a rather restricted

Figure 11.5. Southern ridge sandhill community near the Archbold Biological Station, central Florida. The longleaf pine in the center is approximately 10 m tall (photo by P. Peroni).

Figure 11.6. Sand pine scrub communities, (A) rosemary phase and (B) oak understory phase, near the Archbold Biological Station, central Florida (photo by P. Peroni).

region of northwest Florida (usually on soils derived from recent littoral deposits). This variety has nonserotinous cones, often occurs in mixed-age stands, and may be found in several different community types. Ocala sand pine (*Pinus clausa* var. *clausa*) grows exclusively in the sand pine scrub communities of peninsular Florida. This variety has serotinous cones and successfully establishes primarily from seeds following fire; as would be expected, it forms very nearly even-aged stands (Little and Dorman 1952; Burns 1973). Recent observations by Ronald Myers (pers. commun.) clearly show that both races occur in sand pine scrub.

Mulvania (1931) observed that water availability was one of the major selective factors for shrub morphology in sand pine scrub. For example, several shrub species (especially *Ceratiola ericoides*) have short sclerophyllous needle-like leaves. Leaves of other species are often heavily cutinized, revolute, and sometimes tomentose underneath. Mulvania noted that shrub leaves often are not deployed horizontally in these species and that stomates are smaller and less dense than in nonscrub species. Given the low nutrient concentrations in these sandy soils, it is possible that sclerophylly may also be a response to nutrient limitations (Loveless 1961, 1962). Shrubs and palms reproduce vigorously from underground rhizomes and burls; most shrub reproduction following fire is vegetative (Wade et al. 1980).

Community dynamics and succession. On the driest sand ridges, the rate of fuel accumulation is very slow, and the spatial patterning of vegetation and detritus is discontinuous. Fires in this vegetation type are infrequent and often are localized near the source of ignition. Historical accounts, such as that of Bartram (1791), suggest that during precolonial times these ridges were dominated by longleaf pines, with a relatively open understory. Pine litter would have provided sufficient fuel for occasional fires, which in turn would have favored pine reproduction and suppressed hardwood growth (Christensen 1979b, 1981). These generally misshapen pines were not suitable for saw timber, but were an important source of turpentine, and many of the relict trees on these sites still bear the distinctive scars of this enterprise. Most pines were eventually harvested for tar extraction, allowing such hardwoods as turkey oak and persimmon to invade and initiating a vastly different fire regime. Owing to the sterility of these soils and an altered fire regime, longleaf pine has been very slow to reinvade.

Low-intensity surface fires occur at a very high frequency in wiregrass-dominated pine woods. Fire return intervals may be as brief as 1 yr in areas that are intentionally burned for fuel reduction, but the natural interval probably is 3–10 yr (Heyward 1939; Wells 1942; Garren 1943; Parrott 1967; Christensen 1981). Approximately 3–4 yr are required for sufficient accumulation of dry fuel to carry a surface fire, after which time the probability

Figure 11.7. Stages in the life history of Pinus palustris, *long leaf pine A: : "Grass" or seedling stage. B: After 4–6 yr, often following fire, rapid apical growth is initiated, and the apical bud, which is vulnerable to fire, is carried above the reach of flames. C: A pole-size tree, approximately 10 yr old; survivorship among trees of this size is very high.*

of fire is determined by the availability of ignition sources (Parrott 1967; Christensen 1981).

Fire has the effect of briefly increasing nutrient availability and, thereby, post-fire herb production (Christensen 1977). However, reproduction during the first post-fire growing season is largely vegetative (Hodgkins 1958; Arata 1959; Parrott 1967; Lewis and Harshbarger 1976). Flowering is stimulated by burning in many sandhill species. For example, flowering in *Aristida stricta* is confined to the first post-fire growing season (Parrott 1967). Christensen (1977) found that simply clipping leaves would initiate some flowering. However, flower production was always greater in burned than in clipped areas, presumably owing to fire-caused nutrient enrichment.

The life history of *Pinus palustris* is ideally suited to this high-frequency, low-intensity fire regime. Longleaf pines establish most successfully on bare mineral soil following fire (Chapman 1932; Wahlenberg 1946). During the "grass" stage (Fig. 11.7), seedlings are fire-resistant and allocate most of their photosynthate to the production of an extensive root system, including a massive taproot. After 3–5 yr, apical growth is initiated, carrying the apical bud above the zone of most surface fires. Unlike pines adapted to less frequent crown fires, longleaf pine does not sprout or produce serotinous cones. If fire frequency is reduced, other pine species, such as *Pinus taeda* and *P. elliottii* may replace *Pinus palustris* (Chapman 1926, Little and Dorman 1954; Christensen 1981).

The natural return interval for fires in sand pine scrub is 30–60 yr (Webber 1935; Harper 1940; Laessle 1965; Christensen 1981). These crown fires ordinarily kill the pines and aboveground portions of the shrubs. Seed rain from the sand pine may exceed 250 seeds per square meter (Cooper et al. 1959), and young stands often have sand pine densities exceeding 10,000 trees per hectare (Price 1973; T. S. Coile, unpublished data). As their crowns close, pines begin to thin, converging to densities of 400–800 trees per hectare after 40 yr. In the absence of fire, the pine canopy becomes sufficiently broken to allow establishment of seedlings in the understory, and the stand may become uneven-aged (Peroni 1983; Coile, unpublished data). However, stands more than 70 yr of age are rare. Dry conditions, frequent lightning storms, and accumulation of highly flammable fuels make fire an inevitable event in these communities (Hough 1973).

Laessle (1958, 1967) argued that the Florida sandhill and sand pine scrub communities are segregated with respect to variations in soil characteristics associated with the distribution of Pleistocene shorelines. Recent work suggests that these communities may succeed one another as a consequence of variations in fire regime. Kalisz and Stone (1984) found that sites currently occupied by scrub were, in the recent past, dominated by sandhill vegetation, and vice versa. Myers (1985; Myers and Deyrup 1983) showed that exclusion of fire from sandhill vegetation results in invasion of scrub, which tends to favor intense crown fires at 30–60-yr intervals, which in turn maintains scrub vegetation. In the complete absence of fire, these sites are presumed to be seral to xeric hardwoods (Fig. 11.8).

Mesic Pine Communities

With increasing moisture availability, xeric sandhill communities grade into pine-dominated flatwoods and savannas. Flatwoods and savannas are distinguished primarily on structural grounds (Fig. 11.9). Savannas are typified by an open canopy (pine density usually <150 ha^{-1}), with a graminoid-dominated understory (Penfound and Watkins 1937). The overstory density is greater in flatwoods, and the understory is composed of a diverse array of shrubs and subcanopy trees. However, intermediate situations are common, and some workers (Laessle 1942; Abrahamson et al. 1984) refer to all these communities as flatwoods.

There is no generally accepted regional classification of these mesic pine woods. As with more xeric pine woods, research has been concentrated in the Carolinas, central Florida, and east Texas. Although species composition along the flatwood-savanna gradient varies among these areas, the compositional change along the gradient seems to be due to the interaction of fire frequency and moisture availability in each area.

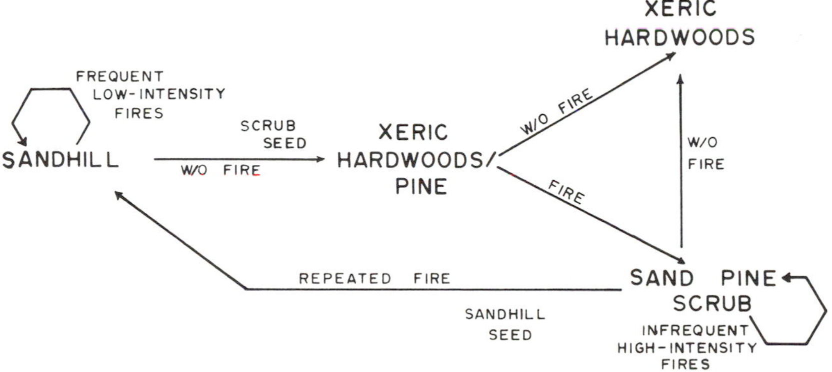

Figure 11.8. A model for successional relations among sandhill, sand pine scrub, and xeric hardwood forest communities. Given the same environmental conditions, the most stable community will depend on the fire regime. Once established, fuel and environmental conditions within a particular community tend to perpetuate that community type (after Myers 1985).

Figure 11.9. Pinus elliottii *flatwoods near the Okefenokee Swamp, Georgia, with a palmetto-gallberry understory (A) 3 mo following fire and (B) unburned for >5 yr (photograph taken near Fargo, Georgia).* Pinus palustris *mesic savanna in the Green Swamp, North Carolina, (C) burned annually for the past 20 yr and (D) nearby area on the same soil type unburned for >20 yr.*

Flatwoods. The term "flatwoods" is used for just about any Coastal Plain pine forest with a well-developed woody understory (Christensen 1979b). Many such forests, particularly on relatively fertile soils, are successional from cropland abandonment and are similar in composition to the successional pine stands of the Piedmont (Oosting 1942) (see Chapter 10). Given the multiplicity of disturbance histories and site conditions, these successional forests virtually defy classification.

Throughout the Gulf Coastal Plain, forests dominated by longleaf and/or slash pine, with an understory of shrubs and saw palmetto, occur on poorly drained Ultisols. Pessin (1933) divided Gulf flatwoods communities into xerophytic forests, dominated by longleaf pine, scrub oaks, and *Serenoa repens*, and meso-hydrophytic forests, in which *Pinus elliottii* codominates in the canopy, and the scrub oaks are replaced with more hydric shrubs such as *Ilex glabra* and *Myrica cerifera*. Edmisten (1963) recognized three flatwoods types based on the dominant canopy tree species (Monk 1968). His

Table 11.4. *Cover of dominant species in Florida flatwoods communities*

	Cover (%)			
Species	SF[a]	FLW	FLP	FLG
Pinus elliottii	—	1	83	24
Pinus clausa	1	—	—	—
Quercus geminata	4	<1	3	—
Quercus chapmanii	16	—	<1	—
Quercus inopina	37	—	—	—
Sabal etonia	8	<1	—	—
Aristida stricta	<1	54	—	1
Quercus minima	—	15	—	—
Hypericum reductum	—	5	—	<1
Serenoa repens	22	36	77	46
Ilex glabra		<1	27	17
Persea borbonia	—	—	12	—
Lyonia lucida	3	3	7	20
Panicum abscissum	—	—	—	86

[a]SF=scrubby flatwoods, FLW=flatwoods/wiregrass, FLP=flatwoods/palmetto, FLG=flatwoods/galberry.
Source: Data from Abrahamson et al. (1984).

longleaf pine and slash pine flatwoods categories correspond to Pessin's xerophytic and meso-hydrophytic forests, respectively. He also described forests dominated by pond pine (*Pinus serotina*) that occur in very wet sites and appear to be successional to swamp forest or wet hammock (Monk 1968).

In central Florida, Abrahamson and associates (1984) differentiated two flatwoods types: scrubby flatwoods and flatwoods proper (Table 11.4). Scrubby flatwoods are transitional between scrub communities and more mesic ecosystems. Unlike other flatwoods communities, these ecosystems occur on well-drained soils (but with a higher water table than pine scrub communities). The canopy is dominated by scattered individuals of slash pine and sand pine. On drier sites, *Quercus inopina* dominates the understory, whereas *Q. virginiana* and *Q. chapmanii* are dominant on moist sites. The flatwoods association occurs on poorly drained soils that vary between very dry during drought periods to inundated during rainy periods. Abrahamson and associates (1984) divided these flatwoods into several phases. The wiregrass phase is dominated by *Aristida stricta* in the understory. It differs from sandhill communities by the abundance of *Serenoa repens*, *Quercus minima*, and other shrubby species. On somewhat more favorable sites, cutthroat grass (*Panicum abscissum*) replaces wiregrass. The palmetto flatwoods phase is characterized by an overstory of *Pinus elliottii* var. *densa* and a dense 1–2-m-tall sward of saw palmetto. In the gallberry-fetterbush flatwoods phase, *Ilex glabra* and *Lyonia lucida* share dominance in the understory with saw palmetto. The fern phase understory is dominated by chain fern (*Woodwardia virginica*), as well as *Osmunda regalis*, *Andropogon* species, and *Panicum hemitomon*. In addition, woody shrubs typical of hardwood swamps occur in the understory, suggesting that fire exclusion from this phase might result in succession to swamp forest or bayhead (Abrahamson et al. 1984).

The shrub oaks *Quercus chapmanii* and *Q. minima*, slash pine, and saw palmetto reach their northernmost limits in southeastern South Carolina. Nevertheless, pine-dominated communities with a shrubby hardwood understory are found on heavier soils north into Virginia. *Pinus taeda* is often the dominant pine, although longleaf pine may be found on sandier soils, and pond pine occurs in moist habitats. The prevalence of loblolly pine in such forests today may be a consequence of historical factors (Snyder 1980; Christensen 1981). Ashe (1915) suggested that loblolly-pine-dominated forests were common only in extreme northeastern North Carolina. Extensive use of this species as a plantation tree has increased its seed rain throughout the Southeast, and fire exclusion from many areas has given it a competitive edge against longleaf pine.

Wells (1928) classified mid-Atlantic flatwoods as part of a *Quercus-Carya-Pinus* association that he said "could be found on virtually any upland site." The subcanopy of these forests is well developed, particularly where fire has been excluded. Trees, such as several gum and oak species, may reach 5–10 m in height. Shrubs such as *Ilex glabra*, *I. coriacea*, *Myrica cerifera*, *Lyonia lucida*, and *Clethra alnifolia* indicate floristic affinity to more southern flatwoods. Herb cover is generally less than 5% in unburned woods, but in areas that have been burned frequently, herb cover may exceed 20%, and the vegetation may become savanna-like (Lewis and Harshbarger 1976). In moist locations, switch cane (*Arundinaria gigantea*) may form dense canebrakes (Hughes 1966).

The "upland pine forests" and the "upper-slope pine-oak forests" described by Marks and Harcombe (1981) and Matos and Rudolph (1985) for east Texas roughly correspond to the flatwoods of the mid-Atlantic states. Longleaf or shortleaf pine dominates the overstory, with scattered individuals of loblolly pine, several oaks, and *Liquidambar styraciflua*. Understory species include *Callicarpa americana*, *Ilex vomitoria*, and *Cornus florida*.

Savannas. Savanna vegetation of the Coastal Plain is often transitional between xeric pine communities and wetland pocosins and bayheads (Harper 1906, 1922; Wells 1928, 1946; Wells and Shunk 1928; Woodwell 1956; Eleuterius and Jones 1969;

Kologiski 1977; Wilson 1978; Christensen 1979b; Snyder 1980; Walker and Peet 1983; Jones and Gresham 1985). References to savannas as "southern upland grass-sedge bogs" (Wells and Shunk 1928), "pine-barren bogs" (Harper 1922), "pitcher plant bogs" (Eleuterius 1968; Eleuterius and Jones 1969), and "wetland pine savannas" (Marks and Harcombe 1981) indicate their seasonally wet condition.

Woodwell (1956) divided savannas in the Carolinas into those dominated by longleaf pine and those dominated by pond pine (*Pinus serotina*). Walker and Peet (1983) recognized a gradient from xeric savannas into xeric sandhill vegetation and into mesic savannas and wet savannas. In xeric savannas, the tree canopy contains only longleaf pine, and *Aristida stricta* dominates the herb layer. Legumes, including *Cassia fasciculata, Lespedeza capitata, Clitoria mariana*, and *Amorpha herbacea*, are common in these areas. Characteristic woody species include *Myrica cerifera* and *Smilax glauca*.

In mesic savannas (Fig. 11.9), the graminoid diversity, indeed herb diversity in general, is very high. Common graminoids include *Sporobolus teretifolius, Muhlenbergia expansa, Ctenium aromaticum, Andropogon* species, and *Rhynchospora plumosa*. Other indicative species include *Lycopodium carolinianum, Lachnocaulon anceps*, and *Xyris smalliana*.

Perhaps the most distinctive occupants of mesic savannas are insectivorous plants, including *Drosera capillaris, Pinguicula* species, *Sarracenia* species, and the celebrated Venus flytrap (*Dionaea muscipula*). The Venus flytrap is found only in savannas of the outer Coastal Plain of southeastern North Carolina and at a few localities in South Carolina (Roberts and Oosting 1958).

Wet savannas occupy depressions and the ecotones between mesic savannas and shrub bogs. Walker and Peet (1983) pointed out that they are most likely to be found where frequent fires on mesic savannas have spread into adjacent pocosins, eliminating the shrubs. *Drosera intermedia, Coreopsis falcata, Rhynchospora chalarocephala, Oxypolis filiformis, Iris tridentata, Aristida affinis*, and *Anthaenantia rufa* are indicative of this portion of the gradient. Sprouts of shrub bog species such as *Cyrilla racemiflora* and *Vaccinium corymbosum* are common. Longleaf pine may be missing from this zone, and one finds only scattered individuals of pond pine, pond cypress (*Taxodium ascendens*), and tupelo (*Nyssa sylvatica* var. *biflora*).

Streng and Harcombe (1982) differentiated two savanna types in the Big Thicket region of east Texas. Wet meadow savannas occur on poorly drained soils and contain sparsely scattered individuals of longleaf and loblolly pine, as well as *Nyssa sylvatica, Liquidambar styraciflua*, and *Magnolia virginiana*. This savanna type floristically resembles other southeastern savannas, save that *Andropogon scoparius* is the dominant grass. The second type, pine-bluestem savanna, is also dominated by *A. scoparius* and is floristically most similar to nearby prairies (Vogl 1973).

Community dynamics and succession. The successional status of flatwoods communities appears to depend largely on site conditions. On comparatively fertile and well-drained sites, where broadleaved trees invade the understory, stands will almost certainly be seral to southern mixed hardwood forest in the absence of fire (Quarterman and Keever 1962; Monk 1968; Blaisdell et al. 1974; Hebb and Clewell 1976; Delcourt and Delcourt 1977b). On sandy soils and in poorly drained areas, these communities display little evidence of compositional change, even when fire has been excluded for long periods (Abrahamson 1984a, 1984b). Givens and associates (1984), studying permanent plots in flatwoods from which fire had been excluded for nearly 50 yr, observed substantial thinning of herb and shrub layers, with the loss of a few shade-intolerant species, increases in tree and shrub species richness, an increase in canopy coverage, and increases in litter coverage and depth. However, they noted no appreciable succession toward hardwood forest (Veno 1976). Streng and Harcombe (1982) found that Texas pine-bluestem savannas that occur on well-drained sites will succeed to forest in the absence of fire, but that poor drainage in wet meadow savannas prevents tree and shrub invasion, regardless of fire regime.

Savannas in North Carolina from which fires have been excluded are consistently less rich than their frequently burned counterparts (Walker and Peet 1983). Roberts and Oosting (1958) found that fire exclusion resulted in exclusion of shade-intolerant savanna species, including the Venus flytrap (*Dionaea muscipula*). But there was no evidence that these savannas would succeed to some other forest type. Normally the ecotone between savanna and adjacent pocosin is very abrupt because frequent surface fires in the savanna burn to the edge of the pocosin, thus sharpening the line (Christensen 1981; Christensen et al. 1981).

The consequences of fire exclusion vary along the xeric-to-wet-savanna gradient. On dry sites, sandhill species, such as *Quercus laevis* and *Diospyros virginiana*, may invade (Christensen 1979b), whereas mesic savannas may succeed to flatwoods in the absence of fire (Kologiski 1977). Wet savannas are quickly invaded by pocosin shrubs if not frequently burned (Kologiski 1977; Walker and Peet 1983).

Flatwoods and savannas demonstrate the nearly circular relationship among vegetation structure and composition and fire regime on the Coastal Plain. Moist fuel conditions on heavy soils in flatwoods diminish the probabilty of ignition, but this is offset by the flammability of plants such as *Myrica cerifera*, *Ilex* species, and scrub oaks (Shafizadeh et al. 1977; Hough and Albini 1978). During interfire years, understory growth produces an increasingly continuous vertical distribution of flammable fuel. Thus, the longer the interfire interval, the more likely that ignition will result in a severe crown-killing fire that initiates a prolonged successional sere. Frequent fires create a distinctly discontinuous vertical fuel distribution, and low-intensity surface fires are the rule (Heyward 1939; Wahlenberg et al. 1939; Lemon 1949; Komarek 1974; Christensen 1981). Lewis and Harshbarger (1976) examined vegetation responses to different fire regimes in South Carolina pine woods and found that frequent (1–3 yr) summer fires favored a herb-dominated savanna-like community, whereas less frequent (5–10 yr) fires resulted in a dense shrubby understory. Komarek (1977) documented similar relationships between community structure and fire frequency in north Florida.

Table 11.5. *Relative densities of canopy (C) and understory (U) trees in two Louisiana coastal plain mesic hardwood stands*

Species	Stand 1		Stand 2	
	C (%)	U (%)	C (%)	U (%)
Magnolia grandiflora	33.8	8.1	14.9	17.6
Fagus grandifolia	6.5	1.6	51.1	17.6
Quercus nigra	2.6	3.2	10.6	11.7
Quercus michauxii	18.2	8.1	4.2	5.9
Pinus taeda	13.0	32.3	—	—
Liquidambar styraciflua	10.4	11.3	—	—
Liriodendron tulipifera	—	—	6.4	5.9
Tilia species	—	—	6.4	—
Quercus falcata var. *pagodaefolia*	6.5	1.6	—	—
Quercus shumardii	—	—	2.1	11.7
Quercus alba	2.6	—	—	—
Quercus virginiana	1.3	—	—	—
Carya species	1.3	3.2	2.1	—
Morus rubra	1.3	6.4	—	—
Ulmus alata	1.3	—	—	—
Ilex opaca	1.3	1.6	—	—
Ostrya virginiana	—	—	2.1	11.7
Carpinus caroliniana	—	17.7	—	11.7
Cornus florida	—	1.6	—	—
Persea borbonia	—	1.6	—	—
Prunus caroliniana	—	—	—	5.9
Nyssa sylvatica	—	1.6	—	—

Source: Data from Braun (1950).

UPLAND HARDWOOD FORESTS

The broad-leaved deciduous forest is described in Chapter 10. Nevertheless, the structure and successional status of deciduous forests on the Coastal Plain have been a matter of considerable debate, and I therefore include a somewhat abbreviated discussion of them here.

Composition and Structure

Harper (1906, 1911) considered that the climax for most of north Florida and southern Georgia was mixed deciduous and evergreen forest. Pessin (1933) concluded that oak and hickory forests would replace the longleaf pine forests of the Gulf states, but would themselves be replaced by forests dominated by beech (*Fagus grandifolia*) and magnolia (*Magnolia grandiflora*). This view was parallel to Wells's view (1928) that beech-maple forests would, in the absence of disturbance, dominate the Coastal Plain of North Carolina where *Magnolia grandiflora* was absent. As data accumulated on the extremes of soil and hydrology, most ecologists later conceded that a variety of edaphic hardwood climaxes might exist (Wells 1942; Harper 1943).

Data on stand structure in mixed hardwood stands have been provided by Pessin (1933), Laessle (1942), Kurz (1944), Braun (1950), Quarterman and Keever (1962), Monk (1965), Beckwith (1967), Ware (1970, 1978), Nesom and Treiber (1977), Marks and Harcombe (1981), and Matos and Rudolph (1985). Representative data for such forests in Louisiana appear in Table 11.5. The diversity of tree species is highest in north Florida and the central Gulf states and diminishes to the north and west (Monk 1967). Although this forest type is often designated as beech-magnolia (SAF 1980; Marks and Harcombe 1981), beech is absent from southeast Georgia and eastern Florida (Monk 1965), and *Magnolia grandiflora* is not abundant in the Carolinas and Virginia (Ware 1978).

There are considerable variations in community structure and composition within particular regions related to gradients of moisture and nutrient availability (Monk 1965; Ware 1978). Monk noted that species diversity was highest on mesic calcareous soils and decreased on wet and dry sites and on less fertile sites. The relative importance of evergreen species varied as a function of nutrients and water availability. On the driest, most sterile sites, evergreen plants accounted for more than 80% of total importance, whereas on the most fertile sites they accounted for only 10%–30% (Monk 1965). It was this pattern that led Monk (1966a) to suggest that evergreen plants were more fit on nutrient-poor sites, because year-round leaf fall in evergreens, coupled with sclerophylly, resulted in constant rates of nutrient turnover and reduced nutrient loss. In the Carolinas, Ware (1978) noted beech and white oak (*Quercus alba*) at one end of a vegetational gradient, with *Quercus laurifolia*, *Liquidambar styraciflua*, and *Carya glabra* at the other. The environmenal factors responsible for this gradient are not clear.

Community Dynamics and Succession

The upland mixed hardwoods are best developed on fertile soils, often derived from limestone, phosphatic deposits, or finer-textured sediments (Monk 1965; Nesom and Treiber 1977; Ware 1978). They are also found on somewhat more sterile soils where fire has been excluded. Nesom and Treiber (1977) considered such communities in North Carolina to be unique "topoedaphic" climaxes confined primarily to river bluffs, but Ware (1970, 1978) suggested that the present restricted distribution of these forests is an artifact of 350 yr of agriculture. DeWitt and Ware (1979) and Monette (1975) observed considerable invasion of mixed hardwood forest species into sites that had been previously farmed.

Although the mixed hardwood forest is considered by many to be the climatic climax for this region, little work has been done on the successional dynamics within these forests. *Magnolia grandiflora* seedlings do not survive beneath adult magnolias (Kurz 1944; Quarterman and Keever 1962; Blaisdell et al. 1974), and most of the associated oaks and hickories are shade-intolerant. Consequently, these tree populations become depleted in sapling and subcanopy size classes (Blaisdell et al. 1974; Harcombe and Marks 1978; Marks and Harcombe 1981), and it is presumed that gap-phase replacement plays a significant role in continued stand maintenance.

Fires are infrequent in these forests (Christensen 1981), and it is generally assumed that such a disturbance would initiate a prolonged successional sere, with an early herbaceous stage, followed by an even-aged pine stage, and eventually reestablishing mixed hardwood forest (Laessle 1942; Kurz 1944; Monk 1968). However, in the only study of fire response in such a community, Blaisdell and associates (1974) found that most of the larger trees survived a canopy fire, although beech suffered considerable trunk scarring. There was a large increase in beech seedling and sprout density following fire, suggesting that these forests are self-replicating, even in the face of major disturbance.

WETLANDS

As a consequence of subdued topography and complex drainage patterns, wetlands cover over 10^7 ha of the southeastern Coastal Plain, or 15% of the total land surface in the region (Turner et al. 1981; Wharton et al. 1982). These wetlands vary widely with respect to virtually every community characteristic, and various classification schemes have been devised to account for that variation (Shaler 1885; Wells 1928; Penfound 1952; Cowardin et al. 1979; Kologiski 1977; Wharton 1978; Huffman and Forsythe 1981; Wharton et al. 1982).

It has long been recognized that much of the variation in the wetland vegetation community is attributable to variations in the inundation regime. Wells (1928, 1942) considered that the length of time each year that soil is saturated is the primary environmental factor underlying wetland vegetation gradients. However, it is now recognized that other characteristics of the inundation regime, such as water quality, velocity, and periodicity, affect the relative success of the diverse collection of southeastern wetland species (Whitlow and Harris 1979; Wharton et al. 1982).

Wetlands are here divided into alluvial and nonalluvial communities. The former group has received intensive study, and vegetation gradients along river courses have been described in detail for many southeastern river systems; consequently, there is considerable agreement on the classification of riverine communities (Huffman and Forsythe 1981; Wharton et al. 1982). Nonalluvial wetlands, including freshwater marshes, shrub bogs, white cedar swamps, bayheads, and wet hammocks, have received considerably less study, and their vegetational relationships are more obscure.

Alluvial Wetlands

Coastal Plain rivers are divisible into three types: (1) Alluvial rivers have their headwaters in the mountains and piedmont and compose the major river systems of this region. Winter rainfalls on these watersheds, coupled with low evapotranspiration rates during these months, result in peak flows and flooding of backswamp areas during the winter and spring months. During the summer months, evapotranspiration limits flows, even during periods of heavy rain (Wharton et al. 1982). (2) Blackwater rivers arise on the Coastal Plain and may discharge into the sea or into other rivers.

These streams often drain shrub bogs and bay swamps and generally have less discharge than other river types. Flow in these rivers is highly variable, being highest when rainfall exceeds the water storage capacity of the bogs they drain. Flooding may occur during any season because of heavy storms. (3) Spring-fed streams originate where Tertiary limestone outcrops occur on the Coastal Plain. Such outcrops are especially common in northwest Florida and at scattered locations in Georgia and the Carolinas. Water in these streams is quite clear and usually alkaline. Because they are fed by groundwater, flows are quite constant.

Many southeastern Coastal Plain rivers are "underfitted"; that is, the modern floodplain is considerably smaller than the historical floodplain. The result is a terraced landscape in which higher terraces represent "paleo-floodplains" (Dury 1977; Wharton 1978). For example, the second terrace for many large rivers on the lower Coastal Plain of the Carolinas and Georgia corresponds to a floodplain formed during a fluvial period at the close of the Wisconsinan ice age (14,000 BP), when river discharge was as much as 18 times that of today (Dury 1977). Alluvial features and soils thus appear well above the present dominion of the river.

As rivers flow onto the Coastal Plain, their floodplains become more sinuous and complex. Often the concave bank of a meander loop will erode into relatively undisturbed floodplain forest, whereas sandbars will form on a convex bank, creating new habitat for colonization. Numerous oxbow lakes along many southeastern rivers are the remains of abandoned meanders. See Leopold and Wolman (1957), Leopold and associates (1964), and Dury (1977) for discussion of floodplain geomorphology.

When a river breaches its banks, sediment is deposited across the floodplain. Such deposition is greatest near the low-water channel, resulting in the formation of upraised levees. Levee formation varies with sediment load and river velocity. Blackwater rivers, for example, carry very little sediment and have poorly formed levees. The areas between the valley walls and natural levees are called flats or backswamps (Wharton et al. 1982). Sediments in flats are generally composed of fine silts and clays or alternating layers of coarse or fine sediment deposited during periods of high and low flood states, respectively. Topographic relief is barely perceptible in many flat areas, but small variations in surface topography may greatly influence plant distribution.

Nearly all of the classification schemes devised for floodplain plant communities are based on changes in inundation regime or the "anaerobic gradient" (Wharton et al. 1982). I use the zonal classification proposed by Huffman and Forsythe (1981) and adopted by the National Wetlands Technical Council. Any such classification is rather arbitrary, and most species occur in several zones.

Zone I: Permanent water courses. Herbs dominate this zone, which includes river channels, oxbow lakes, and other permanently inundated areas. Species composition is greatly affected by water flow. In areas of high water velocity, submerged aquatics, often with streamlined leaves, predominate. Where flow is more sluggish, plants with floating or emergent leaves are more common. If there is no current, floating mats of duckweed (*Lemna* and *Spirodela* species) and water fern (*Azolla caroliniana*) may occur.

Several non-native species have become important "weeds" in this zone. Alligator weed (*Alternanthera philoxeroides*) is a submerged aquatic that dominates and, in some cases, chokes many river channels throughout the Southeast. In southern Georgia and Florida, water hyacinth (*Eichornia crassipes*) forms dense floating mats that clog drainage in many slow-forming streams.

Zone II: River swamp forest. This zone, which includes the wettest flats and sloughs, is generally dominated by cypress-gum swamp forest. Although the silty to sandy soils in this zone may be occasionally exposed, they are saturated with water and typically anoxic.

Bald cypress (*Taxodium distichum*) is the most typical tree species of this habitat. Its buttressed boles and "knees" are a central feature of everyone's concept of a swamp forest (Fig. 11.10). Pond cypress (*Taxodium ascendens*) may replace bald cypress on sandy substrates or in impounded areas. The cypresses often grow in association with one of three species of gum or tupelo. Water tupelo (*Nyssa aquatica*) is most common where water is relatively deep and inundation periods are long. In shallower, less frequently inundated areas, swamp tupelo (*N. sylvatica* var. *biflora*) is abundant. Ogeechee tupelo (*N. ogeche*) is found along blackwater rivers and sloughs of Georgia and Florida. Eastern white cedar (*Chamaecyparis thyoides*) grows along some blackwater rivers, particularly on organic substrates underlain by sand (Wharton et al. 1982). Several other tree species that typically occur in less frequently inundated zones, such as laurel oak (*Quercus laurifolia*), red maple (*Acer rubrum*), American elm (*Ulmus americana*), and sweet gum (*Liquidambar styraciflua*), may become established here on

Figure 11.10. Old-growth zone II alluvial forest near Merchant's Mill Pond, North Carolina. The large tree at the center of the photo is an ancient specimen of Taxodium distichum. *Today trees of this stature are very rare anywhere on the Coastal Plain.*

the stumps of cypress or tupelo or on slightly elevated hummocks.

Trees of smaller stature are common in the understory of these forests. On mineral soils, water elm (*Planera aquatica*), pop ash (*Fraxinus caroliniana*), and pumpkin ash (*F. profunda*) are particularly common. On peaty soils, sweet bay (*Magnolia virginiana*), red bay (*Persea borbonia*), and titi (*Cyrilla racemiflora*) predominate. Shrubs around the bases of trees and on hummocks include swamp leucothoe (*Leucothoe racemosa*), fetterbush (*Lyonia lucida*), sweet pepperbush (*Clethra alnifolia*), and several *Ilex* species.

Zone III: Lower hardwood swamp forest. This is a relatively restricted zone on southeastern floodplains, transitional to the less frequently inundated backwater swamp forests of zone IV. Soils in this zone are saturated for 40%–50% of the year, but may become quite dry during the late summer (Leitman et al. 1981). Here, plants must be able to tolerate inundation in the early part of the growing season and drydown in the late summer. Zone III habitats include wet flats, low levees, and depressions in higher zones (Wharton et al. 1982).

Overcup oak (*Quercus lyrata*) and water hickory (*Carya aquatica*) are the most typical tree species in this zone. These trees remain dormant well into the spring, which may account in part for their tolerance of flooding in this part of the growing season. Common understory trees and shrubs include winterberry (*Ilex verticillata*), water locust (*Gleditsia aquatica*), Virginia "willow" (*Itea virginica*), American snowbell (*Styrax americanum*), and stiff dogwood (*Cornus foemina*). In disturbed areas, black willow (*Salix nigra*), may haw (*Crataegus aestivalis*), and water elm are often common.

Zone IV: Forests of backwaters and flats. This zone encompasses the greatest areas of most southeastern floodplains. These areas are inundated for most of the winter and spring, but only briefly during the growing season; thus, soils are saturated only 20%–30% of the year (Leitman et al. 1981). There is little variation in surface topography over large areas, aside from minibasins, hammocks, and scour channels, but this microtopographic relief contributes significantly to patterning of species. Impenetrable soil layers are common and may cause ponding of water.

Quercus laurifolia is the dominant tree species over much of this zone, with willow oak (*Q. phellos*), *Fraxinus pennsylvanica*, *Ulmus americana*, and *Liquidambar styraciflua* as common associates. Several shrubs and small trees are indicative of this inundation zone, including possum haw (*Ilex decidua*), *Crataegus* species, *Viburnum obovatum*, ironwood (*Carpinus caroliniana*), and several *Rhododendron* species.

Dwarf palm (*Sabal minor*) often forms dense thickets here. Vine diversity is quite high in this community and includes such taxa as poison ivy (*Rhus radicans*), greenbrier (*Smilax* species), supplejack (*Berchemia scandens*), and cross vine (*Anisostichus capriolata*).

Zone V: Transition to upland. Zone V comprises the highest locations of the active floodplain and includes natural levees, higher terraces and flats, and Pleistocene ridges and dunes (Wharton et al. 1982). Inundation is infrequent here, and soils are saturated less than 15% of the year, and not at all during the growing season. Zone V soils are often more sandy and less fertile than those of lower zones.

On flats and old levee ridges, basket oak (*Quercus michauxii*) and cherry bark oak (*Quercus falcata* var. *pagodaefolia*) are usually dominant. Occasional individuals of water oak (*Quercus nigra*) and *Quercus virginiana* become established here on local high spots, such as logs, stumps, or mounds of soil created by tree falls. Several hickory species grow into the canopy in this area. Spruce pine (*Pinus glabra*), one of the few truly shade-tolerant pines, is typically found in wetter portions of this zone and sometimes into zone IV. Loblolly pine is common in dryer areas. Understory trees and shrubs include American holly (*Ilex opaca*), pawpaw (*Asimina triloba*), and spicebush (*Lindera benzoin*). In Florida and across the Gulf states, *Sabal palmetto* and *Serenoa repens* are common in this zone. Zone V is bordered by a variety of upland forest types; however, beech and magnolia forests frequently mark the upper edge of potential inundation. On sandy soils, xeric hammock vegetation, dominated by live oak, may form this border.

Even where elevation increases uniformly away from a single channel, the distribution of vegetation is often more complex than the simple classification suggested earlier. Species such as bald cypress and pop ash that are characteristic of frequently inundated zones are often growing at the same elevation as inundation-intolerant species, such as live oak and pignut hickory. On the scale of just a few meters, seeps or localized clay lenses may create wet soil environments at elevations that are not frequently inundated, and small hummocks, tip-up mounds, and tree stumps provide very localized "dry" habitats in zones that are frequently inundated.

Community dynamics and succession. Fire is not an important disturbance factor in alluvial communities. The most important disturbances are associated with flooding and the geomorphological processes described earlier.

On sandbars and newly formed levees, *Salix nigra* is an important pioneer. Once stabilized, these habitats may support *Populus deltoides* and *Acer saccharinum*. Eventually these comparatively short-lived trees are replaced by zone IV species, such as *Quercus laurifolia* and *Carya aquatica* (Wharton et al. 1982).

Succession following oxbow-lake formation fits the classic model for a hydrarch sere (Clements 1916; Weaver and Clements 1938). Vegetation invades the oxbow in a series of zones that gradually converge as the lake fills in with sediment and organic debris. In the deepest water, submerged aquatics such as *Ceratophyllum, Myriophyllum, Cabomba, Najas,* and *Potamogeton* dominate. Toward shore, floating-leaved aquatics, such as *Brasenia, Castalia, Nelumbo,* and *Numphaea,* are important. In some areas, floating aquatics such as *Alternanthera philoxeroides, Jussiaea grandiflora,* and *Eichornia crassipes* may form a floating mat. Penfound and Earle (1948) observed that such mats of *Eichornia* could form a floating prairie, with a diverse assemblage of aquatic herbs, in less than 25 yr. Along the shore, cattail (*Typha latifolia*), *Salix* species, and *Cephalanthus occidentalis* form a shrubby zone, often with emergent *Taxodium ascendens, Planera aquatica,* and *Persea* species. Occasional alluvial flooding may subsidize the sedimentation process, and oxbow lakes may fill in a few hundred years (Penfound 1952). Nonetheless, the organic soils and poor drainage associated with these old oxbows often result in an assemblage of species distinct from the adjacent backswamp.

Another natural disturbance factor along southeastern rivers is the beaver (*Castor canadensis*). Beavers were nearly extirpated from southeastern rivers by 1900 owing to habitat destruction and trapping, but protection and restocking programs have resulted in a rebound for the beaver population. In the 10 southeastern states, nearly 200,000 ha of floodplain are impounded by beavers (Hill 1976). By damming sluggish streams, beavers inundate portions of a floodplain, killing many plants in mesic zones not adapted to prolonged anoxia. They then girdle and cut trees along the water, creating open habitat that favors the invasion of early successional shrubs and small trees, such as *Alnus serrulata, Cephalanthus occidentalis,* and willows (Hair et al. 1979). It is these shrubs that are the mainstay of the beaver diet.

The history of human disturbance in wetlands is long and extensive. Floodplains were among the first habitats to be farmed by the American Indian, and because of the navigability of the rivers, alluvial forests were among the first southeastern forests to be heavily cut (Pinchot and Ashe 1897; Cowdrey 1983). Thus, most forested wetlands in the Southeast are in some stage of succession from human disturbance. Perhaps the most far-reaching and least understood of human perturbations to alluvial wetlands have been alterations in hydrology. Flows along most of the major river systems of the Southeast have been altered by dams, control structures, and artificial levees. These have greatly altered inundation regimes, in terms of both duration and variability. Thus, the primary determinant of vegetation composition for many successional bottomlands has changed, and the course of succession has undoubtedly been altered.

Nonalluvial Wetlands

Graminoid-dominated wetlands. Shallow marshes dominated by a variety of grass, sedge, and rush species are common across the lower Coastal Plain on a diverse assemblage of sites. Dominant genera include *Panicum, Muhlenbergia, Carex, Rhynchospora, Cladium, Scirpus*, and *Juncus* (Penfound 1952). *Sphagnum* is represented by a range of species, but is rarely dominant in these habitats.

Graminoid-dominated wetlands are most often early seral stages following disturbance of forested wetlands. For example, such communities are common following cutting or burning of bay forests (Wells 1942) or wet slash pine flatwoods (Penfound and Watkins 1937). Severe fires in peatlands also create grass-sedge-dominated communities (Kologiski 1977; Hamilton 1984) and are clearly responsible for maintenance of the "prairies" of the Okefenokee Swamp (Weight and Wright 1932; Cypert 1961; Duever and Riopelle 1983).

The Everglades in south Florida are certainly the best known and most extensive graminoid-dominated wetlands in the Southeast. A massive water track flowing south from Lake Okeechobee creates an extensive wetland dominated over large areas by a single sedge species, *Cladium jamaicense*. See Davis (1943) for a detailed description of Everglades vegetation patterns. Succession to forested wetland is prevented by frequent fire, and spatial variability in fire behavior has long been recognized as a major determinant of landscape patterns (Harper 1911; Wade et al. 1980).

Pocosins. Shrub-dominated wetlands or pocosins are common features of the Atlantic limb of the Coastal Plain (Richardson et al. 1981; Sharitz and Gibbons 1982). These peatlands are dominated by a dense, nearly impenetrable cover of evergreen and deciduous shrubs, with scattered emergent trees (Fig. 11.11). Four types of pocosins are recognized, based on physiographic location (Woodwell 1956; Christensen et al. 1981; Ash et al. 1983).

Pocosin complexes may extend over thousands of hectares in flat interstream "uplands." These bogs began forming in clogged stream channels approximately 6000–8000 yr ago, coincident with a shift toward cooler climates and the rise of sea level to near its present elevation. Beginning as primary mires, these bog complexes "grew," owing to slow decomposition rates and paludification, into tertiary mires or domed bogs (Whitehead 1972; Daniel 1981; Ashe et al 1983). The plant communities of these peatlands are biogeochemically separated from the mineral soil substrate and receive all nutrient inputs from rain and dryfall (i.e., they are ombrotrophic) (Wilbur and Christensen 1983). It is these raised bogs that gave rise to the term "pocosin," an Algonquin Indian word meaning "swamp on a hill" (Tooker 1899).

Pocosins also occur in many of the Carolina bays (Sharitz and Gibbons 1982). These bays are quite varied vegetationally and appear in many cases to be undergoing a typical hydrarch succession from lake to bog to forest.

Ridge and bay pocosins are found in swales of relict dune fields on the lower Coastal Plain. Pocosins may occur near seeps, springs, and the margins of slow-flowing streams, particularly in sandy areas. Such seeps are often associated with clay horizons. Although bay, ridge and bay, and seep types are not ombrotrophic, the water they receive is nonetheless nutrient-poor, and all pocosin types are quite nutrient-limited, especially with respect to phosphorus (Woodwell 1958; Simms 1983, 1985; Wilbur 1985).

Kologiski (1977) divided pocosin vegetation into two classes: conifer-hardwood and pine-ericalean. Conifer-hardwood communities generally occur on mineral soils or shallow peats and are successional to other forest types such as flatwoods or swamp forest (Kologiski 1977; Christensen et al. 1981). Within the pine-ericalean class, Kologiski delimited three types, as briefly described in the following paragraphs.

The *Pinus serotina–Cyrilla racemiflora–Zenobia pulverulenta* type corresponds to Wells's (1946) low pocosin (Table 11.6). These communities are typical of the most nutrient-limited sites with the deepest peats, such as occur at the very centers of bog complexes. *Sphagnum* is not the primary contributor to peat formation in most of these bogs, but may account for as much as 50% of total cover in some places. In bog centers, *Sphagnum magellanicum* and *S. bartlettianum* form extensive hummocks, and *S.*

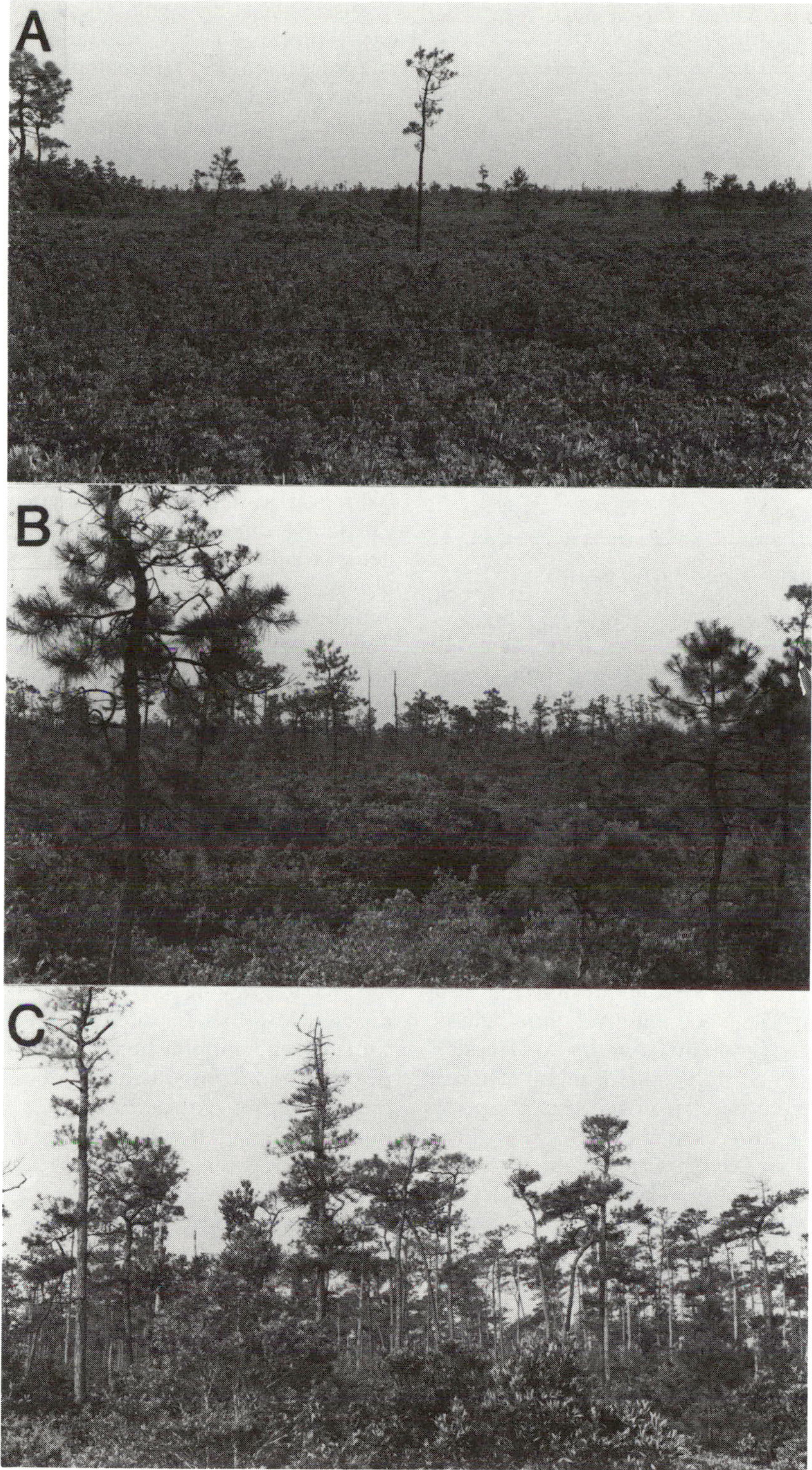

Figure 11.11. Top to bottom: Low-, medium-, and high-stature pocosin communities from the Green Swamp, North Carolina. Left unburned, high pocosin forest will succeed to bay forest, but low pocosin communities show little evidence of successional change.

Table 11.6. *Percentage cover of shrub species from pocosins in Dare County, North Carolina*

Species	Cover (%)
Pinus serotina	19
Ilex glabra	56
Lyonia lucida	26
Chamaedaphne calyculata	17
Zenobia pulverulenta	3
Persea borbonia	2
Gordonia lasianthus	2
Kalmia angustifolia var. *carolina*	<1
Acer rubrum	<1
Clethra alnifolia	<1
Myrica cerifera	<1
Cyrilla racemiflora	<1
Woodwardia virginica	2
Smilax laurifolia	3
Sphagnum species	6
Carex walteriana	<1
Andropogon species	<1

Source: Data from Laney and Noffsinger (1985).

cuspidatum is common in depressions. As shrub cover increases in more productive sites, the *Sphagnum* abundance decreases to scattered clumps.

The *Pinus serotina – Gordonia lasianthus – Lyonia lucida* community type is characteristic of elevated areas or "islands" within the preceding type. These somewhat more productive areas are about 5 – 20 m across and have a very regular distribution through the low pocosin matrix. Their history is uncertain, but they appear to be formed by accumulation of litter and *Sphagnum* around tree boles and stumps.

The *Pinus serotina – Cyrilla racemiflora – Lyonia lucida* community type is frequently referred to as "high pocosin." This community type occurs on shallower peats (generally 0.5 m deep). Trees are 10 – 15 m tall, and shrubs often 5 m tall. There is virtually complete overlap in the lists of species found in these three community types, (*Pinus serotina* with either *Zenobia*, *Gordonia*, or *Lyonia*), but they differ considerably in relative abundance. Species evenness is highest in the low pocosin communities and decreases steadily as productivity increases (Woodwell 1956; Christensen 1979a).

The successional status of pocosins varies with site conditions. On shallow peat soils and in some of the Carolina bays, white cedar and pond cypress are invading; if left unburned, these communities will probably succeed to a variety of swamp forest types (Kologiski 1977; Christensen et al. 1981). Most of these sites were heavily cut over during the nineteenth century and have been frequently burned since, thus maintaining pocosin (Lilly 1981). On deeper peats and truly ombrotrophic sites, pocosins show no sign of invasion by swamp forest species (Christensen et al. 1981). This is consonant with Otte's proposal (1981) that these ecosystems are end points for a long-term paludification process beginning with a grass-sedge bog that is subsequently invaded by swamp forest. Peat accumulation separates the ecosystem from the mineral substrate and drives the succession to pocosin.

The effects of fire are highly variable in these peatland ecosystems. Wilbur and Christensen (1983) found that levels of most nutrients, including phosphorus, were increased in the first post-fire year. However, in most cases the increases lasted only a single growing season. Furthermore, they found that there was considerable spatial variance in enrichment, probably associated with small-scale variations in fire behavior and ash deposition. Thus, phosphorus might be enriched 200-fold at one location and unchanged 10 m away. Christensen and associates (1981) noted that analogous variations occur between fires. Hot summer fires may burn deep into the peat, whereas spring or autumn burns may burn only the surface litter. Burning causes severe drying of surface peat layers, rendering the surface nonwettable for several years (Wilbur 1985). Thus, growing conditions in burned pocosins after the first year may actually be poorer than in unburned areas.

The vegetation response to moderate-intensity spring fires is quite rapid and almost entirely due to vegetative reproduction. All pocosin shrubs sprout vigorously; by the end of the first growing season, the community recovers nearly half of its pre-fire biomass and has a LAI of 2.2, compared with a pre-fire index of 3.9. Production is even greater in the second growing season, but in years 3 and 4 it drops below that for unburned pocosins (N. L. Christensen, unpublished data). This decline appears to be a consequence of decreased nutrient capital coupled with unfavorable moisture conditions. By year 10, however, the peat surface recovers, a thick litter mat develops, and production returns to pre-fire levels.

Species responses to burning are not uniform. Wells (1946) and Woodwell (1956) noted that *Zenobia pulverulenta* tends to be more common on recently burned sites and is subsequently overtopped by *Lyonia lucida* and *Cyrilla racemiflora*. Christensen (unpublished data) observed that production of *Zenobia*, *Sorbus arbutifolia*, and *Ilex glabra* was enhanced by burning and that these species flowered in pocosins only after fire. These species account for large portions of the first- and second-year increases in production. Production of *Cyrilla racemiflora*, the pre-fire dominant, was actually depressed by burning.

Christensen and associates (1981) proposed that species richness in these shrub bogs is as much dependent on the variability created by the fire as on the release from competition. They noted that although reproduction is primarily from vegetative sprouts, the pre-fire composition in permanent plots frequently was not related to post-fire species success. They proposed that a high post-fire variance in limiting nutrients exists and that species are differentially successful along this nutrient gradient.

The peat soils of these ecosystems create the possibility for very intense "ground" fires. During the summer months, evapotranspiration dries out the surface peat layers, creating very flammable conditions. Fires under these conditions may burn several decimeters of peat, killing all subterranean vegetative structures. Succession on such sites is by invasion of herbaceous plants, forming a grass-sedge bog, often with abundant *Woodwardia virginica* (Kologiski 1977). Where peat has burned to mineral soil, shallow lakes may be formed when the water table rises. Such peat-burn lakes are a common feature of the lower terraces of the North Carolina Coastal Plain (Whitehead 1972).

Atlantic white cedar swamp forest. Atlantic white cedar (*Chamaecyparis thyoides*) ranges from New England to northern Florida and Alabama (Clewell and Ward 1985; Belling 1986) and is found in a variety of wetland habitats (Moore and Carter 1985). However, it forms extensive stands in only a few scattered areas, including the New Jersey pine barrens, lower terraces of the North Carolina and Virginia Coastal Plains, and northern Florida (Korstian and Brush 1931; Little 1950; Frost 1986). White-cedar-dominated swamp forests are most frequently associated with deep peats, often over sandy substrates. Very little is known about the factors restricting the distribution of this community type, and Frost (1986) cited historical evidence that such forests may have been more widespread prior to human development of the Coastal Plain.

In the Carolinas and Virginia, white cedar forms dense, even-aged populations, often with a very closed canopy (Fig. 11.12). Compositional data for a typical white cedar swamp forest in North Carolina are shown in Table 11.7. In the Middle Atlantic states, white cedar may account for more than 95% of canopy cover (Korstian and Brush 1931; Buell and Cain 1943), but in the Gulf states, white cedar shares dominance with a wider diversity of tree species and rarely accounts for more than 50% of cover or relative stem density (Harper 1926; Korstian and Brush 1931; Dunn et al. 1985). Shrub cover in the understory may exceed 80% and includes such species as *Lyonia lucida, Ilex coriacea, I. glabra, Clethra alnifolia*, and *Persea borbonia*. *Sphagnum* and *Woodwardia virginica* are the two most important constituents of the herb layer (Laney and Noffsinger 1985). The compositions of the white cedar swamps of New Jersey and the Delmarva peninsula are quite similar (Little 1950; McCormick 1979; Hyll and Whigham 1985).

*Figure 11.12. White cedar (*Chamaecyparis thyoides*) swamp forest on deep peat soils in the Dismal Swamp, North Carolina. This photo was taken at the forest edge; shrub cover beneath the dense canopy is much more sparse.*

Although white cedar is comparatively shade-tolerant, it does not establish in the dark shade of mature stands (Korstian and Brush 1931). Successful seedling establishment occurs when wet conditions follow intense crown-killing fires (Korstian 1924, Buell and Cain 1943). Korstian (1924) found that the top layer of peat in mature white cedar stands contained enormous quantities of viable seeds that could yield very high seedling densities ($>4 \times 10^6$ ha^{-1}); 8-yr-old stands with more than 60,000 trees per hectare are not uncommon. Buell and Cain (1943) documented later thinning in such stands resulting in white cedar stem densities of 2500 and 1700 ha^{-1} in 35- and 85-yr-old stands, respectively. They noted that nearly 70% of the standing stems in stands older than 80 yr were dead, and as stands matured, a considerable quantity of dead fuel accumulated. The broken canopy of these older stands permits invasion of many bay forest species, including *Ilex cassine, Persea borbonia, Gordonia lasianthus*, and *Magnolia virginiana*; in the absence of fire, succession would lead to dominance by these species. However, the accumulated dead fuel of the white cedar increases the flammability of the older stands, increasing the likelihood of crown fire and returning the system to its initial

Table 11.7. *Cover of overstory and understory species in white cedar swamp forests of Dare County, North Carolina*

Species	Cover (%)
Overstory (>5 m high)	
Chamaecyparis thyoides	64
Nyssa sylvatica var. *biflora*	21
Persea borbonia	7
Acer rubrum	7
Magnolia virginiana	4
Pinus taeda	4
Pinus serotina	2
Gordonia lasianthus	2
Understory (<5 m high)	
Lyonia lucida	43
Ilex coriacea	42
Persea borbonia	24
Ilex glabra	17
Smilax laurifolia	15
Sphagnum species	11
Woodwardia virginica	12
Nyssa sylvatica var. *biflora*	7
Clethra alnifolia	6
Amelanchier arborea	2
Myrica species	1
Magnolia virginiana	<1
Gordonia lasianthus	<1

Source: Data from Laney and Noffsinger (1985).

condition. Frost (1986) noted that the present range of white cedar swamp forest may be restricted owing to alteration of fire regimes. Short fire return intervals tend to favor pocosin vegetation on these sites, whereas fire exclusion may result in replacement of white cedar by bay forest species.

Bay forests, bayheads, and baygalls. The term "bay" is widely used in the Southeast to refer to a number of evergreen trees, including sweet bay (*Magnolia virginiana*), red bay (*Persea borbonia*), and loblolly bay (*Gordonia lasianthus*). Communities dominated by these species are referred to as bay forests in the Carolinas and Georgia (Kologiski 1977; McCaffrey and Hamilton 1984; Gresham and Lipscomb 1985) and bayheads in Florida (Gano 1917; Laessle 1942; Monk 1966b; Abrahamson et al. 1984). Related wetlands in east Texas, in which *Gordonia* is absent, are referred to as baygalls (Marks and Harcombe 1981), a term used widely throughout the Southeast during the nineteenth century (Wright and Wright 1932). These ecosystems are often associated with shallow depressions or poorly-drained interstream areas in Carolina bays.

The species found in association with the bay trees listed earlier vary regionally. In the Carolinas they include *Cyrilla racemiflora, Ilex cassine* var. *myrtifolia, Chamaecyparis thyoides, Pinus serotina, Nyssa sylvatica* var. *bilfora, Acer rubrum,* and occasional individuals of *Taxodium distichum* (Table 11.8). Kologiski (1977) differentiated between evergreen and deciduous bay forests in the Green Swamp. The latter community occurs on shallow peats and mineral soils and is probably successional to wet deciduous forest. The shrub stratum of bay forests may be quite diverse, including such species as *Cyrilla racemiflora, Lyonia lucida, Ilex coriacea, I. glabra, Myrica heterophylla, Vaccinium atrococcum, V. corymbosum,* and *Zenobia pulverulenta*. *Woodwardia virginica* is the most abundant herb (Kologiski 1977; Gresham and Lipscomb 1985). In Florida bayheads, *Pinus elliottii* is found with *P. serotina*, and cinnamon fern (*Osmunda cinnamomea*) is an important understory herb (Laessle 1942; Monk 1966b; Abrahamson et al. 1984). Vines, including *Smilax* species, *Gelsemium sempervirens,* and *Parthenocissus quinquifolia,* are important components of bay forest communities across the Coastal Plain. The overstory dominants of the Texas baygalls include *Quercus laurifolia, Nyssa sylvatica, Magnolia virginiana, Ilex vomitoria,* and *Acer rubrum,* with a shrub understory that includes *Cyrilla racemiflora* and *Ilex coriacea*.

Buell and Cain (1943) and Penfound (1952) considered bay forests as the climax of a succession from pocosin and white cedar swamp. Monk (1966b) suggested that these communities might also arise by means of paludification in certain flatwoods types, particularly those dominated by *Pinus serotina*, or by means of hydrarch succession from open-water ponds to bogs to bayheads (Laessle 1942; Davis 1946; Hamilton 1984). Kologiski (1977) and Hamilton (1984) noted that selective cutting of cypress, white cedar, and swamp hardwoods has favored bay species and increased the areal extent of bay forest.

The effects of fire in bay forest are dependent on fire behavior and post-fire hydrology (Wells 1946; Kologiski 1977; Hamilton 1984). Shallow peat burns give rise to pocosin or white cedar stands, which in the absence of further disturbance presumably succeed back to bay forest. Deep peat burns with a high post-burn water table may produce sedge bogs or "prairies" dominated by *Carex walteriana* and *Woodwardia virginica*. These areas are subject to frequent, low-intensity fires; thus, tree invasion may require decades (Cypert 1961, 1972; Kologiski 1977). A deep peat fire when the water table is low can favor the establishment of deciduous species, such as *Nyssa sylvatica* and *Acer rubrum* (Kologiski 1977).

Cypress domes, heads, and islands. Cypress domes (also called cypress heads or wet hammocks) are

Table 11.8. *Density (for overstory and shrub-layer species) and basal area (BA, for overstory species) of bay forest stands in South Carolina*

Species	Density (ha-1)	BA (m^2 ha^{-1})
Overstory (>3 cm dbh)		
Gordonia lasianthus	322	4.64
Persea borbonia	343	2.15
Nyssa sylvatica var. *biflora*	211	3.91
Pinus taeda	134	5.32
Pinus serotina	134	4.93
Liquidambar styraciflua	102	0.77
Vacinium species	130	0.22
Myrica cerifera	156	0.46
Cyrilla racemiflora	72	0.21
Magnolia virginiana	75	0.26
Totals	1912	24.81
Shrub layer (>1 m high, <3 cm dbh)		
Lyonia lucida	26,915	
Clethra alnifolia	6,789	
Vaccinium species	2,709	
Ilex glabra	2,409	
Persea borbonia	1,324	
Smilax species	1,118	
Sorbus arbutifolia	684	
Myrica cerifera	298	
Ilex cassine	675	
Gordonia lasianthus	553	
Total	46,278	

Source: Data from Gresham and Lipscomb (1985).

small, forested wetlands that occur in poorly drained depressions throughout the southeastern Coastal Plain (Kurz and Wagner 1953; Monk and Brown 1965; Marois and Ewel 1983). Similar communities also develop on higher points of land or on accumulations of peat in the midst of the grass-sedge wetlands or on "prairies" of the large southern swamp complexes, such as the Everglades, Big Cypress Swamp, and Okefenokee Swamp (Wright and Wright 1932; Davis 1943; McCaffrey and Hamilton 1984). In the Okefenokee Swamp, such "islands" are referred to as "houses."

Compared with bayheads, cypress domes usually have deeper water and longer periods of inundation (Monk 1966b). Monk and Brown (1965) found that they also had somewhat higher soil pH values (4.1–4.6) and cation concentrations. However, compared with floodplain swamps, these forests are profoundly nutrient-limited. Brown (1981), for example, found that phosphorus inputs into cypress domes amounted to only 0.11 g m^{-2} yr^{-1}, compared with 1625 g m^{-2} yr^{-1} in a flood plain cypress swamp (Schlesinger 1978b).

Compositional data for 17 cypress domes in the Okefenokee Swamp appear in Table 11.9. These data are similar to those for domes in the Carolinas (Kologiski 1977), Florida (Monk and Brown 1965; Marois and Ewel 1983), and the Gulf states (Penfound 1952). Whereas *Taxodium distichum* dominates alluvial floodplain forests, *T. ascendens* is the canopy dominant in the domes (Neufeld 1983). Monk (1968) ascribed their distribution to differences in tolerance to low pH, but Brown (1981) suggested that *T. ascendens* is considerably more drought-tolerant owing to its reduced leaf area and sunken stomates. Neufeld (1984) found that mid-day xylem pressure potentials were consistently lower in *T. distichum* than in *T. ascendens* grown in equivalent moisture regimes. Because cypress domes may dry down periodically, drought tolerance may be important. The reduced leaf area of *T. ascendens* could also be an adaptation to nutrient stress (Brown 1981).

The understory of cypress domes is dominated by a diverse mixture of shrubs, many of which are evergreen sclerophyllous and are found in other nutrient-limited wetlands such as pocosins. Schlesinger (1978a) found that the diversity (richness and evenness) of these understory shrub communities was enhanced by relatively frequent light surface fires.

Spanish moss (*Tillandsia usneoides*) is a ubiquitous epiphyte in this community, often having a biomass in excess of that for total herbs in many

Table 11.9. *Density and basal area (BA) for species in the tree and shrub strata in 18 cypress forest stands in the Okefenokee Swamp*

Species	Density (stems ha^{-1})	BA (m^2 ha^{-1})
Trees (>4cm dbh)		
Taxodium ascendens	1467.0	69.4
Ilex cassine	189.3	1.0
Nyssa sylvatica var. *biflora*	134.1	0.85
Cyrilla racemiflora	59.6	0.16
Magnolia virginiana	43.0	0.31
Lyonia lucida	9.0	0.01
Persea borbonia	11.7	0.02
Clethra alnifolia	1.5	0.004
Vaccinium species	0.8	0.001
Totals	1916.0	71.90
Shrub stratum (>1 high, <4 cm dbh)		
Itea virginica	6840	56.3
Lyonia lucida	7840	38.5
Leucothoe racemosa	3340	35.8
Clethra alnifolia	2080	29.3
Ilex cassine	1180	17.8
Smilax walteri	1620	14.3
Cyrilla racemiflora	1290	13.0
Pieris phillyreifolia	760	15.8
Taxodium ascendens	380	10.3
Decodon verticillatus	260	5.8
Nyssa sylvatica var. *biflora*	190	3.0
Rhus radicans	270	1.3
Vaccinium species	80	1.0
Magnolia virginiana	40	0.8
Ilex glabra	70	0.5
Cephalanthus occidentalis	20	0.8
Ilex coriacea	10	0.5
Smilax laurifolia	10	0.3
Gordonia lasianthus	10	0.3
Acer rubrum	10	0.3

Source: Data from Schlesinger (1978a).

upland forests (Schlesinger 1978b). Schlesinger noted that this species extracts considerable nutrients from incident rainfall and may play an important role in nutrient cycling in these communities. Other common epiphytes include the resurrection fern (*Polypodium polypodioides*) and, in central and southern Florida, other species of *Tillandsia*.

The successional fate of cypress domes appears to be dependent on hydrology. In areas that remain permanently inundated, cypress may form stable, dominant populations. If, however, the area receives nutrient subsidy, such as sediment from a river, cypress may be replaced by broad-leaved deciduous species such as *Acer rubrum*, *Quercus* species, and *Fraxinus caroliniana*. On drier sites, relatively shade-tolerant broad-leaved evergreen species, such as *Pesea* species, *Magnolia virginiana*, and *Gordonia lasianthus*, replace the less shade-tolerant cypress, producing a bayhead forest (Putnam et al. 1960; Monk 1968). This is consistent with the observations of Marois and Ewel (1983) that artificially drained cypress domes have an increased importance of bayhead species.

Cypress seeds do not germinate in flooded soils (Matoon 1916; Demaree 1932, Dickson and Broyer 1972), and it is therefore likely that cypress stands become established on unoccupied sites during periods of drought (Kologiski 1977; Schlesinger 1978a)—conditions that exist following crown-killing fires that do not consume the surface peat layer (Hamilton 1984). Ewel and Mitsch (1978) found that cypress trees survived surface fires better than competing hardwoods. Thus, such fires may favor long-term cypress dominance on sites that might otherwise succeed to bayhead forest. Gunderson (1984) found that *Salix caroliniana* had invaded recently burned cypress stands and that cypress was regenerating successfully from seeds and sprouts.

Recovery from severe fires that burn into the peat may be very slow. Seed dispersal is very limited, particularly in stagnant water (Hamilton 1984). Younger cypress trees (<200 yr) may sprout if the fire is not too severe; however, older trees produce few sprouts. Consequently, where severe fires have burned old-growth cypress stands, there is little sign that the communities are returning to their predisturbance composition (Cypert 1961, 1973; Hamilton 1984).

Succession leading to the formation of cypress islands in the midst of extensive grass-sedge wetlands (such as occur in the Okefenokee Swamp) is illustrated in Fig. 11.13. Peat masses or "batteries" break loose from the substrate and float to the water surface, forming a relatively dry habitat for colonization (Cypert 1972; Spackman et al. 1976; Duever and Riopelle 1983). Batteries may then be invaded by the shrub species *Cephalanthus occidentalis*, *Lyonia lucida*, and *Ilex cassine*, which tend to stabilize this habitat and increase the rate of peat accumulation, thus favoring subsequent invasion of bay species and cypress. The resulting community is similar in structure and composition to cypress domes found in closed depressions. The resulting islands ("houses") are often elliptical, perhaps streamlined by water flow (Wright and Wright 1932). Because they are raised above the surrounding minerotrophic water, they may be ombrotrophic and nutrient-limited.

Much of the cypress in the Southeast has been lost to logging (Pinchot and Ashe 1897; Izlar 1984). Stands that have been high-graded for cypress may

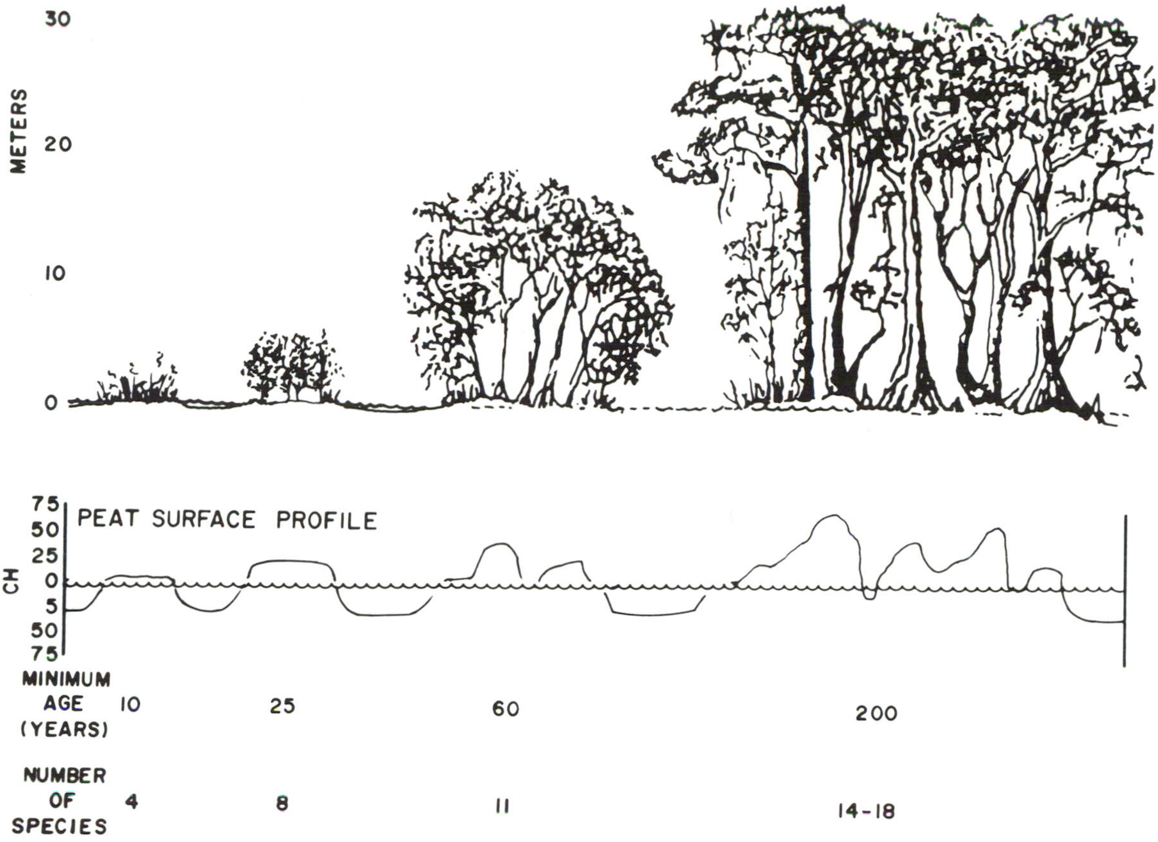

Figure 11.13. Patterns of succession on peat batteries in the Okefenokee Swamp, Georgia (from Deuver and Riopelle 1983).

succeed to bay forest (Kologiski 1977; Hamilton 1984). However, where cypress was the primary canopy constituent, logging opened the canopy sufficiently to allow successful seedling and sprout establishment (Gunderson 1984; Hamilton 1984). In many locations, the combination of drainage and logging has created ideal conditions for intense fires. The combination of fire and logging has a devastating effect on community composition, and most sites so disturbed in the past 50 yr show no sign of returning to their predisturbance composition (Cypert 1961, 1973; Kologiski 1977; Gunderson 1984; Hamilton 1984). The varied effects of fire and logging on community succession in these wetlands are illustrated in Fig. 11.14.

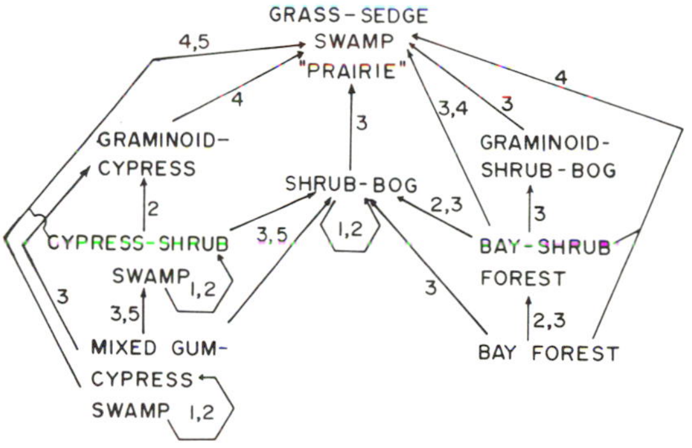

Figure 11.14. Patterns of succession among various wetland communities in the Okefenokee Swamp, with various intensities of fire and logging: 1, light fire; 2, moderate fire; 3, severe fire; 4, very severe or frequent fire; 5, logging followed by fire (after Hamilton 1984).

Because these ecosystems are often hydrologically isolated from open surface water and hydrologic residence times are relatively long, they have received considerable study as potential depositories for wastewater (Dierberg and Brezonik 1984). Such nutrient enrichment results in an immediate response among the herbs and floating aquatics: Ewel (1984) found that growth of duckweed (*Lemna* species) was greatly enhanced, as was dry-matter production of other herbs. However, establishment and growth of *Nyssa sylvatica* var. *biflora* and *Taxodium ascendens* seedlings are inhibited by wastewater (Deghi 1984). Whereas net photosynthesis was increased in cypress growing in sewage-treated domes (Brown et al. 1984), there was no sign of increased tree growth even after 6 yr (Straub 1984). The long-term successional effects of such treatments have yet to be determined.

Marois and Ewel (1983) investigated the impact of artificial drainage on cypress domes and found that cypress tree growth was enhanced by drainage in some, but not all, cases. However, they warned that changes in soil factors and vegetation composition in drained domes inhibit cypress regeneration.

MARITIME VEGETATION GRADIENTS

Composition and Structure

In addition to river deltas, the southeastern coast is bordered by long ribbons of barrier or sea islands that shelter estuaries, lagoons, and sounds. Over very short distances, there are striking gradients of salinity, soil, and climate, and therefore steep gradients in community composition, structure, and physiognomy. The spatial arrangement of geomorphological features and plant communities along the strand varies considerably from location to location. However, the relationship of communities to underlying environmental factors and the composition of those communities are remarkably constant throughout the Southeast (Shaler 1885; Kearney 1900; Coker 1905; Lewis 1917; Penfound and O'Neill 1934; Kurz 1942; Penfound 1952; Brown 1959; Adams 1963; Au 1974; Godfrey and Godfrey 1976). I shall therefore organize this discussion around a hypothetical transect from the sea strand on a typical barrier island, across that island, into the sound, and onto the adjacent mainland.

Strand vegetation consists of an assemblage of short-lived plants whose spatial distribution shifts from season to season and year to year. Many of these species are salt-tolerant and have life-history characteristics that allow them to invade suitable habitat when it becomes available.

The principal grass that stabilizes dunes is sea oats, *Uniola paniculata* (Fig. 11.15). Once established, this grass produces numerous adventitious roots from the culm. Continued apical growth and lateral rhizome and root production allow the plant to stabilize the sand that accumulates at its base. In the absence of inundation or overwash, a dune is formed. *Iva imbricata*, *Physalis maritima*, *Croton punctatus*, and *Euphorbia polygonifolia* grow among the sea-oat culms.

In the lee of the foredune, species richness increases. Several grass species dominate this area,

Figure 11.15. Foredune community dominated by Uniola paniculata *on Shackleford Bank, North Carolina.*

including *Spartina patens*, *Andropogon scoparius* var. *littoralis*, *Distichlis spicata*, *Panicum* species, and *Cenchrus tribuloides*. Among the most common herbs are *Hydrocotyle bonariensis*, *Oenothera humifusa*, and *Opuntia compressa*. Boyce (1954) noted that many species commonly found in old fields and ruderal habitats in nonmaritime locations also occur here; these include *Andropogon virginicus*, *Erigeron canadensis*, *Heterotheca subaxillaris*, *Solidago* species, and *Aster* species.

Oosting and Billings (1942), Oosting (1945), and Boyce (1954) demonstrated experimentally that dune ridge species, such as *Uniola paniculata*, were much more tolerant of salt aerosol impaction than species typically found in dune slack areas.

Freshwater habitats occur at the heads of tidal creeks and in areas where peat or clay lenses result in perched water tables. Taxa such as *Fimbristylis* species, *Cladium jamaicense*, *Juncus* species, and *Typha latifolia* are dominant. However, these communities often include a diverse assemblage of herbs (Odum et al. 1984). Such habitats are important resources for barrier-island animals (Engels 1952; Rubenstein 1981).

Near the foredunes, stunted salt-pruned shrubs may be found in the most protected swales. The pattern of establishment is closely tied to salt-spray tolerance. *Iva imbricata* or *Baccharis halimifolia* are often early invaders. Other species, such as *Ilex*

Figure 11.16. Maritime scrub vegetation in dune slack region of Shackleford Bank. Note the streamlining of the vegetation as a consequence of salt-spray pruning. This particular clump of shrubs includes four species: Baccharis halimifolia, Ilex vomitoria, Juniperus virginiana, *and* Pinus taeda. *The relatively salt-tolerant* Baccharis *probably established first, creating a refuge from salt spray and allowing invasion of the other species. The dominant grass in the surrounding area is* Andropogon scoparius *var.* littoralis.

Table 11.10. *Basal area (BA) of woody species in an old-growth maritime forest on Fort Smith Island, North Carolina*

Species	BA (m^2 ha^{-1})
Quercus virginiana	25.19
Persea borbonia	3.45
Carpinus caroliniana	2.23
Juniperus virginiana	1.78
Pinus taeda	1.50
Ilex vomitoria	1.30
Osmanthus americana	1.09
Ilex opaca	0.85
Sabal palmetto	0.69
Cornus florida	0.65
Prunus caroliniana	0.65
Morus rubra	0.56
Quercus nigra	<0.50
Callicarpa americana	<0.50
Nyssa sylvatica	<0.50
Oxydendrum arboreum	<0.50
Myrica cerifera	<0.50
Aralia spinosa	<0.50
Acer rubrum	<0.50
Total	40.64

Source: Data from Bourdeau and Oosting (1959).

vomitoria, Juniperus virginiana, Myrica species, and *Quercus virginiana* may then become established in the low-aerosol environment on the lee side of early invaders (Fig. 11.16) (Boyce 1954; Oosting 1954).

With reduced salt impaction and increasingly stable environments toward the mainland side of the island, these maritime heaths expand first into dwarf woodland and then into closed-canopy forests (Table 11.10). *Pinus taeda, Persea borbonia,* and *Magnolia virginiana* are particularly common; for regional differences, see Wells (1939), Bourdeau and Oosting (1959), Kurz (1942), and Penfound and O'Neill (1934). A number of taxa with subtropical affinities that are widespread in Florida and the Gulf states extend north into the Carolinas and Virginia only in this community. These include *Sabal palmetto, Sabal minor,* and *Osmanthus americanus.*

Wells (1939) referred to the maritime forest as the "salt spray climax," suggesting that salt aerosol prevented the usual inland sere from taking place. Wells and Shunk (1938) found that when protected from such aerosol, sandhill species such as *Aristida stricta* could thrive in these habitats. Bourdeau and Oosting (1959) found that (aside from modest increases in organic matter and water retention with successional age) soils from young dunes, shrub zones, and mature maritime forest were physically and chemically similar.

Considerable attention has been paid to the ecology of lagoons and marshes (Teal 1962; Gosselink et al. 1979; Peterson and Peterson 1979; Bahr and Lanier 1981). Space limitations permit only a brief overview of vegetation patterns abstracted from Wells (1932), Penfound and Hathaway (1938), Penfound (1952), Hinde (1954), Adams (1963), and Johnson and associates (1974).

Vascular plants are generally absent from those areas of the lagoon that are constantly inundated. However, in shallow, low-energy depositional environments, eel grass (*Zostera marina*) may form dense stands. Salt marshes cover vast stretches of the intertidal portions of these sounds, often dominated by dense swards of a single species, *Spartina alterniflora* (Fig. 11.17). This plant has remarkable tolerance to the widely varying salinity environment of the marsh (Longstreth and Strain 1977). On less frequently inundated flats or pannes, where evaporation may concentrate salts, strict halophytes such as *Salicornia virginica, Sueda linearis,* and *Distichlis spicata* are dominant. Bordering the edge of the salt marsh, in an area inundated at only the highest tides, is a zone dominated by black needle rush (*Juncus roemerianus*). Plants in this zone must deal with considerable soil salinity, but benefit from better soil aeration (Hinde 1954; Adams 1963). Between the *Juncus* zone and the maritime forest,

Figure 11.17. Salt marsh near Beaufort, North Carolina. Spartina-dominated zone is in the lower left. Bordering it is a narrow Juncus zone, and then an infrequently inundated zone dominated by salt-tolerant grasses and shrubs.

further inland, one finds a diverse assemblage of herbs, many of which are also found in dune slack habitats.

Community Dynamics and Succession

Natural disturbances and succession in these ecosystems are closely tied to coastal geomorphic processes. A rising sea level (or coastal subsidence) has resulted in a general erosion of island foredunes and sediment deposition across barrier islands. Thus, in many localities, active wind-driven dunes are devouring maritime scrub, and islands are advancing into lagoons; see Godfrey and Godfrey (1976), Dolan and associates (1980), and Pilkey and associates (1980) for reviews of the dynamics of these processes. As evidence of this island movement, it is not uncommon to find dead snags of *Juniperus virginiana* emerging from the foredunes or strand, or to find strata of lagoonal peat being eroded along the beach strand. The rate of island turnover may be as brief as 200–400 yr in many places (Au 1974; Godfrey and Godfrey 1976).

Among the lasting reminders of human habitation of barrier islands during the seventeenth and eighteenth centuries are large populations of feral animals, including pigs, goats, cattle, and horses. There is little doubt that these animals have altered the structure of the herb communities, particularly in the dune slack and near freshwater marshes (Engels 1952; Au 1974; Rubenstein 1981), but Baron (1982) has suggested that their impact is small compared with that of natural disturbance processes.

Human development of these coastal habitats during the past century, and especially in the past three decades, has had an important impact on barrier-island ecosystems (Pilkey et al. 1980). Although much of this development is concentrated on the strand side of an island, where it has directly resulted in dune erosion, increased salt aerosol deposition owing to dune degradation often results in deterioration of otherwise undisturbed maritime forest (Boyce 1954).

Recent attention has focused on colonization and subsequent succession of river delta sediments (Johnson et al. 1985, Neill and Deegan 1986). Mudflats are initially invaded by saltwater or freshwater marsh, depending on tidal flows and location relative to a river. Such areas accumulate additional sediments and may be invaded by woody vegetation. These areas subsequently undergo erosion and subsidence and, if abandoned by the river, may be invaded by brackish marsh vegetation. With continued erosion and subsidence, the delta lobe may deteriorate completely. This cycle may be completed in about 4000 yr (Neill and Deegan 1986).

SYNTHESIS AND AREAS FOR FUTURE RESEARCH

It should come as no surprise that, to date, there has been no attempt to synthesize data on vegetation composition for the entire Coastal Plain. With the single exception of Quarterman and Keever (1962), studies in this province have been confined to communities and gradients within specific regions, such as the fall-line sandhills, the central Florida highlands, and the Texas Big Thicket. Considerable amounts of unstudied territory separate localities that have received intensive study. Comparisons are made more difficult because of variations in sampling methods, measures of abundance, and community taxonomy.

With these difficulties in mind, I analyzed community data taken from reports of studies performed throughout the Coastal Plain (Table 11.11). I included in this analysis only studies that reported data on species abundances, and I excluded studies from strictly herb-dominated communities, such as grass-sedge marshes and salt marshes. Because sampling procedures and abundance measures varied among studies, I included only the 12 most abundant species from each study and assigned each an importance value (IV) based on the percentage of total community abundance accounted for by that species. For species accounting for more than 10%, 1%–10%, and less than 1% of total community abundance, IV was set equal to 3, 2, and 1, respectively. These data were then ordinated using the DECORANA program for detrended correspondence analysis (DCA) (Hill 1979; Hill and Gauch 1980; Gauch 1982).

The ordinations of sites and species are shown in Fig. 11.18. The first axis corresponds in a general fashion to successional schemes proposed by Wells (1928), Laessle (1942), and Kurz (1944). Mixed hardwoods have low first-axis scores, whereas chronically disturbed (burned) communities, such as sandhills, sand pine scrubs, flatwoods, and pocosins, have high scores. However, this axis also corresponds to a gradient in soil fertility. Communities on coarse sterile sands or nutrient-deficient wetlands have high scores, whereas those on fertile soils or in floodplains that receive nutrient subsidy have low scores. As would be expected from Monk's (1968) demonstration of the correlation between soil infertility and the importance of evergreen plants, the proportion of evergreen relative to deciduous plants increases with increasing first-axis score. The second axis separates wetlands from uplands. Thus, beginning in the lower right corner of the ordination diagram, flatwoods and pocosins grade into bayheads, cypress heads, and white cedar swamps, and these in turn grade into zone II, III, and IV alluvial wetlands. Zone IV alluvial wetlands are intermixed with communities classified as southern mixed hardwoods, although these are clearly separated on the third ordination axis (not shown here). Starting in the upper right-hand corner of the ordination, xeric sand ridges grade into sandhills and savannas, then into xeric oak woods, mixed pine-oak stands, and finally southern mixed hardwoods.

What conclusions can we draw from this ordination? First—at least comparing the relative importances of dominant species—community types identified with specific habitats are quite similar, even though some sites were over 2000 km from each other. This provides hope for the development of a more uniform system of classification for the region. Second, this ordination also verifies the gradient nature of compositional change on this landscape. However, the ordination masks our ignorance of the details of variation within particular community types, either within a particular locality or across the region. We can begin to appreciate the nature and significance of these variations only with a more systematic effort to connect the intensively studied areas with extensive sampling. Given the rate of economic development of this region, time is running out for this effort.

I have discussed the foregoing relationships between vegetation and environmental gradients as though we really understand the mechanisms underlying these relations. Nothing could be further from the truth. For example, we assume that vegetation varies considerably with respect to gradients of soil fertility, but few data are available to indicate exactly which nutrients are limiting. Correlations between soil calcium or pH and evergreenness convince us that nutrients are regulating the composition of communities, but calcium and pH are not the proximal causes of variation in plant performance. Similarly, we can identify obvious moisture gradients, but few studies have either characterized water availability as it might be perceived by plants or compared plant water use efficiency.

The southeastern Coastal Plain is an underused laboratory for testing theories of disturbance, succession, and landscape development. Coastal Plain ecologists were among the first to forsake the notion of the climatic monoclimax (Harper 1911; Laessle 1942; Wells 1942), realizing that it was useless to consider succession in the absence of natural disturbance cycles, especially fire. Furthermore, they were among the first ecologists to realize that such disturbance cycles might be regulated by features of the community itself. This is not to say that the seral nature of various communities has been settled beyond debate! The Coastal Plain offers community ecologists a wide range of fire regimes, from short-return, low-intensity fires of savannas to unpredictable conflagrations of swamp forests. Future studies might well focus on the factors that

Table 11.11. *References, locations, and community types for studies included in Fig. 11.18; community type refers to the nomenclature used in each study*

Sample Number	Reference	Location	Community type
1	Blaisdell et al. (1974)	North Florida	Mesic Hammock
2	"	"	Mesic Hammock
3	Ehrenfeld and Gulick (1981)	New Jersey	Swamp Hardwoods
4	Gemborys and Hodgkins (1971)	Alabama	Stream-bottom Hardwoods
5	"	"	Upland-margin Hardwoods
6	Christensen (unpublished)	North Carolina	Pocosin
7	Quarterman and Keever (1962)	Southeast	Group I Hardwoods
8	Snyder (1980)	North Carolina	Xeric Flatwoods
9	"	"	Pine Savanna
10	"	"	Pocosin
11	"	"	Bottomland Hardwoods
12	Laessle (1942)	North Florida	River Swamp
13	"	"	Sand Pine Scrub
14	"	"	Sandhill
15	"	"	Xeric Hammock
16	"	"	Mesic Hammock
17	"	"	Hydric Hammock
18	"	"	Bayhead
19	Edmisten (1963)	Central Florida	Longleaf Pine Flatwoods
20	"	"	Pond Pine Flatwoods
21	"	"	Slash Pine Flatwoods
22	White (1983)	South Louisiana	Cypress-tupelo Swamp
23	"	"	Bottomland Forest
24	Bourdeau and Oosting (1959)	North Carolina	Maritime Forest
25	Kurz (1942)	Central Florida	Scrub
26	Porcher (1981)	South Carolina	Swamp Forest
27	"	"	Hardwood Bottom
28	"	"	Ridge Bottom
29	"	"	Mixed Mesophytic
30	Monk (1965)	North Florida	Cypress Dome
31	"		Pocosin
32	Monk (1966b)	North Florida	Bayhead
33	"	"	Mixed Hardwood Swamp
34	Marks and Harcombe (1981)	East Texas	Sandhill Pine
35	"	"	Upland Pine-Oak
36	"	"	Wetland Pine Savanna
37	"	"	Upper slope Oak-pine
38	"	"	Mid-slope Oak-pine
39	"	"	Lower slope HW-pine
40	"	"	Floodplain Hardwoods
41	"	"	Flatwood Hardwoods
42	"	"	Baygall Thicket
43	"	"	Cypress-tupelo Swamp
44	Schlesinger (1978a)	Georgia	Cypress Swamp Forest
45	Veno (1976)	North Florida	Pine Oak Forest
46	"	"	Xeric Hammock
47	"	"	Mesic Hammock
48	"	"	Sandhill
49	Hall and Penfound (1943)	Alabama	Cypress-gum Swamp
50	Parsons and Ware (1982)	Virginia	"Dry" Alluvial Swamp
51	"	"	"Wet" Alluvial Swamp
52	Schlesinger (1976)	Georgia	Cypress Swamp
53	Kologiski (1977)	North Carolina	Pocosin
54	"	"	Pine Savanna
55	"	"	White Cedar Swamp
56	"	"	Evergreen Bay Forest
57	"	"	Deciduous Bay Forest
58	Abrahamson et al. (1984)	"	Southern Ridge Sandhill

Table 11.11. *(Cont.)*

Sample Number	Reference	Location	Community type
59	Coile (unpublished)	North Florida	Sand Pine Scrub
60	Abrahamson et el. (1984)	Central Florida	Southern Ridge Sandhill
61	"	"	Sand Pine Scrub
62	"	"	Sand Pine Scrub
63	"	"	Scrubby Flatwoods
64	"	"	Wiregrass Flatwoods
65	"	"	Palmetto Flatwoods
66	"	"	Gallberry Flatwoods
67	"	"	Bayhead
68	Wilbur (1985)	"	Pocosin
69	Whipple et al. (1981)	South Carolina	Oak Hickory Forest
70	"	"	Gum–red bay Forest
71	"	"	Gum–red Maple Forest
72	"	"	Black Oak Forest
73	"	"	Laurel Oak Forest
74	"	"	Gum-ash Forest
75	"	"	Cypress-gum Swamp
76	Hilmon (1968)	Florida	Palmetto Flatwoods
77	Weaver (1969)	North Carolina	Mesic Sandhill
78	"	"	Ridge Sandhill
79	Allen (1958)	North Carolina	Swamp Tupelo–Cypress
80	"	"	Water Tupelo–Cypress
81	"	"	Bar Forest
82	Applequist (1959)	North Carolina	Cypress-gum Swamp
83	"	"	Tupelo-gum Swamp
84	Cypert (1972)	Georgia	Cypress Swamp
85	Hebb and Clewell (1976)	North Florida	Slash Pine Forest
86	McAlister (unpublished)	North Carolina	Sand ridge
87	Caplenor (1968)	Mississippi	Hardwoods, Thick Loess
88	"	"	Hardwoods, Creek Bottom
89	"	"	Hardwoods, Non Loess
90	"	"	Hardwoods, Thin Loess
91	Ware (1970)	Virginia	Mesophytic Hardwoods
92	Christensen (unpublished)	North Carolina	Sand Ridge
93	"	South Carolina	Sand Ridge
94	Quarterman and Keever (1962)	Georgia	Southern Mixed Hardwoods
95	"	Georgia	"
96	"	South Carolina	"
97	"	Georgia	"
98	"	Georgia	"
99	"	Mississippi	"
100	"	Alabama	"
101	"	"	"
102	"	Lousiana	"
103	"	South Carolina	Southern Mixed Hardwoods
104	"	"	"
105–24	DeWitt and Ware (1979)	Virginia	Southern Mixed Hardwoods
125	Laney and Noffsinger (1985)	North Carolina	Pocosin
126	"	"	"
127	"	"	Bay
128	"	"	"
129	"	"	White Cedar Swamp
130	"	"	"
131	"	"	"
132	Conner and Day (1976)	Louisiana	Alluvial Swamp
133	Muzika et al. (1987)	South Carolina	Alluvial Swamp

(Cont.)

Table 11.11. *(Cont.)*

Sample Number	Reference	Location	Community type
134	Jones and Gresham (1985)	South Carolina	Savanna
135	"	"	Pocosin
136	"	"	Bay
137	"	"	Sandy Alluvial Swamp
138	"	"	Red Water Swamp
139	"	"	Alluvial Swamp
140	Matos and Rudolph (1985)	Texas	Floodplain Forest
141	"	"	Alluvial Swamp
152	"	"	Baygall
143	"	"	"
144	"	"	Wet Transition Forest
145	"	"	Dry Transition Forest
146	Golley et al. (1965)	South Carolina	Southern Mixed Hardwoods

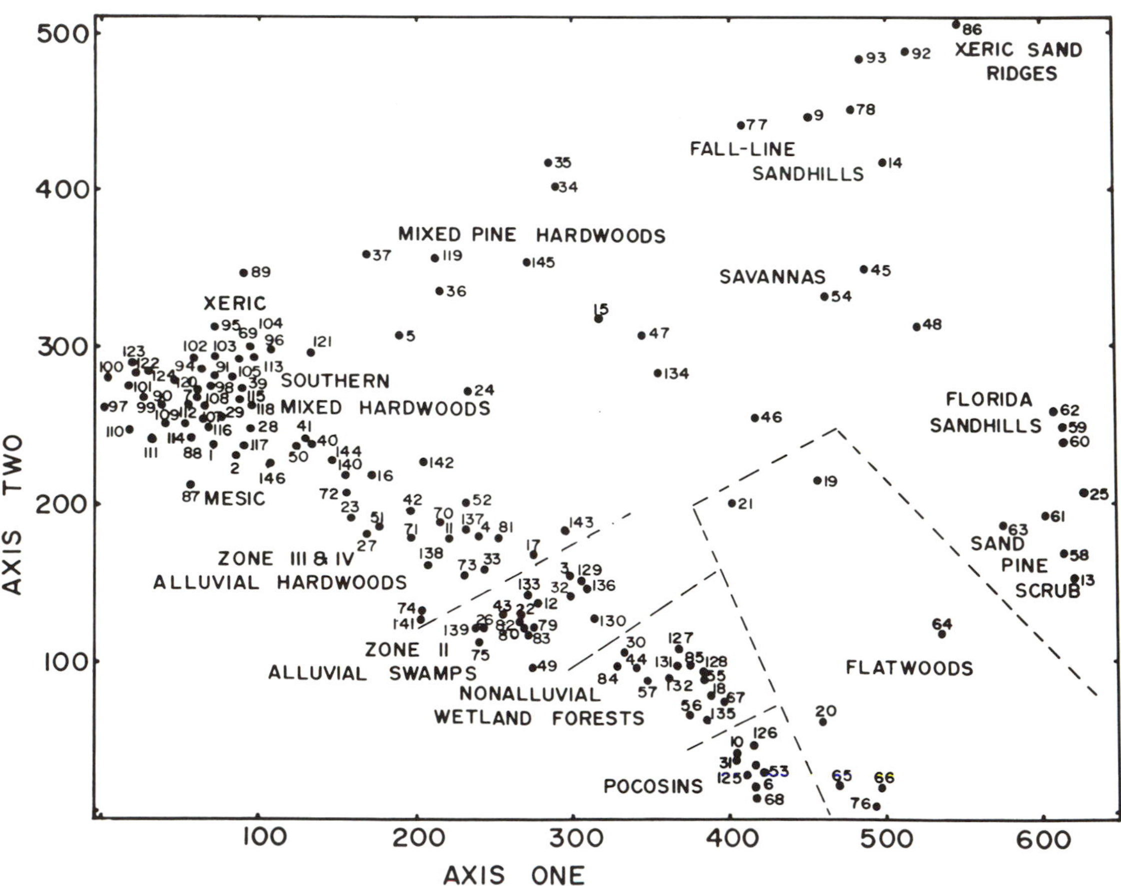

Figure 11.18. An ordination of vegetation samples taken from throughout the Southeast. The numbers refer to specific studies described in Table 11.11.

regulate such fire cycles, on comparative studies of the consequences of differences in fire regimes, and on a consideration of the relation of variations in disturbances to the evolution of plant life-history characteristics.

Delcourt and Delcourt (1981) pointed out that, regionally, the Coastal Plain flora has been relatively constant for the past 9000 yr. We are beginning to appreciate that this regional stability is the consequence of a dynamic equilibrium between local successional trends, constrained by the unique variety of Coastal Plain environments and chronic disturbances.

ACKNOWLEDGMENTS

My interest in Coast Plain ecology was first kindled and has subsequently been encouraged by Lewis E. Anderson. Although he might disagree with some of the notions presented here, he has shaped my understanding of this region more than any other person. However, I suspect that many of the ideas that I thought were mine were stimulated by discussions with Dwight Billings, Robert Peet, Bill Ralston, and Curtis Richardson. Warren Abrahamson, Ralph Good, Catherine Keever, Suzanne McAlister, Pat Peroni, Elsie Quarterman, Richard Schneider, Joan Walker, Stewart Ware, and Rebecca Wilbur provided original data and suggestions for which I am grateful. Various iterations of this manuscript benefitted from reviews and criticism by Ralph Good and Carl Monk. I especially thank Joan Walker for her comments on the structure and content of this chapter.

REFERENCES

Abrahamson, W. G. 1984a. Post-fire recovery of Florida Lake Wales Ridge vegetation. Amer. J. Bot. 71:9–21.

Abrahamson, W. G. 1984b. Species responses to fire on the Florida Lake Wales Ridge. Amer. J. Bot. 71:35–43.

Abrahamson, W. G., A. F. Johnson, J. N.Layne, and P. A. Peroni, 1984. Vegetation of the Archbold Biological Station, Florida: an example of the southern Lake Wales Ridge. Florida Scientist 47:209–250.

Adams, D. A. 1963. Factors affecting vascular plant zonation in North Carolina salt marshes. Ecology 44:445–455.

Allen, P. H. 1958. A tidewater swamp forest and succession after clearcutting. Master's thesis, Duke University, Durham, N.C.

Andresen, J. W. 1959. A study of pseudo-nanism in *Pinus rigida* Mill. Ecol. Monogr. 29:309–322.

Applequist, M. B. 1959. A study of soil and site factors affecting growth and development of swamp blackgum and tupelogum stands in southeastern Georgia. Doctor of Forestry thesis, Duke University, Durham, N.C.

Arata, A. A. 1959. Effects of burning on vegetation and rodent populations in a longleaf pine turkey oak association in north central Florida. Q. J. Florida Acad. Sci. 22:94–104.

Archard, H. O., and M. F. Buell. 1954. Life form spectra of four New Jersey pitch pine communities. Bull. Torrey Botanical Club 81:169–175.

Art, H. W. 1976. Ecological studies of the sunken forest, Fire Island National Seashore, New York. U.S. National Park Service Scientific Monograph no. 7.

Ash, A. N., C. B. McDonald, E. S. Kane, and C. A. Pories. 1983. Natural and modified pocosins: literature synthesis and management options. U.S. Fish and Wildlife Service report FWS/OBS-83/04.

Ashe, W. W. 1915. Loblolly or North Carolina pine. North Carolina Geological and Economic Survey bulletin no. 24.

Au, S. 1974. Vegetation and ecological processes on Shackleford Bank, North Carolina. U.S. National Park Service Scientific Monograph No. 6.

Axelrod, D. I., and H. P. Bailey. 1969. Paleotemperature analysis of Tertiary floras. Paleogeogr. Palaeoclimatol. Paleoecol. 6:163–195.

Bahr, L. M., and W. P. Lanier. 1981. The ecology of intertidal oyster reefs of the south Atlantic coast: a community profile. U.S. Fish and Wildlife Service report FWS/OBS-81/15.

Baron, J. 1982. Effects of feral hogs (*Sus scrofa*) on the vegetation of Horn Island, Mississippi. Amer. Midl. Nat. 107:202–205.

Bartram, W. 1791. Travels through North and South Carolina, Georgia, east and west Florida, the Cherokee country, the extensive territories of the Muscogulges or Creek confederacy, and the country of the Choctaws: containing an account of the soil and natural productions of those regions, together with observations on the manners of the Indians. James and Johnson, Philadelphia.

Beaven, G. F., and H. J. Oosting. 1939. Pocomoke Swamp: a study of a cypress swamp on the Eastern Shore of Maryland. Bull. Torrey Botanical Club 66:367–389.

Beckwith, S. L.1967. Chinsegut Hill-McCarty Woods, Hernando County, Florida. Q. J. Florida Acad. Sci. 30:250–268.

Belling, A. J. 1986. Postglacial history of Atlantic white cedar in the northeastern United States. Presented at the Atlantic white cedar wetlands symposium.

Blaisdell, R. S., J. Wooten, and R. K. Godfrey. 1974. The role of magnolia and beech in forest processes in the Tallahassee, Florida, Thomasville, Georgia area. Proc. Tall Timbers Fire Ecol. Conf. 13:363–397.

Boerner, R. E. J. 1981. Forest structure dynamics following wildfire and prescribed burning in the New Jersey pine barrens. Amer. Midl. Nat. 105:321–333.

Boerner, R. E. J. 1983. Nutrient dynamics of vegetation and detritus following two intensities of fire in the New Jersey pine barrens. Oecologia 59:129–134.

Boerner, R. E. J., and R. T. T. Forman. 1982. Hydrologic and mineral budgets of New Jersey Pine Barrens uplant forests following two intensities of fire. Can. J. Forest Res. 12:503–510.

Bourdeau, P. F., and H. J. Oosting. 1959. The maritime live oak forest in North Carolina. Ecology 40:148–152.

Boyce, S. G. 1954. The salt spray community. Ecol. Monogr. 24:29–67.

Braun, E. L. 1950. Deciduous forests of eastern North America. Blakiston, Philadelphia.

Braun, E. L. 1955. The phytogeography of unglaciated eastern United States and its interpretation. Bot. Rev. 21:297–375.

Brown, C. A. 1959. Vegetation of the Outer Banks of North Carolina. Louisiana State University Coastal Studies Series no. 4.

Brown, S. 1981. A comparison of the structure, primary productivity, and transpiration of cypress ecosystems in Florida. Ecol. Monogr. 51:403–427.

Brown, S. L., E. W. Flohrschutz, and H. T. Odum. 1984. Structure, productivity, and phosphorus cycling of the scrub cypress ecosystem. pp. 304–317 in K. C. Ewel and H. T. Odum (eds.), Cypress swamps. University of Florida Press, Gainesville.

Buchholz, K., and R. E. Good. 1982a. Compendium of New Jersey Pine Barrens literature. Center for Coastal and Environmental Studies, Rutgers–The State University, New Brunswick, N.J.

Buchholz, K., and R. E. Good. 1982b. Density, age structure, biomass, and net annual aboveground productivity of dwarfed *Pinus rigida* Mill. from the New Jersey barren plains. Bull. Torrey Botanical Club 109:24–34.

Buell, M. F. 1939. Peat formation in the Carolina bays. Bull. Torrey Botanical Club 66:483–487.

Buell, M. F. 1946. Jerome Bog, a peat-filled "Carolina Bay." Bull. Torrey Botanical Club 73:24–33.

Buell, M. F., and R. L. Cain. 1943. The successional role of southern white cedar, *Chamaecypris thyoides*, in southeastern North Carolina. Ecology 24:85–93.

Buell, M. F., and J. E. Cantlon. 1950. A study of two communities of the New Jersey pine barrens and a comparison of methods. Ecology 31:567–586.

Buol, S. W. 1973. Soils of the southern states and Puerto Rico. Agricultural Experiment Stations of the Southern States and Puerto Rico Land-Grant Universities, Southern Cooperative Series, bulletin no. 174.

Burns, R. M. 1973. Sand pine: distinguishing characteristics and distribution. pp. 13–22 in Sand pine symposium proceedings. USDA Forest Service general technical report SE-2.

Byrd, W. 1728. Histories of the dividing line betwixt Virginia and North Carolina (facsimile ed.). Dover, New York.

Caplenor, D. 1968. Forest composition on loessal and nonloessal soils in west-central Mississippi. Ecology 49:322–331.

Catesby, M. 1654. The natural history of Carolina, Florida and the Bahama Islands. C. March et al., London.

Chapman, H. H. 1926. Factors determining natural reproduction of longleaf pine on cut-over lands in LaSalle Parish, La. Yale University School of Forestry, bulletin no. 16.

Chapman, H. H. 1932. Is the longleaf type a climax? Ecology 13:328–334.

Christensen, N. L. 1976. The role of carnivory in *Sarracenia flava* L. with regard to specific nutrient deficiencies. J. Elisha Mitchell Scientific Society 92:144–147.

Christensen, N. L. 1977. Fire and soil-plant nutrient relations in a pine wiregrass savanna on the Coastal Plain of North Carolina. Oecologia 31:27–44.

Christensen, N. L. 1979a. Shrublands of the southeastern United States. pp. 441–449 in R. L. Specht (ed.), Heathlands and related shrublands of the world, A. Descriptive studies. Elsevier, Amsterdam.

Christensen, N. L. 1979b. The xeric sandhill and savanna ecosystems of the southeastern Atlantic Coastal Plain, U.S.A. Veroffentlichungen des geobotanischen Institutes der eidgenoessische technische Hochshule Stiftung Rubel, in Zurich 68:246–262.

Christensen, N. L. 1981. Fire regimes in southeastern ecosystems. pp. 112–136 in H. A. Mooney, T. M. Bonnicksen, N. L. Christensen, J. E. Lotan, and W. A. Reinsers (eds.), Fire regimes and ecosystem properties. USDA Forest Service general technical report WO-26.

Christensen, N. L. 1985. Shrubland fire regimes and their evolutionary consequences. pp. 85–100 in S.T.A. Pickett and P.S. White (eds.), The ecology of natural disturbance and patch dynamics. Academic Press, New York.

Christensen, N. L., R. B. Burchell, A. Liggett, and E. L. Simms. 1981. The structure and development of pocosin vegetation. pp. 43–61 in C. J. Richardson (ed.), Pocosin wetlands. Hutchinson Ross Publishing, Stroudsburg, Pa.

Clements, F. E. 1916. Plant succession: an analysis of the development of vegetation. Carnegie Institution of Washington publication no. 242.

Clewell, A. F., and D. B. Ward. 1985. White cedar forests in Florida and Alabama. Presented at the Atlantic white cedar wetlands symposium.

Cohen, A. D. 1973. Petrology of some Holocene peat sediments from the Okefenokee Swamp–marsh complex of southern Georgia. Geol. Soc. Amer. Bull. 84:3867–3878.

Cohen, A. D. 1974. Petrography and paleoecology of Holocene peats from the Okefenokee Swamp–marsh complex at Georgia. J. Sedimentary Petrology 44:716–720.

Cohen, A. D., M. J. Andrejko, W. Spackman, and D. Corrinus. 1984. Peat deposits of the Okefenokee Swamp. pp. 493–553 in A. D. Cohen, D. J. Casagrande, M. J. Andrejko, and G. R. Best (eds.), The Okefenokee Swamp. Wetland Surveys, Los Alamos, N. Mex.

Coker, W. C. 1905. Observations on the flora of the Isle of Palms, Charleston, S.C. Torreya 5:135–145.

Colquhoun, D. J. 1969. Geomorphology of the lower coastal plain of South Carolina. State of South Carolina Division of Geology, State Development Board, Ms-15.

Conner, W. H., and J. W. Day, Jr. 1976. Productivity and composition of a baldcypress–water tupelo site and a bottomland hardwood site in a Louisiana swamp. Amer. J. Bot. 63:1354–1364.

Cooke, C. W. 1931. Seven coastal terraces in the southeastern states. Washington Acad. Sci. J. 21:503–513.

Cooke, C. W. 1933. Origin of the so-called meteorite scars of South Carolina. Washington Acad. Sci. J. 23:569–570.

Cooper, R. W., C. S. Schopmeyer, W. H. D. McGregor. 1959. Sand pine regeneration on the Ocala National Forest. USDA Prod. Res. Rep. 30.

Cowardin, L. M., V. Carter, F. C. Golet, and E. T. LaRoe. 1979. Classification of wetlands and deepwater habitats of the United States. U.S. Fish and Wildlife Service report FWS/OBS-79/31.

Cowdrey, A. E. 1983. This land this South: an environmental history. University Press of Kentucky, Lexington.

Cronin, T. M., B. J. Szabo, T. A. Ager, J. E. Hazel, and J. P. Owens, 1981. Quaternary climates and sea levels of the U.S. Atlantic Coastal Plain. Science 211:233–240.

Croom, J. M. 1978. Sandhills–turkey oak (*Quercus laevis*) ecosystem: community analysis and a model of radiocesium cycling. Ph.D. dissertation, Emory University, Atlanta.

Crum, H. A., and L. E. Anderson. 1981. Mosses of eastern North America. Columbia University Press, New York.

Cypert, E. 1961. The effects of fires in the Okefenokee Swamp in 1954 and 1955. Amer. Midl. Nat. 66:485–503.

Cypert, E. 1972. The origin of houses in the Okefenokee prairies. Amer. Midl. Nat. 87:448–458.

Cypert, E. 1973. Plant succession on burned areas in Okefenokee Swamp following fires of 1954 and 1955. Proc. Tall Timbers Fire Ecol. Conf. 12:199–217.

Dabel, C. V., and F. P. Day, Jr. 1977. Structural comparisons of four plant communities in the Great Dismal Swamp, Virginia. Bull. Torrey Botanical Club 104:352–360.

Dachnowski-Skokes, A. P., and B. W. Wells. 1929. The vegetation, stratigraphy, and age of the "Open Land" peat area in Carteret County, North Carolina. Washington Acad. Sci. J. 19:1–11.

Daniel, C. C., III. 1981. Hydrology, geology and soils of pocosins: a comparison of natural and altered systems. pp. 69–108 in C. J. Richardson (ed.), Pocosin wetlands. Hutchinson Ross Publishing, Stroudsburg, Pa.

Daniels, R. B., E. E. Gamble, Wheeler, W. H., and C. S. Holzhey. 1977. The stratigraphy and geomorphology of the Hofman Forest Pocosin. Soil Sci. Soc. Amer. J. 41:1175–1180.

Daniels, R. B., H. J. Kleiss, S. W. Buol, H. J. Byrd, and J. A. Phillips. Soil systems in North Carolina. North Carolina Agricultural Research Service bulletin 467.

Davis, J. H. 1943. The natural features of southern Florida, especially the vegetation, and the Everglades. Florida Geological Survey bulletin no. 25.

Davis, J. H. 1946. The peat deposits of Florida: their occurrence, development, and uses. Florida Geological Survey bulletin no. 30.

Davis, M. B. 1981. Quaternary history and the stability of deciduous forests. pp. 132–177 in D. C. West, H. H. Shugart, and D. B. Botkin (eds.), Forest succession. Springer-Verlag, Berlin.

Davis, M. B. 1983. Holocene vegetational history of the eastern United States. pp. 166–181 in H. E. Wright, Jr. (ed.), Late Quaternary environments of the United States, vol. 2, The Holocene. University of Minnesota Press, Minneapolis.

Deghi, G. S. 1984. Seedling survival and growth rates in experimental cypress domes. pp. 141–144 in K. C. Ewel and H. T. Odum (eds.), Cypress swamps. University of Florida Press, Gainesville.

Delcourt, H. R. 1976. Presettlement vegetation of the North of Red River Land District, Louisiana. Castanea 41:122–139.

Delcourt, H. R., and P. A. Delcourt. 194. Primeval magnolia-holly-beech climax in Lousiana. Ecology 55:638–644.

Delcourt, H. R., and P. A. Delcourt. 1977a. The Tunica Hills, Louisiana-Mississippi: late glacial locality for spruce and deciduous forest species. Quat. Res. 7:218–237.

Delcourt, H. R., and P. A. Delcourt. 1977b. Presettlement magnolia-beech climax of the Gulf Coastal Plain: quantitative evidence from the Apalachicola River Bluffs, north-central Florida. Ecology 58:1085–1093.

Delcourt, P. A. 1980. Goshen Springs: late Quaternary vegetation record for southern Alabama. Ecology 61:371–386.

Delcourt, P. A., and H. R. Delcourt. 1979. Late Pleistocene and Holocene distributional history of the deciduous forest in the southeastern United States. Veroffentlichungen des geobotanischen Institute der eidgenoessische technische Hochshule Stiftung Rubel, in Zurich 68:79–107.

Delcourt, P. A., and H. R. Delcourt. 1980. Pollen preservation and Quaternary environmental history in the southeastern United States. Palynology 4:215–231.

Delcourt, P. A., and H. R. Delcourt. 1981. Vegetation maps for eastern North America: 40,000 yr B.P. to the present. pp. 123–165 in R. C. Romans (ed.), Geobotany II. Plenum, New York.

Delcourt, P. A., H. R. Delcourt, R. C. Brister, and L. E. Lackey. 1980. Quaternary vegetation history of the Mississippi embayment. Quat. Res. 13:111–132.

Demaree, D. 1932. Submerging experiments with *Taxodium*. Ecology 13:258–262.

DeWitt, R., and S. Ware. 1979. Upland hardwood forests of the central Coastal Plain of Virginia. Castanea 44:163–174.

Dickson, R. E., and T. C. Broyer. 1972. Effects of aeration, water supply, and nitrogen source on growth and development of tupelo gum and bald cypress. Ecology 53:626–634.

Dierberg, F. E., and P. L. Brezonik. 1984. Nitrogen and phosphorus mass balances in a cypress clone receiving waste water. pp. 112–118 in K. C. Ewel and H. T. Odum (eds.), Cypress swamps. University of Florida Press, Gainesville.

Dilcher, D. L. 1973a. Revision of the Eocene flora of the southeastern North America. Paleobotanist 20:7–18.

Dilcher, D. A. 1973b. A paleoclimatic interpretation of

the Eocene floras of southeastern North America. pp. 39–59 in A. Graham (ed.), Vegetation and vegetational history of north Latin America. Elsevier, Amsterdam.

Doering, J. 1958. Citronelle age problem. Amer. Assoc. Petroleum Geologists Bull. 42:764–786.

Dolan, R., B. Hayden, and H. Lins. 1980. Barrier islands. Amer. Sci. 68:16–25.

Duever, M. J., and L. A. Riopelle. 1983. Successional sequences and rates on tree islands in the Okefenokee Swamp. Amer. Midl. Nat. 110:186–193.

Duke, J. A. 1961. The psammophytes of the Carolina fall-line sandhills. J. Elisha Mitchell Scientific Society 77:3–25.

Dunn, W. J., L. N. Schwartz, and G. R. Best. 1985. Structure and water relations of the white cedar forests of north central Florida. Presented at the Atlantic white cedar wetlands symposium.

Dury, G. H. 1977. Underfit streams: retrospect, perspect, and prospect. pp. 281–293 in K. J. Gregory (ed.), River channel changes. Wiley, New York.

Edmisten, J. A. 1963. The ecology of the Florida pine flatwoods. Ph.D. dissertation, University of Florida, Gainesville.

Ehrenfeld, J. G., and M. Gulick. 1981. Structure and dynamics of hardwood swamps in the New Jersey Pine Barrens: contrasting patterns in trees and shrubs. Amer. J. Bot. 68:471–481.

Eleuterius, L. N. 1968. Floristics and ecology of coastal bogs in Mississippi. Master's thesis, University of Southern Mississippi, Hattiesburg.

Eleuterius, L. N., and S. B. Jones, Jr. 1969. A floristic and ecological study of pitcher plant bogs in south Mississippi. Rhodora 71:29–34.

Engels, W. L. 1952. Vertebrate fauna of North Carolina coastal islands, II, Shackleford Banks. Amer. Midl. Nat. 47:702–742.

Ewel, K. C. 1984. Effects of fire and wastewater on understory vegetation in cypress domes. pp. 119–126 in K. C. Ewel and H. T. Odum (eds.), Cypress swamps. University of Florida Press, Gainesville.

Ewel, K. C., and W. J. Mitsch. 1978. The effects of fire on species composition in cypress dome ecosystems. Florida Scientist 41:25–31.

Forman, R. T. T. (ed.). 1979. Pine Barrens: ecosystem and landscape. Academic Press, New York.

Forman, R. T. T., and R. E. Boerner. 1981. Fire frequency and the pine barrens of New Jersey. Bull. Torrey Botanical Club 108:34–50.

Frasco, B., and R. E. Good. 1976. Cone, seed and germination characteristics of pitch pine (*Pinus rigida* Mill.). Bartonia 44:50–57.

Frost, C. C. 1986. Historical overview of Atlantic white cedar in the Carolinas. Presented at the Atlantic white cedar wetlands Symposium.

Gallagher, M. G., and R. E. Good. 1985. Two decades of vegetation change in the New Jersey USA pine barrens. Amer. J. Bot 72:844.

Gano, L. 1917. A study in physiographic ecology in northern Florida. Bot. Gaz. 63:337–372.

Garren, K. H. 1943. Effects of fire on vegetation of the southeastern United States. Bot. Rev. 9:617–654.

Gauch, H. G. 1982. Multivariate analysis in community ecology. Cambridge University Press.

Gemborys, S. R., and E. J. Hodgkins. 1971. Forests of small stream bottoms in the coastal plain of southwestern Alabama. Ecology 52:70–84.

Givens, K. T., J. N. Layne, W. G. Abrahamson, and S. C. White-Schuler. 1984. Structural changes and successional relationships of five Florida Lake Wales ridge plant communities. Bull. Torrey Botanical Club 111:8–18.

Godfrey, P. J., and M. M. Godfrey. 1976. Barrier island ecology of Cape Lookout National Seashore and vicinity, North Carolina. U.S. National Park Service Scientific Monograph No. 9.

Gohlz, H. L., and R. F. Fisher, 1982. Organic matter production and distribution in slash pine (*Pinus elliottii*) plantations. Ecology 63:1827–1839.

Golley, F. B., G. A. Petrides, and J. F. McCormick. 1965. A survey of the vegetation of the Boiling Spring Natural Area, South Carolina. Bull. Torrey Botanical Club 92:355–363.

Good, R. E., and N. F. Good. 1975. Growth characteristics of two populations of *Pinus rigida* Mill. from the Pine Barrens of New Jersey. Ecology 56:1215–1220.

Good, R. E., N. F. Good, and J. W. Andresen. 1979. The Pine Barren plains. pp. 283–295 in R. T. T. Forman (ed.), Pine Barrens: ecosystem and landscape. Academic Press, New York.

Gosselink, J. G., C. L. Cordes, and J. W. Parsons. 1979. An ecological characterization of the Chenier Plain Coastal Ecosystem of Louisiana and Texas. U.S. Fish and Wildlife Service Reports FWS/OBS 79/9, 79/10, and 79/11.

Gosselink, J. G., S. E. Bailey, W. H. Conner, and R. E. Turner. 1981. Ecological factors in the determination of reparian wetland boundaries. pp. 197–219 in J. R. Clark and J. Benforado (eds.), Wetlands of bottomland hardwood forests. Elsevier, Amsterdam.

Gresham, C. A., and D. J. Lipscomb. 1985. Selected ecological characteristics of *Gordonia lasianthus* in coastal South Carolina. Bull. Torrey Botanical Club 112:53–58.

Grubb, P. J. 1977. The maintenance of species richness in plant communities: the importance of the regeneration niche. Biol. Rev. Cambridge Philosoph. Soc. 52:107–145.

Gunderson, L. H. 1984. Regeneration of cypress in logged and burned stands at Corkscrew Swamp Sanctuary, Florida. pp. 349–357 in K. C. Ewel and H. T. Odum (eds.), Cypress Swamps. University of Florida Press, Gainesville.

Hair, J. D., G. T. Hepp, L. M. Luckett, K. P. Reese, and D. K. Woodward. 1979. Beaver pond ecosystems and their multiuse natural resource management. pp. 80–92 in R. R. Johnson and J. F. McCormick (eds.), Strategies for protection and management of floodplain wetlands and other riparian ecosystems. USDA Forest Service general technical report WO-12.

Hall, T. F., and W. T. Penfound. 1939. A phytosociological study of a cypress-gum swamp in southeastern Louisiana. Amer. Midl. Nat. 21:378–395.

Hall, T. F., and W. T. Penfound. 1943. Cypress-gum communities in the Blue Girth Swamp near Selma, Alabama. Ecology 24:208–217.

Hamilton, D. B. 1984. Plant succession and the influence of disturbance in the Okefenokee Swamp.

pp. 86–111 in A. D Cohen, D. J. Casagrande, M. J. Andrejko, and G. R. Best (eds.), The Okefenokee Swamp. Wetland Surveys, Los Alamos, N. Mex.

Harcombe, P. A., and P. L. Marks. 1978. Tree diameter distributions and replacement processes in southeast Texas forests. Forest Sci. 24:153–166.

Harcombe, P. A., and P. L. Marks. 1983. Five years of tree death in a *Fagus Magnolia* forest, southeast Texas (USA). Oecologia 57:49–54.

Harper, R. M. 1906. A phytogeographical sketch of the Altamaka Grit Region of the Coastal Plain of Georgia. Ann. N.Y. Acad. Sci. 7:1–415.

Harper, R. M. 1911. The relation of climax vegetation to islands and peninsulas. Bull. Torrey Botanical Club 38:515–525.

Harper, R. M. 1914a. The "pocosin" of Pike Co., Ala., and its bearing on certain problems of succession. Bull. Torrey Botanical Club 41:209–220.

Harper, R. M. 1914b. The geography and vegetation of northern Florida. Florida Geological Survey Annual Report no. 6.

Harper, R. M. 1922. Some pine-barren bogs in central Alabama. Torreya 22:57–60.

Harper, R. M. 1926. A middle Florida white cedar swamp. Torreya 26:81–84.

Harper, R. M. 1940. Fire and forests. American Botanist 46:5–7.

Harper, R. M. 1943. Forests of Alabama. Geological Survey of Alabama monograph no. 10.

Harshberger, J. W. 1916. The vegetaion of the New Jersey Pine Barrens: an ecological investigation. Christopher Sower Co., Philadelphia.

Hebb, E. A., and A. F. Clewell. 1976. A remnant stand of old-growth slash pine in the Florida panhandle. Bull. Torrey Botanical Club 103:1–9.

Heyward, F. 1939. The relation of fire to stand composition of longleaf pine forests. Ecology 20:287–304.

Hill, E. P. 1976. Control methods for nuisance beaver in the southeastern United States. Vertebrate Pest Control Conference 7:85–98.

Hill, M. O. 1979. DECORANA–a FORTRAN Program for detrended correspondence analysis and reciprocal averaging. Cornell University, Ithaca, N.Y.

Hill, M. O., and H. G. Gauch. 1980. Detrended correspondence analysis, an improved ordination technique. Vegetatio 42:47–58.

Hilmon, J. B. 1968. Autecology of palmetto [*Serenoa repens* (Bartr.) Small]. Ph.D. dissertation, Duke University, Durham, N.C.

Hinde, H. P. 1954. The vertical distribution of phanerogams in relation to tide level. Ecol. Monogr. 24:209–225.

Hodgkins, E. J. 1958. Effects of fire on undergrowth vegetation in upland southern pine forests. Ecology 58:36–46.

Hough, W. A. 1973. Fuel and weather influence wildfires in sand pine forests. USDA Forest Service research paper SE-106.

Hugh, W. A., and F. A. Albini. 1978. Predicting fire behavior in palmetto gallberry fuel complexes. USDA Forest Service research paper SE-174.

Hudson, C. 1976. The southeastern Indians. University of Tennessee Press, Knoxville.

Huffman, R. T., and S. W. Forsythe. 1981. Bottomland hardwood forest communities and their relation to anaerobic soil conditions. pp. 197–196 in J. R. Clark and J. Benforads (eds.), Wetlands of bottomland hardwood forests. Elsevier, Amsterdam.

Hughes, R. H. 1966. Fire ecology of canebrakes. Proc. Tall Timbers Fire Ecol. Conf. 5:149–158.

Hull, J. C., and D. F. Whigham. 1985. Atlantic white cedar in the Maryland inner Coastal Plain and the Delmarva Peninsula. Presented at the Atlantic white cedar wetlands symposium.

Izlar, R. L. 1984. Some comments on fire and climate in the Okefenokee Swamp–marsh complex. pp. 70–85 in A. D. Cohen, D. J. Casagrande, M. J. Andrejko, and G. R. Best (eds.), The Okefenokee Swamp. Wetland Surveys, Los Alamos, N. Mex.

Johnson, A. S., H. O. Hillsted, S. Shanholtzer, and G. F. Shanholtzer. 1974. Ecological survey of the coastal region of Georgia. National Park Service scientific monograph no. 3.

Johnson, D. 1942. The origin of the Carolina bays. Columbia University Press, New York.

Johnson, W. B., C. E. Sasser, and J. G. Gosselink. 1985. Succession of vegetation in an evolving river delta, Atchafalaya Bay, Louisana. J. Ecol. 73:973–986.

Jones, R. H., and C. A. Gresham. 1985. Analysis of composition, environmental gradients, and structure in the Coastal Plain lowland forests of South Carolina. Castanea 50:207–227.

Kalisz, P. J., and E. L. Stone. 1984. The longleaf pine islands of Ocala National Forest, Florida: a soil study. Ecology 65:1743–1754.

Kearney, T. H. 1900. The plant covering of Ocracoke Island; a study of the ecology of the North Carolina strand vegetation. U.S. National Herbarium, Contributions 5:261–319.

Kearney, T. H. 1901. Report on a botanical survey of the Dismal Swamp region. U.S. National Herbarium, Contributions 5:321–585.

Kologiski, R. L. 1977. The phytosociology of the Green Swamp, North Carolina. North Carolina Agricultural Experiment Station technical bulletin no. 250.

Komarek, E. V., Sr. 1968. Lightning and lightning fires as ecological forces. Proc. Tall Timbers Fire Ecol. Conf. 9:169–197.

Komarek, E. V., Sr. 1974. Effects of fire on temperate forests and related ecosystems: southeastern United States. pp. 251–277 in T. T. Kozlowski and C. E. Ahlgren (eds.), Fire and ecosystems. Academic Press, New York.

Komarek, E. V., Sr. 1977. A quest for ecological understanding. Tall Timbers Research Station miscellaneous publication no. 5.

Korstian, C. F. 1924. Natural regeneration of southern white cedar. Ecology 5:188–191.

Korstian, C. F., and W. D. Brush. 1931. Southern white cedar. USDA technical bulletin no. 251.

Küchler, A. W. 1964. Potential natural vegetation of the coterminous United States. American Geographical Society Special Publication No. 36.

Kurz, H. 1938. A physiographic study of the tree associations of the Apalachicola River. Proc. Florida Acad. Sci. 3:78–90.

Kurz, H. 1942. Florida dunes and scrub, vegetation and geology. State of Florida Department of Conservation, geological bulletin no. 23.

Kurz, H. 1944. Secondary forest succession in the

Tallahassee Red Hills. Proc. Florida Acad. Sci. 7:59–100.
Kurz, H. and K. A. Wagner. 1953. Factors in cypress dome development. Ecology 34:157–164.
Laessle, A. M. 1942. The plant communities of the Welaka area. University of Florida publications, biological science series 4:5–141.
Laessle, A. M. 1958. The origin and successional relationship of sandhill vegetation and sand-pine scrub. Ecol. Monogr. 28:361–387.
Laessle, A. M. 1965. Spacing and competition in natural stands at sand pine. Ecology 46:65–72.
Laessle, A. M. 1967. Relationship of sand pine scrub to former shore lines. Q. J. Florida Acad. Sci. 30:269–286.
Lanley, R. W.,and R. E. Noffsinger. 1985. Vegetative composition of Atlantic white cedar [*Chamaecyparis thyoides* (L.) B.S.P.] swamps in Dare County, North Carolina. Presented at the Atlantic white cedar wetlands symposium.
Lawson, J. 1714. Lawson's history of North Carolina. Garrett and Massil, Richmond, Va. (reprinted 1952).
Ledig, F. T., and S. Little. 1979. Pitch pine (*Pinus rigida* Mill.): ecology, physiology, and genetics. pp. 347–372 in R. T. T. Forman (ed.), Pine Barrens: ecosystem and landscape. Academic Press, New York.
Leitman, H. M., J. E. Sohm, and M. A. Franklin. 1981. Wetland hydrology and tree distribution of the Apalachicola River flood-plain, Florida. U.S. Geological Survey water supply paper no. 2196-A.
Lemon, P. C. 1949. Successional responses of herbs in the longleaf–slash pine forest after fire. Ecology 30:135–145.
Leopold, L. B., and M. G. Wolman. 1957. River channel patterns: braided, meandering and straight. U.S. Geological Survey professional paper 282-B.
Leopold, L., M. G. Wolman, and J. Miller. 1964. Alluvial processes in geomorphology. W. H. Freeman, San Francisco.
Lewis, C. E., and T. J. Harshbarger. 1976. Shrub and herbaceous vegetation after 20 years of prescribed burning in the South Carolina Coastal plain. J. Range Management 29:13–18.
Lewis, I. F. 1917. The vegetation of Shackleford Bank. North Carolina Geological and Economic Survey paper no. 46.
Lilly, J. P. 1981. A history of swamp development in North Carolina. pp. 20–39 in C. J. Richardson (ed.), Pocosin wetlands. Hutchinson Ross Publishing, Stroudsburg, Pa.
Little, E. L., Jr., and K. W. Dorman. 1952. Geographic differences in cone-opening in sand pine. J. Forestry 50:204–205.
Little, E. L., Jr., and K. W. Dorman. 1954. Slash pine (*Pinus elliottii*), including South Florida slash pine. USDA Forest Service paper SE-36.
Little, S. 1950. Ecology and silviculture in the white cedar and associated hardwoods in southern New Jersey. Yale Univesity School of Forestry bulletin 56.
Litle, S. 1979. Fire and plant succession in the New Jersey pine barrens. pp. 297–314 in R. T. T Forman (ed.), Pine Barrens: ecosystem and landscape. Academic Press, New York.
Longstreth, D. J., and B. R. Strain. 1977. Effects of salinity and illumination on photosynthesis and water balance of *Spartina alterniflora* Loisel. Oecologia 31:191–199.
Loveless, A. R. 1961. A nutritional interpretation of sclerophylly based on differences in the chemical composition of sclerophyllous and meophytis leaves. Ann. Bot. (London) 25:168–184.
Loveless, A. R. 1962. Further evidence to support a nutritional interpretation of sclerophylly. Ann. Bot. (London) 26:551–561.
Lutz, H. J. 1934. Ecological relations in the pine pitch plains of southern New Jersey. Yale University School of Forestry bulletin no. 38.
McCaffrey, C. A., and D. B. Hamilton. 1984. Vegetation mapping of the Okefenokee ecosystem. pp. 201–211 in A. D. Cohen, D. J. Casagrande, M. J. Andrejko, and G. R. best (eds.), The Okefenokee Swamp. Wetland Surveys, Los Alamos, N. Mex.
McCormick, J. 1979. The vegetation of the New Jersey pine barrens. pp. 229–244 in R. T. T. Forman (ed.), Pine Barrens: ecosystem and landscape. Academic Press, New York.
McCormick, J., and M. F. Buell. 1968. The plains: pigmy forests of the New Jersey Pine barrens, a review and annotated bibliography. Bull. N.J. Acad. Sci. 13:20–34.
Marks, P. L., and P. A. Harcombe. 1975. Community diversity of Coastal Plain forests in southern East Texas. Ecology 56:1004–1008.
Marks, P. L., and P. A. Harcombe. 1981. Forest vegetation of the Big Thicket, southeast Texas. Ecol. Monogr. 51:287–305.
Marois, K. C., and K. C. Ewel. 1983. Natural and management-related variation in cypress domes. Forest Sci. 29:627–640.
Matoon, W. R. 1916. Water requirements and growth of young cypress. Proc. Soc. Amer. Foresters11: 192–197.
Matos, J. A., and D. C. Rudolph. 1985. The vegetation of the Roy E. Larsen Sandylands Sancturary in the Big Thicket of Texas. Castanea 50:228–249.
Mitsch, W. J., and K. C. Ewel. 1979. Comparative biomass and growth of cypress in Florida wetlands. Amer. Midl. Nat. 101:417–426.
Monette, R. 1975. Early forest succession in the southeastern Virginia Coastal Plain. Virginia J. Sci. 26:65.
Monk, C. D. 1960. A preliminary study on the relationships between the vegetation of a mesic hammock community and a sandhill community. Q. J. Florida Acad. Sci. 23:1–12.
Monk, C. D. 1965. Southern mixed hardwood forest of northcentral Florida. Ecol. Monogr. 35:335–354.
Monk, C. D. 1966a. An ecological significance of evergreenness. Ecology 47:504–505.
Monk, C. D. 1966b. An ecological study of hardwood swamps in northcentral Florida. Ecology 47: 649–654.
Monk, C. D. 1967. Tree species diversity in the eastern deciduous forest with particular reference to north central Florida. American Naturalist 101:173–187.
Monk, C. D. 1968. Successional and environmental relationships of the forest vegetation of north central Florida. Amer. Midl. Nat. 79:441–457.
Monk, C. D., and T. W. Brown, 1965. Ecological consideration of cypress heads in northcentral Florida. Amer. Midl. Nat. 74:127–140.
Moore, J. H., and J. H. Carter III. 1985. The range and

habitats of Atlantic white cedar in North Carolina. Presented at the Atlantic white cedar wetlands symposium.

Moore, W. H., B. F. Swindel, and W. S. Terry. 1982. Vegetative response to prescribed fire in a north Florida flatwoods forest. J. Range Management 35:386–389.

Mulvania, M. 1931. Ecological survey of a Florida scrub. Ecology 12:528–540.

Murray, G. E. 1961. Geology of the Atlantic and Gulf Coastal Province of North America. Harper, New York.

Muzika, R. M., J. B. Gladden, and J. D. Haddock. 1987. Structural and functional aspects of recovery in southeastern floodplain forests following a major disturbance. Amer. Midl. Nat. (in press).

Myers, R. L. 1985. Fire and the dynamic relationship between Florida sandhill and sand pine scrub vegetation. Bull. Torrey Botanical Club 112:241–252.

Myers, R. L., and N. D. Deyrup. 1983. The dynamic relationship between Florida sandhill and sand pine scrub vegetation. Bull. Ecol. Soc. Amer. 64:62.

Neill, C., and L. A. Deegan. 1986. The effect of Mississippi River delta lobe development on the habitat composition and diversity of Louisana coastal wetlands. Amer. Midl. Nat. 116:296–303.

Nesom, G. L., and M. Treiber. 1977. Beech-mixed hardwoods communities: a topo-edaphic climax on the North Carolina Coastal Plain. Castanea 42:119–140.

Neufeld, H. S. 1983. Effects of light on growth, morphology, and photosynthesis in bald cypress [*Taxodium distichum* (L. Rich.)] and pond cypress (*T. ascendens* Brongn.) seedlings. Bull. Torrey Botanical Club 110:43:54.

Neufeld, H. S. 1984. Comparative ecophysiology of baldcypress [*Taxodium distichum* (L.) Rich.] and pondcypress (*Taxodium ascendens* Brongn.). Ph.D. dissertation, University of Georgia, Athens.

Odum, W. E., T. J. Smith III, J. K. Hoover, and C. C. McIvor. 1984. The ecology of tidal freshwater marshes of the United States east coast: a community profile. U.S. Fish and Wildlife Service, FWS/OBS 83/17.

Olsson, H. 1979. Vegetation of the New Jersey Pine Barrens: a phytosociological classification. pp. 245–264 in R. T. T. Forman (ed.), Pine Barrens: ecosystem and landscape. Academic Press, New York.

Olsvig, L. S., J. F. Cryan, and R. H. Whittaker. 1979. Vegetational gradients of the pine plains and barrens of Long Island, New York, pp. 265–282 in R. T. T. Forman (ed.), Pine Barrens: ecosystem and landscape. Academic Press, New York.

Oosting, H. J. 1942. An ecological analysis of the plant communities of piedmont, North Carolina. Amer. Midl. Nat. 28:1–126.

Oosting, H. J. 1945. Tolerance to salt spray of plants of coastal dunes. Ecology 26:85–89.

Oosting, H. J. 1954. Ecological processes and vegetation of the maritime strand in the southeastern United States. Bot. Rev. 20:226–262.

Oosting, H. J., and W. D. Billings, 1942. Factors effecting vegetational zonation on coastal dunes. Ecology 23:131–142.

Otte, L. J. 1981. Origin, development, and maintenance of the pocosin wetlands of North Carolina. Unpublished report of North Carolina Department of Natural Resources and Community Development Natural Heritage Program, Raleigh, N.C.

Parrish, F. K., and E. J. Rykiel, Jr. 1979. Okefenokee Swamp origin: review and reconsideration. J. Elisha Mitchell Scientific Society 95:17–31.

Parrott, R. T. 1967. A study of wiregrass (*Aristida stricta*) with particular reference to fire. Master's thesis, Duke University, Durham, N.C.

Parsons, S. E., and S. Ware. 1982. Edaphic factors and vegetation in Virginia Coastal Plain swamps. Bull. Torrey Botanical Club 109:365–370.

Penfound, W. T. 1952. Southern swamps and marshes. Bot. Rev. 18:413–446.

Penfound, W. T., and T. T. Earle. 1948. The biology of the water hyacinth. Ecol. Monogr. 18:447–472.

Penfound, W. T., and E. S. Hathaway. 1938. Plant communities in the marshlands of southeastern Louisiana. Ecol. Monogr. 8:1–56.

Penfound, W. T., and J. A. Howard. 1940. A phytosociological study of an evergreen oak forest in the vicinity of New Orleans, Louisiana. Amer. Midl. Nat. 23:165–174.

Penfound, W. T., and M. E. O'Neill. 1934. The vegetation of Cat Island, Mississippi. Ecology 15:1–16.

Penfound, W. T., and A. G. Watkins. 1937. Phytosociological studies in the pinelands of southeastern Louisiana. Amer. Midl. Nat. 18:661–682.

Peroni, P. A. 1983. Vegetation history of the southern Lake Wales ridge, Highlands County, Florida. Master's thesis, Bucknell University, Lewisburg, Pa.

Pessin, L. J. 1933. Forest associations in the uplands of the lower Gulf Coastal Plain. Ecology. 14:1–14.

Peterson, C. H., and N. M. Peterson. 1979. The ecology of intertidal flats of North Carolina: a community profile. U.S. Fish and Wildlife Service, FWS/OBS-79-39.

Pilkey, O. H., Jr., W. J. Neal, and O. H. Pilkey, Sr. 1980. From currituck to calabash. North Carolina Science and Technology Center, Research Triangle Park, N.C.

Pinchot. G., and W. W. Ashe. 1897. Timber trees and forests of North Carolina. North Carolina Geological Survey bulletin no. 6.

Porcher, R. D. 1981. The vascular flora of the Francis Beidler Forest in Four Holes Swamp, Berkeley and Dorchester Counties, South Carolina. Castanea 46:248–280.

Price, M. B. 1973. Management of natural stands of Ocala sand pine. pp. 153–163 in Sand Pine symposium proceedings. USDA Forest Service general technical report SE-2.

Prouty, W. F. 1952. Carolina bays and their origin. Bull. Geol. Soc. Amer. 63:187–224.

Putnam, J. A., G. M. Furnival, and J. S. McKnight. 1960. Management and inventory of southern hardwoods. USDA Forest Service Agricultural handbook no. 181.

Pyne, S. J. 1982. Fire in America. A cultural history of wildland and rural fire. Princeton University Press, Princeton, N.J.

Quarterman, E., and C. Keever. 1962. Southern mixed hardwood forest: climax in the southeastern coastal plain: U.S.A. Ecol. Monogr. 32:167–185.

Radford, A. E., H. E. Ahles, and C. R. Bell. 1968. Manual of the vascular flora of the Carolinas. University of North Carolina Press, Chapel Hill.

Raff, P. J. 1954. Aspects of the ecological life-history of turkey oak (*Quercus laevis* Walter). Master's thesis, Duke University, Durham, N.C.

Richardson, C. J., R. Evans, and D. Carr. 1981. Pocosins: an ecosystem in transition. pp. 3–19. in C. J. Richardson (ed.), Pocosin wetlands. Hutchinson Ross Publishing, Stroudsburg, Pa.

Roberts, P. R., and H. J. Oosting. 1958. Responses of venus fly trap (*Dionaea muscipula*) to factors involved in its endemism. Ecol. Monogr. 28:193–218.

Rubenstein, D. I. 1981. Behavioral ecology of island feral horses. Equine Veterinary Journal 13:27–34.

SAF. 1980. Forest cover types of the United States and Canada. Society of American Foresters, Washington, D.C.

Savage, H., Jr. 1982. The mysterious Carolina bays. University of South Carolina Press, Columbia.

Schlesinger, W. H. 1976. Biogeochemical limits on two levels of plant community organization in the cypress forest of Okefenokee Swamp. Ph.D. dissertation, Cornell University, Ithaca, N.Y.

Schlesinger, W. H. 1978a. Community structure, dynamics and nutrient cycling in the Okefenokee cypress swamp-forest. Ecol. Monogr. 48:43–65.

Schlesinger, W. H. 1978b. On the relative dominance of shrubs in Okefenokee Swamp. American Naturalist 112:949–954.

Shafizadeh, F., P. P. S. Chin, and W. F. De Groot. 1977. Effective heat content of forest fuels. Forest Sci. 23:81–89.

Shaler, N. S. 1885. Seacost swamps of the eastern United States. U.S. Geological Survey Annual Report 6:353–398.

Sharitz, R. R., and J. W. Gibbons. 1982. The ecology of southeastern shrub bogs (pocosins) and Carolina bays: a community profile. U.S. Fish and Wildlife Service report FWS/OBS-82/04.

Simms, E. L. 1983. The growth, reproduction, and nutrient dynamics of two pocosin shrubs, the evergreen *Lyonia lucida* and the deciduous *Zenobia pulverulenta*. Ph.D. dissertation, Duke University, Durham, N.C.

Simms, E. L. 1985. Growth response to clipping and nutrient addition to *Lyonia lucida* and *Zenobia pulverulenta*. Amer. Mid. Nat. 114:44–50.

Snyder, J. R. 1980. Analysis of coastal plain vegetation, Croatan National Forest, North Carolina. Veroffentlichongen des geobotanischen Institutes der eidgenoessiche technische Hochschule Stiftung Rubel, in Zurich 69:40–113.

Spackman, W., A. D. Cohen, P. H. Given, and D. J. Casagrande. 1976. Comparative study of the Okefenokee Swamp and the Everglades-Mangrove Complex of southern Florida. Coal Research Section, Pennsylvania State University, State College, Pa.

Stephenson, S. N. 1965. Vegetation change in the Pine Barrens of New Jersey. Bull. Torrey Botanical Club 92:102–114.

Stone, W. 1911. The plants of southern New Jersey, with especial reference to the flora of the Pine Barrens and the geographical distribution of the species. New Jersey State Museum Annual Report 1910:3–828.

Straub, P. A. 1984. Effects of wastewater and inorganic fertilizer on growth rates and nutrient concentrations in dominant tree species in cypress domes. pp. 127–140 in K. C. Ewel and H. T. Odum (eds.), Cypress swamps. University of Florida Press, Gainesville.

Streng, D. R., and P. A. Harcombe. 1982. Why don't east Texas savannas grow up to forest. Amer. Midl. Nat. 108:278–294.

Teal, J. M. 1962. Energy flow in the salt marsh ecosystem of Georgia. Ecology 43:614–624.

Tedrow, J. D. F. 1979. Development of pine barrens soils. pp. 61–80 in R. T. T. Forman (ed.), Pine Barrens: ecosystem and landscape. Academic Press, New York.

Thornthwaite, C. W., J. R. Mather, and D. B. Carter. 1958. Three water balance maps of eastern North America. Resources for the future, Washington, D.C.

Tooker, W. W. 1899. The adapted Algonquin term "poguosin." American Anthropology January: 162–170.

Trewartha, G. T. 1968. An introduction to climate, 4th ed. McGraw-Hill, New York.

Turner, R. E., S. W. Forsythe, and N. J. Craig. 1981. Bottomland hardwood forest land resources of the southeastern United States. pp. 13–28 in J. R. Clark and J. Benforado (eds.), Wetlands of bottomland hardwood forests. Elsevier, Amsterdam.

Veno, P. A. 1976. Successional relationships of five Florida plant communities. Ecology 57:498–508.

Vitousek, P. M. 1982. Nutrient cycling and nutrient use efficiency. American Naturalist 119:553–572.

Vogl, R. J. 1973. Fire in the southeastern grasslands. Proc. Tall Timbers Fire Ecol. Conf. 12:175–198.

Wade, D., J. Ewel, and R. Hofstetter. 1980. Fire in south Florida ecosystems. USDA Forest Service general technical report SE-17.

Wagner, R. H. 1964. The ecology of *Uniola paniculata* L. in the dune-strand habitat of North Carolina. Ecol. Monogr. 34:79–96.

Wahlenberg, W. G. 1946. Longleaf pine. Charles Lathrop Pack Forest Foundation, Washington, D.C.

Wahlenberg, W. G., S. W. Greene, and H. R. Reed. 1939. Effect of fire and cattle grazing on longleaf pine lands studied at McNeill, Mississippi. USDA agricultural technical bulletin no. 683.

Walker, J. 1984. Species diversity and production in pine-wiregrass savannas of the Green Swamp, North Carolina. Ph.D. dissertation, University of North Carolina, Chapel Hill.

Walker, J., and R. K. Peet. 1983. Composition and species diversity of pine–wire grass savannas of the Green Swamp, North Carolina. Vegetatio 55:163–179.

Walter, H. 1979. Vegetation of the earth and ecological systems of the geo-biosphere. Springer-Verlag, Berlin.

Walter, H., and H. Lieth. 1967. Klimadiagramm-Weltatlas. Gustav Fischer-Verlag, Jena.

Wang, D. 1984. Fire and nutrient dynamics in a pine-oak forest ecosystem in the New Jersey pine barrens. Ph.D. disseration, Yale University, New Haven, Conn.

Ware, S. 1970. Southern mixed hardwood forest in the Virginia Coastal Plain. Ecology 51:921–924.

Ware, S. 1978. Vegetational role of beech in the

southern mixed hardwood forest and the Virginia Coastal Plain. Virginia J. Sci. 29:231–235.

Watts, W. A. 1971. Postglacial and interglacial vegetation history of southern Georgia and central Florida. Ecology 52:676–690.

Watts, W. A. 1975. A late Quaternary record of vegetation from Lake Annie, south-central Florida. Geology 3:344–346.

Watts, W. A. 1980a. The late Quaternary vegetation history of the southeastern United States. Ann. Rev. Ecol. Syst. 11:387–409.

Watts, W. A. 1980b. Late Quaternary vegetation history of White Pond on the inner Coastal Plain of South Carolina. Quat. Res. 13:187–199.

Weaver, J. E., and F. E. Clements. 1938. Plant ecology. McGraw-Hill, New York.

Weaver, T. W, III. 1969. Gradients in the Carolina fall line sandhills: environment, vegetation, and comparative ecology of the oaks. Ph.D. dissertation, Duke University, Durham, N.C.

Webber, J. 1935. Florida scrub, a fire fighting association. Amer. J. Bot. 22:344–361.

Wells, B. W. 1928. Plant communities of the Coastal Plain of North Carolina and their successional relations. Ecology 9:230–242.

Wells, B. W. 1932. The natural gardens of North Carolina. University of North Carolina Press, Chapel Hill.

Wells, B. W. 1939. A new forest climax: the salt spray climax of Smith Island, N.C. Bull. Torrey Botanical Club 66:629–634.

Wells, B. W. 1942. Ecological problems of the southeastern United States Coastal Plain. Bot. Rev. 8:533–561.

Wells, B. W. 1946. Vegetation of Holly Shelter Wildlife Management area. North Carolina Department of Conservation and Development, Division of Game and Inland Fisheries, bulletin no. 2.

Wells, B. W., and S. G. Boyce. 1953. Carolina bays: additional data on their origin, age and history. J. Elisha Mitchell Scientific Society 69:119–141.

Wells, B. W., and I. V. Shunk. 1928. A southern upland grass-sedge bog. North Carolina State College Agricultural Experimental Station technical bulletin no. 32.

Wells, B. W., and I. V. Shunk. 1931. The vegetation and habitat factors of coarser sands of the North Carolina Coastal Plain: an ecological study. Ecol. Monogr. 1:465–520.

Wells, B. W., and I. V. Shunk. 1938. Salt spray: an important factor in coastal ecology. Bull. Torrey Botanical Club 65:485–492.

Wells, B. W., and L. A. Whitford, 1976. History of stream-head swamp forests, pocosins, and savannahs in the Southeast. J. Elisha Mitchell Scienfitic Society 92:148–150.

Wharton, C. H. 1978. The natural environments of Georgia. Georgia Department of Natural Resources, Atlanta.

Wharton, C. H., W. M. Kitchens, E. C. Pendleton, and T. W. Sipe. 1982. The ecology of bottomland hardwood swamps of the Southeast: a community profile. U.S. Fish and Wildlife Service, FWS/OBS 81/37.

Whipple, S. A., L. H. Wellman, and B. J. Good. 1981. A classification of hardwood and swamp forests on the Savannah River Plant, South Carolina. U.S. Department of Energy, Savannah River Plant publication SRO-NERP-6.

White, D. A. 1983. Plant communities of the lower Pearl River basin, Louisiana. Amer. Midl. Nat. 110:381–396.

Whitehead, D. R. 1972. Development and environmental history of the Dismal Swamp. Ecol. Monogr. 42:301–315.

Whitehead, D. R. 1981. Late-Pleistocene vegetational changes in northeastern North Carolina. Ecol. Monogr. 51:451–471.

Whitlow, T. H., and R. W. Harris. 1979. Flood tolerance in plants: a state of the art review. U.S. Army Corps of Engineers, Environmental and Water Quality Operational Studies, technical report E-79-2.

Whittaker, R. H. 1979. Vegetational relationships of the pine barrens. pp. 315–332 in R. T. T. Forman (ed.), Pine Barrens: ecosystem and landscape. Academic Press, New York.

Wilbur, R. B. 1985. Effects of fire on nitrogen and phosphorus availability in a North Carolina Coastal Plain pocosin. Ph.D. dissertation, Duke University, Durham, N.C.

Wilbur, R. B., and N. L. Christensen. 1983. Effects of fire on nutrient availability in a North Carolina coastal plain pocosin. Amer. Midl. Nat. 110:54–61.

Wilson, J. E. 1978. A floristic study of the "savannahs" on pine plantations in the Croatan National Forest. Master's thesis, University of North Carolina, Chapel Hill.

Wolfe, J. A. 1978. A paleobotanical interpretation of Tertiary climates in the Northern Hemisphere. Amer. Sci. 66:694–704.

Woodwell, G. M. 1956. Phytosociology of Coastal Plain wetlands in the Carolinas. Master's thesis, Duke University, Durham, N.C.

Woodwell, G. M. 1958. Factors controlling growth of pond pine seedlings in organic soils of the Carolinas. Ecol. Monogr. 28:219–236.

Woodwell, G. M. 1979. Leaky ecosystems: nutrient fluxes and succession in the Pine Barrens vegetation. pp. 333–343 in R. T. T. Forman (ed.), Pine Barrens: ecosystem and landscape. Academic Press, New York.

Wright, A. H., and A. A. Wright. 1932. The habitats and composition of the vegetation of Okefinokee Swamp, Georgia. Ecol. Monogr. 2:110–232.

Wright, H. E., Jr. 1976. The dynamic nature of Holocene vegetation, a problem in paleoclimatology, biogeography and stratigraphic nomenclature. Quat. Res. 6:581–596.

Wright, N. O. 1984. A cultural history of the Okefenokke. pp. 58–69 in A. D. Cohen, D. J. Casagrande, M. J. Andrejko, and G. R. Best (eds.), The Okefenokee Swamp. Wetland Surveys, Los Alamos, N. Mex.

Wunderlin, R. P. 1982. Guide to the vascular plants of central Florida. University of Florida Press, Gainesville.

Chapter 12

Tropical and Subtropical Vegetation of Meso-America

GARY S. HARTSHORN

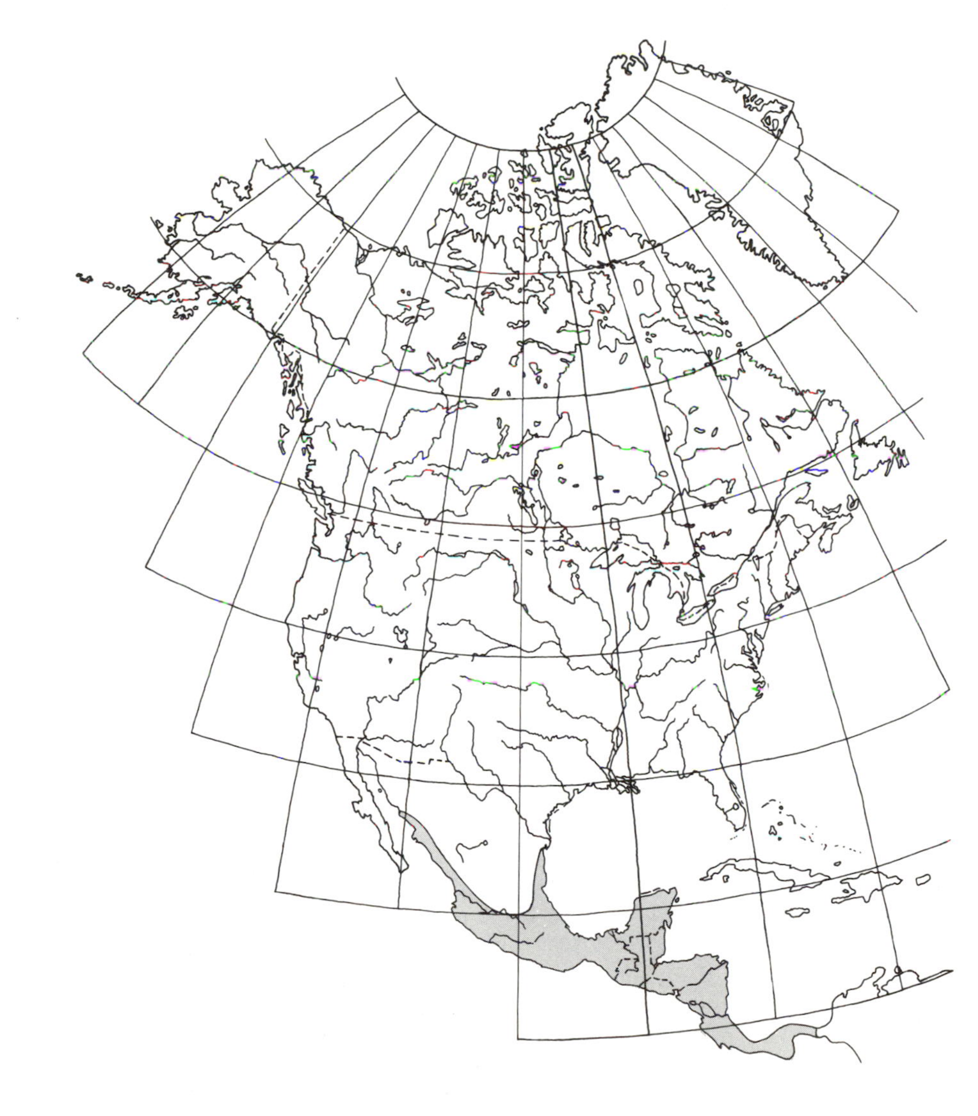

SETTING

Meso-America is a convenient, nonpolitical term for the region between South America and North America, but where do we draw the southern and northern limits for purposes of this chapter? It makes no sense ecologically to separate the Darién forests of eastern Panama from the very similar forests of the Colombian Chocó. A more reasonable geographical boundary is the isthmus of Panama, preferably east of the Panama Canal, if for no other reason than that the renowned Barro Colorado Island has been the site of much research on tropical vegetation and plants.

The northern ecological limit of Meso-America is not so easily defined. One possibility is the Tropic of Cancer (23° 30′ N), that is, the astronomic northern limit of the tropics. The narrow isthmus of Tehuántepec (225 km wide) in southern Mexico is an attractive candidate for the northern limit, but that would exclude the lowland subtropical forests of Veracruz. A more meaningful ecological boundary for Meso-America is the northern limit of mangrove forests at 26–27° N. However, the preponderance of temperate vegetation in the Mexican highlands makes it impossible to draw a northern latitudinal boundary to Meso-America. This chapter focuses on the major types of tropical and subtropical vegetation prevalent in Meso-America, without attempting to define precise latitudinal or altitudinal limits. Vegetation of the Caribbean islands (Lugo et al. 1981) is not covered in this chapter.

The division of Meso-America into tropical and subtropical regions occurs at approximately 12–13° N latitude (Holdridge 1967). This latitudinal division corresponds to the inner tropics and outer tropics of geographers. Others prefer to consider the entire region as tropical. Less understandable is the designation of all of Meso-America as subtropical (*Encyclopaedia Britannica* 1977). Tropical Meso-America and subtropical Meso-America differ in several ways. Seasonal climate is more pronounced farther from the equator. Tropical storms and hurricanes often strike subtropical Meso-America, whereas tropical Meso-America is usually south of hurricane paths. More important for this chapter, forest structure and composition differ between subtropical and tropical Meso-America.

Physical Features

Physiography. The remarkable diversity of vegetation types highlighted in this chapter is due not only to Meso-America's low latitude but also to the fundamental influences that mountains and volcanoes have on the region's climate, geology, and soils. Mountain ranges are dominant physiographic features in every country of the region, ranging from the 5700-m Mount Orizaba in Mexico (the highest North American peak south of Alaska) to the low Maya Mountains in Belize. Several active volcanoes (Mexico's El Chichón, Guatemala's Fuego, Nicaragua's Masaya, Costa Rica's Arenal, et al.) are part of the Pacific "ring of fire" that forms the spectacularly beautiful backbone of Meso-America.

The rugged Sierra Madre Occidental and Sierra Madre Oriental dominate the physiography of much of Mexico, with vast, warm-temperate tablelands between these two parallel ranges. The transversal Cordillera Neo-Volcánica, which forms the southern rim of the Mexico City basin, is one of several west-east mountain ranges prominent in southern Mexico and in Guatemala. A dominant mountain range is absent east of Guatemala; nevertheless, rugged highlands occupy 65% of Honduras. These highlands extend into northern and central Nicaragua, finally disappearing in the Nicaraguan Depression (Lake Managua, Lake Nicaragua, and the Río San Juan). Costa Rica differs from the pattern of parallel mountain ranges in northern Meso-America in that its four major ranges are aligned from northwest to southeast. The Cordillera de Talamanca includes the highest peaks of Costa Rica as well as of Panama, where it also forms the mountainous backbone of western Panama.

Mexico's Yucatán Peninsula is part of an extensive limestone platform that also underlies northern Belize and the Guatemalan Petén. Another prominent physiographic feature along the Caribbean coast is the Honduran-Nicaraguan bulge terminating at Cape Gracias de Diós. The Pacific coast of Meso-America has some prominent peninsulas (Nicoya, Osa, Azuero) and gulfs (Tehuántepec, Fonseca, Nicoya, Dulce, Chiriquí, Montijo, Panamá). The coastal plain is more extensive on the Caribbean side than on the Pacific side of Meso-America.

Climate. The Meso-American climate is characterized by predictable temperature regimes and unpredictable rainfall patterns. The daily variation in temperature (on a sunny day) exceeds the difference between average temperatures for the coolest and warmest months. Because of minimal variation in temperature patterns during the year, seasonal differences are based on rainfall. When the "heat equator" or Intertropical Convergence Zone (ITCZ) is south of the region, weather depends on the northeast tradewinds that blow incessantly from

November-December to March-April. Picking up moisture in their passage over the Caribbean Sea, the tradewinds are the source of the famous cloud forests characteristic of the windward mountain slopes in many parts of the circum-Caribbean region. Stripped of moisture by the mountainous cloud forests and rapidly heated as they descend the hot Pacific slopes, the northeast tradewinds have an opposite, rain-shadow, effect on the leeward side of the mountains. Most of the Pacific lowlands and slopes, as well as many of the intermountain valleys, have a monsoon type of climate, with a very pronounced dry season of up to 6 mo without significant rainfall.

The rain season is associated with the presence of the ITCZ over the southern part of Meso-America. Late in the dry season, the northeast tradewinds diminish some 6–8 wk before the abrupt start of the rains. Because of proximity to large bodies of water along both coasts, the region's climate is strongly maritime, with on-shore breezes usually bringing afternoon rains. Convectional air currents produce abundant thunderstorms, with typically heavy rainfall. The predictable pattern of daily zenithal rains associated with the ITCZ is overlain by the unpredictable occurrence of storms caused by northern polar air masses penetrating far to the south. These "northers" occur between November and March, bringing cooler-than-normal and rainy weather to the highlands and Caribbean lowlands during the normal dry season.

Tropical cyclones also bring heavy rains to the region, usually during the traditional rainy season. Pacific storms often arrive late in the rainy season (August to October), not only resulting in very heavy rainfall but also producing a modest rain-shadow effect on the Caribbean side. Tropical cyclones known as hurricanes occasionally come across the Caribbean to strike northern Meso-America, with devastating consequences. Although hurricane paths are usually absent from the inner tropics (Coen 1983), high winds and heavy rains do affect the eastern lowlands of Nicaragua and Costa Rica.

Geology and soils. The geological history of Meso-America has been characterized by episodes of mountain building and oceanic submergence that first provided islands in the region and finally the present land bridge between continents (Raven and Axelrod 1975). The Quaternary has been characterized by intense volcanic activity throughout much of the region. Even in Belize, which has no geological vestiges of volcanoes and lies far from known volcanoes, some of the soils include volcanic ash in the marine sediments (Wright et al. 1959). Volcanic eruptions have blanketed extensive areas with ash; the Central Valley of Costa Rica has multiple layers of volcanic ash as deep as 50 m.

In addition to lava flows, volcanoes eject two principal types of ash: andesitic and rhyolitic. Andesitic ash is mostly plagioclase feldspar, whereas rhyolitic ash is high in silica. As soil parent material, these two different types of volcanic ash have influenced both soil genesis and vegetation, as well as land use patterns. Andesitic ash is the parent material for a well-known group of soils called "andepts" that characterize the major coffee-growing areas of the world. Andepts are common in southern Mexico, western Guatemala, El Salvador, Costa Rica, and western Panama, and they also characterize other regions supporting high population densities, such as highland Colombia, Kenya, Java, and Bali. In contrast to the fertile andepts, rhyolitic ash and pumice lava flows are parent materials of relatively infertile soils.

The combination of rugged topography and high rainfall results in appreciable erosion of soils, often exacerbated by inappropriate land use. The high sediment loads carried by rivers are deposited on the active floodplains. Because most of the region's mountainous areas are geologically young, frequent floods enrich the lowlands with minerals from the eroded slopes. These extensive alluvial floodplains have long been prized for intensive agriculture, such as banana plantations in eastern Costa Rica and northern Honduras, or sugarcane fields in seasonally dry valleys of the Pacific lowlands.

The Yucatán Peninsula is very different geologically and ecologically from the rest of Meso-America because of the preponderance of limestone and soils derived from marine sediments. Where limestone rock is exposed, the topography tends to be abrupt, and drainage is internal through the porous limestone. Although soil fertility is generally good, seasonal drought has a strong influence on natural vegetation and often limits land use options. Other locally important soil types will be mentioned in the treatments of vegetation that follow.

Vegetation Classification Systems

Though many classification systems have been developed to categorize vegetation, it is beyond the scope of this overview of Meso-American vegetation to evaluate or compare them. Rather, a few major classification systems relevant to Meso-American vegetation will be mentioned (Shimwell 1971; Mueller-Dombois and Ellenberg 1974). The vegetation classification systems used in Meso-

America can be grouped into three types: floristic, physiognomic (structural), and bioclimatic.

Most of the early descriptions of Meso-American vegetation were based on floristic criteria (Stevenson 1928; Lundell 1937; Standley 1937; Allen 1956; Gómez-Pompa et al. 1964; Wagner 1964; Rzedowski 1978). The descriptions often included substantial lists of plant species, but there were few specifics concerning abundance, stature, dominance, or ease of taxonomic identification. Such species lists mean very little to the reader unfamiliar with most of the species. The great species richness of tropical forests, the wealth of life forms, and the frequent lack of dominance make it exceedingly difficult to classify tropical vegetation solely based on floristics.

The classic work of Davis and Richards (1933–4) was the first to include physiognomic criteria in the classification of neotropical vegetation. These authors pioneered the use of profile diagrams to illustrate the complex vertical structures of tropical forests. Key physiognomic features include the number of layers of tree crowns, the canopy height, the proportion of deciduous species in the canopy, the abundance of leaf forms, and the types of synusiae or life forms.

Beard's (1944) physiognomic classification of Meso-American vegetation distinguished six formations: rain; seasonal; dry evergreen; montane; swamp forests; and marsh or seasonal swamp. The UNESCO physiognomic classification system (Mueller-Dombois and Ellenberg 1974) uses a confusing plethora of adjectives (e.g., deciduous, drought-deciduous, evergreen, ombrophilous, seasonal, semi-deciduous) to define vegetation formations. Except for a few superficial trials (Küchler and Montoya Maquín 1971), the UNESCO system has not been used in Meso-America.

A fundamental criterion for physiognomic classification of vegetation is that it be applied to climax communities. Because of natural dynamic processes such as tree-fall gaps (Hartshorn 1980), the validity of the "climatic climax" concept for tropical forests has been seriously questioned (Aubréville 1938; Hewetson 1956). Where a dry season permits burning, slash-and-burn agriculture has profoundly and perhaps permanently altered the natural vegetation (Gómez-Pompa et al. 1972). The use of fire by indigenous cultures for hunting game or shifting cultivation of subsistence crops may have affected the composition and structure of present forests (Sauer 1958; Budowski 1959).

Holdridge (1947) devised a bioclimatic classification system using precipitation and temperature, plus the ratio of potential evapotranspiration (PET) to precipitation. These parameters are arranged logarithmically in an isogonal diagram (Fig. 12.1). Each hexagon is called a life zone, and the small triangles around the periphery of a hexagon are called transitions. Because of the geometry of the life-zone diagram, tropical transitional areas can occur in the subtropical latitudinal region. Some examples occur in the lower Aguan Valley of Honduras (cool transition of tropical moist forest life zone) and in southeastern Belize (cool transition of tropical wet forest life zone). Holdridge's life-zone classification system has been used extensively in Meso-America to map natural vegetation (Tosi 1969, 1971; Holdridge 1975a; De la Cruz 1976; Tosi and Hartshorn 1978; González et al. 1983; Hartshorn et al. 1984).

In the life-zone diagram (Fig. 12.1), note that a latitudinal basal belt does not have a corresponding altitudinal belt. For example, the subtropical latitudinal region does not have a premontane altitudinal belt; rather, altitudinal belts in the subtropics start with the lower montane. The lowest row of complete hexagons (Fig. 12.1) is divided by a frost or critical-temperature line into subtropical and warm-temperate latitudinal regions or premontane and lower-montane altitudinal belts. The arrangement of latitudinal regions and altitudinal belts is clearer in the third dimension (Fig. 12.2). At the equator (right margin of Fig. 12.2), the theoretical limits of altitudinal belts are shown (e.g., 3000–4000 m for tropical montane life zone). Moving poleward, the altitudinal limits decrease; in Costa Rica (8–11° N), the change from tropical basal belt to tropical premontane altitudinal belt occurs at only 500–700 m elevation.

MAJOR VEGETATION TYPES

Meso-America is exceptionally rich in plant species that occur in an impressive number of plant communities. No comprehensive survey of Meso-American vegetation has been done, and given the extensive deforestation that has occurred, it is difficult to classify the region's natural vegetation. Excluding Mexico, Meso-America is estimated to have about 18,000 species of vascular plants (D'Arcy 1977). Even tiny Belize lists about 4000 species of higher plants (Spellman et al. 1975; Dwyer and Spellman 1981). The tree flora in some lowland sites (Table 12.1) is indicative of the plant species richness in local areas of Meso-America.

Protected Meso-American forests are major attractions for research on tropical vegetation (Leigh et al. 1982; Estrada and Coates-Estrada 1983; Janzen 1983a; Clark et al. in press). It is no surprise that much of the information and many of the references of this chapter are from a few research sites: La Selva,

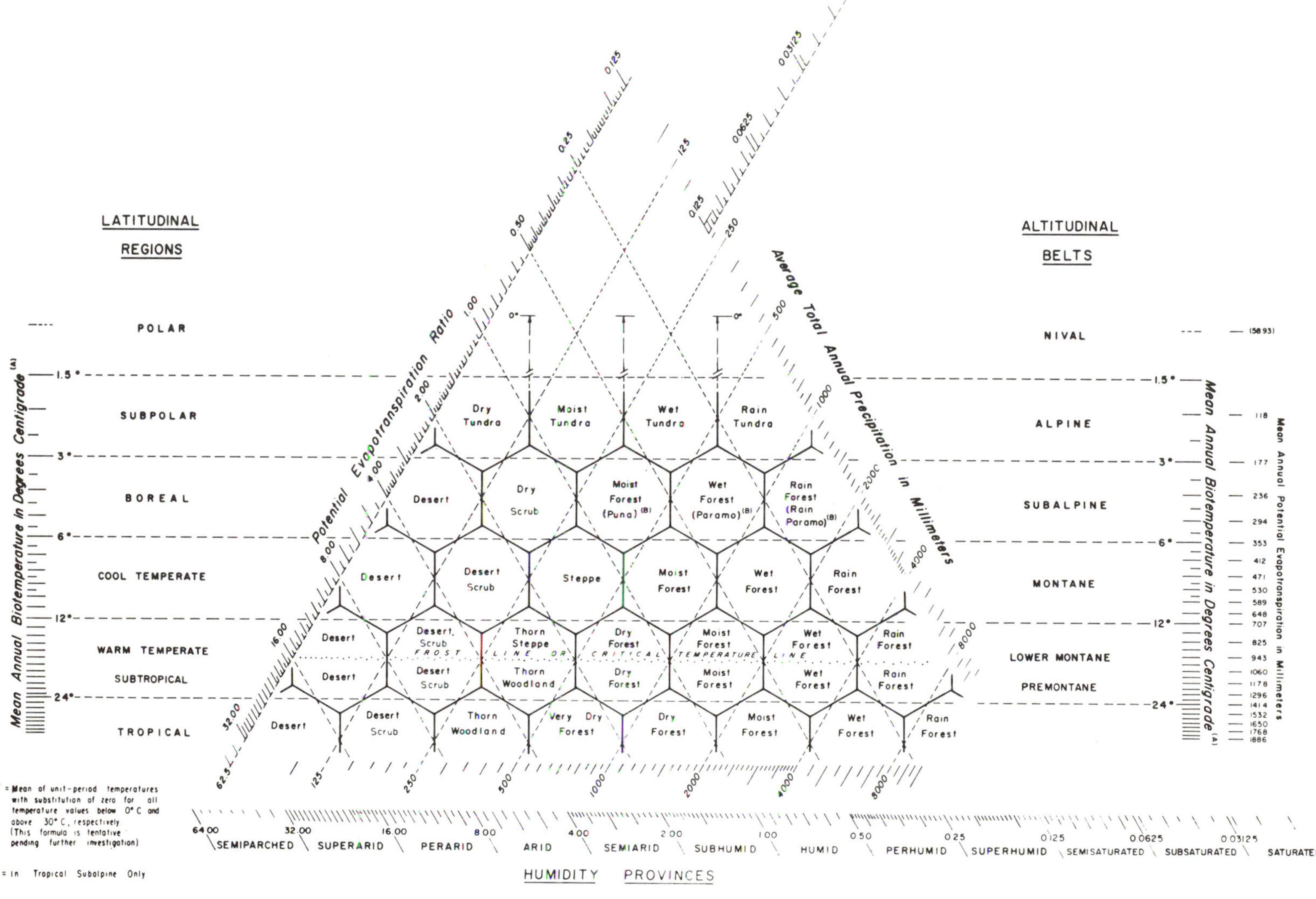

Figure 12.1. Diagram for the Holdridge classification of world life zones or plant formation (used with permission of L. R. Holdridge).

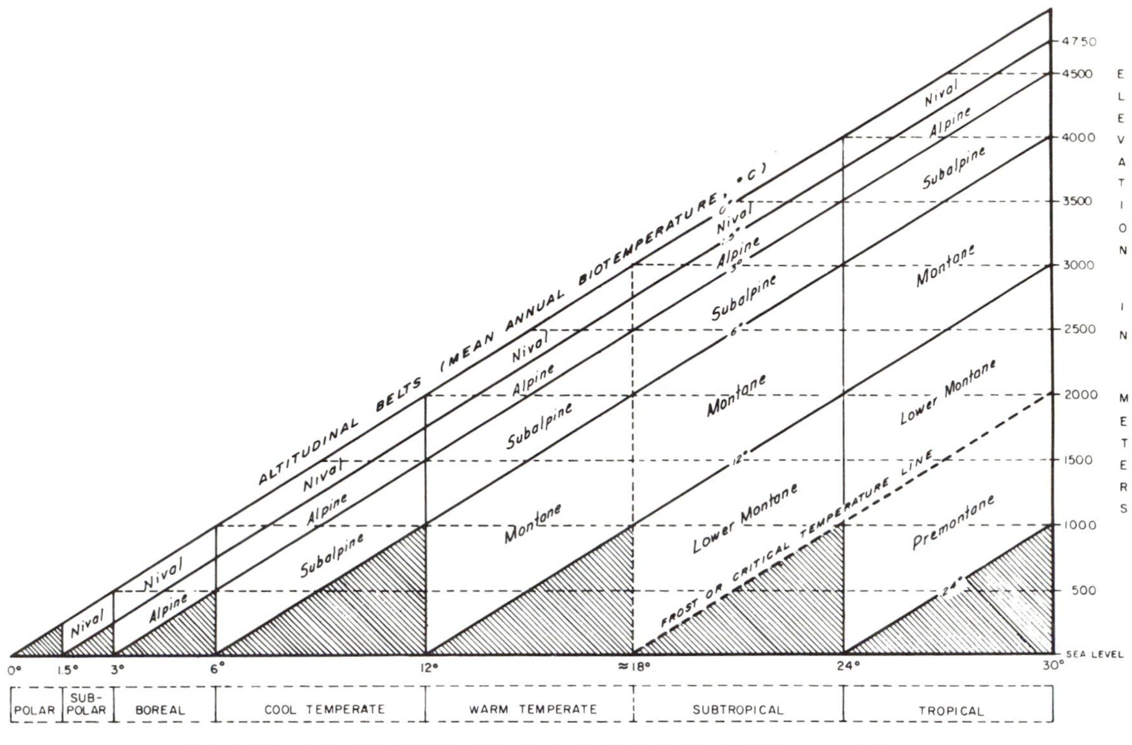

Figure 12.2. Approximate guideline positions of latitudinal regions and altitudinal belts of Holdridge's world life-zone system (used with permission of L. R. Holdridge). Based on lapse rate of 6°C per 1000m. Shaded areas represent basal belt positions.

Table 12.1. *Tree flora richness in Meso-American lowland forests*

Site and country	Life zone	Area (ha)	Species per hectare[a]	Species[b]	Genera	Families	Reference
Deininger National Park, El Salvador	Subtropical moist	732	45	143	105	48	Witsberger et al. (1982)
Palo Verde Wildlife Refuge, Costa Rica	Tropical dry	4,757	52	152	116	51	Hartshorn and Poveda (1983)
Santa Rosa National Park, Costa Rica	Tropical dry	10,700	—	206	162	60	Hartshorn and Poveda (1983)
Barro Colorado Island, Panama	Tropical moist	1,560	—	362	221	67	Croat (1978)
La Selva Biological Station, Costa Rica	Tropical wet	1,330	98	448	224	65	Hartshorn and Hammel (1982)

[a]Species per hectare based on an inventory of trees ≥10 cm dbh on one or more hectares.
[b]For the species tally, a tree is defined as ≥5 m tall or ≥10 cm dbh. The Leguminosae are treated as three families.

Monteverde, Palo Verde, and Santa Rosa in Costa Rica; Barro Colorado Island (BCI) in Panama; and lowland Veracruz in Mexico. Increasing use of the Mexican biological stations at Chamela and Los Tuxtlas should generate much new information from studies of different types of vegetation.

Even through Mexico has not been mapped using Holdridge's life-zone classification system, it can be assumed that all of the subtropical life zones represented in Guatemala (which have been mapped) also occur in Mexico. Thus, approximately 25 life zones occur in the tropical and sub-

Table 12.2. *Relation of Meso-American vegetation types to Holdridge life zones*

This chapter	Holdridge life zones
Lowlands (0–500 m)	
Mangrove forests	Tropical dry forest Tropical moist forest Tropical wet forest Subtropical dry forest Subtropical moist forest Subtropical wet forest
Freshwater swamp forests	Tropical moist forest Tropical wet forest Tropical premontane wet forest Subtropical wet forest
Perhumid forests	Tropical wet forest Tropical premontane wet forest Subtropical wet forest
Humid forests	Tropical moist forest Tropical premontane moist forest Subtropical moist forest
Subhumid forests	Tropical dry forest Subtropical dry forest Subtropical thorn woodland
Savannas	Tropical dry forest Tropical moist forest Tropical premontane moist forest Subtropical moist forest
Low mountains (500–2000 m)	
Subhumid forests	Subtropical thorn woodland Subtropical dry forest Subtropical lower-montane dry forest
Humid forests	Tropical premontane moist forest Tropical lower-montane moist forest Subtropical moist forest Subtropical lower-montane moist forest
Perhumid forests	Tropical premontane wet forest Tropical premontane rain forest Tropical lower-montane wet forest Tropical lower-montane rain forest Subtropical lower-montane wet forest Subtropical lower-montane rain forest
High mountains (>2000 m)	
Humid forests	Tropical montane moist forest Subtropical montane moist forest Subtropical subalpine moist forest
Perhumid forests	Tropical montane wet forest Tropical montane rain forest Subtropical montane wet forest
Páramo	Tropical subalpine rain páramo

tropical regions of Meso-America. For comparison, the eastern and central United States has 11 life zones (Sawyer and Lindsey 1964). Rather than treat each life zone of Meso-America in this short chapter, major vegetation types (Table 12.2) are grouped into three humidity regimes – perhumid (including superhumid), humid, and subhumid (including semiarid) – within three altitudinal zones: lowlands (0–500 m), low mountains (500–2000 m), and high mountains (>2000 m). These three altitudinal zones are used strictly as a framework to assist the reader, rather than as an indication of specific altitudinal limits for any vegetation type. I agree with Standley (1937) and Holdridge (1967) that it is impossible to assign regional altitudinal limits to vegetation types.

Lowlands

For our purposes, lowlands are arbitrarily defined as occurring from sea level to roughly 500 m in elevation. This includes the tropical and subtropical basal latitudinal regions (Fig. 12.2), plus some of the lower-elevation tropical premontane altitudinal belt.

Mangrove forests. Nothing is more typical of the tropics than mangroves along the coasts of many low-latitude regions of the world. Mangroves typify low-energy coasts, deltas, and estuaries, but they do not occur along beaches with high wave action (Lugo and Snedaker 1974). Because of much higher daily fluctuations of tides on the Pacific coast of Meso-America and the low base flow of rivers during the long dry season, mangrove forests extend inland for several kilometers and form extensive stands in the major Pacific coast deltas (Golfo de San Miguel, Río Grande de Térraba, Río Grande de Tárcoles, Golfo de Nicoya, Golfo de Fonseca, Bahía de Jiquilisco). In contrast, the Caribbean coast has lower daily tidal fluctuations, much greater input of freshwater, and a less severe dry season. Hence, on the Caribbean coast, mangrove forests occur in a narrower band and penetrate only a few hundred meters inland along rivers. As rainfall decreases and seasonality increases, there is a marked decrease in mangrove height (Lot-Helgueras et al. 1975; Pool et al. 1977; Hartshorn et al. 1984). The tallest (35–40 m) mangrove forests occur in the tropical wet forest life zone, with rainfall greater than 4000 mm yr^{-1}.

Meso-American mangroves, characterized by eight tree species in five genera and four families, are not nearly as species-rich as are Old World mangroves (Chapman 1975). Best known is the red mangrove, *Rhizophora mangle* (Rhizophoraceae), with its impressive arching stilt roots forming an almost impenetrable barrier at the sea edge of mangrove forests. On reaching the substrate, the aerial roots branch profusely; up to 50% of the lower root volume is gas space (Gill and Tomlinson 1977), which facilitates aeration of the roots in the anaerobic mud (Scholander et al. 1955).

Rhizophora mangle, *Avicennia germinans* (Avicenniaceae), *Pelliciera rhizophorae* (Theaceae), *Conocarpus erecta*, and *Laguncularia racemosa* (both Combretaceae) usually form distinct zones along a salinity gradient (Cintron et al. 1978). Ecologists generally interpret mangrove zonation as representing a successional sequence, but Lugo (1980) argued that mangroves are steady-state ecosystems that undergo cyclic succession. Rabinowitz (1978) found that the dispersal properties of mangrove propagules correlate well with the spatial distribution of adult mangroves. Mangrove seedlings grow equally well or better when planted under different mangrove species than when under conspecifics (Rabinowitz 1975); thus, tidal sorting of available propagules may influence mangrove zonation.

Freshwater swamp forests. In the wet lowlands, freshwater swamp forests often occur inland of mangrove forests, where salinity no longer restricts floristic composition. The frequency, depth, and duration of flooding appear to be important determinants of species composition and dominance in swamp forests (Budowski 1966; Holdridge et al. 1971). As with mangrove forests, some freshwater swamp forests are nearly pure stands of one tree species (Table 12.3). Two characteristic swamp trees (Fig. 12.3), *Mora oleifera* (Caesalpiniaceae) and

Table 12.3. *Stand table for lowland freshwater swamp forest at Osa, Costa Rica*[a]; *relative basal area (BA), density (D), frequency (F), and importance value (IV)*

Tree species	BA	D	F	IV
Mora oleifera	95	70	48	71
Amphitecna latifolia	1	15	19	12
Avicennia germinans	2	4	10	5
Tabebuia rosea	1	4	10	5
Luehea seemannii	0.2	2	5	2
Subtotal top five species	100	96	90	95

[a]Site 8B: 2 m elevation; 0.2 ha; 7 species and 47 trees (≥10 cm dbh); BA = 7.0 m^2; stand height = 26 m.
Source: Relative data from Holdridge et al. 1971:243).

Figure 12.3. *Freshwater swamp forest in the Darién of Panama. The buttressed trees are* Pterocarpus officinalis, *and the nonbuttressed trees in the foregound are* Prioria copaifera.

Table 12.4. *Stand table for lowland freshwater swamp forest at Río Colorado, Costa Rica*[a]*; abbreviations as in Table 12.3*

Tree species	BA	D	F	IV
Prioria copaifera	95	80	50	75
Stemmadenia donnell-smithii	1	13	29	14
Pithecellobium latifolium	3	3	11	6
Grias fendleri	0.1	2	7	3
Ixora finlaysoniana	0.1	1	4	2
Subtotal top five species	100	100	100	100

[a]Site 19C: 5 m elevation; 0.3 ha; 5 species and 87 trees (≥10 cm dbh); BA = 16.5 m^2; stand height = 47 m. *Source:* Relative data from Holdridge et al. (1971:314).

Pterocarpus officinalis (Fabaceae), also occur occasionally in mangrove forests. *P. officinalis* may form a pure stand of large trees, such as near the mouth of the Río Llorona in Costa Rica's Corcovado National Park (Janzen 1978). It is unclear how the Llorona stand became established, but it probably was not induced by humans.

Swamp forests of cativo, *Prioria copaifera* (Caesalpiniaceae), are truly impressive because of the abundance of large trees (Table 12.4). Unfortunately, all pure cativo stands have been logged for plywood core stock (Mayo Meléndez 1965). Cativo seems to do best where there are strong seasonal pulses of flooding, with substantial intervals of moderately good drainage. Along the Caribbean, cativo occurs from Colombia to Nicaragua, whereas it is restricted to Panama and Colombia on the Pacific coast. Cativo is surprisingly absent from southwestern Costa Rica.

Even more restricted in occurrence is the oré tree, *Campnosperma panamensis* (Anacardiaceae), which forms pure stands in the Laguna de Chiriquí region of Panama's Bocas del Toro province (Holdridge and Budowski 1956), as well as along coastal Costa Rica and Nicaragua. The oré swamp forests have exceptionally high volume, averaging 382 m^3 ha^{-1} for stems greater than 40 cm dbh (Falla 1978).

Palms dominate some swamp forests on the coastal plains (Fig. 12.4). The yolillo palm, *Raphia taedigera* (Palmae), is common in southwestern and northeastern Costa Rica (Table 12.5). In a detailed study of the Tortuguero swamp forests, Myers (1981) showed that *Manicaria saccifera* (Palmae) dominates the more interior palm swamps. In northeastern Costa Rica, *Raphia* lines the natural levees along slow-moving rivers. It also occurs in

Figure 12.4. Freshwater swamp forest in Corcovado National Park, Costa Rica. The dominant palm is Raphia taedigera.

Table 12.5. *Stand table for lowland freshwater swamp forest at Tortuguero, Costa Rica[a]; abbreviations as in Table 12.3*

Tree species	BA	D	F	IV
Raphia taedigera	98	84	65	82
Pentaclethra macroloba	1	4	9	5
Grias fendleri	0.3	4	9	4
Crudia acuminata	0.1	3	7	3
Prioria copaifera	0.1	1	4	2
Subtotal top five species	99	96	93	96

[a]0.3 ha; 9 species and 191 trees (≥10 cm dbh); BA = 103.2 m^2; stand height = 15.3 m.
Source: Relative data from Myers (1981:65).

natural sloughs farther inland (e.g., in the Sarapiquí region of Costa Rica) that are inundated during heavy rains with water backed up by flooding rivers (Anderson and Mori 1967).

More heterogeneous swamp forests occur farther inland or peripheral to the pure stands described earlier. At the La Selva Biological Station in northeastern Costa Rica, the primary swamp forest is dominated by *Pentaclethra macroloba* (Mimosaceae) and *Carapa guianensis* (Meliaceae). A 1-ha plot had 79 species of trees and lianas (≥10 cm dbh), which is only about 20 species per hectare less than in nearby firm-ground forest (Table 12.1). In the same La Selva swamp forest there are dramatic reductions in stem density and species richness on sites with poorer drainage because of the exclusion of tree species intolerant of waterlogged soils (Lieberman et al. 1985). Several typical swamp tree species also occur in lower densities in non-swamp habitats. The absence of an effective dry season may permit swamp species like *Pentaclethra* to grow on well-drained ridges (Hartshorn 1983b). The occurrence of cativo on seasonally dry BCI (Croat 1978) is more difficult to interpret.

Most of the dominant tree species in swamp forests have large seeds that float (Janzen 1983b; McHargue and Hartshorn 1983). Many large seeds or fruits are common drift debris on tropical beaches (Gunn and Dennis 1976). The coconut palm, *Cocos nucifera*, is so widely dispersed along tropical coasts that it is difficult to determine its geographical origin (Gruezo and Harries 1984). The large cotyledonary reserves of swamp species seeds not only enhance viability while floating but also may facilitate better anchoring of the seedling in soft soil, or raise the first leaves above the typical level of floodwaters.

Perhumid forests. Lowland perhumid forests are often called tropical rain forest (*sensu* Richards 1952). A tropical rain forest life zone (*sensu* Holdridge 1967) does not occur in Meso-America; hence, our lowland perhumid forests fall mostly in the tropical wet forest, tropical premontane wet forest, and subtropical wet forest life zones. These lowland perhumid life zones are extensive along the Caribbean coast of western Panama, northeastern Costa Rica, much of southeastern Nicaragua, northern Honduras, from the lower Motagua Valley of Guatemala northwest to the Transversal, and the Caribbean lowlands of Veracruz, Mexico. Perhumid life zones also occur in the Pacific lowlands of Guatemala and southern Costa Rica. These two isolated perhumid areas may be caused by tall mountains (>3000 m) partially blocking the northeast tradewinds and creating a vortex that brings moisture-laden air inland off the Pacific Ocean.

Lowland perhumid forests (Fig. 12.5) are the most species-rich plant communities in Meso-America (Table 12.1). Even where there is moderate dominance by one species (Table 12.6), the heterogeneous forest averages about 100 tree species per hectare (Lieberman et al. 1985). Palms, which may be subcanopy trees (Fig. 12.6), understory treelets, or dwarf species only a meter tall, are a conspicuous component of lowland perhumid forests. Seven palm species constitute 25.5% of the stems (≥10 cm dbh) on 12.4 ha of permanent inventory plots at La Selva (Lieberman et al. 1985).

Lowland perhumid forests are considered to have three tree strata – canopy, subcanopy, and understory – though it is often difficult or impossible to recognize distinct layers. Even if discrete tree strata do exist, the functional significance of layers for light interception, pollination, seed dispersal, and other processes has not yet been studied. Some canopy tree species are facultatively deciduous for a few weeks in response to a short-term moisture deficit (Borchert 1980; Reich and Borchert 1982). A few canopy species like *Ceiba pentandra* (Bombacaceae) and *Tabebuia guayacan* (Bignoniaceae) are deciduous for a few months, during which time massive flowering occurs for a few days.

Dependent life forms such as epiphytes (Fig. 12.7), woody climbers (lianas), herbaceous vines, and stranglers are abundant in lowland perhumid forests. Epiphytic orchids, ferns, bromeliads, gesneriads, and so forth, are most prevalent on the branches of canopy trees (Perry 1978). Even cacti occur as epiphytes in lowland perhumid forests. Lianas are well represented in the families Apocynaceae, Bignoniaceae, Dilleniaceae, and Marcgraviaceae. The largest may attain 35–40 cm dbh and be more than 100 m long as they spread from crown to crown through the canopy. Climbing

Figure 12.5 Lowland perhumid forest at La Selva, Costa Rica. The emergent crown (center is Dipteryx panamensis.

Table. 12.6. *Stand table for lowland perhumid forest at La Selva, Costa Rica*[a]*; abbreviations as in Table 12.3*

Tree species	BA	D	F	IV
Pentaclethra macroloba	28	16	11	19
Welfia georgii	3	13	11	9
Protium species	4	6	7	6
Guarea species	2	7	4	4
Carapa guianensis	9	1	2	4
Subtotal top five species	45	44	34	41

[a]Site 11: 60 m elevation; 0.8 ha; 65 species and 328 trees (≥10 cm dbh); BA = 41.1 m^2; stand height = 46 m.
Source: Relative data from Holdridge et al. (1971:367).

vines are epitomized by the aroids, with dozens of species of *Anthurium*, *Monstera*, *Philodendron*, and *Syngonium* (Strong and Ray 1975; Ray 1983a, 1983b).

Though some stranglers belong to the genus *Clusia* (Guttiferae), most are figs (*Ficus*, Moraceae). A strangler fig begins life as an epiphyte whose aerial roots clasp the host tree and eventually form a stout lattice of coalesced aerial roots that can stand as an independent tree long after the host tree has died and decomposed. The popular notion of "strangling" has never been proved scientifically; rather, I have noted that stranglers usually have as hosts shade-intolerant gap species and that shad-

Figure 12.6. Stilt roots supporting the subcanopy palm, Socrates durissima, *at La Selva, Costa Rica.*

Figure 12.7. Tremendous epiphyte load on an emergent Tabebuia guayacan *at La Selva, Costa Rica. The 45-m-tall tree is deciduous during the brief dry season.*

ing by the strangler crown may cause death of the shade-intolerant host tree.

Tree-fall gaps in the primary forest canopy are essential for successful regeneration of up to 50% of the tree species (Hartshorn 1980). Shade intolerance is more common (63%) among canopy tree species at La Selva, decreasing in subcanopy (43%) and understory (38%) tree species. Based on a long-term study of gap occurrence in three permanent inventory plots, Hartshorn (1978) calculated an average turnover rate of 118 yr for the La Selva primary forest. Gap-phase dynamics appears to be important in the maintenance of species-rich perhumid forests in Meso-America (Hartshorn 1980; Brokaw 1982, 1985; Lang and Knight 1983; Pickett 1983; Hubbell and Foster 1985a), as in other tropical areas (Whitmore 1984).

Humid forests. Lowland humid forests in the tropical moist forest, tropical premontane moist forest, and subtropical moist forest life zones (Fig. 12.1) were originally the most extensive vegetation type in Meso-America. Large areas of these life zones occur in eastern, central, and western Panama, northwestern and northern Costa Rica, northeastern Nicaragua, eastern Honduras, northern Petén in Guatemala, northern Belize, and the southern Yucatán Peninsula in Mexico.

Table 12.7. *Stand table for lowland humid forest at Barranca, Costa Rica[a]; abbreviations as in Table 12.3*

Tree species	BA	D	F	IV
Scheelea rostrata	35	33	23	30
Luehea seemannii	22	11	10	14
Bravaisia integerrima	3	8	7	6
Enterolobium cyclocarpum	8	1	1	3
Spondias mombin	1	3	4	3
Subtotal top five species	70	56	44	57

[a]Site 2A: 40 m elevation; 0.6 ha; 49 species and 304 trees (≥10 cm dbh); BA = 28 m^2; stand height = 45 m.
Source: Relative data from Holdridge et al. (1971:147).

Tropical humid forests are nearly as tall and impressive as tropical wet forests (Table 12.7), whereas subtropical humid forests are shorter, with only two tree strata. Lowland humid forests have an appreciable proportion (25%−50%) of the canopy deciduous during the dry season. Epiphytic orchids and gesneriads are less abundant than in lowland perhumid forests; however, supple, thin, woody vines are more abundant in lowland humid forests (Holdridge et al. 1971). Putz (1984) counted 1597 climbing lianas on 1 ha of old-growth forest on BCI. Liana recruitment to the forest canopy is usually on the edge of tree-fall gaps and by vegetative sprouting (Peñalosa 1984; Putz 1984). Robust palms, especially *Scheelea rostrata*, are common subcanopy trees. Because adults are fire-tolerant and juveniles build a substantial trunk below ground, *Scheelea* is an aggressive colonizer of deforested land (Sousa 1964). It is a conspicuous component of pastures in western Panama, Costa Rica's El General Valley and Nicoya Peninsula, the northern foothills of Honduras, and southeastern Mexico.

The seasonality of rainfall and drought not only determines many community functions and ecosystem processes in lowland humid forests but also has profound effects on faunal populations (Leigh et al. 1982). Particularly striking is the strong synchronization of massive flowering with the dry season (Janzen 1967; Frankie et al. 1974; Croat 1975; Opler et al. 1976; Foster 1982a). Dry-season flowering while tree crowns are leafless facilitates foraging by pollinators such as large bees (Janzen 1971; Frankie 1975; Frankie et al. 1976). Borchert (1983) found that seasonal moisture stress is the proximal cause of dry-season flowering. An unusually wet 1970 dry season on BCI limited the normal pulse of tree flowering at the start of the rainy season, causing a threefold decrease in rainy-season fruiting and a severe famine among frugivores (Foster 1982b).

The pioneering studies on spatial distributions of Meso-American tree species by Hubbell (1979) and Hubbell and Foster (1983, 1985a, 1985b) indicated that very few species have the classic hyperdispersion so often suggested in the literature (e.g., (Richards 1952). Many tree species have a random dispersion of adults, whereas some species are clumped. Tree species dispersion patterns have important consequences for pollination syndromes (Janzen 1971; Frankie 1975; Feinsinger 1976; Heithaus 1979; Bawa et al. 1985a), tree breeding systems (Dobzhansky 1950; Baker 1970; Bawa 1974, 1980; Bawa and Opler 1975, 1977; Frankie et al. 1976; Opler and Bawa 1978; Bawa and Beach 1983; Bawa et al. 1985b), herbivory (Janzen 1969, 1970a), and the gap-phase regeneration of shade-intolerant trees (Hartshorn 1978, 1980; Denslow 1980; Brokaw 1982, 1985; Augspurger 1984).

Mahogany, *Swietenia macrophylla* (Meliaceae), is largely restricted to lowland humid forests, where it is more abundant in the subtropical moist forest life zone than in the tropical moist forest life zone. South of Nueva Guinea, Nicaragua, mahogany abruptly drops out of natural forests as the life zone changes to tropical wet forest. Mahogany was once quite common in southern Yucatán and northern Guatemala and Belize, where it was the principal object of logging efforts for more than two centuries (Lamb 1966; Hartshorn et al. 1984).

The Mayan civilization that developed in the lowland humid areas of Belize, Guatemala, Honduras, and Mexico is believed to have favored useful plants such as mahogany, zapote (*Manilkara zapota*, Sapotaceae), and ramón (*Brosimum alicastrum*, Moraceae) (Lundell 1933; Lambert and Arnason 1978). The demise of the classic Mayan civilization occurred about A.D. 900 (Deevey et al. 1979; Turner and Harrison 1983; Driever and Hoy 1984); thus, Mayan fields have been abandoned for more than a millennium. Recent studies of tropical forest dynamics suggest that 1000 yr should be ample time for succession to have attained climax status in the region. I observed no major differences in mahogany abundance or stand structure between mature subtropical moist forests of northern Belize and those on the Beni Plain at the northern base of the Bolivian Andes, which suggests that the effects of Mayan agriculture have long since disappeared. Future demographic studies such as those by Hartshorn (1975) and Sarukhán (1978) in the protected forests at the Tikal World Heritage Site could clarify the successional status of mahogany in these humid forests.

Subhumid forests. Lowland subhumid forests and woodland occur in the tropical dry forest (Fig. 12.8), subtropical dry forest, and subtropical thorn woodland life zones. This vegetation type was once common in a wide arc bordering a large area: Panama's Bahía de Parita; Costa Rica's lower Tempisque Valley; north and west of Lake Managua; extensive areas around the Golfo de Fonseca shared by Nicaragua, Honduras, and El Salvador; much of El Salvador's La Unión department, eastern San Vicente and Cabañas departments, and the upper Río Lempa Valley in Chalatenango and Santa Ana departments; substantial areas in Guatemala's Baja Verapaz, El Progreso, Chiquimula, and Zacapa departments; and Mexico's northern Yucatán Peninsula and along the central Pacific lowlands. Because of a long history of extensive grazing, frequent burning, and some agriculture, nearly all subhumid forests have been eliminated or severely degraded. Costa Rica's Santa Rosa National park, Palo Verde Wildlife Refuge and National Park, and Mexico's Chamela Biological Station are the only three protected areas with substantial representative vegetation of lowland subhumid forests in Meso-America. Tree species richness is substantially lower in these forests than in lowland humid and perhumid forests (Tables 12.1 and 12.8, p. 379).

Lowland subhumid forests are seasonally deciduous (Fig. 12.9), with only two tree strata, whose canopy seldom exceeds 20 m in height. The shrub layer is often dense, with multiple stems and thorns. Thin, woody vines are common, but epiphytes are sparse. Columnar cacti (e.g., *Lemaireocereus aragonii*) are occasional on xeric sites. *Jacquinia pungens* (Theophrastaceae) is a small understory tree that is deciduous during the rainy season, but in the dry season it takes advantage of sunlight passing through the barren overstory to store appreciable photosynthate, as well as to flower and fruit (Janzen 1970b). While leafless during the rainy season, an individual can lose up to 50% of the starch stored in the stem (Janzen and Wilson 1974).

Taller, evergreen forests occur along streams or in limestone basins where groundwater is available. Riparian forests are distinct floristically from the adjoining deciduous forest, often with species from nonriparian habitats of wetter life zones (Table 12.9). Some characteristic tree species of hillside subhumid forests have the opposite habitat pattern: *Tabebuia rosea* (Bignoniaceae) and *Bursera simaruba* (Burseraceae) are three times larger on alluvial soils in perhumid forests, where they attain 100 cm dbh and 40 m in height, than in subhumid forests.

In contrast to the monospecific stands in some freshwater swamps mentioned earlier, pure stands

Figure 12.8. Lowland subhumid forest at Palo Verde, Costa Rica. At the beginning of the dry season, the light-colored crowns are Calycophyllum candidissimum *in flower.*

Figure 12.9. Lowland subhumid forest interior at Palo Verde, Costa Rica. Late in the dry season, most trees and shrubs are deciduous. Note the accumulation of leaf litter on the forest floor.

Table 12.8. *Stand table for lowland subhumid forest at Palo Verde, Costa Rica*[a]; *abbreviations as in Table 12.3*

Tree species	BA	D	F	IV
Calycophyllum candidissimum	21	6	6	11
Licania arborea	7	6	6	6
Brosimum alicastrum	9	4	3	5
Spondias mombina	5	5	5	5
Guazuma ulmifolia	3	6	5	5
Subtotal top five species	46	28	24	33

[a]10 m elevation; 4.0 ha; 68 species and 691 trees (≥10 cm dbh); BA = 79.4m^2; stand height = 20 m.
Source: Relative data from Hartshorn (1983a:131).

Table 12.9. *Stand table for riparian lowland subhumid forest at Taboga, Costa Rica*[a]; *abbreviations as in Table 12.3*

Tree species	BA	D	F	IV
Luehea seemannii	22	15	16	18
Guarea excelsa	10	21	15	15
Scheelea rostrata	11	16	15	14
Terminalia oblonga	5	10	11	9
Anacardium excelsum	13	4	5	8
Subtotal top five species	61	67	62	63

[a]Site IF: 10 m elevation; 0.4 ha; 18 species and 67 trees (≥10 cm dbh); BA = 14.0 m^2; stand height = 33 m.
Source: Relative data from Holdridge et al. (1971:123).

Table 12.10 *Stand table for lowland subhumid oak forest at Bagaces, Costa Rica*[a]; *abbreviations as in Table 12.3*

Tree species	BA	D	F	IV
Quercus oleoides	58	34	18	37
Byrsonima crassifolia	6	12	11	10
Apeiba tibourbou	6	11	11	9
Spondias mombin	7	8	8	7
Cordia alliodora	2	5	7	5
Subtotal top five species	79	70	55	68

[a]4.0 ha; 44 species and 814 trees (≥10 cm dbh); BA = 50.7 m^2; stand height = 17 m.
Source: Relative data from Hartshorn (1983a:127).

in subhumid and some humid lowlands are usually attributable to edaphic factors, such as pumice or montmorillonite clay. The calabash tree (*Crescentia alata*, Bignoniaceae), *Erythrina fusca* (Fabaceae), palo verde (*Parkinsonia aculeata*, Caesalpiniaceae), and a white oak (*Quercus oleoides*, Fagaceae) often form pure stands or dominate forests on restrictive sites (Holdridge et al. 1971; Hartshorn 1983a). *Quercus oleoides* is the only lowland oak in Meso-America; it occurs in a series of disjunct populations that extend from Guanacaste, Costa Rica, to Tamaulipas, Mexico (Montoya Maquín 1966). In northwestern Costa Rica it once dominated (Table 12.10) the soils derived from pumice, but most of these forests have been converted to rangeland of the naturalized African savanna grass *Hyparrhenia rufa* (Daubenmire 1972). *Q. oleoides* has thick bark that effectively protects it from dry-season fires that sweep the rangeland. Seedlings can also survive fire because the seed reserves are translocated to a tuberous radicle extending 4–8 cm below the soil surface (Boucher 1983).

Savannas. By definition, savanna has a continuous grass cover (Beard 1953). According to the UNESCO classification system, trees may form up to 30% of the plant cover in savannas (Mueller-Dombois and Ellenberg 1974:479). Considerable areas of Meso-American vegetation meet these criteria for savanna; however, it is not clear if they are natural savannas or are derived from forest or woodland through repetitive burning by humans (Cook 1909; Budowski 1956; Sauer 1958). Regardless of origin, Meso-American savannas are not nearly as vast or as ecologically important as the Colombian-Venezuelan llanos and the pampas of the Bolivian Beni.

The extensive rangelands in the Pacific lowlands of Meso-America can be considered derived savannas created by almost annual burning to remove the rank old growth of grass and to suppress woody invasion. Meso-American savanna habitats are most prevalent in the tropical dry forest, tropical moist forest, subtropical dry forest, and subtropical moist forest life zones. Many of these derived savannas are dotted with fire-resistant tree species such as *Acrocomia vinifera* (Palmae), *Byrsonima crassifolia* (Malpighiaceae), *Crescentia alata*, and *Curatella americana* (Dilleniaceae). The first three can be found occasionally in remnant subhumid forests, but *Curatella* is restricted to xeric savanna habitats where the soil is very shallow over pumice or a plinthic hardpan.

Lowland pine (*Pinus caribaea* var. *hondurensis*) savannas occur in northeastern Nicaragua, much of Honduras, part of the Guatemalan Petén, and east-central Belize. In contrast to the derived savannas of the Pacific lowlands, pine savannas appear to be natural associations maintained by frequent fire. The flat pine savannas of Belize occur on sandy soils that usually flood during the rainy season (Wright et al. 1959). The severe seasonal drying of sandy soils make the vegetation highly flammable. The palmetto, *Paurotis wrightii* (Palmae), is a conspicuous component of the frequently burned and flooded savannas of Belize (Anderson and Fralish 1975).

Inland from the Misquito coast of Nicaragua and adjoining Honduras, extensive pine savannas occur on gravelly outwash and finer quartz sediments (Parsons 1955; Taylor 1962, 1963; Alexander 1973). The infertile soil, in conjunction with frequent burning, delimits pine savanna in northeastern Nicaragua (Alexander 1973). The presence of pine savanna and broad-leaved savanna in humid and subhumid life zones indicates that climate is not the primary determinant of Meso-American savannas. Rather, edaphic factors such as soil infertility and shallowness, coupled with frequent burning, have determined the distribution of savanna. Kellman (1984) suggested that vegetation on infertile soils is more susceptible to fire than on fertile soils, because slower growth delays canopy closure, thus permitting persistence of herbaceous grasses.

Low Mountains

For purposes of this chapter, low mountains have been given arbitrary altitudinal limits of roughly 500–2000 m. It is important to remember that each pair of life zones in subtropical and warm temperate latitudinal regions, or in premontane and lower-montane altitudinal belts (Fig. 12.1), has similar vegetation physiognomy. For example, subtropical wet forest and subtropical lower-montane wet forest life zones have similar stand structures, but differ floristically. The critical-temperature line (Fig. 12.1) separates species that can tolerate occasional frost from species intolerant of near-freezing temperatures. The upper limit for coffee plantations in Meso-America coincides well with the critical-temperature line separating the tropical premontane belt from the lower-montane altitudinal belt, and the subtropical basal belt from the subtropical lower-montane altitudinal belt (Holdridge 1967).

The similarity in vegetation physiognomy and the high species richness within a hexagon make it difficult to detect the change between two life zones separated by the critical-temperature line (Fig. 12.1). In northern Meso-America, the presence of temperate forest genera (*Alnus, Carpinus, Cornus, Nyssa, Ostrya, Platanus, Ulmus*) is a good indication of the subtropical lower-montane altitudinal belt. In southern Meso-America, several tree genera have a few closely related species that are geographically separated by the critical-temperature line: *Billia* (Hippocastanaceae); *Brunellia* (Brunelliaceae); *Byrsonima* (Malpighiaceae); *Cedrela* (Meliaceae); *Didymopanax* (Araliaceae); *Hedyosmum* (Chloranthaceae); *Pithecellobium* (Mimosaceae); *Rapanea* (Myrsinaceae); *Symplocos* (Symplocaceae); and *Turpinia* (Staphylaeaceae).

Subhumid forests. This vegetation type was once extensive on the lower- to middle-elevation tablelands and valleys that occur over much of northern Meso-America. Intermountain valleys usually are in the subtropical thorn woodland or subtropical dry forest life zone, whereas the higher slopes are in the subtropical lower-montane dry forest life zone. These cooler tablelands have been inhabited by humans for centuries and have been severely degraded by overgrazing and fire. The relatively open forests are generally of short stature (<10 m tall), with an abundance of spiny, mimosoid legumes, such as *Pithecellobium dulce, P. flexicaule,* and *Prosopis laevigata*. In Mexico, this vegetation type is called "low, spiny, evergreen forest" or "low, spiny, deciduous forest" (Miranda and Hernández 1963). *Cercidium praecox* is a characteristic tree of the subtropical thorn woodland life zone in Guatemala's Zacapa Valley (Holdridge 1956). On upper slopes around Chimaltenango, Guatemala, *Pinus montezumae* is common (Holdridge 1957a).

Humid forests. Tropical premontane moist forest, subtropical moist forest, and subtropical lower-montane moist forest life zones are widespread on the low mountains of Meso-America. In the northern region there is a complex mosaic of pine forests on the upper slopes or tablelands, and broad-leaved forests in the valleys. On the middle slopes of the Sierra Madre Oriental of Mexico, the latter vegetation type is called "deciduous forest" (Miranda and Hernández 1963). It has a mixture of temperate trees (mentioned earlier) with tropical trees such as *Brunellia* (Brunelliaceae), *Beilschmiedia* and *Phoebe* (both Lauraceae), and *Alchornea* (Euphorbiaceae) (Pennington and Sarukhán 1968).

Low-mountain pine forests of Belize, Guatemala, Honduras, and Nicaragua usually are relatively pure stands of *Pinus oocarpa*, although some stands include *P. caribaea* (Denevan 1961; Johnson et al. 1973). An active fire-control program by the Belize Forest Department has successfully established excellent 20–30-yr-old pine forests in the Mountain Pine Ridge (Hartshorn et al. 1984). Because of fire suppression, there is a vigorous understory of broad-leaved trees, confirming the importance of fire in maintaining pine forests on these poor soils. Kellman (1979) reported that some broad-leaved trees accumulate nutrients on poor sites in the pine forest and suggested that these locally enriched sites may facilitate invasion by broad-leaved species typical of more fertile soils.

Low-mountain humid forests in southern Meso-America are restricted to intermountain valleys and Pacific slopes. This vegetation type has a relatively tall canopy (<40 m tall) that is fairly open and partially deciduous (Tables 12.11 and 12.12). Because of accessibility and hospitable climate, virtually all of this vegetation type has been deforested,

Table 12.11. *Stand table for low-mountain humid forest at Alajuela, Costa Rica[a]; abbreviations as in Table 12.3*

Tree species	BA	D	F	IV
Luehea candida	29	15	9	18
Phoebe mexicana	6	11	11	9
Stemmadenia obovata	3	12	8	8
Roupala complicata	4	10	5	7
Anacardium excelsum	8	4	5	6
Subtotal top five species	50	52	39	47

[a]Site 18: 800 m elevation; 0.5 ha; 44 species and 245 trees (≥10 cm dbh); BA = 16.8 m^2; stand height = 23 m.
Source: Relative data from Holdridge et al. (1971:337).

Table 12.12. *Stand table for low-mountain humid forest at Turrialba, Costa Rica[a]; abbreviations as in Table 12.3*

Tree species	BA	D	F	IV
Brosimum alicastrum	25	17	11	18
Guarea aligera	5	9	7	7
Hymenolobium pulcherrimum	9	6	6	7
Simarouba amara	4	8	7	6
Anacardium excelsum	16	1	1	6
Subtotal top five species	58	41	32	44

[a]Site 3: 590 m elevation; 0.4 ha; 71 species and 236 trees (≥10 cm dbh); BA = 17.7 m^2; stand height = 40 m.
Source: Relative data from Holdridge et al. (1971:178).

mostly for cattle pastures (Hartshorn et al. 1982; Heckadon Moreno 1983).

Perhumid forests. Perhumid (including superhumid) forests of low mountains occur in the tropical premontane wet forest, tropical premontane rain forest, tropical lower-montane wet forest, tropical lower-montane rain forest, subtropical lower-montane wet forest, and subtropical lower-montane rain forest life zones. This vegetation occurs in a fairly broad band along the "front ranges" (Fig. 12.10) exposed to the northeast tradewinds, as well as ringing most of the volcanoes, and on some of the Pacific middle slopes (e.g. Costa Rican Talamancas). Where the physiographic position is more or less perpendicular to the northeast tradewinds, luxuriant cloud forest usually occurs (Vogelmann 1973; Cruz and Erazo Peña 1977; Reyna Vásquez 1979; Lawton and Dryer 1980).

Low-mountain perhumid forests have a well-developed canopy 30–45 m tall (Tables 12.13 and 12.14), with occasional emergents 50–55 m tall, including *Ulmus mexicana* and *Quercus copeyensis*. In the Sierra Negra of Honduras, Vogel (1954) measured a *Pinus pseudostrobus* 231 cm dbh and more than 60 m tall. Canopy trees are mostly evergreen although a few are briefly deciduous. Several species of oaks and other temperate genera (mentioned earlier) are conspicuous components of the lower-montane canopy. These perhumid forests are characterized by a profusion of epiphytic bryophytes and vascular plants. Trunks and

Figure 12.10. Low-mountain perhumid forest in the rugged mountains of Braulio Carrillo National Park, Costa Rica. Quaternary lava flows are deeply dissected by streams, such as the Río Puerto Viejo and its tributaries shown here.

Table 12.13. *Stand table for low-mountain perhumid forest at Volcán, Costa Rica*[a]*; abbreviations as in Table 12.3*

Tree species	BA	D	F	IV
Vantanea barbourii	38	13	12	21
Socretes durissima	11	36	13	20
Brosimum utile	20	3	3	8
Dendropanax arboreus	2	5	7	5
Conostegia globulifera	1	4	5	3
Subtotal top five species	71	61	41	58

[a]Site 7: 620 m elevation; 1.0 ha; 67 species and 503 trees (≥10 cm dbh); BA = 22.7 m^2; stand height = 44 m.
Source: Relative data from Holdridge et al. (1971:355).

Table 12.14. *Stand table for low-mountain perhumid forest at Valle Escondido, Costa Rica*[a]*; abbreviations as in Table 12.3*

Tree species	BA	D	F	IV
Euterpe macrospadix	6	17	9	11
Oreomunnea mexicana	18	4	4	8
Inga species	6	9	6	7
Chrysophyllum mexicanum	9	5	5	6
Clusia pithecobia	7	5	4	5
Subtotal top five species	46	40	29	38

[a]Site 22E: 1100 m elevation; 0.4 ha; 66 species and 281 trees (≥10 cm dbh); BA = 13.12 m^2; stand height = 38 m.
Source: Relative data from Holdridge et al. (1971:457).

branches are festooned with mosses, leaves are covered with epiphylls, and even epiphytic trees and shrubs are abundant.

The variations in vegetation structure and floristic composition on tropical mountains have long attracted the attention of plant ecologists (Shreve 1914; Beard 1949; Holdridge et al. 1971). Nevertheless, considerable controversy still exists over the physical factors controlling the distribution of forest types (Leigh 1975; Grubb and Tanner 1976; Grubb 1977). In a study of low-mountain perhumid forest in Jamaica, Tanner and Kapos (1982; Kapos and Tanner 1985) found that leaves are smaller and thicker in the cloud forest than in the lowlands, but the xeromorphic leaves in cloud forest do not seem to be an adaptation to moisture limitation (Kapos and Tanner 1985).

In the Monteverde Cloud Forest Reserve of Costa Rica (Hartshorn 1983a), the location, structure, and floristic composition of forest types are controlled mainly by the degree of exposure to the northeast tradewinds (Lawton and Dryer 1980). The northeast tradewinds roar up the Peñas Blancas Valley to spill over the low (1500–1600 m) continental divide. Lawton and Dryer (1980) described six forest types for the Monteverde Reserve: cove, leeward cloud, oak ridge, windward cloud, elfin, and swamp communities. Increasing exposure to the tradewinds increases rainfall and wind shear, which result in luxuriant epiphyte loads, shorter forest stature, and a more open canopy. The smoothly sculptured canopy of the elfin forest (Fig. 12.11) is an exception to the pattern of less continuous canopy with increasing exposure to the tradewinds; gusts regularly exceed 100 km hr^{-1} across the Brillante gap (Lawton and Dryer 1980).

The abundance of brownish epiphytic mosses (Fig. 12.12) gives such a distinctive appearance to lower-montane wet forest and rain forest canopies that it is possible to recognize them from a low-flying airplane. Also notable is the abundance of ericaceous epiphytic shrubs (*Cavendishia, Gaultheria, Gonocalyx, Macleania, Psammisia, Satyria*). The profusion of epiphytes in cloud forests results in a considerable amount of organic matter on the branches. Breakage of heavily laden branches is a common source of canopy gaps (Lawton and Dryer 1980). Some host trees produce adventitious roots in the organic debris associated with epiphytes (Nadkarni 1984), thus partially short-circuiting the nutrient cycle.

High Mountains

High-mountain vegetation types generally occur above 2000–2500 m in Meso-America. The tropical and subtropical montane altitudinal belt is restricted to south central Mexico, western Guatemala, Costa Rica's Cordillera Volcánica Central, and the high Talamancas of Costa Rica and Panama. Tropical and subtropical subalpine vegetation occurs only on the high peaks of southern Mexico, western Guatemala, and the Costa Rican Talamancas. Guatemala's highlands are the low-latitude center of diversity for conifers: *Abies, Cupressus, Juniperus* (two species), *Pinus* (nine), *Podocarpus* (two), *Taxodium*, and *Taxus* (Holdridge 1957a, 1975b; Veblen 1976). Guatemala's volcanic history (Williams 1960), rugged terrain, and altitudinal variation have contributed to the richness of pine species.

Humid forests. The high-mountain humid forests of northern Meso-America are characterized by coniferous trees that are not native to southern Meso-America. Subtropical montane humid forests are fairly open stands, usually dominated by *Juniperus standleyi, Pinus ayacahuite*, or *P. rudis* (Holdridge

Figure 12.11. Elfin forest on the continental divide in the Monteverde Cloud Forest Reserve, Costa Rica. Wind shear by the northeast tradewinds contributes to the smoothly sculptured canopy of the elfin forest.

1956). Human uses of natural resources have destroyed or degraded most native forests through logging, slash-and-burn agriculture for maize, and overgrazing by livestock. An interesting exception in the Guatemalan highlands is the Totonicapán furniture industry based on *Pinus ayacahuite* dating to the sixteenth century (Veblen 1978). The remaining white pine forests are held communally and vigorously protected by the communities of woodcutters and carpenters dependent on the forests for their livelihood.

The tropical montane moist forest life zone is restricted to small rain-shadow areas in the high mountains of Costa Rica. These areas are mostly in agriculture, with remnant trees that suggest the original forests were dominated by oaks (Table 12.15).

Pinus hartwegii, *P. rudis*, and *Juniperus standleyi* occur up to 4000–4100 m elevation in northern

Figure 12.12. Moss-laden branches in the Monteverde Cloud Forest Reserve, Costa Rica.

Table 12.15. *Stand table for high-mountain humid forest at Irazú, Costa Rica*[a]*; abbreviations as in Table 12.3*

Tree species	BA	D	F	IV
Quercus copeyensis	66	32	19	39
Rapanea guianensis	4	11	12	9
Citharexylum lankesteri	6	10	10	9
Freziera candicans	5	8	10	8
Ocotea seibertii	3	8	9	7
Subtotal top five species	84	69	61	72

[a]Site 17: 2360 m elevation; 0.3 ha; 20 species and 37 trees (≥10 cm dbh); BA = 11.9 m^2; stand height = 33 m.
Source: Relative data from Holdridge et al. (1971:467).

Table 12.16. *Stand table for high-mountain perhumid forest at Villa Mills, Costa Rica*[a]*; abbreviations as in Table 12.3*

Tree species	BA	D	F	IV
Quercus costaricensis	61	41	20	41
Miconia biperulifera	10	16	15	14
Vaccinium consanguineum	4	10	12	9
Weinmannia pinnata	10	7	9	8
Didymopanax pittieri	5	4	6	5
Subtotal top five species	89	78	62	77

[a]Site 6: 3080 m elevation; 0.4 ha; 22 species and 245 trees (≥10 cm dbh); BA = 17.9 m^2; stand height = 30 m.
Source: Relative data from Holdridge et al. (1971:527).

Meso-America (Holdridge 1956; Veblen 1976). The occurrence of these high-mountain coniferous forests in the subtropical subalpine altitudinal belt (Fig. 12.2) indicates that the treeline is higher in northern than in southern Meso-America. The treeline in the Talamanca Mountains of Costa Rica occurs at about 3300–3500 m, above which is subalpine páramo. Longer days during summer apparently permit tree growth at higher elevations in extra-tropical regions.

Perhumid forests. *Abies guatemalensis* and *A. religiosa* often dominate the high-mountain perhumid forests of western Guatemala and south central Mexico, respectively (Holdridge 1956). Also present in these perhumid forests are *Cupressus lusitanica, Pinus pseudostrobus*, and *Podocarpus oleifolius* (Veblen 1976). Mixed oak-pine forests are also common in northern Meso-America.

The tropical montane wet forest and tropical montane rain forest life zones of the high Talamancas in Costa Rica are dominated by several species of oaks (Table 12.16). Large trees are heavily laden with epiphytes, particularly mosses, large tank bromeliads, and shrubby ericads. A scandent bamboo, *Chusquea tonduzii*, dominates the shrub layer over extensive areas of high-elevation oak forests. Many of these oak forests apparently are not regenerating, and the bamboo is hypothesized to be interfering with oak regeneration.

Páramo. In the Western Hemisphere, páramo is restricted to tropical subalpine regions of the Talamanca Mountains of Costa Rica and the northern Andes of South America (Salgado-Labouriau 1979). The Andean páramo is characterized by abundant *Espeletia* species (Compositae) that are absent from the Costa Rican páramo. Nevertheless, the floristic and physiognomic similarities are sufficient to consider the Costa Rican páramo an outlier of the Andea páramo (Weber 1959).

Tropical subalpine rain páramo of Costa Rica is a low scrub, with a few slightly taller trees. In the Cerro de la Muerte region, fires have lowered the treeline by 300–400 m (Fig. 12.13). Vegetative regeneration following fire is quite slow (Janzen 1973). During Quaternary glacials, the treeline was 700–900 m lower in the Talamanca Mountains (Martin 1964).

Hypericum species (Hypericaceae) and a miniature bamboo, *Swallenochloa subtessellata* (Gramineae), dominate the shrubby páramo. Where drainage is poor, bogs have an abundance of the cycad-like fern, *Blechnum buchtienii*, plus the Andean fern *Jamesonia* species and terrestrial bromeliad *Puya dasylirioides*. The high Talamanca vegetation also has strong floristic affinities with that of North America, with the presence of temperate genera such as *Alchemilla* (Rosaceae), *Castilleja* (Scrophulariaceae), *Cirsium* (Compositae), *Vaccinium* (Ericaceae), and others.

AREAS FOR FUTURE RESEARCH

The scientific research needs of developing countries have been the focus of many studies (NRC 1980a, 1982). It would be easy to list many intriguing questions about tropical vegetation. Rather, I prefer to stress the urgent need for conservation of tropical and subtropical vegetation. We are destroying the world's tropical and subtropical forests at alarming rates (NRC 1980b; FAO 1981). In the moist tropics, nowhere is deforestation occurring more rapidly than in Meso-America (Heckadon Moreno and McKay 1982; Hartshorn 1983c; Nations and Komer 1983). Excluding Belize, no country in the region has more than 30% of its territory still in natural, broad-leaved forests, and the remaining primary forests are being cut down at rates of 2%–5% per year. It seems inevitable that unprotected broad-leaved forests of Meso-America will have been destroyed well before the year 2000. The

Figure 12.13. Treeline near the Cerro de la Muerte, Costa Rica. The treeline is 300–400 m lower because of occasional burning of the subalpine páramo. The montane forest is dominated by oaks.

consequences of tropical deforestation for global climate, food crops, pharmaceutical drugs, extinction of species, soil erosion, hydrologic regimes and projects, desertification, and so forth, have received considerable attention (Parsons 1976; Prance and Elias 1977; Myers 1979, 1984; Ehrlich and Ehrlich 1981; Caufield 1985).

Economic difficulties hinder or threaten the existence of national parks and equivalent reserves in Meso-America (Hartshorn 1983c). In Costa Rica, foreign donations and assistance have played a key role in the establishment of an excellent national parks system (Boza and Mendoza 1981). Foreign financial assistance is essential for expansion and consolidation of conservation units and environmental education programs. Without public appreciation and support, however, few conservation units in Meso-America will survive into the next century.

Basic biological inventories of tropical and subtropical forests are woefully incomplete. Expeditions to remote areas easily turn up dozens of species new to science (Pringle et al. 1984). Without basic inventories, it is impossible to know what species are adequately protected in conservation units or to legitimately list endangered or threatened species. Particularly relevant to conservation is the role of gap-phase dynamics in the regeneration and maintenance of species in primary tropical forests. The stochastic nature of gap occurrence may maintain dozens of ecologically similar, shade-intolerant species in complex tropical forests. If tropical forests are nonequilibrium communities with hundreds of species sharing few ecological guilds determined by unpredictable disturbances, we will need to modify our thinking on how to conserve and manage tropical forests. We know far too little about managing tropical forests for particular species to be able to maximize community diversity or even to produce timber on a sustained-yield basis.

ACKNOWLEDGMENTS

This chapter is dedicated to Dr. L. R. Holdridge. Early drafts were improved by the comments and suggestions of Lynne Hartshorn, Les Holdridge, David Janos, Diana Lieberman, and Milton Lieberman.

REFERENCES

Alexander, E. B. 1973. A comparison of forest and savanna soils in northeastern Nicaragua. Turrialba (Costa Rica) 23:181–191.

Allen, P. H. 1956. The rain forests of Golfo Dulce. University of Florida Press, Gainesville.

Anderson, R., and S. Mori. 1967. A preliminary investigation of *Raphia* palm swamps, Puerto Viejo, Costa Rica. Turrialba 17:221–224.

Anderson, R. C., and J. S. Fralish. 1975. An investigation of palmetto, *Paurotis wrightii* (Griseb. & Wendl.) Britt., communities in Belize, Central America. Turrialba 25:37–44.

Aubréville, A. 1938. La forêt coloniale; les forêts de l'Afrique occidentale française. Ann. Acad. Sci. Colon. (Paris) 9:1–245.

Augspurger, C. K. 1984. Light requirements of neotropical tree seedlings: a comparative study of growth and survival. J. Ecol. 72:777–795.

Baker, H. G. 1970. Evolution in the tropics. Biotropica 2:101–111.

Bawa, K. S. 1974. Breeding systems of tree species of a lowland tropical community. Evolution 28:85–92.

Bawa, K. S. 1980. Evolution of dioecy in flowering plants. Ann. Rev. Ecol. Syst. 11:15–39.

Bawa, K. S., and J. H. Beach. 1983. Self-incompatibility systems in the Rubiaceae of a tropical lowland wet forest. Amer. J. Bot. 70:1281–1288.

Bawa, K. S., S. H. Bullock, D. R. Perry, R. E. Colville, and M. H. Grayum. 1985a. Reproductive biology of tropical lowland rain forest trees, II, Pollination systems. Amer. J. Bot. 72:346–356.

Bawa, K. S., and P. A. Opler. 1975. Dioecism in tropical forest trees. Evolution 29:167–178.

Bawa, K. S., and P. A. Opler. 1977. Spatial relationships between staminate and pistillate plants of dioecious tropical forest trees. Evolution 31:64–68.

Bawa, K. S., D. R. Perry, and J. H. Beach. 1985b. Reproductive biology of tropical lowland rain forest trees, I, Sexual systems and incompatibility mechanisms. Amer. J. Bot. 72:331–345.

Beard, J. S. 1944. Climax vegetation in tropical America. Ecology 25:127–158.

Beard, J. S. 1949. The natural vegetation of the Windward and Leeward Islands. Oxford Forestry Memoirs no. 21.

Beard, J. S. 1953. The savanna vegetation of northern tropical America. Ecol. Monogr. 23:149–215.

Borchert, R. 1980. Phenology and ecophysiology of tropical trees: *Erythrina poeppigiana* O. F. Cook. Ecology 61:1065–1074.

Borchert, R. 1983. Phenology and control of flowering in tropical trees. Biotropica 15:81–89.

Boucher, D. H. 1983. *Quercus oleoides* (Roble Encino, Oak). pp. 319–320 in D. H. Janzen (ed.), Costa Rican natural history. University of Chicago Press.

Boza, M. A., and R. Mendoza. 1981. The national parks of Costa Rica. INCAFO, Madrid.

Brokaw, N. V. L. 1982. Treefalls: frequency, timing, and consequences. pp. 101–108 in E. G. Leigh, Jr., A. S. Rand, and D. M. Winsor (eds.), The ecology of a tropical forest: seasonal rhythms and long-term changes. Smithsonian Institution Press, Washington, D.C.

Brokaw, N. V. L. 1985. Gap-phase regeneration in a tropical forest. Ecology 66:682–687.

Budowski, G. 1956. Tropical savannas, a sequence of forest felling and repeated burnings. Turrialba 6:23–33.

Budowski, G. 1959. Algunas relaciones entre la presente vegetación y antiguas actividades del hombre en el trópico americano. pp. 259–263 in Actas 33 Congreso Internacional de Americanistas. vol. 1. Lehmann, San José, Costa Rica.

Budowski, G. 1966. Los bosques de los trópicos húmedos de América. Turrialba 16:278–285.

Caufield, C. 1985. In the rainforest. Alfred Knopf, New York.

Chapman, V. J. 1975. Mangrove biogeography. pp. 3–22 in G. E. Walsh, S. C. Snedaker, and H. J. Teas (eds.), Proceedings of the international symposium on biology and management of mangroves. University of Florida Institute of Food and Agricultural Sciences, Gainesville.

Cintron, G., A. E. Lugo, D. J. Pool, and G. Morris. 1978. Mangroves of arid environments in Puerto Rico and adjacent islands. Biotropica 10:110–121.

Clark, D., R. Dirzo, and N. Fetcher (eds.). in press. Simposio sobre plantas mesoamericanas. Rev. Biol. Trop. (Costa Rica) [Suppl.]

Coen, E. 1983. Climate. pp. 35–46 in D. H. Janzen (ed.), Costa Rican natural history. University of Chicago Press.

Cook, O. F. 1909. Vegetation affected by agriculture in Central America. USDA Bureau of Plant Science Industry, bulletin no. 145.

Croat, T. B. 1975. Phenological behavior of habit and habitat classes on Barro Colorado Island (Panama Canal Zone). Biotropica 7:270–277.

Croat, T. B. 1978. Flora of Barro Colorado Island. Stanford University Press, Stanford, Calif.

Cruz, G. A., and M. Erazo Peña. 1977. Análisis de la vegetación del bosque nebuloso "La Tigra" (Reserva Forestal San Juancito). Ceiba (Tegucigalpa, Honduras) 21:19–60.

D'Arcy, W. D. 1977. Endangered landscapes in Panama and Central America: the threat to plant species. pp. 89–104 in G. T. Prance and T. S. Elias (eds.), Extinction is forever: threatened and endangered species of plants in the Americas and their significance in ecosystems today and in the future. New York Botanical Garden.

Daubenmire, R. 1972. Ecology of *Hyparrhenia rufa* (Ness) in derived savanna in north-western Costa Rica. J. Appl. Ecol. 9:11–23.

Davis, T. A. W., and P. W. Richards. 1933–4. The vegetation of Moraballi Creek, British Guiana: an ecological study of a limited area of tropical rain forest, Parts I & II. J. Ecol. 21:350–384, 22:106–155.

Deevey, E. S., D. S. Rice, P. M. Rice, H. H. Vaughan, M. Brenner, and M. S. Flannery. 1979. Mayan urbanism: impact on a tropical karst environment. Science 206:298–306.

De la Cruz, J. R. 1976. Mapa de zonas de vida de Guatemala. INAFOR, Guatemala.

Denevan, W. M. 1961. The upland pine forests of Nicaragua: a study in cultural plant geography. Univ. California Pub. Geogr. 12:251–320.

Denslow, J. S. 1980. Gap partitioning among tropical rainforest trees. Biotropica [Suppl.] 12:47–55.

Dobzhansky, T. 1950. Evolution in the tropics. Amer. Sci. 38:209–221.

Driever, S. L., and D. R. Hoy. 1984. Vegetation productivity and the potential population of the classic Maya. Singapore J. Trop. Geogr. 5:140–153.

Dwyer, J. D., and D. L. Spellman. 1981. A list of the Dicotyledoneae of Belize. Rhodora 83:161–236.

Ehrlich, P., and A. Ehrlich. 1981. Extinction: the causes

and consequences of the disappearance of species. Random House, New York.

Encyclopaedia Britannica. 1977. Jungles and rain forests. pp. 336–346 in Encyclopaedia Britannica, vol. 10, 15th ed. Encyclopaedia Britannica, Chicago.

Estrada, A., and R. Coates-Estrada. 1983. Rain forest in Mexico: research and conservation at Los Tuxtlas. Oryx 17:201–204.

Falla R., A. 1978. Plan de desarrollo forestal, parte II, Estudio de las perspectivas del desarrollo forestal en Panamá. PCT/6/PAN/01/1, no. 2, FAO, Panama.

FAO. 1981. Tropical forest resources assessment project (GC-MS): tropical America. FAO/UNEP, Rome.

Feinsinger, P. 1976. Organization of a tropical guild of nectarivorous birds. Ecol. Monogr. 46:257–291.

Foster, R. B. 1982a. The seasonal rhythm of fruitfall on Barro Colorado Island. pp. 151–172 in E. G. Leigh, Jr., A. S. Rand, and D. M. Windsor (eds.), The ecology of a tropical forest: seasonal rhythms and long-term changes. Smithsonian Institution Press, Washington, D.C.

Foster, R. B. 1982b. Famine on Barro Colorado Island. pp. 201–212 in E. G. Leigh, Jr., A. S. Rand, and D. M. Windsor (eds.), The ecology of a tropical forest: seasonal rhythms and long-term changes. Smithsonian Institution Press, Washington, D.C.

Frankie, G. W. 1975. Tropical forest phenology and pollinator plant coevolution. pp. 192–209 in L. E. Gilbert and P. H. Raven (eds.), Coevolution of animals and plants. University of Texas Press, Austin.

Frankie, G. W., H. G. Baker, and P. A. Opler. 1974. Comparative phenological studies of trees in tropical wet and dry forests in the lowlands of Costa Rica. J. Ecol. 62:881–919.

Frankie, G. W., P. A. Opler, and K. S. Bawa. 1976. Foraging behaviour of solitary bees: implications for outcrossing of a neotropical forest tree species. J. Ecol. 64:1049–1057.

Gill, A. M., and P. B. Tomlinson. 1977. Studies on the growth of red mangrove (*Rhizophora mangle* L.), 4, The adult root system. Biotropica 9:145–155.

Gómez-Pompa, A., P. L. Hernández, and S. M. Sousa. 1964. Estudio fitoecológico de la cuenca intermedia del Río Papaloapán. INIF (Mexico) Publ. Esp. 3:37–90.

Gómez-Pompa, A., C. Vásquez-Yanes, and S. Guevara. 1972. The tropical rainforest: a non-renewable resource. Science 177:762–765.

González, L., M. Ramírez, and R. Peralta. 1983. Estudio ecológico y dendrológico: zonas de vida y vegetación. ACDI-COHDEFOR, Tegucigalpa.

Grubb, P. J. 1977. Control of forest growth and distribution on wet tropical mountains. Ann. Rev. Ecol. Syst. 8:83–107.

Grubb, P. J., and E. V. J. Tanner. 1976. The montane forests and soils of Jamaica: a reassessment. J. Arnold Arboretum 57:313–368.

Gruezo, W. S., and H. C. Harries. 1984. Self-sown, wild-type coconuts in the Philippines. Biotropica 16:140–147.

Gunn, C. R., and J. V. Dennis. 1976. World guide to tropical drift seeds and fruit. Quadrangle, New York.

Hartshorn, G. S. 1975. A matrix model of tree population dynamics. pp. 41–51 in F. B. Golley and E. Medina (eds.), Tropical ecological systems: trends in terrestrial and aquatic research. Springer-Verlag, Berlin.

Hartshorn, G. S. 1978. Tree falls and tropical forest dynamics. pp. 617–638 in P. B. Tomlinson and M. H. Zimmermann (eds.), Tropical trees as living systems. Cambridge University Press.

Hartshorn, G. S. 1980. Neotropical forest dynamics. Biotropica [Suppl.] 12:23–30.

Hartshorn, G. S. 1983a. Plants: introduction. pp. 118–157 in D. H. Janzen (ed.), Costa Rican natural history. University of Chicago Press.

Hartshorn, G. S. 1983b. *Pentaclethra macroloba* (Gavilán). pp. 301–303 in D. H. Janzen (ed.), Costa Rican natural history. University of Chicago Press.

Hartshorn, G. S. 1983c. Wildlands conservation in Central America. pp. 423–444 in S. L. Sutton, T. C. Whitmore, and A. C. Chadwick (eds.), Tropical rain forest: ecology and management. Blackwell, London.

Hartshorn, G. S., and B. Hammel. 1982. Trees of La Selva. Organization for Tropical Studies mimeograph, San José, Costa Rica.

Hartshorn, G., L. Hartshorn, A. Atmella, L. D. Gomez, A. Mata, L. Mata, R. Morales, R. Ocampo, D. Pool, C. Quesada, C. Solera, R. Solorzano, G. Stiles, J. Tosi, Jr., A. Umana, C. Villalobos, and R. Wells. 1982. Costa Rica country environmental profile: a field study. Tropical Science Center, San José.

Hartshorn, G., L. Nicolait, L. Hartshorn, G. Bevier, R. Brightman, J. Cal, A. Cawich, W. Davidson, R. DuBois, C. Dyer, J. Gibson, W. Hawley, J. Leonard, R. Nicolait, D. Weyer, H. White and C. Wright. 1984. Belize country environmental profile: a field study, R. Nicolait & Associates, Belize City.

Hartshorn, G. S., and L. J. Poveda. 1983. Plants: checklist of trees. pp. 158–183 in D. H. Janzen (ed.), Costa Rican natural history. University of

Heckadon Moreno, S. 1983. Cuando se acaban los montes: los campesinos Santeños y la colonización de Tonosí. University of Panama and STRI, Panama.

Heckadon Moreno, S., and A. McKay (eds.). 1982. Colonización y destrucción de bosques en Panamá: ensayos sobre un grave problema ecológico. Associación Panameña de Antropoligía, Panama.

Heithaus, E. R. 1979. Community structure of neotropical flower visiting bees and wasps: diversity and phenology. Ecology 60:190–202.

Hewetson, C. E. 1956. A discussion on the "climax" concept in relation to the tropical rain and deciduous forest. Empire Forestry Review 35:274–291.

Holdridge, L. R. 1947. Determination of world plant formations from simple climatic data. Science 105:367–368.

Holdridge, L. R. 1956. Middle America. pp. 183–200 in S. Haden-Guest, J. K. Wright, and E. M. Teclaff (eds.), A world geography of forest resources. Ronald Press, New York.

Holdridge, L. R. 1957a. The vegetation of mainland Middle America. Eighth Pacific Sci. Congr. Proc. IV:148–161.

Holdridge, L. R. 1957b. Pine and other conifers. pp. 332–338 in Tropical silviculture. FAO

Forestry and Forest Products Studies no. 13, vol. 2. FAO, Rome.

Holdridge, L. R. 1967. Life zone ecology, rev. ed. Tropical Science Center, San José, Costa Rica.

Holdridge, L. R. 1975a. Zonas de vida de El Salvador. PNUD/FAO/ELS/73/004, no. 6, San Salvador.

Holdridge, L. R. 1975b. Las coníferas de Guatemala. PNUD/FAO/FO:DP/GUA/72/006, no. 1, Guatemala City.

Holdridge, L. R., and G. Budowski. 1956. Report of an ecological survey of the Republic of Panama. Carib. Forester 17:92–110.

Holdridge, L. R., W. C. Grenke, W. H. Hatheway, T. Liang, and J. A. Tosi, Jr. 1971. Forest environments in tropical life zones: a pilot study. Pergamon Press, Elmsford, N.Y.

Hubbell, S. P. 1979. Tree dispersion, abundance, and diversity in a tropical dry forest. Science 203:1299–1309.

Hubbell, S. P., and R. B. Foster. 1983. Diversity of canopy trees in a neotropical forest and implications for conservation. pp. 25–41 in S. L. Sutton, T. C. Whitmore, and A. C. Chadwick (eds.), Tropical rain forest: ecology and management. Blackwell, London.

Hubbell, S. P., and R. B. Foster. 1985a. Canopy gaps and the dynamics of a neotropical forest. in M. J. Crawley (ed.), Plant ecology. Blackwell, London.

Hubbell, S. P., and R. B. Foster. 1985b. Biology, chance, and history, and the structure of tropical tree communities. in J. M. Diamond and T. J. Case (eds.), Community ecology. Harper & Row, New York.

Janzen, D. H. 1967. Synchronization of sexual reproduction of trees within the dry season in Central America. Evolution 21:620–637.

Janzen, D. H. 1969. Seed-eaters versus seed size, number, toxicity and dispersal. Evolution 23:1–27.

Janzen, D. H. 1970a. Herbivores and the number of tree species in tropical forests. American Naturalist 104:501–528.

Janzen, D. H. 1970b. *Jacquinia pungens*, a heliophile from the understory of tropical deciduous forest. Biotropica 2:112–119.

Janzen, D. H. 1971. Euglossine bees as long-distance pollinators of tropical plants. Science 171:203–205.

Janzen, D. H. 1973. Rate of regeneration after a tropical high elevation fire. Biotropica 5:117–122.

Janzen, D. H. 1978. Description of a *Pterocarpus officinalis* (Leguminosae) monoculture in Corcovado National Park, Costa Rica. Brenesia (Costa Rica) 14–15:305–309.

Janzen, D. H. (ed.). 1983a. Costa Rican natural history. University of Chicago Press.

Janzen, D. H. 1983b. *Mora megistosperma* (Alcornoque, Mora). pp. 280–281 in D. H. Janzen (ed.), Costa Rican natural history. University of Chicago Press.

Janzen, D. H., and D. E. Wilson. 1974. The cost of being dormant in the tropics. Biotropica 6:260–262.

Johnson, M. S., D. R. Chaffey, and C. J. Birchall. 1973. A forest inventory of part of the mountain pine ridge, Belize. ODA Land Resources Division, land resource study no. 13, Surrey, England.

Kapos, V., and E. V. J. Tanner. 1985. Water relations of Jamaican upper montane rain forest trees. Ecology 66:241–250.

Kellman, M. 1979. Soil enrichment by neotropical savanna trees. J. Ecol. 67:565–577.

Kellman, M. 1984. Synergistic relationships between fire and low soil fertility in neotropical savannas: a hypothesis. Biotropica 16:158–160.

Küchler, A. W., and J. M. ontoya Maquín. 1971. The UNESCO classification of vegetation: some tests in the tropics. Turrialba 21:98–109.

Lamb, F. B. 1966. Mahogany of tropical America: its ecology and management. University of Michigan Press, Ann Arbor.

Lambert, J. D. H., and T. Arnason. 1978. Distribution of vegetation on Maya ruins and its relationship to ancient land-use at Lamanai, Belize. Turrialba 28:33–41.

Lang, G. E., and D. H. Knight. 1983. Tree growth, mortality, recruitment, and canopy gap formation during a 10-year period in a tropical moist forest. Ecology 64:1075–1080.

Lawton, R., and V. Dryer. 1980. The vegetation of the Monteverde Cloud Forest Reserve. Brenesia 18:101–116.

Leigh, E. G., Jr. 1975. Structure and climate in tropical rain forest. Ann. Rev. Ecol. Syst. 6:67–86.

Leigh, E. G., Jr., A. S. Rand, and D. M. Windsor (eds.). 1982. The ecology of a tropical forest: seasonal rhythms and long-term changes. Smithsonian Institution Press, Washington, D.C.

Lieberman, M., D. Lieberman, G. Hartshorn, and R. Peralta. 1985. Small-scale altitudinal variation in lowland wet tropical vegetation. J. Ecol. 73:505–516.

Lot-Helgueras, A., C. Vásquez-Yanes, and L. F. Menéndez. 1975. Physiognomic and floristic changes near the northern limit of mangroves in the Gulf Coast of Mexico. pp. 52–61 in G. E. Walsh, S. C. Snedaker, and H. J. Teas (eds.), Proceedings of the international symposium on biology and management of mangroves. University of Florida Institute of Food and Agricultural Sciences, Gainesville.

Lugo, A. E. 1980. Mangrove ecosystems: successional or steady state? Biotropica [Suppl.] 12:65–72.

Lugo, A. E., R. Schmidt, and S. Brown. 1981. Tropical forests in the Caribbean. Ambio 10:318–324.

Lugo, A. E., and S. C. Snedaker. 1974. The ecology of mangroves. Ann. Rev. Ecol. Syst. 5:39–64.

Lundell, C. L. 1933. The agriculture of the Maya. Southwest Rev. 19:65–77.

Lundell, C. L. 1937. The vegetation of Petén. Carnegie Institution Publication no. 478, Washington, D.C.

McHargue, L. A., and G. S. Hartshorn. 1983. Seed and seedling ecology of *Carapa guianensis*. Turrialba 33:399–404.

Martin, P. S. 1964. Paleoclimatology and a tropical pollen profile. VI Int. Quat. Congr. (Warsaw) Rept. 2:319–323.

Mayo Meléndez, E. 1965. Algunas caracteristicas ecológicas de los bosques inundables de Darién, Panamá, con miras a su posible utilización. Turrialba 15:336–347.

Miranda, F., and X. E. Hernández. 1963. Los tipos de vegetación de México y su clasificación. Bol. Soc. Bot. Mex. 28:29–179.

Montoya Maquín, J. M. 1966. Notas fitogeográficas sobre el *Quercus oleoides* Cham. Schlecht. Turrialba 16:57–66.

Mueller-Dombois, D., and H. Ellenberg. 1974. Aims

and methods of vegetation ecology. Wiley, New York.

Myers, N. 1979. The sinking ark: a new look at the problem of disappearing species. Pergamon Press, Elmsford, N.Y.

Myers, N. 1984. The primary source: tropical forests and our future. W. W. Norton, New York.

Myers, R. L. 1981. The ecology of low diversity palm swamps near Tortuguero, Costa Rica. Ph.D. dissertation, University of Florida, Gainesville.

Nadkarni, N. M. 1984. Epiphyte biomass and nutrient capital of a neotropical elfin forest. Biotropica 16:249–256.

Nations, J. D., and D. I. Komer. 1983. Central America's tropical rainforests: positive steps for survival. Ambio 12:232–238.

NRC (National Research Council). 1980a. Research priorities in tropical biology. National Academy of Science, Washington, D.C.

NRC. 1980b. Conversion of tropical moist forests. National Academy of Science, Washington, D.C.

NRC. 1982. Ecological aspects of development in the humid tropics. National Academy Press, Washington, D.C.

Opler, P. A., and K. S. Bawa. 1978. Sex ratios in tropical forest trees. Evolution 32:812–821.

Opler, P. A., G. W. Frankie, and H. G. Baker. 1976. Rainfall as a factor in the synchronization, release, and timing of anthesis by tropical trees and shrubs. J. Biogeography 3:231–236.

Parsons, J. J. 1955. The Miskito pine savanna of Nicaragua and Honduras. Ann. Assoc. Amer. Geogr. 45:36–63.

Parsons, J. J. 1976. Forest to pasture: development or destruction? Rev. Biol. Trop. [Suppl.] 24:121–138.

Peñalosa, J. 1984. Basal branching and vegetative spread in two tropical rain forest lianas. Biotropica 16:1–9.

Pennington, T. D., and J. Sarukhán. 1968. Arboles tropicales de México. INIF, Mexico.

Perry, D. R. 1978. Factors influencing arboreal epiphytic phytosociology in Central America. Biotropica 10:235–237.

Pickett, S. T. A. 1983. Differential adaptation of tropical tree species to canopy gaps and its role in community dynamics. Trop. Ecology 24:68–84.

Pool, D. J., S. C. Snedaker, and A. E. Lugo. 1977. Structure of mangrove forests in Florida, Puerto Rico, Mexico, and Costa Rica. Biotropica 9:195–212.

Prance, G. T., and T. S. Elias (eds.). 1977. Extinction is forever: the status of threatened and endangered plants of the Americas. New York Botanical Garden.

Pringle, C., I. Chacon, M. Grayum, H. Greene, G. Hartshorn, G. Schatz, G. Stiles, C. Gomez, and M. Rodriguez. 1984. Natural history observations and ecological evaluation of the La Selva Protection Zone, Costa Rica. Brenesia 22:189–206.

Putz, F. E. 1984. The natural history of lianas on Barro Colorado Island, Panama. Ecology 65:1713–1724.

Rabinowitz, D. 1975. Planting experiments in mangrove swamps of Panama. pp. 385–393 in G. E. Walsh, S. C. Snedaker, and H. J. Teas (eds.), Proceedings of the international symposium on biology and management of mangroves. University of Florida Institute of Food and Agricultural Sciences, Gainesville.

Rabinowitz, D. 1978. Dispersal properties of mangrove propagules. Biotropica 10:47–57.

Raven, P. H., and D. I. Axelrod. 1975. History of the flora and fauna of Latin America. Amer. Sci. 63:420–429.

Ray, T. 1983a. *Monstera tenuis* (Chirravaca, Mano de Tigre, Monstera). pp. 278–280 in D. H. Janzen (ed.), Costa Rican natural history. University of Chicago Press.

Ray, T. 1983b. *Syngonium triphyllum* (Mano de Tigre). pp. 333–335 in D. H. Janzen (ed.), Costa Rican natural history. University of Chicago Press.

Reich, P. B., and R. Borchert. 1982. Phenology and ecophysiology of the tropical tree, *Tabebuia neochrysantha* (Bignoniaceae). Ecology 63:294–299.

Reyna Vásquez, M. L. 1979. Vegetación arborea del bosque nebuloso de Montecristo. Thesis, University of El Salvador, San Salvador.

Richards, P. W. 1952. The tropical rainforest: an ecological study. Cambridge University Press.

Rzedowski, J. 1978. Vegetación de México. Editorial Limusa, México, D. F.

Salgado-Labouriau, M. L. (ed.). 1979. El medio ambiente páramo. Ediciones Centro de Estudios Avanzados, IVIC, Caracas.

Sarukhán, J. 1978. Studies on the demography of tropical trees. pp. 163–184 in P. B. Tomlinson and M. H. Zimmermann (eds.), Tropical trees as living systems. Cambridge University Press.

Sauer, C. O. 1958. Man in the ecology of tropical America. pp. 104–110 in Ninth Pacific science congress proceedings (Bangkok).

Sawyer, J. O., and A. A. Lindsey. 1964. The Holdridge bioclimatic formations of the eastern and central United States. Indiana Acad. Sci. Proc. 73:105–112.

Scholander, P. F., L. Van Dam, and S. I. Scholander. 1955. Gas exchange in the roots of mangroves. Amer. J. Bot. 42:92–98.

Shimwell, D. W. 1971. The description and classification of vegetation. University of Washington Press, Seattle.

Shreve, F. 1914. A montane rain-forest: a contribution to the physiological plant geography of Jamaica. Carnegie Institution Publication no. 199, Washington, D.C.

Sousa S., M. 1964. Estudio de la vegetación secundaria en la región de Tuxtepec. Oax. INIF (Mexico) Publ. Esp. 3:91–105.

Spellman, D. L., J. D. Dwyer, and G. Davidse. 1975. A list of the Monocotyledoneae of Belize – including a historical introduction to plant collecting in Belize. Rhodora 77:105–140.

Standley, P. C. 1937. Flora of Costa Rica. Field Museum of Natural History, Chicago.

Stevenson, D. 1928. Types of forest growth in British Honduras. Trop. Woods. 14:20–25.

Strong, D. R., Jr., and T. S. Ray, Jr. 1975. Host tree location behavior of a tropical vine *Monstera gigantea* by skototropism. Science 190:804–806.

Tanner, E. V. J., and V. Kapos. 1982. Leaf structure of Jamaican upper montane rain-forest trees. Biotropica 14:16–24.

Taylor, B. W. 1962. The status and development of the Nicaraguan pine savannas. Carib. Forester 23:21–26.

Taylor, B. W. 1963. An outline of the vegetation of Nicaragua. J. Ecol. 51:27–54.

Tosi, J. A., Jr. 1969. Mapa Ecológico, República de

Costa Rica. Centro Científico Tropical, San José.
Tosi, J. A., Jr. 1971. Zonas de vida: una base ecológica para investigaciones silvícolas e inventariación forestal en la República de Panamá. FAO/FO:SF/PAN/6, no. 2, Panama.
Tosi, J., Jr., and G. Hartshorn. 1978. Mapa ecológico de El Salvador. MAG & CATIE, San Salvador.
Turner, B. L., II, and P. D. Harrison (eds.). 1983. Pulltrouser Swamp: ancient Maya habitat, agriculture, and settlement in northern Belize. University of Texas Press, Austin.
Veblen, T. T. 1976. The urgent need for forest conservation in highland Guatemala. Biol. Conserv. 9:141–154.
Veblen, T. T. 1978. Forest preservation in the western highlands of Guatemala. Geogr. Rev. 68:417–434.
Vogel, F. H. 1954. Los bosques de Honduras. Ceiba 4:85–121.
Vogelmann, H. W. 1973. Fog precipitation in the cloud forests of eastern Mexico. BioScience 23:96–100.
Wagner, P. L. 1964. Natural vegetation of Middle America. pp. 216–264 in R. C. West (ed.), Natural environment and early cultures. University of Texas Press, Austin.
Weber, H. 1959. Los Páramos de Costa Rica y su concatenación fitogeográfica con los Andes Suramericanos. Instituto Geográfico Nacional, San José.
Whitmore, T. C. 1984. Tropical rain forests of the Far East, 2nd ed. Oxford University Press.
Williams, H. 1960. Volcanic history of the Guatemalan highlands. University of California Publications in Geological Science, no. 38, Berkeley.
Witsberger, D., D. Current, and E. Archer. 1982. Arboles del Parque Deininger. Ministerio de Educación, San Salvador.
Wright, A. C. S., D. H. Romney, R. H. Arbuckle, and V. E. Vial. 1959. Land in British Honduras. Colonial Research Publication no. 24, London.

Chapter 13

Alpine Vegetation

WILLIAM DWIGHT BILLINGS

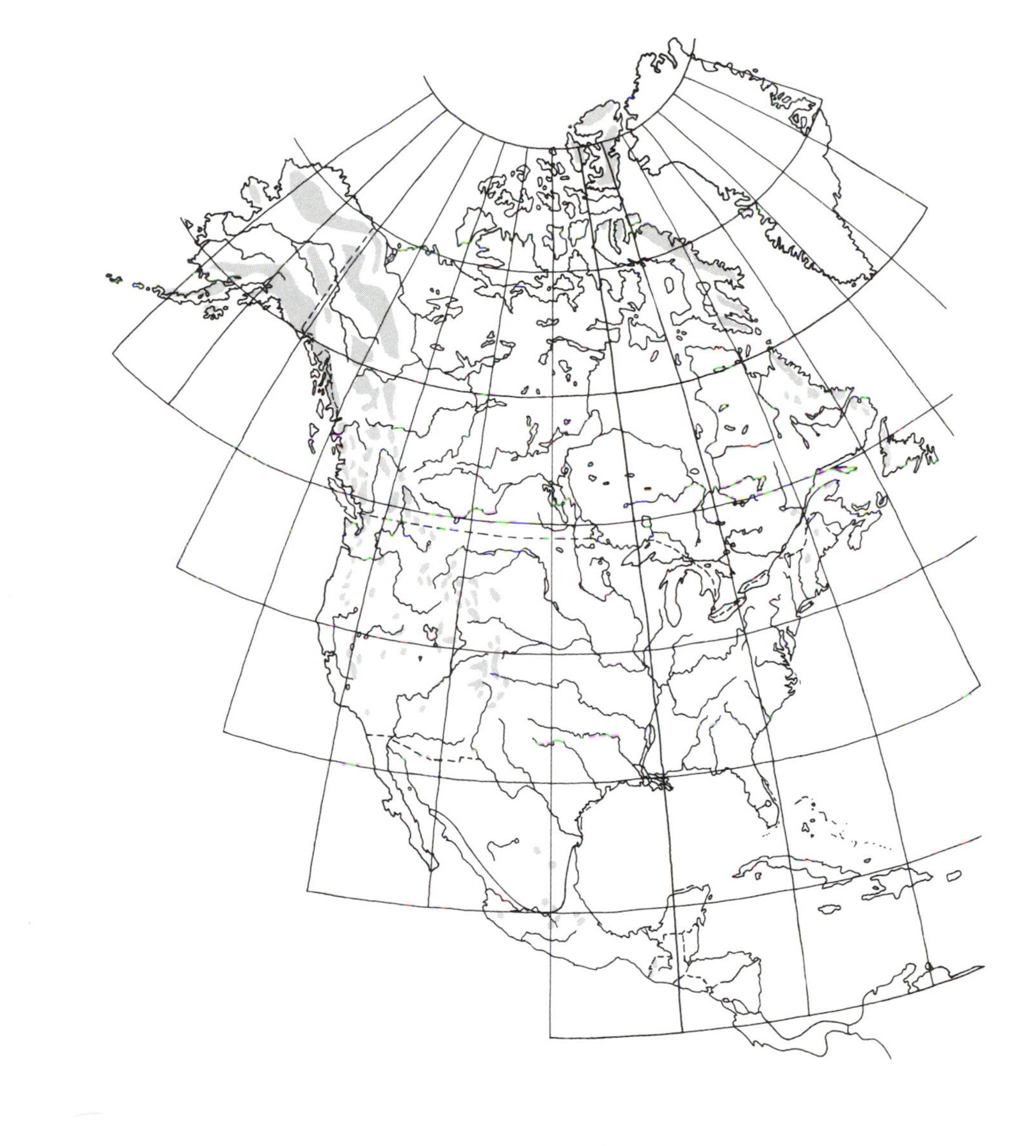

INTRODUCTION

The alpine biome, one of the smaller in area of the major North American ecosystem complexes, occupies high-mountain summits, slopes, and ridges above the upper limits of forests. In a world of intense radiation, wind, cold, snow, and ice, alpine vegetation is close to the ground and consists mainly of dwarf shrubs and perennial herbaceous plants less than a few decimeters tall. Vegetational types vary from lawn-like meadows in the moister and wetter sites to relatively dry fellfields on windward slopes, and nothing more than scattered cushion plants and lichens on rocky ridges. Places with persistent or long-lasting snowdrifts, screes, and cliffs are almost plantless.

Except for isolated summits such as Mt. Washington and Mt. Katahdin in the northern Appalachians, the Long Range of Newfoundland, and the Torngat Mountains of Labrador, for example, alpine vegetation in North America exists only on the high mountains in the western third of the continent. Here, as seen on the map, it occurs from the Brooks Range in the Arctic and the coastal Alaska Range southward as a series of montane "islands" along two great cordilleras: (1) the Coastal Cordillera from the Alaska Range to the Sierra Nevada and (2) the Rocky Mountains from the Brooks Range of northern Alaska to the higher peaks of northern Arizona and New Mexico. South of latitude 35° N, there is a gap of more than 1500 km to a few alpine summits in the Sierra Madre Oriental. Deeper in the tropics, high volcanoes rise from the Mexican Plateau. Vascular plants on these go to at least 4570 m (Webster 1961). The southernmost mountains in North America with vegetation that can be called "North American alpine" are in Guatemala. The richest alpine floras in Guatemala are on the older mountains, not the young volcanoes, and are related to the mountain floras of the southern Rocky Mountains and Mexico (Steyermark 1950). Southward, the high-mountain floras on Cerro de la Muerte and Cerro Chirripó (Costa Rica) and Volcan de Chiriquí (Panama) form "páramos," in the sense of Weber (1958), that are more closely related to the páramo floras and vegetation of the northern Andes of Venezuela and Colombia than they are to alpine vegetation farther north.

On a map of this scale, it is not possible to indicate the extent of permanent ice and snow that exist under present climatic conditions. Therefore, in the mountains of coastal Alaska, considerable glacial ice is included in the alpine ecosystems as mapped. The same is true of the ice caps on Ellesmere, Devon, and Baffin islands in the eastern Canadian Arctic. No attempt has been made to map alpine areas in Greenland, where more than 95% of the land is covered with ice cap. Although glaciers exist in the mountains of the middle latitudes in western North America, they are too small to affect the areal distribution of alpine ecosystems as depicted. It should be noted that glacial ice in alpine situations was much more extensive at times in the past, and with the postulated climatic warming due to the CO_2 effect and other environmental changes, glaciers may be less extensive in the future.

There is a tendency among some ecologists in North America to speak of "alpine tundra" and to equate this with the true tundra of the Arctic. Although it is tempting to accept this equivalence, from the standpoints of environment, flora, and vegetation it is only partly true. As shown in Fig. 13.1, environmentally, only low temperatures during the growing season are held in common by the Arctic tundra of the Far North and the alpine ecosystems of the Rocky Mountains near 40° N latitude. All else in their respective physical environments differs in considerable degree: intensity and wavelengths of solar radiation, ultraviolet radiation, daylength, wind, soils, snow cover, and topography (Billings 1973, 1979).

However, with changing climates during the Pleistocene and Holocene, there have been considerable north and south migrations of plants along the cordillera of the Rocky Mountains. Also, in eastern North America, migration southward of arctic species in front of the Wisconsinan ice and migration back northward after the melting of the ice sheets have resulted in the alpine floras of the New England mountains having a strong arctic component. On Mt. Washington, this amounts to about 70% of the alpine flora, using the 63 alpine vascular plant species mentioned by Bliss (1963) as a sample (W. D. Billings, unpublished data). Similarly, about 50% of the alpine flora at the species level of the Beartooth Mountains in the central Rocky Mountains also occurs in the Arctic (Johnson and Billings 1962; Billings 1978). However, the younger Sierra Nevada, with a richer and more diverse alpine flora, has only about a 20% floristic relationship with the Arctic (Billings 1978).

ALPINE ENVIRONMENTS

In alpine regions, the physical aspects of the mountain massif dominate: bare rocky crests, thin atmosphere, low temperatures, intense solar radiation,

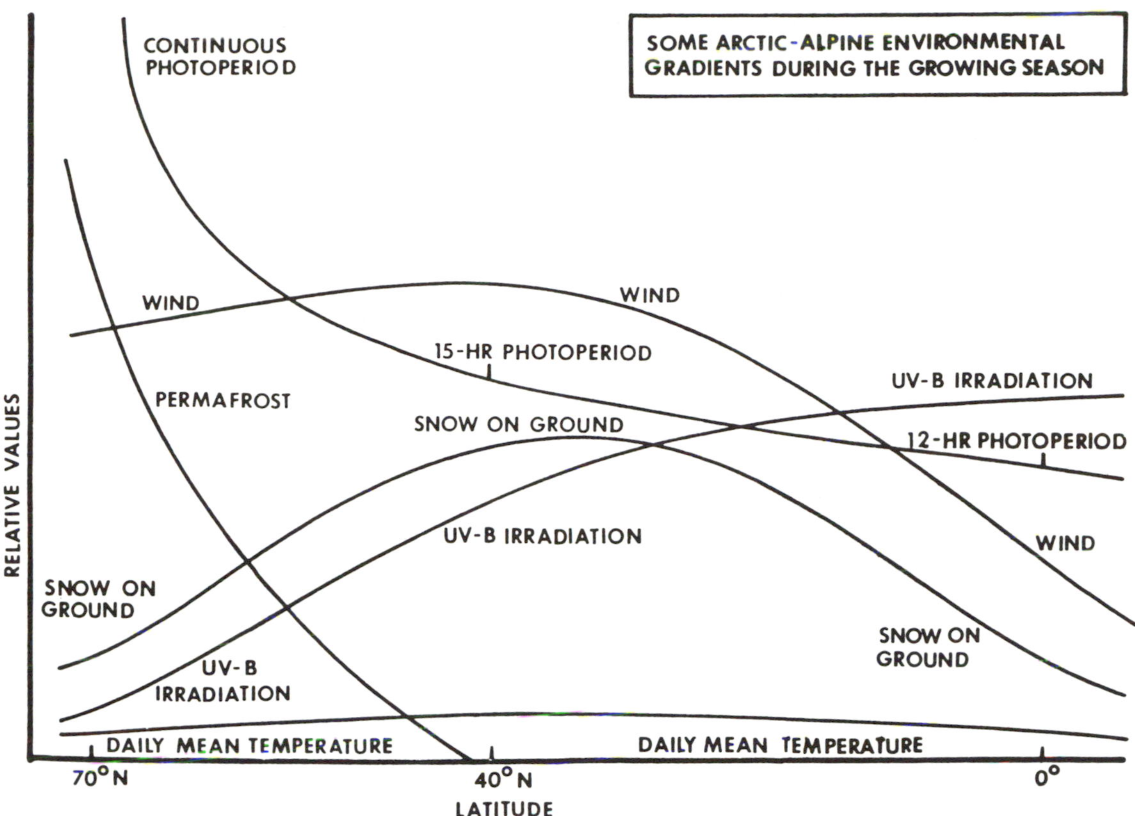

Figure 13.1. Relative values for six principal environmental factors in arctic and alpine ecosystems along a latitudinal gradient from the equator to the Arctic. The factor that changes the least along the gradient is daily mean air temperature. Factor combinations differ considerably at any given latitude (from Billings 1979).

blowing snow, long-lasting drifts, wind. However, these are not separable from the principal biological factor: the presence of tree cover below timberline and its absence above. The sharpest environmental gradient on the mountain is that crossing timberline from relative peace within the forest to the elemental dangers of the exposed alpine ridges. Even scattered trees and scrubby krummholz reduce the intensity of solar radiation received at ground level, capture drifting snow, provide protection from wind, and ameliorate day and night temperatures. Timberline, in turn, is controlled largely by the physical aspects of the mountain: weather, temperature, wind, snowdrifts, rocks, and soils. As these change through long periods of time, trees die or become established even higher, so that timberline retreats or advances. The forest gives way, and alpine vegetation replaces it, or, conversely, the new forest shades out the fellfield plants as tree seedlings establish upward during warmer times.

Geological and Geomorphological Components

At the risk of oversimplification, one can recognize four large mountain systems in North America north of Mexico:

1. In the east, the Appalachians trend northeast from Alabama to Newfoundland, a distance of almost 3000 km.
2. Across the Straits of Belle Isle from the Long Range of Newfoundland, the mountains of the eastern rim of the Canadian Shield extend into the American Arctic for 3600 km from Labrador to Ellesmere and Axel Heiberg islands, and to within 750 km of the North Pole.
3. The longest and most massive of the systems is the Cordillera of the continental divide, the Rocky Mountains. These mountains stretch for more than 5100 km from southern New Mexico to the western end of the Brooks Range in Alaska.

4. The Western or Coastal Cordillera is almost as long (4000 km) as the Rocky Mountains. With the highest summits on the continent, it extends from southwestern Alaska along the Pacific Coast through the Cascades and Sierra Nevada to southern California

South of the desert gap in northern Mexico, the two Sierra Madres (Occidental and Oriental) frame the Mexican Plateau until they meet the transverse belt of high volcanoes south of latitude 20° N near Mexico City. Beyond the Tehuantepec Isthmus, the North American mountains continue in a southeasterly direction from Guatemala to Panama, with much physical and biological diversity.

In general, among the four main mountain systems, their tectonic ages decrease markedly from east to west. The Appalachians are very old (Cambrian to Pennsylvanian), the eastern mountains of the Canadian Arctic not quite as old, the Rocky Mountains much less so (upper Cretaceous to Oligocene), and the various members of the Coastal Cordillera from the Alaska Range to the Sierra Nevada are relatively young, having risen to their present heights during the Pliocene and Pleistocene, and including the present time. Not all ranges as components in each system are equally old, however.

Also, the present elevations of the highest peaks increase from east to west. Mt. Mitchell in the Black Mountains of North Carolina is the highest in the Appalachians at 2037 m. Under present climatic conditions, Mt. Mitchell is capped to the top with spruce-fir forest. It probably had a timberline and alpine vegetation when the continental ice of the Wisconsinan reached its terminal moraines only 375 km to the north. The same would be true of the other peaks over 2000 m in the southern Appalachians: LeConte, Clingman's Dome, and Guyot, all in the Great Smoky Mountains. Conversely, the higher peaks of the northern Appalachians, such as Mt. Washington (1909 m), were covered by the continental ice. Their present climate is such that there are timberlines considerably below their summits that are capped, therefore, with alpine vegetation.

The Rocky Mountains and Sierra Nevada each have several peaks over 4300 m in elevation, with the highest being Mt. Whitney in the Sierra at 4418 m. The Rocky Mountains are geologically complex and consist of folded strata and uplifted blocks. The Sierra Nevada is a large fault block of granitoid rocks tilted to the west and capped, particularly in the north, by andesitic flows. Between the Rocky Mountains and the Sierra are many mountain ranges and isolated peaks. In the Great Basin, these ranges trend north-south, with some peaks over 4000 m and as high as 4341 m. The general crest level of the Cascades Mountains is modest (1500–1800 m), but this ridge system is punctuated from northern California to northwestern Washington by high volcanoes between 3200 and 4400 m in elevation.

The highest mountains in North America are in Alaska and along the Yukon-Alaska boundary near the sea. There is a sizable suite of peaks in the Alaska Range, the Wrangells, and the St. Elias Range over 5000 m in elevation. Denali (Mt. McKinley) in the Alaska Range, at 6187 m, and Mt. Logan in the St. Elias Range, at 6050 m, are the highest, with several others not much lower. Only Pico de Orizaba (5700 m) and Popocatepetl (5452 m) in central Mexico are in this class. The high Alaskan mountains, being close to the Pacific, are laden with alpine ice and valley glaciers, extending, in some cases, to the sea.

All of the higher mountains in North America north of Mexico were glacially carved during the Pleistocene, with the exception of the southern Appalachians. Alpine glaciers and ice fields still characterize many mountains, especially those near the Pacific from California to Alaska. Smaller glaciers occur as far inland as Colorado and Wyoming. Under present climatic conditions, most of these smaller glaciers are wasting away. There are extensive glaciers and ice caps on the mountains of the eastern Canadian Arctic from Baffin to Ellesmere.

Weather and Climate

The common bond among high mountains anywhere is cold weather. This holds for the tropical mountains as well as for those of Alaska. Absolute alpine temperatures should be considered also relative to those in the adjacent lowlands and to those along latitudinal gradients on the mountain crests.

Because high mountains rise into thin, clear atmosphere, solar radiation received at any given latitude will be more intense at the high elevations than in the valleys. This is true not only across the visible spectrum but especially so in the ultraviolet (UV). Caldwell and associates (1980) measured global UV-B (280–320 nm) in the Arctic-alpine life zone from sea level in Alaska at 70° 28′ N latitude to Laguna Mucubají (3560 m) at 8° 50′ N latitude in the Venezuelan Andes. Their results show that integrated effective UV-B irradiances can differ by a full order of magnitude along this Arctic-alpine gradient. Robberecht and associates (1980) found that alpine plant species along this same gradient attenuate the high UV-B radiation mainly by absorption

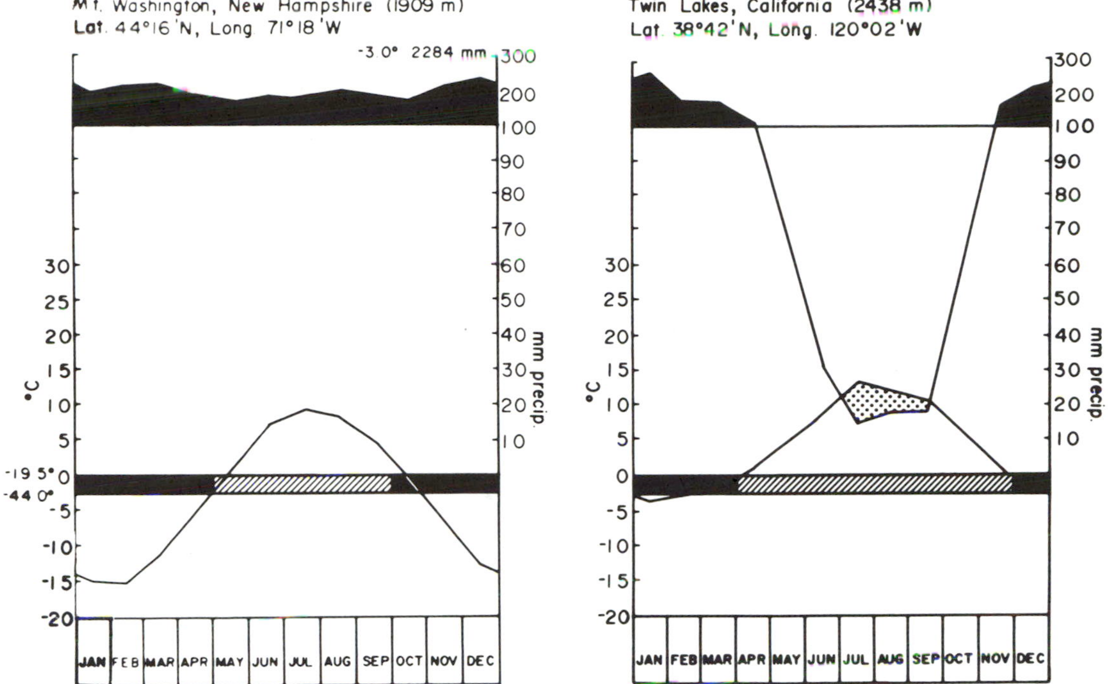

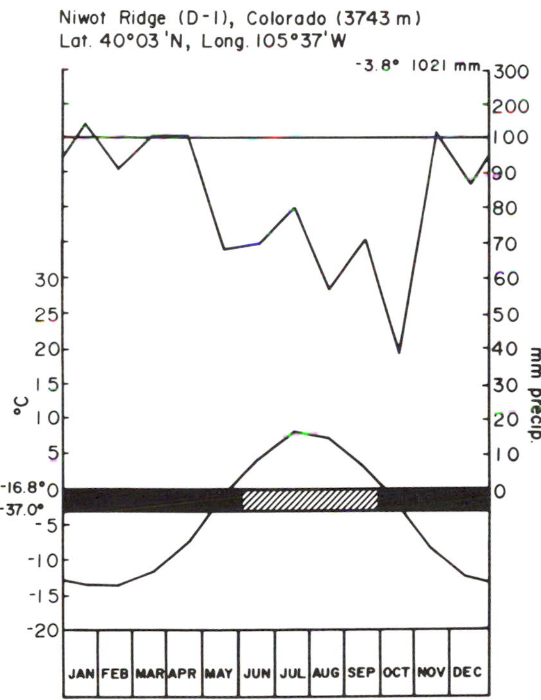

Figure 13.2. Walter-type climate diagrams for three different kinds of alpine climates: northern Appalachians, Rocky Mountains, Sierra Nevada. For explanation of data and axes, see p. 323.

in the epidermis, whereas the leaves of certain species reflect some of the UV radiation.

The normal temperature lapse rate in the atmosphere with increasing altitude assures that temperatures at high elevations will be cooler at midday than those on the lower mountain slopes and in the valleys. However, on a micro scale, absorption of intense solar radiation by rocks, soils, and plants above timberline allows air temperatures near the ground to heat up during the day. Alpine air temperatures, therefore, are much higher near the soil surface than they are in the wind at 1 m above. At timberline, small plants are at an advantage as compared with trees.

If one takes long-term weather data from standard stations at or above timberline, the general climatic background of alpine vegetation in middle-latitude North America can be seen (Fig. 13.2). These Walter-type climate diagrams are representative of alpine sites on Mt. Washington in the northern Appalachians, on Niwot Ridge in the Rocky Mountains of Colorado (Fig. 13.3), and at Twin Lakes near timberline in the Sierra Nevada of California. The Appalachian and Rocky Mountain alpine zones have much in common in regard to temperature. The mean temperature of the warmest month is less than 10°C in each, and the season dur- which the mean air temperature is above 0°C is

Figure 13.3. Aerial photograph looking southwest over Niwot Ridge alpine area in the Indian Peaks area of the Colorado Front Range, Rocky Mountains (photograph by Patrick J. Webber).

short: 4 to 5 mo at most. On the other hand, the sunny Sierra warms up more during its longer summer after snowmelt, with the mean temperature of the warmest month reaching 13°C. In the high-radiation alpine environment of the Sierra, the climate near the ground can be quite warm and dry as compared with foggy and damp Mt. Washington; Niwot Ridge is intermediate.

From the standpoint of precipitation, Mt. Washington receives a lot of snow and rain rather evenly throughout the year; drought is absent there. Niwot Ridge receives only about half as much precipitation as Mt. Washington, but most of that as snow in the winter; its summer thundershowers keep it generally green until the gentle drought of late summer brings the brown of dormancy. In contrast, the high Sierra, in its semi-Mediterranean climate, receives over 1.25 m of precipitation during the year (25% more than Niwot), but almost all of it as snow in the winter and spring months. The summers are dry on the alpine ridges; moist alpine meadows are confined to meltwater slopes below long-lasting snowdrifts. In places, Sierran vegetation has a desert-like aspect, and there are clear floristic relationships with the arid ecosystems at the eastern base of these mountains. Notably absent here are many of those arctic plant species so characteristic of the Rocky Mountains and the northern Appalachians.

Wind is a strong modifying factor in alpine environments: It chills during the day, it warms at night, and it moves water around, especially in the form of snow. There are few quiet days on the alpine scene. This is particularly true during the winter, and during storms. The highest wind speed recorded at the surface of the earth was in the alpine zone at the summit of Mt. Washington, New Hampshire, on April 12, 1934: 231 miles per hour.

Strong wind action and its movement of snow off the peaks and ridges result in topographic moisture gradients in alpine regions outside the tropics. Such a mesotopographic gradient (Billings 1973) is illustrated in Fig. 13.4. On any mountain, such repeatable gradients determine to a large extent the patterning of alpine vegetation from dry ridges to the wet meadows and bogs below large snowdrifts. This vegetational patterning results from complex combinations of thickness of snowpack, time of snowmelt, speed and direction of wind, steepness of slopes, and diversity of the extant flora. In the high mountains of the North American middle latitudes, the most common wind direction is from the

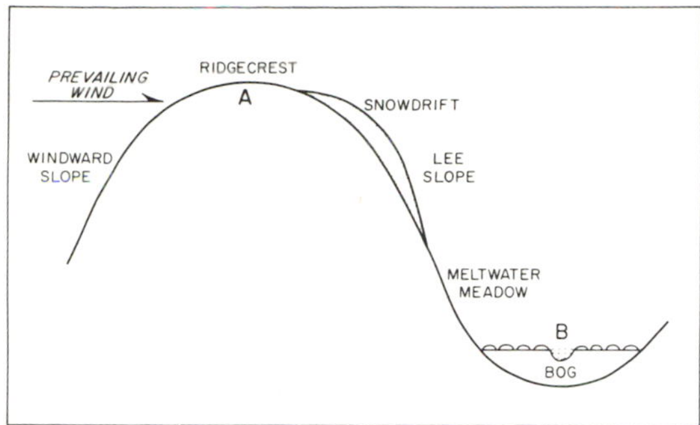

Figure 13.4. Diagram of a typical alpine mesotopographic gradient (adapted from Billings 1973).

southwest. This causes the uppermost slopes that face west or southwest to be relatively snow-free, or at least to have a shallower pack than the great drifts on eastern or northeastern slopes across the bare ridge. Spring and summer come early where snow lies thinly or not at all. Drought also comes early on these upper windward slopes and bare ridges.

Such steep local gradients of snow cover and soil moisture result in similarly steep vegetational gradients. Within the cold alpine zone, it is often the availability of water that governs productivity and the distribution of species. But long-lasting snowbanks shorten the growing season so much that only a few species can tolerate the aridity and cold of late summer after 10 to 11 mo of darkness under the snow. So the productivity of alpine vegetation that lies under big drifts is quite low compared with that on intermediate sites (Billings and Bliss 1959).

By measuring leaf water potentials and leaf conductances within the array of species along an alpine mesotopographic gradient throughout the short growing season, one can evaluate the roles of soil water and soil drought on the growth and local distributions of the different species. Oberbauer and Billings (1981) did this across an alpine ridge at 3300 m in the Medicine Bow Mountains of Wyoming (Table 13.1). As expected, leaf water potentials were generally lowest on the ridge top and highest in the meltwater wet meadow. Highest leaf conductances were found in the wet meadow, and lowest on the upper windward slope. All plants growing on the ridge top had very small, tightly overlapping leaves that were too small for the porometer. Each of the 29 species measured at frequent intervals had its own individualistic distribution along the gradient. Some were very localized. For example, *Paronychia pulvinata* occurred only on the upper windward slope and ridge top, whereas *Caltha leptosepala* was present only in the wet meadow. A few species, such as *Geum rossii* and *Trifolium parryi*, were present almost universally along the gradient. At each site, plants of *Geum rossii* showed diurnal and seasonal courses of leaf water potentials typical of the other species characteristic of that particular site. Whether this is due to phenotypic acclimation in *Geum* or to local ecotypes is not known, but both are probably involved. Certainly, rooting depth has considerable effect. For example, the deep-rooted *Trifolium parryi* maintained steady leaf water potentials at midday of approximately −2.00 MPa, no matter what the site and no matter that some of its more shallow-rooted neighbors were showing very low potentials near −3.5 to −4.0 MPa at the same time.

These data on the seasonal courses of plant water potentials and stomatal conductances translate well into alpine vegetational patterning along a topographic gradient. The floristic composition along this same vegetational transect in the Medicine Bow Mountains is shown in Table 13.2 (W. D. Billings, unpublished data). Whereas the vegetational gradient shows rather gradual floristic shifts from the lower windward slope to the lee slope, there are some fairly abrupt spatial changes in species compositions. The most conspicuous of these occur just to the lee of the ridge top and near the bottom of the lee slope where the large rocks abut the level wet meadow and temporary pond. Also, the ground under the longest-lasting snowbanks is rather barren and is occupied only by an open vegetation of distinct floristic composition that includes *Sibbaldia procumbens*, *Juncus drummondii*, and *Deschampsia caespitosa*. So it is possible to recognize, in an arbitrary way, certain plant communities along the transect:

Windward slope from −150 m to ridge top at 0 m: open, rocky fellfield
Upper lee slope from 0 m to 40 m: modified fellfield
Upper lee slope from 40 m to 60 m: transitional from fellfield to early snowbed community
Middle lee slope from 60 m to 90 m: early snowbed community
Middle lee slope from 90 m to 120 m: late snowbed community
Middle lee slope from 120 m to 150 m: moist meadow
Lower lee slope from 150 m to 260 m: mostly late snowbed community
Bottom of lee slope from 260 m to 300 m: very wet meadow

This complex gradient of plant communities is depicted in Fig. 13.5, in conjunction with Table 13.2. From these data, the effects of water distribution on

Table 13.1. *Lowest mean midday (P_{min}) and lowest mean predawn (P_{max}) leaf water potentials (MPa) for the season, change in mean predawn water potential from the 5th to the 6th week following light rain (dP_{max}), and the maximum leaf conductance for the season (C_{max}, cm sec^{-1}). Genera appear in T. 13.2*

Site	P_{min}	P_{max}	dP_{max}	C_{max}
Windward (−60 m)				
M. obtusiloba	−3.10	−1.75	+0.79	—
H. grandiflora	−3.72	−2.90	+1.43	—
A. scopulorum	−4.00	−2.33	+1.17	—
C. jamesii	−3.42	−2.07	+0.95	0.92
T. parryi	−2.18	−1.57	−0.06	0.81
T. dasyphyllum	−2.54	−1.46	−0.14	—
G. rossii	−2.43	−1.82	+0.86	—
Ridge (0 m)				
M. obtusiloba	−4.00	−2.81	+1.29	—
T. dasyphyllum	−2.96	−1.65	−0.04	—
G. rossii	−3.12	−2.15	+0.61	—
P. pulvinata	−4.00	−3.70	+2.14	—
A. scribneri	−4.00	−4.00	+2.82	—
Lee (60 m)				
T. parryi	−2.00	−0.86	+0.03	0.89
G. rossii	−2.76	−1.69	−0.18	—
A. scopulorum	−3.28	−1.79	+0.46	—
B. bistortoides	−2.28	−1.46	−0.05	1.37
T. alpestre	—	—	—	—
P. rupicola	−4.00	−2.30	−0.02	—
P. diversifolia	−2.84	−1.72	−0.20	0.90
A. lanulosa	—	—	—	—
Snowdrift (100 m)				
T. parryi	−2.34	−1.04	−0.29	0.78
A. scopulorum	−4.00	−1.95	+0.43	—
B. bistortoides	−2.02	−0.77	−0.10	1.86
T. alpestre	−4.00	−4.00	+0.07	—
P. diversifolia	−2.68	−1.33	+0.02	1.28
P. alpinum	−4.00	−1.64	+0.09	1.06
D. caespitosa	−3.25	−1.35	−0.14	—
E. peregrinus	−3.39	−1.33	+0.04	0.91
140m				
T. parryi	−1.92	−0.99	+0.04	0.81
G. rossii	−2.04	−0.89	+0.19	—
A. scopulorum	−3.15	−1.83	+0.50	—
B. bistortoides	−1.65	−0.82	−0.03	1.44
P. diversifolia	−2.81	−0.78	+0.17	0.73
C. rupestris	−2.92	−1.96	+0.57	—
C. jamesii	−2.84	−1.37	+0.38	0.56
Rocks (190 m)				
G. rossii	−1.30	−0.60	—	—
E. peregrinus	−1.38	−0.62	—	0.51
S. dimorphophyllus	−1.08	−0.56	—	0.83
S. procumbens	−1.60	−0.52	—	0.63
A. mollis	−1.41	−0.32	—	0.81
R. montigenum	−1.89	−0.54	—	0.67
A. caerulea	−1.65	−0.96	—	0.42
P. engelmannii	−1.50	−0.59	—	—
Pond (230 m)				
B. bistortoides	−1.11	−0.42	—	1.20
D. caespitosa	−1.89	−0.28	—	—
S. dimorphophyllus	−1.36	−0.59	—	0.99
S. brachycarpa	−1.21	−0.54	—	1.85
S. rhomboidea	−1.10	−0.30	—	0.87
S. rodanthum	−0.88	−0.34	—	—
Meadow (280 m)				
G. rossii	−1.41	−0.51	−0.08	—
P. diversifolia	−1.78	−0.36	−0.17	0.98
B. bistortoides	−1.10	−0.28	−0.04	2.17
A. scopulorum	−1.95	−0.68	+0.02	—
C. leptosepala	−1.35	−0.38	−0.28	1.36
D. caespitosa	−2.10	−0.48	−0.06	—
C. chalciolepsis	−1.76	−0.25	−0.12	1.29

Source: From Oberbauer and Billings (1981).

vegetational composition in an alpine site are obvious. An ordination by May and Webber (1982) shows a strikingly similar set of vegetational noda on Niwot Ridge, Colorado; the Medicine Bow transect lacks only one node of the Niwot set, the dry *Kobresia* meadow. Isard (1986), in a notable recent vegetation-environmental study on Niwot Ridge, came to the same conclusion: that the alpine plant communities are controlled primarily by snow cover and soil moisture relations.

Soil Frost Action

Alpine soils are cold and often wet. These environmental circumstances, when combined with the strong diurnal cycling of soil surface temperatures, lead to considerable freezing and thawing of the upper soil. Because these soils are shallow and overlie bedrock, rocky glacial till, or permafrost, the consequent formation of ice pushes boulders, rocks, and gravel toward the surface – and thence outward along the surface. The result is an array of patterned ground made up of sorted polygons of various types: stone nets, frost boils or sorted circles, sorted stripes, sorted steps, and hummocks, as described by Washburn (1956).

Such frost-churned soils provide an unstable soil environment through much, but not all, of the alpine zone. The principal exceptions are the rocky outcrops on ridge tops, glacially scoured rocks, cliffs, and wet meadows covered with deep peat. The most active sites of cryopedogenic activity are barren gravelly places in which the water table fluctuates between just above and just below the surface. Such an active substratum is not easily invaded by plants, and active polygon centers are quite barren; plants are restricted to the rocky borders of such polygons. It is only after the water

Table 13.2. *Floristic composition of alpine vegetational gradient at about 3300 m in the Medicine Bow Mountains, Wyoming*

Species	−140	−130	−120	−110	−100	−90	−80	−70	−60	−50	−40
Geum rossii	2	2	7	5	8	9	5	13	8	8	7
Minuartia obtusiloba	2	4	1	1	5	4	16	6	13	4	10
Potentilla diversifolia	1	p	2	2	9	1	—	p	—	—	—
Trifolium parryi	1	p	p	1	7	2	8	2	6	5	7
Carex elynoides	1	3	1	3	1	p	1	1	1	1	1
Silene acaulis	1	2	9	3	1	p	4	1	—	1	—
Trifolium dasyphyllum	1	2	2	3	1	16	2	7	p	p	5
Artemisia scopulorum	p[b]	—	—	—	—	—	—	—	p	—	1
Bistorta bistortoides	—	1	—	1	—	—	—	—	—	1	1
Selaginella densa	—	1	1	1	p	—	p	3	1	4	2
Phlox pulvinata	—	3	3	2	—	8	1	1	p	1	2
Saxifraga rhomboidea	—	1	—	—	—	—	—	—	—	—	—
Poa alpina	—	—	1	—	—	—	—	—	—	—	—
Paronychia pulvinata	—	—	1	—	—	—	p	5	3	1	6
Lichens	—	1	—	1	—	p	1	p	—	5	4
Calamagrostis purpurascens	—	—	—	—	—	—	—	—	—	—	—
Achillea lanulosa	—	—	—	—	—	—	—	—	—	—	—
Juncus drummondii	—	—	—	—	—	—	—	—	—	—	—
Deschampsia caespitosa	—	—	—	—	—	—	—	—	—	—	—
Sibbaldia procumbens	—	—	—	—	—	—	—	—	—	—	—
Senecio dimorphophyllus	—	—	—	—	—	—	—	—	—	—	—
Caltha leptosepala	—	—	—	—	—	—	—	—	—	—	—
Carex scopulorum	—	—	—	—	—	—	—	—	—	—	—
Sedum rhodanthum	—	—	—	—	—	—	—	—	—	—	—
Dead and litter	79	49	27	44	14	16	9	16	10	14	12
Rock	8	20	15	20	16	20	35	28	30	43	26
Bare soil	0	1	22	1	37	20	10	17	25	11	15

Species	−30	−20	−10	0	+10	+20	+30	+40	+50	+60	+70
Geum rossii	5	9	14	6	7	9	2	8	7	p	2
Minuartia obtusiloba	5	13	8	6	5	10	11	4	2	1	3
Potentilla diversifolia	—	—	—	—	—	—	—	1	1	p	3
Trifolium parryi	3	—	—	—	—	7	8	—	8	3	28
Carex elynoides	p	1	p	p	1	1	1	1	—	—	—
Silene acaulis	—	1	p	1	1	2	5	5	3	2	—
Trifolium dasyphyllum	10	17	17	17	13	5	4	11	9	—	—
Artemisia scopulorum	—	—	—	—	—	—	—	1	4	4	13
Bistorta bistortoides	—	—	—	—	—	—	—	2	—	1	2
Selaginella densa	1	1	3	3	1	1	p	1	1	1	—
Phlox pulvinata	—	—	p	p	1	6	p	1	—	—	—
Saxifraga rhomboides	—	—	—	—	—	—	—	—	—	—	—
Poa alpina	—	—	—	—	—	—	—	—	—	—	p
Paronychia pulvinata	5	8	6	5	2	1	1	2	1	—	—
Lichens	1	1	1	1	2	2	8	p	4	1	2
Calamagrostis purpurascens	—	—	—	1	1	1	1	—	—	1	1
Achilles lanulosa	—	—	—	—	—	—	—	—	—	p	6
Juncus drummondii	—	—	—	—	—	—	—	—	—	15	—
Deschampsia caespitosa	—	—	—	—	—	—	—	—	—	—	1
Sibbaldia procumbens	—	—	—	—	—	—	—	—	—	—	—
Senecio dimorphophyllus	—	—	—	—	—	—	—	—	—	—	—
Caltha leptosepala	—	—	—	—	—	—	—	—	—	—	—
Carex scopulorum	—	—	—	—	—	—	—	—	—	—	—
Sedum rhodanthum	—	—	—	—	—	—	—	—	—	—	—
Dead and litter	12	13	28	19	29	24	21	25	18	17	14
Rock	47	30	20	35	27	25	28	14	21	20	14
Bare soil	11	11	5	7	13	3	6	15	13	30	8

(Cont.)

Table 13.2. *(Cont.)*

Species	+80	+90	+100	+110	+120	+130	+140	+150	+280
Geum rossii	5	—	—	—	—	2	7	20	6
Minuartia obtusiloba	5	—	—	—	—	—	1	1	—
Potentilla diversifolia	3	—	2	1	4	7	3	5	7
Trifolium parryi	22	12	8	3	6	6	16	19	10
Carex elynoides	—	—	—	—	—	—	—	1	—
Silene acaulis	2	—	—	—	—	—	—	—	—
Trifolium dasyphyllum	—	—	—	—	—	—	—	—	—
Artemisia scopulorum	10	2	p	—	—	1	p	—	2
Bistorta bistortoides	3	p	p	—	11	8	1	1	8
Selaginella densa	—	—	—	—	—	—	1	1	—
Phlox pulvinata	—	—	—	—	—	—	—	—	—
Saxifraga rhomboidea	—	1	—	—	—	—	—	—	3
Poa alpina	—	—	1	—	—	—	—	—	5
Paronychia pulvinata	—	—	—	—	—	—	—	—	—
Lichens	—	3	—	—	—	5	1	1	—
Calamagrostis purpurascens	1	1	—	—	—	1	p	—	—
Achillea lanulosa	7	14	5	2	7	—	—	—	—
Juncus drummondii	—	—	—	—	—	—	—	—	—
Deschampsia caespitosa	—	13	6	8	29	18	2	—	35
Sibbaldia procumbens	—	—	23	8	9	5	—	—	—
Senecio dimorphophyllus	—	—	1	3	1	3	—	—	—
Caltha leptosepala	—	—	—	—	—	—	—	—	4
Carex scopulorum	—	—	—	—	—	—	—	—	4
Sedum rhodanthum	—	—	—	—	—	—	—	—	9
Dead and litter	23	20	13	10	17	18	18	8	water
Rock	5	17	24	46	5	8	8	3	water
Bare soil	5	17	24	46	5	8	8	3	water

[a]The figures are percentage coverages for the most important plant species at 10-m intervals from the lowermost windward slope at −140 m across the ridge at 0 m to the wet meadow at +280 m. From −140 m through +60 m, the data are from late June 1978. The remainder of the data (+70 m to +280 m) are from August of that year; these were snow-covered in June. Data are averages of three plots of 20 × 50 cm across the line of the transect. Compare with map and slope diagram in Fig. 13.4
[b]p=present, but <1% cover.

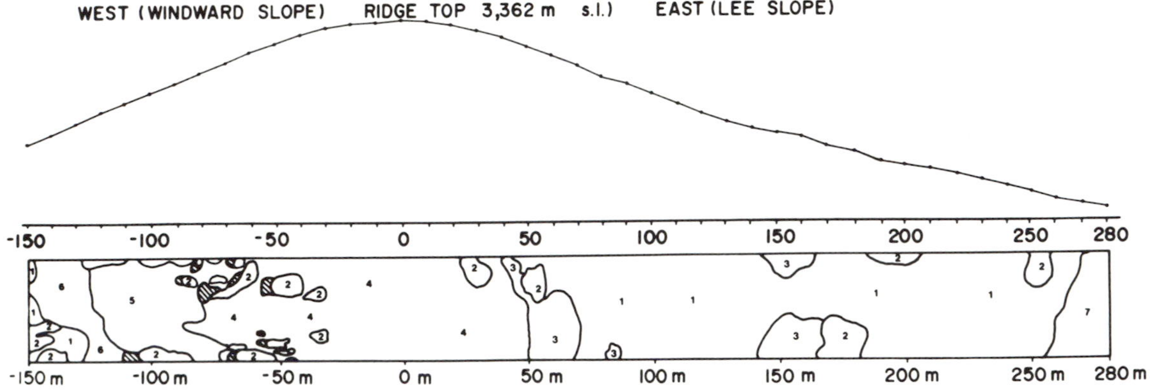

Figure 13.5. Vegetational map and profile diagram along a mesotopographic gradient at 3300 m elevation in the Medicine Bow Mountains, Wyoming.

table falls that plants can invade the polygon centers and help to build up peat layers. Then vegetational succession can proceed. The peat acts as insulation to temperature changes, ice lenses appear in the peat, and the permafrost table may rise (Billings and Mooney 1959; Johnson and Billings 1962).

In wetter and colder times, on ridges above the Rocky Mountain glaciers, stone nets of large size with very large rocks were formed. Today, these large "fossil" stone polygons with turf-covered soils in the centers of the polygons are characteristics of many high alpine sites in the Rocky Mountains. These include Niwot Ridge and the Beartooth Mountains.

ALPINE VEGETATION TYPES

Plant Life Forms

The vegetation of alpine environments is short in stature, with a tendency toward a perennial herbaceous habit and low or prostrate shrubs. Such shrubs may be either deciduous or evergreen. Perennial herbs dominate the alpine landscape; they have much more root and rhizome biomass than that of shoots, leaves, and flowers. The roots and rhizomes not only function in water and nutrient absorption but also play a very important role in overwinter carbohydrate storage (Mooney and Billings 1960; Fonda and Bliss 1966; Rochow 1969; Wallace and Harrison 1978). Annual plants are rare in this vegetation and usually are only a few centimeters tall, with weak root systems (Reynolds 1984). Annuals constitute only about 1%–2% of the flora in most alpine regions.

Whereas most alpine plants have the perennial herb (with preformed flower buds) or dwarf shrub life form, other life forms do exist above timberline. These include lichens, mosses, giant rosette plants on columnar stems, and even succulents. The giant rosette plants and succulents, such as a few cacti, are almost entirely restricted to the tropical mountains of South America and Africa. However, there are such plants in the páramos of Costa Rica in North America (Weber 1958).

Table 13.3 shows a scheme that I believe is realistic for classification of the growth forms of alpine plants. It could be useful in the mountains of both North America and South America.

Table 13.3. *Scheme for classification of alpine plant growth forms. Many combinations are possible.*

- Perennial herbs
  - Ferns
  - Dicots
    - Acaulescent rosettes
    - Caulescent forbs
  - Graminoids
    - Sod-formers
    - Tussocks
    - Dwarf mat types
  - Lycopodioids
    - True lycopods
    - Lycopodioid angiosperms
- Perennial cushion plants
  - Small
  - Large (giant)
- Giant rosette plants
  - Acaulescent
  - Tall-stemmed
- Shrubs (dwarf or prostrate)
  - Deciduous
  - Evergreen (also wintergreen)
    - Flat-leaved
    - Scale-leaved, including lycopodioid angiosperms
- Succulents
  - Stem (cacti)
  - Leaf (Sedum)
- Annual herbs
  - Graminoids
  - Dicots
    - Terminal flowering
    - Lateral flowering
- Lichens
  - Crustose
  - Thallose
  - Fruticose
- Mosses
  - Acrocarpous
  - Pleurocarpous

Geographical Coverage

Reference to the alpine map and to the Introduction for this chapter will convey some sense of the scattered array of alpine terrain in North America. The vegetations and floras of these widely separated places have been studied in a modern sense only here and there. Also, some mountaintops, especially in the Arctic and Subarctic, have emerged from glaciation only in the last millennia and centuries of the Holocene. Others, such as the alpine slopes of the St. Elias and Alaska ranges, are still covered by ice. Existing glaciers also occupy many of the mountains of the eastern Canadian Arctic on Ellesmere, Devon, and Baffin islands. Conversely, glaciation has touched only lightly some of the alpine areas in the Southwest, and has long since departed.

It is my purpose here to include brief information on the alpine floras and vegetations of those mountain systems most prominent and relatively ice-free. Even among these, information can be given only for certain places that seem representative in alpine vegetation for relatively large geographical regions. Unfortunately, little is known of the vegetation for some large mountainous regions, particularly in the Subarctic. However, the following sections will serve as an introduction to the plant communities of the higher mountains of the continent.

Mountains of Labrador

These low, heavily glaciated mountains are botanical bridges between the truly arctic mountains of the Canadian Archipelago and the northern Appalachians. As with so many isolated mountain ranges in the Subarctic, the floras of the Labrador mountains are better known than the vegetation. Hustich (1962), using his own collections and those of L. A. Viereck, has compared the flora of Gerin Mountain (940 m) with that of Pallas-Ounastunturi in arctic Finland. The alpine area of Gerin Mountain is about 39 km^2.

Of the 151 species of plants in the alpine part of Gerin Mountains, 94 (62%) also occur on Ounastunturi, 4800 km across the Atlantic. Such "circumpolarity" in the nature of the flora is characteristic, of course, of many places in the Arctic. But it is also prominent in the vegetations of subarctic mountain ranges, as in Labrador. Because of close genetic relationships between some vicarious species, the arctic element is certainly greater than 62%. We can conclude that on Gerin, and probably on other mountains of Labrador, the flora is primarily of arctic derivation; one could classify it as truly "arctic-alpine." A few of the species held in common between mountainous Labrador and arctic Finland are *Poa alpina*, *Eriophorum vaginatum*, *Carex aquatilis*, *C. bigelowii*, *Oxyria digyna*, *Sibbaldia procumbens*, *Rubus chamaemorus*, *Arctostaphylos alpina*, *Diapensia lapponica*, *Bartsia alpina*, and *Phyllodice coerulea*. The vegetation is typically low, dry tundra of a fellfield type.

Northern Appalachian Mountains

The alpine summits of the White Mountains of New Hampshire were the earliest collected and studied botanically of vegetation above timberline in North America. On Mt. Washington alone, the topography of the alpine zone bears the imprint of these early nineteenth-century plant explorers: Tuckerman's Ravine, Boott Spur, Bigelow Lawn. And in the twentieth-century came the plant ecologists: R. F. Griggs, R. S. Monahan, H. I. Baldwin, and L. C. Bliss, among others. For the last century and a half, such people have been drawn to this rather small (1917 m) deglaciated mountain by the presence of an island of arctic plants surrounded by the forests of a temperate climate. But the top of this mountain is not temperate. The weather there can be as severe as any on earth: the world's record wind speed (231 miles per hour), cloud fog, rime ice, extreme drops in temperature. Even the averages in the climate diagram (Fig. 13.2) reflect its nature. As Griggs (1956) said, "It is nothing much of a mountain . . . yet . . . Mt. Washington has one of the worst of climates." The changeability and severity of the summer weather on the summit contribute to its danger for the unprepared casual visitor. There have been many fatalities even during the warmer months.

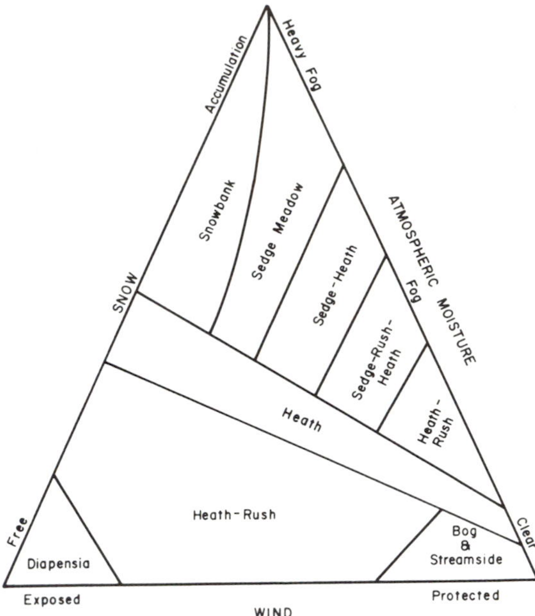

Figure 13.6. Environmental relationships of alpine plant communities of the Presidential Range, White Mountains, New Hampshire (adapted from Bliss 1963).

The alpine plant communities of Mt. Washington and the other peaks in the Presidential Range of the White Mountains have been described thoroughly by Bliss (1963). The environmental relationships of the principal alpine vegetation types are shown in Fig. 13.6 as these communities exist along gradients of snow, wind, and fog. The most exposed, snow-free sites are characterized by communities of *Diapensia lapponica*, whereas sedge meadows of *Carex bigelowii* occupy the snowy and foggy places. The floristic compositions of the eight main community types are listed as importance values in Table 13.4.

The alpine vegetation of the northern Appalachian Mountains is more closely related to the vegetation of the eastern Canadian Arctic than to that of the Rocky Mountains or the Sierra Nevada. This is partly because of the cold and foggy climate of

Table 13.4. *Relative importance values of vascular plants in the alpine plant communities of the Presidential Range, New Hampshire*

Species	Sedge meadow	Sedge–dwarf shrub heath	Sedge–rush–dwarf shrub heath	shrub heath–rush	Dwarf shrub heath	Diapensia	Snow–bank	Stream–side	Bog
Vascular plants									
Carex bigelowii	93.0	60.6	31.7	6.0	4.2	1.9	11.4	4.6	46.9
Vaccinium uliginosum incl. var. *alpinum*		2.6	3.6	11.0	18.8	8.8	11.3	9.8	8.5
Arenaria groenlandica	7.0	17.7	2.2	1.4	0.5	2.5		0.1	0.4
Juncus trifidus		3.3	23.8	25.7	1.1	21.8	1.0	0.7	
Potentilla tridentata		2.7	9.3	23.6	4.7	4.6	0.2	6.5	
Vaccinium vitis-idaea var. *minus*		12.8	19.7	21.6	12.7	0.4		4.6	8.2
Diapensia lapponica		0.2	5.4	4.6		31.3		0.1	
Agrostis borealis			0.3	1.1	0.3	5.5		1.3	
Rhododendron lapponicum			3.0	0.8		6.8		0.2	
Solidago cutleri			0.8	1.0		8.3		1.0	
Prenanthes nana				0.1	0.1	0.1	0.1	4.4	
Scirpus caespitosus var. *callosus*				0.7		0.2		10.2	11.0
Loiseleuria procumbens				0.5		7.9			0.2
Hierochloe alpina			0.1	0.6				2.1	
Carex canescens				0.5			4.0		
Betula minor				0.7					
Geum peckii				0.2				16.2	5.2
Ledum groenlandicum					15.4		1.2		0.5
Cornus canadensis					11.8		5.1		
Vaccinium angustifolium					10.7		1.0		
Maianthemum canadense					6.8		1.4		
Betula glandulosa var. *rotundifolia*					4.5				6.6
Empetrum earnesii ssp. *hermaphroditum*					2.1				
Deschampsia flexuosa					2.0		14.0		
Trientalis borealis					1.5		0.3		0.8
Vaccinium caespitosum					1.5		16.3		
Lycopodium annotinum var. *pungens*					1.6		0.2		
Solidago macrophylla var. *thyrsoidea*							11.0		
Clintonia borealis							7.4		
Coptis groenlandica							4.7		
Veratrum viride							3.8		
Houstonia caerulea var. *faxonorum*							2.9		
Juncus filiformis							1.3		
Calamagrostis canadensis var. *scabra*							1.0	0.1	
Cassiope hypnoides							0.2		
Luzula parviflora var. *melanocarpa*							0.2		
Spiraea latifolia var. *septentrionalis*							0.2		
Dryopteris spinulosa							0.1		
Salix uva–ursi								16.8	
Carex scirpoidea								10.4	
Polygonum viviparum								6.2	

(Cont.)

Table 13.4. *(Cont.)*

Species	Sedge meadow	Sedge–dwarf shrub heath	Sedge–rush–dwarf shrub heath	shrub heath–rush	Dwarf shrub heath	Diapensia	Snow–bank	Stream–side	Bog
Campanula rotundifolia var. *arctica*								3.0	
Salix planifolia								0.8	
Carex capitata								0.4	
Poa alpigena								0.3	
Carex capillaris								0.1	
Kalmia polifolia									6.3
Vaccinium oxycoccos									5.5
Total number of species	2	7	11	17	18	13	25	23	12
Average number of species	2	5	7	9	13	10	14	14	9
Number of species restricted to community type	0	0	0	0	1	0	10	8	2

Source: From Bliss (1963).

places such as Mt. Washington. But also it seems to be due to some extent to the relatively recent continental deglaciation of the New England mountains. There was tundra vegetation south of the continental ice on the mid-Atlantic East Coast at full glacial. The taxa in this vegetation were of arctic derivation. The present alpine plants of the summit of Mt. Washington are derived from these and show arctic characteristics. For example, the plants of *Oxyria digyna* from there have rhizomes, as do arctic oxyrias. Plants of this species from the mountains of western North America do not have rhizomes.

In summary, according to my calculations, using the data of Bliss, there are at least 63 species that are truly alpine in the Presidential Range. Of these, 44 also occur in the Arctic, a percentage of 69.8%. The proportion of arctic species in this alpine flora is much greater than occurs anywhere in the western mountains.

Brooks Range

Ecological publications on the alpine vegetation of this east-west mountain system (68°–69° N) in northern Alaska are rare. This is in contrast to the relatively abundant literature on the arctic tundra to the north and the taiga on the south. The Brooks Range is the very real barrier to the northern limit of forest in Alaska. The spruce taiga comes up from the south to reach its limits at between 675 m and 800 m elevation on the southern slopes and valleys of this range (Densmore 1980). There are no forests (in a real sense) north of these mountains in Alaska; it is all arctic tundra.

Spetzman (1959), in his vegetational survey of the Arctic Slope, did visit the alpine zone of the Brooks Range in several places, but he reported only short lists of species, and general descriptions. At 1860 m elevation near Killik River, he listed *Luzula confusa, Potentilla elegans, Saxifraga bronchialis, Selaginella sibirica,* and *Smelowskia calycina* as growing sparsely on the north face of a summit ridge of quartzite, shale, and sandstone rubble. At the time of his publication, this was the highest elevation from which seed plants had been collected on the North Slope. Vegetation of this type might be termed "alpine desert." All of these species also grow at lower elevations in the main alpine zone. For example, I have seen *Saxifraga bronchialis* growing with *Geum rossii* in talus above Atigun Pass at 1500 m.

Spetzman and others have also collected around above Anaktuvuk Pass (550 m) on the North Slope. The most widespread community in that part of the range is the *Dryas*-lichen dry meadow, only a few centimeters high; about 90% of the plant cover is *Dryas octopetala*. Associated species are *Lupinus arcticus, Hierochloe alpina, Silene acaulis, Kobresia myrosuroides,* and *Polygonum viviparum*. In wetter places are sedge meadows and tussock vegetation, with *Carex aquatilis, C. bigelowii, C. membranacea, Dryas integrifolia, Eriophorum angustifolium, E. vaginatum,* and *Salix reticulata*. Intermediate places in regard to moisture are occupied by low shrubs with herbaceous undergrowth.

Around Lake Peters and Lake Schrader at elevations of approximately 1070 m, Spetzman described a complex vegetational cover of about 150 species of vascular plants. Dry meadows are the most extensive type, with minor amounts of tussock meadow, wet sedge meadows, and willows. Most of the vegetation above 1225 m is confined to rock crevices and the edges of small mountain streams. On north-facing slopes with late snow cover, *Cassiope tetragona* is the dominant species. But the most widespread and conspicuous genus in this elevational zone is *Saxifraga*, with the following species being important: *S. bronchialis*, *S. cespitosa*, *S. eschscholtzii*, *S. flagellaris*, *S. oppositifolia*, *S. punctata*, *S. reflexa*, *S. serpyllifolia*, *S. tricuspidata*, and *S. davurica*. The presence or absence of water plays an important role in plant distribution and community development in this part of the Brooks Range.

Mountains of Central Alaska

Between the Brooks Range of northern Alaska and the high and icy Alaska Range there are some relatively low mountains of about 1800 m elevation with varied geologic composition. These White Mountains lie between the Tanana and Yukon rivers. Only the summits are alpine. General descriptions of the alpine vegetation and detailed lists of the flora have been provided by Gjaerevoll (1958, 1963, 1980). However, quantitative vegetational compositions are lacking. On the other hand, the alpine zone of these mountains has been rather thoroughly investigated around Eagle Summit by P. C. Miller and his colleagues (Miller 1982). Included in the latter publication are many tables that describe the vegetation in the quantitative terms of physiological plant ecology. One gets a very good idea of the structure and operation of this arctic-alpine vegetation and its communities from this modern and useful viewpoint. Unfortunately, space will not allow a thorough description here; I refer the reader to Miller's work.

Rocky Mountains

The most intensively studied alpine vegetations in North America are those of the long Cordillera: the Rocky Mountains. This system extends from central Alaska to Colorado and New Mexico. Along such a great distance there are marked differences in geology and climate; and, of course, there is the increasing elevation of timberline southward. Surprisingly, there is a rather strong coherence in floristic composition of the alpine vegetations along this latitudinal gradient. This seems to be due to the fact that the cordilleran pathway connecting the Arctic with the continental midlands has been there for a long time – at least from the beginning of the Cenozoic. As Billings (1974) and Axelrod and Raven (1985) have suggested, however, most of the upward and latitudinal migrations of plant taxa occurred during the wildly fluctuating climates of the Pleistocene. There were many times in the last million years or so when land bridges of relatively low elevations connected separate ranges of the Rockies and provided avenues of dispersal to and from Colorado, Wyoming, and the Arctic. Nor did these all have to be open simultaneously; plants can be patient in their migratory movements.

Apart from the rocky cliffs and peaks, the typical alpine vegetation on the gentler slopes and somewhat level areas of the central and southern Rocky Mountains is very much like that mapped in Fig. 13.5 and listed in Table 13.2. In other words, it is essentially a repeatable gradient from open fellfields on upper windward slopes, across ridges to moist meadows, and to wet meadows and bogs at the bases of the lee slopes. From north to south, these may differ somewhat in floristic composition, but the community physiognomy does not vary much along the topographic water gradients from one latitude to another. Such moisture gradients are primarily functions of the redistribution of snow by winter winds. However, there is considerable summer precipitation on the crests of these mountains. This increases toward the south, particularly in the form of thunderstorms that reach a maximum in southern Colorado and northern New Mexico. This summer mix of rain, hail, and snow has little effect on the moist and wet meadows, but I believe it is a strong factor in maintaining certain arctic species on the ridges and fellfields.

In this regard, Billings (1978), using 194 species growing in the alpine zone of the Beartooth Mountains at 45° N on the boundary between Wyoming and Montana, has found that at least 91 of them (47%) also occur in the Arctic. This is a lower percentage of arctic species than on Mt. Washington in New England, but higher than one finds to the west in the mountains of the Great Basin or in the Sierra Nevada. However, more important, the absolute number of arctic species is higher than that on Mt. Washington or on other American mountains at similar latitudes. Such a high number of arctic species is also characteristic of other Rocky Mountain ranges, particularly the high peaks of Colorado. The pathway of the Cordillera to the Arctic may be partly responsible for arctic species getting to the central Rockies, but I believe that the summer rains help to maintain them there.

Because environment plays such an important role in the present distribution and composition of the Rocky Mountain alpine vegetation types, it is not surprising that there has been emphasis in the last 30 yr on the effects of environmental factors on these communities. Bliss (1956), for example, compared plant growth at four alpine sites in Wyoming with similar sites in the Arctic in regard to temperature, water, and soils. Billings and Bliss (1959) made similar studies on alpine productivity through the growing season around a semipermanent snowbank. Scott and Billings (1964) carried this temperature-water approach even further with their studies of productivity and net photosynthesis both in the field and in the laboratory. All three of these studies were done in the Medicine Bow Mountains of Wyoming.

The nature and activity of soils and geologic substrata have also been subjects of attention in regard to vegetation-environment interactions above the timberline in the Rocky Mountains. For example, from much previous work in the mountains of Europe it has been known for a long time that alpine plant distribution and communities are influenced to a great extent by limestone and other calcareous substrata. Bamberg and Major (1968), in a thorough study of such limy sites in three mountain ranges in Montana, found that the effects of these calcareous parent materials are long persistent in cold, frost-churned soil. Such cryopedogenic activity and its effects on the nature and dynamics of Rocky Mountain alpine vegetation have been subjects of intensive research by Billings and Mooney (1959), Johnson and Billings (1962), and Bryant and Scheinberg (1970).

The most thorough work on alpine vegetation in the Rocky Mountains has been done in the Indian Peaks area of the Front Range above Boulder, Colorado. This includes the excellent work of Komárková (1979, 1980) using standard Braun-Blanquet methods and ordination. Niwot Ridge (Fig. 13.3) is the site of most of the intensive work, including a detailed vegetation map in color at a scale of 1 : 10,000 (Komárková and Webber 1978). This map details the distribution of 21 associations, alliances, orders, and classes of vegetation occurring on Niwot above timberline. The compositions of these communities are given in detail by Komárková (1979).

In the central and southern Rocky Mountains, there are several alpine vegetation types that appear over and over again along dry-to-wet moisture gradients. Among these are relatively dry fellfields (Fig. 13.7), moist meadows (Fig. 13.8), wet meltwater meadows (Fig. 13.9), and rather barren areas beneath long-lasting snowbanks. These appear also in Fig. 13.5 and Table 13.2 earlier in this chapter. Stand data from the most productive of these (moist meadow = *Geum* turf; wet meadow = *Deschampsia* meadow) have been published by Johnson and Billings (1962). These data are included here in Tables 13.5 and 13.6 (pp. 409, 410) as representative of the compositions of such alpine communities in the Rocky Mountains.

Mountains of the Great Basin and the Southwest

From the Wasatch Mountains of central Utah to the Sierra Nevada, the rivers drain inward to the old lake basins of the Pleistocene. Rimming the long desert valleys of Nevada and western Utah are almost two hundred tall but narrow mountain ranges. These montane "islands" trend north and south above the cold desert "seas" (Billings 1978). Southward, other montane islands rise above the "seas" of the warm deserts of Arizona (Billings 1980; Gehlbach 1981). One of these latter is of particular importance as an alpine refugium: San Francisco Mountain (3857 m) in northern Arizona.

This isolated Pleistocene volcano rises above the pines and junipers of the Coconino Plateau, with no other alpine peaks visible even in the distance. Baldy Peak (3533 m) in the White Mountains, 260 km to the southeast, is the nearest candidate. But Baldy is marginally alpine at best; it is essentially a subalpine "bald." Charleston Mountain (3630 m), in southern Nevada, 380 km to the northwest, has only a small alpine flora on limestone.

The remarkable thing about the top of San Francisco Mountain is that within an alpine area of only about 5.18 km^2 is an aggregation of alpine species with close relationships to the circumpolar arctic flora and alpine floras of the Rocky Mountains. Little (1941) was the first to list these species, to describe the communities, and to speculate on their geographical origins. His figures show 41% arctic-alpine, 49% Rocky Mountain alpine, 6% southwestern, and 4% endemic to the mountain. The total alpine flora, according to Little, was 49 vascular plant species.

Figure 13.7. Alpine fellfield vegetation on glacially deposited quartzite at 3300 m in the Medicine Bow Mountains, Wyoming. The prominent plant in the Compositae is Hymenoxys grandiflora *(photograph by W. D. Billings).*

Figure 13.8. Moist alpine meadow at 3275 m in the Beartooth Mountains, Wyoming. The prominent plant is Bistorta bistortoides. *In the middleground is a semipermanent snowbank near the end of summer showing the barren gravels recently exposed by snowmelt. Only a few species of dwarf plants can grow below such snowbanks, and they may not be exposed every year (photograph by W. D. Billings).*

Figure 13.7.

Figure 13.8.

Figure 13.9. Meltwater pond and hummocky wet meadow of sedges at 3675 m near Independence Pass, Colorado (photograph by W. D. Billings).

Only two associations were recognized by Little: alpine rock field and alpine meadow. The rock field association has nearly all the species growing above timberline. The soil surface is generally covered with volcanic rocks of various sizes, but there is fine material beneath the dry-looking rocks; the fine material is surprisingly moist in the summer, with water derived from the frequent thundershowers that gather over the peak. The principal species are *Geum rossii, Silene acaulis, Cerastium beeringianum, Festuca brachyphylla, Poa rupicola, Trisetum spicatum, Primula parryi,* and *Oxyria digyna*. These all have deep, well-developed root systems in the porous soil. The alpine meadow community also is dominated by *Geum rossii,* but present are *Sibbaldia procumbens, Silene acaulis, Trisetum spicatum, Carex albonigra, Pedicularis parryi,* and *Luzula spicata*.

Recently, Schaak (1983) has restudied the alpine vegetation of this mountain. With additional collections, he now lists the alpine flora as consisting of 80 taxa of vascular plants. Most of these are rare or uncommon, but their presence is significant phytogeographically. This is particularly true of the arctic saxifrages *S. cespitosa* and *S. flagellaris,* which here reach their southernmost limits on the continent. I concur with Schaak's hypothesis that the dispersal pathway for these taxa to San Francisco Mountain was from the Rocky Mountains by way of the mountains of western New Mexico, the White Mountains of Arizona, and the nearby Mogollon Rim. In sum, it is a remarkable assemblage of alpine plants, most of them at their southern and most isolated limits in North America.

Relatively little is known about the alpine floras and vegetation of the mountains of the Great Basin per se. There are some notable exceptions. Two of these are the Deep Creek Mountains in western Utah and the Ruby Mountains in northeastern Nevada. McMillan (1948) listed 80 species as being present above timberline in the relatively small Deep Creeks. In Loope's classic work (1969) on the Ruby Mountains, a much larger range than the Deep Creeks, 189 alpine species were reported. In both cases, the floras have a distinctly Rocky Mountain affinity. This is shown in Table 13.7 (p. 411) which lists the ground coverage for plant species in six alpine turf communities in the Ruby Mountains as compared with the vegetation in four dry alpine sites in the same range. I do not know of similar studies on Wheeler Peak in the Snake Range, but in

Table 13.5. *Point quadrat data from Geum turf expressed as percentage of ground covered*

Species	Stands																					
	24	36	23	38	31	6	15	14	32	5	13	17	18	19	37	33	12	11	34	10	4	Mean
Geum rossii	21	26	23	26	22	20	16	12	12	10	15	15	5	6	13	7	10	5	5	3		11.7
Carex drummondiana		8	9	14	13	10	10	7	8	8	14	15	5	8	17	11	18	11	19	20	16	16.1
Phlox caespitosa	4		5	8	9	6				3		5	4	3	5		10	5	7	3	3	4.0
Artemisia scopulorum	10		13	4	7		6	4			3								2			2.8
Trifolium parryi	8	13	5																5			1.5
Luzula spicata	2	4		5		4	4	3	2				3		2	2						1.5
Polygonum bistortoides	4		6			4	6	4		3	5	3		10	5	4	4		3	2		3.0
Potentilla diversifolia	2	3	4	3			8		3													1.0
Deschampsia caespitosa	2	3	4	3			8		3								2					1.5
Mosses	4		5				6	2		2												0.9
Carex scopulorum	3							5														0.4
Festuca ovina	3	3	3				2	4	4	3		10		4			2	2	2		2	2.1
Trifolium nanum			2		8				25				6	20					4			3.3
Erigeron simplex		2							7	2	2		7		2		2		5	4	7	2.0
Poa species	5	6	6	2									2	3				2				1.2
Senecio fuscatus	2																					0.1
Gentiana algida						4																0.2
Polemonium viscosum						2						2		3								0.3
Cerastium beeringianum						2						3										0.2
Mertensia alpina						2											2					0.2
Bupleurum americanum						4			3	3		10		5				3				1.3
Antennaria species					6							3								2		0.5
Sedum rosea							2															0.1
Carex elynoides					3																	0.1
Lupinus monticola										3		3								3		0.4
Dodecatheon radicatum										2		2										0.2
Smelowskia calycina				2	4				5	4			2				4	4	5	2	3	0.5
Silene acaulis				4					3				11	3			14	7				2.1
Arenaria obtusiloba				2		4	2				4			12	6		4			6	4	1.9
Trisetum spicatum		3				2			6	6		7						2	4	3	7	1.9
Lomatium montanum				2		4				6				5				8		4	3	1.5
Selaginella densa						4		2	5					3	3	5	2	3				1.3
Eritrichium elongatum													2	2		2		3	5	2	2	0.9
Oxytropis species													2				4	5	5		5	1.0
Potentilla nivea																					2	0.1
Senecio canus																					4	0.2
Lichens	2			2		10	4	8		15			18					5		14	19	4.5
Litter	6		2	4	34	14	16	18	6	10	35	6	8	9	20	61	60	19	6	20	10	14.6
Bare soil or rock	10	18	8	14	1	4	14	11	8	9		6	8	9	7		6	10	6		6	7.4

Source: From Johnson and Billings (1962).

Table 13.6. *Point quadrat data from* Deschampsia *meadow vegetation expressed as percentage of ground covered*

	Stands													
Species	16	35	25	27	26	3	40	42	41	30	28	29	43	Mean
Deschampsia caespitosa	11	10	44	33	43		13		2	10	4	6		13.5
Geum rossii	5		11	9	9	11								3.5
Trifolium parryi	31		3											2.6
Luzula spicata	2	2				2								0.5
Trisetum spicatum		2				2								0.3
Arenaria obtusiloba	3													0.2
Salix nivalis	12	5												1.3
Salix planifolia		19												1.5
Calamagrostis purpurascens		7												0.5
Polygonum viviparum		5												0.4
Artemisia scopulorum	11	2	4	5										1.7
Potentilla diversifolia		2	2	3	2	9						8		2.0
Festuca ovina		2		3	4	5							2	1.2
Erigeron simplex			4			7								0.9
Mertensia alpina			3											0.2
Cerastium beeringianum			3	4	5									0.9
Polygonum bistortoides	7		6	5					4		2	9		2.5
Gentiana algida				2										0.2
Lichens	3	4						2	11					1.5
Poa species				2	22				3	3	3			2.5
Smelowskia calycina						5								0.4
Mosses		3		13		2		4	8	4	8	11	3	4.2
Carex scopulorum		3				44	44	37	26	32	25	7	3	16.8
Antennaria species							5	9	20	3				2.8
Carex species		2		2			4		4	19	2	8	5	3.5
Caltha leptosepala		5		5						9	24	24		5.2
Sibbaldia procumbens							5			3	4	14		2.0
Senecio cymbalariodes										3	6	4		1.0
Juncus drummondii										2	7		2	0.8
Arabis species													3	0.2
Water										2	10	7		1.5
Litter	3	6	15	7	10		12	4	5	2	3			5.2
Bare soil or rock	8	22	5	2		9	14	40	12	5			81	15.2

Source: From Johnson and Billings (1962).

that higher and more extensive alpine area there should also be strong floristic ties with the Rocky Mountains.

There have been scattered studies on the mountain ranges of central Nevada, but as yet nothing definitive in regard to alpine vegetation to the extent of the studies by McMillan and Loope. The Toiyabe, Toquima, Grant, and other high ranges there await more work.

The alpine floras of some high mountains in the western Great Basin are better known. These are all west of the Pleistocene Lake Lahontan shoreline. The alpine zone on Mt. Grant (3426 m) in the Wassuk Range has an area of only 2.6 km^2, but a flora of 70 species (Bell and Johnson 1980). The affinities are mainly with the Sierra Nevada and with mountainous western North America in general. However, 12 species (17%) are widespread Arctic-alpine. The same authors (Bell Hunter and Johnson 1983) also studied the larger alpine zone of the Sweetwater Mountains (3558 m) about 45 km southwest of Mt. Grant and only 33 km east of the crest of the Sierra. The alpine area is about 16 km^2. The alpine flora consists of 173 species of vascular plants. Of these, 94% also occur in the Sierra, 75% in the Great Basin, 52% in the southern Rocky Mountains, and 18% in the Arctic. It is clearly Sierran in derivation.

The White Mountains of California stand tall at the very western edge of the Great Basin, a figurative stone's throw east of the Sierra Nevada, about 40 km from crest to crest. White Mountain Peak reaches 4341 m, and much of the range is above the timberline of *Pinus longaeva*. Of all the Great Basin mountains, the White Mountains are the best known ecologically.

Lloyd and Mitchell (1973) made a thorough study of the flora of this whole range. In that volume, Mitchell discussed the phytogeography and comparative floristics. Half of the endemic taxa in the White Mountains are found in the alpine zone. The flora of the whole alpine zone is about 200 taxa,

Table 13.7. *Coverage values (% ground cover) (line-intercept method) for species in alpine turf (columns 1–6) and dry-site alpine communities (columns 7–10) in North Ruby Mountains*

Species	1[a]	2	3	4	5	6	7	8	9	10
Erigeron peregrinus	18.9	25.3	13.2	13.9	12.7	3.1	4.5	1.1	—	0.4
Salix arctica	12.0	23.9	7.1	1.9	22.5	33.2	—	5.6	—	—
Carex elynoides	—	8.5	5.4	21.3	—	42.7	—	—	—	—
Caltha leptosepala	10.2	22.0	—	10.4	10.5	—	—	—	—	—
Sibbaldia procumbens	0.7	2.1	6.4	2.7	4.7	8.4	—	2.2	18.6	4.5
Polygonum bistortoides	6.9	3.9	8.9	10.1	3.3	—	—	2.6	3.6	0.7
Pedicularis groenlandica	1.1	1.1	2.5	6.6	5.1	—	—	—	—	—
Festuca brachyphylla	2.2	0.7	4.6	0.8	3.3	—	—	—	—	—
Geum rossii	3.3	—	—	—	—	1.1	76.5	5.6	14.7	5.2
Epilobium alpinum	2.2	—	0.4	—	—	—	—	—	—	—
Epilobium latifolium	1.8	—	—	—	—	—	1.5	—	—	—
Veronica wormskjoldii	2.2	—	6.4	2.3	1.1	—	—	—	—	—
Astragalus alpinus	5.1	6.7	5.0	0.4	0.4	2.3	—	0.7	—	—
Antennaria umbrinella	4.4	—	2.1	—	—	1.1	—	—	—	—
Potentilla diversifolia	4.4	0.7	—	2.3	7.6	—	—	0.7	1.1	0.7
Senecio cymbalarioides	0.4	—	—	16.3	13.4	1.1	7.1	0.7	0.7	0.4
Viola adunca	2.5	—	1.4	—	—	—	—	—	—	—
Gentiana calycosa	1.5	—	0.4	1.2	2.9	—	—	—	—	—
Carex pseudoscirpoidea	3.3	—	—	—	—	—	—	4.8	—	4.5
Carex festivella	0.4	—	—	—	—	—	—	—	—	—
Ranunculus eschscholtzii	—	1.1	—	—	—	4.2	—	—	0.4	—
Thalictrum alpinum	—	—	6.1	1.9	—	—	—	—	—	—
Taraxacum officinale	—	—	2.1	—	—	—	—	—	—	—
Dodecatheon alpinum	—	—	0.7	—	—	—	—	—	—	—
Kalmia polifolia	—	—	—	0.4	1.8	—	—	—	0.4	—
Mimulus primuloides	—	—	1.8	1.9	—	—	—	—	—	—
Juncus mertensianus	—	—	—	—	0.4	—	—	—	—	—
Vaccinium caespitosum	—	—	1.4	0.8	—	—	—	27.4	—	23.6
Parnassia fimbriata	—	—	3.9	—	1.1	—	—	—	—	—
Potentilla fruticosa	—	—	0.4	—	—	—	—	—	—	—
Castilleja linariaefolia	—	—	—	—	—	—	1.9	—	—	—
Mertensia ciliata	—	—	—	—	—	—	0.7	—	—	—
Hackelia jessicae	—	—	—	—	—	—	0.4	—	—	—
Festuca ovina	—	—	—	—	—	0.4	—	5.2	36.5	—
Phleum alpinum	—	—	0.4	—	0.4	—	—	—	—	—
Juncus drummondii	—	—	1.1	—	—	0.4	—	1.5	3.6	1.4
Deschampsia caespitosa	—	—	—	—	—	—	—	—	3.6	—
Salix orestera	—	—	—	1.9	—	—	—	—	—	—
Penstemon procerus	—	—	—	—	—	—	—	0.4	—	1.1
Trifolium monanthum	—	—	—	—	—	—	—	1.1	—	—
Antennaria rosea	—	—	—	—	—	—	—	5.6	4.6	3.7
Agrostis rossae	—	—	—	—	—	—	—	—	0.4	—
Trisetum spicatum	—	—	—	—	—	—	—	—	—	1.4
Carex nova	—	—	5.0	—	4.4	—	—	—	—	—
Carex species	—	—	—	—	—	—	—	—	1.1	—
Poa palustris	—	—	—	—	—	—	1.5	—	—	—
Poa epilis	—	—	—	—	—	—	—	3.8	0.7	4.5
Carex pseudoscirpoidea	—	—	—	—	—	—	—	—	—	2.2
Poa fendleriana	—	—	—	—	—	—	—	—	—	4.1
Soil	17.1	1.1	6.8	2.7	2.5	—	4.1	14.8	1.4	26.2
Litter	—	2.1	6.4	—	2.2	1.9	—	1.9	4.6	9.7
Rock	—	1.1	—	—	—	—	1.9	—	3.8	5.2

[a]Locations of sampling sites: 1, 2, and 7, 9750 ft, south of Lamoilie Lake (7/30/68); 3, 10,000 ft, right fork of Lamoille Canyon (8/4/68); 4–6, 8–10, 10,000 ft, Island Lake cirque (8/22/68) and (8/23/68).
Source: From Loope (1967).

of which 125 are characteristically alpine. About 62% of these also occur in the nearby Sierra, and only 28% in the Rocky Mountains.

Mooney (1973) described the plant communities of the White Mountains. The alpine zone extends from about 3500 m to the summit at 4341 m. The range is dry and very cold, so that the vegetation around the peak itself is extremely sparse. But even so, three species are restricted to the high alpine zone above 4000 m: *Phoenicaulis eurycarpa*, *Polemon-*

Table 13.8. *Plant cover (%) in three alpine communities in the White Mountains, California*

Species	White Mt. Peak pyramid, 4176 m	Dolomite barrens, 3597 m	Fellfield granite, 3871 m
Herbs			
Erigeron vagus	0.40	—	0.52
Festuca brachyphylla	0.56	—	—
Calyptridium umbellatum	0.24	—	—
Polemonium chartaceum	0.28	—	—
Eriogonum gracilipes	—	1.32	—
Poa rupicola	—	2.00	—
Sitanion hystrix	—	0.36	0.96
Phlox covillei	—	6.52	—
Erigeron pygmaeus	—	0.92	—
Draba sierrae	—	0.28	0.48
Arenaria kingii	—	0.64	—
Castilleja nana	—	0.20	—
Eriogonum ovalifolium	—	—	6.16
Carex helleri	—	—	0.24
Trifolium monoense	—	—	27.64
Selaginella watsoni	—	—	0.68
Haplopappus apargioides	—	—	1.04
Koeleria cristata	—	—	11.12
Lewisia pygmaea	—	—	0.08
Potentilla pensylvanica	—	—	0.20
Total cover, all plants	1.48	12.24	50.08

Source: From Mooney (1973).

ium chartaceum, and *Erigeron vagus*. There is a strong relationship to geologic substratum. The dolomites have a rather barren appearance and low plant cover as compared with the granites and quartzites. Table 13.8 compares the plant cover at three sites in the alpine zone. Additional data on the alpine vegetation and flora in relation to active soil frost features are provided by Mitchell and associates (1966).

Billings (1978) calculated floristic similarity (Sørensen's index) between alpine floras across the Great Basin along two transects from the Rocky Mountains to the Sierra. The northern transect extends from the Beartooth Mountains on the Wyoming-Montana line to Piute Pass in the Sierra. The southern transect is from San Francisco Mountain in Arizona to Olancha Peak in the Sierra. These percentages are shown in Table 13.9, which demonstrates that the alpine floras east of a hypothetical line extending southward from near Elko, Nevada, to Las Vegas, Nevada, are strongly related to those of the Rocky Mountains. West of that line, alpine floras of the Great Basin are unique. However, in the White Mountains there is some floristic relationship of the alpine flora to that of the Sierra, but vegetationally the two ranges are distinct.

The Sierra Nevada

This large mountain range in eastern California extends almost 650 km from northwest to southeast. Its base is a large granitic batholith tilted toward the west and capped by andesitic flows in its northern reaches. In the south, Mt. Whitney (4418 m) is the highest peak in the lower 48 states. It is a young range, uplifted mainly during the Pliocene and Pleistocene.

The alpine flora is the largest and richest such flora on the continent. Both it and the alpine vegetation are not as well known as they ought to be. However, Major and Taylor (1977) have done an excellent job in synthesizing the available information on the alpine vegetation. They presented long and detailed tables on the compositions of the main alpine communities. The reader is referred to them for details, particularly concerning the close relationships between geologic substratum and present vegetation.

Chabot and Billings (1972) studied a series of elevational sites on the steep eastern escarpment of the Sierra from the desert near Bishop at 1400 m to above Piute Pass at 3540 m on the glacially scoured granites of the Sierran crest (Fig. 13.10). Emphasis was on the question "How does an alpine flora originate?" Most of the research in the field was devoted to measuring the microenvironments along this steep transect in regard to the distribution of different species along the gradient. Physiological measurements were made on these species under controlled conditions in the Duke University Phytotron. These included water relations, photosynthesis, respiration, storage products, and acclimation.

Only 19% of the alpine species on the granites above Piute Pass also occur in the Arctic. This is in strong contrast to the almost 50% in the Beartooth Mountains. Alpine species here endemic to the Sierra (17%) are in genera predominantly present in California or Great Basin floras at lower elevations. A number of species near Piute Pass have populations also in the desert below. Desert taxa certainly have contributed to the Sierran alpine gene pool. Most of the alpine flora consists of perennial herbaceous species, but there are more annual species

Table 13.9. *Sørensen's indices of floristic similarity (%) between alpine floras for all combinations of alpine areas in both transects; numbers in parentheses are distances (km) between alpine regions*

	Northern transect						Southern transect		
	Bear-tooth	Deep Creeks	Ruby Mts.	Toiyabe Range	Pellisier Flats (White Mts.)	Piute Pass (Sierra)	Olancha Pk. group (Sierra)	Spring Mts.	San Francisco Pks.
Beartooth	—	24	33	14	14	9	14	8	23
Deep Creeks	(676)	—	39	22	19	10	15	13	31
Ruby Mts.	(692)	(153)	—	20	14	11	14	8	19
Toiyabe Range	(917)	(322)	(257)	—	21	16	13	18	21
Pellisier Flats (White Mts.)	(1102)	(451)	(418)	(145)	—	34	36	16	19
Piute Pass (Sierra)	(1167)	(515)	(483)	(217)	(72)	—	32	10	4
Olancha Pk. group (Sierra)	(1223)	(547)	(539)	(290)	(169)	(129)	—	13	13
Spring Mts.	(1110)	(434)	(475)	(325)	(298)	(306)	(225)	—	14
San Francisco Pks.	(1086)	(531)	(668)	(636)	(660)	(668)	(603	(378)	—

Source: From Billings (1978).

Figure 13.10. Alpine zone and glacially scoured granite at 3700 m, above Piute Pass, Sierra Nevada (from Chabot and Billings 1972).

than in other alpine floras, where they are unusually rare. The alpine vegetation on these granites is very sparse. It is characterized by low plants of *Phlox caespitosa*, *Penstemon davidsonii*, *Ivesia pygmaea*, *Ivesia lycopodioides*, *Draba lemmonii*, *Arenaria nuttallii*, *Carex helleri*, *Eriogonum ochrocephalum*, and *Potentilla brewerii*. At slightly higher elevations, on the same substratum, *Oxyria digyna*, *Polemonium eximium*,

Figure 13.11. Very open fellfield on glacially scoured granite at 3540 m, above Piute Pass, Sierra Nevada. The scattered plants are Lupinus breweri, Potentilla breweri, Antennaria rosea, *and* Calyptridium umbellatum; *Mt. Humphreys (4263 m) in background (photograph by W. D. Billings).*

and *Hulsea algida* are constant members of the alpine vegetation wherever soil has accumulated in the granite. The most extremely arid sites of this type have almost no vascular plants except for small individuals of *Calyptridium umbellatum* and *Polygonum minimum*; the latter is a miniature annual. Figure 13.11 is typical of the sparse vegetation in these dry granitic sites. The composition of the alpine community at this site appears in Table 13.10.

Carson Pass, south of Lake Tahoe, is the northern limit for many alpine species in the Sierra, according to Major and Taylor (1977). Here, the granodiorites are overlain by volcanic flows, mainly andesites. The granitic rocks have a better representation of truly alpine plant species, particularly on north-facing slopes. Even on the andesites, however, in places where snow stays late into the summer there are alpine plants such as the rare but ubiquitous *Oxyria digyna*. However, where the andesites are swept clear of winter snow on ridge tops, there are communities of plant species from the cold deserts of western Nevada: *Balsamorrhiza sagittata*, *Viola beckwithii*, *Wyethia mollis*, *Eriogonum umbellatum*, *E. ovalifolium*, *Crepis occidentalis*, and *Lygodesmia spinosa*.

Cascade Range

The most conspicuous mountains of the Cascades Range are the high ice-covered volcanoes that are scattered along the crest of the range, from Mt. Lassen and Mt. Shasta in northern California to Mt. Baker in northwestern Washington. Above timberline on ridges, but below the ice, these "island" mountains support small, open communities of herbaceous plants in the alpine zone. These show considerable relationships with similar communities on ridges in the Sierra.

However, in northern Washington, the North Cascades provide a rugged set of young mountains, geologically complex, and heavily glaciated in the past. The alpine vegetation is well developed in this snowy, cool environment. Douglas and Bliss (1977) have sampled 128 of these communities thoroughly and have related the alpine vegetation to soils and microenvironments using methods of ordination. The results show a large array of plant community types across these mountains from west to east. Their detailed tables of the compositions of these communities are too long to present here, but I refer the reader to their study.

Table 13.10. *Composition and coverage of the alpine plant community on granite, above Piute Pass, at an elevation of 3,540 m in the central Sierrra Nevada*

Species	Coverage (%)
Lupinus breweri	7.0
Antennaria rosea	6.5
Carex helleri	5.1
Carex phaeocephala	2.1
Sedum rosea ssp. *integrifolium*	1.7
Ivesia pygmaea	1.5
Solidago multiradiata	1.2
Selaginella watsoni	1.1
Lewisia pygmaea	0.5
Phlox covillei	0.3
Agrostis variabilis	0.3
Gentiana newberryi	0.2
Draba lemmonii	0.2
Trisetum spicatum	0.1
Calyptridium umbellatum	0.1
Castilleja nana	p[a]
Antennaria alpina var. *media*	p
Arenaria nuttallii ssp. *gracilis*	p
Eriogonum ovalifolium	p
Dodecatheon jeffreyi	p
Potentilla breweri	p
Poa hansenii	p
Sitanion hystrix	p
Polygonum minimum	p
Penstemon davidsonii	p
Poa nervosa	p
Carex nigricans	p
Total plant cover	27.9
Rock	40.7
Soil	31.4

Source: Exerpted from Table 1 of Chabot and Billings (1972), as determined by intercept on a 20-m line.
[a]p=cover <0.1%

The alpine flora of the North Cascades has floristic affinities with many regions in western North America. Unlike that of the Sierra, the strongest relationships are with the Arctic and cordilleran regions to the north and east. A smaller part of the flora is restricted to the mountains from British Columbia to California. The number of Cascadian endemic species in the North Cascades is very low (Douglas and Bliss 1977).

The High Mountains of Mexico

Webster (1961) listed the upper elevational limits of alpine plants on the higher mountains over the earth. He included Citlaltepetl (Pico de Orizaba) at 19° N, whose summit elevation in central Mexico is about 5700 m. The highest vascular plant is apparently *Castilleja tolucensis* at 4570 m, over 1000 m below the summit. As Webster noted, vascular plants in the Himalayas and the southern Andes reach their upper limits almost 1500 m higher. Why should plants not go higher in Mexico? In my opinion, this could be a result of the relatively young age of these high Mexican volcanoes and their isolated geographical position between the Rocky Mountains to the north and the Andes to the south. The arctic taxa at the species level apparently have had difficulties getting to these high mountains. The same thing could be said for the members of the antarctic element, some of which have come up the Andes to Colombia and Venezuela. *Colobanthus quitensis* reaches the high mountains of central Mexico, occurring at 3900 m on Pico de Orizaba. Its nearest station to the south is on Pichincha, the volcano overlooking Quito, Ecuador. How did it make this jump, if indeed it did? No one has yet collected one of its arctic-antarctic counterparts, *Saxifraga cespitosa*, on the Mexican volcanoes. The gap in the saxifrage distribution is wide: San Francisco Mountain, Arizona, to Pichincha, an airline distance of 5500 km (Mulroy 1979). Why is it not in Mexico?

Beaman and Andresen (1966) have described alpine meadow vegetation from the summit of Cerro Potosí (ca. 3650 m) at 24° 53′ N in the Sierra Madre Oriental. In this alpine meadow, presence and percentage plant cover by species were recorded on 100 quadrats that were 50 cm on a side along a line 1 km long extending from the north to the south end of the meadow. The results are shown in Table 13.11. *Potentilla leonina*, a species of *Arenaria*, and *Bidens muelleri* together compose 49.3% of the total plant cover. These dominant species are all perennial herbs with a depressed or prostrate habit and small or finely divided leaves. In sixth and seventh places are the familiar *Linum lewisii* and *Trisetum spicatum*, the first a Rocky Mountain species, and the second a true arctic-alpine. Inspection of the list in Table 13.11 shows that other species also add to the feeling that there is an influence from the north on the alpine summit of the Cerro.

SUCCESSION

Vegetational change and soil formation are slow in any cold climate. These processes are slower on glacially polished rock than on glacial till or alluvium. Succession is extremely slow on the Medicine Bow Peak quartzite in Wyoming, for example. This rock is extremely hard, and it has been subject to glaciation. The same holds true for the Sierran granites west of Lake Tahoe in Desolation Valley and the granites above Piute Pass farther south (Figs. 13.10 and 13.11). Volcanic rocks in the Sierra,

Table 13.11. *Frequency cover, importance values (scale of 0–200), and life forms for 38 species encountered in quadrats in the alpine meadow on Cerro Potosí, Mexico*

Species	(%) Frequency	(%) Cover	Importance value	Life form[a]
Potentilla leonina	81	8.63	33.54	Ch
Arenaria species (Beaman 2664)	82	4.84	23.14	Ch
Bidens muelleri	54	4.31	18.36	H
Astragalus purpusii	72	1.36	12.31	Ch
Lupinus cacuminis	32	2.76	11.45	H
Festuca hephaestophila	55	1.28	10.79	H
Linum lewisii	69	0.93	10.77	H
Trisetum spicatum	62	0.83	9.65	H
Grindelia inuloides	38	1.74	9.34	H
Pinus culminicola	11	2.35	7.82	N
Astranthium beamanii	42	0.55	6.39	H
Thlaspi mexicanum	41	0.43	6.05	H
Senecio loratifolius	11	1.59	5.71	H
Draba helleriana	32	0.35	4.77	H
Senecio scalaris	26	0.36	4.08	H
Phacelia platycarpa	26	0.31	3.94	H
Hymenoxys insignis	7	0.56	2.38	H
Senecio sanguisorbae	7	0.50	2.22	H
Castilleia bella	13	0.20	2.10	H
Juniperus monticola	2	0.60	1.90	N
Villadia species (Beaman 4461)	13	0.03	1.62	H
Senecio carnerensis	7	0.25	1.52	Ch
Trifolium species	9	0.15	1.47	H
Tauschia madrensis	4	0.30	1.30	H
Sedum species (Beaman 4462)	8	0.09	1.20	Ch
Campanula rotundifolia	8	0.08	1.17	H
Achillea lanulosa	7	0.09	1.08	H
Delphinium valens	1	0.30	0.95	H
Sisyrinchium species (Schneider 939)	5	0.05	0.73	H
Stellaria cuspidata	3	0.06	0.52	H
Blepharoneuron tricholepis	3	0.06	0.52	H
Erysimun species (Beaman 2648)	3	0.03	0.39	H
Gnaphalium species (Beaman 2655)	2	0.02	0.30	H
Penstemon leonensis	2	0.02	0.30	H
Smilacina stellata	2	0.02	0.30	G
Allium species	1	0.01	0.15	G
Bromus frondosus	1	0.01	0.15	H
Cerastium brachypodum	1	0.01	0.15	H

Source: From Beaman and Andreson (1966).
[a]Ch=chamaephyte, H=hemicryptophyte, N=nanophanerophyte, G=geophyte

however, are more readily covered by vegetation and soils. Pockets of till or alluvium in the lower parts of the topographic gradient are invaded more readily and often become moist or wet meadows given enough time. Flow diagrams for succession in alpine ecosystems have been published; most of these are hypothetical. It would be difficult to use them for prediction except in a general way.

Two successional processes, however, are worthy of note, one biological and one physical. Griggs (1956) presented some interesting evidence for the biological process of plant competition in regard to invasion and succession in alpine fellfields in the Front Range of Colorado. He found that cushion plants such as *Silene acaulis, Minuartia obtusiloba, Paronychia pulvinata, Trifolium nanum*, and *Trifolium dasyphyllum* are readily invaded by a number of species. As many as 33 other species were observed invading *Silene acaulis* alone. From such observations, Griggs constructed a tentative competitive ladder. Up near the top of this ladder are *Geum rossii* and *Trifolium parryi*. A look at Table 13.2 shows that these two species do indeed prevail in the Medicine Bow alpine zone – but their success is primarily in the moist parts of the gradient rather than in the dry or wet sections. Surprisingly, the ideas of Griggs in regard to succession have not been avidly pursued; they deserve better.

Not all succession in alpine ecosystems is linear, as in the foregoing examples. Wherever soil frost polygonization is concerned, succession is likely to be cyclic, with no end point of stability. This is true

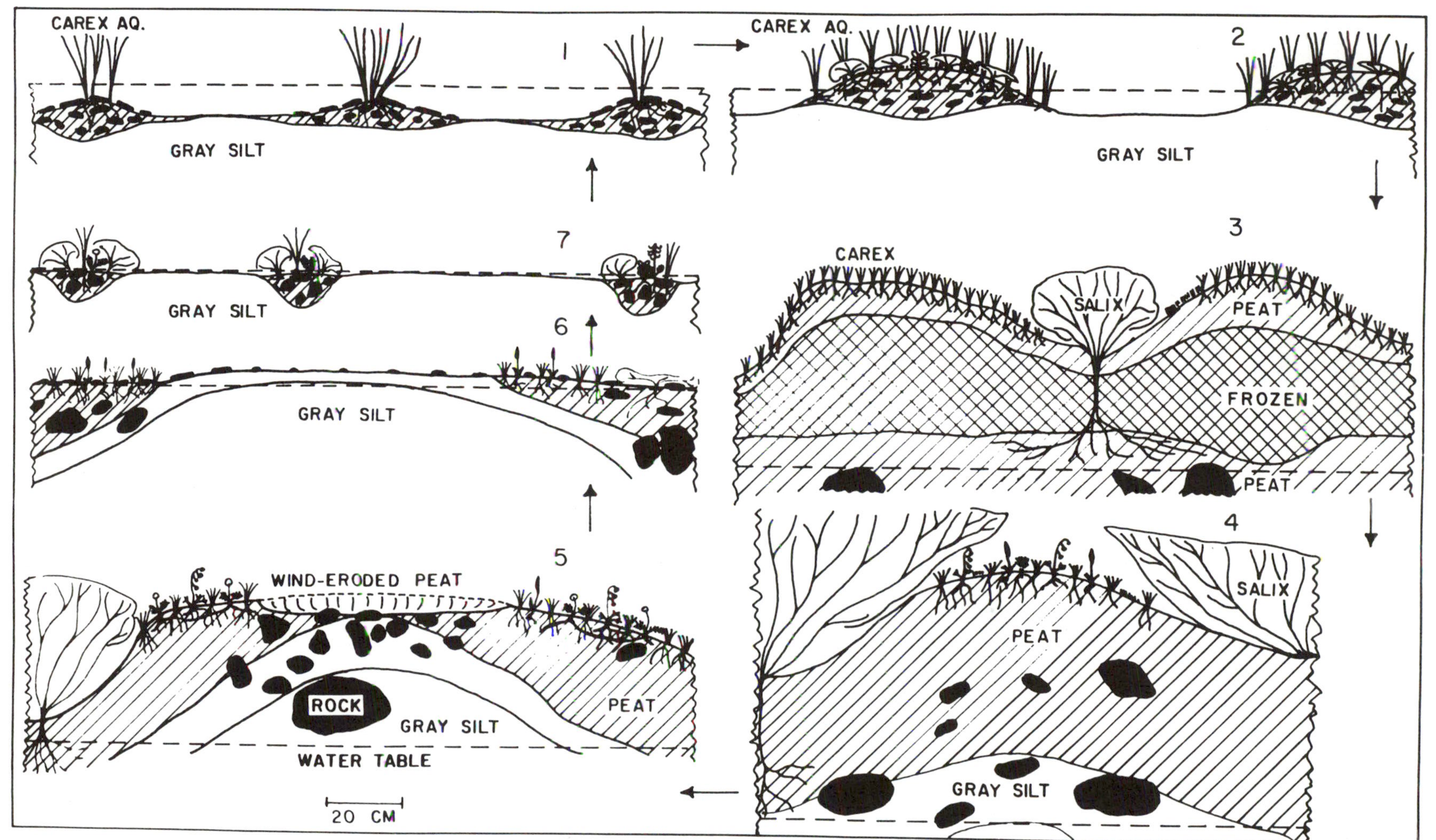

Figure 13.12. Initiation of peat hummocks by vegetation and frost action (stages 1, 2, 3, 4) in an alpine meadow of the Medicine Bow Mountains and subsequent disintegration of the hummocks into frost scars (stage 5) and sorted polygons (stages 6 and 7). Cross sections traced from field drawings of W. D. Billings (from Billings and Mooney, 1959).

on a large scale in the case of the thaw-lake cycle in the arctic wet tundra (Billings and Peterson 1980). But smaller-scale cyclic successions can be seen in the alpine zone. Billings and Mooney (1959) studied such a cycle in wet meadows in the Medicine Bow Mountains. As sedge peat hummocks build up in the meadow, they eventually get too high for complete winter snow protection. Strong winter winds carrying sharp snow and ice crystals blast off the tops of the higher hummocks. This allows soil frost action to thrust up rocks from the geologic substratum. The hummocks degrade into a stone net at or below the water table, with the stones as boundaries between the cells of the net. The net becomes barren and covered with shallow water. In time, mosses and vascular plants get established on the border rocks. These eventually form embryonic hummocks that grow higher than the water table, then form peat, and the process is renewed (Fig. 13.12).

PEOPLE AND ALPINE VEGETATION

Traditionally, in Europe, alpine ecosystems have been used for centuries as pastures for summer grazing and for production of hay. During the last century these practices became standard in the American West as the mountains were opened up for settlement. The greatest use has been as summer range for sheep. Toward the end of the nineteenth century and during the first half of the twentieth century, hundreds of thousands of sheep were driven up into the alpine meadows of the Rocky Mountains, the Great Basin ranges, and the Sierra Nevada. The effects of this heavy, relatively uncontrolled grazing are still apparent even in the national parks where grazing is not allowed now. Grazing is also closely controlled in the national forests and other public lands. The present composition of alpine vegetation reflects this impact of grazing. Was the unpalatable *Geum rossii* always as common in the high Rocky Mountains? Or has it increased because of grazing pressure on other, more choice species?

Mining also has had strong local impacts on mountain vegetation during the last century or more. This is particularly true of the very acid and nutrient-poor spoil dumps, which usually have high concentrations of heavy metals. Many of these are still barren of plants even after more than a century. The USDA Forest Service and also the companies operating the large modern mines have active research in progress on the use of native alpine plants in revegetation processes.

The twentieth century has seen a curvilinear increase in tourist use of the western American mountains, including the alpine zones. Millions of people are involved, with heavy concentrations in the national parks and national forests. The impacts are primarily those of foot traffic on trails, and backpacking and camping in the back country. But there have been other ecosystemic impacts, as, for example, the accidental introduction of *Giardia*, a protozoan, into alpine streams and lakes by back-country hikers. This causes the disease "giardiasis." Such water supplies are no longer safe.

As walkers have increased in number, casual trails through alpine vegetation have proliferated into extensive networks as the vegetation has been damaged or destroyed. Hartley (1976) made a thorough study of these impacts in Glacier National Park, Montana. Using a "sensitivity index," he found that 35 alpine plant species decreased in abundance as the trail system grew, and only 7 species increased their numbers. Some of the latter are semi-weedy species from lower elevations. He found some of the reasons for decreased vigor of these plants. Using cross-trail transects and carbohydrate analyses, Hartley found that late-summer total utilizable stored carbohydrates decreased by 20% to almost 50% within 0.5 m of the trail, as compared with the amounts stored by plants of the same species just beyond 2 m from the trail. Certainly the effects of off-road vehicles are similar.

SUGGESTIONS FOR FUTURE RESEARCH

I do not wish to restrict the scope of future research on alpine vegetation by focusing only on certain aspects and procedures. There are enough problems mentioned in this chapter to get people to thinking imaginatively. It is apparent that there is much to learn from vegetation and ecosystems at their upper limits. I am confident that we shall know a great deal more about them within the foreseeable future.

REFERENCES

Axelrod, D. I., and P. H. Raven. 1985. Origins of the cordilleran flora. J. Biogeography 12:21–47.

Bamberg, S. A., and J. Major. 1968. Ecology of the vegetation and soils associated with calcareous parent materials in three alpine regions of Montana. Ecol. Monogr. 38:127–167.

Beaman, J. H., and J. W. Andresen. 1966. The vegetation, floristics and phytogeography of the summit of Cerro Potosí, Mexico. Amer. Midl. Nat. 75:1–33.

Bell, K. L., and R. E. Johnson. 1980. Alpine flora of the Wassuk Range, Mineral County, Nevada. Madroño 27:25–35.

Bell Hunter, K., and R. E. Johnson. 1983. Alpine flora

of the Sweetwater Mountains, Mono County, California. Madroño 30:89–105.

Billings, W. D. 1973. Arctic and alpine vegetations: similarities, differences, and susceptibility to disturbance. BioScience 23:697–704.

Billings, W. D. 1974. Adaptations and origins of alpine plants. Arct. Alp. Res. 6:129–142.

Billings, W. D. 1978. Alpine phytogeography across the Great Basin. Great Basin Naturalist Memoirs 2:105–117.

Billings, W. D. 1979. Alpine ecosystems of western North America. pp. 6–21 in D. A. Johnson (ed.), Special management needs of alpine ecosystems. Society for Range Management, Denver.

Billings, W. D. 1980. American deserts and their mountains: an ecological frontier. Bull. Ecol. Soc. Amer. 61:203–209.

Billings, W. D., and L. C. Bliss. 1959. An alpine snowbank environment and its effects on vegetation, plant development, and productivity. Ecology 40:388–397.

Billings, W. D., and H. A. Mooney. 1959. An apparent frost hummock-sorted polygon cycle in the alpine tundra of Wyoming. Ecology 40:16–20.

Billings, W. D., and K. M. Peterson. 1980. Vegetational change and ice-wedge polygons through the thaw-lake cycle in arctic Alaska. Arct. Alp. Res. 12:413–432.

Bliss, L. C. 1956. A comparison of plant development in microenvironments of arctic and alpine tundras. Ecol. Monogr. 26:303–337.

Bliss, L. C. 1963. Alpine plant communities of the Presidential Range, New Hampshire. Ecology 44:678–697.

Bryant, J. P., and E. Scheinberg. 1970. Vegetation and frost activity in an alpine fellfield on the summit of Plateau Mountain, Alberta. Can. J. Bot. 48:751–771.

Caldwell, M. M., R. Robberecht, and W. D. Billings. 1980. A steep latitudinal gradient of solar ultraviolet-B radiation in the arctic-alpine life zone. Ecology 61:600–611.

Chabot, B. F., and W. D. Billings. 1972. Origins and ecology of the Sierran alpine flora and vegetation. Ecol. Monogr. 42:163–199.

Densmore, D. 1980. Vegetation and forest dynamics of the upper Dietrich River valley, Alaska. M.Sc. thesis, North Carolina State University, Raleigh.

Douglas, G. W., and L. C. Bliss. 1977. Alpine and high subalpine plant communities of the North Cascades Range, Washington and British Columbia. Ecol. Monogr. 47:113–150.

Fonda, R. W., and L. C. Bliss. 1966. Annual carbohydrate cycle of alpine plants on Mt. Washington, New Hampshire. Bull. Torrey Botanical Club 93:268–277.

Gehlbach, F. R. 1981. Mountain islands and desert seas: a natural history of the U.S.-Mexican borderlands. Texas A&M University Press, College Station.

Gjaerevoll, O. 1958. Botanical investigations in Central Alaska, especially in White Mountains, part I, Pteridophytes and monocotyledones. Det Kgl Norske Videnskabers Selskabs Skrifter no. 5, Trondheim.

Gjaerevoll, O. 1963. Botanical investigations in Central Alaska, especially in White Mountains, part II, Dicotyledones. Salicaceae-Umbelliferae. Det Kgl Norske Videnskabers Selskabs Skrifter no. 4, Trondheim.

Gjaerevoll, O. 1980. A comparison between the alpine plant communities of Alaska and Scandinavia. Acta Phytogeographica Suecica 68:83–88.

Griggs, R. F. 1956. Competition and succession on a Rocky Mountain fellfield. Ecology 37:8–20.

Hartley, E. A. 1976. Man's effects on the stability of alpine and subalpine vegetation in Glacier National Park, Montana. Ph.D. dissertation, Duke University, Durham, N.C.

Hustich, I. 1962. A comparison of the floras of subarctic mountains of Labrador and in Finnish Lapland. Acta Geographica 17:1–24.

Isard, S. A. 1986. Factors influencing soil moisture and plant community distributions on Niwot Ridge, Front Range, Colorado, U.S.A. Arct. Alp. Res. 18:83–96.

Johnson, P. L., and W. D. Billings. 1962. The alpine vegetation of the Beartooth Plateau in relation to cryopedogenic processes and patterns. Ecol. Monogr. 32:105–135.

Komárková, V. 1979. Alpine vegetation of the Indian Peaks area, Front Range, Colorado Rocky Mountains. 2 vols. J. Cramer, Vaduz, Liechtenstein.

Komárková, V. 1980. Classification and ordination in the Indian Peaks area, Colorado Rocky Mountains. Vegetatio 42:149–163.

Komárková, V., and P. J. Webber. 1978. An alpine vegetation map of Niwot Ridge, Colorado. Arct. Alp. Res. 10:1–29.

Little, E. L., Jr. 1941. Alpine flora of San Francisco Mountain, Arizona. Madroño 6:65–96.

Lloyd, R. M., and R. S. Mitchell. 1973. A flora of the White Mountains, California and Nevada. University of California Press, Berkeley.

Loope, L. L. 1969. Subalpine and alpine vegetation of northeastern Nevada. Ph.D. dissertation, Duke University, Durham, N.C.

McMillan, C. 1948. A taxonomic and ecological study of the flora of the Deep Creek Mountains of central western Utah. M.S. thesis, University of Utah, Salt Lake City.

Major, J., and D. W. Taylor. 1977. Alpine. pp. 601–675 in M. G. Barbour and J. Major (eds.), Terrestrial vegetation of California. Wiley, New York.

May, D. E., and P. J. Webber. 1982. Spatial and temporal variation of the vegetation and its productivity, Niwot Ridge, Colorado. pp. 35–62 in J. Halfpenny (ed.), Ecological studies in the Colorado alpine, festschrift for John W. Marr. Institute of Arctic and Alpine Research, University of Colorado, occasional paper no. 37.

Miller, P. C. (ed.). 1982. The availability and utilization of resources in tundra ecosystems. Holarctic Ecol. 5:81–220.

Mitchell, R. S., V. C. LaMarche, Jr., and R. M. Lloyd. 1966. Alpine vegetation and active frost features of Pellisier Flats, White Mountains, California. Amer. Midl. Nat. 75:516–525.

Mooney, H. A. 1973. Plant communities and vegetation. pp. 7–17 in R. M. Lloyd, and R. S. Mitchell (eds.), A flora of the White Mountains, California and Nevada. University of California Press, Berkeley.

Mooney, H. A., and W. D. Billings. 1960. The annual carbohydrate cycle of alpine plants as related to growth. Amer. J. Bot. 47:594–598.

Mulroy, J. C. 1979. Contributions to the ecology and biogeography of the *Saxifraga cespitosa* L. complex in the Americas. Ph.D. dissertation, Duke University, Durham, N.C.

Oberbauer, S. F., and W. D. Billings. 1981. Drought tolerance and water use by plants along an alpine topographic gradient. Oecologia 50:325–331.

Reynolds, D. N. 1984. Alpine annual plants: phenology, germination, photosynthesis, and growth of three Rocky Mountain species. Ecology 65:759–966.

Robberecht, R., M. M. Caldwell, and W. D. Billings. 1980. Leaf ultraviolet optical properties along a latitudinal gradient in the arctic-alpine life zone. Ecology 61:612–619.

Rochow, T. F. 1969. Growth, caloric content, and sugars in *Caltha leptosepala* in relation to alpine snowmelt. Bull. Torrey Botanical Club 96:689–698.

Schaak, C. G. 1983. The alpine vascular flora of Arizona. Madroño 30:79–88.

Scott, D., and W. D. Billings. 1964. Effects of environmental factors on standing crop and productivity of an alpine tundra. Ecol. Monogr. 34:243–270.

Spetzman, L. A. 1959. Vegetation of the arctic slope of Alaska. Geological Survey professional paper 302-B. U.S. Government Printing Office, Washington, D.C.

Steyermark, J. A. 1950. Flora of Guatemala. Ecology 31:368–372.

Wallace, L. L., and A. T. Harrison. 1978. Carbohydrate mobilization and movement in alpine plants. Amer. J. Bot. 65:1035–1040.

Washburn, A. L. 1956. Classification of patterned ground and review of suggested origins. Bull. Geol. Soc. Amer. 67:823–865.

Weber, H. 1958. Die Páramos von Costa Rica und ihre pflanzen-geographische Verkettung mit den Hochanden Südamerikas. pp. 123–194 plus 105 figures and plates in Mathematisch-Naturwissenschaftlichen Klasse. Jahrgang 3. Akademie der Wissenschaften und der Literatur, Mainz.

Webster, G. L. 1961. The altitudinal limits of vascular plants. Ecology 42:587–590.

Index